T0335421

ELEMENTS OF ∞–CATEGORY THEORY

The language of ∞-categories provides an insightful new way of expressing many results in higher-dimensional mathematics but can be challenging for the uninitiated. To explain what exactly an ∞-category is requires various technical models, raising the question of how they might be compared. To overcome this, a model-independent approach is desired, so that theorems proven with any model would apply to them all. This text develops the theory of ∞-categories from first principles in a model-independent fashion using the axiomatic framework of an ∞-cosmos, the universe in which ∞-categories live as objects. An ∞-cosmos is a fertile setting for the formal category theory of ∞-categories, and in this way the foundational proofs in ∞-category theory closely resemble the classical foundations of ordinary category theory. Equipped with exercises and appendices with background material, this first introduction is meant for students and researchers who have a strong foundation in classical 1-category theory.

Emily Riehl is an associate professor of mathematics at Johns Hopkins University. She received her PhD from the University of Chicago and was a Benjamin Peirce and NSF postdoctoral fellow at Harvard University. She is the author of *Categorical Homotopy Theory* (Cambridge University Press, 2014) and *Category Theory in Context* (Dover, 2016), and a coauthor of *Fat Chance: Probability from 0 to 1* (Cambridge University Press, 2019). She and her present coauthor have published ten articles over the course of the past decade that develop the new mathematics appearing in this book.

Dominic Verity is a professor of mathematics at Macquarie University in Sydney and is a director of the Centre of Australian Category Theory. While he is a leading proponent of "Australian-style" higher category theory, he received his PhD from the University of Cambridge and migrated to Australia in the early 1990s. Over the years he has pursued a career that has spanned the academic and nonacademic worlds, working at times as a computer programmer, quantitative analyst, and investment banker. He has also served as Chair of the Academic Senate of Macquarie University, the principal academic governance and policy body.

CAMBRIDGE STUDIES IN ADVANCED MATHEMATICS

All the titles listed below can be obtained from good booksellers or from Cambridge University Press. For a complete series listing, visit www.cambridge.org/mathematics.

Elements of ∞–Category Theory

EMILY RIEHL
The Johns Hopkins University

DOMINIC VERITY
Macquarie University, Sydney

CAMBRIDGE
UNIVERSITY PRESS

CAMBRIDGE
UNIVERSITY PRESS

University Printing House, Cambridge CB2 8BS, United Kingdom

One Liberty Plaza, 20th Floor, New York, NY 10006, USA

477 Williamstown Road, Port Melbourne, VIC 3207, Australia

314–321, 3rd Floor, Plot 3, Splendor Forum, Jasola District Centre,
New Delhi – 110025, India

79 Anson Road, #06–04/06, Singapore 079906

Cambridge University Press is part of the University of Cambridge.

It furthers the University's mission by disseminating knowledge in the pursuit of
education, learning, and research at the highest international levels of excellence.

www.cambridge.org
Information on this title: www.cambridge.org/9781108837989
DOI: 10.1017/9781108936880

First published 2022

A catalogue record for this publication is available from the British Library.

Library of Congress Cataloging-in-Publication Data
Names: Riehl, Emily, author. | Verity, Dominic, 1966- author.
Title: Elements of ∞–category theory / Emily Riehl, Dominic Verity.
Description: Cambridge ; New York, NY : Cambridge University Press, 2022. |
In title "[infinity]" appears as the infinity symbol. | Includes
bibliographical references and index.
Identifiers: LCCN 2021024580 (print) | LCCN 2021024581 (ebook) |
ISBN 9781108837989 (hardback) | ISBN 9781108936880 (epub)
Subjects: LCSH: Categories (Mathematics) | Infinite groups. |
BISAC: MATHEMATICS / Logic
Classification: LCC QA169 .R55 2022 (print) | LCC QA169 (ebook) | DDC 512/.55–dc23
LC record available at https://lccn.loc.gov/2021024580
LC ebook record available at https://lccn.loc.gov/2021024581

ISBN 978-1-108-83798-9 Hardback

Peter –

My debt to you is evident in everything that I write. I found it astounding that, after each of three or four consecutive close-readings of my thesis in which you suggested complete structural revisions, your advice remained entirely correct every time. You taught me to derive immense pleasure from each comprehensive reappraisal that is demanded whenever deepened mathematical understanding allows a more fundamental narrative to emerge. And you always encouraged me to pursue my own esoteric interests while challenging me to consider competing mathematical perspectives. I'm hoping you like the result.

– Emily

Martin –

In writing these words, I realise that I am truly unable to articulate the deep and profound positive influence that you have had on my life and mathematical career. It was you who taught me that in mathematics, as in life, it is not good enough simply to state truths or blindly follow convention. Proofs don't only serve to demonstrate propositions; they are a palimpsest, there to be painted and repainted in the pursuit of aesthetic perfection and pragmatic clarity. In short, it was you who inspired me to see the beauty in our art and to appreciate that every mathematical odyssey must be philosophically grounded and driven by a desire to communicate the elegance of our craft. It is for others to judge how closely we have met that ideal in these pages. All I can say is that in every page we have done our best to live up to your blueprint.

– Dom

Contents

Preface

Mathematical objects of a certain sophistication are frequently accompanied by higher homotopical structures: Maps between them might be connected by homotopies that witness the weak commutativity of diagrams, which might then be connected by higher homotopies expressing coherence conditions among these witnesses, which might then be connected by even higher homotopies ad infinitum. The natural habitat for such mathematical objects is not an ordinary 1-category but instead an ∞-*category* or, more precisely, an $(\infty, 1)$-*category*, with the index "1" referring to the fact that the morphisms above the lowest dimension – the homotopies just discussed – are weakly invertible.

Here the homotopies defining the higher morphisms of an ∞-category are to be regarded as *data* rather than as mere witnesses to an equivalence relation borne by the 1-dimensional morphisms. This shift in perspective is illustrated by the relationship between two algebraic invariants of a topological space: the fundamental groupoid, an ordinary 1-category, and the fundamental ∞-groupoid, an ∞-category in which all of the morphisms are weakly invertible. The objects in both cases are the points of the ambient topological space, but in the former, the 1-morphisms are homotopy classes of paths, while in the latter, the 1-morphisms are the paths themselves and the 2-morphisms are explicit endpoint-preserving homotopies. To encompass examples such as these, all of the categorical structures in an ∞-category are weak. Even at the base level of 1-morphisms, composition is not necessarily uniquely defined but is instead witnessed by a 2-morphism and associative up to a 3-morphism whose boundary data involves specified 2-morphism witnesses. Thus, diagrams valued in an ∞-category cannot be said to *commute* on the nose but are instead interpreted as *homotopy coherent*, with explicitly specified higher data.

A fundamental challenge in defining ∞-categories has to do with giving a precise mathematical meaning of this notion of a weak composition law, not just for the 1-morphisms but also for the morphisms in higher dimensions. Indeed, there is a sense in which our traditional set-based foundations for mathematics are not really suitable for reasoning about ∞-categories: Sets do not feature prominently in ∞-categorical data, especially when ∞-categories are considered in a morally correct fashion as objects that are only well-defined up to equivalence. When considered up to equivalence, ∞-categories, like ordinary categories, do not have a well-defined set of objects. In addition, the morphisms between a fixed pair of objects in an ∞-category assemble into an ∞-groupoid, which describes a well-defined homotopy type, though not a well-defined space.[1]

Precision is achieved through a variety of *models* of (∞, 1)-categories, which are Bourbaki-style mathematical structures that represent infinite-dimensional categories with a weak composition law in which all morphisms above dimension 1 are weakly invertible. In order of appearance, these include *simplicial categories*, *quasi-categories* (née *weak Kan complexes*), *relative categories*, *Segal categories*, *complete Segal spaces*, and *1-complicial sets* (née *saturated 1-trivial weak complicial sets*), each of which comes with an associated array of naturally occurring examples. The proliferation of models of (∞, 1)-categories begs the question of how they might be compared. In the first decades of the twenty-first century, Julia Bergner, André Joyal and Myles Tierney, Dominic Verity, Jacob Lurie, and Clark Barwick and Daniel Kan built various bridges that prove that each of the models listed above "has the same homotopy theory" in the sense of defining the fibrant objects in Quillen equivalent model categories.[2]

In parallel with the development of models of (∞, 1)-categories and the construction of comparisons between them, Joyal pioneered and Lurie and many others expanded a wildly successful project to extend basic category theory from ordinary 1-categories to (∞, 1)-categories modeled as quasi-categories in such a way that the new quasi-categorical notions restrict along the standard embedding $\mathcal{C}at \hookrightarrow \mathcal{QC}at$ to the classical 1-categorical concepts. A natural question is then, does this work extend to other models of (∞, 1)-categories? And to what extent are basic ∞-categorical notions invariant under change of model? For instance, (∞, 1)-categories of manifolds are most naturally constructed as complete Segal spaces, so Kazhdan–Varshavsky [65], Boavida de Brito [34], and Rasekh [95, 96, 98] have recently endeavored to redevelop some of the category theory of quasi-categories using complete Segal spaces instead in order to have direct access to constructions and definitions that had previously been introduced only in the quasi-categorical model.

For practical, aesthetic, and moral reasons, the ultimate desire of practitioners

---

[1] Grothendieck's homotopy hypothesis posits that ∞-groupoids up to equivalence correspond to homotopy types.

[2] A recent book by Bergner surveys all but the last of these models and their interrelationships [15]. For a more whirlwind tour, see [3].

is to work "model independently," meaning that theorems proven with any of the models of $(\infty, 1)$-categories would apply to them all, with the technical details inherent to any particular model never entering the discussion. Since all models of $(\infty, 1)$-categories "have the same homotopy theory," the general consensus is that the choice of model should not matter greatly, but one obstacle to proving results of this kind is that, to a large extent, precise versions of the categorical definitions that have been established for quasi-categories had not been given for the other models. In cases where comparable definitions do exist in different models, an ad hoc heuristic proof of model invariance of the categorical notion in question can typically be supplied, with details to be filled in by experts fluent in the combinatorics of each model, but it would be more reassuring to have a systematic method of comparing the category theory of $(\infty, 1)$-categories in different models via arguments that are somewhat closer to the ground.

Aims

In this text we develop the theory of ∞-categories from first principles in a model independent fashion using a common axiomatic framework that is satisfied by a variety of models. In contrast with prior "analytic" treatments of the theory of ∞-categories – in which the central categorical notions are defined in reference to the coordinates of a particular model – our approach is "synthetic," proceeding from definitions that can be interpreted simultaneously in many models to which our proofs then apply. While synthetic, our work is not schematic or hand-wavy, with the details of how to make things fully precise left to "the experts" and turtles all the way down.[3] Rather, we prove our theorems starting from a short list of clearly enumerated axioms, and our conclusions are thus valid in any model of ∞-categories satisfying these axioms.

The synthetic theory is developed in any ∞-*cosmos*, which axiomatizes the universe in which ∞-categories live as objects. So that our theorem statements suggest their natural interpretation, we recast ∞-*category* as a technical term, to mean an object in some (typically fixed) ∞-cosmos. Several common models of $(\infty, 1)$-categories[4] are ∞-categories in this sense, but our ∞-categories also

---

[3] A less rigorous "model independent" presentation of ∞-category theory might confront a problem of infinite regress, since infinite-dimensional categories are themselves the objects of an ambient infinite-dimensional category, and in developing the theory of the former one is tempted to use the theory of the latter. We avoid this problem by using a very concrete model for the ambient $(\infty, 2)$-category of ∞-categories that arises frequently in practice and is designed to facilitate relatively simple proofs. While the theory of $(\infty, 2)$-categories remains in its infancy, we are content to cut the Gordian knot in this way.

[4] Quasi-categories, complete Segal spaces, Segal categories, and 1-complicial sets (naturally marked quasi-categories) all define the ∞-categories in an ∞-cosmos.

include certain models of (∞, n)-categories[5] as well as fibered versions of all of the above. Thus each of these objects are ∞-categories in our sense and our theorems apply to all of them.[6] This usage of the term "∞-categories" is meant to interpolate between the classical one, which refers to any variety of weak infinite-dimensional categories, and the common one, which is often taken to mean quasi-categories or complete Segal spaces.

Much of the development of the theory of ∞-categories takes place not in the full ∞-cosmos but in a quotient that we call the *homotopy 2-category*, the name chosen because an ∞-cosmos is something like a category of fibrant objects in an enriched model category and the homotopy 2-category is then a categorification of its homotopy category. The homotopy 2-category is a strict 2-category – like the 2-category of categories, functors, and natural transformations[7] – and in this way the foundational proofs in the theory of ∞-categories closely resemble the classical foundations of ordinary category theory except that the universal properties they characterize, e.g., when a functor between ∞-categories defines a cartesian fibration, are slightly weaker than in the familiar case of strict 1-categories.

There are many alternate choices we could have made in selecting the axioms of an ∞-cosmos. One of our guiding principles, admittedly somewhat contrary to the setting of homotopical higher category theory, was to allow us to work as strictly as possible, with the aim of shortening and simplifying proofs. As a consequence of these choices, the ∞-categories in an ∞-cosmos and the functors and natural transformations between them assemble into a 2-category rather than a bicategory. To help us achieve this counterintuitive strictness, each ∞-cosmos comes with a specified class of maps between ∞-categories called *isofibrations*. The isofibrations have no homotopy-theoretic meaning, as any functor between ∞-categories is equivalent to an isofibration with the same codomain. However, isofibrations permit us to consider strictly commutative diagrams between ∞-categories and allow us to require that the limits of such diagrams satisfy a universal property up to simplicially enriched isomorphism. Neither feature is

---

[5] n-quasi-categories, Θ_n-spaces, iterated complete Segal spaces, and n-complicial sets also define the ∞-categories in an ∞-cosmos, as do saturated (née weak) complicial sets, a model for (∞, ∞)-categories.

[6] There is a sense, however, in which many of our definitions are optimized for those ∞-cosmoi whose objects are $(\infty, 1)$-categories. A good illustration is provided by the notion of *discrete ∞-category* introduced in Definition 1.2.26. In the ∞-cosmoi of $(\infty, 1)$-categories, the discrete ∞-categories are the ∞-groupoids. While this is not true for the ∞-cosmoi of (∞, n)-categories, we nevertheless put this concept to use in certain exotic ∞-cosmoi (see, for instance, Definition 7.4.1).

[7] In fact this is another special case: there is an ∞-cosmos whose objects are ordinary categories and its homotopy 2-category is the usual category of categories, functors, and natural transformations. This 2-category is as old as category theory itself, introduced in Eilenberg and Mac Lane's foundational paper [42].

essential for the development of ∞-category theory. Similar proofs carry through to a weaker setting, at the cost of more time spent considering coherence of higher cells.

In Part I, we define and develop the notions of equivalence and adjunction between ∞-categories, limits and colimits in ∞-categories, and cartesian and cocartesian fibrations and their discrete variants, for which we prove a version of the Yoneda lemma. The majority of these results are developed from the comfort of the homotopy 2-category. In an interlude, we digress into abstract ∞-cosmology to give a more careful account of the full class of limit constructions present in any ∞-cosmos. This analysis is used to develop further examples of ∞-cosmoi, whose objects are pointed ∞-categories, or stable ∞-categories, or cartesian or cocartesian fibrations in a given ∞-cosmos.[8]

What is missing from this basic account of the category theory of ∞-categories is a satisfactory treatment of the "hom" bifunctor associated to an ∞-category, which is the prototypical example of what we call a *module*. An instructive exercise for a neophyte is the challenge of defining the ∞-groupoid-valued hom bifunctor in a preferred model. What is edifying is to learn that this construction, so fundamental to ordinary category theory, is prohibitively difficult.[9] In our axiomatization, any ∞-category in an ∞-cosmos has an associated ∞-category of arrows, equipped with domain and codomain projection functors that respectively define cartesian and cocartesian fibrations in a compatible manner. Such modules, which themselves assemble into an ∞-cosmos, provide a convenient vehicle for encoding universal properties as fibered equivalences. In Part II, we develop the calculus of modules between ∞-categories and apply this to define and study pointwise Kan extensions. This will give us an opportunity to repackage universal properties proven in Part I as part of the "formal category theory" of ∞-categories.

This work is all "model-agnostic" in the sense of being blind to details about the specifications of any particular ∞-cosmos. In Part III we prove that the category theory of ∞-categories is also "model independent" in a precise sense: all categorical notions are preserved, reflected, and created by any "change-of-model" functor that defines what we call a *cosmological biequivalence*. This model independence theorem is stronger than our axiomatic framework might initially suggest in that it also allows us to transfer theorems proven using analytic techniques to all biequivalent ∞-cosmoi. For instance, the four ∞-

[8] The impatient reader could skip this interlude and take on faith that any ∞-cosmos begets various other ∞ without compromising their understanding of what follows – though they would miss out on some fun.
[9] Experts in quasi-category theory know to use Lurie's straightening–unstraightening construction [78, 2.2.1.2] or Cisinski's universal left fibration [28, 5.2.8] and the twisted arrow quasi-category.

cosmoi whose objects model $(\infty, 1)$-categories are all biequivalent.[10] It follows that the analytically-proven theorems about quasi-categories from [78] hold for complete Segal spaces, and vice versa. We conclude with several applications of this transfer principle. For instance, in the ∞-cosmoi whose objects are $(\infty, 1)$-categories, we demonstrate that various universal properties are "pointwise-determined" by first proving these results for quasi-categories using analytical techniques and then appealing to model independence to extend these results to biequivalent ∞-cosmoi.

The question of the model invariance of statements about ∞-categories is more subtle than one might expect. When passing an ∞-category from one model to another and then back, the resulting object is typically equivalent but not identical to the original, and certain "evil" properties of ∞-categories fail to be invariant under equivalence: the assertion that an ∞-category has a single object is a famous example. A key advantage to our systematic approach to understanding the model independence of ∞-category theory is that it allows us to introduce a formal language and prove that statement about ∞-categories expressible in that language are model independent. This builds on work of Makkai that resolves a similar question about the invariance of properties of a 2-category under biequivalence [82].

Regrettably, space considerations have prevented us from exploring the homotopy coherent structures present in an ∞-cosmos. For instance, a companion paper [109] proves that any adjunction between ∞-categories in an ∞-cosmos extends homotopically uniquely to a homotopy coherent adjunction and presents a monadicity theorem for homotopy coherent monads as a mechanism for ∞-categorical universal algebra. The formal theory of homotopy coherent monads is extended further by Sulyma [124] who develops the corresponding theory of monadic and comonadic descent and Zaganidis [133] who defines and studies homotopy coherent monad maps. Another casualty of space limitations is an exploration of a "macrocosm principle" for cartesian fibrations, which proves that the codomain projection functor from the ∞-cosmos of cartesian fibrations to the base ∞-cosmos defines a "cartesian fibration of ∞-cosmoi" in a suitable sense [111]. We hope to return to these topics in a sequel.

The ideal reader might already have some acquaintance with enriched category theory, 2-category theory, and abstract homotopy theory so that the constructions and proofs with antecedents in these traditions will be familiar. Because ∞-categories are of interest to mathematicians with a wide variety of backgrounds,

---

[10] A closely related observation is that the Quillen equivalences between quasi-categories, complete Segal spaces, and Segal categories constructed by Joyal and Tierney in [64] can be understood as equivalences of $(\infty, 2)$-categories not just of $(\infty, 1)$-categories by making judicious choices of simplicial enrichments (see §E.2).

we review all of the material we need on each of these topics in Appendices A, B, and C, respectively. Some basic facts about quasi-categories first proven by Joyal are needed to establish the corresponding features of general ∞-cosmoi in Chapter 1. We state these results in §1.1 but defer the proofs that require lengthy combinatorial digressions to Appendix D, where we also introduce *n-complicial sets*, a model of (∞, n)-categories for any $0 \leq n \leq \infty$. The examples of ∞-cosmoi that appear "in the wild" can be found in Appendix E, where we also present general techniques that the reader might use to find ∞-cosmoi of their own. The final appendix addresses a crucial bit of unfinished business. Importantly, the synthetic theory developed in the ∞-cosmos of quasi-categories is fully compatible with the analytic theory developed by Joyal, Lurie, and many others. This is the subject of Appendix F.

We close with the obligatory disclaimer on sizes. To apply the theory developed here to the ∞-categories of greatest interest, one should consider three infinite inaccessible cardinals $\alpha < \beta < \gamma$, as is the common convention [5, 2]. Colloquially, α-small categories might be called "small," while β-small categories are the default size for ∞-categories. For example, the ∞-categories of (small) spaces, chain complexes of (small) abelian groups, or (small) homotopy coherent diagrams are all β-small. These normal-sized ∞-categories are then the objects of an ∞-cosmos that is γ-small – "large" in colloquial terms. Of course, if one is only interested in small simplicial sets, then the ∞-cosmos of small quasi-categories is β-small, rather than γ-small, and the theory developed here equally applies. For this reason, we set aside the Grothendieck universes and do not refer to these inaccessible cardinals elsewhere.

Acknowledgments

The first draft of much of this material was written over the course of a semester-long topics course taught at Johns Hopkins in the spring of 2018 and benefitted considerably from the perspicuous questions asked during lecture by Qingci An, Thomas Brazelton, tslil clingman, Daniel Fuentes-Keuthan, Aurel Malapani-Scala, Mona Merling, David Myers, Apurv Nakade, Martina Rovelli, and Xiyuan Wang.

Further revisions were made during the 2018 MIT Talbot Workshop on the model independent theory of ∞-categories, organized by Eva Belmont, Calista Bernard, Inbar Klang, Morgan Opie, and Sean Pohorence, with faculty sponsor Haynes Miller. We were inspired by compelling lectures given at that workshop by Kevin Arlin, Timothy Campion, Kyle Ferendo, Daniel Fuentes-Keuthan, Joseph Helfer, Paul Lessard, Lyne Moser, Emma Phillips, Nima Rasekh, Martina

Rovelli, Maru Sarazola, Matthew Weatherley, Jonathan Weinberger, Laura Wells, and Liang Ze Wong as well as by myriad discussions with Elena Dimitriadis Bermejo, Luciana Basualdo Bonatto, Olivia Borghi, Tai-Danae Bradley, Tejas Devanur, Aras Ergus, Matthew Feller, Sina Hazratpour, Peter James, Zhulin Li, David Myers, Maximilien Péroux, Mitchell Riley, Luis Scoccola, Brandon Shapiro, Pelle Steffens, Raffael Stenzel, Paula Verdugo, and Marco Vergura.

John Bourke suggested Lemma 2.1.11, which unifies the proofs of several results concerning the 2-category theory of adjunctions. Denis-Charles Cisinski drew our attention to an observation of André Joyal that appears as Exercise 4.2.iii. Omar Antolín-Camarena told us about condition (iii) of Theorem 4.4.12, which turns out to be the most expeditious characterization of stable ∞-categories with which to prove that they assemble into an ∞-cosmos in Proposition 6.3.16. Gabriel Drummond-Cole encouraged us to do some much-needed restructuring of Chapter 5. Anna Marie Bohmann suggested Exercise 8.2.i. tslil clingman pointed us to an observation by Tom Leinster, which inspired Exercise 8.3.ii. The presentation in Chapter 9 was greatly improved by observations of Kevin Arlin, who inspired a reorganization, and David Myers, who first stated the result appearing as Proposition 9.1.8. The material on the groupoid core of an $(\infty, 1)$-category was informed by discussions with Alexander Campbell and Yuri Sulyma, the latter of whom developed much of this material independently (and first) while working on his PhD thesis. The results of §11.3 were inspired by a talk by Simon Henry in the Homotopy Type Theory Electronic Seminar Talks [52]. Timothy Campion, Yuri Sulyma, and Dimitri Zaganidis each made excellent suggestions concerning material that was ultimately cut from the final version of this text.

Alexander Campbell was consulted several times during the writing of Appendix A, in particular regarding the subtle interaction between change-of-enrichment functors and underlying categories. He also supplied the proof of the ∞-cosmos of n-quasi-categories appearing in Proposition E.3.3. Naruki Masuda caught several typos and inconsistencies in Appendix D, while Christian Sattler supplied a simpler proof of the implication (iii)$\Rightarrow$(i) in Proposition D.5.6.

tslil clingman suggested a more aesthetic way to typeset proarrows in the virtual equipment, greatly improving the displayed diagrams in Chapter 8. Doug Ravenel has propagated the use of the character " $ょ$ " for the Yoneda embedding. Anna Marie Bohmann, John Bourke, Alexander Campbell, Antoine Chambert-Loir, Arun Debray, Gabriel Drummond-Cole, David Farrell, Harry Gindi, Philip Hackney, Peter Haine, Dodam Ih, Stephen Lack, Chen-wei (Milton) Lin, Naruki Masuda, Mark Myers, Viktoriya Ozornova, Jean Kyung Park, Emma Phillips, Maru Sarazola, Yuri Sulyma, Paula Verdugo, Mira Wattal, Jonathan Weinberger, and Hu Xiao wrote to point out typos.

Peter May suggested the name "∞-cosmos," a substantial improvement upon previous informal terminology. Mike Hopkins proposed the title of this volume. We are grateful for the stewardship of Kaitlin Leach and Amy He at Cambridge University Press, for the technical support provided by Suresh Kumar, for the superlative copyediting services of Anbumani Selvam and production management by Dhivya Elavazhagan, and for the perspicuous suggestions shared by the anonymous reviewers.

Finally, the authors are grateful for the financial support provided by the National Science Foundation via grants DMS-1551129 and DMS-1652600, the Australian Research Council via the grants DP160101519 and DP190102432, and the Johns Hopkins Catalyst and President's Frontier Award programs. We also wish to thank the Department of Mathematics at Johns Hopkins University, the Centre of Australian Category Theory at Macquarie University, and the Mathematical Sciences Research Institute for hosting our respective visits. And most of all, we are grateful to our families for their love, support, patience, and understanding, without which this would have never been possible.

PART ONE

BASIC ∞-CATEGORY THEORY

It is difficult and time-consuming to learn a new language. The standard advice to "fake it til you make it" is disconcerting in mathematical contexts, where the validity of a proof hinges upon the correctness of the statements it cites. The aim in Part I of this text is to develop a substantial portion of the theory of ∞-categories from first principles, as rapidly and painlessly as possible – at least assuming that the reader finds classical abstract nonsense to be relatively innocuous.[11]

The axiomatic framework that justifies this is introduced in Chapter 1, but the impatient or particularly time-constrained reader might consider starting directly in Chapter 2 with the study of adjunctions, limits, and colimits. In adopting this approach, one must take for granted that there is a well-defined 2-category of ∞-categories, ∞-functors between them, and ∞-natural transformations between these. This 2-category is constructed in Chapter 1, where we see that any ∞-cosmos has a homotopy 2-category and that the familiar models of (∞, 1)-categories define biequivalent ∞-cosmoi, with biequivalent homotopy 2-categories. To follow the proofs in Chapter 2, it is necessary to understand the general composition of natural transformations by *pasting diagrams*. This and other concepts from 2-category theory are reviewed in Appendix B, which should be consulted as needed.

The payoff for acquainting oneself with some standard 2-category theory is that numerous fundamental results concerning equivalences and adjunctions and limits and colimits can be proven quite expeditiously. We prove one such theorem, that right adjoint functors between ∞-categories preserve any limits found in those ∞-categories, via a formal argument that is arguably even simpler than the classical one.

The definitions of adjunctions, limits, and colimits given in Chapter 2 are optimized for ease of use in the homotopy 2-category of ∞-categories, ∞-functors, and ∞-natural transformations in an ∞-cosmos, but especially in the latter cases, these notions are not expressed in their most familiar forms. To encode a limit of a diagram valued in an ∞-category as a terminal cone, we introduce the powerful and versatile construction of the *comma ∞-category* built from a cospan of functors in Chapter 3. We then prove various "representability theorems" that characterize those comma ∞-categories that are equivalent to ones defined by a single functor. These general results specialize in Chapter 4 to the expected equivalent definitions of adjunctions, limits, and colimits. This theory is then applied to study limits and colimits of particular diagram

---

[11] Dan Freed defines the category number of a mathematician to be the largest integer n so that they may ponder n-categories for half an hour without developing a migraine. Here we require a category number of 2, which we note is much smaller than ∞!

shapes, which in turn is deployed to establish an equivalence between various presentations of the important notion of a *stable ∞-category*.

The basic theory of ∞-categories is extended in Chapter 5 to encompass *cocartesian* and *cartesian fibrations*, which can be understood as indexed families of ∞-categories acted upon covariantly or contravariantly by arrows in the base ∞-category. After developing the theory of the various classes of categorical fibrations, we conclude by proving a fibrational form of the Yoneda lemma that will be used to further develop the formal category theory of ∞-categories in Part II.

1

∞-Cosmoi and Their Homotopy 2-Categories

In this chapter, we introduce a framework to develop the formal category theory of ∞-categories, which goes by the name of an ∞-*cosmos*. Informally, an ∞-cosmos is an (∞, 2)-category – a category enriched over (∞, 1)-categories – that is equipped with (∞, 2)-categorical limits. In the motivating examples of ∞-cosmoi, the objects are ∞-categories in some model. To focus this abstract theory on its intended interpretation, we recast "∞-category" as a technical term, reserved to mean an object of some ∞-cosmos.

Unexpectedly, the motivating examples permit us to use a quite strict interpretation of "(∞, 2)-category with (∞, 2)-categorical limits": an ∞-cosmos is a particular type of simplicially enriched category and the (∞, 2)-categorical limits are modeled by simplicially enriched limits. More precisely, an ∞-cosmos is a category enriched over quasi-categories, these being one of the models of (∞, 1)-categories defined as certain simplicial sets. The (∞, 2)-categorical limits are defined as limits of diagrams involving specified maps called *isofibrations*, which have no intrinsic homotopical meaning – since any functor between ∞-categories is equivalent to an isofibration – but allow us to consider strictly commuting diagrams.

In §1.1, we introduce quasi-categories, reviewing the classical results that are needed to show that quasi-categories themselves assemble into an ∞-cosmos – the prototypical example. General ∞-cosmoi are defined in §1.2, where several examples are given and their basic properties are established. In §1.3, we turn our attention to *cosmological functors* between ∞-cosmoi, which preserve all of the defining structure. Cosmological functors serve dual purposes, on the one hand providing technical simplifications in many proofs, and then later on serving as the "change of model" functors that establish the model independence of ∞-category theory.

Finally, in §1.4, we introduce a strict 2-category whose objects are ∞-categories, whose 1-cells are the ∞-functors between them, and whose 2-cells define

5

∞-natural transformations between these. Any ∞-cosmos has a 2-category of this sort, which we refer to as the *homotopy 2-category* of the ∞-cosmos. In fact, the reader who is eager to get on to the development of the category theory of ∞-categories can skip this chapter on first reading, taking the existence of the homotopy 2-category for granted, and start with Chapter 2.

1.1 Quasi-Categories

Before introducing an axiomatic framework that allows us to develop ∞-category theory in general, we first consider one model in particular: *quasi-categories*, which were introduced in 1973 by Boardman and Vogt [21] in their study of homotopy coherent diagrams. Ordinary 1-categories give examples of quasi-categories via the construction of Definition 1.1.4. Joyal first undertook the task of extending 1-category theory to quasi-category theory in [61] and [63] and in several unpublished draft book manuscripts. The majority of the results in this section are due to him.

NOTATION 1.1.1 (the simplex category). Let Δ denote the **simplex category** of finite nonempty ordinals $[n] = \{0 < 1 < \cdots < n\}$ and order-preserving maps. These include in particular the

$$\text{elementary face operators} \qquad [n-1] \overset{\delta^i}{\rightarrowtail} [n] \qquad 0 \leq i \leq n$$

$$\text{elementary degeneracy operators} \qquad [n+1] \overset{\sigma^i}{\twoheadrightarrow} [n] \qquad 0 \leq i \leq n$$

whose images, respectively, omit and double up on the element $i \in [n]$. Every morphism in Δ factors uniquely as an epimorphism followed by a monomorphism; these epimorphisms, the **degeneracy operators**, decompose as composites of elementary degeneracy operators, while the monomorphisms, the **face operators**, decompose as composites of elementary face operators.

The category of **simplicial sets** is the category $s\mathcal{S}et := \mathcal{S}et^{\Delta^{op}}$ of presheaves on the simplex category. We write $\Delta[n]$ for the **standard n-simplex** the simplicial set represented by $[n] \in \Delta$, and $\Lambda^k[n] \subset \partial\Delta[n] \subset \Delta[n]$ for its k-**horn** and **boundary sphere**, respectively. The sphere $\partial\Delta[n]$ is the simplicial subset generated by the codimension-one faces of the n-simplex, while the horn $\Lambda^k[n]$ is the further simplicial subset that omits the face opposite the vertex k.

Given a simplicial set X, it is conventional to write X_n for the set of n-**simplices**, defined by evaluating at $[n] \in \Delta$. By the Yoneda lemma, each n-simplex $x \in X_n$ corresponds to a map of simplicial sets $x \colon \Delta[n] \to X$. Accordingly, we write $x \cdot \delta^i$ for the ith face of the n-simplex, an $(n-1)$-simplex classified by the composite map

$$\Delta[n-1] \xrightarrow{\delta^i} \Delta[n] \xrightarrow{x} X.$$

The right action of the face operator defines a map $X_n \xrightarrow{\cdot \delta^i} X_{n-1}$. Geometrically, $x \cdot \delta^i$ is the "face opposite the vertex i" in the n-simplex x.

DEFINITION 1.1.2 (quasi-category). A **quasi-category** is a simplicial set A in which any **inner horn** can be extended to a simplex, solving the displayed lifting problem:

$$
\begin{array}{ccc}
\Lambda^k[n] & \longrightarrow & A \\
\downarrow & \nearrow & \\
\Delta[n] & &
\end{array}
\qquad \text{for } n \geq 2,\, 0 < k < n. \qquad (1.1.3)
$$

Quasi-categories were first introduced by Boardman and Vogt [21] under the name "weak Kan complexes," a **Kan complex** being a simplicial set admitting extensions as in (1.1.3) along all horn inclusions $n \geq 1, 0 \leq k \leq n$. Since any topological space can be encoded as a Kan complex,[1] in this way spaces provide examples of quasi-categories.

Categories also provide examples of quasi-categories via the nerve construction.

DEFINITION 1.1.4 (nerve). The category $\mathcal{C}at$ of 1-categories embeds fully faithfully into the category of simplicial sets via the **nerve** functor. An n-simplex in the nerve of a 1-category C is a sequence of n composable arrows in C, or equally a functor $\mathfrak{n}+\mathbb{1} \to C$ from the ordinal category $\mathfrak{n}+\mathbb{1} := [n]$ with objects $0, \ldots, n$ and a unique arrow $i \to j$ just when $i \leq j$.

The map $[n] \mapsto \mathfrak{n}+\mathbb{1}$ defines a fully faithful embedding $\Delta \hookrightarrow \mathcal{C}at$. From this point of view, the nerve functor can be described as a "restricted Yoneda embedding" which carries a category C to the restriction of the representable functor $\hom(-, C)$ to the image of this inclusion. More general "nerve-type constructions" are described in Exercise 1.1.i.

REMARK 1.1.5. The nerve of a category C is **2-coskeletal** as a simplicial set, meaning that every sphere $\partial\Delta[n] \to C$ with $n \geq 3$ is filled uniquely by an n-simplex in C (see Definition C.5.2). Note a sphere $\partial\Delta[2] \to C$ extends to a

---

[1] The total singular complex construction defines a functor from topological spaces to simplicial sets that is an equivalence on their respective homotopy categories – weak homotopy types of spaces correspond to homotopy equivalence classes of Kan complexes [93, §II.2]. The left adjoint constructed by Exercise 1.1.i "geometrically realizes" a simplicial set as a topological space.

2-simplex if and only if that arrow along its diagonal edge is the composite of the arrows along the edges in the inner horn $\Lambda^1[2] \subset \partial\Delta[2] \to C$. The simplices in dimension 3 and above witness the associativity of the composition of the path of composable arrows found along their **spine**, the 1-skeletal simplicial subset formed by the edges connecting adjacent vertices. In fact, as suggested by the proof of Proposition 1.1.6, any simplicial set in which inner horns admit *unique* fillers is isomorphic to the nerve of a 1-category (see Exercise 1.1.iv).

We decline to introduce explicit notation for the nerve functor, preferring instead to identify 1-categories with their nerves. As we shall discover the theory of 1-categories extends to ∞-categories modeled as quasi-categories in such a way that the restriction of each ∞-categorical concept along the nerve embedding recovers the corresponding 1-categorical concept. For instance, the standard simplex $\Delta[n]$ is isomorphic to the nerve of the ordinal category $\mathbb{n} + 1$, and we frequently adopt the latter notation – writing $\mathbb{1} := \Delta[0]$, $\mathbb{2} := \Delta[1]$, $\mathbb{3} := \Delta[2]$, and so on – to suggest the correct categorical intuition.

To begin down this path, we must first verify the implicit assertion that has just been made:

PROPOSITION 1.1.6 (nerves are quasi-categories). *Nerves of categories are quasi-categories.*

Proof Via the isomorphism $C \cong \mathrm{cosk}_2 C$ from Remark 1.1.5 and the adjunction $\mathrm{sk}_2 \dashv \mathrm{cosk}_2$ of C.5.2, the required lifting problem displayed below-left transposes to the one displayed below-right:

The functor sk_2 replaces a simplicial set by its **2-skeleton**, the simplicial subset generated by the simplices of dimension at most two. For $n \geq 4$, the inclusion $\mathrm{sk}_2 \Lambda^k[n] \hookrightarrow \mathrm{sk}_2 \Delta[n]$ is an isomorphism, in which case the lifting problems on the right admit (unique) solutions. So it remains only to solve the lifting problems on the left in the cases $n = 2$ and $n = 3$.

To that end consider

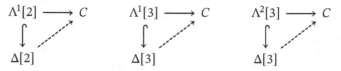

An inner horn $\Lambda^1[2] \to C$ defines a composable pair of arrows in C; an extension to a 2-simplex exists precisely because any composable pair of arrows admits a (unique) composite.

An inner horn $\Lambda^1[3] \to C$ specifies the data of three composable arrows in C, as displayed in the following diagram, together with the composites gf, hg, and $(hg)f$.

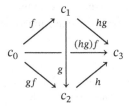

Because composition is associative, the arrow $(hg)f$ is also the composite of gf followed by h, which proves that the 2-simplex opposite the vertex c_1 is present in C; by 2-coskeletality, the 3-simplex filling this boundary sphere is also present in C. The filler for a horn $\Lambda^2[3] \to C$ is constructed similarly. □

DEFINITION 1.1.7 (homotopy relation on 1-simplices). A parallel pair of 1-simplices f, g in a simplicial set X are **homotopic** if there exists a 2-simplex whose boundary takes either of the following forms[2]

$$
\begin{array}{ccc}
& y & \\
f \nearrow & & \searrow \\
x \xrightarrow[g]{} y & &
\end{array}
\qquad
\begin{array}{ccc}
& x & \\
\diagup\diagup & & \searrow f \\
x \xrightarrow[g]{} y & &
\end{array}
\tag{1.1.8}
$$

or if f and g are in the same equivalence class generated by this relation.

In a quasi-category, the relation witnessed by either of the types of 2-simplex on display in (1.1.8) is an equivalence relation and these equivalence relations coincide.

LEMMA 1.1.9 (homotopic 1-simplices in a quasi-category). *Parallel 1-simplices f and g in a quasi-category are homotopic if and only if there exists a 2-simplex of any or, equivalently all of the forms displayed in* (1.1.8).

Proof Exercise 1.1.ii. □

DEFINITION 1.1.10 (the homotopy category [44, §2.4]). By 1-truncating, any simplicial set X has an underlying reflexive directed graph with the 0-simplices of X defining the objects and the 1-simplices defining the arrows:

$$
X_1 \underset{\cdot\delta^0}{\overset{\cdot\delta^1}{\underset{\longrightarrow}{\overset{\longrightarrow}{\leftarrow \cdot\sigma^0 -}}}} X_0,
$$

---

[2] The symbol "$=$" is used in diagrams to denote a degenerate simplex or an identity arrow.

By convention, the source of an arrow $f \in X_1$ is its 0th face $f \cdot \delta^1$ (the face opposite 1) while the target is its 1st face $f \cdot \delta^0$ (the face opposite 0). The **free category** on this reflexive directed graph has X_0 as its object set, degenerate 1-simplices serving as identity morphisms, and nonidentity morphisms defined to be finite directed paths of nondegenerate 1-simplices. The **homotopy category** hX of X is the quotient of the free category on its underlying reflexive directed graph by the congruence[3] generated by imposing a composition relation $h = g \circ f$ witnessed by 2-simplices

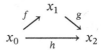

This relation implies in particular that homotopic 1-simplices represent the same arrow in the homotopy category.

The homotopy category of the nerve of a 1-category is isomorphic to the original category, as the 2-simplices in the nerve witness all of the composition relations satisfied by the arrows in the underlying reflexive directed graph. Indeed, the natural isomorphism h$C \cong C$ forms the counit of an adjunction, embedding $\mathcal{C}at$ as a reflective subcategory of $s\mathcal{S}et$.

PROPOSITION 1.1.11. *The nerve embedding admits a left adjoint, namely the functor which sends a simplicial set to its homotopy category:*

$$\mathcal{C}at \overset{h}{\underset{}{\leftrightarrows}} \perp \; s\mathcal{S}et$$

The adjunction of Proposition 1.1.11 exists for formal reasons (see Exercise 1.1.i), but nevertheless, a direct proof can be enlightening.

Proof For any simplicial set X, there is a natural map from X to the nerve of its homotopy category hX; since nerves are 2-coskeletal, it suffices to define the map $\mathrm{sk}_2 X \to \mathrm{h}X$, and this is given immediately by the construction of Definition 1.1.10. Note that the quotient map $X \to \mathrm{h}X$ becomes an isomorphism upon applying the homotopy category functor and is already an isomorphism whenever X is the nerve of a category. Thus the adjointness follows from Lemma B.4.2 or by direct verification of the triangle equalities. □

The homotopy category of a quasi-category admits a simplified description.

LEMMA 1.1.12 (the homotopy category of a quasi-category). *If A is a quasi-category then its **homotopy category** hA has*

---

[3] A binary relation $\sim$ on parallel arrows of a 1-category is a **congruence** if it is an equivalence relation that is closed under pre- and post-composition: if $f \sim g$ then $hfk \sim hgk$.

- *the set of 0-simplices A_0 as its objects*
- *the set of homotopy classes of 1-simplices A_1 as its arrows*
- *the identity arrow at $a \in A_0$ represented by the degenerate 1-simplex $a \cdot \sigma^0 \in A_1$*
- *a composition relation $h = g \circ f$ in hA between the homotopy classes of arrows represented by any given 1-simplices $f, g, h \in A_1$ if and only if there exists a 2-simplex with boundary*

Proof Exercise 1.1.iii. □

DEFINITION 1.1.13 (isomorphism in a quasi-category). A 1-simplex in a quasi-category is an **isomorphism**[4] just when it represents an isomorphism in the homotopy category. By Lemma 1.1.12 this means that $f : a \to b$ is an isomorphism if and only if there exists a 1-simplex $f^{-1} : b \to a$ together with a pair of 2-simplices

The properties of the isomorphisms in a quasi-category are most easily proved by arguing in a closely related category where simplicial sets have the additional structure of a "marking" on a specified subset of the 1-simplices; maps of these so-called *marked simplicial sets* must then preserve the markings (see Definition D.1.1). For instance, each quasi-category has a *natural marking*, where the marked 1-simplices are exactly the isomorphisms (see Definition D.4.5). Since the property of being an isomorphism in a quasi-category is witnessed by the presence of 2-simplices with a particular boundary, every map between quasi-categories preserves isomorphisms, inducing a map of the corresponding naturally marked quasi-categories. Because marked simplicial sets seldom appear outside of the proofs of certain combinatorial lemmas about the isomorphisms in quasi-categories, we save the details for Appendix D.

Let us now motivate the first of several results proven using marked techniques. A quasi-category A is defined to have extensions along all *inner horns*. But when the initial or final edges, respectively, of an outer horn $\Lambda^0[2] \to A$ or

---

[4] Joyal refers to these maps as "isomorphisms" while Lurie refers to them as "equivalences." We prefer, wherever possible, to use the same term for ∞-categorical concepts as for the analogous 1-categorical ones.

$\Lambda^2[2] \to A$ map to isomorphisms in A, then a filler

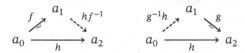

should intuitively exist. The higher-dimensional "special outer horns" behave similarly:

PROPOSITION 1.1.14 (special outer horn filling). *Any quasi-category A admits fillers for those outer horns*

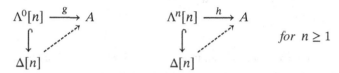

in which the edges $g|_{\{0,1\}}$ and $h|_{\{n-1,n\}}$ are isomorphisms.[5]

The proof of Proposition 1.1.14 requires clever combinatorics, due to Joyal, and is deferred to Proposition D.4.6. Here, we enjoy its myriad consequences. Immediately:

COROLLARY 1.1.15. *A quasi-category is a Kan complex if and only if its homotopy category is a groupoid.*

Proof If the homotopy category of a quasi-category is a groupoid, then all of its 1-simplices are isomorphisms, and Proposition 1.1.14 then implies that all inner and outer horns have fillers. Thus, the quasi-category is a Kan complex. Conversely, in a Kan complex, all outer horns can be filled and in particular fillers for the horns displayed in Definition 1.1.13 can be used to construct left and right inverses for any 1-simplex, which can be rectified to a single two-sided inverse by Lemma 1.1.12. □

A quasi-category contains A a canonical **maximal sub Kan complex** $A^{\simeq}$, the simplicial subset spanned by those 1-simplices that are isomorphisms. Just as the arrows in a quasi-category A are represented by simplicial maps $2 \to A$ whose domain is the nerve of the free-living arrow, the isomorphisms in a quasi-category can be represented by diagrams $\mathbb{I} \to A$ whose domain, called the **homotopy coherent isomorphism**, is the nerve of the free-living isomorphism:

---

[5] In the case $n = 1$, no condition is needed on the horns; degenerate 1-simplices define the required lifts.

COROLLARY 1.1.16. *An arrow f in a quasi-category A is an isomorphism if and only if it extends to a homotopy coherent isomorphism*

Proof If f is an isomorphism, the map $f : 2 \to A$ lands in the maximal sub Kan complex contained in A:

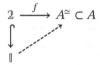

By Exercise 1.1.v, the inclusion $2 \hookrightarrow \mathbb{I}$ can be expressed as a sequential composite of pushouts of outer horn inclusions. Since $A^\simeq$ is a Kan complex, this shows that the required extension exists and in fact lands in $A^\simeq \subset A$. □

The category of simplicial sets, like any category of presheaves, is cartesian closed. By the Yoneda lemma and the defining adjunction, an n-simplex in the exponential Y^X corresponds to a simplicial map $X \times \Delta[n] \to Y$, and its faces and degeneracies are computed by precomposing in the simplex variable. Our next aim is to show that the quasi-categories define an exponential ideal in the simplicially enriched category of simplicial sets: if X is a simplicial set and A is a quasi-category, then A^X is a quasi-category. We deduce this as a corollary of the "relative" version of this result involving certain maps called isofibrations that we now introduce.

DEFINITION 1.1.17 (isofibration). A simplicial map $f : A \to B$ between quasi-categories is an **isofibration** if it lifts against the inner horn inclusions, as displayed below-left, and also against the inclusion of either vertex into the free-living isomorphism $\mathbb{I}$.

To notationally distinguish the isofibrations, we depict them as arrows "$\twoheadrightarrow$" with two heads.

Proposition 1.1.14 is subsumed by its relative analogue, proven as Theorem D.5.1:

PROPOSITION 1.1.18 (special outer horn lifting). *Any isofibration between quasi-categories* $f : A \twoheadrightarrow B$ *admits lifts against those outer horns*

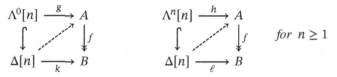

in which the edges $g|_{\{0,1\}}$, $h|_{\{n-1,n\}}$, $k|_{\{0,1\}}$, *and* $\ell|_{\{n-1,n\}}$ *are isomorphisms.*

OBSERVATION 1.1.19.

   (i) For any simplicial set X, the unique map $X \to 1$ whose codomain is the terminal simplicial set is an isofibration if and only if X is a quasi-category.
  (ii) Any collection of maps, such as the isofibrations, that is characterized by a right lifting property is automatically closed under composition, product, pullback, retract, and (inverse) limits of towers (see Lemma C.2.3).
 (iii) Combining (i) and (ii), if $A \twoheadrightarrow B$ is an isofibration, and B is a quasi-category, then so is A.
 (iv) The isofibrations generalize the eponymous categorical notion. The nerve of any functor $f : A \to B$ between categories defines a map of simplicial sets that lifts against the inner horn inclusions. This map then defines an isofibration if and only if given any isomorphism in B and specified object in A lifting either its domain or codomain, there exists an isomorphism in A with that domain or codomain lifting the isomorphism in B.

Much harder to establish is the stability of the isofibrations under the formation of "Leibniz[6] exponentials" as displayed in (1.1.21). This is proven in Proposition D.5.2.

PROPOSITION 1.1.20. *If* $i : X \hookrightarrow Y$ *is a monomorphism and* $f : A \twoheadrightarrow B$ *is an*

---

[6] The name alludes to the Leibniz rule in differential calculus, or more specifically to the identification of the domain of the Leibniz product of Lemma D.3.1 with the boundary of the prism (see Definition C.2.8 and Remark D.3.2).

isofibration, then the induced Leibniz exponential map i $\widehat{\pitchfork}$ f

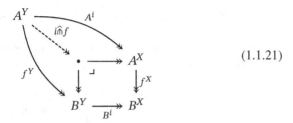

$$(1.1.21)$$

is again an isofibration.[7]

CorOLLARY 1.1.22. *If X is a simplicial set and A is a quasi-category, then A^X is a quasi-category. Moreover, a 1-simplex in A^X is an isomorphism if and only if its components at each vertex of X are isomorphisms in A.*

Proof The first statement is a special case of Proposition 1.1.20 (see Exercise 1.1.vii), while the second statement is proven similarly by arguing with marked simplicial sets (see Corollary D.4.19). □

DEFINITION 1.1.23 (equivalences of quasi-categories). A map $f : A \to B$ between quasi-categories is an **equivalence** if it extends to the data of a "homotopy equivalence" with the free-living isomorphism $\mathbb{I}$ serving as the interval: that is, if there exist maps $g : B \to A$,

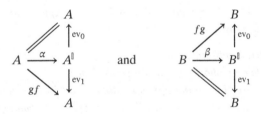

We write "$\xrightarrow{\sim}$" to decorate equivalences and $A \simeq B$ to indicate the presence of an equivalence $A \xrightarrow{\sim} B$.

REMARK 1.1.24. If $f : A \to B$ is an equivalence of quasi-categories, then the functor $hf : hA \to hB$ is an equivalence of categories, where the data displayed above defines an equivalence inverse $hg : hB \to hA$ and natural isomorphisms

---

[7] Degenerate cases of this result, taking $X = \varnothing$ or $B = 1$, imply that the other six maps in this diagram are also isofibrations (see Exercise 1.1.vii).

encoded by the composite[8] functors

$$hA \xrightarrow{h\alpha} h(A^{\mathbb{I}}) \longrightarrow (hA)^{\mathbb{I}} \qquad hB \xrightarrow{h\beta} h(B^{\mathbb{I}}) \longrightarrow (hB)^{\mathbb{I}}$$

DEFINITION 1.1.25. A map $f : X \to Y$ between simplicial sets is a **trivial fibration** if it admits lifts against the boundary inclusions for all simplices

$$
\begin{array}{ccc}
\partial\Delta[n] & \longrightarrow & X \\
\downarrow & \nearrow & \downarrow f \\
\Delta[n] & \longrightarrow & Y
\end{array}
\qquad \text{for } n \geq 0 \qquad (1.1.26)
$$

We write "⤇" to decorate trivial fibrations.

REMARK 1.1.27. The simplex boundary inclusions $\partial\Delta[n] \hookrightarrow \Delta[n]$ "cellularly generate" the monomorphisms of simplicial sets (see Definition C.2.4 and Lemma C.5.9). Hence the dual of Lemma C.2.3 implies that trivial fibrations lift against any monomorphism between simplicial sets. In particular, it follows that any trivial fibration $X \twoheadrightarrow Y$ is a split epimorphism.

The notation "⤇" is suggestive: the trivial fibrations between quasi-categories are exactly those maps that are both isofibrations and equivalences. This can be proven by a relatively standard although rather technical argument in simplicial homotopy theory, appearing as Proposition D.5.6.

PROPOSITION 1.1.28. *For a map $f : A \to B$ between quasi-categories the following are equivalent:*

(i) *f is a trivial fibration*

(ii) *f is both an isofibration and an equivalence*

(iii) *f is a **split fiber homotopy equivalence**: an isofibration admitting a section s that is also an equivalence inverse via a homotopy α from id_A to sf that composes with f to the constant homotopy from f to f.*

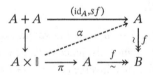

As a class characterized by a right lifting property, the trivial fibrations are also closed under composition, product, pullback, limits of towers, and contain

---

[8] Note that $h(A^{\mathbb{I}}) \not\cong (hA)^{\mathbb{I}}$ in general. Objects in the latter are homotopy classes of isomorphisms in A, while objects in the former are homotopy coherent isomorphisms, given by a specified 1-simplex in A, a specified inverse 1-simplex, together with an infinite tower of coherence data indexed by the nondegenerate simplices in $\mathbb{I}$.

the isomorphisms. The stability of these maps under Leibniz exponentiation is proven along with Proposition 1.1.20 in Proposition D.5.2.

PROPOSITION 1.1.29. *If* $i : X \to Y$ *is a monomorphism and* $f : A \to B$ *is an isofibration, then if either* f *is a trivial fibration or if* i *is in the class cellularly generated by the inner horn inclusions and the map* $\mathbb{1} \hookrightarrow \mathbb{I}$ *then the induced Leibniz exponential map*

$$A^Y \xrightarrow{\;\widehat{i} \widehat{\pitchfork} f\;} B^Y \times_{B^X} A^X$$

a trivial fibration.

To illustrate the utility of these Leibniz stability results, we give an "internal" or "synthetic" characterization of the Kan complexes.

LEMMA 1.1.30. *A quasi-category* A *is a Kan complex if and only if the map* $A^{\mathbb{I}} \twoheadrightarrow A^2$ *induced by the inclusion* $2 \hookrightarrow \mathbb{I}$ *is a trivial fibration.*

Note that Proposition 1.1.20 implies that $A^{\mathbb{I}} \twoheadrightarrow A^2$ is an isofibration.

Proof The lifting property that characterizes trivial fibrations transposes to another lifting property, displayed below-right

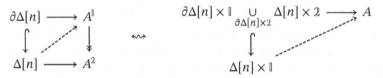

that asserts that A admits extensions along maps formed by taking the *Leibniz product* – also known as the pushout product – of a simplex boundary inclusion $\partial \Delta[n] \hookrightarrow \Delta[n]$ with the inclusion $2 \hookrightarrow \mathbb{I}$. By Exercise 1.1.v(ii) the inclusion $2 \hookrightarrow \mathbb{I}$ is a sequential composite of pushouts of left outer horn inclusions. By Corollary D.3.11, a key step along the way to the proofs of Propositions 1.1.20 and 1.1.29, it follows that the Leibniz product is also a sequential composite of pushouts of left and inner horn inclusions. If A is a Kan complex, then the extensions displayed above right exist, and, by transposing, the map $A^{\mathbb{I}} \twoheadrightarrow A^2$ is a trivial fibration.

Conversely, if $A^{\mathbb{I}} \twoheadrightarrow A^2$ is a trivial fibration then in particular it is surjective on vertices. Thus every arrow in A is an isomorphism, and Corollary 1.1.15 tells us that A must be a Kan complex. □

DIGRESSION 1.1.31 (the Joyal model structure). The category of simplicial sets bears a Quillen model structure, in the sense of Definition C.3.1, whose fibrant objects are exactly the quasi-categories and in which all objects are cofibrant.

Between fibrant objects, the fibrations, weak equivalences, and trivial fibrations are precisely the isofibrations, equivalences, and trivial fibrations just introduced. Proposition 1.1.28 proves that the trivial fibrations are the intersection of the fibrations and the weak equivalences. Propositions 1.1.20 and 1.1.29 reflect the fact that the Joyal model structure is a *cartesian closed model category*, satisfying the additional axioms of Definition C.3.10.

We decline to elaborate further on the Joyal model structure for quasi-categories since we have highlighted all of the features that we need. The results enumerated here suffice to show that the category of quasi-categories defines an ∞-*cosmos*, a concept to which we now turn.

Exercises

EXERCISE 1.1.i ([103, §1.5]). Given any cosimplicial object $C : \Delta \to \mathcal{E}$ valued in any category $\mathcal{E}$, there is an associated **nerve functor** N_C defined by:

$$\mathcal{E} \xrightarrow{\ N_C\ } s\mathcal{S}et$$

$$E \longmapsto \hom(C^-, E)$$

By construction n-simplices in $N_C E$ correspond to maps $C^n \to E$ in $\mathcal{E}$. Show that if $\mathcal{E}$ is cocomplete, then N_C has a left adjoint defined as the left Kan extension of the functor C along the Yoneda embedding $ よ : \Delta \hookrightarrow s\mathcal{S}et$. This gives a second proof of Proposition 1.1.11.

EXERCISE 1.1.ii (Boardman–Vogt [21]). Consider the set of 1-simplices in a quasi-category with initial vertex a and final vertex b.

(i) Prove that the relation defined by $f \sim g$ if and only if there exists a

2-simplex with boundary

 is an equivalence relation.

(ii) Prove that the relation defined by $f \sim g$ if and only if there exists a

2-simplex with boundary

 is an equivalence relation.

(iii) Prove that the equivalence relations defined by (i) and (ii) are the same.

This proves Lemma 1.1.9.

EXERCISE 1.1.iii. Consider the free category on the reflexive directed graph

$$A_1 \underset{\cdot\delta^0}{\overset{\cdot\delta^1}{\underset{\longrightarrow}{\longleftarrow}}} A_0,$$

underlying a quasi-category A.

(i) Consider the binary relation that identifies sequences of composable 1-simplices with common source and common target whenever there exists a simplex of A in which the sequences of 1-simplices define two paths from its initial vertex to its final vertex. Prove that this relation is stable under pre- and post-composition with 1-simplices and conclude that its transitive closure is a **congruence**: an equivalence relation that is closed under pre- and post-composition.[9]

(ii) Consider the congruence relation generated by imposing a composition relation $h = g \circ f$ witnessed by 2-simplices

$$\begin{array}{ccc} & a_1 & \\ {}^{f}\nearrow & & \searrow^{g} \\ a_0 & \xrightarrow{\quad h \quad} & a_2 \end{array}$$

and prove that this coincides with the relation considered in (i).

(iii) In the congruence relations of (i) and (ii), prove that every sequence of composable 1-simplices in A is equivalent to a single 1-simplex. Conclude that every morphism in the quotient of the free category by this congruence relation is represented by a 1-simplex in A.

(iv) Prove that for any triple of 1-simplices f, g, h in A, $h = g \circ f$ in the homotopy category hA of Definition 1.1.10 if and only if there exists a 2-simplex with boundary

$$\begin{array}{ccc} & a_1 & \\ {}^{f}\nearrow & & \searrow^{g} \\ a_0 & \xrightarrow{\quad h \quad} & a_2 \end{array}$$

This proves Lemma 1.1.12.

EXERCISE 1.1.iv. Show that any quasi-category in which inner horns admit unique fillers is isomorphic to the nerve of its homotopy category.

EXERCISE 1.1.v. Let $\mathbb{I}$ be the nerve of the free-living isomorphism.

---

[9] Given a congruence relation on the hom-sets of a 1-category, the quotient category can be formed by quotienting each hom-set (see [81, §II.8]).

(i) Prove that 𝕀 contains exactly two nondegenerate simplices in each dimension.

(ii) Inductively build 𝕀 from 2 by expressing the inclusion $2 \hookrightarrow 𝕀$ as a sequential composite of pushouts of left outer horn inclusions[10] $\Lambda^0[n] \hookrightarrow \Delta[n]$, one in each dimension starting with $n = 2$.[11]

EXERCISE 1.1.vi. Prove the relative version of Corollary 1.1.16: for any isofibration $p : A \twoheadrightarrow B$ between quasi-categories and any $f : 2 \to A$ that defines an isomorphism in A any homotopy coherent isomorphism in B extending pf lifts to a homotopy coherent isomorphism in A extending f.

EXERCISE 1.1.vii. Specialize Proposition 1.1.20 to prove the following:

(i) If A is a quasi-category and X is a simplicial set then A^X is a quasi-category.

(ii) If A is a quasi-category and $X \hookrightarrow Y$ is a monomorphism then $A^Y \twoheadrightarrow A^X$ is an isofibration.

(iii) If $A \twoheadrightarrow B$ is an isofibration and X is a simplicial set then $A^X \twoheadrightarrow B^X$ is an isofibration.

EXERCISE 1.1.viii. Anticipating Lemma 1.2.17:

(i) Prove that the equivalences defined in Definition 1.1.23 are closed under retracts.

(ii) Prove that the equivalences defined in Definition 1.1.23 satisfy the 2-of-3 property.

EXERCISE 1.1.ix. Prove that if $f : X \xrightarrow{\sim\!\!\!\twoheadrightarrow} Y$ is a trivial fibration between quasi-categories then the functor $hf : hX \xrightarrow{\sim\!\!\!\twoheadrightarrow} hY$ is a surjective equivalence of categories.

1.2 ∞-Cosmoi

In §1.1, we presented "analytic" proofs of a few of the basic facts about quasi-categories. The category theory of quasi-categories can be developed in a similar

---

[10] By the duality described in Definition 1.2.25, the right outer horn inclusions $\Lambda^n[n] \hookrightarrow \Delta[n]$ can be used instead.

[11] This decomposition of the inclusion $2 \hookrightarrow 𝕀$ reveals which data extends homotopically uniquely to a homotopy coherent isomorphism. For instance, the 1- and 2-simplices of Definition 1.1.13 together with a single 3-simplex that has these as its outer faces with its inner faces degenerate. Homotopy type theorists refer to this data as a **half adjoint equivalence** [125, §4.2].

style, but we aim instead to develop the "synthetic" theory of infinite-dimensional categories, so that our results apply to many models at once. To achieve this, our strategy is not to axiomatize what infinite-dimensional categories *are*, but rather to axiomatize the categorical "universe" in which they *live*.

The definition of an ∞-*cosmos* abstracts the properties of the category of quasi-categories together with the isofibrations, equivalences, and trivial fibrations introduced in §1.1.[12] First, the category of quasi-categories is *enriched* over the category of simplicial sets – the set of morphisms from A to B coincides with the set of vertices of the simplicial set B^A – and moreover these hom spaces are all quasi-categories. Second, certain limit constructions that can be defined in the underlying unenriched category of quasi-categories satisfy universal properties relative to this simplicial enrichment, with the usual isomorphism of sets extending to an isomorphism of simplicial sets. And finally, the isofibrations, equivalences, and trivial fibrations satisfy properties that are familiar from abstract homotopy theory, forming a *category of fibrant objects* à la Brown [23] (see §C.1). In particular, the use of isofibrations in diagrams guarantees that their strict limits are equivalence invariant, so we can take advantage of up-to-isomorphism universal properties and strict functoriality of these constructions while still working "homotopically."

As explained in Digression 1.2.13, there are a variety of models of infinite-dimensional categories for which the category of "∞-categories," as we call them, and "∞-functors" between them is enriched over quasi-categories and admits classes of isofibrations, equivalences, and trivial fibrations satisfying analogous properties. This motivates the following axiomatization:

DEFINITION 1.2.1 (∞-cosmos). An ∞-**cosmos** $\mathcal{K}$ is a category that is enriched over quasi-categories,[13] meaning in particular that

- its morphisms $f : A \to B$ define the vertices of a quasi-category denoted $\mathrm{Fun}(A, B)$ and referred to as a **functor space**,

that is also equipped with a specified collection of maps that we call **isofibrations** and denote by "↠" satisfying the following two axioms:

(i) (completeness) The quasi-categorically enriched category $\mathcal{K}$ possesses a terminal object, small products, pullbacks of isofibrations, limits of countable towers of isofibrations, and cotensors with simplicial sets, each

---

[12] Metaphorical allusions aside, our ∞-cosmoi resemble the fibrational cosmoi of Street [117].
[13] This is to say $\mathcal{K}$ is a simplicially enriched category (see Digression 1.2.4) whose hom spaces are all quasi-categories.

of these limit notions satisfying a universal property that is enriched over simplicial sets.[14]

(ii) (isofibrations) The isofibrations contain all isomorphisms and any map whose codomain is the terminal object; are closed under composition, product, pullback, forming inverse limits of towers, and Leibniz cotensors with monomorphisms of simplicial sets; and have the property that if $f : A \twoheadrightarrow B$ is an isofibration and X is any object then $\mathsf{Fun}(X, A) \twoheadrightarrow \mathsf{Fun}(X, B)$ is an isofibration of quasi-categories.

For ease of reference, we refer to the simplicially enriched limits of diagrams of isofibrations enumerated in (i) as the **cosmological limit notions**.

DEFINITION 1.2.2. In an ∞-cosmos $\mathcal{K}$, a morphism $f : A \to B$ is

- an **equivalence** just when the induced map $f_* : \mathsf{Fun}(X, A) \xrightarrow{\sim} \mathsf{Fun}(X, B)$ on functor spaces is an equivalence of quasi-categories for all $X \in \mathcal{K}$, and
- a **trivial fibration** just when f is both an isofibration and an equivalence.

These classes are denoted by "$\xrightarrow{\sim}$" and "$\twoheadrightarrow{\sim}$", respectively.

Put more concisely, one might say that an ∞-cosmos is a "quasi-categorically enriched category of fibrant objects" (see Definition C.1.1 and Example C.1.3).

CONVENTION 1.2.3 (∞-category, as a technical term). Henceforth, we recast ∞-**category** as a technical term to refer to an object in an arbitrary ambient ∞-cosmos. Similarly, we use the term ∞-**functor** – or more commonly the elision "**functor**" – to refer to a morphism $f : A \to B$ in an ∞-cosmos. This explains why we refer to the quasi-category $\mathsf{Fun}(A, B)$ between two ∞-categories in an ∞-cosmos as a "functor space": its vertices are the (∞-)functors from A to B.

DIGRESSION 1.2.4 (simplicial categories, §A.2). A **simplicial category** $\mathcal{A}$ is given by categories $\mathcal{A}_n$, with a common set of objects and whose arrows are called n-**arrows**, that assemble into a diagram $\mathbf{\Delta}^{\mathrm{op}} \to \mathcal{C}at$ of identity-on-objects functors

$$\cdots \mathcal{A}_3 \underset{\substack{\xleftarrow{\;-\,.\delta^2\,-\;} \\ \xrightarrow{\;-\,.\delta^2\,-\;} \\ \xleftarrow{\;-\,.\sigma^1\,-\;} \\ \xrightarrow{\;-\,.\delta^1\,-\;} \\ \xleftarrow{\;-\,.\sigma^0\,-\;} \\ \xrightarrow{\;-\,.\delta^0\,-\;}}}{\overset{\substack{\xrightarrow{\;-\,.\delta^3\,-\;}}}{}} \mathcal{A}_2 \underset{\substack{\xleftarrow{\;-\,.\sigma^1\,-\;} \\ \xrightarrow{\;-\,.\delta^1\,-\;} \\ \xleftarrow{\;-\,.\sigma^0\,-\;} \\ \xrightarrow{\;-\,.\delta^0\,-\;}}}{\overset{\substack{\xrightarrow{\;-\,.\delta^2\,-\;}}}{}} \mathcal{A}_1 \underset{\substack{\xleftarrow{\;-\,.\sigma^0\,-\;} \\ \xrightarrow{\;-\,.\delta^0\,-\;}}}{\overset{\substack{\xrightarrow{\;-\,.\delta^1\,-\;}}}{}} \mathcal{A}_0 =: \mathcal{A} \qquad (1.2.5)$$

The category $\mathcal{A}_0$ of 0-arrows is the **underlying category** of the simplicial category $\mathcal{A}$, which forgets the higher dimensional simplicial structure.

---

[14] We elaborate on these simplicially enriched limits in Digression 1.2.6.

The data of a simplicial category can equivalently be encoded by a **simplicially enriched category** with a set of objects and a simplicial set $A(x, y)$ of morphisms between each ordered pair of objects: an n-arrow in A_n from x to y corresponds to an n-simplex in $A(x, y)$ (see Exercise 1.2.i). Each endo-hom space contains a distinguished identity 0-arrow (the degenerate images of which define the corresponding identity n-arrows) and composition is required to define a simplicial map

$$A(y, z) \times A(x, y) \xrightarrow{\ \circ\ } A(x, z)$$

the single map encoding the compositions in each of the categories A_n and also the functoriality of the diagram (1.2.5). The composition is required to be associative and unital, in a sense expressed by the commutative diagrams of simplicial sets

$$A(y, z) \times A(x, y) \times A(w, x) \xrightarrow{\circ \times \mathrm{id}} A(x, z) \times A(w, x)$$

$$\mathrm{id} \times \circ \downarrow \qquad\qquad\qquad\qquad \downarrow \circ$$

$$A(y, z) \times A(w, y) \xrightarrow{\qquad \circ \qquad} A(w, z)$$

$$A(x, y) \xrightarrow{\mathrm{id}_y \times \mathrm{id}} A(y, y) \times A(x, y)$$

$$\mathrm{id} \times \mathrm{id}_x \downarrow \qquad\searrow^{\mathrm{id}}\qquad \downarrow \circ$$

$$A(x, y) \times A(x, x) \xrightarrow{\qquad \circ \qquad} A(x, y)$$

On account of the equivalence between these two presentations, the terms "simplicial category" and "simplicially enriched category" are generally taken to be synonyms.[15]

In particular, the underlying category $\mathcal{K}_0$ of an ∞-cosmos $\mathcal{K}$ is the category whose objects are the ∞-categories in $\mathcal{K}$ and whose morphisms are the 0-arrows, i.e., the vertices in the functor spaces. In all of the examples to appear in what follows, this recovers the expected category of ∞-categories in a particular model and functors between them.

DIGRESSION 1.2.6 (simplicially enriched limits, §A.4-A.5). Let A be a simplicial category. The **cotensor** of an object $A \in A$ by a simplicial set U is characterized by a natural isomorphism of simplicial sets

$$A(X, A^U) \cong A(X, A)^U \tag{1.2.7}$$

---

[15] The phrase "simplicial object in $\mathcal{C}at$" is reserved for the more general yet less common notion of a diagram $\Delta^{\mathrm{op}} \to \mathcal{C}at$ that is not necessarily comprised of identity-on-objects functors.

Assuming such objects exist, the simplicial cotensor defines a bifunctor

$$s\mathcal{S}et^{\mathrm{op}} \times \mathcal{A} \longrightarrow \mathcal{A}$$

$$(U, A) \longmapsto A^U$$

in a unique way making the isomorphism (1.2.7) natural in U and A as well.

The other simplicial limit notions postulated by axiom 1.2.1(i) are **conical**, which is the term used for ordinary 1-categorical limit shapes that satisfy an enriched analog of the usual universal property (see Definition A.5.2). Such limits also define limits in the underlying category, but the usual universal property is strengthened. By applying the covariant representable functor $\mathcal{A}(X, -) \colon \mathcal{A}_0 \to s\mathcal{S}et$ to a limit cone $(\lim_{j \in J} A_j \to A_j)_{j \in J}$ in $\mathcal{A}_0$, we obtain a natural comparison map

$$\mathcal{A}(X, \lim_{j \in J} A_j) \longrightarrow \lim_{j \in J} \mathcal{A}(X, A_j). \tag{1.2.8}$$

We say that $\lim_{j \in J} A_j$ defines a **simplicially enriched limit** if and only if (1.2.8) is an isomorphism of simplicial sets for all $X \in \mathcal{A}$.

The general theory of enriched categories is reviewed in Appendix A.

PREVIEW 1.2.9 (flexible weighted limits in ∞-cosmoi). The axiom 1.2.1(i) implies that any ∞-cosmos $\mathcal{K}$ admits all *flexible limits*, a much larger class of simplicially enriched "weighted" limits (see Definition 6.2.1 and Proposition 6.2.8).

We quickly introduce the three examples of ∞-cosmoi that are most easily absorbed, deferring more sophisticated examples to the end of this section. The first of these is the prototypical ∞-cosmos.

PROPOSITION 1.2.10 (the ∞-cosmos of quasi-categories). *The full subcategory $\mathcal{QC}at \subset s\mathcal{S}et$ of quasi-categories defines an ∞-cosmos in which the isofibrations, equivalences, and trivial fibrations coincide with the classes already bearing these names.*

Proof The subcategory $\mathcal{QC}at \subset s\mathcal{S}et$ inherits its simplicial enrichment from the cartesian closed category of simplicial sets: by Proposition 1.1.20, whenever A and B are quasi-categories, $\mathsf{Fun}(A, B) := B^A$ is again a quasi-category.

The cosmological limits postulated in 1.2.1(i) exist in the ambient category of simplicial sets.[16] For instance, the defining universal property of the simplicial cotensor (1.2.7) is satisfied by the exponentials of simplicial sets. Moreover,

---

[16] Any category of presheaves is cartesian closed, complete, and cocomplete – a "cosmos" in the sense of Bénabou.

since the category of simplicial sets is cartesian closed, each of the conical limits is simplicially enriched in the sense discussed in Digression 1.2.6 (see Exercise 1.2.ii and Proposition A.5.4).

We now argue that the full subcategory of quasi-categories inherits all these limit notions and at the same time establish the stability of the isofibrations under the formation of these limits. In fact, this latter property helps to prove the former. To see this, note that a simplicial set is a quasi-category if and only if the map from it to the point is an isofibration. More generally, if the codomain of any isofibration is a quasi-category then its domain must be as well. So if any of the maps in a limit cone over a diagram of quasi-categories are isofibrations, then it follows that the limit is itself a quasi-category.

Since the isofibrations are characterized by a right lifting property, Lemma C.2.3 implies that the isofibrations contains all isomorphism and are closed under composition, product, pullback, and forming inverse limits of towers. In particular, the full subcategory of quasi-categories possesses these limits. This verifies all of the axioms of 1.2.1(i) and 1.2.1(ii) except for the last two: Leibniz closure and closure under exponentiation $(-)^X$. These last closure properties are established in Proposition 1.1.20, and in fact by Exercise 1.1.vii, the former subsumes the latter . This completes the verification of the ∞-cosmos axioms.

It remains to check that the equivalences and trivial fibrations coincide with those maps defined by 1.1.23 and 1.1.25. By Proposition 1.1.28 the latter coincidence follows from the former, so it remains only to show that the equivalences of 1.1.23 coincide with the **representably defined equivalences**: those maps of quasi-categories $f : A \to B$ for which $A^X \to B^X$ is an equivalence of quasi-categories in the sense of Definition 1.1.23. Taking $X = \Delta[0]$, we see immediately that representably defined equivalences are equivalences, and the converse holds since the exponential $(-)^X$ preserves the data defining a simplicial homotopy. $\qquad\Box$

Two further examples fit into a common paradigm: both arise as full subcategories of the ∞-cosmos of quasi-categories and inherit their ∞-cosmos structures from this inclusion (see Lemma 6.1.4). But it is also instructive, and ultimately takes less work, to describe the resulting ∞-cosmos structures directly.

PROPOSITION 1.2.11 (the ∞-cosmos of categories). *The category 𝒞at of 1-categories defines an ∞-cosmos whose isofibrations are the* **isofibrations:** *functors* *satisfying the displayed right lifting property:*

*The equivalences are the equivalences of categories and the trivial fibrations are **surjective equivalences***: *equivalences of categories that are also surjective on objects.*

Proof   It is well-known that the 2-category of categories is complete (and in fact also cocomplete) as a $\mathcal{C}at$-enriched category (see Definition A.6.17 or [67]). The categorically enriched category of categories becomes a quasi-categorically enriched category by applying the nerve functor to the hom-categories (see §A.7). Since the nerve functor is a right adjoint, it follows formally that these 2-categorical limits become simplicially enriched limits. In particular, as proscribed in Proposition A.7.8, the cotensor of a category A by a simplicial set U is defined to be the functor category A^{hU}. This completes the verification of axiom (i).

Since the class of isofibrations is characterized by a right lifting property, Lemma C.2.3 implies that the isofibrations are closed under all of the limit constructions of 1.2.1(ii) except for the last two, and by Exercise 1.1.vii, the Leibniz closure subsumes the closure under exponentiation.

To verify that isofibrations of categories $f : A \twoheadrightarrow B$ are stable under forming Leibniz cotensors with monomorphisms of simplicial sets $i : U \hookrightarrow V$, we must solve the lifting problem below-left

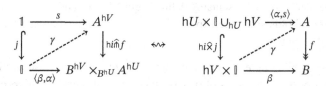

which transposes to the lifting problem above-right, which we can solve by hand. Here the map β defines a natural isomorphism between $fs\colon hV \to B$ and a second functor. Our task is to lift this to a natural isomorphism γ from s to another functor that extends the natural isomorphism α along $hi\colon hU \to hV$. Note this functor hi need not be an inclusion, but it is injective on objects, which is enough.

We define the components of γ by cases. If an object $v \in hV$ is equal to $i(u)$ for some $u \in hU$ define $\gamma_{i(u)} := \alpha_u$; otherwise, use the fact that f is an isofibration to define γ_v to be any lift of the isomorphism β_v to an isomorphism in A with domain $s(v)$. The data of the map $\gamma : hV \times \mathbb{I} \to A$ also entails the specification of the functor $hV \to A$ that is the codomain of the natural isomorphism γ. On objects, this functor is given by $v \mapsto \text{cod}(\gamma_v)$. On morphisms, this functor defined in the unique way that makes γ into a natural transformation:

$$(k : v \to v') \mapsto \gamma_{v'} \circ s(k) \circ \gamma_v^{-1}.$$

This completes the proof that $\mathcal{C}at$ defines an ∞-cosmos. Since the nerve of a functor category, such as $A^{\mathbb{I}}$, is isomorphic to the exponential between their nerves, the equivalences of categories coincide with the equivalences of Definition 1.1.23. It follows that the equivalences in the ∞-cosmos of categories coincide with equivalences of categories, and since the surjective equivalences are the intersection of the equivalences and the isofibrations, this completes the proof. □

PROPOSITION 1.2.12 (the ∞-cosmos of Kan complexes). *The category $\mathcal{K}an$ of Kan complexes defines an ∞-cosmos whose isofibrations are the **Kan fibrations**: maps that lift against all horn inclusions $\Lambda^k[n] \hookrightarrow \Delta[n]$ for $n \geq 1$ and $0 \leq k \leq n$.*

The proof proceeds along the lines of Lemma 6.1.4. We show that the subcategory of Kan complexes inherits an ∞-cosmos structure by restricting structure from the ∞-cosmos of quasi-categories.

Proof By Proposition 1.1.18, an isofibration between Kan complexes is a Kan fibration. Conversely, since the homotopy coherent isomorphism $\mathbb{I}$ can be built from the point $\mathbb{1}$ by attaching fillers to a sequence of outer horns, all Kan fibrations define isofibrations. This shows that between Kan complexes, isofibrations and Kan fibrations coincide. So to show that the category of Kan complexes inherits an ∞-cosmos structure by restriction from the ∞-cosmos of quasi-categories, we need only verify that the full subcategory $\mathcal{K}an \hookrightarrow \mathcal{Q}\mathcal{C}at$ is closed under all of the limit constructions of axiom 1.2.1(i). For the conical limits, the argument mirrors the one given in the proof of Proposition 1.2.10, while the closure under cotensors is a consequence of Corollary D.3.11, which implies that the Kan complexes also define an exponential ideal in the category of simplicial sets. The remaining axiom 1.2.1(ii) is inherited from the analogous properties established for quasi-categories in Proposition 1.2.10. □

We mention a common source of ∞-cosmoi found in nature to build intuition for readers familiar with Quillen's model categories, a popular framework for abstract homotopy theory, but reassure newcomers that model categories are not needed outside of Appendix E where these results are proven.

DIGRESSION 1.2.13 (a source of ∞-cosmoi in nature). As explained in §E.1, certain easily described properties of a model category imply that the full subcategory of fibrant objects defines an ∞-cosmos whose isofibrations, equivalences, and trivial fibrations are the fibrations, weak equivalences, and trivial fibrations between fibrant objects. Namely, any model category that is enriched as such over the Joyal model structure on simplicial sets in which all fibrant objects are cofibrant presents an ∞-cosmos (see Proposition E.1.1). This model-categorical enrichment over quasi-categories can be defined when the model category is cartesian closed and equipped with a right Quillen adjoint to the Joyal model structure on simplicial sets whose left adjoint preserves finite products (see Corollary E.1.4). In this case, the right adjoint becomes the underlying quasi-category functor (see Proposition 1.3.4(ii)) and the ∞-cosmoi so-produced is cartesian closed (see Definition 1.2.23). The ∞-cosmoi listed in Example 1.2.24 all arise in this way.

The following results are consequences of the axioms of Definition 1.2.1. To begin, observe that the trivial fibrations enjoy the same stability properties satisfied by the isofibrations.

LEMMA 1.2.14 (stability of trivial fibrations). *The trivial fibrations in an ∞-cosmos define a subcategory containing the isomorphisms and are stable under product, pullback, and forming inverse limits of towers. Moreover, the Leibniz cotensors of any trivial fibration with a monomorphism of simplicial sets is a trivial fibration as is the Leibniz cotensor of an isofibration with a map in the class cellularly generated by the inner horn inclusions and the map $\mathbb{1} \hookrightarrow \mathbb{I}$, and if $E \twoheadrightarrow B$ is a trivial fibration then so is* $\mathrm{Fun}(X, E) \twoheadrightarrow \mathrm{Fun}(X, B)$.

Proof   We prove these statements in the reverse order. By axiom 1.2.1(ii) and the definition of the trivial fibrations in an ∞-cosmos, we know that if $E \twoheadrightarrow B$ is a trivial fibration then $\mathrm{Fun}(X, E) \twoheadrightarrow \mathrm{Fun}(X, B)$ is both an isofibration and an equivalence, and hence by Proposition 1.1.28 a trivial fibration. For stability under the remaining constructions, we know in each case that the maps in question are isofibrations in the ∞-cosmos; it remains to show only that the maps are also equivalences. The equivalences in an ∞-cosmos are defined to be the maps that $\mathrm{Fun}(X, -)$ carries to equivalences of quasi-categories, so it suffices to verify that trivial fibrations of quasi-categories satisfy the corresponding stability properties. For the Leibniz stability properties, this is established in Proposition 1.1.29, while the remaining properties are covered by Lemma C.2.3.   □

By a Yoneda-style argument, the "homotopy equivalence" characterization of the equivalences in the ∞-cosmos of quasi-categories of Definition 1.1.23 extends to an analogous characterization of the equivalences in any ∞-cosmos:

LEMMA 1.2.15 (equivalences are homotopy equivalences). *A map $f : A \to B$ between ∞-categories in an ∞-cosmos $\mathcal{K}$ is an equivalence if and only if it extends to the data of a "homotopy equivalence" with the free-living isomorphism $\mathbb{I}$ serving as the interval: that is, if there exist maps $g : B \to A$*

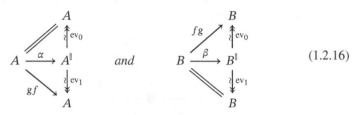

$$(1.2.16)$$

in the ∞-cosmos.

Proof By hypothesis, if $f : A \to B$ defines an equivalence in the ∞-cosmos $\mathcal{K}$ then the induced map on post-composition $f_* : \mathrm{Fun}(B, A) \xrightarrow{\sim} \mathrm{Fun}(B, B)$ is an equivalence of quasi-categories in the sense of Definition 1.1.23. Evaluating the inverse equivalence $\tilde{g} : \mathrm{Fun}(B, B) \xrightarrow{\sim} \mathrm{Fun}(B, A)$ and homotopy $\tilde{\beta} : \mathrm{Fun}(B, B) \to \mathrm{Fun}(B, B)^{\mathbb{I}}$ at the 0-arrow $\mathrm{id}_B \in \mathrm{Fun}(B, B)$, we obtain a 0-arrow $g : B \to A$ together with an isomorphism $\beta : \mathbb{I} \to \mathrm{Fun}(B, B)$ from the composite fg to id_B. By the defining universal property of the cotensor (1.2.7), this isomorphism internalizes to define the map $\beta : B \to B^{\mathbb{I}}$ in $\mathcal{K}$ displayed on the right of (1.2.16).

Now the hypothesis that f is an equivalence also provides an equivalence of quasi-categories $f_* : \mathrm{Fun}(A, A) \xrightarrow{\sim} \mathrm{Fun}(A, B)$, and the map $\beta f : A \to B^{\mathbb{I}}$ represents an isomorphism in $\mathrm{Fun}(A, B)$ from fgf to f. Since f_* is an equivalence, we conclude from Remark 1.1.24 that id_A and gf are isomorphic in the quasi-category $\mathrm{Fun}(A, A)$: explicitly, such an isomorphism may be defined by applying the inverse equivalence $\tilde{h} : \mathrm{Fun}(A, B) \to \mathrm{Fun}(A, A)$ and composing with the components at $\mathrm{id}_A, gf \in \mathrm{Fun}(A, A)$ of the isomorphism $\tilde{\alpha} : \mathrm{Fun}(A, A) \to \mathrm{Fun}(A, A)^{\mathbb{I}}$ from $\mathrm{id}_{\mathrm{Fun}(A,A)}$ to $\tilde{h} f_*$. Now by Corollary 1.1.16 this isomorphism is represented by a map $\mathbb{I} \to \mathrm{Fun}(A, A)$ from id_A to gf, which internalizes to a map $\alpha : A \to A^{\mathbb{I}}$ in $\mathcal{K}$ displayed on the left of (1.2.16).

The converse is easy: the simplicial cotensor construction commutes with $\mathrm{Fun}(X, -)$, so a homotopy equivalence (1.2.16) induces a homotopy equivalence of quasi-categories as in Definition 1.1.23. $\square$

LEMMA 1.2.17. *The equivalences in an ∞-cosmos are closed under retracts and satisfy the **2-of-3 property**: given a composable pair of functors and their composite, if any two of these are equivalences so is the third.*

By the representable definition of equivalences and functoriality, Lemma 1.2.17 follows easily from the corresponding results for equivalences between quasi-categories (see Exercise 1.1.viii). But for sake of completeness, we prove the general cosmological result without relying on this base case, subsuming Exercise 1.1.viii.

Proof   Let $f : A \overset{\sim}{\to} B$ be an equivalence equipped with the data of (1.2.16) and consider a retract diagram

$$
\begin{array}{ccccc}
C & \overset{u}{\longrightarrow} & A & \overset{v}{\longrightarrow} & C \\
h \downarrow & & f \downarrow \wr & & \downarrow h \\
D & \overset{s}{\longrightarrow} & B & \overset{t}{\longrightarrow} & D
\end{array}
$$

By Lemma 1.2.15, to prove that $h : C \to D$ is an equivalence, it suffices to construct the data of an inverse homotopy equivalence. To that end define $k : D \to C$ to be the composite vgs and then observe from the commutative diagrams

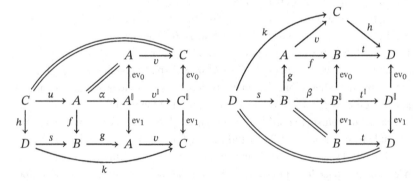

that $v^{\mathbb{I}}\alpha u : C \to C^{\mathbb{I}}$ and $t^{\mathbb{I}}\beta s : D \to D^{\mathbb{I}}$ define the required homotopy coherent isomorphisms.

Via Lemma 1.2.15, the 2-of-3 property for equivalences follows from the fact that the set of isomorphisms in a quasi-category is closed under composition. Homotopy coherent isomorphisms in a quasi-category represent isomorphisms in the homotopy category, whose composite in the homotopy category is then an isomorphism, which can be lifted to a representing homotopy coherent isomorphism by Corollary 1.1.16.[17] We now apply this to the homotopy coherent isomorphisms in the functor spaces of an ∞-cosmos that form part of the data of an equivalence of ∞-categories.

---

[17] In fact, by Example D.5.5, homotopy coherent isomorphisms can be composed directly, but we do not need this here.

To prove that equivalences are closed under composition, consider a composable pair of equivalences with their inverse equivalences

$$A \underset{\underset{k}{\sim}}{\overset{f}{\underset{\sim}{\longrightarrow}}} B \underset{\underset{h}{\sim}}{\overset{g}{\underset{\sim}{\longrightarrow}}} C$$

The equivalence data of Lemma 1.2.15 defines isomorphisms $\alpha : \operatorname{id}_A \cong kf \in$ Fun(A, A) and $\gamma : \operatorname{id}_B \cong hg \in$ Fun(B, B), the latter of which whiskers to define $k\gamma f : kf \cong khgf \in$ Fun(B, B). Composing these, we obtain an isomorphism $\operatorname{id}_A \cong khgf \in$ Fun(A, A), witnessing that kh defines a left equivalence inverse of gf. The other isomorphism is constructed similarly.

To prove that the equivalences are closed under right cancelation, consider a diagram

$$A \underset{\underset{k}{\sim}}{\overset{f}{\underset{\sim}{\longrightarrow}}} \overset{\overset{\ell}{\overset{\sim}{\frown}}}{B} \overset{g}{\longrightarrow} C$$

with k an inverse equivalence to f and ℓ and inverse equivalence to gf. We claim that $f\ell$ defines an inverse equivalence to g. One of the required isomorphisms $\operatorname{id}_C \cong gf\ell$ is given already. The other is obtained by composing three isomorphisms in Fun(B, B)

$$\operatorname{id}_B \overset{\beta^{-1}}{\underset{\approx}{\longrightarrow}} fk \overset{f\delta k}{\underset{\approx}{\longrightarrow}} f\ell gfk \overset{f\ell g\beta}{\underset{\approx}{\longrightarrow}} f\ell g.$$

The proof of stability of equivalence under left cancelation is dual. □

The trivial fibrations admit a similar characterization as split fiber homotopy equivalences.

LEMMA 1.2.18 (trivial fibrations split). *Every trivial fibration admits a section*

$$\begin{array}{ccc} & & E \\ & \overset{s}{\nearrow} & \downarrow p \\ B & = \!\!\!= \!\!\!= & B \end{array}$$

that defines a split fiber homotopy equivalence

$$\begin{array}{ccc} E & \overset{(\operatorname{id}_E, sp)}{\overset{\longrightarrow}{\underset{\alpha}{\longrightarrow}}} E^{\mathbb{I}} \overset{}{\underset{(\operatorname{ev}_0, \operatorname{ev}_1)}{\longrightarrow\!\!\!\longrightarrow}} E \times E \\ p \downarrow & \quad p^{\mathbb{I}} \downarrow & \\ B & \overset{\longrightarrow}{\underset{\Delta}{\longrightarrow}} B^{\mathbb{I}} & \end{array}$$

and conversely any isofibration that defines a split fiber homotopy equivalence is a trivial fibration.

Proof If $p : E \twoheadrightarrow B$ is a trivial fibration, then by the final stability property of Lemma 1.2.14, so is $p_* : \mathrm{Fun}(X, E) \twoheadrightarrow \mathrm{Fun}(X, B)$ for any ∞-category X. By Definition 1.1.25, we may solve the lifting problem below-left

$$
\begin{array}{ccc}
\varnothing = \partial\Delta[0] & \longrightarrow & \mathrm{Fun}(B, E) \\
\Big\downarrow & \nearrow^{s} & \Big\downarrow{\scriptstyle p_*} \\
\mathbb{1} = \Delta[0] & \xrightarrow[\mathrm{id}_B]{} & \mathrm{Fun}(B, B)
\end{array}
\qquad
\begin{array}{ccc}
\mathbb{1} + \mathbb{1} & \xrightarrow{(\mathrm{id}_E, sp)} & \mathrm{Fun}(E, E) \\
\Big\downarrow & \nearrow^{\alpha} & \Big\downarrow{\scriptstyle p_*} \\
\mathbb{I} \xleftarrow{!} \mathbb{1} & \xrightarrow{p} & \mathrm{Fun}(E, B)
\end{array}
$$

to find a map $s : B \to E$ so that $ps = \mathrm{id}_B$, and then solve the lifting problem above-right to construct the desired fibered homotopy. The converse is immediate from Lemma 1.2.15. □

A classical construction in abstract homotopy theory proves the following:

LEMMA 1.2.19 (Brown factorization lemma). *Any functor $f : A \to B$ in an ∞-cosmos may be factored as an equivalence followed by an isofibration, where this equivalence is constructed as a section of a trivial fibration.*

$$
\begin{array}{ccc}
 & Pf & \\
{}^{q}\nearrow \;{}_{s}\Big\uparrow\; \searrow^{p} & & \\
A & \xrightarrow[f]{} & B
\end{array}
\tag{1.2.20}
$$

Moreover, f is an equivalence if and only if the isofibration p is a trivial fibration.

Proof The displayed factorization is constructed by the pullback of an isofibration formed by the simplicial cotensor of the inclusion $\mathbb{1} + \mathbb{1} \hookrightarrow \mathbb{I}$ into the ∞-category B.

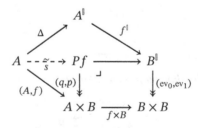

Note the map q is a pullback of the trivial fibration $\mathrm{ev}_0 : B^{\mathbb{I}} \twoheadrightarrow B$ and is hence a trivial fibration. Its section s, constructed by applying the universal property of the pullback to the displayed cone with summit A, is thus an equivalence by the

2-of-3 property. Again by 2-of-3, it follows that f is an equivalence if and only if p is. □

REMARK 1.2.21 (equivalences satisfy the 2-of-6 property). In fact the equivalences in any ∞-cosmos satisfy the stronger **2-of-6 property**: for any composable triple of functors

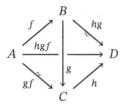

if gf and hg are equivalences then f, g, h, and hgf are too. An argument of Blumberg and Mandell [20, 6.4] reproduced in Proposition C.1.8 uses Lemmas 1.2.17, 1.2.18, and 1.2.19 to prove that the equivalences have the 2-of-6 property (see Corollary C.1.9).

One of the key advantages of the ∞-cosmological approach to abstract category theory is that there are a myriad varieties of "fibered" ∞-cosmoi that can be built from a given ∞-cosmos, which means that any theorem proven in this axiomatic framework specializes and generalizes to those contexts. The most basic of these derived ∞-cosmoi is the ∞-cosmos of isofibrations over a fixed base, which we introduce now. Other examples of ∞-cosmoi are developed in Chapter 6, once we have a deeper understanding of the cosmological limits of axiom 1.2.1(i).

PROPOSITION 1.2.22 (sliced ∞-cosmoi). *For any ∞-cosmos $\mathcal{K}$ and any ∞-category $B \in \mathcal{K}$ there is an ∞-cosmos $\mathcal{K}_{/B}$ of isofibrations over B whose*

(i) objects are isofibrations $p : E \twoheadrightarrow B$ with codomain B
(ii) functor spaces, say from $p : E \twoheadrightarrow B$ to $q : F \twoheadrightarrow B$, are defined by pullback

$$
\begin{array}{ccc}
\mathrm{Fun}_B(p : E \twoheadrightarrow B, q : F \twoheadrightarrow B) & \longrightarrow & \mathrm{Fun}(E,F) \\
\downarrow & \lrcorner & \downarrow{\scriptstyle q_*} \\
\mathbb{1} & \xrightarrow{\quad p \quad} & \mathrm{Fun}(E,B)
\end{array}
$$

and abbreviated to $\mathrm{Fun}_B(E, F)$ when the specified isofibrations are clear from context

(iii) *isofibrations are commutative triangles of isofibrations over B*

$$E \xrightarrow{\;\;r\;\;} F$$
$$p \searrow \quad \swarrow q$$
$$B$$

(iv) *terminal object is* id : $B \twoheadrightarrow B$ *and products are defined by the pullback along the diagonal*

$$\times_i^B E_i \longrightarrow \prod_i E_i$$
$$\downarrow \qquad\quad \downarrow \Pi_i p_i$$
$$B \xrightarrow{\;\Delta\;} \prod_i B$$

(v) *pullbacks and limits of towers of isofibrations are created by the forgetful functor* $\mathcal{K}_{/B} \to \mathcal{K}$

(vi) *simplicial cotensor of* $p : E \twoheadrightarrow B$ *with* $U \in sSet$ *is constructed by the pullback*

$$U \pitchfork_B p \longrightarrow E^U$$
$$\downarrow \qquad\qquad \downarrow p^U$$
$$B \xrightarrow{\;\Delta\;} B^U$$

(vii) *and in which a map over B*

$$E \xrightarrow{\;\;f\;\;} F$$
$$p \searrow \quad \swarrow q$$
$$B$$

is an equivalence in the ∞-*cosmos* $\mathcal{K}_{/B}$ *if and only if f is an equivalence in* $\mathcal{K}$.

Proof   The functor spaces are quasi-categories since axiom 1.2.1(ii) asserts that for any isofibration $q : F \twoheadrightarrow B$ in $\mathcal{K}$ the map $q_* : \mathrm{Fun}(E, F) \twoheadrightarrow \mathrm{Fun}(E, B)$ is an isofibration of quasi-categories. Other parts of this axiom imply that each of the limit constructions – such as the products and cotensors constructed in (iv) and (vi) – define isofibrations over B. The closure properties of the isofibrations in $\mathcal{K}_{/B}$ follow from the corresponding ones in $\mathcal{K}$. The most complicated of these is the Leibniz cotensor stability of the isofibrations in $\mathcal{K}_{/B}$, which follows from the corresponding property in $\mathcal{K}$, since for a monomorphism of simplicial sets $i : X \hookrightarrow Y$ and an isofibration r over B as in (iii) above, the map $i \widehat{\pitchfork}_B r$ is constructed by pulling back $i \widehat{\pitchfork} r$ along $\Delta : B \to B^Y$.

The fact that the above constructions define simplicially enriched limits in a simplicially enriched slice category are standard from enriched category theory.

It remains only to verify that the equivalences in the ∞-cosmos of isofibrations are created by the forgetful functor $\mathcal{K}_{/B} \to \mathcal{K}$. Suppose first that the map f displayed in (vii) defines an equivalence in $\mathcal{K}$. Then for any isofibration $s : A \twoheadrightarrow B$ the induced map on functor spaces in $\mathcal{K}_{/B}$ is defined by the pullback:

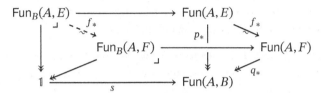

Since f is an equivalence in $\mathcal{K}$, the map $f_* : \mathrm{Fun}(A, E) \to \mathrm{Fun}(A, F)$ is an equivalence, and so it follows that the induced map on fibers over s is an equivalence as well.[18]

For the converse implication, we appeal to Lemma 1.2.15. If $f : E \to F$ is an equivalence in $\mathcal{K}_{/B}$ then it admits a homotopy inverse in $\mathcal{K}_{/B}$. The inverse equivalence $g : F \to E$ also defines an inverse equivalence in $\mathcal{K}$ and the required simplicial homotopies in $\mathcal{K}$

$$E \xrightarrow{\alpha} \mathbb{1} \pitchfork_B p \longrightarrow E^{\mathbb{I}} \qquad F \xrightarrow{\beta} \mathbb{1} \pitchfork_B q \to F^{\mathbb{I}}$$

are defined by composing with the top horizontal leg of the pullback defining the cotensor in $\mathcal{K}_{/B}$. □

As mentioned in Digression 1.2.13, many of the ∞-cosmoi we encounter "in the wild" satisfy an additional axiom. Note, however, that this axiom is not inherited by the sliced ∞-cosmos of Proposition 1.2.22, which is one of the reasons it was not included in Definition 1.2.1.

DEFINITION 1.2.23 (cartesian closed ∞-cosmoi). An ∞-cosmos $\mathcal{K}$ is **cartesian closed** if the product bifunctor $- \times - : \mathcal{K} \times \mathcal{K} \to \mathcal{K}$ extends to a simplicially enriched two-variable adjunction

$$\mathrm{Fun}(A \times B, C) \cong \mathrm{Fun}(A, C^B) \cong \mathrm{Fun}(B, C^A)$$

in which the right adjoints $(-)^A : \mathcal{K} \to \mathcal{K}$ preserve isofibrations for all $A \in \mathcal{K}$.

For instance, the ∞-cosmos of quasi-categories is cartesian closed, with the exponentials defined as (special cases of) simplicial cotensors. This is one of the reasons that we use the same notation for cotensor and for exponential.[19]

---

[18] The stability of equivalences between isofibrations under pullback can be proven either as a consequence of Lemmas 1.2.14 and 1.2.19 using standard techniques from simplicial homotopy theory (see Lemma C.1.11) or by arguing 2-categorically (see Proposition 3.3.4).

[19] Other advantages of this convenient notational conflation are discussed in §2.3 and in Proposition 10.3.5.

Note in this case the functor spaces and the exponentials coincide. The same is true for the cartesian closed ∞-cosmoi of categories and of Kan complexes. In general, the functor space from A to B is the "underlying quasi-category" of the exponential B^A whenever it exists (see Remark 1.3.11).

EXAMPLE 1.2.24 (∞-cosmoi of $(\infty, 1)$-categories; §E.2). The following models of $(\infty, 1)$-categories define cartesian closed ∞-cosmoi:

(i) Rezk's **complete Segal spaces** define the objects of an ∞-cosmos $\mathcal{CSS}$, in which the isofibrations, equivalences, and trivial fibrations are the corresponding classes of the model structure of [100].[20]

(ii) The **Segal categories** defined by Dwyer, Kan, and Smith [38] and developed by Hirschowitz and Simpson [56] define the objects of an ∞-cosmos $\mathcal{Segal}$, in which the isofibrations, equivalences, and trivial fibrations are the corresponding classes of the model structure of [13, 90].[21]

(iii) The **1-complicial sets** of [129], equivalently the "naturally marked quasi-categories" of [78], define the objects of an ∞-cosmos $1\text{-}\mathcal{Comp}$ in which the isofibrations, equivalences, and trivial fibrations are the corresponding classes of the model structure from either of these sources.

In §E.3, we show that certain models of (∞, n)-categories or even (∞, ∞)-categories define ∞-cosmoi: n-quasi-categories, Θ_n-spaces, iterated complete Segal spaces, and n-complicial sets.

DEFINITION 1.2.25 (co-dual ∞-cosmoi). There is an identity-on-objects involutive functor $(-)^\circ : \Delta \to \Delta$ that reverses the ordering of the elements in each ordinal $[n] \in \Delta$. In the notation of 1.1.1, the functor $(-)^\circ$ sends a face map $\delta^i : [n-1] \rightarrowtail [n]$ to the face map $\delta^{n-i} : [n-1] \rightarrowtail [n]$ and sends the degeneracy map $\sigma^i : [n+1] \twoheadrightarrow [n]$ to the degeneracy map $\sigma^{n-i} : [n+1] \twoheadrightarrow [n]$. Precomposition with this involutive automorphism induces an involution $(-)^{\mathrm{op}} : s\mathcal{Set} \to s\mathcal{Set}$ that sends a simplicial set X to its **opposite simplicial set** X^{op}, with the orientation of the vertices in each simplex reversed. This construction preserves all conical limits and colimits and induces an isomorphism $(Y^X)^{\mathrm{op}} \cong (Y^{\mathrm{op}})^{X^{\mathrm{op}}}$ on exponentials.

---

[20] Warning: the model category of complete Segal spaces is enriched over simplicial sets in two distinct "directions" – one enrichment makes the simplicial set of maps between two complete Segal spaces into a Kan complex that probes the "spacial" structure while another enrichment makes the simplicial set of maps into a quasi-category that probes the "categorical" structure [64]. It is this latter enrichment that we want.

[21] Here we reserve the term "Segal category" for those simplicial objects with a discrete set of objects that are Reedy fibrant and satisfy the Segal condition. The traditional definition does not include the Reedy fibrancy condition because it is not satisfied by the simplicial object defined as the nerve of a Kan complex enriched category. Since Kan complex enriched categories are not among our preferred models of $(\infty, 1)$-categories this does not bother us.

For any ∞-cosmos $\mathcal{K}$, there is a **dual ∞-cosmos** $\mathcal{K}^{co}$ with the same objects but with functor spaces defined by:

$$\mathsf{Fun}_{\mathcal{K}^{co}}(A, B) := \mathsf{Fun}_{\mathcal{K}}(A, B)^{op}.$$

The isofibrations, equivalences, and trivial fibrations in $\mathcal{K}^{co}$ coincide with those of $\mathcal{K}$.

Conical limits in $\mathcal{K}^{co}$ coincide with those in $\mathcal{K}$, while the cotensor of $A \in \mathcal{K}$ with $U \in s\mathcal{S}et$ is defined to be $A^{U^{op}}$.

A 2-categorical justification for this notation is given in Exercise 1.4.ii.

DEFINITION 1.2.26 (discrete ∞-categories). An ∞-category E in an ∞-cosmos $\mathcal{K}$ is **discrete** just when for all $X \in \mathcal{K}$ the functor space $\mathsf{Fun}(X, E)$ is a Kan complex.

In the ∞-cosmos of quasi-categories, the discrete ∞-categories are exactly the Kan complexes. Similarly, in the ∞-cosmoi of Example 1.2.24 whose ∞-categories are $(\infty, 1)$-categories in some model, the discrete ∞-categories are the ∞-groupoids. Importantly for what follows, the discrete ∞-categories can be characterized "internally" to the ∞-cosmos as follows:

LEMMA 1.2.27. *An ∞-category E is discrete if and only if $E^{\mathbb{I}} \twoheadrightarrow E^2$ is a trivial fibration.*

Proof By Definition 1.2.2, the isofibration $E^{\mathbb{I}} \twoheadrightarrow E^2$ is a trivial fibration if and only if for all ∞-categories X the induced map on functor spaces

$$\begin{array}{ccc}
\mathsf{Fun}(X, E^{\mathbb{I}}) & \longtwoheadrightarrow & \mathsf{Fun}(X, E^2) \\
\rotatebox{90}{$\cong$} & & \rotatebox{90}{$\cong$} \\
\mathsf{Fun}(X, E)^{\mathbb{I}} & \longtwoheadrightarrow & \mathsf{Fun}(X, E)^2
\end{array}$$

is a trivial fibration of quasi-categories. Via the universal property of the simplicial cotensor, Lemma 1.1.30 tells us that this map is a trivial fibration if and only if $\mathsf{Fun}(X, E)$ is a Kan complex. $\square$

The reader may check that the discrete ∞-categories in any ∞-cosmos assemble into an ∞-cosmos $\mathcal{K}^{\simeq}$. A proof appears in Proposition 6.1.6 where general techniques for producing new ∞-cosmoi from given ones are developed.

Exercises

EXERCISE 1.2.i. Define an equivalence between the categories of:

(i) simplicial categories, as in (1.2.5), and

(ii) categories enriched over simplicial sets.

EXERCISE 1.2.ii. Elaborate on the proof of Proposition 1.2.10 by proving that the simplicially enriched category $\mathcal{QC}at$ admits conical products satisfying the universal property of Digression 1.2.6. That is:

(i) Define the cartesian product $A \times B$ and the projection maps $\pi_A : A \times B \to A$ and $\pi_B : A \times B \to B$ for a pair of quasi-categories A and B and prove that this data satisfies the usual (unenriched) universal property.

(ii) Given another quasi-category X, use (i) and the Yoneda lemma to show that the projection maps induce an isomorphism of quasi-categories

$$(A \times B)^X \xrightarrow{\ \cong\ } A^X \times B^X.$$

(iii) Explain how this relates to the universal property of Digression 1.2.6.

(iv) Express the usual 1-categorical universal property of (i) as the "0-dimensional aspect" of the universal property of (ii).

EXERCISE 1.2.iii. Prove that any object in an ∞-cosmos has a **path object**

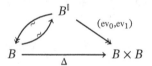

constructed by cotensoring with the free-living isomorphism.

EXERCISE 1.2.iv. Show that if $\mathcal{K}$ is a cartesian closed ∞-cosmos then $\mathcal{K}^{co}$ is as well.

EXERCISE 1.2.v (6.1.6). Use Proposition 1.2.12 to show that the discrete ∞-categories in any ∞-cosmos define an ∞-cosmos whose functor spaces are all Kan complexes.

1.3 Cosmological Functors

Certain "right adjoint type" constructions define maps between ∞-cosmoi that preserve all of the structures axiomatized in Definition 1.2.1. The simple observation that such constructions define *cosmological functors* between ∞-cosmoi streamlines many proofs.

DEFINITION 1.3.1 (cosmological functor). A **cosmological functor** is a simplicial functor (see Definition A.2.6) between ∞-cosmoi that preserves the specified isofibrations and all of the cosmological limits.

In general, cosmological functors preserve any ∞-categorical notion that can be characterized *internally* to the ∞-cosmos – for instance, as a map equipped with additional structure – as opposed to *externally* – for instance, by a statement that involves a universal or existential quantifier. For example, the equivalences in an ∞-cosmos are characterized externally in Definition 1.2.2, which might lead one to suspect that a nonsurjective cosmological functor could fail to preserve them. However, Lemma 1.2.15 characterizes equivalences in terms of the presence of structures defined internally to an ∞-cosmos, so as a result:

LEMMA 1.3.2. *Any cosmological functor also preserves equivalences and trivial fibrations.*

Proof By Lemma 1.2.15 the equivalences in an ∞-cosmos coincide with the "homotopy equivalences" defined by cotensoring with the free-living isomorphism. Since a cosmological functor preserves simplicial cotensors, it preserves the data displayed in (1.2.16) and hence carries equivalences to equivalences. The preservation of trivial fibrations follows. □

REMARK 1.3.3. Similarly, arguing from Definition 1.2.26 it would not be clear whether cosmological functors preserve discrete ∞-categories, but using the internal characterization of Lemma 1.2.27 – an ∞-category A is discrete if and only if $A^{\mathbb{I}} \twoheadrightarrow A^2$ is a trivial fibration – this follows from the fact that cosmological functors preserve simplicial cotensors and trivial fibrations.

We now demonstrate that cosmological functors are abundant:

PROPOSITION 1.3.4. *The following constructions define cosmological functors for any ∞-cosmos $\mathcal{K}$:*

(i) *The functor space* $\mathrm{Fun}(X, -) : \mathcal{K} \to \mathcal{Q}\mathcal{C}at$*, for any ∞-category X.*
(ii) *The **underlying quasi-category** functor*

$$(-)_0 := \mathrm{Fun}(1, -) : \mathcal{K} \to \mathcal{Q}\mathcal{C}at,$$

specializing (i) to the terminal ∞-category 1.
(iii) *The simplicial cotensor* $(-)^U : \mathcal{K} \to \mathcal{K}$*, for any simplicial set U.*
(iv) *The exponential* $(-)^A : \mathcal{K} \to \mathcal{K}$*, for any ∞-category A in a cartesian closed ∞-cosmos $\mathcal{K}$.*
(v) *Pullback of isofibrations* $f^* : \mathcal{K}_{/B} \to \mathcal{K}_{/A}$ *along any functor $f : A \to B$ in an ∞-cosmos $\mathcal{K}$.*
(vi) *Moreover, for any cosmological functor $F : \mathcal{K} \to \mathcal{L}$ and any ∞-category $A \in \mathcal{K}$, the induced map on slices $F : \mathcal{K}_{/A} \to \mathcal{L}_{/FA}$ defines a cosmological functor.*

Proof The first four of these statements are nearly immediate, the preservation of isofibrations being asserted explicitly as a hypothesis in each case and the preservation of limits following from familiar arguments.

For (v), pullback in an ∞-cosmos $\mathcal{K}$ is a simplicially enriched limit construction; one consequence of this is that $f^* : \mathcal{K}_{/B} \to \mathcal{K}_{/A}$ defines a simplicial functor. The action of the functor f^* on a 0-arrow g in $\mathcal{K}_{/B}$ is also defined by a pullback square: since the front and back squares in the displayed diagram are pullbacks the top square is as well

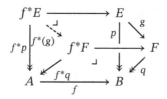

Since isofibrations are stable under pullback, it follows that $f^* : \mathcal{K}_{/B} \to \mathcal{K}_{/A}$ preserves isofibrations. It remains to prove that this functor preserves the simplicial limits constructed in Proposition 1.2.22, which is fundamentally a consequence of the commutativity of limit constructions. In each case, this can be verified explicitly. We illustrate this computation for simplicial cotensors by constructing the commutative cube:

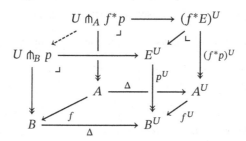

Since the front, back, and right faces are pullbacks, the left is as well.

The final statement (vi) is left as Exercise 1.3.i.   □

EXAMPLE 1.3.5. By Propositions 1.2.11 and 1.2.12, the full subcategory inclusions $\mathcal{C}at \hookrightarrow \mathcal{Q}\mathcal{C}at$ and $\mathcal{K}an \hookrightarrow \mathcal{Q}\mathcal{C}at$ both define cosmological functors (see also Lemma 6.1.4). These cosmological embeddings explicate the intuition that the formal category theory of 1-categories or of ∞-groupoids can be recovered as a special case of the formal category theory of $(\infty, 1)$-categories.

Non-examples of cosmological functors are also instructive:

NON-EXAMPLE 1.3.6. The forgetful functor $\mathcal{K}_{/B} \to \mathcal{K}$ is simplicial and preserves isofibrations but does *not* define a cosmological functor, failing to preserve cotensors and products. However, by Proposition 1.3.4(v), its right adjoint $- \times B : \mathcal{K} \to \mathcal{K}_{/B}$ does define a cosmological functor.

NON-EXAMPLE 1.3.7. The cosmological embedding $\mathcal{K}an \hookrightarrow \mathcal{Q}\mathcal{C}at$ has a right adjoint $(-)^{\simeq} : \mathcal{Q}\mathcal{C}at \to \mathcal{K}an$ that carries each quasi-category to its "∞-groupoid core" or maximal sub Kan complex, the simplicial subset containing those n-simplices whose edges are all isomorphisms. This core functor preserves isofibrations and 1-categorical limits but is *not* cosmological since it is not simplicially enriched: any functor $K \to Q$ whose domain is a Kan complex and whose codomain is a quasi-category factors through the inclusion $Q^{\simeq} \hookrightarrow Q$ via a unique map $K \to Q^{\simeq}$ but in general $\mathrm{Fun}(K, Q) \ncong \mathrm{Fun}(K, Q^{\simeq})$, since a natural transformation $K \times \Delta[1] \to Q$ only factors through $Q^{\simeq} \hookrightarrow Q$ in the case where its components are invertible (see Lemma 12.1.12 however).

Certain cosmological functors are especially well-behaved:

DEFINITION 1.3.8 (cosmological biequivalence). A cosmological functor defines a **cosmological biequivalence** $F : \mathcal{K} \overset{\simeq}{\to} \mathcal{L}$ if it additionally

(i) is **essentially surjective on objects up to equivalence**: for all $C \in \mathcal{L}$ there exists $A \in \mathcal{K}$ so that $FA \simeq C$ and

(ii) defines a **local equivalence**: for all $A, B \in \mathcal{K}$, the action of F on functor spaces defines an equivalence of quasi-categories

$$\mathrm{Fun}(A, B) \overset{\sim}{\longrightarrow} \mathrm{Fun}(FA, FB).$$

Cosmological biequivalences are studied more systematically in Chapter 10, where we think of them as "change-of-model" functors. Crucially for our proof of the "model independence" of $(\infty, 1)$-category theory in Chapter 11, there are a variety of cosmological biequivalences between the ∞-cosmoi of $(\infty, 1)$-categories:

EXAMPLE 1.3.9 (§E.2).

(i) The underlying quasi-category functors defined on the ∞-cosmoi of complete Segal spaces, Segal categories, and 1-complicial sets

$$\mathcal{C}\mathcal{S}\mathcal{S} \xrightarrow[\sim]{(-)_0} \mathcal{Q}\mathcal{C}at \quad \mathcal{S}egal \xrightarrow[\sim]{(-)_0} \mathcal{Q}\mathcal{C}at \quad 1\text{-}\mathcal{C}omp \xrightarrow[\sim]{(-)_0} \mathcal{Q}\mathcal{C}at$$

are all biequivalences. In the first two cases these are defined by "evaluating at the 0th row" and in the last case this is defined by "forgetting the markings."

(ii) There are also cosmological biequivalences nerve : $\mathcal{QC}at \rightsquigarrow \mathcal{CSS}$ and nerve : $\mathcal{QC}at \rightsquigarrow \mathcal{S}egal$ defined by Joyal and Tierney [64].

(iii) The functor disc : $\mathcal{CSS} \rightsquigarrow \mathcal{S}egal$ defined by Bergner [14] that "discretizes" a complete Segal spaces also defines a cosmological biequivalence.

(iv) Another cosmological biequivalence $(-)^{\natural} : \mathcal{QC}at \rightsquigarrow 1\text{-}\mathcal{C}omp$ that gives each quasi-category its "natural marking."

In terminology justified by Proposition 10.2.1:

DEFINITION 1.3.10. An ∞-cosmos $\mathcal{K}$ is an **∞-cosmos of** $(\infty, 1)$**-categories** just when the underlying quasi-category $(-)_0 : \mathcal{K} \to \mathcal{QC}at$ is a cosmological biequivalence.

REMARK 1.3.11. The underlying quasi-category functor $(-)_0 : \mathcal{K} \to \mathcal{QC}at$ carries the internal homs of a cartesian closed ∞-cosmos $\mathcal{K}$ to the corresponding functor spaces: for any ∞-categories A and B in $\mathcal{K}$, we have

$$(B^A)_0 := \mathrm{Fun}(1, B^A) \cong \mathrm{Fun}(A, B).$$

In the case where the ∞-cosmos $\mathcal{K}$ is biequivalent to $\mathcal{QC}at$, we see in Chapters 10 and 11 that this entails no essential loss of categorical information.

Cosmological biequivalences not only preserve equivalences but also reflect and create them.

LEMMA 1.3.12. *Let* $F : \mathcal{K} \rightsquigarrow \mathcal{L}$ *be a cosmological biequivalence. Then:*

(i) *A functor* $f : A \to B$ *between* ∞-*categories in* $\mathcal{K}$ *is an equivalence if and only if* $Ff : FA \to FB$ *is an equivalence in* $\mathcal{L}$.

(ii) *A pair of* ∞-*categories in* $\mathcal{K}$ *are equivalent if and only if their images in* $\mathcal{L}$ *are equivalent.*

Proof Lemma 1.3.2 implies that cosmological functors preserve equivalences and thus also the existence of an equivalence between a pair of ∞-categories in $\mathcal{K}$. To see that equivalences are also reflected, suppose $f : A \to B$ is a functor in $\mathcal{K}$ with the property that $Ff : FA \rightsquigarrow FB$ is an equivalence in $\mathcal{L}$. Now for any ∞-category X, simplicial functoriality provides a commutative diagram

$$
\begin{array}{ccc}
\mathrm{Fun}(X, A) & \xrightarrow{\ f_* \ } & \mathrm{Fun}(X, B) \\
\downarrow{\wr} & & \downarrow{\wr} \\
\mathrm{Fun}(FX, FA) & \xrightarrow[\ \approx \ ]{Ff_*} & \mathrm{Fun}(FX, FB)
\end{array}
$$

so from the 2-of-3 property we conclude that $f_* : \mathrm{Fun}(X, A) \twoheadrightarrow \mathrm{Fun}(X, B)$ is an equivalence, proving that f is an equivalence in $\mathcal{K}$.

To see that equivalences are created, suppose now that A and B are ∞-categories in $\mathcal{K}$ equipped with an equivalence:

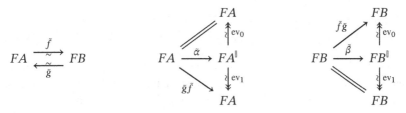

in $\mathcal{L}$. Since $\mathrm{Fun}(A, B) \twoheadrightarrow \mathrm{Fun}(FA, FB)$ and $\mathrm{Fun}(B, A) \twoheadrightarrow \mathrm{Fun}(FB, FA)$ are equivalences of quasi-categories the induced functors of homotopy categories $h(\mathrm{Fun}(A, B)) \twoheadrightarrow h(\mathrm{Fun}(FA, FB))$ and $h(\mathrm{Fun}(B, A)) \twoheadrightarrow h(\mathrm{Fun}(FB, FA))$ are equivalences of categories, by Remark 1.1.24, and in particular essentially surjective. So we may lift $\tilde{f}$ and $\tilde{g}$ to functors $f : A \to B$ and $g : B \to A$ in $h(\mathrm{Fun}(A, B))$ and $h(\mathrm{Fun}(B, A))$, respectively, so that $Ff \cong \tilde{f}$ and $Fg \cong \tilde{g}$. The commutative diagram of quasi-categories

$$
\begin{array}{ccc}
\mathrm{Fun}(B, A) \times \mathrm{Fun}(A, B) & \xrightarrow{\ \circ\ } & \mathrm{Fun}(A, A) \\
\downarrow{\scriptstyle\wr} & & \downarrow{\scriptstyle\wr} \\
\mathrm{Fun}(FB, FA) \times \mathrm{Fun}(FA, FB) & \xrightarrow{\ \circ\ } & \mathrm{Fun}(FA, FA)
\end{array}
$$

induces a commutative diagram between their homotopy categories. In particular, by applying the composition bifunctor to the isomorphisms $Ff \cong \tilde{f}$ and $Fg \cong \tilde{g}$, we see that

$$
F(\mathrm{id}_A) = \mathrm{id}_{FA} \cong \tilde{g} \circ \tilde{f} \cong Fg \circ Ff = F(g \circ f)
$$

in $h(\mathrm{Fun}(A, A))$. By fully faithfulness of $h(\mathrm{Fun}(A, A)) \twoheadrightarrow h(\mathrm{Fun}(FA, FA))$, this isomorphism lifts to an isomorphism $\mathrm{id}_A \cong g \circ f$ in $h(\mathrm{Fun}(A, A))$. By Corollary 1.1.16, this isomorphism can be represented by a homotopy coherent isomorphism $\mathbb{I} \to \mathrm{Fun}(A, A)$, which internalizes to define a map $\alpha : A \to A^{\mathbb{I}}$ as required. The construction of the homotopy coherent isomorphism $\beta : B \to B^{\mathbb{I}}$ from $f \circ g$ to id_B proceeds similarly. $\qquad\square$

The proof of the creation of equivalences in Lemma 1.3.12 is surprisingly delicate, passing to the homotopy categories of the functor spaces to avoid lifting and composing homotopy coherent isomorphisms; an argument along those lines is also possible, and left to the reader as Exercise 1.3.ii. The next section provides context for the argument just given by introducing the *homotopy*

2-*category of an* ∞-*cosmos.* The reader is then invited to revisit the creation of equivalences in Exercise 1.4.vi.

Exercises

EXERCISE 1.3.i. Prove that for any cosmological functor $F\colon \mathcal{K} \to \mathcal{L}$ and any $A \in \mathcal{K}$, the induced map $F\colon \mathcal{K}_{/A} \to \mathcal{L}_{/FA}$ defines a cosmological functor.

EXERCISE 1.3.ii. Sketch a proof that cosmological biequivalences create equivalences between ∞-categor ies without passing to homotopy categories, by lifting and composing the homotopy coherent isomorphisms given as part of the data of the hypothesized equivalences.

EXERCISE 1.3.iii. Suppose $F\colon \mathcal{K} \to \mathcal{L}$, $G\colon \mathcal{L} \to \mathcal{M}$, and $H\colon \mathcal{M} \to \mathcal{N}$ are cosmological functors, and assume that GF and HG are cosmological biequivalences. Show that F, G, H, and HGF are cosmological biequivalences.

1.4 The Homotopy 2-Category

Small 1-categories define the objects of a strict 2-category[22] $\mathcal{C}at$ of categories, functors, and natural transformations. Many basic categorical notions – those defined in terms of categories, functors, and natural transformations – can be defined internally to the 2-category $\mathcal{C}at$. This suggests a natural avenue for generalization: reinterpreting these same definitions in a generic 2-category using its objects in place of small categories, its 1-cells in place of functors, and its 2-cells in place of natural transformations.

In Chapter 2, we develop a significant portion of the theory of ∞-categories in any fixed ∞-cosmos following exactly this outline, working internally to a 2-category that we refer to as the *homotopy 2-category* that we associate to any ∞-cosmos. The homotopy 2-category of an ∞-cosmos is a quotient of the full ∞-cosmos, replacing each quasi-categorical functor space by its homotopy category. Surprisingly, this rather destructive quotienting operation preserves quite a lot of information. Indeed, essentially all of the development of the

---

[22] Appendix B introduces 2-categories and 2-functors, reviewing the 2-category theory needed here. Succinctly, in parallel with Digression 1.2.4, 2-categories (see Definition B.1.1) can be understood equally as:

- "two-dimensional" categories, with objects; **1-cells**, whose boundary are given by a pair of objects; and **2-cells**, whose boundary are given by a parallel pair of 1-cells between a pair of objects – together with partially defined composition operations governed by this boundary data
- or as categories enriched over $\mathcal{C}at$.

theory of ∞-categories in Part I takes place in the homotopy 2-category of an ∞-cosmos. This said, we caution the reader against becoming overly seduced by homotopy 2-categories, which are more of a technical convenience for reducing the complexity of our arguments than a fundamental notion of ∞-category theory.

The homotopy 2-category for the ∞-cosmos of quasi-categories was first introduced by Joyal in his work on the foundations of quasi-category theory [63].

DEFINITION 1.4.1 (homotopy 2-category). Let $\mathcal{K}$ be an ∞-cosmos. Its **homotopy 2-category** is the 2-category $\mathfrak{h}\mathcal{K}$ whose

- objects are the the objects A, B of $\mathcal{K}$, i.e., the ∞-categories;
- 1-cells $f : A \to B$ are the 0-arrows in the functor space $\mathsf{Fun}(A, B)$, i.e., the ∞-functors; and
- 2-cells $A \underset{g}{\overset{f}{\rightrightarrows}}{}^{\Downarrow \alpha} B$ are homotopy classes of 1-simplices in $\mathsf{Fun}(A, B)$, which we call ∞-**natural transformations**.

Put another way $\mathfrak{h}\mathcal{K}$ is the 2-category with the same objects as $\mathcal{K}$ and with hom-categories defined by

$$\mathsf{hFun}(A, B) := \mathsf{h}(\mathsf{Fun}(A, B)),$$

that is, $\mathsf{hFun}(A, B)$ is the homotopy category of the quasi-category $\mathsf{Fun}(A, B)$.

The **underlying category** of a 2-category is defined by simply forgetting its 2-cells. Note that an ∞-cosmos $\mathcal{K}$ and its homotopy 2-category $\mathfrak{h}\mathcal{K}$ share the same underlying category $\mathcal{K}_0$ of ∞-categories and ∞-functors in $\mathcal{K}$.

DIGRESSION 1.4.2 (change of base, §A.7). The homotopy category functor preserves finite products, as of course does its right adjoint. It follows that the adjunction of Proposition 1.1.11 induces a change-of-base adjunction

$$2\text{-}\mathcal{C}at \underset{\longrightarrow}{\overset{h_*}{\underset{\perp}{\longleftarrow}}} s\mathcal{S}et\text{-}\mathcal{C}at$$

whose left and right adjoints change the enrichment by applying the homotopy category functor or the nerve functor to the hom objects of the enriched category. Here $2\text{-}\mathcal{C}at$ and $s\mathcal{S}et\text{-}\mathcal{C}at$ can each be understood as 2-categories – of enriched categories, enriched functors, and enriched natural transformations – and both change of base constructions define 2-functors (see Propositions A.7.3 and A.7.5). Since the nerve embedding is fully faithful, 2-categories can be identified as a full subcategory comprised of those simplicial categories whose hom spaces are nerves of categories.

The proof of Lemma 1.3.12 uses an observation worth highlighting:

LEMMA 1.4.3.

(i) *Every 2-cell* $A \underset{g}{\overset{f}{\underset{\Downarrow\alpha}{\rightrightarrows}}} B$ *in the homotopy 2-category of an ∞-cosmos is represented by a map of quasi-categories as below-left or equivalently by a functor as below-right*

and two such maps represent the same 2-cell if and only if they are homotopic as 1-simplices in $\mathsf{Fun}(A, B)$.

(ii) *Every invertible 2-cell* $A \underset{g}{\overset{f}{\underset{\cong\Downarrow\alpha}{\rightrightarrows}}} B$ *in the homotopy 2-category of an ∞-cosmos is represented by a map of quasi-categories as below-left or equivalently by a functor as below-right*

$$\begin{array}{ccc} & \mathbb{1} + \mathbb{1} & \\ & \diagup \quad \searrow^{(f,g)} & \\ \mathbb{I} & \xrightarrow[\alpha]{} \mathsf{Fun}(A,B) & \end{array} \quad \rightsquigarrow \quad \begin{array}{ccc} A & \xrightarrow{\ulcorner\alpha\urcorner} & B^{\mathbb{I}} \\ {\scriptstyle(g,f)}\searrow & & \swarrow{\scriptstyle(p_1,p_0)} \\ & B \times B & \end{array}$$

and two such maps represent the same invertible 2-cell if and only if their common restrictions along $\mathbb{2} \hookrightarrow \mathbb{I}$ *are homotopic as 1-simplices in* $\mathsf{Fun}(A, B)$.

The notion of homotopic 1-simplices referenced here is defined in Lemma 1.1.9. Since the 2-cells in the homotopy 2-category are referred to as ∞-natural transformations, we refer to the invertible 2-cells in the homotopy 2-category as ∞-**natural isomorphisms**.

Proof The statement (i) records the definition of the 2-cells in the homotopy 2-category and the universal property (1.2.7) of the simplicial cotensor. For (ii), a 2-cell in the homotopy 2-category is **invertible** if and only if it defines an isomorphism in the appropriate hom-category $\mathrm{hFun}(A, B)$. By Corollary 1.1.16 it follows that each invertible 2-cell α is represented by a homotopy coherent isomorphism $\alpha : \mathbb{I} \to \mathsf{Fun}(A, B)$, which similarly internalizes to define a functor $\ulcorner\alpha\urcorner : A \to B^{\mathbb{I}}$. □

An upshot of Digression 1.4.2 is that change of base is an operation that applies to enriched functors as well as enriched categories, as can be directly verified in the case of greatest interest.

LEMMA 1.4.4. *Any simplicial functor* $F: \mathcal{K} \to \mathcal{L}$ *between* ∞*-cosmoi induces a 2-functor* $F: \mathfrak{h}\mathcal{K} \to \mathfrak{h}\mathcal{L}$ *between their homotopy 2-categories.*

Proof The action of the induced 2-functor $F: \mathfrak{h}\mathcal{K} \to \mathfrak{h}\mathcal{L}$ on objects and 1-cells is given by the corresponding action of $F: \mathcal{K} \to \mathcal{L}$; recall an ∞-cosmos and its homotopy 2-category have the same underlying 1-category. Each 2-cell in $\mathfrak{h}\mathcal{K}$ is represented by a 1-simplex in $\mathsf{Fun}(A, B)$ which is mapped via

$$\mathsf{Fun}(A, B) \xrightarrow{\quad F \quad} \mathsf{Fun}(FA, FB)$$

$$A \underset{g}{\overset{f}{\Longrightarrow}} B \longmapsto FA \underset{Fg}{\overset{Ff}{\Longrightarrow}} FB$$

to a 1-simplex representing a 2-cell in $\mathfrak{h}\mathcal{L}$. Since the action $F: \mathsf{Fun}(A, B) \to \mathsf{Fun}(FA, FB)$ on functor spaces defines a morphism of simplicial sets, it preserves faces and degeneracies. In particular, homotopic 1-simplices in $\mathsf{Fun}(A, B)$ are carried to homotopic 1-simplices in $\mathsf{Fun}(FA, FB)$ so the action on 2-cells just described is well-defined. The 2-functoriality of these mappings follows from the simplicial functoriality of the original mapping. □

We now begin to relate the simplicially enriched structures of an ∞-cosmos to the 2-categorical structures in its homotopy 2-category by proving that homotopy 2-categories inherit products from their ∞-cosmoi that satisfy a 2-categorical universal property. To illustrate, recall that the terminal ∞-category $1 \in \mathcal{K}$ has the universal property $\mathsf{Fun}(X, 1) \cong \mathbb{1}$ for all $X \in \mathcal{K}$. Applying the homotopy category functor we see that $1 \in \mathfrak{h}\mathcal{K}$ has the universal property $\mathsf{hFun}(X, 1) \cong \mathbb{1}$ for all $X \in \mathfrak{h}\mathcal{K}$, which is expressed by saying that the ∞-category 1 defines a **2-terminal object** in the homotopy 2-category. This 2-categorical universal property has both a 1-dimensional and a 2-dimensional aspect. Since $\mathsf{hFun}(X, 1) \cong \mathbb{1}$ is a category with a single object, there exists a unique morphism $X \to 1$ in $\mathcal{K}$, and since $\mathsf{hFun}(X, 1) \cong \mathbb{1}$ has only a single morphism, the only 2-cells in $\mathfrak{h}\mathcal{K}$ with codomain 1 are identities.

PROPOSITION 1.4.5 (cartesian (closure)).

(i) *The homotopy 2-category of any* ∞*-cosmos has 2-categorical products.*

(ii) *The homotopy 2-category of a cartesian closed* ∞*-cosmos is cartesian closed as a 2-category.*

Proof   While the functor $h : sSet \to Cat$ only preserves finite products, the restricted functor $h : QCat \to Cat$ preserves *all* products on account of the simplified description of the homotopy category of a quasi-category given in Lemma 1.1.12. Thus for any set I and family of ∞-categories $(A_i)_{i \in I}$ in $\mathcal{K}$, the homotopy category functor carries the isomorphism of functor spaces to an isomorphism of hom-categories

$$\mathrm{Fun}(X, \textstyle\prod_{i \in I} A_i) \xrightarrow{\;\cong\;} \prod_{i \in I} \mathrm{Fun}(X, A_i)$$
$$\Big\downarrow_h$$
$$\mathrm{hFun}(X, \textstyle\prod_{i \in I} A_i) \xrightarrow{\;\cong\;} \prod_{i \in I} \mathrm{hFun}(X, A_i).$$

This proves that the homotopy 2-category $\mathfrak{h}\mathcal{K}$ has products whose universal properties have both a 1- and 2-dimensional component, as described in the empty case for terminal objects above.

If $\mathcal{K}$ is a cartesian closed ∞-cosmos, then for any triple of ∞-categories $A, B, C \in \mathcal{K}$ there exist exponential objects $C^A, C^B \in \mathcal{K}$ characterized by natural isomorphisms

$$\mathrm{Fun}(A \times B, C) \cong \mathrm{Fun}(A, C^B) \cong \mathrm{Fun}(B, C^A).$$

Passing to homotopy categories we have natural isomorphisms

$$\mathrm{hFun}(A \times B, C) \cong \mathrm{hFun}(A, C^B) \cong \mathrm{hFun}(B, C^A),$$

which demonstrates that $\mathfrak{h}\mathcal{K}$ is cartesian closed as a 2-category: functors $A \times B \to C$ transpose to define functors $A \to C^B$ and $B \to C^A$, and natural transformations transpose similarly.   □

There is a standard definition of *isomorphism* between two objects in any 1-category, preserved by any functor. Similarly, there is a standard definition of *equivalence* between two objects in any 2-category, preserved by any 2-functor:

DEFINITION 1.4.6 (equivalence).   An **equivalence** in a 2-category is given by

- a pair of objects A and B;
- a pair of 1-cells $f : A \to B$ and $g : B \to A$; and
- a pair of invertible 2-cells

$$A \underset{gf}{\overset{\frown}{\longrightarrow}} A \quad\Big(\cong \Downarrow \alpha\Big) \qquad \text{and} \qquad B \overset{\frown}{\longrightarrow} B \quad\Big(\cong \Downarrow \beta\Big)$$

When A and B are **equivalent**, we write $A \simeq B$ and refer to the 1-cells f and g as **equivalences**, denoted by "$\xrightarrow{\sim}$."

In the case of the homotopy 2-category of an ∞-cosmos we have a competing definition of equivalence from 1.2.1: namely a 1-cell $f : A \overset{\sim}{\to} B$ that induces an equivalence $f_* : \mathsf{Fun}(X, A) \overset{\sim}{\to} \mathsf{Fun}(X, B)$ on functor spaces – or equivalently, by Lemma 1.2.15, a homotopy equivalence defined relative to the interval $\mathbb{I}$. Crucially, all three notions of equivalence coincide:

THEOREM 1.4.7 (equivalences are equivalences). *In any ∞-cosmos $\mathcal{K}$, the following are equivalent and characterize what it means for a functor $f : A \to B$ between ∞-categories to define an* **equivalence**.

(i) *For all $X \in \mathcal{K}$, the post-composition map $f_* : \mathsf{Fun}(X, A) \overset{\sim}{\to} \mathsf{Fun}(X, B)$ defines an equivalence of quasi-categories.*

(ii) *There exists a functor $g : B \to A$ and natural isomorphisms $\alpha : \mathrm{id}_A \cong gf$ and $\beta : fg \cong \mathrm{id}_B$ in the homotopy 2-category.*

(iii) *There exists a functor $g : B \to A$ and maps*

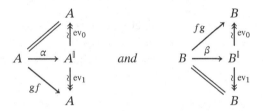

in the ∞-cosmos in $\mathcal{K}$.

As an illustrative comparison of 2-categorical and quasi-categorical techniques, rather than appealing to Lemma 1.2.15 to prove (i)⇔(iii), we re-prove it.

Proof For (i)⇒(ii), if the induced map $f_* : \mathsf{Fun}(X, A) \overset{\sim}{\to} \mathsf{Fun}(X, B)$ defines an equivalence of quasi-categories then the functor $f_* : \mathsf{hFun}(X, A) \overset{\sim}{\to} \mathsf{hFun}(X, B)$ defines an equivalence of categories, by Remark 1.1.24. In particular, the equialence $f_* : \mathsf{hFun}(B, A) \overset{\sim}{\to} \mathsf{hFun}(B, B)$ is essentially surjective so there exists $g \in \mathsf{hFun}(B, A)$ and an isomorphism $\beta : fg \cong \mathrm{id}_B \in \mathsf{hFun}(B, B)$. Now since $f_* : \mathsf{hFun}(A, A) \overset{\sim}{\to} \mathsf{hFun}(A, B)$ is fully faithful, the isomorphism $\beta f : fgf \cong f \in \mathsf{hFun}(A, B)$ can be lifted to define an isomorphism $\alpha^{-1} : gf \cong \mathrm{id}_A \in \mathsf{hFun}(A, A)$. This defines the data of a 2-categorical equivalence in Definition 1.4.6.

To see that (ii)⇒(iii) recall from Lemma 1.4.3 that the natural isomorphisms $\alpha : \mathrm{id}_A \cong gf$ and $\beta : fg \cong \mathrm{id}_B$ in $\mathfrak{h}\mathcal{K}$ are represented by maps $\alpha : A \to A^{\mathbb{I}}$ and $\beta : B \to B^{\mathbb{I}}$ in $\mathcal{K}$ as in (1.2.16).

Finally, (iii)⇒(i) since $\mathsf{Fun}(X, -)$ carries the data of (iii) to the data of an equivalence of quasi-categories as in Definition 1.1.23. □

It is hard to overstate the importance of Theorem 1.4.7 for the work that follows. The categorical constructions that we introduce for ∞-categories, ∞-functors, and ∞-natural transformations are invariant under 2-categorical equivalence in the homotopy 2-category and the universal properties we develop similarly characterize 2-categorical equivalence classes of ∞-categories. Theorem 1.4.7 then asserts that such constructions are "homotopically correct": both invariant under equivalence in the ∞-cosmos and precisely identifying equivalence classes of objects.

The equivalence invariance of the functor space in the codomain variable is axiomatic, but equivalence invariance in the domain variable is not.[23] Nor is it evident how this could be proven from either (i) or (iii) of Theorem 1.4.7. But using (ii) and 2-categorical techniques, there is now a short proof.

COROLLARY 1.4.8. *Equivalences of* ∞-*categories* $A' \xrightarrow{\sim} A$ *and* $B \xrightarrow{\sim} B'$ *induce an equivalence of functor spaces* $\mathrm{Fun}(A, B) \xrightarrow{\sim} \mathrm{Fun}(A', B')$.

Proof The representable simplicial functors $\mathrm{Fun}(A, -) : \mathcal{K} \to \mathcal{QC}at$ and $\mathrm{Fun}(-, B) : \mathcal{K}^{\mathrm{op}} \to \mathcal{QC}at$ induce 2-functors $\mathrm{Fun}(A, -) : \mathfrak{h}\mathcal{K} \to \mathfrak{h}\mathcal{QC}at$ and $\mathrm{Fun}(-, B) : \mathfrak{h}\mathcal{K}^{\mathrm{op}} \to \mathfrak{h}\mathcal{QC}at$, which preserve the 2-categorical equivalences of Definition 1.4.6. By Theorem 1.4.7 this is what we wanted to show.   □

There is also a standard 2-categorical notion of an isofibration, defined in the statement of Proposition 1.4.9 and elaborated upon in Definition B.4.4. We now show that any isofibration in an ∞-cosmos defines an isofibration in its homotopy 2-category.

PROPOSITION 1.4.9 (isofibrations are isofibrations). *An isofibration* $p : E \twoheadrightarrow B$ *in an* ∞-*cosmos* $\mathcal{K}$ *also defines an* **isofibration** *in the homotopy 2-category* $\mathfrak{h}\mathcal{K}$: *given any invertible 2-cell as displayed below-left abutting to B with a specified lift of one of its boundary 1-cells through p, there exists an invertible 2-cell abutting to E with this boundary 1-cell as displayed below-right that whiskers with p to the original 2-cell.*

$$
\begin{array}{ccc}
X \xrightarrow{\ e\ } E & & X \underset{\bar{e}}{\overset{e}{\rightrightarrows}} E \\
\ \cong\Downarrow\beta \ \Big\downarrow p & = & \quad\cong\Downarrow\gamma \quad \Big\downarrow p \\
b\searrow B & & B
\end{array}
$$

Proof The universal property of the statement says that the functor

$$p_* : \mathrm{hFun}(X, E) \twoheadrightarrow \mathrm{hFun}(X, B)$$

---

[23] Lemma 1.3.2 does not apply since $\mathrm{Fun}(-, B)$ is not cosmological.

is an isofibration of categories in the sense defined in Proposition 1.2.11. By axiom 1.2.1(ii), since $p : E \twoheadrightarrow B$ is an isofibration in $\mathcal{K}$, the induced map $p_* : \mathrm{Fun}(X, E) \twoheadrightarrow \mathrm{Fun}(X, B)$ is an isofibration of quasi-categories. So it suffices to show that the functor $\mathrm{h} : \mathcal{QCat} \to \mathcal{Cat}$ carries isofibrations of quasi-categories to isofibrations of categories.

So let us now consider an isofibration $p : E \twoheadrightarrow B$ between quasi-categories. By Corollary 1.1.16, every isomorphism β in the homotopy category $\mathrm{h}B$ of the quasi-category B is represented by a simplicial map $\beta : \mathbb{I} \to B$. By Definition 1.1.17, the lifting problem

$$
\begin{array}{ccc}
\mathbb{1} & \xrightarrow{\ e\ } & E \\
{\scriptstyle\iota}\downarrow & {\scriptstyle\gamma}\nearrow & \downarrow{\scriptstyle p} \\
\mathbb{I} & \xrightarrow[\ \beta\ ]{} & B
\end{array}
$$

can be solved, and the map $\gamma : \mathbb{I} \to E$ so produced represents a lift of the isomorphism from $\mathrm{h}B$ to an isomorphism in $\mathrm{h}E$ with domain e. □

CONVENTION 1.4.10 (on isofibrations in homotopy 2-categories). Since the converse to Proposition 1.4.9 does not hold, there is a potential ambiguity when using the term "isofibration" to refer to a map in the homotopy 2-category of an ∞-cosmos. We adopt the convention that when we declare a map in $\mathfrak{h}\mathcal{K}$ to be an isofibration we always mean this is the stronger sense of defining an isofibration in $\mathcal{K}$. This stronger condition gives us access to the 2-categorical lifting property of Proposition 1.4.9 and also to homotopical properties axiomatized in Definition 1.2.1, which ensure that the strictly defined limits of 1.2.1(i) are automatically equivalence invariant constructions (see §C.1 and Proposition 6.2.8).

We conclude this chapter with a final definition that can be extracted from the homotopy 2-category of an ∞-cosmos. The 1- and 2-cells in the homotopy 2-category from the terminal ∞-category $1 \in \mathcal{K}$ to a generic ∞-category $A \in \mathcal{K}$ define the objects and morphisms in the homotopy category of the ∞-category A.

DEFINITION 1.4.11 (homotopy category of an ∞-category). The **homotopy category** of an ∞-category A in an ∞-cosmos $\mathcal{K}$ is defined to be the homotopy category of its underlying quasi-category, that is:

$$\mathrm{h}A := \mathrm{hFun}(1, A) := \mathrm{h}(\mathrm{Fun}(1, A)).$$

As we shall discover, homotopy categories generally inherit "derived" analogues of structures present at the level of ∞-categories. An early example of this appears in Proposition 2.1.7(ii).

Exercises

EXERCISE 1.4.i.

(i) What is the homotopy 2-category of the ∞-cosmos $\mathcal{C}at$ of 1-categories?
(ii) Prove that the nerve defines a 2-functor $\mathcal{C}at \hookrightarrow \mathfrak{h}\mathcal{Q}\mathcal{C}at$ that is locally fully faithful.

EXERCISE 1.4.ii. Demonstrate that the homotopy 2-category of the dual cosmos $\mathcal{K}^{co}$ of an ∞-cosmos $\mathcal{K}$ is the co-dual of the homotopy 2-category $\mathfrak{h}\mathcal{K}$ – in symbols $\mathfrak{h}(\mathcal{K}^{co}) \cong (\mathfrak{h}\mathcal{K})^{co}$ – with the domains and codomains of 2-cells but not 1-cells reversed (see Definition B.1.6).

EXERCISE 1.4.iii. Consider a natural isomorphism $A \underset{g}{\overset{f}{\rightrightarrows}} B$ between a
parallel pair of functors in an ∞-cosmos. Give two proofs that if either f or g is an equivalence then both functors are, either by arguing entirely in the homotopy 2-category or by appealing to Lemma 1.4.3.

EXERCISE 1.4.iv. Extend Lemma 1.2.27 to show that the following four conditions are equivalent, characterizing the discrete objects E in an ∞-cosmos $\mathcal{K}$:

(i) E is a discrete object in the homotopy 2-category $\mathfrak{h}\mathcal{K}$, that is, every 2-cell with codomain E is invertible.
(ii) For each $X \in \mathcal{K}$, the hom-category $\mathsf{hFun}(X, E)$ is a groupoid.
(iii) For each $X \in \mathcal{K}$, the mapping quasi-category $\mathsf{Fun}(X, E)$ is a Kan complex.
(iv) The isofibration $E^{\mathbb{I}} \twoheadrightarrow E^2$, induced by the inclusion of simplicial sets $2 \hookrightarrow \mathbb{I}$, is a trivial fibration.

EXERCISE 1.4.v (10.3.1). Extend Lemma 1.4.4 to show that if $F: \mathcal{K} \to \mathcal{L}$ is a cosmological biequivalence then $F: \mathfrak{h}\mathcal{K} \to \mathfrak{h}\mathcal{L}$ is a 2-categorical **biequivalence**, a 2-functor that is essentially surjective on objects up to equivalence that locally defines an equivalence of hom-categories.

EXERCISE 1.4.vi. Let $F: \mathcal{K} \xrightarrow{\sim} \mathcal{L}$ be a cosmological biequivalence and let $A, B \in \mathcal{K}$. Re-prove part of the statement of Lemma 1.3.12: that if $FA \simeq FB$ in $\mathcal{L}$ then $A \simeq B$ in $\mathcal{K}$.

EXERCISE 1.4.vii (3.6.2). Let B be an ∞-category in the ∞-cosmos $\mathcal{K}$ and let $\mathfrak{h}\mathcal{K}_{/B}$ denote the 2-category whose

• objects are isofibrations $E \twoheadrightarrow B$ in $\mathcal{K}$ with codomain B;

- 1-cells are 1-cells in $\mathfrak{h}\mathcal{K}$ over B; and

- 2-cells are 2-cells α in $\mathfrak{h}\mathcal{K}$

that lie over B in the sense that $q\alpha = \mathrm{id}_p$.

Argue that the homotopy 2-category $\mathfrak{h}(\mathcal{K}_{/B})$ of the sliced ∞-cosmos has the same underlying 1-category but different 2-cells. How do these compare with the 2-cells of $\mathfrak{h}\mathcal{K}_{/B}$?

2

Adjunctions, Limits, and Colimits I

Heuristically, ∞-categories generalize ordinary 1-categories by adding in higher dimensional morphisms and weakening the composition law. One could imagine "∞-tizing" other types of categorical structure similarly, by adding in higher dimension and weakening properties. The naïve hope is that proofs establishing the theory of 1-categories might similarly generalize to give proofs for ∞-categories, just by adding a prefix "∞-" everywhere. In this chapter, we make this dream a reality – at least for a library of basic propositions concerning equivalences, adjunctions, limits, and colimits and the interrelationships between these notions.

Recall that categories, functors, and natural transformations assemble into a 2-category $\mathcal{C}at$. Similarly, the ∞-categories, ∞-functors, and ∞-natural transformations in any ∞-cosmos assemble into a 2-category, namely the *homotopy 2-category* of the ∞-cosmos, introduced in §1.4. In fact, $\mathcal{C}at$ can be regarded as a special case of a homotopy 2-category (by Exercise 1.4.i). In this chapter, we use 2-categorical techniques to define *adjunctions* between ∞-categories and *limits* and *colimits* of diagrams valued in an ∞-category and prove that these notions interact in the expected ways. In the homotopy 2-category of categories, this recovers classical results from 1-category theory, and in some cases even specializes to the standard proofs. As these arguments are equally valid in any homotopy 2-category, our proofs also establish the desired generalizations by simply appending the prefix "∞-."

In §2.1, we define an adjunction between ∞-categories to be an adjunction in the homotopy 2-category of ∞-categories, ∞-functors, and ∞-natural transformations. While it takes some work to justify the moral correctness of this simple definition, it has the great advantage that proofs of a number of results concerning the calculus of adjunctions and equivalences can be taken "off the shelf" in the sense that anyone who is sufficiently well-acquainted with 2-categories might know them already. In §2.2, we specialize the theory of adjunctions between ∞-categories to define and study initial and terminal elements inside an ∞-category. This section also serves as a warmup for the more subtle general theory of limits and colimits of diagrams valued in an ∞-category, which is the subject of §2.3. Finally, in §2.4, we study the interactions between these notions, proving that right adjoints preserve limits and left adjoints preserve colimits.

Missing from this discussion is an account of the universal properties associated to the unit of an adjunction or to a limit cone. These will be incorporated when we return to these topics in Chapter 4 after introducing an appropriate "hom ∞-category" with which to state them.

2.1 Adjunctions and Equivalences

In §1.4, we encounter the definition of an *equivalence* between a pair of objects in a 2-category. In the case where the ambient 2-category is the homotopy 2-category of an ∞-cosmos, Theorem 1.4.7 observes that the 2-categorical notion of equivalence precisely recaptures the notion of equivalence between ∞-categories in the full ∞-cosmos. In each of the examples of ∞-cosmoi we have considered, the representably defined equivalences in the ∞-cosmos coincide with the standard notion of equivalences between ∞-categories as presented in that particular model.[1] Thus, the 2-categorical notion of equivalence is the "correct" notion of equivalence between ∞-categories.

Similarly, there is a standard definition of an *adjunction* between a pair of objects in a 2-category, which, when interpreted in the homotopy 2-category of ∞-categories, functors, and natural transformations in an ∞-cosmos, will define the correct notion of adjunction between ∞-categories.

DEFINITION 2.1.1 (adjunction). An **adjunction** between ∞-categories is comprised of:

- a pair of ∞-categories A and B;
- a pair of ∞-functors $u : A \to B$ and $f : B \to A$; and
- a pair of ∞-natural transformations $\eta : \text{id}_B \Rightarrow uf$ and $\epsilon : fu \Rightarrow \text{id}_A$, called the **unit** and **counit** respectively,

---

[1] For instance, as outlined in Digression 1.2.13, the equivalences in the ∞-cosmoi of Example 1.2.24 recapture the weak equivalences between fibrant–cofibrant objects in the usual model structure.

so that the triangle equalities hold:[2]

$$
\begin{array}{ccccccc}
B == B & B & & B == B & & B \\
\end{array}
$$

$$
\xymatrix{
B \ar@{=}[rr] \ar@/^/[rd]^{u} & & B \\
A \ar@/^/[ru]^{u} \ar@{=}[rr] & & A
} \quad = u\left(=\right)u \qquad
\xymatrix{
B \ar@{=}[rr] \ar@/_/[rd]_{f} & & B \\
A & & A \ar@/_/[lu]
} \quad = f\left(=\right)f
$$

The functor f is called the **left adjoint** and u is called the **right adjoint**, a relationship that is denoted symbolically in text by writing $f \dashv u$ or in a displayed diagram such as[3]

$$
A \underset{u}{\overset{f}{\underset{\rightarrow}{\overset{\leftarrow}{\perp}}}} B
$$

We typically drop the prefix "∞" from the functors and natural transformations between ∞-categories.

DIGRESSION 2.1.2 (justifying the 2-categorical definition of an adjunction). We offer a few words of justification for those who find Definition 2.1.1 implausible – perhaps too simple to be trusted. Joyal was the first to propose using the standard 2-categorical definition to define an adjunction between ∞-categories, defining an adjunction between quasi-categories to be an adjunction in the homotopy 2-category $\mathfrak{h}\mathcal{QC}at$ in the preface to [63]. However, this definition was not widely adopted, with most practitioners instead using Lurie's definition of adjunction between quasi-categories [78, §5.2], which takes a quite different form. In §F.5, we prove that in the ∞-cosmos of quasi-categories, Joyal's 2-categorical definition of adjunction precisely recovers Lurie's. As explained in Part III, each of the models of $(\infty, 1)$-categories described in Example 1.2.24 "has the same category theory," so Definition 2.1.1 agrees with the community consensus notion of adjunction between $(\infty, 1)$-categories.

In the ∞-cosmoi whose objects model (∞, n)- or (∞, ∞)-categories, the adjunctions defined in the homotopy 2-category are the "pseudo-style" adjunctions. While these are not the most general adjunctions that might be considered – for instance, one might have the triangle equality relations satisfied only up to coherent noninvertible 3-cells – they are an important class of adjunctions. One reason for the relevance of Definition 2.1.1 in all ∞-cosmoi is its relationships to the equivalences, which Theorem 1.4.7 establishes are morally "correct," and to the notions of limits and colimits to be introduced.

---

[2] The left-hand equality of pasting diagrams asserts the composition relation $u\epsilon \cdot \eta u = \mathrm{id}_u$ in the hom-category $\mathsf{hFun}(A, B)$, while the right-hand equality asserts that $\epsilon f \cdot f\eta = \mathrm{id}_f$ in $\mathsf{hFun}(B, A)$. The calculus of pasting diagrams is surveyed in §B.1.

[3] Some authors contort adjunction diagrams so that the left adjoint is always oriented in a particular direction; we instead use the turnstile symbol "$\perp$" to indicate which adjoint is the left adjoint.

Finally, a reasonable objection is that Definition 2.1.1 appears too "low dimensional," comprised of data found entirely in the homotopy 2-category and ignoring the higher dimensional morphisms in an ∞-cosmos. In fact, any adjunction between ∞-categories extends to a *homotopy coherent adjunction* involving data in all dimensions, and moreover such extensions are homotopically unique [109].

The definition of an adjunction given in Definition 2.1.1 is "equational" in character: stated in terms of the objects, 1-cells, and 2-cells of a 2-category and their composites. Immediately:

LEMMA 2.1.3. *An adjunction in a 2-category is preserved by any 2-functor.* □

EXAMPLE 2.1.4 (adjunctions between 1-categories). Via the nerve embedding $\mathcal{C}at \hookrightarrow \mathfrak{h}\mathcal{Q}\mathcal{C}at$, any adjunction between 1-categories induces an adjunction between their nerves regarded as quasi-categories.

EXAMPLE 2.1.5 (adjunctions between topological categories). Cordier's *homotopy coherent nerve* [29, 30] defines a 2-functor $\mathfrak{N} : \mathcal{K}an\text{-}\mathcal{C}at \to \mathfrak{h}\mathcal{Q}\mathcal{C}at$ from the 2-category of Kan complex enriched categories, simplicially enriched functors, and simplicial natural transformations, to the homotopy 2-category $\mathfrak{h}\mathcal{Q}\mathcal{C}at$. In this way, topologically enriched adjunctions define adjunctions between quasi-categories.

EXAMPLE 2.1.6 (Quillen adjunctions). Topologically enriched adjunctions are relatively rare. More prevalent are "up-to-homotopy" topologically enriched adjunctions, such as those presented by Quillen adjunctions between (simplicial) model categories. These also define adjunctions between quasi-categories (see Mazel-Gee [85] or [108, §6.2]).

These examples define adjunctions between quasi-categories, but Lemma 2.1.3 applies to the 2-functors underlying the cosmological functors of Example 1.3.9 to transfer adjunctions defined in one model of $(\infty, 1)$-categories to adjunctions defined in each of the other models. The preservation of adjunctions by 2-functors, such as those given by Lemma 1.4.4, also proves:

PROPOSITION 2.1.7. *Given an adjunction* $A \underset{u}{\overset{f}{\underset{\perp}{\rightleftarrows}}} B$ *between* ∞-*categories:*

(i) for any ∞-*category* X,

$$\mathsf{Fun}(X,A) \underset{u_*}{\overset{f_*}{\underset{\perp}{\rightleftarrows}}} \mathsf{Fun}(X,B)$$

defines an adjunction between quasi-categories;

(ii) for any ∞-category X,

$$\mathsf{hFun}(X,A) \underset{u_*}{\overset{f_*}{\rightleftarrows}} \mathsf{hFun}(X,B)$$

defines an adjunction between categories;
(iii) for any simplicial set U,

$$A^U \underset{u^U}{\overset{f^U}{\rightleftarrows}} B^U$$

defines an adjunction between ∞-categories; and
(iv) if the ambient ∞-cosmos is cartesian closed, then for any ∞-category C,

$$A^C \underset{u^C}{\overset{f^C}{\rightleftarrows}} B^C$$

defines an adjunction between ∞-categories.

For instance, taking $X = 1$ in (ii) yields a "derived" adjunction between the homotopy categories of the ∞-categories A and B (see Definition 1.4.11):

$$\mathsf{h}A \underset{u_*}{\overset{f_*}{\rightleftarrows}} \mathsf{h}B$$

Proof Any adjunction $f \dashv u$ in the homotopy 2-category $\mathfrak{h}\mathcal{K}$ is preserved by each of the 2-functors $\mathsf{Fun}(X,-): \mathfrak{h}\mathcal{K} \to \mathfrak{h}\mathcal{Q}\mathcal{C}at$, $\mathsf{hFun}(X,-): \mathfrak{h}\mathcal{K} \to \mathcal{C}at$, $(-)^U: \mathfrak{h}\mathcal{K} \to \mathfrak{h}\mathcal{K}$, and $(-)^C: \mathfrak{h}\mathcal{K} \to \mathfrak{h}\mathcal{K}$. □

REMARK 2.1.8. There are contravariant versions of each of the adjunction preservation results of Proposition 2.1.7, the first of which we explain in detail (see Exercise 2.1.i for further discussion). Fixing the codomain variable of the functor space at any ∞-category $C \in \mathcal{K}$ defines a 2-functor

$$\mathsf{Fun}(-,C): \mathfrak{h}\mathcal{K}^{\mathrm{op}} \longrightarrow \mathfrak{h}\mathcal{Q}\mathcal{C}at$$

that is contravariant on 1-cells and covariant on 2-cells.[4] Such 2-functors preserve adjunctions, but exchange left and right adjoints: for instance, given $f \dashv u$

---

[4] On a 2-category, the superscript "op" is used to signal that the 1-cells should be reversed but not the 2-cells, the superscript "co" is used to signal that the 2-cells should be reversed but not the 1-cells, and the superscript "coop" is used to signal that both the 1- and 2-cells should be reversed (see Definition B.1.6).

in $\mathcal{K}$, we obtain an adjunction

$$\mathrm{Fun}(A,C) \underset{f^*}{\overset{u^*}{\rightleftarrows}} \mathrm{Fun}(B,C)$$

between the functor spaces.

The next five results have standard proofs that can be taken "off the shelf" by querying any 2-category theorist who may happen to be standing nearby. The only novelty is the observation that these standard arguments can be applied to the theory of adjunctions between ∞-categories.

PROPOSITION 2.1.9. *Adjunctions compose: given adjoint functors*

$$C \underset{u'}{\overset{f'}{\rightleftarrows}} B \underset{u}{\overset{f}{\rightleftarrows}} A \qquad \rightsquigarrow \qquad C \underset{u'u}{\overset{ff'}{\rightleftarrows}} A$$

the composite functors are adjoint.

Proof Writing $\eta : \mathrm{id}_B \Rightarrow uf, \epsilon : fu \Rightarrow \mathrm{id}_A, \eta' : \mathrm{id}_C \Rightarrow u'f'$, and $\epsilon' : f'u' \Rightarrow \mathrm{id}_B$ for the respective units and counits, the pasting diagrams

define the unit and counit of $ff' \dashv u'u$ so that the triangle equalities hold:

$\square$

An adjoint to a given functor is unique up to natural isomorphism:

PROPOSITION 2.1.10 (uniqueness of adjoints).

(i) *If $f \dashv u$ and $f' \dashv u$, then $f \cong f'$.*
(ii) *Conversely, if $f \dashv u$ and $f \cong f'$, then $f' \dashv u$.*

Proof Writing $\eta : \mathrm{id}_B \Rightarrow uf$, $\epsilon : fu \Rightarrow \mathrm{id}_A$, $\eta' : \mathrm{id}_B \Rightarrow uf'$, and $\epsilon' : f'u \Rightarrow \mathrm{id}_A$ for the respective units and counits, the pasting diagrams

define 2-cells $f \Rightarrow f'$ and $f' \Rightarrow f$. The composites $f \Rightarrow f' \Rightarrow f$ and $f' \Rightarrow f \Rightarrow f'$ are computed by pasting these diagrams together horizontally on one side or on the other. Applying the triangle equalities for the adjunctions $f \dashv u$ and $f' \dashv u$ both composites are easily seen to be identities. Hence $f \cong f'$ as functors from B to A.

Part (ii) is left as Exercise 2.1.ii.   □

The following result weakens the hypotheses of Definition 2.1.1.

LEMMA 2.1.11 (minimal adjunction data). *A pair of functors $f : B \to A$ and $u : A \to B$ form an adjoint pair $f \dashv u$ if and only if there exist natural transformations $\mathrm{id}_B \Rightarrow uf$ and $fu \Rightarrow \mathrm{id}_A$ so that the triangle equality composites $f \Rightarrow fuf \Rightarrow f$ and $u \Rightarrow ufu \Rightarrow u$ are both invertible.*

Proof The unit and counit of an adjunction certainly satisfy these hypotheses. For the converse, consider natural transformations $\eta : \mathrm{id}_B \Rightarrow uf$ and $\epsilon' : fu \Rightarrow \mathrm{id}_A$ so that the triangle equality composites

$$\phi := f \xrightarrow{\; f\eta \;} fuf \xrightarrow{\; \epsilon' f \;} f \qquad\qquad \psi := u \xrightarrow{\; \eta u \;} ufu \xrightarrow{\; u\epsilon' \;} u$$

are isomorphisms. We construct an adjunction $f \dashv u$ with unit η by modifying ϵ' to form the counit ϵ.[5] To explain the idea of the construction, note that for a fixed pair of generalized elements $b : X \to B$ and $a : X \to A$, pasting with η and

---

[5] By the co-dual of this construction, we could alternatively take ϵ' to be the counit at the cost of modifying η to form the unit (see Exercise 2.1.iii).

with ϵ' defines functions between the displayed sets of natural transformations:

$$
\begin{array}{ccccc}
& & \left\{ \begin{array}{c} B \\ b\nearrow \ \Downarrow \ \searrow f \\ X \xrightarrow{\ a\ } A \end{array} \right\} & & \\
& \xrightarrow{\epsilon' \cdot f(-)} & & \searrow -\cdot\phi & \\
\left\{ \begin{array}{c} X \xrightarrow{\ b\ } B \\ a\searrow \ \Downarrow \ \nearrow u \\ A \end{array} \right\} & \xrightarrow{\hspace{2cm}} & \Big\downarrow u(-)\cdot\eta & \cong & \left\{ \begin{array}{c} B \\ b\nearrow \ \Downarrow \ \searrow f \\ X \xrightarrow{\ a\ } A \end{array} \right\} \\
& \searrow \psi\cdot- & & \nearrow \epsilon' \cdot f(-) & \\
& \cong & \left\{ \begin{array}{c} X \xrightarrow{\ b\ } B \\ a\searrow \ \Downarrow \ \nearrow u \\ A \end{array} \right\} & & \\
\end{array}
$$

From the hypothesis that the triangle equality composites are isomorphisms, two of these functions are invertible, and then by the 2-of-6 property for isomorphisms all six maps are bijections.

Define the "corrected" counit to be the composite:

$$
\epsilon := \ \begin{array}{c} B \\ u\nearrow \ \Downarrow\epsilon' \ {\cong}{\Downarrow}\phi^{-1} \searrow f \\ A \xRightarrow[\ \ f\ \ ]{} A \end{array}
$$

so that one of the triangle equality composites reduces to the identity:

$$
\begin{array}{c} B =\!=\!=\!= B \\ f\searrow \ \Downarrow\eta \ \nearrow u \ \Downarrow\epsilon \ \searrow f \\ A =\!=\!=\!= A \end{array} = \begin{array}{c} B =\!=\!=\!= B \\ f\searrow \ \Downarrow\eta \ \nearrow u \ \Downarrow\epsilon' \ {\cong}{\Downarrow}\phi^{-1} \searrow f \\ A \xRightarrow{} A \end{array} = f\left(\begin{array}{c} B \\ = \\ A \end{array}\right) f
$$

Now from the pasting equality

$$
\begin{array}{c} B =\!=\!= B \xrightarrow{f} B \\ u\nearrow \ \Downarrow\epsilon' \ f\searrow \ \Downarrow\eta \ \nearrow u \ \Downarrow\epsilon' \ {\cong}{\Downarrow}\phi^{-1} \ \Downarrow\eta \ \nearrow u \\ A =\!=\!= A \xrightarrow{f} A \end{array} = \begin{array}{c} B =\!=\!= B \\ u\nearrow \ \Downarrow\epsilon' \ f\searrow \ \Downarrow\eta \ \nearrow u \\ A =\!=\!= A \end{array}
$$

we see that $(u\epsilon \cdot \eta u) \cdot \psi = \psi$. Since ψ is invertible, we may cancel to conclude that $u\epsilon \cdot \eta u = \mathrm{id}_u$. □

A standard 2-categorical result is that any equivalence in a 2-category can be promoted to an equivalence that also defines an adjunction:

PROPOSITION 2.1.12 (adjoint equivalences). *Any equivalence can be promoted to an **adjoint equivalence** by modifying one of the 2-cells. That is, the invertible 2-cells in an equivalence can be chosen so as to satisfy the triangle equalities. Hence, if f and g are inverse equivalences then $f \dashv g$ and $g \dashv f$.*

Proof   Consider an equivalence comprised of functors $f : A \to B$ and $g : B \to A$ and invertible 2-cells

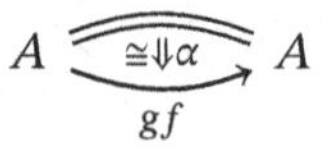

$$A \xrightarrow[gf]{\overset{}{\cong \Downarrow \alpha}} A \qquad \text{and} \qquad B \overset{fg}{\underset{}{\cong \Downarrow \beta}} B$$

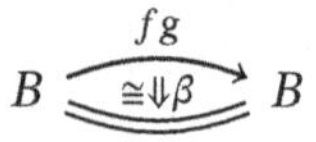

Since α and β are both invertible, the triangle equality composites are as well, and the construction of Lemma 2.1.11 applies.   □

One use of Proposition 2.1.12 is to show that adjunctions are equivalence invariant:

PROPOSITION 2.1.13. *A functor $u : A \to B$ between ∞-categories admits a left adjoint if and only if, for any pair of equivalent ∞-categories $A' \simeq A$ and $B' \simeq B$, the equivalent functor $u' : A' \to B'$ admits a left adjoint.*

As we shall discover, all of ∞-category theory is equivalence invariant in this way.

Proof   If $u : A \to B$ admits a left adjoint then by composing $f \dashv u$ with the adjoint equivalences $A' \simeq A$ and $B \simeq B'$ we obtain an equivalent adjunction:

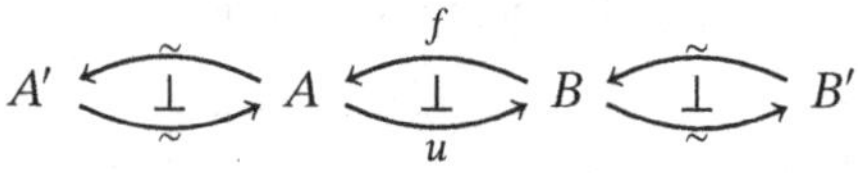

$$A' \overset{f}{\underset{u}{\cong \perp \cong}} A \underset{}{\perp} B \overset{}{\cong \perp \cong} B'$$

Conversely, if the equivalent functor $u' : A' \overset{\sim}{\to} A \overset{u}{\to} B \overset{\sim}{\to} B'$ admits a left adjoint f' then again we obtain a composite adjunction:

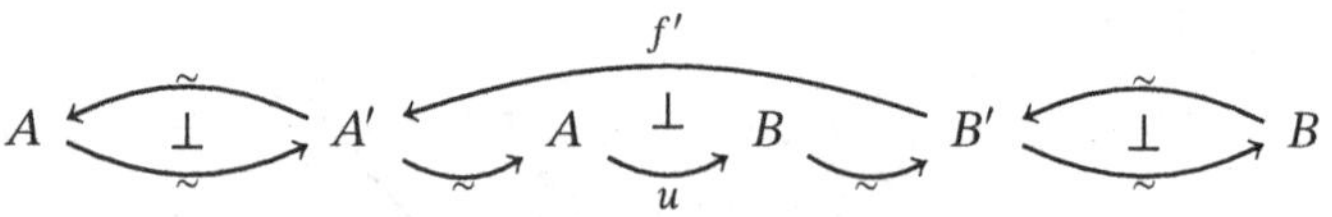

$$A \overset{f'}{\underset{}{\perp}} A' \overset{}{\perp} A \overset{}{\underset{u}{\perp}} B \overset{}{\perp} B' \overset{}{\underset{}{\perp}} B$$

whose right adjoint is naturally isomorphic to the original functor u. By Proposition 2.1.10 the displayed left adjoint is then a left adjoint to u.   □

For later use, we close with an example of an abstractly defined adjunction that can be constructed for any ∞-category in any ∞-cosmos via the results proven in this section.

LEMMA 2.1.14. *For any ∞-category A, the "composition" functor*

$$A^2 \times_A A^2 \xleftrightarrow[\substack{(\mathrm{id}_{\mathrm{cod}(-)},-)}]{\substack{(-,\mathrm{id}_{\mathrm{dom}(-)}) \\ \perp \\ \circ \\ \perp}} A^2 \qquad (2.1.15)$$

admits left and right adjoints, which extend an arrow into a composable pair by pairing it with the identities at its domain or its codomain, respectively.

Proof There is a dual adjunction in $\Delta \subset \mathcal{C}at$ whose functors we describe using notation for simplicial operators introduced in 1.1.1:

$$3 \xleftrightarrow[\substack{\sigma^1}]{\substack{\sigma^0 \\ \top \\ \delta^1 \\ \top}} 2 \qquad \rightsquigarrow \qquad A^3 \xleftrightarrow[\substack{A^{\sigma^1}}]{\substack{A^{\sigma^0} \\ \perp \\ A^{\delta^1} \\ \perp}} A^2$$

For any ∞-category A in an ∞-cosmos $\mathcal{K}$, Exercise 2.1.i describes a 2-functor $A^{(-)} : \mathcal{C}at^{\mathrm{op}} \to \mathfrak{h}\mathcal{K}$ carrying the adjoint triple displayed above-left to the one displayed above-right.

Now we claim there is a trivial fibration $A^3 \twoheadrightarrow A^2 \times_A A^2$ constructed as follows. The pushout diagram of simplicial sets displayed below-left is carried by the simplicial cotensor $A^{(-)} : sSet^{\mathrm{op}} \to \mathcal{K}$ to a pullback diagram of ∞-categories below-right; since the legs of the pushout square are monomorphisms, the legs of the pullback square are isofibrations by 1.2.1(ii):

$$\begin{array}{ccc} \Lambda^1[2] & \longleftarrow & 2 \\ \Big\uparrow & \lrcorner & \Big\uparrow {\scriptstyle\delta^1} \\ 2 & \xleftarrow{\delta^0} & \mathbb{1} \end{array} \qquad\qquad \begin{array}{ccc} A^{\Lambda^1[2]} & \longrightarrow & A^2 \\ \Big\downarrow & \lrcorner & \Big\downarrow {\scriptstyle\mathrm{ev}_0} \\ A^2 & \xrightarrow{\mathrm{ev}_1} & A \end{array}$$

By Lemma 1.2.14, the cotensor of the inner horn inclusion $\Lambda^1[2] \hookrightarrow \Delta[2] \cong 3$ with the ∞-category A defines a trivial fibration $q : A^3 \twoheadrightarrow A^{\Lambda^1[2]} \cong A^2 \times_A A^2$. By Lemma 1.2.18, the trivial fibration $q : A^3 \twoheadrightarrow A^2 \times_A A^2$ admits a section s, which defines an equivalence inverse. By Proposition 2.1.12, these functors are both left and right adjoints. The desired adjoint triple may then be constructed as the composite adjunction:

$$A^2 \times_A A^2 \xleftrightarrow[\substack{q}]{\substack{q \\ \perp \\ s \\ \perp}} A^3 \xleftrightarrow[\substack{A^{\sigma^1}}]{\substack{A^{\sigma^0} \\ \perp \\ A^{\delta^1} \\ \perp}} A^2 \qquad\qquad \square$$

Note that the adjoint functors of (2.1.15) commute with the "endpoint evaluation" functors to $A \times A$. In fact, the units and counits can similarly be fibered over $A \times A$ (see Example 3.6.13).

Exercises

EXERCISE 2.1.i. The aim of this exercise is to spell out the most subtle of the dual adjunction-preservation results discussed in Remark 2.1.8.

(i) Let A be an ∞-category is an ∞-cosmos $\mathcal{K}$. Show that the simplicial cotensor restricts to define a 2-functor $A^{(-)} : \mathfrak{h} \mathcal{Q} \mathcal{C} at^{op} \to \mathfrak{h} \mathcal{K}$.

(ii) Argue that the 2-functor of (i) restricts further along the nerve embedding to define a 2-functor $A^{(-)} : \mathcal{C} at^{op} \to \mathfrak{h} \mathcal{K}$.

(iii) Conclude that for any adjunction between 1-categories as below-left there is an induced adjunction between ∞-categories as below-right:

$$I \underset{u}{\overset{f}{\rightleftarrows}} \bot J \qquad \leadsto \qquad A^I \underset{f^*}{\overset{u^*}{\rightleftarrows}} \bot A^J$$

EXERCISE 2.1.ii. Prove Proposition 2.1.10(ii).

EXERCISE 2.1.iii. Dualize the proof of Lemma 2.1.11 so that it applies in the context of Proposition 2.1.12 to show that any equivalence can be promoted into an adjoint equivalence in which the counit is part of the originally specified data.

EXERCISE 2.1.iv. Prove that an adjoint equivalence between ∞-categories descends to an adjoint equivalence between their homotopy categories.

2.2 Initial and Terminal Elements

Employing the tactic used in Definition 1.4.11 to define the homotopy category of an ∞-category, we use the terminal ∞-category 1 to probe inside an ∞-category A. An object a in the homotopy category hA is defined to be a map of ∞-categories $a : 1 \to A$. To avoid the proliferation of the term "objects," and in deference to Lawvere's notion of (generalized) elements [74], we refer to maps $a : 1 \to A$ as **elements**[6] of the ∞-category A henceforth. This terminology will help us keep track of the "category level" under discussion: elements a

---

[6] A **generalized element** of A is a functor $f : X \to A$. By the Yoneda lemma, an ∞-category is determined by its generalized elements.

live inside ∞-categories A, which are the objects of ∞-cosmoi $\mathcal{K}$ – which themselves define "infinite-dimensional categories," albeit of a different sort.

DEFINITION 2.2.1 (initial/terminal element). An **initial element** in an ∞-category A is a left adjoint to the unique functor $! : A \to 1$, as displayed below-left, while a **terminal element** in an ∞-category A is a right adjoint, as displayed below-right.

$$1 \underset{!}{\overset{i}{\rightleftarrows}} \perp\, A \qquad\qquad 1 \underset{t}{\overset{!}{\rightleftarrows}} \perp\, A$$

Let us unpack the definition of an initial element; dual remarks apply to terminal elements.

LEMMA 2.2.2 (minimal data). *To define an initial element in an ∞-category A, it suffices to specify*

- *an element $i : 1 \to A$ and*

- *a natural transformation* $\begin{array}{c} 1 \\ {}^{!}\nearrow \;{\Downarrow\epsilon}\; \searrow^{i} \\ A =\!=\!=\!= A \end{array}$ *from the constant functor at i to the identity functor*

so that the component $\epsilon i : i \Rightarrow i$, an arrow from i to i in hA, is invertible.

Proof Proposition 1.4.5, whose proof starts in the paragraph before its statement, demonstrates that the ∞-category $1 \in \mathcal{K}$ is 2-terminal in the homotopy 2-category $\mathfrak{h}\mathcal{K}$. The 1-dimensional aspect of this universal property implies that any element $i : 1 \to A$ defines a section of the unique map $! : A \to 1$, while the 2-dimensional aspect asserts that there exist no nonidentity 2-cells with codomain 1. In particular, the unit of the adjunction $i \dashv !$ is necessarily an identity and one of the triangle equalities comes for free. What remains of Definition 2.1.1 in this setting is the data of a counit natural transformation $\epsilon : i! \Rightarrow \mathrm{id}_A$ together with the condition that its component $\epsilon i = \mathrm{id}_i$. But in fact we can prove that this natural transformation must be the identity from the weaker and more natural assumption that $\epsilon i : i \cong i$ is invertible.

To see this consider, the horizontal composite

$$1 \overset{i}{\to} A \begin{array}{c} {}^{!}\nearrow {\Downarrow\epsilon} \searrow^{i} \\ =\!=\!=\!= \end{array} A \begin{array}{c} {}^{!}\nearrow {\Downarrow\epsilon} \searrow^{i} \\ =\!=\!=\!= \end{array} A \quad \rightsquigarrow \quad \begin{array}{ccc} i!i!i & \overset{\epsilon i!i}{\Longrightarrow} & i!i \\ {}_{i!\epsilon i}\Downarrow & & \Downarrow{\epsilon i} \\ i!i & \underset{\epsilon i}{\Longrightarrow} & i \end{array}$$

By naturality of whiskering,[7] we can evaluate this composite as a vertical composite in two ways. Since 1 is 2-terminal, the whiskered cell $!\epsilon = \mathrm{id}_1$, so the composition relation reduces to $\epsilon i \cdot \epsilon i = \epsilon i$. Thus ϵi is an idempotent isomorphism, and hence, by cancelation, an identity.   □

Put more concisely, an initial element defines a *left adjoint right inverse* to the functor $! : A \twoheadrightarrow 1$, while a terminal element defines a *right adjoint right inverse* (see §B.4).

LEMMA 2.2.3 (uniqueness). *Any two initial elements in an ∞-category A are isomorphic in hA and any element of hA that is isomorphic to an initial element is initial.*

Proof By Proposition 2.1.10, any two left adjoints i and i' to the functor $! : A \to 1$ are naturally isomorphic, and any $a : 1 \to A$ that is isomorphic to a left adjoint to $! : A \to 1$ is itself a left adjoint. A natural isomorphism between a pair of functors $i, i' : 1 \to A$ gives exactly the data of an isomorphism $i \cong i'$ between the corresponding elements of the homotopy category hA.   □

REMARK 2.2.4. Applying the 2-functor $\mathsf{Fun}(X, -) : \mathfrak{h}\mathcal{K} \to \mathfrak{h}\mathfrak{Q}\mathcal{C}at$ to an initial element $i : 1 \to A$ of an ∞-category $A \in \mathcal{K}$ yields an adjunction

$$1 \cong \mathsf{Fun}(X, 1) \underset{!}{\overset{i_*}{\rightleftarrows}} \mathsf{Fun}(X, A)$$

Via the isomorphism $\mathsf{Fun}(X, 1) \cong 1$ that expresses the universal property of the terminal ∞-category 1, the constant functor at an initial element

$$X \xrightarrow{\ !\ } 1 \xrightarrow{\ i\ } A$$

defines an initial element of the functor space $\mathsf{Fun}(X, A)$. This observation can be summarized by saying that initial elements are representably initial at the level of the ∞-cosmos.

Conversely, if $i : 1 \to A$ is representability initial, then i defines an initial element of A. This is most easily seen by passing to the homotopy 2-category, where we can show that an initial element $i : 1 \to A$ is initial among all *generalized elements* $f : X \to A$ in the following precise sense.

---

[7] "Naturality of whiskering" refers to the observation of Lemma B.1.3 that any horizontal-composite of 2-cells in a 2-category can be expressed as a vertical composite of whiskerings of those cells in two different ways, in this case giving rise to the commutative diagram in $hA := hFun(1, A)$ displayed above-right.

LEMMA 2.2.5. *An element $i: 1 \to A$ is initial if and only if for all $f: X \to A$ there exists a unique 2-cell with boundary*

$$
\begin{array}{ccc}
 & 1 & \\
{}^{!}\nearrow & \Downarrow \exists! & \searrow^{i} \\
X & \xrightarrow{\quad f \quad} & A
\end{array}
$$

Proof If $i: 1 \to A$ is initial, then the adjunction of Definition 2.2.1 is preserved by the 2-functor $\mathrm{hFun}(X, -): \mathfrak{h}\mathcal{K} \to \mathcal{C}at$, defining an adjunction

$$
\mathbb{1} \cong \mathrm{hFun}(X, 1) \underset{!}{\overset{i_*}{\underset{\perp}{\rightleftarrows}}} \mathrm{hFun}(X, A)
$$

Via the isomorphism $\mathrm{hFun}(X, 1) \cong \mathbb{1}$, this adjunction proves that the constant functor $i!: X \to A$ is initial in the category $\mathrm{hFun}(X, A)$ and thus has the universal property of the statement.

Conversely, if $i: 1 \to A$ satisfies the universal property of the statement, applying this to the generic element of A (the identity map $\mathrm{id}_A: A \to A$) produces the data of Lemma 2.2.2. $\qquad\square$

Lemma 2.2.5 says that initial elements are representably initial in the homotopy 2-category. Specializing the generalized elements to ordinary elements, we see that initial and terminal elements in A respectively define initial and terminal elements in its homotopy category:

$$
\mathbb{1} \underset{\underset{t}{\overset{\perp}{\longleftarrow}}}{\overset{\overset{i}{\overset{\perp}{\longrightarrow}}}{\underset{!}{\rightleftarrows}}} \mathrm{h}A \tag{2.2.6}
$$

In general the property of being "homotopy initial," i.e., initial in the homotopy category, is weaker than being initial in the ∞-category. However Nguyen, Raptis, and Schrade observe that a homotopy initial element in a complete $(\infty, 1)$-category necessarily defines an initial element [88, 2.2.2].

Continuing the theme of the equivalence invariance of ∞-categorical notions:

LEMMA 2.2.7. *If A has an initial element and $A \simeq A'$ then A' has an initial element and these elements are preserved up to isomorphism by the equivalences.*

Proof By Proposition 2.1.12, the equivalence $A \simeq A'$ can be promoted to an adjoint equivalence, which can immediately be composed with the adjunction characterizing an initial element i of A:

$$
1 \underset{!}{\overset{i}{\underset{\perp}{\rightleftarrows}}} A \underset{\sim}{\overset{\sim}{\underset{\perp}{\rightleftarrows}}} A'
$$

The composite adjunction provided by Proposition 2.1.9 proves that the image of i defines an initial element of A', which by construction is preserved by the equivalence $A \xrightarrow{\sim} A'$. By the uniqueness of initial elements established in Lemma 2.2.3, this argument also shows that the equivalence $A' \xrightarrow{\sim} A$ preserves initial elements.   □

We now turn to the general theory of limits and colimits of diagrams valued in an ∞-category. The theory of initial elements previews this material well since in fact an initial element can be understood as an example of both notions: an initial element is the colimit of the empty diagram and also the limit of the diagram encoded by the identity functor, as we explain in Example 2.3.11.

Exercises

EXERCISE 2.2.i. Use Lemma 2.2.5 to show that a representably initial element, as described in Remark 2.2.4, necessarily defines an initial element in A.

EXERCISE 2.2.ii. Prove that initial elements are preserved by left adjoints and terminal elements are preserved by right adjoints.

2.3 Limits and Colimits

We now introduce limits and colimits of diagram valued *inside* an ∞-category A in some ∞-cosmos. We consider two varieties of diagrams:

- diagrams indexed by a simplicial set J and valued in an ∞-category A in a generic ∞-cosmos and
- diagrams indexed by an ∞-category J and valued in an ∞-category A in a cartesian closed ∞-cosmos.[8]

DEFINITION 2.3.1 (diagram ∞-category). For an ∞-category A and a simplicial set J – or possibly, in the case of a cartesian closed ∞-cosmos, an ∞-category J – we refer to A^J as the ∞-**category of J-shaped diagrams in** A. A **diagram** of shape J in A is an element $d : 1 \to A^J$.[9]

---

[8] For the ∞-cosmoi of $(\infty, 1)$-categories of Example 1.2.24, there is no essential difference between these notions: in $\mathcal{QC}at$ they are tautologically the same, and in all biequivalent ∞-cosmoi the ∞-category of diagrams indexed by an ∞-category J is equivalent to the ∞-category of diagrams indexed by its underlying quasi-category, regarded as a simplicial set (see Proposition 10.3.5).

[9] When A^J is the exponential of a cartesian closed ∞-cosmos, diagrams stand in bijection with functors $d : J \to A$.

Both constructions of the ∞-category of diagrams in an ∞-cosmos $\mathcal{K}$ define simplicial bifunctors

$$sSet^{op} \times \mathcal{K} \longrightarrow \mathcal{K} \qquad\qquad \mathcal{K}^{op} \times \mathcal{K} \longrightarrow \mathcal{K}$$

$$(J, A) \longmapsto A^J \qquad\qquad\qquad (J, A) \longmapsto A^J$$

In either indexing context, there is a terminal object 1 with the property that $A^1 \cong A$ for any ∞-category A. Restriction along the unique map $!: J \to 1$ induces the **constant diagram functor** $\Delta: A \to A^J$.

We deliberately conflate the notation for ∞-categories of diagrams indexed by a simplicial set or by another ∞-category because all of the results we prove in Part I about the former case also apply to the latter. For economy of language, we refer only to simplicial set indexed diagrams for the remainder of this section.

DEFINITION 2.3.2 (limit and colimit functor). An ∞-category A **admits all colimits** of shape J if the constant diagram functor $\Delta: A \to A^J$ admits a left adjoint, while A **admits all limits** of shape J if the constant diagram functor admits a right adjoint:

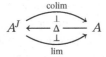

In the ∞-cosmos of categories, Definition 2.3.2 reduces to the classically defined limit and colimit functors, but in a general ∞-category limits and colimits should be thought of as analogous to the classical notions of "homotopy limits" and "homotopy colimits." In certain cases, this correspondence can be made precise. Every quasi-category is equivalent to the homotopy coherent nerve of a Kan complex enriched category [111, 7.2.2], and homotopy limit or homotopy colimit cones in the Kan complex enriched category correspond exactly to limit or colimit cones in the homotopy coherent nerve (see Lurie's [78, 4.2.4.1] or [113, 6.1.4, 6.2.7]). In the ∞-categorical context, no stricter notion of limit or colimit is available, so the "homotopy" qualifier is typically dropped.

Limits or colimits of set-indexed diagrams – the case where the indexing shape is a coproduct of the terminal object 1 indexed by a set J – are called **products** or **coproducts**, respectively.

LEMMA 2.3.3. *Products or coproducts in an ∞-category A also define products or coproducts in its homotopy category* hA.

Proof When J is a set, the ∞-category of diagrams itself decomposes as a product $A^J \cong \prod_J A$. Since the 2-functor that carries an ∞-category to its homotopy category

$$\begin{array}{ccc} \mathfrak{h}\mathcal{K} & \xrightarrow{\mathsf{hFun}(1,-)} & \mathcal{C}at \\ A & \longmapsto & \mathsf{h}A \end{array}$$

preserves products, there is a chain of isomorphisms

$$\mathsf{h}(A^J) \cong \mathsf{h}(\textstyle\prod_J A) \cong \prod_J \mathsf{h}A \cong (\mathsf{h}A)^J$$

when J is a set. Thus, in this special case, the adjunctions of Definition 2.3.2 that define products or coproducts in an ∞-category descend to the adjunctions that define products or coproducts in its homotopy category:

$$(\mathsf{h}A)^J \cong \mathsf{h}(A^J) \underset{\underset{\lim}{\longrightarrow}}{\overset{\overset{\mathrm{colim}}{\longrightarrow}}{\xleftarrow[\bot]{\;\bot\;\Delta}}} \mathsf{h}A$$

This remains true in the case $J = \varnothing$, explaining the observation made in (2.2.6). □

Warning 2.3.4. This argument does *not* extend to more general limit or colimit notions, and such ∞-categorical limits or colimits do not typically descend to limits or colimits in the homotopy category.[10] In §3.2, we observe that the homotopy category construction fails to preserve more complicated cotensors, even in the relatively simple case of $J = \mathbb{2}$.

The problem with Definition 2.3.2 is that it is insufficiently general: many ∞-categories have certain, but not all, limits of diagrams of a particular indexing shape. So it would be desirable to re-express Definition 2.3.2 in a form that allows us to consider the limit of a single diagram $d\colon 1 \to A^J$ or of a family of diagrams. To achieve this, we make use of the following 2-categorical notion that op-dualizes the more familiar absolute (Kan) extension diagrams.

Definition 2.3.5 (absolute lifting diagram). Given a cospan $C \xrightarrow{g} A \xleftarrow{f} B$ in a 2-category, an **absolute left lifting** of g through f is given by a 1-cell ℓ and 2-cell λ as displayed below-left

$$\begin{array}{ccc} & & B \\ & {}^{\ell}\nearrow \;\Uparrow\lambda & \downarrow f \\ C & \xrightarrow[g]{} & A \end{array} \qquad \begin{array}{ccc} X & \xrightarrow{b} & B \\ c\downarrow & \Uparrow\chi & \downarrow f \\ C & \xrightarrow[g]{} & A \end{array} \quad = \quad \begin{array}{ccc} X & \xrightarrow{b} & B \\ c\downarrow & \exists!\Uparrow\zeta \;\; {}^{\ell}\nearrow \;\Uparrow\lambda & \downarrow f \\ C & \xrightarrow[g]{} & A \end{array}$$

[10] This sort of behavior is familiar from abstract homotopy theory: homotopy limits and colimits are not generally limits or colimits in the homotopy category.

so that any 2-cell as displayed above-center factors uniquely through (ℓ, λ) as displayed above-right.

Dually, an **absolute right lifting** of g through f is given by a 1-cell r and 2-cell ρ as displayed below-left

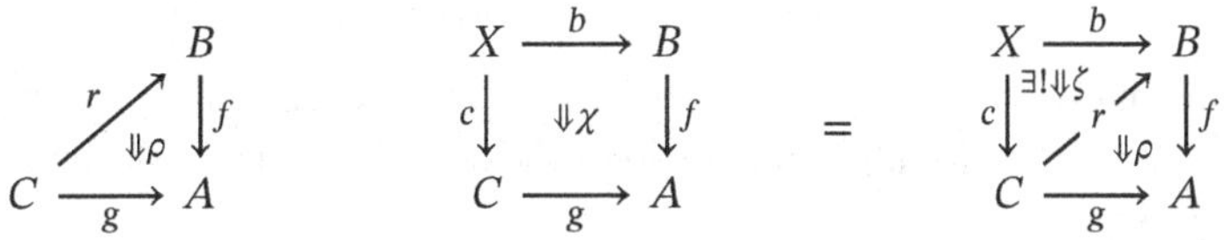

$$
\begin{array}{ccc}
\begin{array}{c}
B \\
\nearrow{\scriptstyle r} \quad \Downarrow{\scriptstyle f} \\
\Downarrow\rho \\
C \xrightarrow[g]{} A
\end{array}
&
\begin{array}{c}
X \xrightarrow{b} B \\
{\scriptstyle c}\downarrow \quad \Downarrow\chi \quad \downarrow{\scriptstyle f} \\
C \xrightarrow[g]{} A
\end{array}
=
&
\begin{array}{c}
X \xrightarrow{b} B \\
{\scriptstyle c}\downarrow \; {}^{\exists!\Downarrow\zeta}\nearrow \; \downarrow{\scriptstyle f} \\
\nearrow{\scriptstyle r} \; \Downarrow\rho \\
C \xrightarrow[g]{} A
\end{array}
\end{array}
$$

so that any 2-cell as displayed above-center factors uniquely through (r, ρ) as displayed above-right.

When these exist, left and right liftings respectively define left and right adjoints to the composition functor $f_* : \mathsf{hFun}(C, B) \to \mathsf{hFun}(C, A)$, with the 2-cells defining the components of the unit and counit of these adjunctions, respectively, at the object g. The adjective "absolute" refers to the following stability property.

LEMMA 2.3.6. *Absolute left or right lifting diagrams are stable under restriction of their domain object: if (ℓ, λ) defines an absolute left lifting of g through f, then for any $c : X \to C$, the restricted diagram $(\ell c, \lambda c)$ defines an absolute left lifting of gc through f.*

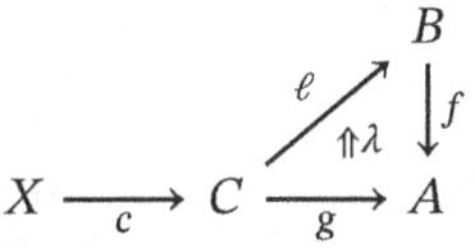

$$
\begin{array}{c}
B \\
{\scriptstyle \ell}\nearrow \quad \downarrow{\scriptstyle f} \\
\Uparrow\lambda \\
X \xrightarrow[c]{} C \xrightarrow[g]{} A
\end{array}
$$

Proof Exercise 2.3.ii. □

Units and counits of adjunctions provide important examples of absolute left and right lifting diagrams, respectively:

LEMMA 2.3.7. *A 2-cell $\eta : \mathrm{id}_B \Rightarrow uf$ defines the unit of an adjunction $f \dashv u$ if and only if (f, η) defines an absolute left lifting diagram, displayed below-left.*

$$
\begin{array}{cc}
\begin{array}{c}
A \\
{\scriptstyle f}\nearrow \quad \downarrow{\scriptstyle u} \\
\Uparrow\eta \\
B =\!=\!= B
\end{array}
&
\begin{array}{c}
B \\
{\scriptstyle u}\nearrow \quad \downarrow{\scriptstyle f} \\
\Downarrow\epsilon \\
A =\!=\!= A
\end{array}
\end{array}
$$

Dually a 2-cell $\epsilon : fu \Rightarrow \mathrm{id}_A$ defines the counit of an adjunction if and only if (u, ϵ) defines an absolute right lifting diagram, displayed above-right.

Proof The universal property of the absolute right lifting diagram

$$
\begin{array}{ccc}
X \xrightarrow{\ b\ } B & & X \xrightarrow{\ b\ } B \\
{\scriptstyle a}\downarrow \quad \Downarrow\alpha \quad \downarrow{\scriptstyle f} & = & {\scriptstyle a}\downarrow \ {}^{\exists!\Downarrow\beta}\nearrow_{u}\ \downarrow{\scriptstyle f} \\
A =\!\!=\!\!= A & & A =\!\!=\!\!= A
\end{array}
$$

asserts that every natural transformation $\alpha : fb \Rightarrow a$ has a unique transpose $\beta : b \Rightarrow ua$ across the adjunction between the hom-categories of the homotopy 2-category:

$$
\mathrm{hFun}(X, B) \underset{u_*}{\overset{f_*}{\rightleftarrows}} \bot \ \mathrm{hFun}(X, A)
$$

Thus if $f \dashv u$ with counit ϵ, Proposition 2.1.7(ii) supplies this induced adjunction and (u, ϵ) defines an absolute right lifting of id_A through f.

Conversely, the unit and triangle equalities of an adjunction can extracted from the universal property of the absolute right lifting diagram. The details are left as Exercise 2.3.iii. □

In particular, the unit and counit of the adjunctions $\mathrm{colim} \dashv \Delta \dashv \mathrm{lim}$ of Definition 2.3.2 define absolute left and right lifting diagrams:

$$
\begin{array}{cc}
\begin{array}{c}
A \\
{}^{\mathrm{colim}}\nearrow \ {\scriptstyle\Uparrow\eta} \ \downarrow{\scriptstyle\Delta} \\
A^J =\!\!=\!\!= A^J
\end{array}
&
\begin{array}{c}
A \\
{}^{\mathrm{lim}}\nearrow \ {\scriptstyle\Downarrow\epsilon} \ \downarrow{\scriptstyle\Delta} \\
A^J =\!\!=\!\!= A^J
\end{array}
\end{array}
$$

By Lemma 2.3.6, these universal properties are retained upon restricting to any subobject of the ∞-category of diagrams. This motivates the following definition:

DEFINITION 2.3.8 (limit and colimit). A **colimit** of a family of diagrams $d : D \to A^J$ of shape J in an ∞-category A is given by an absolute left lifting diagram

$$
\begin{array}{c}
A \\
{}^{\mathrm{colim}\,d}\nearrow \ {\scriptstyle\Uparrow\eta} \ \downarrow{\scriptstyle\Delta} \\
D \xrightarrow{\ d\ } A^J
\end{array}
$$

comprised of a generalized element $\mathrm{colim}\,d : D \to A$ and a **colimit cone** $\eta : d \Rightarrow \Delta\,\mathrm{colim}\,d$.

Dually, a **limit** of a family of diagrams $d : D \to A^J$ of shape J in an ∞-

category A is given by an absolute right lifting diagram

$$\begin{array}{ccc} & & A \\ & \overset{\lim d}{\nearrow} \quad \Big\downarrow \Delta & \\ & {\scriptstyle\Downarrow\epsilon} & \\ D & \xrightarrow{\quad d \quad} & A^J \end{array}$$

comprised of a generalized element $\lim d : D \to A$ and a **limit cone** $\epsilon :$ $\Delta \lim d \Rightarrow d$.

REMARK 2.3.9. If A has all limits of shape J, then Lemma 2.3.6 implies that any family of diagrams $d : D \to A^J$ has a limit, defined by composing the limit functor $\lim : A^J \to A$ with d. In an ∞-cosmos of $(\infty, 1)$-categories, if every diagram $d : 1 \to A^J$ has a limit, then A admits all limits of shape J (see Corollary 12.2.10), but in general families of diagrams cannot be reduced to single diagrams.

EXAMPLE 2.3.10. An initial element $i : 1 \to A$ can be regarded as a colimit of the empty diagram. The ∞-category $A^\varnothing \simeq 1$ of empty diagrams in A is terminal, so the constant diagram functor reduces to $! : A \to 1$. To show that initial elements are colimits in the sense of Definition 2.3.8, we must verify that an initial element defines an absolute left lifting diagram whose 2-cell is the identity:

$$\begin{array}{ccc} & A & \\ \overset{i}{\nearrow} & \Big\uparrow{\scriptstyle !} & \\ 1 & =\!\!=\!\!= & 1 \end{array} \qquad \begin{array}{ccc} X & \xrightarrow{\ f\ } & A \\ {\scriptstyle !}\Big\downarrow & {\scriptstyle\Uparrow\chi} & \Big\downarrow{\scriptstyle !} \\ 1 & =\!\!=\!\!= & 1 \end{array} \;=\; \begin{array}{ccc} X & \xrightarrow{\ f\ } & A \\ {\scriptstyle !}\Big\downarrow{\scriptstyle\exists!\Uparrow\zeta} & \overset{i}{\nearrow} & \Big\downarrow{\scriptstyle !} \\ 1 & =\!\!=\!\!= & 1 \end{array}$$

Since the ∞-category 1 is 2-terminal, there is a unique 2-cell χ inhabiting the central square above, namely the identity. Thus, the universal property of the absolute left lifting diagram asserts the existence of a unique 2-cell $\zeta : i! \Rightarrow f$ for any $f : X \to A$, exactly as provided by Lemma 2.2.5.

EXAMPLE 2.3.11. In a cartesian closed ∞-cosmos, an initial element $i : 1 \to A$ can also be regarded as a limit of the identity functor $\mathrm{id}_A : A \to A$.[11] The counit $\epsilon : i! \Rightarrow \mathrm{id}_A$ of the adjunction $i \dashv !$ transposes across the 2-adjunction $A \times - \dashv (-)^A$ of Proposition 1.4.5 to define the limit cone displayed below-left:

$$\begin{array}{ccc} & A & \\ \overset{i}{\nearrow} & \Big\downarrow \Delta & \\ {\scriptstyle\Downarrow\bar\epsilon} & & \\ 1 & \xrightarrow{\ \mathrm{id}_A\ } & A^A \end{array} \qquad \begin{array}{ccc} X & \xrightarrow{\ f\ } & A \\ {\scriptstyle !}\Big\downarrow & {\scriptstyle\Downarrow\chi} & \Big\downarrow \Delta \\ 1 & \xrightarrow{\ \mathrm{id}_A\ } & A^A \end{array} \;=\; \begin{array}{ccc} X & \xrightarrow{\ f\ } & A \\ {\scriptstyle !}\Big\downarrow{\scriptstyle\exists!\Downarrow\zeta} & \overset{i}{\nearrow}{\scriptstyle\Downarrow\bar\epsilon} & \Big\downarrow \Delta \\ 1 & \xrightarrow{\ \mathrm{id}_A\ } & A^A \end{array}$$

[11] This result is extended to ∞-cosmoi that are not cartesian closed in Proposition 9.4.10.

The universal property displayed above-right is easiest to verify by transposing across the 2-adjunction $A \times - \dashv (-)^A$ again, where we must establish the pasting equality

$$
\begin{array}{c}
X \times A \quad \Downarrow \hat{\chi} \quad \Big| f \; = \; \cdots \; \cdots
\end{array}
\tag{2.3.12}
$$

Observe that when we restrict the right-hand side of (2.3.12) along the functor $\mathrm{id}_X \times i : X \cong X \times 1 \to X \times A$ we recover the 2-cell ζ, since $\epsilon i = \mathrm{id}_i$. This tells us that given χ, we must necessarily define the 2-cell $\zeta : f \Rightarrow i!$ to be the restriction of $\hat{\chi}$ along the functor $\mathrm{id}_X \times i : X \to X \times A$.

From this definition of ζ and the 2-functoriality of the cartesian product – which tells us that $\epsilon \pi_A = \pi_A (X \times \epsilon)$ – we have

By "naturality of whiskering" (see Lemma B.1.3), the right-hand pasted composite can be computed as the vertical composite of $\pi_X(X \times \epsilon)$ followed by $\hat{\chi}$, but $\pi_X(X \times \epsilon)$ is the identity 2-cell, so this composite is just $\hat{\chi}$. This verifies the desired pasting equality (2.3.12).

Certain limits and colimits in ∞-categories exist for formal reasons. For example, an abstract 2-categorical lemma enables a formal proof of a classical result from homotopy theory that computes the colimits, typically called *geometric realizations*, of "split" simplicial objects. Before proving this, we introduce the indexing shapes involved.

DEFINITION 2.3.13 (split augmented (co)simplicial object). The simplex category Δ of finite nonempty ordinals and order-preserving maps introduced in 1.1.1 defines a full subcategory of the category Δ_+ of finite ordinals and order-preserving maps, which freely appends the empty ordinal "$[-1]$" as an initial object. The category Δ_+ in turn defines a wide subcategory of a category $\Delta_\perp$, which adds an "extra" degeneracy $\sigma^{-1} : [n+1] \twoheadrightarrow [n]$ between each pair of consecutive ordinals, including $\sigma^{-1} : [0] \twoheadrightarrow [-1]$. The category Δ_+ also defines a wide subcategory of a category $\Delta_\top$, which adds an "extra" degeneracy

$\sigma^{n+1} : [n+1] \twoheadrightarrow [n]$ on the other side between each pair of consecutive ordinals, including $\sigma^0 : [0] \twoheadrightarrow [-1]$. The categories $\Delta_\perp$ and $\Delta_\top$ can be described in another way: there are faithful embeddings of these categories into Δ that act on objects by $[n] \mapsto [n+1]$ and identify $\Delta_\perp$ and $\Delta_\top$ with the subcategories of finite nonempty ordinals and order-preserving maps that preserve the bottom and top elements respectively.

Covariant diagrams indexed by $\Delta \subset \Delta_+ \subset \Delta_\perp, \Delta_\top$ are, respectively, called **cosimplicial objects, coaugmented cosimplicial objects**, and **split coaugmented cosimplicial objects** (in the case of either $\Delta_\perp$ or $\Delta_\top$), while contravariant diagrams are respectively called **simplicial objects, augmented simplicial objects**, and **split augmented simplicial objects**. When it is useful to disambiguate between $\Delta_\perp$ and $\Delta_\top$ we refer to the former category as a "bottom splitting" and the latter category as a "top splitting," but this terminology is not standard.

A cosimplicial object $d : 1 \to A^\Delta$ in an ∞-category A **admits a coaugmentation** or **admits a splitting** if it lifts along the restriction functors

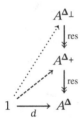

where in the case of a top splitting, $\Delta_\perp$ is replaced by $\Delta_\top$. The family of cosimplicial objects admitting a coaugmentation and splitting is represented by the generalized element $\text{res} : A^{\Delta_\perp} \twoheadrightarrow A^\Delta$. In any augmented cosimplicial object, there is a cone over the underlying cosimplicial object whose summit is obtained by evaluating at $[-1] \in \Delta_+$. This cone is defined by cotensoring with the unique natural transformation

$$\begin{array}{ccc} \Delta & \hookrightarrow & \Delta_+ \\ {\scriptstyle !} \searrow & {\scriptstyle \exists! \Uparrow \nu} & \nearrow {\scriptstyle [-1]} \\ & \mathbb{1} & \end{array} \qquad (2.3.14)$$

that exists because $[-1] : \mathbb{1} \to \Delta_+$ is initial (see Lemma 2.2.5).

PROPOSITION 2.3.15 (totalization/geometric realization). *Let A be any ∞-category. Every cosimplicial object in A that admits a coaugmentation and a splitting has a limit, whose limit cone is defined by the coaugmentation. Dually, every simplicial object in A that admits an augmentation and a splitting has a colimit, whose colimit cone is defined by the augmentation. That is, there exist* *absolute right and left lifting diagrams*

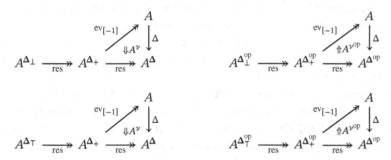

in which the 2-cells are obtained as restrictions of the cotensor of the 2-cell (2.3.14) into A. Moreover, such limits and colimits are **absolute**, *preserved by any functor* $f : A \rightarrow B$ *of ∞-categories.*

Proof By Example B.5.2, the inclusion $\Delta \hookrightarrow \Delta_\perp$ admits a right adjoint, which can automatically be regarded as an adjunction "over $\mathbb{1}$" since $\mathbb{1}$ is 2-terminal in $\mathcal{C}at$. The initial element $[-1] \in \Delta_+ \subset \Delta_\perp$ defines a left adjoint to the constant functor:

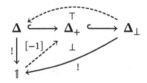

and the counit of this adjunction restricts along the inclusions $\Delta \subset \Delta_+ \subset \Delta_\perp$ to the 2-cell (2.3.14). For any ∞-category A in an ∞-cosmos $\mathcal{K}$, these adjunctions are preserved by the 2-functor $A^{(-)} : \mathcal{C}at^{\mathrm{op}} \rightarrow \mathfrak{h}\mathcal{K}$, yielding a diagram

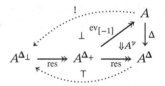

By Lemma B.5.1 these adjunctions witness the fact that evaluation at $[-1]$ and the 2-cell from (2.3.14) define an absolute right lifting of the canonical restriction functor $A^{\Delta_\perp} \twoheadrightarrow A^\Delta$ through the constant diagram functor, as claimed. The colimit case is proven similarly by applying the composite 2-functor

$$\mathcal{C}at^{\mathrm{coop}} \xrightarrow{(-)^{\mathrm{op}}} \mathcal{C}at^{\mathrm{op}} \xrightarrow{A^{(-)}} \mathfrak{h}\mathcal{K}$$

A similar argument, starting from Example B.5.3, constructs the absolute lifting diagrams from the top splitting.

Finally, by 2-functoriality of the simplicial cotensor, any functor $f : A \to B$ commutes with the 2-cells defined by cotensoring with ν or its opposite:

$$
A^{\Delta_\perp} \xrightarrow[\text{res}]{} A^{\Delta_+} \xrightarrow[\text{res}]{} A^{\Delta} \xrightarrow{f^{\Delta}} B^{\Delta}
\quad = \quad
A^{\Delta_\perp} \xrightarrow[\text{res} \cdot f^{\Delta_\perp}]{} B^{\Delta_+} \xrightarrow[\text{res}]{} B^{\Delta}
$$

Since the right-hand composite is an absolute right lifting diagram by Lemma 2.3.6, so is the left-hand composite, and thus $f : A \to B$ preserves the totalization of any split coaugmented cosimplicial object in A. □

Exercises

EXERCISE 2.3.i. Prove that if an ∞-category A has binary products then it also has ternary products (and in fact n-ary products for all $n \geq 1$). Show further that the ternary product functor can be defined from the binary product functor $- \times - : A \times A \to A$ either as the composite $(- \times -) \times -$ or as the composite $- \times (- \times -)$; that is, show that these composites are naturally isomorphic and both satisfy the universal property that characterizes ternary products.

EXERCISE 2.3.ii. Prove Lemma 2.3.6.

EXERCISE 2.3.iii. Re-prove the forwards implication of Lemma 2.3.7 by a pasting diagram calculation and prove the converse similarly.

EXERCISE 2.3.iv. Let $\lrcorner$ be the category defined by gluing two arrows along their codomain. Diagrams of shape $\lrcorner$ are called **cospans**. Consider either of the surjective functors $\pi : \lrcorner \to 2$ that send one of these arrows to an identity. Show that for any ∞-category A, the corresponding functor $A^\pi : A^2 \to A^\lrcorner$ admits an absolute right lifting through the constant diagram functor $\Delta : A \to A^\lrcorner$. That is, show that any ∞-category admits pullbacks of cospans in which one of the two arrows is an identity.

EXERCISE 2.3.v (3.5.6). Show that for any functor $f : A \to B$, the identity

$$
\begin{array}{ccc}
 & & A \\
 & \nearrow & \downarrow f \\
A & \xrightarrow{f} & B
\end{array}
$$

defines an absolute *right* lifting of f through itself if and only if the identity defines an absolute *left* lifting of f through itself by proving that each of these conditions is equivalent to the assertion that for any ∞-category X the induced functor

$$f_* : \mathsf{hFun}(X, A) \to \mathsf{hFun}(X, B)$$

is fully faithful. A fourth equivalent characterization of what it means for a functor between ∞-categories to be **fully faithful** appears in Corollary 3.5.6.

EXERCISE 2.3.vi. Show that diagrams that are isomorphic to absolute right lifting diagrams are themselves absolute right lifting: given an absolute right lifting diagram and natural isomorphisms

show that the pasted composite is an absolute right lifting diagram.

Conclude that limits and colimits are invariant under natural isomorphism.

2.4 Preservation of Limits and Colimits

A functor $f : A \to B$ *preserves* limits if the image of a limit cone in A also defines a limit cone in B. In the other direction, a functor $f : A \to B$ *reflects* limits if a cone in A that defines a limit cone in B is also a limit cone in A. A functor $f : A \to B$ *creates* limits if whenever a diagram in A admits a limit cone in B, then there must exist a limit cone in A whose image under f is isomorphic to the given limit cone in B.

Famously, right adjoint functors preserve limits and left adjoints preserve colimits. Our aim in this section is to prove this in the ∞-categorical context and exhibit the first examples of initial and final functors, in the sense introduced in Definition 2.4.5. We conclude with a result about the reflection of limits whose proof relies in a crucial way on a result – that cosmological functors preserve absolute lifting diagrams – that motivates Chapter 3.

The commutativity of right adjoints and limits is very easily established in the case where the ∞-categories in question admit *all* limits of a given shape: under these hypotheses, the limit functor is right adjoint to the constant diagram functor, which commutes with all functors between the base ∞-categories. Since the left adjoints commute, the uniqueness of adjoints (Proposition 2.1.10) implies that the right adjoints commute up to isomorphism. This outline gives a hint for Exercise 2.4.i.

A more delicate argument is needed in the general case, involving, say, the preservation of a single limit diagram without a priori assuming that any other limits exist. We appeal to a general lemma about composition and cancelation of absolute lifting diagrams:

LEMMA 2.4.1 (composition and cancelation of absolute lifting diagrams). *Suppose* (r, ρ) *defines an absolute right lifting of* h *through* f:

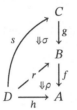

Then (s, σ) *defines an absolute right lifting of* r *through* g *if and only if* $(s, \rho \cdot f\sigma)$ *defines an absolute right lifting of* h *through* fg.

Proof Exercise 2.4.ii. □

THEOREM 2.4.2. *Right adjoints preserve limits and left adjoints preserve colimits.*

The usual argument that right adjoints preserve limits is this: a cone over a *J*-shaped diagram in the image of a right adjoint u transposes across the adjunction $f^J \dashv u^J$ to a cone over the original diagram, which factors uniquely through the designated limit cone. This factorization transposes across the adjunction $f \dashv u$ to define the sought-for unique factorization through the image of the limit cone. An ∞-categorical proof along these lines can be given as well (see Exercise 2.4.iii), but instead we present a slicker packaging of the standard argument. We use absolute lifting diagrams to express the universal properties of limits and colimits (Definition 2.3.8) and adjoint transposition (Lemma 2.3.7), allowing us to suppress consideration of a generic test cone that must be shown to uniquely factor through the limit cone.

Proof We prove that right adjoints preserve limits. By taking co-duals the same argument demonstrates that left adjoints preserve colimits.

Suppose a functor $u : A \to B$ in an ∞-cosmos $\mathcal{K}$ admits a left adjoint $f : B \to A$ with counit $\epsilon : fu \Rightarrow \mathrm{id}_A$. Our aim is to show that any absolute right lifting diagram as displayed below-left is carried to an absolute right lifting diagram as displayed below-right:

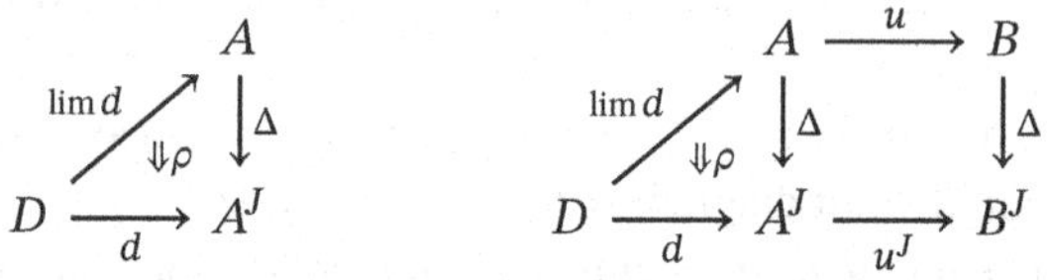

$$
\begin{array}{cc}
\begin{array}{ccc}
& A & \\
\lim d \nearrow {\scriptstyle \Downarrow \rho} & & \downarrow \Delta \\
D \xrightarrow{\quad d \quad} & & A^J
\end{array}
&
\begin{array}{ccccc}
& A & \xrightarrow{\;u\;} & B & \\
\lim d \nearrow {\scriptstyle \Downarrow \rho} & \downarrow \Delta & & \downarrow \Delta & \\
D \xrightarrow{\; d \;} & A^J & \xrightarrow{\; u^J \;} & B^J &
\end{array}
\end{array}
\qquad (2.4.3)
$$

By Proposition 2.1.7, the cotensor $(-)^J : \hbar\mathcal{K} \to \hbar\mathcal{K}$ carries the adjunction $f \dashv u$ to an adjunction $f^J \dashv u^J$ with counit ϵ^J. In particular, by Lemma 2.3.7, (u^J, ϵ^J) defines an absolute right lifting of the identity through f^J, which is then preserved by restriction along the functor d. Thus, by Lemma 2.4.1, the diagram on the right of (2.4.3) is an absolute right lifting diagram if and only if the pasted composite displayed below-left defines an absolute right lifting diagram:

$$
\begin{array}{ccc}
\begin{array}{l}
\begin{array}{ccccc}
& A & \xrightarrow{\;u\;} & B & \\
\lim d \nearrow {\scriptstyle \Downarrow \rho} & \downarrow \Delta & & \downarrow \Delta & \\
D \xrightarrow{\; d \;} & A^J & \xrightarrow{\; u^J \;} & B^J & \\
& & {\scriptstyle \Downarrow \epsilon^J} & \downarrow f^J & \\
& & & A^J &
\end{array}
\end{array}
& = &
\begin{array}{l}
\begin{array}{ccc}
& B & \\
u \nearrow {\scriptstyle \Downarrow \epsilon} & \downarrow f & \\
A & == & A \\
\lim d \nearrow {\scriptstyle \Downarrow \rho} & \downarrow \Delta & \downarrow \Delta \\
D \xrightarrow{\; d \;} & A^J & == A^J
\end{array}
\end{array}
& =
\begin{array}{l}
\begin{array}{ccc}
& & B \\
u \lim d \nearrow {\scriptstyle \Downarrow \epsilon \lim d} & & \downarrow f \\
& & A \\
\lim d \nearrow {\scriptstyle \Downarrow \rho} & \downarrow \Delta & \\
D \xrightarrow{\; d \;} & A^J &
\end{array}
\end{array}
\end{array}
$$

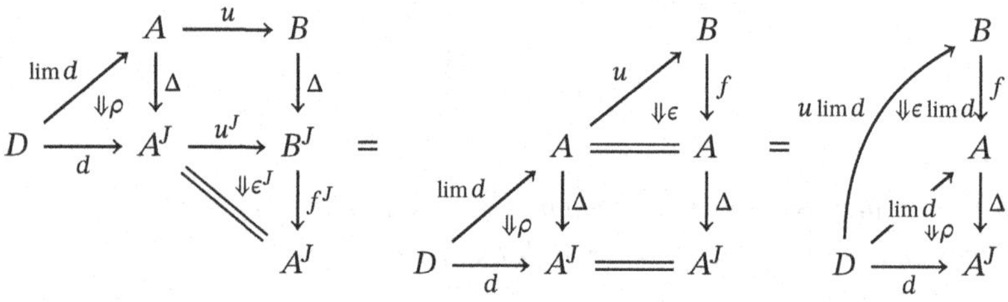

As noted in the proof of Lemma 2.3.7, pasting the 2-cell on the right of (2.4.3) with the counit in this way amounts to transposing the cone $u^J \rho$ across the adjunction $f^J \dashv u^J$.

We now argue that this transposed cone above-left factors through the limit cone $(\lim d, \rho)$ in a canonical way. From the 2-functoriality of the simplicial cotensor in its exponent variable, $f^J \Delta = \Delta f$ and $\epsilon^J \Delta = \Delta \epsilon$. Hence, the pasting diagram displayed above-left equals the one displayed above-center, which equals the diagram above-right. This latter diagram is a pasted composite of two absolute right lifting diagrams, and is then an absolute right lifting diagram in its own right by Lemma 2.4.1; this universal property says that any cone over d whose summit factors through f factors uniquely through the limit cone $(\lim d, \rho)$ through a map that then transposes along the adjunction $f \dashv u$. Hence the diagram on the right-hand side of (2.4.3) is an absolute right lifting diagram as claimed. $\square$

PROPOSITION 2.4.4. *An equivalence $f : A \xrightarrow{\sim} B$ preserves, reflects, and creates limits and colimits.*

Proof By Proposition 2.1.12, equivalences define adjoint functors, so Theorem 2.4.2 implies that equivalences preserve limits. To see that limits are reflected, consider a J-shaped cone ρ in A whose image $f^J\rho$ is a limit cone in B. The inverse equivalence $g \colon B \rightsquigarrow A$ carries this to a limit cone $g^J f^J\rho$ in A, which is naturally isomorphic to the original cone ρ. By Exercise 2.3.vi, ρ must also define a limit cone. Finally to see that limits are created, consider a diagram $d \colon D \to A^J$ so that fd has a limit cone ν in B. Then $g^J\nu$ defines a limit cone for the diagram gfd in A, and by Exercise 2.3.vi, a limit cone for d may be defined by composing with the isomorphism $gfd \cong d$. $\qquad\square$

We turn now to a limit-preservation result of another sort, which can be used to simplify the calculation of limits or colimits of diagrams with particular shapes. This simplification comes about by reindexing the diagrams, by restricting along a functor $k \colon I \to J$. For certain functors, called "initial" or "final," this reindexing preserves and reflects limits or colimits, respectively.

At present, we give a teleological, rather than an intrinsic, description of these functors. The following definition makes sense for an arbitrary functor in a cartesian closed ∞-cosmos or for a map between simplicial sets serving as indexing shapes in an arbitrary ∞-cosmos. In Definition 9.4.11 we extend the adjectives "initial" and "final" to functors between ∞-categories in an arbitrary ∞-cosmos and prove that the functors characterized there satisfy the property described here.

DEFINITION 2.4.5 (initial and final functor). A functor $k \colon I \to J$ is **final** if a J-shaped cone defines a colimit cone if and only if the restricted I-shaped cone is a colimit cone and **initial** if any J-shaped cone defines a limit cone if and only if the restricted I-shaped cone is a limit cone. That is, $k \colon I \to J$ is final if and only if for any ∞-category A, the square

$$
\begin{array}{ccc}
A & =\!=\!= & A \\
\Delta\downarrow & & \downarrow\Delta \\
A^J & \xrightarrow[A^k]{} & A^I
\end{array}
$$

preserves and reflects all absolute left lifting diagrams, and initial if and only if this squares preserves and reflects all absolute right lifting diagrams.

Historically, final functors were called "cofinal" with no obvious name for the dual notion. Our preferred terminology hinges on the following mnemonic: the inclusion of an initial element defines an initial functor, while the inclusion of a terminal (aka final) element defines a final functor. These facts are special cases of a more general result we now establish, using exactly the same tactics as deployed to prove Theorem 2.4.2.

PROPOSITION 2.4.6. *Left adjoints define initial functors and right adjoints define final functors.*

Proof If $k \dashv r$ with counit $\epsilon : kr \Rightarrow \mathrm{id}_J$, then cotensoring into A yields an adjunction

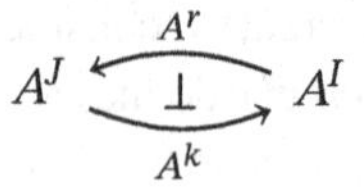

$$A^J \underset{A^k}{\overset{A^r}{\rightleftarrows}} \bot \; A^I \qquad \text{with counit } A^\epsilon : A^r A^k \Rightarrow \mathrm{id}_{A^J}.$$

To prove that k is initial we must show that for any cone $\rho : \Delta\ell \Rightarrow d$ as displayed below-left,

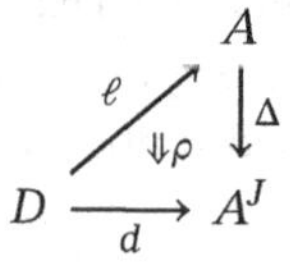

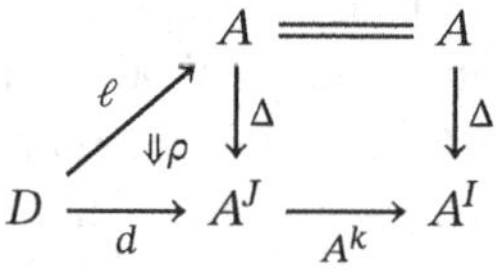

$$
\begin{array}{ccc}
& A & \\
\ell \nearrow \;\Downarrow\rho\; & \downarrow \Delta & \\
D \xrightarrow{\;\;d\;\;} & A^J &
\end{array}
\qquad
\begin{array}{ccccc}
& A & = & A & \\
\ell \nearrow \;\Downarrow\rho\; & \downarrow \Delta & & \downarrow \Delta & \\
D \xrightarrow{\;\;d\;\;} & A^J & \xrightarrow{\;A^k\;} & A^I &
\end{array}
$$

the left-hand diagram is an absolute right lifting diagram if and only if the right-hand diagram is an absolute right lifting diagram.

By Lemmas 2.3.7 and 2.4.1, the right-hand diagram is an absolute right lifting diagram if and only if the pasted composite displayed below-left is also an absolute right lifting diagram.

$$
\begin{array}{ccc}
& A = A & A \\
\ell \nearrow \Downarrow\rho \downarrow\Delta & \downarrow\Delta & \ell \nearrow \Downarrow\rho \downarrow\Delta \\
D \xrightarrow{d} A^J \xrightarrow{A^k} A^I & & = \quad D \xrightarrow{d} A^J \\
\searrow \Downarrow A^\epsilon \downarrow A^r & & \\
A^J & &
\end{array}
$$

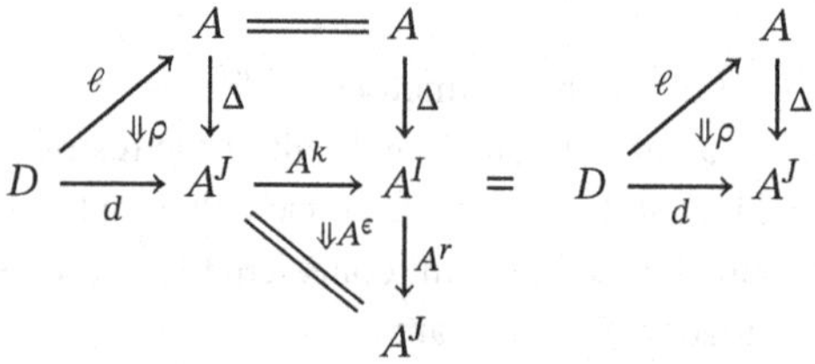

Since $A^r\Delta = \Delta$ and $A^\epsilon\Delta = \mathrm{id}_\Delta$, the left-hand side reduces to the right-hand side, which proves the claim. $\qquad\square$

Exercise 2.3.v defines a functor $f : A \to B$ between ∞-categories to be **fully faithful** just when

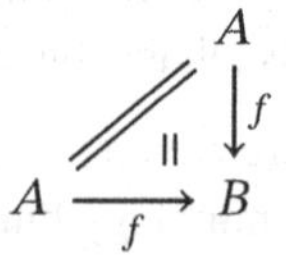

$$
\begin{array}{ccc}
& A & \\
\nearrow \;\|\; & \downarrow f & \\
A \xrightarrow{\;\;f\;\;} & B &
\end{array}
$$

defines absolute right lifting diagram or equivalently an absolute left lifting diagram. Modulo a result we borrow from Chapter 3, we show:

PROPOSITION 2.4.7. *A fully faithful functor* $f : A \to B$ *reflects any limits or colimits that exist in B.*

Proof The statement for limits asserts that for any family of diagrams $d : D \to A^J$ of shape J in A, any functor $\ell : D \to A$, and any cone $\rho : \Delta \ell \Rightarrow d$ so that the whiskered composite with $f^J : A^J \to B^J$ is an absolute right lifting diagram

$$
\begin{array}{ccc}
 & A \xrightarrow{\ f\ } B & \\
\ell \nearrow\ \Downarrow\rho\ \downarrow\Delta & & \downarrow\Delta \\
D \xrightarrow[d]{} A^J \xrightarrow[f^J]{} B^J
\end{array}
$$

then (ℓ, ρ) defines an absolute right lifting of $d : D \to A^J$ through $\Delta : A \to A^J$. By Exercise 2.3.v, to say that f is fully faithful is to say that $\mathrm{id}_A : A \to A$ defines an absolute right lifting of f through itself. So by Lemma 2.4.1, the composite diagram below-left is an absolute right lifting diagram, and by 2-functoriality of the simplicial cotensor with J, the diagram below-left coincides with the diagram below-right:

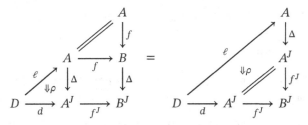

Now if we knew that $\mathrm{id}_{A^J} : A^J \to A^J$ defines an absolute right lifting of f^J through itself – that is, if we know that $f^J : A^J \to B^J$ is also fully faithful – then we could apply Lemma 2.4.1 again to conclude that (ℓ, ρ) is an absolute right lifting of d through Δ as required. And indeed this is the case: by Corollary 3.5.7, any cosmological functor, such as $(-)^J$, preserves absolute lifting diagrams. $\square$

It is worth asking why we have not already proven that cosmological functors preserve absolute lifting diagrams, since after all, by Lemma 1.4.4, cosmological functors induce 2-functors between homotopy 2-categories, which is where absolute lifting diagrams are defined. But unlike adjunctions, which are defined by pasting equations in a 2-category, absolute lifting diagrams are defined using universal quantifiers and hence are not preserved by all 2-functors. However, the 2-functors that underlie cosmological functors *do* preserve absolute lifting diagrams, even when the cosmological functor is "forgetful" or fails to be essentially surjective. This is because the universal property of absolute lifting diagrams can be re-expressed internally to the ambient ∞-cosmos by deploying the axiomatized limits of 1.2.1(i), at which point their preservation by cosmological functors is a direct corollary (see Theorem 3.5.3 and Corollary 3.5.7). In pursuit of results such as these, we now turn our attention to the 2-categorical properties of the cosmological limits.

Exercises

EXERCISE 2.4.i. Show that any left adjoint $f : B \to A$ between ∞-categories admitting all J-shaped colimits preserves them in the sense that the square of functors commutes up to isomorphism.

$$
\begin{array}{ccc}
B^J & \xrightarrow{\;f^J\;} & A^J \\
{\scriptstyle\text{colim}}\downarrow & \cong & \downarrow{\scriptstyle\text{colim}} \\
B & \xrightarrow[\;f\;]{} & A
\end{array}
$$

EXERCISE 2.4.ii. Prove Lemma 2.4.1.

EXERCISE 2.4.iii. Give a proof of Theorem 2.4.2 that does not appeal to Lemma 2.4.1 by directly verifying that the diagram on the right of (2.4.3) is an absolute right lifting diagram.

EXERCISE 2.4.iv. Use Lemma 2.4.1 to give a new proof that adjunctions compose (Proposition 2.1.9).

EXERCISE 2.4.v. For any composable pair of maps $k : I \to J$ and $\ell : J \to K$, show that if k and ℓk are final, then so is ℓ.

3

Comma ∞-Categories

In Chapter 2, we introduce adjunctions between ∞-categories and limits of diagrams valued within an ∞-category through definitions that are particularly expedient for establishing the expected interrelationships, as illustrated by the proof that right adjoints preserve limits. These definitions are 2-categorical in nature – stated in reference to the ∞-categories, ∞-functors, and ∞-natural transformations of the homotopy 2-category – but neither clearly articulates the universal properties of these notions. Definition 2.3.8 does not obviously express the expected universal property of the limit cone: namely, that a limit cone over a diagram d defines a terminal element in some ∞-*category of cones over d*. Nor does Definition 2.1.1 explain how an adjunction $f \dashv u$ induces an equivalence between *hom-spaces* $\mathrm{Hom}_A(fb, a) \simeq \mathrm{Hom}_B(b, ua)$.[1] In this chapter, we make use of the axiomatized limits in an ∞-cosmos to exhibit a general construction that specializes to define both this ∞-category of cones and also these hom-spaces. This construction also permits us to represent a functor between ∞-categories *as* an ∞-category, in dual "left" or "right" fashions, so that an adjunction consists of a pair of functors $f : B \to A$ and $u : A \to B$ so that the left representation of f is equivalent to the right representation of u over $A \times B$ (see Proposition 4.1.1).

Our vehicle for all of these new definitions is the *comma* ∞-*category* associated to a cospan:

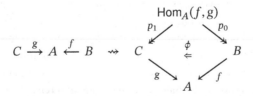

---

[1] A 2-categorical version of this result – exhibiting a bijection between *sets* of 2-cells – appears as Lemma 2.3.7, but in an ∞-category one would expect a similar equivalence of hom-spaces.

Our aim in this chapter is to develop the general theory of comma constructions from the point of view of the homotopy 2-category of an ∞-cosmos. Our first payoff for this work appears in Chapter 4 where we study the universal properties of adjunctions, limits, and colimits along these lines. The comma construction also provides the essential vehicle in Part III for establishing the model independence of the categorical notions we introduce throughout this text.

There is a standard definition of a "comma object" that can be stated in any 2-category, defined as a particular weighted limit (see Example A.6.14). Comma ∞-categories do *not* satisfy this universal property in the homotopy 2-category, however. Instead, they satisfy a somewhat peculiar "weak" variant of the usual 2-categorical universal property that to our knowledge has not appeared elsewhere in the literature. The weak universal property is encoded by something we call a *smothering functor*, which relates homotopy coherent and homotopy commutative diagrams of suitable shapes. To introduce these universal properties in a concrete rather than abstract framework, we start in §3.1 by considering smothering functors involving homotopy categories of quasi-categories.

In §3.2, we use a smothering functor to encode the weak universal property of the ∞-category of arrows A^2 associated to an ∞-category A, considered as an object in the homotopy 2-category. In §3.3, we briefly study the analogous weak universal properties associated to the pullback of an isofibration, which we exploit to prove that the pullback of an equivalence along an isofibration is an equivalence.

Comma ∞-categories are introduced in §3.4 where we describe both their strict universal properties as simplicially enriched limits as well as their weak universal properties in the homotopy 2-category. Each have their uses, for instance in describing the induced actions on comma ∞-categories of various types of morphisms between their generating cospans. The weak 2-categorical universal property is deployed in §3.5 to prove a general representability theorem that characterizes those comma ∞-categories that are right or left represented by a functor. In Chapter 4, we reap the payoff for this work, achieving the desired representable characterizations of adjunctions, limits, and colimits as special cases of these general results.

In §3.6, we tighten the main theorem of §3.5 to say that a comma ∞-category is right represented by a functor if and only if its codomain-projection functor admits a terminal element, when considered as an object in the sliced ∞-cosmos. This result requires a careful analysis of the subtle difference between the homotopy 2-category of a sliced ∞-cosmos and the sliced 2-category of the homotopy 2-category of an ∞-cosmos. Those readers who would rather stay out of the weeds are invited to take note of Definition 3.6.5 and Corollary 3.6.10 but otherwise skip this section.

3.1 Smothering Functors

Let Q be a quasi-category. Recall from Lemma 1.1.12 that its homotopy category hQ has

- elements of Q as its objects;
- homotopy classes of 1-simplices of Q as its arrows, where parallel 1-simplices are homotopic just when they bound a 2-simplex whose remaining outer edge is degenerate; and
- a composition relation if and only if any chosen 1-simplices representing the three arrows bound a 2-simplex.

For a 1-category J, it is well-known in classical homotopy theory that the homotopy category of diagrams $h(Q^J)$ is not equivalent to the category $(hQ)^J$ of diagrams in the homotopy category – except in very special cases, such as when J is a set (see Lemma 2.3.3). The objects of $h(Q^J)$ are *homotopy coherent* diagrams of shape J in Q, while the objects of $(hQ)^J$ are mere *homotopy commutative* diagrams. There is, however, a canonical comparison functor

$$h(Q^J) \longrightarrow (hQ)^J$$

defined by applying $h : \mathcal{QC}at \to \mathcal{C}at$ to the evaluation functor $Q^J \times J \to Q$ and then transposing; a homotopy coherent diagram is in particular homotopy commutative.

Our first aim in this section is to better understand the relationship between the arrows in the homotopy category hQ and the arrows of Q, meaning the 1-simplices in the quasi-category. To study this, we consider the quasi-category Q^2 in which the arrows of Q live as elements, where $2 = \Delta[1]$ is the nerve of the walking arrow. Our notation deliberately imitates the notation commonly used for the **category of arrows**: if C is a 1-category, then C^2 is the category whose objects are arrows in C and whose morphisms are commutative squares, regarded as a morphism from the arrow displayed vertically on the left-hand side to the arrow displayed vertically on the right-hand side. This notational conflation suggests our first question: how does the homotopy category of Q^2 relate to the category of arrows in the homotopy category hQ?

LEMMA 3.1.1. *The canonical functor* $h(Q^2) \to (hQ)^2$ *is*

  (i) surjective on objects,

(ii) full, and

*(iii) **conservative**, i.e., reflects invertibility of morphisms,*

but not necessarily injective on objects nor faithful.

Proof　Surjectivity on objects asserts that every arrow in the homotopy category hQ is represented by a 1-simplex in Q. This is the conclusion of Exercise 1.1.iii(iii) which outlines the proof of Lemma 1.1.12.

To prove fullness, consider a pair of arrows f and g in Q that form the source and target of a commutative square in hQ. By (i), we may choose arbitrary 1-simplices representing each morphism in hQ and their common composite:

$$
\begin{array}{ccc}
\bullet & \xrightarrow{\ h\ } & \bullet \\
{\scriptstyle f}\downarrow & \searrow^{\ell} & \downarrow{\scriptstyle g} \\
\bullet & \xrightarrow[\ k\ ]{} & \bullet
\end{array}
$$

By Lemma 1.1.12, every composition relation in hQ is witnessed by a 2-simplex in Q; choosing a pair of such 2-simplices defines a diagram $2 \times 2 \to Q$, which represents a morphism from f to g in $h(Q^2)$, proving fullness.

Surjectivity on objects and fullness of the functor $h(Q^2) \to (hQ)^2$ are special properties having to do with the diagram shape 2, while conservativity holds for generic diagram shapes by Corollary 1.1.22. The construction of counterexamples illustrating the general failure of injectivity on objects and faithfulness is left to Exercise 3.1.i, with a hint.　□

The properties of the canonical functor $h(Q^2) \to (hQ)^2$ frequently reappear, so we bestow them with a suggestive name:

DEFINITION 3.1.2 (smothering functor).　A functor $f : A \to B$ between 1-categories is **smothering** if it is surjective on objects, full, and conservative. That is, a functor is smothering if and only if it has the right lifting property with respect to the set of functors:

$$
\left\{
\begin{array}{ccc}
\varnothing & 1+1 & 2 \\
\downarrow & \downarrow & \downarrow \\
1 & 2 & 1
\end{array}
\right\}
$$

Various elementary properties of smothering functors are established in Exercise 3.1.ii; here we highlight one worthy of particular attention:

LEMMA 3.1.3 (smothering fibers).　*Each fiber of a smothering functor is a nonempty connected groupoid.*

Proof Suppose $f : A \to B$ is smothering and consider the fiber

$$
\begin{array}{ccc}
A_b & \longrightarrow & A \\
\downarrow & \lrcorner & \downarrow f \\
\mathbb{1} & \xrightarrow{\;b\;} & B
\end{array}
$$

over an object b of B. By surjectivity on objects, the fiber is nonempty. Its morphisms are defined to be arrows between objects in the fiber of b that map to the identity on b. By fullness, any two objects in the fiber are connected by a morphism, indeed, by morphisms pointing in both directions. By conservativity, all the morphisms in the fiber are necessarily invertible. □

The argument used to prove Lemma 3.1.1 generalizes to:

LEMMA 3.1.4. *If J is a 1-category that is free on a reflexive directed graph and Q is a quasi-category, then the canonical functor* $h(Q^J) \to (hQ)^J$ *is smothering.*

Proof Exercise 3.1.iii. □

Cotensors are one of the cosmological limits axiomatized in Definition 1.2.1. Other limit constructions listed there also give rise to smothering functors.

LEMMA 3.1.5. *For any pullback diagram of quasi-categories in which p is an isofibration*

$$
\begin{array}{ccc}
A \underset{B}{\times} E & \longrightarrow & E \\
\downarrow & \lrcorner & \downarrow p \\
A & \xrightarrow{\;f\;} & B
\end{array}
$$

the canonical functor $h(A \underset{B}{\times} E) \to hA \underset{hB}{\times} hE$ *is smothering.*

Proof As $h : \mathcal{Q}\mathcal{C}at \to \mathcal{C}at$ does not preserve pullbacks, the canonical comparison functor of the statement is not an isomorphism. It is however bijective on objects since the composite functor

$$
\mathcal{Q}\mathcal{C}at \xrightarrow{\;h\;} \mathcal{C}at \xrightarrow{\;\mathrm{obj}\;} \mathcal{S}et
$$

passes to the underlying set of vertices of each quasi-category, and this functor *does* preserve pullbacks.

For fullness, note that a morphism in $hA \times_{hB} hE$ is represented by a pair of 1-simplices $\alpha : a \to a'$ in A and $\epsilon : e \to e'$ in E whose images are homotopic in B, a condition that implies in particular that $f(a) = p(e)$ and $f(a') = p(e')$.

By Lemma 1.1.9, we can configure this homotopy however we like, and thus we choose a 2-simplex witness β so as to define a lifting problem

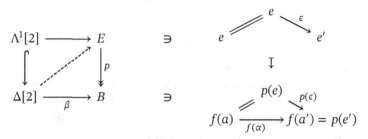

Since p is an isofibration, a solution exists, defining an arrow $\tilde{\epsilon}\colon e \to e'$ in E in the same homotopy class as ϵ so that $p(\tilde{\epsilon}) = f(\alpha)$. The pair $(\alpha, \tilde{\epsilon})$ now defines the lifted arrow in $\mathrm{h}(E \times_B A)$.

Finally, consider an arrow $2 \to A \times E$ whose image in $\mathrm{h}A \underset{\mathrm{h}B}{\times} \mathrm{h}E$ is an isomorphism, which is the case just when the projections to E and A define isomorphisms. By Corollary 1.1.16, we may choose a homotopy coherent isomorphism $\mathbb{I} \to A$ extending the given isomorphism $2 \to A$. This data presents us with a lifting problem

$$
\begin{array}{ccc}
2 & \longrightarrow A \times E & \longrightarrow E \\
\downarrow & {}_{{}^{\urcorner}} \,\, B \,\, {}_{\lrcorner} & \downarrow p \\
\mathbb{I} & \xrightarrow{\quad\sim\quad} A & \xrightarrow{\quad f \quad} B
\end{array}
$$

which Exercise 1.1.vi tells us we can solve. This proves that $\mathrm{h}(A \underset{B}{\times} E) \to \mathrm{h}A \underset{\mathrm{h}B}{\times} \mathrm{h}E$ is conservative and hence also smothering. $\qquad\square$

A similar argument proves:

LEMMA 3.1.6. *For any tower of isofibrations between quasi-categories*

$$\cdots \longrightarrow E_n \longrightarrow E_{n-1} \longrightarrow \cdots \longrightarrow E_2 \longrightarrow E_1 \longrightarrow E_0$$

the canonical functor $\mathrm{h}(\lim_n E_n) \to \lim_n \mathrm{h}E_n$ *is smothering.*

Proof Exercise 3.1.iv. $\qquad\square$

LEMMA 3.1.7. *For any cospan between quasi-categories* $C \xrightarrow{g} A \xleftarrow{f} B$ *consider the quasi-category defined by the pullback*

$$
\begin{array}{ccc}
\mathrm{Hom}_A(f,g) & \longrightarrow & A^2 \\
\downarrow & {}^{\lrcorner} & \downarrow {\scriptstyle(\mathrm{cod},\mathrm{dom})} \\
C \times B & \xrightarrow{g \times f} & A \times A
\end{array}
$$

The canonical functor $\mathsf{hHom}_A(f,g) \to \mathsf{Hom}_{\mathsf{h}A}(\mathsf{h}f,\mathsf{h}g)$ *is smothering.*

Proof The codomain of this functor is the category defined by an analogous pullback in $\mathcal{C}at$

$$
\begin{array}{ccc}
\mathsf{Hom}_{\mathsf{h}A}(\mathsf{h}f,\mathsf{h}g) & \longrightarrow & (\mathsf{h}A)^2 \\
\downarrow \quad \lrcorner & & \downarrow {\scriptstyle(\mathrm{cod},\mathrm{dom})} \\
\mathsf{h}C \times \mathsf{h}B & \xrightarrow[\mathsf{h}g\times\mathsf{h}f]{} & \mathsf{h}A \times \mathsf{h}A
\end{array}
$$

and the canonical functor factors as

$$
\mathsf{hHom}_A(f,g) \longrightarrow \mathsf{h}(A^2) \times_{\mathsf{h}A\times\mathsf{h}A} (\mathsf{h}C \times \mathsf{h}B) \longrightarrow (\mathsf{h}A)^2 \times_{\mathsf{h}A\times\mathsf{h}A} (\mathsf{h}C \times \mathsf{h}B)
$$

By Lemma 3.1.5 the first of these functors is smothering. By Lemma 3.1.1 the second is a pullback of a smothering functor. By Exercise 3.1.ii(i) it follows that the composite functor is smothering. □

In the sections that follow, we discover that the smothering functors just constructed express weak universal properties of arrow, pullback, and comma constructions in the homotopy 2-category of any ∞-cosmos.

Exercises

EXERCISE 3.1.i. Find an explicit example of a quasi-category Q for which the canonical smothering functor $\mathsf{h}(Q^2) \to (\mathsf{h}Q)^2$ fails to be injective on objects and faithful for instance by defining Q to be the total singular complex of a suitable topological space.

EXERCISE 3.1.ii. Prove that:

(i) Smothering functors are closed under composition, retract, product, pullback, and limits of towers.
(ii) Surjective equivalences of categories are smothering functors.
(iii) Smothering functors are isofibrations, that is, maps that have the right lifting property with respect to $\mathbb{1} \hookrightarrow \mathbb{I}$.
(iv) Prove that if f and gf are smothering functors, then g is a smothering functor.[2]

EXERCISE 3.1.iii. Prove Lemma 3.1.4.

EXERCISE 3.1.iv. Prove Lemma 3.1.6.

[2] In fact, it suffices to merely assume that f is surjective on objects and arrows.

3.2 ∞-Categories of Arrows

In this section, we replicate the discussion from the start of §3.1 using an arbitrary ∞-category A in place of the quasi-category Q. The analysis of the previous section could have been developed natively in this general setting but at the cost of an extra layer of abstraction and more confusing notation – with a functor space $\mathsf{Fun}(X, A)$ replacing the quasi-category Q.

Recall an **element** of an ∞-category is defined to be a functor $a : 1 \to A$. Tautologically, the elements of A are the vertices of the **underlying quasi-category** $\mathsf{Fun}(1, A)$ of A. In this section, we define and study an ∞-category A^{2} whose elements are the 1-simplices in the underlying quasi-category of A. We refer to A^{2} as the *∞-category of arrows* in A and call its elements simply *arrows* of A. In fact, we have tacitly introduced this construction already. Recall 2 is our preferred notation for the quasi-category $\Delta[1]$, the nerve of the category 2 with a single nonidentity morphism $0 \to 1$.

DEFINITION 3.2.1 (arrow ∞-category). Let A be an ∞-category. The **∞-category of arrows** in A is the simplicial cotensor A^{2} together with the canonical endpoint evaluation isofibration

$$A^{2} := A^{\Delta[1]} \xrightarrow{(p_1, p_0)} A^{\partial\Delta[1]} \cong A \times A$$

induced by the inclusion $\partial\Delta[1] \hookrightarrow \Delta[1]$. For conciseness, we write $p_0 : A^{2} \twoheadrightarrow A$ for the domain evaluation induced by the inclusion $0 : 1 \hookrightarrow 2$ and write $p_1 : A^{2} \twoheadrightarrow A$ for the codomain evaluation induced by $1 : 1 \hookrightarrow 2$.

As an object of the homotopy 2-category, the ∞-category of arrows comes equipped with a canonical 2-cell that we now construct.

LEMMA 3.2.2 (generic arrow). *For any ∞-category A, the ∞-category of arrows A^{2} comes equipped with a canonical 2-cell*

$$A^{2} \underset{p_1}{\overset{p_0}{\Longrightarrow}} A \qquad (3.2.3)$$

*that we refer to as the **generic arrow** with codomain A.*

Proof The simplicial cotensor has a strict universal property described in Digression 1.2.6: namely A^{2} is characterized by the natural isomorphism

$$\mathsf{Fun}(X, A^{2}) \cong \mathsf{Fun}(X, A)^{2}. \qquad (3.2.4)$$

By the Yoneda lemma, the data of the natural isomorphism (3.2.4) is encoded by its "universal element", which is defined to be the image of the identity at the representing object. Here the identity functor id : $A^2 \to A^2$ is mapped to an element of $\mathsf{Fun}(A^2, A)^2$, a 1-simplex in $\mathsf{Fun}(A^2, A)$, which by Lemma 1.4.3 represents a 2-cell κ in the homotopy 2-category.

To see that the source and target of κ must be the domain evaluation and codomain evaluation functors, defined by cotensoring with the endpoint inclusion $\mathbb{1} + \mathbb{1} \hookrightarrow \mathbb{2}$, we use the naturality of the isomorphism (3.2.4) in the cotensor variable:

$$
\begin{array}{ccc}
\mathsf{Fun}(X, A^2) & \cong & \mathsf{Fun}(X, A)^2 \\
{\scriptstyle (p_1, p_0)_*} \downarrow & & \downarrow {\scriptstyle (\mathrm{cod}, \mathrm{dom})} \\
\mathsf{Fun}(X, A \times A) & \cong & \mathsf{Fun}(X, A) \times \mathsf{Fun}(X, A)
\end{array}
$$

The identity functor maps around the top-right composite to the pair of functors $(\mathrm{cod}\,\kappa, \mathrm{dom}\,\kappa)$ and around the left-bottom composite to the pair (p_1, p_0). □

There is a 2-categorical limit notion that is analogous to Definition 3.2.1, which constructs, for any object A, the universal 2-cell with codomain A: namely the (categorical) cotensor with the 1-category $\mathbb{2}$. Its universal property is analogous to (3.2.4) but with the hom-categories of the 2-category in place of the functor spaces (see Definition A.4.1). In the 2-category of categories, the $\mathbb{2}$-cotensor defines the arrow category.

In the homotopy 2-category, by the Yoneda lemma again, the data (3.2.3) encodes a natural transformation

$$
\mathsf{hFun}(X, A^2) \to \mathsf{hFun}(X, A)^2
$$

of categories but this is *not* a natural isomorphism, nor even a natural equivalence of categories. However, it does furnish the ∞-category of arrows with a "weak" universal property of the following form:

PROPOSITION 3.2.5 (the weak universal property of the arrow ∞-category). *The generic arrow* (3.2.3) *with codomain A has a weak universal property in the homotopy 2-category given by three operations:*

(i) ***1-cell induction**: Given a natural transformation over A as below-left* *there exists a functor* $\ulcorner\alpha\urcorner : X \to A^2$ *so that* $s = p_0\ulcorner\alpha\urcorner$, $t = p_1\ulcorner\alpha\urcorner$, *and* $\alpha = \kappa\ulcorner\alpha\urcorner$.

(ii) **2-cell induction:** *Given functors* $a, a' : X \to A^2$ *and natural transformations* τ_1 *and* τ_0 *so that*

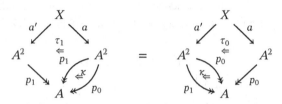

there exists a natural transformation $\tau : a \Rightarrow a'$ *so that* $p_1\tau = \tau_1$ *and* $p_0\tau = \tau_0$.

(iii) **2-cell conservativity:** *For any natural transformation* $X \underset{a'}{\overset{a}{\rightrightarrows}}{}_{\Downarrow\tau} A^2$

if both $p_1\tau$ *and* $p_0\tau$ *are isomorphisms then* τ *is an isomorphism.*

Proof   Let $Q = \mathrm{Fun}(X, A)$ and apply Lemma 3.1.1 to observe that the natural map of hom-categories

$$\mathrm{hFun}(X, A^2) \xrightarrow{\hspace{5cm}} \mathrm{hFun}(X, A)^2$$

$$((p_1)_*, (p_0)_*) \searrow \qquad \swarrow (\mathrm{cod}, \mathrm{dom})$$
$$\mathrm{hFun}(X, A) \times \mathrm{hFun}(X, A)$$

over $\mathrm{hFun}(X, A \times A) \cong \mathrm{hFun}(X, A) \times \mathrm{hFun}(X, A)$ is a smothering functor. Surjectivity on objects is expressed by 1-cell induction, fullness by 2-cell induction, and conservativity by 2-cell conservativity.   □

Note that the functors $\ulcorner\alpha\urcorner : X \to A^2$ that represent a given natural transformation α with domain X and codomain A are not unique. However, they are unique up to "fibered" isomorphisms that whisker with $(p_1, p_0) : A^2 \twoheadrightarrow A \times A$ to identities:

PROPOSITION 3.2.6. *Whiskering with* (3.2.3) *induces a bijection between natural transformations with domain* X *and codomain* A *as displayed below-left*

$$\left\{\begin{array}{c} X \\ t\left(\underset{\alpha}{\overset{\Leftarrow}{}}\right)s \\ A \end{array}\right\} \quad\longleftrightarrow\quad \left\{\begin{array}{c} X \\ t \swarrow \ \big\downarrow\ulcorner\alpha\urcorner\ \searrow s \\ A \ \nwarrow \ \underset{A^2}{} \ \nearrow A \\ p_1 \qquad\qquad p_0 \end{array}\right\}\bigg/{\cong}$$

and fibered isomorphism classes of functors $X \to A^2$ *as displayed above-right, where the fibered isomorphisms are given by invertible 2-cells*

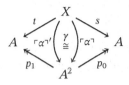

so that $p_1 \gamma = \mathrm{id}_t$ *and* $p_0 \gamma = \mathrm{id}_s$.

Proof Lemma 3.1.3 proves that the fibers of the smothering functor of Proposition 3.2.5 are connected groupoids. The objects of the fiber over α are functors $X \to A^2$ that whisker with the generic arrow κ to α, and the morphisms are invertible 2-cells that whisker with $(p_1, p_0) : A^2 \twoheadrightarrow A \times A$ to the identity 2-cell $(\mathrm{id}_t, \mathrm{id}_s)$. The action of the smothering functor defines a bijection between the objects of its codomain and their corresponding fibers. □

Our final task is to observe that the universal property of Proposition 3.2.5 is also enjoyed by any object $(e_1, e_0) : E \twoheadrightarrow A \times A$ that is equivalent to the ∞-category of arrows $(p_1, p_0) : A^2 \twoheadrightarrow A \times A$ in the slice ∞-cosmos over $A \times A$. We have special terminology to allow us to concisely express the type of equivalence we have in mind.

DEFINITION 3.2.7 (fibered equivalence). A **fibered equivalence** over an ∞-category B in an ∞-cosmos $\mathcal{K}$ is an equivalence

$$
\begin{array}{ccc}
E & \xrightarrow{\quad \sim \quad} & F \\
& \searrow \quad \swarrow & \\
& B &
\end{array}
\tag{3.2.8}
$$

in the sliced ∞-cosmos $\mathcal{K}_{/B}$. We write $E \simeq_B F$ to indicate that the specified isofibrations with these domains are equivalent **over** B.

By Proposition 1.2.22(vii), a fibered equivalence is just a map between a pair of isofibrations over a common base that defines an equivalence in the underlying ∞-cosmos: the forgetful functor $\mathcal{K}_{/B} \to \mathcal{K}$ preserves and reflects equivalences. Note, however, that it does not create them: It is possible for two ∞-categories E and F to be equivalent without there existing any equivalence compatible with a pair of specified isofibrations $E \twoheadrightarrow B$ and $F \twoheadrightarrow B$.

WARNING 3.2.9. At this point, there is some ambiguity about the 2-categorical data that presents a fibered equivalence in an ∞-cosmos $\mathcal{K}_{/B}$ related to the question posed in Exercise 1.4.vii about the relationship between the 2-categories

$\mathfrak{h}(\mathcal{K}_{/B})$ and $(\mathfrak{h}\mathcal{K})_{/B}$. But since Proposition 1.2.22(vii) tells us that a mere equivalence in $\mathfrak{h}\mathcal{K}$ involving a functor of the form (3.2.8) is sufficient to guarantee that this as-yet-unspecified 2-categorical data exists, we defer a careful analysis of this issue to Proposition 3.6.4.

PROPOSITION 3.2.10 (uniqueness of arrow ∞-categories). *For any isofibration* $(e_1, e_0) : E \twoheadrightarrow A \times A$ *that is fibered equivalent to* $(p_1, p_0) : A^2 \twoheadrightarrow A \times A$ *the 2-cell*

$$E \underset{e_1}{\overset{e_0}{\rightrightarrows}} \Downarrow\epsilon \; A$$

encoded by the equivalence $e : E \xrightarrow{\sim} A^2$ *satisfies the weak universal property of Proposition 3.2.5. Conversely, if the isofibrations* $(d_1, d_0) : D \twoheadrightarrow A \times A$ *and* $(e_1, e_0) : E \twoheadrightarrow A \times A$ *are equipped with 2-cells*

$$D \underset{d_1}{\overset{d_0}{\rightrightarrows}} \Downarrow\delta \; A \qquad and \qquad E \underset{e_1}{\overset{e_0}{\rightrightarrows}} \Downarrow\epsilon \; A$$

satisfying the weak universal property of Proposition 3.2.5, then $D \simeq_{A \times A} E$.

Proof   We prove the first statement. By the defining equation of 1-cell induction $\epsilon = \kappa e$, where κ is the generic arrow (3.2.3). Hence, the functor induced by pasting with ϵ factors as a composite

$$\mathsf{hFun}(X, E) \xrightarrow[\sim]{e_*} \mathsf{hFun}(X, A^2) \longrightarrow \mathsf{hFun}(X, A)^2$$
$$((p_1)_*, (p_0)_*) \searrow \qquad \swarrow (\mathrm{cod}, \mathrm{dom})$$
$$\mathsf{hFun}(X, A) \times \mathsf{hFun}(X, A)$$

and our task is to prove that this composite functor is smothering. The first functor, defined by postcomposing with the equivalence $e : E \xrightarrow{\sim} A^2$, is an equivalence of categories, and the second functor is smothering. Thus, the composite is clearly full and conservative. To see that it is also surjective on objects, note first that by 1-cell induction any 2-cell

$$X \underset{t}{\overset{s}{\rightrightarrows}} \Downarrow\alpha \; A$$

is represented by a functor $\ulcorner\alpha\urcorner : X \to A^2$ over $A \times A$. Composing with any fibered inverse equivalence e' to e yields a functor

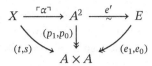

$$X \xrightarrow{\ulcorner\alpha\urcorner} A^2 \xrightarrow{\underset{\sim}{e'}} E$$

whose image after postcomposing with e is isomorphic to $\ulcorner\alpha\urcorner$ over $A \times A$. Because this isomorphism is fibered in the sense of Proposition 3.2.6, the image of $e'\ulcorner\alpha\urcorner$ under the functor $\mathsf{hFun}(X, E) \to \mathsf{hFun}(X, A)^2$ returns the 2-cell α. This proves that this mapping is surjective on objects and hence defines a smothering functor as claimed.

The converse is left to Exercise 3.2.ii and proven in a more general context in Proposition 3.4.11. □

CONVENTION 3.2.11. On account of Proposition 3.2.10, we extend the appellation "∞-category of arrows" from the strict model constructed in Definition 3.2.1 to any ∞-category that is fibered equivalent to it.

Via Lemma 3.1.4, the results of this section extend to corresponding weak universal properties for the cotensors A^J of an ∞-category A with a free category J, as the reader is invited to explore.

Exercises

EXERCISE 3.2.i. This exercise revisits the result of Proposition 3.2.6.

(i) Prove that a parallel pair of 1-simplices $f, g : x \to y$ in a quasi-category Q are homotopic if and only if they are isomorphic as elements of Q^2 via an isomorphism that projects to an identity along $(p_1, p_0) : Q^2 \twoheadrightarrow Q \times Q$.

(ii) Conclude that a parallel pair of 1-arrows in the functor space between two ∞-categories X and A in any ∞-cosmos represent the same natural transformation if and only if they are isomorphic as elements of $\mathsf{Fun}(X, A)^2 \cong \mathsf{Fun}(X, A^2)$ via an isomorphism whose domain and codomain components are an identity.

(iii) Conclude that a parallel pair of 1-arrows in $\mathsf{Fun}(X, A)$, which may be encoded as functors $X \to A^2$, represent the same natural transformation if and only if they are connected by a fibered isomorphism:

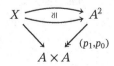

EXERCISE 3.2.ii. Prove the converse implication of Proposition 3.2.10.

EXERCISE 3.2.iii. Extend the results of Propositions 3.2.5 and Proposition 3.2.6 to describe the weak universal property of the cotensor A^J of an ∞-category A by a category J that is freely generated from some reflexive directed graph.

3.3 Pullbacks of Isofibrations

Pullbacks and limits of towers of isofibrations in an ∞-cosmos also have weak 2-dimensional universal properties in the homotopy 2-category, though we generally exploit the strict universal properties of the simplicially enriched limits instead. However, the weak 2-dimensional universal property of pullbacks can be used to prove that equivalences pull back along isofibrations to equivalences, which in turn is used to establish the equivalence invariance of pullbacks in an ∞-cosmos.

PROPOSITION 3.3.1 (the weak universal property of the pullback). *The pullback of an isofibration along a functor in an ∞-cosmos*

$$
\begin{array}{ccc}
A \underset{B}{\times} E & \xrightarrow{\;g\;} & E \\
{\scriptstyle q}\downarrow & \lrcorner & \downarrow{\scriptstyle p} \\
A & \xrightarrow[\;f\;]{} & B
\end{array}
$$

has a weak universal property in the homotopy 2-category given by three operations:

(i) **1-cell induction**: *Commutative squares $pe = fa$ over the cospan underlying a pullback diagram factor uniquely through the pullback square*

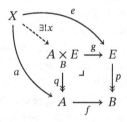

(ii) **2-cell induction**: *Given functors $x, x' : X \to A \underset{B}{\times} E$ and natural transformations $\alpha : qx \Rightarrow qx'$ and $\epsilon : gx \Rightarrow gx'$ so that $p\epsilon = f\alpha$, there exists a* *natural transformation* $\tau : x \Rightarrow x'$ *so that* $q\tau = \alpha$ *and* $g\tau = \epsilon$.

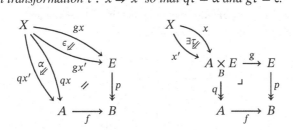

(iii) **2-cell conservativity:** *For any* $X \overset{x}{\underset{x'}{\rightrightarrows}} A \underset{B}{\times} E$ *if both* $q\tau$ *and* $g\tau$
are isomorphisms then τ *is an isomorphism.*

Proof Apply Lemma 3.1.5 to the pullback diagram of quasi-categories

$$\begin{array}{ccc}
\mathsf{Fun}(X, A \underset{B}{\times} E) & \overset{g_*}{\longrightarrow} & \mathsf{Fun}(X, E) \\
q_* \downarrow & \lrcorner & \downarrow p_* \\
\mathsf{Fun}(X, A) & \underset{f_*}{\longrightarrow} & \mathsf{Fun}(X, B)
\end{array}$$

to observe that the natural map of hom-categories

$$\mathsf{hFun}(X, A \underset{B}{\times} E) \longrightarrow \mathsf{hFun}(X, A) \underset{\mathsf{hFun}(X,B)}{\times} \mathsf{hFun}(X, E)$$

is a bijective-on-objects smothering functor. Bijectivity on objects is expressed by 1-cell induction, fullness by 2-cell induction, and conservativity by 2-cell conservativity. □

DIGRESSION 3.3.2 (weakly cartesian squares). A commutative square between parallel isofibrations is **weakly cartesian** if the induced map to the pullback is an equivalence:

$$\begin{array}{ccc}
F & \overset{g}{\longrightarrow} & E \\
q \downarrow & \lrcorner & \downarrow p \\
A & \underset{f}{\longrightarrow} & B
\end{array}$$

Weakly cartesian squares also satisfy 2-cell induction and 2-cell conservativity as well as a modified form of the 1-cell induction property, where the essentially unique induced functor commutes strictly over B and up to an isomorphism over E that projects along p to the identity [110, 3.5.4].

It follows from the weak 2-categorical universal property of the pullback that ∞-cosmoi are **right proper**, meaning that the pullback of any equivalence along an isofibration defines an equivalence.

PROPOSITION 3.3.3. *In any ∞-cosmos, the pullback of an equivalence along an isofibration is an equivalence.*

$$
\begin{array}{ccc}
F & \xrightarrow{\ \sim\ g\ } & E \\
q \downarrow & \quad\lrcorner & \downarrow p \\
A & \xrightarrow[\ f\ ]{\ \sim\ } & B
\end{array}
$$

Proof By Proposition 2.1.12, we may choose an adjoint equivalence inverse to f and pick invertible 2-cells $\alpha \colon \mathrm{id}_A \cong f^{-1}f$ and $\beta \colon ff^{-1} \cong \mathrm{id}_B$ satisfying the triangle equalities in the homotopy 2-category.[3] Now since the map p is an isofibration, we may use Proposition 1.4.9 to lift the isomorphism $\beta p \colon ff^{-1}p \cong p$ along p to define an isomorphism $\epsilon \colon e \cong \mathrm{id}_E$ with codomain $\mathrm{id}_E \colon E \to E$. By construction $pe = ff^{-1}p$, so by 1-cell induction the pair $(f^{-1}p, e)$ induces a map $g^{-1} \colon E \to F$ so that $qg^{-1} = f^{-1}p$ and $gg^{-1} = e$. In this way we obtain an isomorphism $\epsilon \colon gg^{-1} \cong \mathrm{id}_E$ with $p\epsilon = \beta p$.

Now by 2-cell induction and conservativity of Proposition 3.3.1, to define an isomorphism $\mathrm{id}_F \cong g^{-1}g$, it suffices to exhibit a pair of isomorphisms

$$
\alpha q \colon q \cong f^{-1}fq = f^{-1}pg = qg^{-1}g \qquad \text{and} \qquad \epsilon^{-1}g \colon g \cong gg^{-1}g
$$

so that $f\alpha q = p\epsilon^{-1}g$. This latter equation holds because $p\epsilon^{-1}g = \beta^{-1}pg = \beta^{-1}fq = f\alpha q$ by the triangle equality $\beta f \cdot f\alpha = \mathrm{id}_f$ for the adjoint equivalence $f \dashv f^{-1}$. Thus, we may lift the data of an inverse equivalence to f to define an inverse equivalence to its pullback g.   □

As a consequence of right properness, pullback is an equivalence invariant construction in any ∞-cosmos.

PROPOSITION 3.3.4. *Given a diagram of isofibrations and equivalences in any ∞-cosmos*

$$
\begin{array}{ccccc}
C & \xrightarrow{\ g\ } & A & \xleftarrow{\ f\ } & B \\
r \downarrow & & \downarrow p & & \downarrow q \\
\bar{C} & \xrightarrow[\ \bar{g}\ ]{} & \bar{A} & \xleftarrow[\ \bar{f}\ ]{} & \bar{B}
\end{array}
$$

the induced map $C \times_A B \to \bar{C} \times_{\bar{A}} \bar{B}$ between the pullbacks of the horizontal rows is again an equivalence.

---

[3] It is for this reason that we work with the weak 2-categorical universal property of the pullback rather than the simplicially enriched universal property.

Proof By factoring via Lemma 1.2.19, we can replace the map $\bar{g}$ by an isofibration. By the 2-of-3 property and the right properness of Proposition 3.3.3, the pullback of this isofibration along the equivalence p is equivalent to the map g:

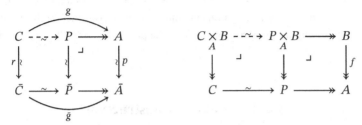

By right properness again, the pullback of $P \twoheadrightarrow A$ along f is equivalent to the pullback of $g : C \to A$ along f and similarly for the lower-horizontal maps. So without loss of generality, we may assume that the maps g and $\bar{g}$ of the statement are isofibrations and the left-hand square is a pullback.

Under these new hypothesis, the top, bottom, and front faces of the cube are pullback squares:

$$
\begin{array}{ccc}
C \times_A B & \longrightarrow\!\!\!\twoheadrightarrow & B \\
& g & f & q \\
C & \longrightarrow\!\!\!\twoheadrightarrow & A \\
r & \bar{C} \times_{\bar{A}} \bar{B} & \longrightarrow & \bar{B} \\
& & p & \bar{f} \\
\bar{C} & \xrightarrow{\ \bar{g}\ } & \bar{A}
\end{array}
$$

so by pullback composition and cancelation, the back face is a pullback square as well. Now the induced map $C \times_A B \to \bar{C} \times_{\bar{A}} \bar{B}$ is the pullback of the equivalence q along an isofibration and hence is an equivalence by Proposition 3.3.3. □

Exercises

EXERCISE 3.3.i. Use Proposition 3.3.1 to prove that for any isofibration and parallel pair of isomorphic functors

$$
\begin{array}{c}
E \\
\downarrow p \\
A \underset{f'}{\overset{f}{\underset{\cong\Downarrow\alpha}{\rightleftarrows}}} B
\end{array}
$$

their pullbacks are equivalent over A.

EXERCISE 3.3.ii. State and prove an analogous result to Proposition 3.3.1 that describes the weak 2-categorical universal property of limits of towers of isofibrations.

EXERCISE 3.3.iii. Use the result of Exercise 3.3.ii to prove that a natural equivalence between towers of isofibrations induces an equivalence between their limits by adapting the construction given in the proofs of Propositions 3.3.3 and 3.3.4.

3.4 The Comma Construction

The *comma ∞-category* is defined by restricting the domain and codomain of the ∞-category of arrows A^2 along a pair of specified functors with codomain A.

DEFINITION 3.4.1 (comma ∞-category). Let $C \xrightarrow{g} A \xleftarrow{f} B$ be a diagram of ∞-categories. The **comma ∞-category** is constructed as a pullback of the simplicial cotensor A^2 along $g \times f$

$$
\begin{array}{ccc}
\mathrm{Hom}_A(f,g) & \xrightarrow{\ulcorner \phi \urcorner} & A^2 \\
{\scriptstyle (p_1,p_0)}\big\downarrow \quad \lrcorner & & \big\downarrow{\scriptstyle (p_1,p_0)} \\
C \times B & \xrightarrow[g \times f]{} & A \times A
\end{array}
\tag{3.4.2}
$$

This construction equips the comma ∞-category with a specified isofibration $(p_1, p_0) \colon \mathrm{Hom}_A(f,g) \twoheadrightarrow C \times B$ and a canonical natural transformation

$$
\begin{array}{c}
\mathrm{Hom}_A(f,g) \\
{\scriptstyle p_1}\swarrow \qquad \searrow{\scriptstyle p_0} \\
C \quad \overset{\phi}{\underset{\Leftarrow}{}} \quad B \\
{\scriptstyle g}\searrow \qquad \swarrow{\scriptstyle f} \\
A
\end{array}
\tag{3.4.3}
$$

in the homotopy 2-category called the **comma cone**.

By the universal property (3.4.2), an element $(\alpha, b, c) \colon 1 \to \mathrm{Hom}_A(f,g)$ of the comma ∞-category is a triple where b and c are elements of B and C, respectively, and $\alpha \colon fb \to gc$ is an arrow in A with domain fb and codomain gc.

EXAMPLE 3.4.4. The ∞-category of arrows arises as a special case of the comma construction applied to the identity span. This provides us with alternate notation for the generic arrow of (3.2.3), which may be regarded as a particular instance of a comma cone.

$$
\begin{array}{ccc}
 & \mathsf{Hom}_A & \\
{\scriptstyle p_1}\swarrow & & \searrow{\scriptstyle p_0} \\
A & \overset{\phi}{\Leftarrow} & A \\
& \searrow \quad \swarrow & \\
& A &
\end{array}
$$

The following proposition encodes the homotopical properties of the comma construction.

PROPOSITION 3.4.5 (maps between commas). *A commutative diagram*

$$
\begin{array}{ccccc}
C & \overset{g}{\longrightarrow} & A & \overset{f}{\longleftarrow} & B \\
{\scriptstyle r}\downarrow & & \downarrow{\scriptstyle p} & & \downarrow{\scriptstyle q} \\
\bar{C} & \overset{\bar{g}}{\longrightarrow} & \bar{A} & \overset{\bar{f}}{\longleftarrow} & \bar{B}
\end{array}
$$

induces a map between the comma ∞-categories

$$
\begin{array}{ccc}
\mathsf{Hom}_A(f,g) & \overset{\mathsf{Hom}_p(q,r)}{\dashrightarrow} & \mathsf{Hom}_{\bar{A}}(\bar{f},\bar{g}) \\
{\scriptstyle (p_1,p_0)}\downarrow & & \downarrow{\scriptstyle (p_1,p_0)} \\
C \times B & \overset{r \times q}{\longrightarrow} & \bar{C} \times \bar{B}
\end{array}
$$

Moreover, if p, q, and r are all isofibrations, all trivial fibrations, or all equivalences then the induced map is again an isofibration, trivial fibration, or equivalence, respectively.

Proof The map of cospans gives rise to a commutative diagram

$$
\begin{array}{ccccc}
C \times B & \overset{g \times f}{\longrightarrow} & A \times A & \overset{(p_1,p_0)}{\longleftarrow} & A^2 \\
{\scriptstyle r \times q}\downarrow & & \downarrow{\scriptstyle p \times p} & & \downarrow{\scriptstyle p^2} \\
\bar{C} \times \bar{B} & \overset{\bar{g} \times \bar{f}}{\longrightarrow} & \bar{A} \times \bar{A} & \overset{(p_1,p_0)}{\longleftarrow} & \bar{A}^2
\end{array}
$$

in which the dotted map is the Leibniz tensor of the monomorphism $\mathbb{1} + \mathbb{1} \hookrightarrow \mathbb{2}$ with p. If p, q, and r are isofibrations or trivial fibrations, then this map and the four other downwards pointing maps are all isofibrations or trivial fibrations, respectively, by axiom 1.2.1(ii) and Lemma 1.2.14. By Proposition C.1.12, the map $\mathsf{Hom}_p(q,r)$ is again a isofibration or trivial fibration (see Exercise 3.4.i). In the case where p, q, and r are equivalences, Lemma 1.2.15 implies that the maps $r \times q$, $p \times p$, and p^2 are as well, so Proposition 3.3.4 applies to prove that $\mathsf{Hom}_p(q,r)$ is an equivalence. $\square$

There is a 2-categorical limit notion that is analogous to Definition 3.4.1, which constructs the universal 2-cell inhabiting a square over a specified cospan. In $\mathcal{C}at$ this universal property characterizes the *comma category*, from which we borrow the name. As with the case of ∞-categories of arrows, comma ∞-categories do *not* satisfy this 2-universal property strictly. Instead:

PROPOSITION 3.4.6 (the weak universal property of the comma ∞-category). *The comma cone* (3.4.3) *has a weak universal property in the homotopy 2-category given by three operations:*

(i) **1-cell induction:** *Given a natural transformation over* $C \xrightarrow{g} A \xleftarrow{f} B$ *as below-left*

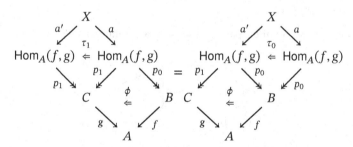

there exists a functor $\ulcorner\alpha\urcorner : X \to \mathrm{Hom}_A(f,g)$ so that $b = p_0\ulcorner\alpha\urcorner$, $c = p_1\ulcorner\alpha\urcorner$, and $\alpha = \phi\ulcorner\alpha\urcorner$.

(ii) **2-cell induction:** *Given functors* $a, a' : X \to \mathrm{Hom}_A(f,g)$ *and natural transformations* τ_1 *and* τ_0 *so that*

$$
\begin{array}{ccc}
& X & \\
a' \swarrow & \ {\tau_1} & \searrow a \\
\mathrm{Hom}_A(f,g) & \Leftarrow & \mathrm{Hom}_A(f,g) \\
& p_1 \swarrow \ \searrow p_0 \\
p_1 \searrow & C \ \ \underset{\Leftarrow}{\phi} \ \ B & \swarrow p_0 \\
& g \searrow \ \swarrow f & \\
& A &
\end{array}
=
\begin{array}{ccc}
& X & \\
a' \swarrow & \ {\tau_0} & \searrow a \\
\mathrm{Hom}_A(f,g) & \Leftarrow & \mathrm{Hom}_A(f,g) \\
p_1 \swarrow \ \searrow p_0 & & \swarrow p_0 \\
C \ \ \underset{\Leftarrow}{\phi} \ \ B & & \\
g \searrow \ \swarrow f & & \\
A & &
\end{array}
$$

there exists a natural transformation $\tau : a \Rightarrow a'$ *so that* $p_1\tau = \tau_1$ *and* $p_0\tau = \tau_0$.

(iii) **2-cell conservativity:** *For any* $X \underset{a'}{\overset{a}{\rightrightarrows}} \Downarrow\tau \ \mathrm{Hom}_A(f,g)$ *if both* $p_1\tau$ *and* $p_0\tau$ *are isomorphisms then* τ *is an isomorphism.*

Proof The cosmological functor $\mathrm{Fun}(X,-) : \mathcal{K} \to \mathcal{QC}at$ carries the pullback

(3.4.2) to a pullback

$$\begin{array}{ccc} \mathsf{Fun}(X,\mathsf{Hom}_A(f,g)) & & \\ \wr\Vert & & \\ \mathsf{Hom}_{\mathsf{Fun}(X,A)}(\mathsf{Fun}(X,f),\mathsf{Fun}(X,g)) & \xrightarrow{\ \phi\ } & \mathsf{Fun}(X,A)^{\mathbb{2}} \\ {\scriptstyle (p_1,p_0)}\big\downarrow & \lrcorner & \big\downarrow{\scriptstyle (p_1,p_0)} \\ \mathsf{Fun}(X,C)\times\mathsf{Fun}(X,B) & \xrightarrow[\mathsf{Fun}(X,g)\times\mathsf{Fun}(X,f)]{} & \mathsf{Fun}(X,A)\times\mathsf{Fun}(X,A) \end{array}$$

of quasi-categories. Now Lemma 3.1.7 demonstrates that the canonical 2-cell (3.4.3) induces a natural map of hom-categories

$$\begin{array}{ccc} \mathsf{hFun}(X,\mathsf{Hom}_A(f,g)) & \longrightarrow & \mathsf{Hom}_{\mathsf{hFun}(X,A)}(\mathsf{hFun}(X,f),\mathsf{hFun}(X,g)) \\ & {\scriptstyle ((p_1)_*,(p_0)_*)}\searrow \quad \swarrow{\scriptstyle (\mathrm{cod},\mathrm{dom})} & \\ & \mathsf{hFun}(X,C)\times\mathsf{hFun}(X,B) & \end{array}$$

over $\mathsf{hFun}(X,C\times B)\cong\mathsf{hFun}(X,C)\times\mathsf{hFun}(X,B)$ that is a smothering functor. The properties of 1-cell induction, 2-cell induction, and 2-cell conservativity follow from surjectivity on objects, fullness, and conservativity of this smothering functor, respectively. □

The functors $\ulcorner\alpha\urcorner\colon X\to\mathsf{Hom}_A(f,g)$ induced by a fixed natural transformation $\alpha\colon fb\Rightarrow gc$ are unique up to fibered isomorphism over $C\times B$.

Proposition 3.4.7. *Whiskering with the comma cone* (3.4.3) *induces a bijection between natural transformations as displayed below-left*

$$\left\{\begin{array}{ccc} & X & \\ {\scriptstyle c}\swarrow & & \searrow{\scriptstyle b} \\ C & \overset{\alpha}{\Leftarrow} & B \\ {\scriptstyle g}\searrow & & \swarrow{\scriptstyle f} \\ & A & \end{array}\right\} \quad \leftrightsquigarrow \quad \left\{\begin{array}{ccc} & X & \\ {\scriptstyle c}\swarrow & \big\downarrow{\scriptstyle \ulcorner\alpha\urcorner} & \searrow{\scriptstyle b} \\ C & & B \\ {\scriptstyle p_1}\nwarrow & & \nearrow{\scriptstyle p_0} \\ & \mathsf{Hom}_A(f,g) & \end{array}\right\}_{/\cong}$$

and fibered isomorphism classes of maps of spans from C to B as displayed above-right, where the fibered isomorphisms are given by invertible 2-cells

$$\begin{array}{ccc} & X & \\ {\scriptstyle c}\swarrow & {\scriptstyle \ulcorner\alpha\urcorner'}\big\downarrow \overset{\gamma}{\underset{\cong}{\Rightarrow}} \big\downarrow{\scriptstyle \ulcorner\alpha\urcorner} & \searrow{\scriptstyle b} \\ C & & B \\ {\scriptstyle p_1}\nwarrow & & \nearrow{\scriptstyle p_0} \\ & \mathsf{Hom}_A(f,g) & \end{array}$$

so that $p_1\gamma=\mathrm{id}_c$ and $p_0\gamma=\mathrm{id}_b$.

Proof Lemma 3.1.3 proves that the fibers of the smothering functor of Proposition 3.4.6 are connected groupoids. The objects of the fiber over α are functors $X \to \mathrm{Hom}_A(f, g)$ that whisker with the comma cone ϕ to α, and the morphisms are invertible 2-cells that whisker with

$$(p_1, p_0) \colon \mathrm{Hom}_A(f, g) \longrightarrow\!\!\!\!\!\rightarrow C \times B$$

to the identity 2-cell $(\mathrm{id}_c, \mathrm{id}_b)$. The action of the smothering functor defines a bijection between the objects of its codomain and their corresponding fibers. $\square$

Oplax maps of cospans in the homotopy 2-category also induce maps of comma ∞-categories:

OBSERVATION 3.4.8. By 1-cell induction a diagram

$$
\begin{array}{ccccc}
C & \xrightarrow{g} & A & \xleftarrow{f} & B \\
r\downarrow & \Downarrow\gamma & \downarrow p & \Downarrow\beta & \downarrow q \\
\bar{C} & \xrightarrow[\bar{g}]{} & \bar{A} & \xleftarrow[\bar{f}]{} & \bar{B}
\end{array}
$$

induces a map between comma ∞-categories as displayed below-right:

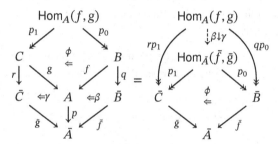

that is well-defined and functorial up to fibered isomorphism (see Exercise 3.4.ii).

One of many uses of comma ∞-categories is to define the internal mapping spaces between two elements of an ∞-category A. This is one motivation for our notation "Hom_A."

DEFINITION 3.4.9 (mapping space). The **mapping space** between two elements $x, y \colon 1 \to A$ of an ∞-category is the comma ∞-category $\mathrm{Hom}_A(x, y)$ defined by the pullback diagram

$$
\begin{array}{ccc}
\mathrm{Hom}_A(x, y) & \xrightarrow{\ulcorner\phi\urcorner} & A^2 \\
(p_1, p_0)\downarrow & \lrcorner & \downarrow(p_1, p_0) \\
1 & \xrightarrow[(y, x)]{} & A \times A
\end{array}
$$

Mapping spaces are discrete in the sense of Definition 1.2.26:

PROPOSITION 3.4.10 (mapping spaces are discrete). *For any pair of elements* $x, y : 1 \to A$ *of an* ∞*-category A, the mapping space* $\mathrm{Hom}_A(x, y)$ *is discrete.*

Proof We must show that the functor space $\mathrm{Fun}(X, \mathrm{Hom}_A(x, y))$ is a Kan complex for any ∞-category X. This is so just when $\mathrm{hFun}(X, \mathrm{Hom}_A(x, y))$ is a groupoid, i.e., when any 2-cell with codomain $\mathrm{Hom}_A(x, y)$ is invertible. By 2-cell conservativity, a 2-cell with codomain $\mathrm{Hom}_A(x, y)$ is invertible just when its whiskered composite with the isofibration $(p_1, p_0) : \mathrm{Hom}_A(x, y) \twoheadrightarrow 1 \times 1$ is an invertible 2-cell, but in fact this whiskered composite is an identity since 1 is 2-terminal. □

The weak universal property of Proposition 3.4.6 characterizes comma ∞-categories up to fibered equivalence (see Definition 3.2.7) over $C \times B$.

PROPOSITION 3.4.11 (uniqueness of comma ∞-categories). *For any isofibration* $(e_1, e_0) : E \twoheadrightarrow C \times B$ *that is fibered equivalent to* $\mathrm{Hom}_A(f, g) \twoheadrightarrow C \times B$ *the 2-cell*

$$
\begin{array}{ccc}
 & E & \\
e_1 \swarrow & & \searrow e_0 \\
C & \overset{\epsilon}{\Leftarrow} & B \\
g \searrow & & \nearrow f \\
 & A &
\end{array}
$$

encoded by the equivalence $e : E \overset{\sim}{\twoheadrightarrow} \mathrm{Hom}_A(f, g)$ *satisfies the weak universal property of Proposition 3.4.6. Conversely, if* $(d_1, d_0) : D \twoheadrightarrow C \times B$ *and* $(e_1, e_0) : E \twoheadrightarrow C \times B$ *are equipped with 2-cells*

$$
\begin{array}{ccc}
 & D & \\
d_1 \swarrow & & \searrow d_0 \\
C & \overset{\delta}{\Leftarrow} & B \\
g \searrow & & \nearrow f \\
 & A &
\end{array}
\qquad and \qquad
\begin{array}{ccc}
 & E & \\
e_1 \swarrow & & \searrow e_0 \\
C & \overset{\epsilon}{\Leftarrow} & B \\
g \searrow & & \nearrow f \\
 & A &
\end{array}
\qquad (3.4.12)
$$

satisfying the weak universal property of Proposition 3.4.6, then $D \simeq_{C \times B} E$.

Proof The proof of the first statement proceeds exactly as in the special case of Proposition 3.2.10. We prove the converse, solving Exercise 3.2.ii.

Consider a pair of 2-cells (3.4.12) satisfying the weak universal properties enumerated in Proposition 3.4.6. 1-cell induction supplies a map of spans

Exchanging the roles of δ and ϵ yields a second map of spans $\ulcorner \epsilon \urcorner : E \to D$ with the property that $\epsilon \ulcorner \delta \urcorner \ulcorner \epsilon \urcorner = \epsilon$ and $\delta \ulcorner \epsilon \urcorner \ulcorner \delta \urcorner = \delta$. By Proposition 3.4.7 it follows that $\ulcorner \delta \urcorner \ulcorner \epsilon \urcorner \cong \mathrm{id}_E$ over $C \times B$ and $\ulcorner \epsilon \urcorner \ulcorner \delta \urcorner \cong \mathrm{id}_D$ over $C \times B$. This defines the data of a fibered equivalence $D \simeq E$.[4] □

As is our convention for ∞-categories of arrows, it is convenient extend the appellation "comma ∞-category" from the strict model constructed in Definition 3.4.1 to any ∞-category that is fibered equivalent to it and refer to its accompanying 2-cell as the "comma cone." For example, in §4.2, we introduce multiple models for the ∞-*category of cones* over a fixed simplicial set indexed diagram, which are useful in developing various equivalent formulations of the universal property of limits.

Exercises

EXERCISE 3.4.i (C.1.12). Complete the proof of Proposition 3.4.5 by observing that the map $\mathrm{Hom}_p(q, r)$ factors as a pullback of the Leibniz cotensor of $\partial \Delta[1] \hookrightarrow \Delta[1]$ with p followed by a pullback of $r \times q$.

EXERCISE 3.4.ii. Use Proposition 3.4.7 to justify the functoriality up to isomorphism of the comma construction in oplax morphisms described in Observation 3.4.8.

EXERCISE 3.4.iii. Exercises 3.4.i and 3.4.ii illustrate the relative advantages and disadvantages of strict simplicial and weak 2-categorical universal properties of the comma ∞-category construction: the former gives a strictly functorial action but only of strictly commutative maps of cospans, while the latter gives an action of oplax transformations of cospans that is only functorial up to isomorphism. Mediating between these two constructions, use Lemma 1.2.19 and Proposition

---

[4] As alluded to in Warning 3.2.9, there is a slight ambiguity in the 2-categorical data that encodes a fibered equivalence. Proposition 3.6.4 provides a small boost to finish this proof.

1.4.9 to rectify a pseudo-commutative diagram

$$
\begin{array}{ccccc}
C & \xrightarrow{g} & A & \xleftarrow{f} & B \\
{\scriptstyle r}\downarrow & {\scriptstyle \cong\Downarrow\gamma} & {\scriptstyle p}\downarrow & {\scriptstyle \cong\Downarrow\beta} & \downarrow{\scriptstyle q} \\
\bar{C} & \xrightarrow[\bar{g}]{} & \bar{A} & \xleftarrow[\bar{f}]{} & \bar{B}
\end{array}
$$

into an equivalent strictly commutative diagram and prove that the induced map $\mathrm{Hom}_p(q,r)$ is equivalent to $\beta \downarrow \gamma$.

Exercise 3.4.iv. Show that the functor between comma ∞-categories induced by a diagram

$$
\begin{array}{ccccc}
C & \xrightarrow{g} & A & \xleftarrow{f} & B \\
{\scriptstyle r}\downarrow & {\scriptstyle \cong\Downarrow\gamma} & {\scriptstyle p}\downarrow & {\scriptstyle \cong\Downarrow\beta} & \downarrow{\scriptstyle q} \\
\bar{C} & \xrightarrow[\bar{g}]{} & \bar{A} & \xleftarrow[\bar{f}]{} & \bar{B}
\end{array}
$$

in which β and γ are isomorphisms and p, q, and r are equivalences defines an equivalence over $r \times q$.

$$
\begin{array}{ccc}
\mathrm{Hom}_A(f,g) & \xrightarrow[\approx]{\beta\downarrow\gamma} & \mathrm{Hom}_{\bar{A}}(\bar{f},\bar{g}) \\
{\scriptstyle (p_1,p_0)}\downarrow & & \downarrow{\scriptstyle (p_1,p_0)} \\
C \times B & \xrightarrow[r\times q]{\sim} & \bar{C} \times \bar{B}
\end{array}
$$

3.5 Representable Comma ∞-Categories

Definition 3.4.1 constructs a comma ∞-category for any cospan. In the special cases where one of the legs of the cospan is an identity, this provides two dual mechanisms to encode a functor between ∞-categories as an ∞-category itself.

Definition 3.5.1 (left and right representations). Any functor $f \colon A \to B$ admits a **left representation** and a **right representation** as a comma ∞-category, displayed below-left and below-right:

$$
\begin{array}{ccc}
 & \mathrm{Hom}_B(f,B) & \\
{\scriptstyle p_1}\swarrow & {\scriptstyle \Leftarrow\not{\phi}} & \searrow{\scriptstyle p_0} \\
B & \xleftarrow[f]{} & A
\end{array}
\qquad
\begin{array}{ccc}
 & \mathrm{Hom}_B(B,f) & \\
{\scriptstyle p_1}\swarrow & {\scriptstyle \Leftarrow\not{\phi}} & \searrow{\scriptstyle p_0} \\
A & \xrightarrow[f]{} & B
\end{array}
$$

where we save space by depicting the **left comma cone over** f displayed above-left and the **right comma cone over** f displayed above-right as inhabiting triangles rather than squares.

By Proposition 3.4.11, the weak universal property of the comma cone characterizes the comma span up to fibered equivalence over the product of the codomain objects. Thus:

DEFINITION 3.5.2. Given a cospan $C \xrightarrow{g} A \xleftarrow{f} B$, the comma ∞-category $\mathsf{Hom}_A(f, g) \twoheadrightarrow C \times B$ is **left representable** if there exists a functor $\ell : B \to C$ so that

$$\mathsf{Hom}_A(f, g) \simeq_{C \times B} \mathsf{Hom}_C(\ell, C)$$

and **right representable** if there exists a functor $r : C \to B$ so that

$$\mathsf{Hom}_A(f, g) \simeq_{C \times B} \mathsf{Hom}_B(B, r).$$

In this section, we prove a representability theorem: a comma ∞-category $\mathsf{Hom}_A(f, g)$ is right representable if and only if $g : C \to A$ admits an absolute right lifting along $f : B \to A$, in which case the representing functor $r : C \to B$ defines the postulated lifting. We prove this result over the course of three theorems, each strengthening the previous statement.

The first theorem characterizes those natural transformations

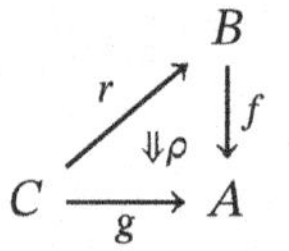

$$
\begin{array}{ccc}
& & B \\
& \overset{r}{\nearrow} {\scriptstyle \Downarrow \rho} & \downarrow {\scriptstyle f} \\
C & \underset{g}{\longrightarrow} & A
\end{array}
$$

that define absolute right lifting diagrams as those that induce fibered equivalences $\mathsf{Hom}_B(B, r) \simeq_{C \times B} \mathsf{Hom}_A(f, g)$ between comma ∞-categories. The second theorem proves that a functor r defines an absolute right lifting of g through f just when $\mathsf{Hom}_A(f, g)$ is right represented by r; the difference is that no natural transformation $\rho : fr \Rightarrow g$ need be provided. The final theorem gives a general right representability criterion that can be applied to construct a right representation to $\mathsf{Hom}_A(f, g)$ without a priori specifying the representing functor r. Dual results characterize left representable comma ∞-categories.

THEOREM 3.5.3. *The triangle below-left defines an absolute right lifting diagram if and only if the induced functor below-right*

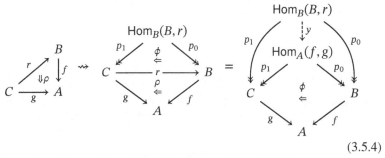

$$(3.5.4)$$

defines a fibered equivalence $\mathrm{Hom}_B(B, r) \simeq_{C \times B} \mathrm{Hom}_A(f, g)$.

In [123], Street and Walters interpret the equivalence $\mathrm{Hom}_B(B, r) \simeq_{C \times B}$ $\mathrm{Hom}_A(f, g)$ encoding an absolute right lifting diagram as asserting that "f is left adjoint to r relative to g." This notion of relative adjunction, first studied by Ulmer [126], should be compared with the definitions of adjunction given in Lemma 2.3.7 and Proposition 4.1.1.

Proof Suppose that (r, ρ) defines an absolute right lifting of g through f, and consider the unique factorization of the comma cone under $\mathrm{Hom}_A(f, g)$ through ρ displayed by the left-hand pasting equality:

$$(3.5.5)$$

By 1-cell induction, the natural transformation ζ factors through the right comma cone over r as displayed above-center. Substituting (3.5.4) into the bottom portion of the above-right diagram, we see that $yz : \mathrm{Hom}_A(f, g) \to \mathrm{Hom}_A(f, g)$ is a functor that factors the comma cone for $\mathrm{Hom}_A(f, g)$ through itself. Applying the universal property of Proposition 3.4.7, it follows that there is a fibered isomorphism $yz \cong \mathrm{id}_{\mathrm{Hom}_A(f, g)}$ over $C \times B$.

To prove that $zy \cong \mathrm{id}_{\mathrm{Hom}_B(B, r)}$ it suffices to argue similarly that the right comma cone over r restricts along zy to itself. Since (r, ρ) is absolute right lifting, it suffices to verify the equality $\phi zy = \phi$ after pasting below with ρ. But now reversing the order of the equalities in (3.5.5) and (3.5.4) we have

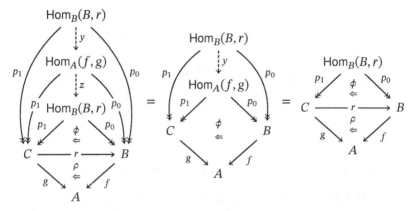

which is exactly what we wanted to show. Thus, we see that if (r, ρ) is an absolute right lifting of g through f, then the induced map (3.5.4) defines a fibered equivalence $\mathrm{Hom}_B(B, r) \simeq \mathrm{Hom}_A(f, g)$.

Now, conversely, suppose the functor y defined by (3.5.4) is a fibered equivalence and let us argue that (r, ρ) is an absolute right lifting of g through f. By Proposition 3.4.11, the natural transformation displayed on the left-hand side of the equality in (3.5.4) inherits the weak universal property of a comma cone from $\mathrm{Hom}_A(f, g)$. So Proposition 3.4.7 supplies a bijection displayed below-left

$$
\left\{
\begin{array}{c}
X \\
{}^{c}\swarrow \quad \searrow^{b} \\
C \quad \overset{\alpha}{\Leftarrow} \quad B \\
{}_{g}\searrow \quad \swarrow_{f} \\
A
\end{array}
\right\}
\longleftrightarrow
\left\{
\begin{array}{c}
X \\
{}^{c}\swarrow \; \downarrow^{a} \; \searrow^{b} \\
C \quad\quad\quad B \\
{}_{p_1}\nwarrow \; \downarrow \; \nearrow_{p_0} \\
\mathrm{Hom}_B(B, r)
\end{array}
\right\}_{/\cong}
\longleftrightarrow
\left\{
\begin{array}{c}
X \\
{}^{c}\swarrow \overset{\xi}{\Leftarrow} \searrow^{b} \\
C \quad \overset{}{\underset{r}{\longrightarrow}} \quad B
\end{array}
\right\}
$$

between 2-cells over the cospan and fibered isomorphism classes of maps of spans that is implemented, from center to left, by whiskering with the 2-cell $\rho p_1 \cdot f \phi : f p_0 \Rightarrow g p_1$ in the center of (3.5.4). Proposition 3.4.7 also applies to the right comma cone ϕ over $r : C \to B$ giving us a second bijection, displayed above-center-right between the same fibered isomorphism classes of maps of spans and 2-cells over r. This second bijection is implemented, from center to right, by pasting with the right comma cone $\phi : p_0 \Rightarrow r p_1$. Combining these yields a bijection between the 2-cells displayed on the right and the 2-cells displayed on the left implemented by pasting with ρ, which is precisely the universal property that characterizes absolute right lifting diagrams. $\quad\square$

As a special case of this result:

COROLLARY 3.5.6. *The following are equivalent, and define what it means for a functor* $f : A \to B$ *between ∞-categories to be **fully faithful**:*

  (i) *The identity defines an absolute right lifting diagram:*

$$
\begin{array}{ccc}
 & & A \\
 & \nearrow\!\!\!\nearrow & \big\downarrow f \\
 & \| & \\
A & \xrightarrow{\ f\ } & B
\end{array}
$$

  (ii) *The identity defines an absolute left lifting diagram:*

$$
\begin{array}{ccc}
 & & A \\
 & \nearrow\!\!\!\nearrow & \big\downarrow f \\
 & \| & \\
A & \xrightarrow{\ f\ } & B
\end{array}
$$

  (iii) *For any ∞-category X the induced functor*

$$f_* : \mathsf{hFun}(X, A) \to \mathsf{hFun}(X, B)$$

  is a fully faithful functor of 1-categories.

  (iv) *The functor induced by the identity 2-cell* id_f *defines a fibered equivalence* $A^2 \simeq_{A \times A} \mathsf{Hom}_B(f, f)$.

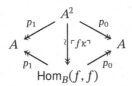

Proof   The statement (iii) is an unpacking of the meaning of both (i) and (ii). Theorem 3.5.3 specializes to prove (i)⇔(iv) or dually (ii)⇔(iv).   □

It is not surprising that postcomposition with a fully faithful functor of ∞-categories induces a fully faithful functor of hom-categories in the homotopy 2-category, and in particular between the homotopy categories (see Definition 1.4.11). What is less apparent is that this condition is strong enough to capture the ∞-categorical notion of fully faithfulness, when certainly it would not be enough to merely require that the functor $hf : hA \to hB$ is fully faithful. The unexpected power of condition (iii) is that its statement quantifies over all generalized elements $a : X \to A$ of A, in contrast to the objects of hA which are limited to the elements $a : 1 \to A$. This provides a retroactive justification for our work in the homotopy 2-category.

Theorem 3.5.3 has another important consequence cited in the proof of Proposition 2.4.7.

COROLLARY 3.5.7. *Cosmological functors preserve absolute lifting diagrams.*

Proof Consider a cosmological functor $F: \mathcal{K} \to \mathcal{L}$ together with an absolute right lifting diagram in $\mathcal{K}$:

$$\begin{array}{ccc} & & B \\ & \overset{r}{\nearrow} & \Big\downarrow f \\ & \overset{\Downarrow \rho}{} & \\ C & \xrightarrow[g]{} & A \end{array}$$

By Theorem 3.5.3, the induced functor of (3.5.4) defines a fibered equivalence $y: \operatorname{Hom}_B(B, r) \xrightarrow{\sim} \operatorname{Hom}_A(f, g)$ over $C \times B$.

Since cosmological functors preserve the simplicial limits and isofibrations of (3.4.2), F carries y to a functor $Fy: \operatorname{Hom}_{FB}(FB, Fr) \to \operatorname{Hom}_{FA}(Ff, Fg)$ over $FC \times FB$. By Lemma 1.3.2, this functor is again a fibered equivalence. Since cosmological functors define 2-functors, this functor satisfies a pasting equation

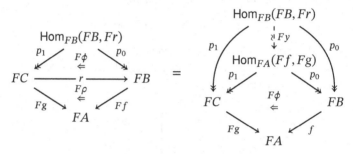

in $\mathcal{L}$. By Theorem 3.5.3, this fibered equivalence witnesses the fact that

$$\begin{array}{ccc} & & FB \\ & \overset{Fr}{\nearrow} & \Big\downarrow Ff \\ & \overset{\Downarrow F\rho}{} & \\ FC & \xrightarrow[Fg]{} & FA \end{array}$$

defines an absolute right lifting diagram in $\mathcal{L}$.   □

Having proven Theorem 3.5.3 our immediate aim is to strengthen it to show that a fibered equivalence $\operatorname{Hom}_B(B, r) \simeq_{C \times B} \operatorname{Hom}_A(f, g)$ implies that $r: C \to B$ defines an absolute right lifting of g through f without a previously specified 2-cell $\rho: fr \Rightarrow g$.

THEOREM 3.5.8. *Given a trio of functors* $r: C \to B$, $f: B \to A$, *and* $g: C \to A$ *there is a bijection between natural transformations as displayed below-left and fibered isomorphism classes of maps of spans as displayed below-right*

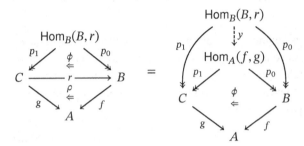

that is constructed by pasting with the right comma cone over r and then applying 1-cell induction to factor through the comma cone for $\mathrm{Hom}_A(f,g)$.

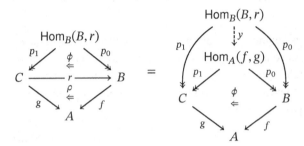

Moreover, a natural transformation $\rho : fr \Rightarrow g$ *displays r as an absolute right lifting of g through f if and only if the corresponding map* $y : \mathrm{Hom}_B(B,r) \to \mathrm{Hom}_A(f,g)$ *is an equivalence.*

The second clause is the statement of Theorem 3.5.3, so it remains only to prove the first. We show the claimed construction is a bijection by exhibiting its inverse, the construction of which involves a rather mysterious lemma whose significance will gradually become apparent.

LEMMA 3.5.9. *For any functor* $f : A \to B$, *the codomain projection functor* $p_1 : \mathrm{Hom}_B(B,f) \twoheadrightarrow A$ *from its right representation admits a right adjoint right inverse*[5] $i := \ulcorner \mathrm{id}_f \urcorner$ *induced from the identity 2-cell* id_f, *defining an adjunction over A whose counit is an identity and whose unit* $\eta : \mathrm{id} \Rightarrow ip_1$ *satisfies the*

---

[5] A functor admits a **right adjoint right inverse** just when it is the left adjoint in an adjunction whose counit is invertible (see §B.4). When the original functor is an isofibration, as is the case here, any right adjoint right inverse adjunction can be upgraded to one in which the counit is an identity, which can then be upgraded further to a *fibered adjunction* (see Definition 3.6.5 and Lemma 3.6.9).

conditions $\eta i = \mathrm{id}_i$, $p_1\eta = \mathrm{id}_{p_1}$, *and* $p_0\eta = \phi$.

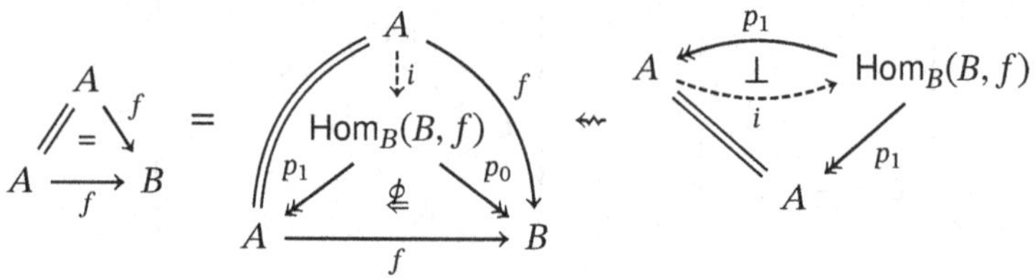

This lemma figures prominently in the proof of the Yoneda lemma in §5.7 and is also the main ingredient in a "cheap" version of the Yoneda lemma appearing as Corollary 3.5.11.

Proof This adjunction is constructed via the weak universal properties of the right comma cone over f. The identity 2-cell id_f induces a functor $i := \ulcorner \mathrm{id}_f \urcorner$ over the right comma cone over f as displayed in the statement. Note that $p_1 i = \mathrm{id}_A$, so we may take the counit to be the identity 2-cell. Since $\phi i = \mathrm{id}_f$, we have a pasting equality:

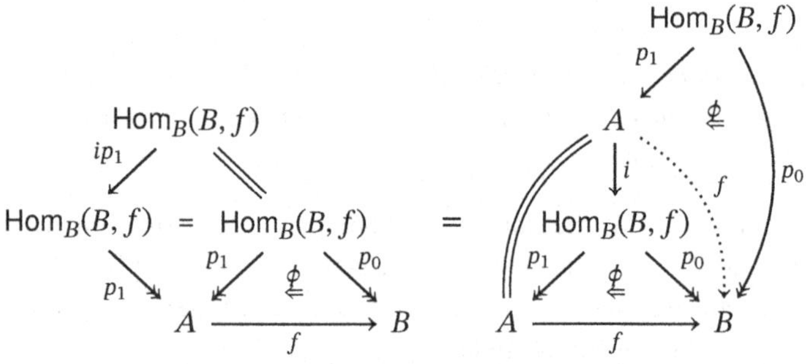

which induces a 2-cell $\eta \colon \mathrm{id} \Rightarrow ip_1$ with defining equations $p_1\eta = \mathrm{id}_{p_1}$ and $p_0\eta = \phi$. The first of these conditions provides one triangle equality; for the other, we must verify that $\eta i = \mathrm{id}_i$. By 2-cell conservativity, ηi is an isomorphism since $p_1\eta i = \mathrm{id}_A$ and $p_0\eta i = \mathrm{id}_f$ are both invertible. By naturality of whiskering, we have

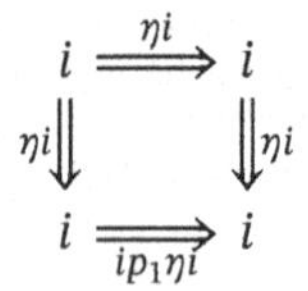

$$
\begin{array}{ccc}
i & \overset{\eta i}{\Longrightarrow} & i \\
\eta i \big\Downarrow & & \big\Downarrow \eta i \\
i & \underset{ip_1\eta i}{\Longrightarrow} & i
\end{array}
$$

and since $p_1\eta = \mathrm{id}_{p_1}$ the bottom edge is an identity. So $\eta i \cdot \eta i = \eta i$ and since ηi is an isomorphism cancelation implies that $\eta i = \mathrm{id}_i$ as required. $\square$

One interpretation of Lemma 3.5.9 is best revealed through a special case:

COROLLARY 3.5.10. *For any element* $b : 1 \to B$, *the identity at* b *defines a terminal element in* $\mathrm{Hom}_B(B, b)$.

Proof By Lemma 3.5.9, the codomain projection – which in this case reduces to the unique functor $! : \mathrm{Hom}_B(B, b) \to 1$ – admits a right adjoint right inverse induced from its identity 2-cell. Thus, by Definition 2.2.1, this right adjoint identifies a terminal element $\ulcorner \mathrm{id}_b \urcorner : 1 \to \mathrm{Hom}_B(B, b)$ corresponding to the identity morphism $\mathrm{id}_b : b \to b$ in the homotopy category hB. □

The general version of Lemma 3.5.9 has a similar interpretation: id_f induces a terminal element in $\mathrm{Hom}_B(B, f)$ "over A," that is, in the sliced ∞-cosmos (see Definition 3.6.8 and Example 3.6.12).

Proof of Theorem 3.5.8 It remains to construct an inverse to the function in the statement that takes a natural transformation $fr \Rightarrow g$ and produces a fibered isomorphism class of functors $\mathrm{Hom}_B(B, r) \to \mathrm{Hom}_A(f, g)$ over $C \times B$. Our construction makes use of the right adjoint right inverse $i : C \to \mathrm{Hom}_B(B, r)$ of Lemma 3.5.9. Given a functor $\mathrm{Hom}_B(B, r) \to \mathrm{Hom}_A(f, g)$, restrict along i, and paste with the comma cone for $\mathrm{Hom}_A(f, g)$ to define a natural transformation $fr \Rightarrow g$.

Starting from a natural transformation $\rho : fr \Rightarrow g$, the composite of these two functions constructs the natural transformation displayed below-left

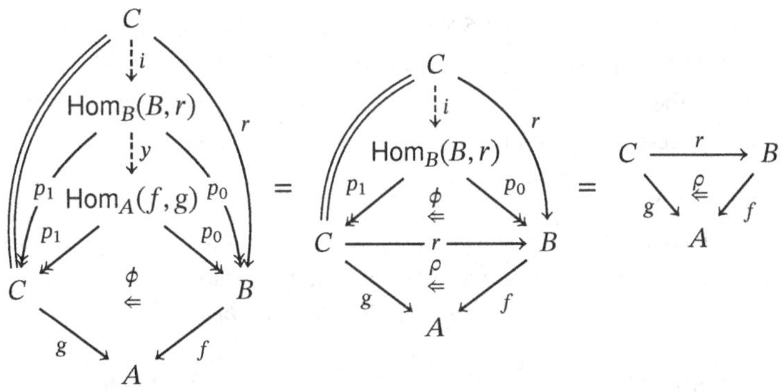

which equals the above-center pasted composite by the definition of y from ρ, and equals the above-right composite since $\phi i = \mathrm{id}_r$. Thus, when a natural transformation $\rho : fr \Rightarrow g$ is encoded as a map $y : \mathrm{Hom}_B(B, r) \to \mathrm{Hom}_A(f, g)$ over $C \times B$, and then re-converted into a natural transformation, the original natural transformation ρ is recovered.

For the converse, starting with a map $z : \mathrm{Hom}_B(B, r) \to \mathrm{Hom}_A(f, g)$ over $C \times B$, the composite of these two functions constructs an isomorphism class of maps of spans w displayed below-left by applying 1-cell induction for the comma cone for $\text{Hom}_A(f, g)$ to the composite natural transformation pasted below-center:

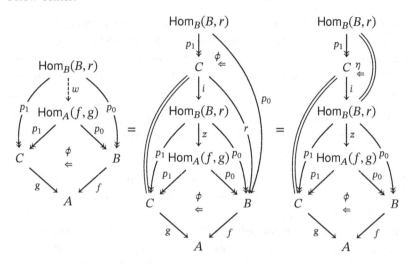

By Lemma 3.5.9, there exists a natural transformation $\eta : \text{id} \Rightarrow ip_1$ so that $p_0\eta = \phi$ – this gives the pasting equality above center – and $p_1\eta = \text{id}$, which tells us that the right-hand pasting diagram reduces to ϕz. Proposition 3.4.7 now implies that $w \cong z$ over $C \times B$.   □

A dual version of Theorem 3.5.8 represents natural transformations $g \Rightarrow f\ell$ as fibered isomorphism classes of maps $\text{Hom}_B(\ell, B) \to \text{Hom}_A(g, f)$ over $B \times C$. Specializing these results to the case where one of f or g is the identity, we immediately recover a "cheap" form of the Yoneda lemma:

COROLLARY 3.5.11. *Given a parallel pair of functors, $f, g : A \to B$, there are bijections between natural transformations as displayed below-center and fibered isomorphism classes of maps between their left and right representations as comma ∞-categories, as displayed below-left and below-right:*

$$\left\{ \begin{array}{c} \text{Hom}_B(g, B) \\ {}^{p_1}\swarrow \quad \downarrow{}^{\ulcorner \alpha^* \urcorner} \quad \searrow^{p_0} \\ B \quad\quad\quad\quad A \\ {}^{p_1}\searrow \quad \downarrow \quad \nearrow^{p_0} \\ \text{Hom}_B(f, B) \end{array} \right\}_{/\cong} \quad\longleftrightarrow\quad \left\{ A \underset{g}{\overset{f}{\underset{\Downarrow\alpha}{\rightrightarrows}}} B \right\} \quad\longleftrightarrow\quad \left\{ \begin{array}{c} \text{Hom}_B(B, f) \\ {}^{p_1}\swarrow \quad \downarrow{}^{\ulcorner \alpha_* \urcorner} \quad \searrow^{p_0} \\ A \quad\quad\quad\quad B \\ {}^{p_1}\searrow \quad \downarrow \quad \nearrow^{p_0} \\ \text{Hom}_B(B, g) \end{array} \right\}_{/\cong}$$

that are constructed by pasting with the left comma cone over g and right comma cone over f, respectively:

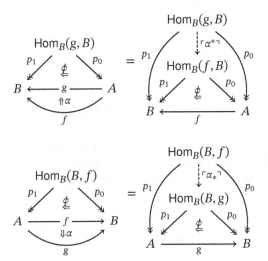

and then applying 1-cell induction to factor through the left comma cone over f in the former case or the right comma cone over g in the latter.   □

Combining the results of this section, we prove one final representability theorem that allows us to recognize when a comma ∞-category is right representable in the absence of a predetermined representing functor. This result specializes to give existence theorems for adjoint functors and for limits and colimits in the next chapter.

THEOREM 3.5.12. *The comma ∞-category* $\mathrm{Hom}_A(f, g)$ *associated to a cospan* $C \xrightarrow{g} A \xleftarrow{f} B$ *is right representable if and only if its codomain projection functor admits a right adjoint right inverse*

$$\mathrm{Hom}_A(f, g)$$

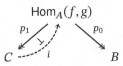

in which case the composite $p_0 i : C \to B$ *defines the representing functor and the natural transformation encoded by the functor* $i : C \to \mathrm{Hom}_A(f, g)$ *defines an absolute right lifting of g through f.*

Proof Suppose that the comma $\mathrm{Hom}_A(f, g)$ is represented on the right by a functor $r : C \to B$. By Lemma 3.5.9, $p_1 : \mathrm{Hom}_B(B, r) \twoheadrightarrow C$ admits a right adjoint right inverse i', which composes with the fibered equivalence to define a right adjoint right inverse for the equivalent functor $p_1 : \mathrm{Hom}_A(f,g) \twoheadrightarrow C$.

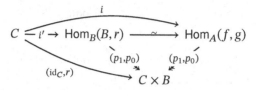

Note that $r = p_0 i$, and by the construction in the proof of Theorem 3.5.8, the functor $i : C \to \mathrm{Hom}_A(f,g)$ encodes an absolute right lifting diagram $\rho : rf \Rightarrow g$. Thus, it remains only to prove the converse.

To that end, suppose we are given a right adjoint right inverse adjunction $p_1 \dashv i$. Unpacking the definition, this provides an adjunction

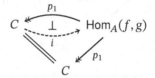

over C whose counit is an identity and whose unit $\eta : \mathrm{id} \Rightarrow ip_1$ satisfies the conditions $\eta i = \mathrm{id}_i$ and $p_1\eta = \mathrm{id}_{p_1}$. By Theorem 3.5.8, to construct the fibered equivalence $\mathrm{Hom}_B(B,r) \simeq_{C\times B} \mathrm{Hom}_A(f,g)$ with $r := p_0 i$, it suffices to demonstrate that the natural transformation defined by restricting the comma cone for $\mathrm{Hom}_A(f,g)$ along i

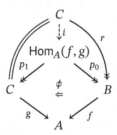

defines an absolute right lifting diagram.

By 1-cell induction any natural transformation $\chi : fb \Rightarrow gc$ induces a functor $\ulcorner\chi\urcorner$ as displayed below-left:

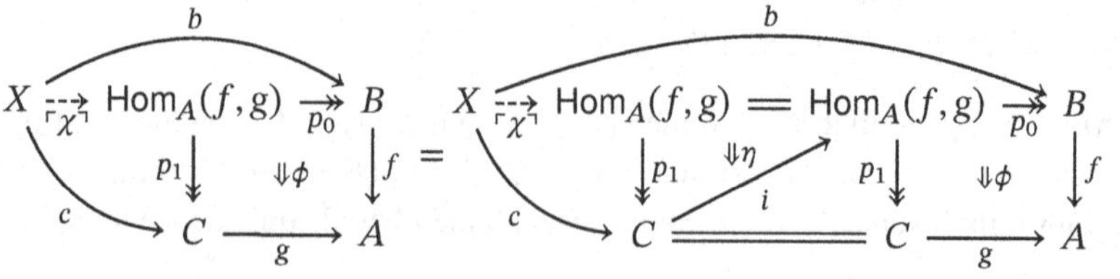

Inserting the triangle equality $p_1\eta = \mathrm{id}_{p_1}$ as displayed above-right constructs the desired factorization $p_0\eta^\ulcorner\chi^\urcorner : b \Rightarrow rc$ of χ through ϕi.

In fact, given any natural transformation $\tau_0 : b \Rightarrow rc$ that defines a factorization of $\chi : fb \Rightarrow gc$ through ϕi, the pair (id_c, τ_0) satisfies the compatibility condition of Proposition 3.4.6(ii), inducing a natural transformation $\tau : \ulcorner\chi^\urcorner \Rightarrow ic$ so that $\mathrm{id}_c = p_1\tau$ and $\tau_0 = p_0\tau$. We argue that the natural transformation τ is unique, proving that the factorization $p_0\tau : b \Rightarrow rc$ is also unique.

To see this, note that the adjunction $p_1 \dashv i$ over C exhibits the right adjoint as a terminal element of the object $p_1 : \mathrm{Hom}_A(f, g) \twoheadrightarrow C$ in the strict slice of the homotopy 2-category over C.[6] It follows, as in Lemma 2.2.5, that for any object $c : X \to C$ and any morphism $\ulcorner\chi^\urcorner : X \to \mathrm{Hom}_A(f, g)$ over C, there exists a unique natural transformation $\ulcorner\chi^\urcorner \Rightarrow ic$ over C. Thus, there is a unique natural transformation $\tau : \ulcorner\chi^\urcorner \Rightarrow ic$ with the property that $p_1\tau = \mathrm{id}_c$, and so the factorization $p_0\tau : b \Rightarrow rc$ of χ through ϕi must also be unique. $\square$

In the next section, we discover that Theorem 3.5.12 may be expressed more concisely as the assertion that a comma ∞-category $\mathrm{Hom}_A(f, g)$ in an ∞-cosmos $\mathcal{K}$ is right representable precisely when its codomain projection functor $p_1 : \mathrm{Hom}_A(f, g) \twoheadrightarrow C$ admits a terminal element as an object of the sliced ∞-cosmos $\mathcal{K}_{/C}$ (see Corollary 3.6.10). Dually, $\mathrm{Hom}_A(f, g)$ is left representable just when its domain projection functor admits an initial element as an object of the sliced ∞-cosmos $\mathcal{K}_{/B}$. There is a small gap between this statement and the version proven in Theorem 3.5.12 having to do with the discrepancy between the homotopy 2-category of the sliced ∞-cosmos $\mathcal{K}_{/C}$ and the slice of the homotopy 2-category $\mathfrak{h}\mathcal{K}$ over C. This is the subject to which we now turn.

Exercises

EXERCISE 3.5.i. Anticipate Proposition 4.1.1 by exploring how one might encode the existence of an adjunction $f \dashv u$ between a given opposing pair of functors using comma ∞-categories.

EXERCISE 3.5.ii. Extend the result of Exercise 2.3.vi to show that for any equivalence of cospans

$$
\begin{array}{ccccc}
C & \xrightarrow{\ g\ } & A & \xleftarrow{\ f\ } & B \\
\downarrow{\wr} & \cong & \downarrow{\wr} & \cong & \downarrow{\wr} \\
\bar{C} & \xrightarrow[\ \bar{g}\ ]{} & \bar{A} & \xleftarrow[\ \bar{f}\ ]{} & \bar{B}
\end{array}
$$

---

[6] An object is a functor of ∞-categories $c : X \to C$, a 1-cell is a functor between the domain ∞-categories defining a strictly commutative triangle, and a 2-cell is a natural transformation between such functors that whiskers to define an identity 2-cell with codomain C.

there exists an absolute right lifting of g through f if and only if there exists an absolute right lifting of $\bar{g}$ through $\bar{f}$.

EXERCISE 3.5.iii ([124, 3.7]). Use Theorem 3.5.3 and Corollary 3.5.6(iv) to prove that a fully faithful functor $f : A \to B$ reflects all limits or colimits that exist in A. Why does this argument not also show that $f : A \to B$ preserves them?

3.6 Fibered Adjunctions and Fibered Equivalences

In Proposition 3.2.10, we discovered that the ∞-category A^2 of arrows in A together with its codomain and domain evaluation functors $(p_1, p_0) : A^2 \twoheadrightarrow A \times A$ satisfies a weak universal property in the homotopy 2-category that characterizes it up to equivalence *over* $A \times A$. Similarly, Proposition 3.4.11 tells us that the comma ∞-category associated to a given pair of functors with common codomain is characterized up to *fibered equivalence*, as defined in Definition 3.2.7.

As noted in Warning 3.2.9 there is some ambiguity regarding the 2-categorical data required to specify a fibered equivalence that we now address head-on. The issue is that, for an ∞-category B in an ∞-cosmos $\mathcal{K}$, the homotopy 2-category $\mathfrak{h}(\mathcal{K}_{/B})$ of the sliced ∞-cosmos (see Proposition 1.2.22 and Definition 1.4.1) is not isomorphic to the 2-category $(\mathfrak{h}\mathcal{K})_{/B}$ of isofibrations, functors, and 2-cells over B in the homotopy 2-category $\mathfrak{h}\mathcal{K}$. However, there is a canonical comparison functor relating these 2-categories that satisfies a property we now introduce:

DEFINITION 3.6.1 (smothering 2-functor). A 2-functor $F : \mathcal{A} \to \mathcal{B}$ is **smothering** if it is

- surjective on objects;
- full on 1-cells: for any pair of objects A, A' in $\mathcal{A}$ and 1-cell $k : FA \to FA'$ in $\mathcal{B}$, there exists $f : A \to A'$ in $\mathcal{A}$ with $Ff = k$;
- full on 2-cells: for any parallel pair of 1-cells $f, g : A \to A'$ in $\mathcal{A}$ and 2-cell

$$FA \underset{Fg}{\overset{Ff}{\underset{\Downarrow\beta}{\rightrightarrows}}} FA'$$

in $\mathcal{B}$, there exists a 2-cell $\alpha : f \Rightarrow g$ in $\mathcal{A}$ with $F\alpha = \beta$;

and

- conservative on 2-cells: for any 2-cell α in $\mathcal{A}$ if $F\alpha$ is invertible in $\mathcal{B}$ then α is invertible in $\mathcal{A}$.

This is to say, a smothering 2-functor is a surjective-on-objects 2-functor that is "locally smothering," meaning that the action on hom-categories is by a smothering functor, as codified in Definition 3.1.2.

The prototypical example of a smothering 2-functor solves Exercise 1.4.vii.

PROPOSITION 3.6.2. *Let B be an ∞-category in an ∞-cosmos $\mathcal{K}$. There is a canonical 2-functor*

$$\mathfrak{h}(\mathcal{K}_{/B}) \longrightarrow (\mathfrak{h}\mathcal{K})_{/B}$$

from the homotopy 2-category of the sliced ∞-cosmos $\mathcal{K}_{/B}$ to the 2-category of isofibrations, functors, and 2-cells over B in $\mathfrak{h}\mathcal{K}$ and this 2-functor is smothering.

This follows more or less immediately from Lemma 3.1.5 but we spell out the details nonetheless.

Proof The 2-categories $\mathfrak{h}(\mathcal{K}_{/B})$ and $(\mathfrak{h}\mathcal{K})_{/B}$ have the same objects – isofibrations with codomain B – and the same 1-cells – functors between the domains that commute with these isofibrations – so the canonical mapping may be defined to act as the identity on underlying 1-categories.

By the definition of the sliced ∞-cosmos given in Proposition 1.2.22, a 2-cell between functors $f, g : E \to F$ from $p : E \twoheadrightarrow B$ to $q : F \twoheadrightarrow B$ is a homotopy class of 1-simplices in the quasi-category defined by the pullback of simplicial sets below-left

$$\begin{array}{ccc}
\mathsf{Fun}_B(E,F) \longrightarrow \mathsf{Fun}(E,F) & & (\mathsf{hFun})_{/B}(E,F) \longrightarrow \mathsf{hFun}(E,F) \\
\downarrow \quad \lrcorner \qquad \downarrow {\scriptstyle q_*} & & \downarrow \quad \lrcorner \qquad \downarrow {\scriptstyle q_*} \\
\mathbb{1} \xrightarrow{\ p\ } \mathsf{Fun}(E,B) & & \mathbb{1} \xrightarrow{\ p\ } \mathsf{hFun}(E,B)
\end{array}$$

Unpacking, a 2-cell $\alpha : f \Rightarrow g$ is represented by a 1-simplex $\alpha : f \to g$ in $\mathsf{Fun}(E,F)$ that whiskers with q to the degenerate 1-simplex on the vertex $p \in \mathsf{Fun}(E,B)$, and two such 1-simplices represent the same 2-cell if and only if they bound a 2-simplex of the form displayed in (1.1.8) that also whiskers with q to the degenerate 2-simplex on p.

By contrast, a 2-cell in $(\mathfrak{h}\mathcal{K})_{/B}$ is a morphism in the category defined by the pullback of categories above-right. Such 2-cells are represented by 1-simplices $\alpha : f \to g$ in $\mathsf{Fun}(E,F)$ that whisker with q to 1-simplices in $\mathsf{Fun}(E,B)$ that are homotopic to the degenerate 1-simplex on p, and two such 1-simplices represent the same 2-cell if and only if they are homotopic in $\mathsf{Fun}(E,F)$.

Applying the homotopy category functor $\mathsf{h} : \mathcal{QC}at \to \mathcal{C}at$ to the above-left pullback produces a cone over the above-right pullback, inducing a canonical map

$$\mathsf{h}(\mathsf{Fun}_B(E,F)) \longrightarrow (\mathsf{hFun})_{/B}(E,F),$$

which is the action on hom-categories of the canonical 2-functor $\mathfrak{h}(\mathcal{K}_{/B}) \to$
$(\mathfrak{h}\mathcal{K})_{/B}$. By Lemma 3.1.5, this canonical map defines a bijective-on-objects
smothering functor. Thus, we have defined a 2-functor $\mathfrak{h}(\mathcal{K}_{/B}) \to (\mathfrak{h}\mathcal{K})_{/B}$ that
is bijective on 0- and 1-cells and locally smothering, as claimed.   □

Smothering 2-functors are not strictly speaking invertible, but nevertheless
2-categorical structures from the codomain can be lifted to the domain.

LEMMA 3.6.3. *Smothering 2-functors reflect and create equivalences.*

Proof For any smothering 2-functor $F: \mathcal{A} \to \mathcal{B}$ and 1-cell $f: A \to B$ in
$\mathcal{A}$, if $Ff: FA \overset{\sim}{\to} FB$ is an equivalence in $\mathcal{B}$, then by fullness on 1-cells, an
equivalence inverse $g': FB \overset{\sim}{\to} FA$ to Ff lifts to a 1-cell $g: B \to A$ in $\mathcal{A}$. By
fullness on 2-cells, the isomorphisms $\mathrm{id}_{FA} \cong g' \circ Ff$ and $Ff \circ g' \cong \mathrm{id}_{FB}$ also
lift to $\mathcal{A}$ and by conservativity on 2-cells these lifted 2-cells are also invertible.
This proves that equivalences are reflected. To see that they are created, note
that any $f': FA \overset{\sim}{\to} FB$ in $\mathcal{B}$ lifts to a 1-cell $f: A \to B$, which is an equivalence
by the construction just given.   □

Applying Lemma 3.6.3 to the smothering 2-functor

$$\mathfrak{h}(\mathcal{K}_{/B}) \longrightarrow (\mathfrak{h}\mathcal{K})_{/B}$$

we resolve the ambiguity about the 2-categorical data of a fibered equivalence.

PROPOSITION 3.6.4 (fibered equivalence data). *Let B be an ∞-category in an*
∞-cosmos $\mathcal{K}$.

  (i) *Any equivalence in $(\mathfrak{h}\mathcal{K})_{/B}$ lifts to an equivalence in $\mathfrak{h}(\mathcal{K}_{/B})$. That is,*
      fibered equivalences over B may be specified by defining an opposing
      pair of 1-cells $f: E \to F$ and $g: F \to E$ over B together with invertible
      2-cells $\mathrm{id}_E \cong gf$ and $fg \cong \mathrm{id}_F$ that lie over B in $\mathfrak{h}\mathcal{K}$.
  (ii) *Moreover, if $f: E \to F$ is a map between isofibrations over B that*
       admits an not-necessarily fibered equivalence inverse $g: F \to E$ with
       not-necessarily fibered 2-cells $\mathrm{id}_E \cong gf$ and $fg \cong \mathrm{id}_F$, then this data is
       isomorphic to a genuine fibered equivalence.

Thus, the forgetful 2-functor $\mathfrak{h}(\mathcal{K}_{/B}) \to (\mathfrak{h}\mathcal{K})_{/B} \to \mathfrak{h}\mathcal{K}$ reflects equivalences.

Proof The first statement is proven by Lemma 3.6.3 and Proposition 3.6.2.
The second statement, which asserts that the forgetful 2-functor $(\mathfrak{h}\mathcal{K})_{/B} \to \mathfrak{h}\mathcal{K}$
reflects equivalences, is left as Exercise 3.6.i, and requires only the 2-categorical
lifting property of isofibrations (see Proposition 1.4.9).   □

This gives a 2-categorical proof of Proposition 1.2.22(vii), that for any ∞-category B in an ∞-cosmos $\mathcal{K}$, the forgetful functor $\mathcal{K}_{/B} \to \mathcal{K}$ preserves and reflects equivalences. The smothering 2-functor $\mathfrak{h}(\mathcal{K}_{/B}) \to (\mathfrak{h}\mathcal{K})_{/B}$ can also be used to lift adjunctions that are fibered 2-categorically over B to adjunctions in the sliced ∞-cosmos $\mathcal{K}_{/B}$.

DEFINITION 3.6.5 (fibered adjunction). A **fibered adjunction** over an ∞-category B in an ∞-cosmos $\mathcal{K}$ is an adjunction in the sliced ∞-cosmos $\mathcal{K}_{/B}$.

We write $f \dashv_B u$ to indicate that specified functors over B are **adjoint over** B.

LEMMA 3.6.6 (pullback and projection of fibered adjunctions).

   (i) A fibered adjunction over B pulls back along any functor $k : A \to B$ to define a fibered adjunction over A.

 (ii) A fibered adjunction over A can be composed with any isofibration[7] $p : A \twoheadrightarrow B$ to define a fibered adjunction over B.

Proof For any ∞-cosmos $\mathcal{K}$, pullback along $k : A \to B$ defines a cosmological functor $k^* : \mathcal{K}_{/B} \to \mathcal{K}_{/A}$, by Proposition 1.3.4(v), which descends to a 2-functor $k^* : \mathfrak{h}(\mathcal{K}_{/B}) \to \mathfrak{h}(\mathcal{K}_{/A})$ that carries fibered adjunctions over B to fibered adjunctions over A. This proves (i).

Composition with an isofibration $p : A \twoheadrightarrow B$ also defines a 2-functor of slices $p_* : \mathfrak{h}(\mathcal{K}_{/A}) \to \mathfrak{h}(\mathcal{K}_{/B})$. Thus, composition with an isofibration carries a fibered adjunction over A to a fibered adjunction over B, proving (ii). $\qquad\square$

In analogy with Lemma 3.6.3:

LEMMA 3.6.7. *If $F : \mathcal{A} \to \mathcal{B}$ is a smothering 2-functor, then any adjunction in $\mathcal{B}$ may be lifted to an adjunction in $\mathcal{A}$. In particular, any adjunction in the slice 2-category $(\mathfrak{h}\mathcal{K})_{/B}$ of an ∞-cosmos $\mathcal{K}$ lifts to a fibered adjunction over B.*

Proof Exercise 3.6.iii. $\qquad\square$

Combining Definitions 3.6.5 and 2.2.1, we obtain notions of *fibered* initial and terminal elements.

---

[7] We require p to be an isofibration due to our convention that the objects in sliced ∞-cosmoi are isofibrations over a fixed base.

DEFINITION 3.6.8. Given an isofibration $p : E \twoheadrightarrow B$, we say that E admits an **initial element over** B or admits a **terminal element over** B if there exists a fibered left or right adjoint, respectively, to the unique functor from p to id_B over B:

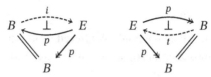

That is, E admits an initial or terminal element over B just when $p : E \twoheadrightarrow B$ admits an initial or terminal element when considered as an object of the sliced ∞-cosmos over B.

The next result shows that fibered initial or terminal elements exist just when the isofibration $p : E \twoheadrightarrow B$ admits a left adjoint right inverse or a right adjoint right inverse, respectively.

LEMMA 3.6.9. *Let* $p : E \twoheadrightarrow B$ *be any isofibration that admits a right adjoint right inverse* $r' : B \to E$. *Then* r' *is isomorphic to a functor* r *that defines a fibered adjunction:*

Thus, an isofibration $p : E \twoheadrightarrow B$ *admits a right adjoint right inverse if and only if* E *admits a terminal element over* B.

Proof Since an isofibration $p : E \twoheadrightarrow B$ in an ∞-cosmos $\mathcal{K}$ defines an isofibration in the homotopy 2-category $\mathfrak{h}\mathcal{K}$, the invertible counit $\epsilon' : pr' \cong \mathrm{id}_B$ of the adjunction $p \dashv r'$ lifts to define a functor $r : B \to E$ together with a natural isomorphism $\gamma : r' \cong r$ so that $p\gamma = \epsilon'$ and $pr = \mathrm{id}_B$:

$$
\begin{array}{ccc}
B \xrightarrow{r'} E & & B \overset{r'}{\underset{r}{\cong\Downarrow\gamma}} E \\
\cong\Downarrow\epsilon' \searrow \downarrow p & = & \hspace{1em} \downarrow p \\
B & & B
\end{array}
$$

By the construction left to the reader in Exercise 2.1.ii, $p \dashv r$ with unit $\eta := \gamma p \cdot \eta'$ defined by composing the original unit η' with γ and with counit $\epsilon := \epsilon' \cdot p\gamma^{-1}$. In particular, since $p\gamma = \epsilon'$, the counit ϵ is the identity 2-cell, and consequently one of the triangle equality composites reduces to the assertion that $p\eta = \mathrm{id}_p$.

This constructs a right adjoint to p considered as a functor in $(\mathfrak{h}\mathcal{K})_{/B}$. By Lemma 3.6.7, this adjunction lifts along the smothering 2-functor of Proposition 3.6.2 to define a fibered adjunction over B of the desired form in $\mathfrak{h}(\mathcal{K}_{/B})$ (see Exercise 3.6.iii). Definition 3.6.8 interprets the fibered adjunction just constructed as defining a terminal element in E over B. □

With this observation, Theorem 3.5.12 may be summarized more compactly as follows:

COROLLARY 3.6.10. *For any cospan* $C \xrightarrow{g} A \xleftarrow{f} B$, *the comma ∞-category* $\mathrm{Hom}_A(f, g)$ *is right representable if and only if* $\mathrm{Hom}_A(f, g)$ *admits a terminal element over* C *– in which case the representing functor defines an absolute right lifting of* g *through* f. □

REMARK 3.6.11. In an ∞-cosmos of $(\infty, 1)$-categories, the representability theorem can be improved still further to say that $\mathrm{Hom}_A(f, g)$ is right representable if and only if, for all elements $c : 1 \to C$, the ∞-category $\mathrm{Hom}_A(f, gc)$ has a terminal element (see Corollary 12.2.8). The proof requires "analytic" techniques, in contrast with the purely synthetic reasoning in this chapter.

EXAMPLE 3.6.12. By Lemmas 3.5.9 and 3.6.9, for any functor $f : A \to B$, there is a fibered adjunction

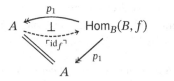

which asserts that $\ulcorner\mathrm{id}_f\urcorner : A \to \mathrm{Hom}_B(B, f)$ defines a terminal element in $\mathrm{Hom}_B(B, f)$ over A.

By Lemma 3.6.6(i), we may pull back the fibered adjunction along any element $a : 1 \to A$ to obtain an adjunction that identifies a terminal element in the fiber

$$
1 \underset{\ulcorner\mathrm{id}_{f_a}\urcorner = \ulcorner\mathrm{id}_{f_a}\urcorner}{\overset{!}{\underset{\bot}{\rightleftarrows}}} \mathrm{Hom}_B(B, fa)
\qquad\qquad
\begin{array}{ccc}
\mathrm{Hom}_B(B, fa) & \longrightarrow & \mathrm{Hom}_B(B, f) \\
\downarrow & \lrcorner & \downarrow{\scriptstyle p_1} \\
1 & \xrightarrow{\quad a \quad} & A
\end{array}
$$

generalizing the result of Corollary 3.5.10.

EXAMPLE 3.6.13 (the fibered adjoints to composition). For any ∞-category A, the adjoints to the "composition" functor $\circ : A^2 \underset{A}{\times} A^2 \to A^2$ constructed in Lemma 2.1.14 are constructed by composing a triple of adjoint functors that are fibered over the endpoint evaluation functors

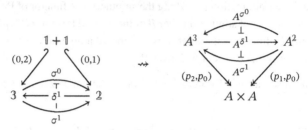

with an adjoint equivalence involving a functor $A^3 \twoheadrightarrow A^2 \underset{A}{\times} A^2$, which also lies over $A \times A$. Lemma 3.6.9 and its dual implies that these adjoint equivalences can be lifted to fibered adjoint equivalences over $A \times A$, and now both adjoint triples and hence the composite adjunctions are also fibered:

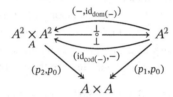

This fibered adjunction, which allows us to work at the ∞-cosmos level rather than purely in the homotopy 2-category, figures in the proof of a result that allows us to convert limit and colimit diagrams into right and left Kan extension diagrams (see Proposition 4.3.4).

PROPOSITION 3.6.14. *A cospan as displayed below-left admits an absolute right lifting if and only if the cospan displayed below-right admits an absolute right lifting*

$$
\begin{array}{ccc}
 & B & \\
{}^{r}\nearrow & \Downarrow\rho & \downarrow f \\
C & \xrightarrow[g]{} & A
\end{array}
\qquad
\begin{array}{ccc}
 & \mathrm{Hom}_A(f, A) & \\
{}^{i}\nearrow & \Downarrow\epsilon & \downarrow p_1 \\
C & \xrightarrow[g]{} & A
\end{array}
$$

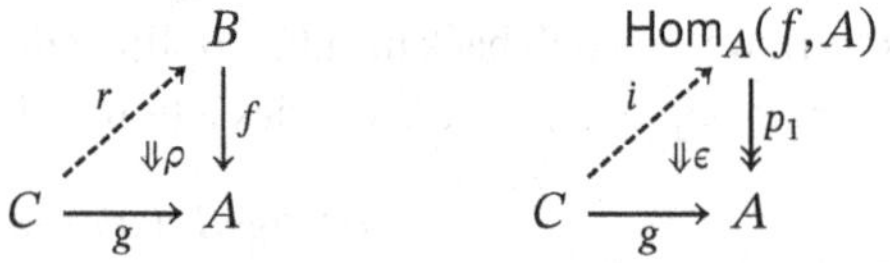

in which case the 2-cell ϵ *is necessarily an isomorphism and the pair* (i, ϵ) *can be chosen to be* $(\ulcorner\rho\urcorner, \mathrm{id}_g)$.

Proof By Theorem 3.5.12 and Corollary 3.6.10, to verify the existence statement it suffices to show that $\mathrm{Hom}_A(f, g)$ admits a terminal element over C if and only if $\mathrm{Hom}_A(p_1, g)$ admits a terminal element over C.

From the defining pullback (3.4.2) that constructs the comma ∞-category

$\mathsf{Hom}_A(p_1,g)$, we see that $\mathsf{Hom}_A(p_1,g) \cong \mathsf{Hom}_A(A,g) \times_A \mathsf{Hom}_A(f,A)$.

$$
\begin{array}{ccccc}
\mathsf{Hom}_A(p_1,g) & \longrightarrow & \mathsf{Hom}_A(A,g) & \longrightarrow & A^2 \\
{\scriptstyle (p_1,p_0)}\downarrow & \lrcorner & {\scriptstyle (p_1,p_0)}\downarrow & \lrcorner & \downarrow{\scriptstyle (p_1,p_0)} \\
C \times \mathsf{Hom}_A(f,A) & \xrightarrow{C \times p_1} & C \times A & \xrightarrow{g \times A} & A \times A \\
{\scriptstyle \pi}\downarrow & \lrcorner & \downarrow{\scriptstyle \pi} & & \\
\mathsf{Hom}_A(f,A) & \xrightarrow{\quad p_1 \quad} & A & &
\end{array}
$$

Thus, by Lemma 3.6.6, the composition-identity fibered adjunction of Example 3.6.13 pulls back along the functors $g \times f : C \times B \to A \times A$ to define a fibered adjunction

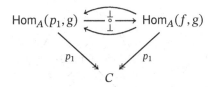

which then composes with the projection $\pi : C \times B \twoheadrightarrow C$ to give a fibered adjunction over C

$$
\begin{array}{ccc}
\mathsf{Hom}_A(p_1,g) & \overset{\perp}{\underset{\perp}{\rightleftarrows}} & \mathsf{Hom}_A(f,g) \\
{\scriptstyle p_1}\searrow & & \swarrow{\scriptstyle p_1} \\
& C &
\end{array}
$$

between the codomain projection for $\mathsf{Hom}_A(p_1,g)$ and the codomain projection for $\mathsf{Hom}_A(f,g)$, considered as objects of the sliced ∞-cosmos over C. Since we have right adjoints pointing in both directions, by Theorem 2.4.2, a terminal element on either side is carried by the appropriate right adjoint to a terminal element on the other side. This proves the equivalence of the absolute right lifting conditions conditions.

Now we assume that either and thus both absolute right liftings exist. Observe that the rightmost adjoint $(\mathrm{id}_{\mathrm{cod}(-)}, -) : \mathsf{Hom}_A(f,g) \to \mathsf{Hom}_A(p_1,g)$ is characterized up to fibered isomorphism by the pasting equality:

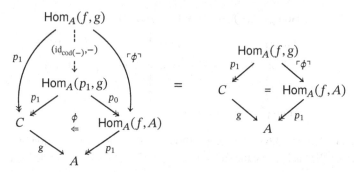

where $\ulcorner \phi \urcorner : \mathrm{Hom}_A(f, g) \to \mathrm{Hom}_A(f, A)$ is the functor that encodes the factorization of the comma cone for $\mathrm{Hom}_A(f, g)$ through the comma cone for $\mathrm{Hom}_A(f, A)$. By Theorem 3.5.12, the functor $\ulcorner \rho \urcorner : C \to \mathrm{Hom}_A(f, g)$ defines a right adjoint right inverse to $p_1 : \mathrm{Hom}_A(f, g) \twoheadrightarrow C$. Thus, by the argument just given, the composite of $\ulcorner \rho \urcorner$ and $(\mathrm{id}_{\mathrm{cod}(-)}, -)$ defines a right adjoint right inverse to $p_1 : \mathrm{Hom}_A(p_1, g) \twoheadrightarrow C$, encoding the data of an absolute right lifting of g through p_1, necessarily isomorphic to the pair (i, ϵ). The pasting equalities

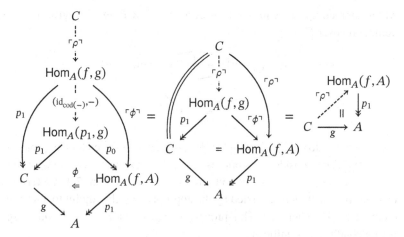

demonstrate that this absolute right lifting diagram is given by $(\ulcorner \rho \urcorner, \mathrm{id}_g)$ as claimed.   □

The following example hints at one application of Proposition 3.6.14.

EXAMPLE 3.6.15. The left representation of a functor $A^f : A^V \to A^U$ induced by cotensoring with a map of simplicial sets $f : U \to V$ is itself definable as a cotensor with the simplicial set formed by attaching V to the domain end of the cylinder $U \times 2$ via the map f:

$$
\begin{array}{ccc}
\operatorname{Hom}_A(A^f, A) \cong A^{\operatorname{cone}(f)} & \xrightarrow{\ \ulcorner\phi\urcorner\ } & A^{U\times 2} \\
{\scriptstyle (p_1,p_0)}\Big\downarrow \quad\quad {}^{\lrcorner} & & \Big\downarrow{\scriptstyle (p_1,p_0)} \\
A^U \times A^V & \xrightarrow[\ \operatorname{id}\times A^f\ ]{} & A^U \times A^U
\end{array}
\qquad
\begin{array}{ccc}
U + U & \xrightarrow{\ f+U\ } & V + U \\
\Big\uparrow & & \Big\downarrow \\
U \times 2 & \longrightarrow & \operatorname{cone}(f)
\end{array}
$$

Proposition 3.6.14 establishes a correspondence between absolute right lifting problems

$$
\begin{array}{ccc}
 & & A^V \\
 & {}^{r}\nearrow & \Big\downarrow{\scriptstyle A^f} \\
 & \Downarrow\rho & \\
D & \xrightarrow[\ d\ ]{} & A^U
\end{array}
\qquad
\begin{array}{ccc}
 & & A^{\operatorname{cone}(f)} \\
 & {}^{i}\nearrow & \Big\downarrow{\scriptstyle p_1} \\
 & \| & \\
D & \xrightarrow[\ d\ ]{} & A^U
\end{array}
$$

under which a single functor $i: D \to A^{\operatorname{cone}(f)}$ is used to encode the data of both the functor

$$
r := D \xrightarrow{\ i\ } A^{\operatorname{cone}(f)} \xrightarrow{\ p_0\ } A^V
$$

and the natural transformation

$$
\ulcorner\rho\urcorner := D \xrightarrow{\ i\ } A^{\operatorname{cone}(f)} \xrightarrow{\ \ulcorner\phi\urcorner\ } A^{U\times 2}
$$

Exercises

EXERCISE 3.6.i. Let B be an object in a 2-category and consider a map

$$
\begin{array}{ccc}
E & \xrightarrow{\ \ f\ \ } & F \\
 & \searrow \quad \swarrow & \\
 & B &
\end{array}
$$

between isofibrations over B. Prove that if f is an equivalence in the ambient 2-category then f is also an equivalence in the slice 2-category of isofibrations over B, 1-cells that form commutative triangles over B, and 2-cells that lie over B in the sense that they whisker with the codomain isofibration to the identity 2-cell on the domain isofibration.

EXERCISE 3.6.ii. Under the correspondence of Corollary 3.5.11, show that the following are equivalent:

(i) $\alpha: f \Rightarrow g$ is an isomorphism.
(ii) The functor $\ulcorner\alpha_*\urcorner: \operatorname{Hom}_B(B, f) \to \operatorname{Hom}_B(B, g)$ defines a fibered equivalence over $A \times B$.

(iii) The functor $\ulcorner \alpha^* \urcorner : \mathrm{Hom}_B(g, B) \to \mathrm{Hom}_B(f, B)$ defines a fibered equivalence over $B \times A$.

EXERCISE 3.6.iii. Let $F : \mathcal{A} \to \mathcal{B}$ be a smothering 2-functor. Use Lemma 2.1.11 to show that any adjunction in $\mathcal{B}$ can be lifted to an adjunction in $\mathcal{A}$. Demonstrate furthermore that if we have previously specified a lift of the objects, 1-cells, and either the unit or counit of the adjunction in $\mathcal{B}$, then there is a lift of the remaining 2-cell that combines with the previously specified data to define an adjunction in $\mathcal{A}$. This proves a more precise version of Lemma 3.6.7.

EXERCISE 3.6.iv. Extend the proof of Proposition 3.6.14 to prove that a square preserves the absolute right lifting (r, ρ) if and only if the induced square preserves the absolute right lifting (i, ϵ):

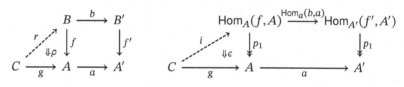

4

Adjunctions, Limits, and Colimits II

In Chapter 2, we develop the basic theory of adjunctions between ∞-categories and limits and colimits of diagrams valued in ∞-categories by characterizing these notions in terms of absolute lifting diagrams in the homotopy 2-category of ∞-categories, functors, and natural transformations in an ∞-cosmos. While absolute lifting diagrams are expedient for proving theorems relating adjunctions, limits, and colimits, they do not obviously express the familiar universal properties associated to these notions. In this chapter, we use the comma ∞-categories of Chapter 3 as a vehicle to give precise expressions to these universal properties and prove that new characterizations of adjunctions, limits, and colimits are equivalent to the previous definitions. In fact, many of the main results in this section are mere special cases of the general theorems characterizing representable comma ∞-categories.

Using the theory of comma ∞-categories, in §4.1 we quickly prove a variety of results describing the universal property of adjunctions. In particular, Theorem 3.5.8 specializes in Proposition 4.1.1 to characterize adjoint pairs of functors $f : B \to A$ and $u : A \to B$ via a "transposing equivalence"

$$\operatorname{Hom}_A(f, A) \simeq_{A \times B} \operatorname{Hom}_B(B, u),$$

while Corollary 3.6.10 specializes in Proposition 4.1.6 to give a criterion that guarantees that a left or right adjoint to a given functor exists.

In an interlude in §4.2, we introduce the ∞-*category of cones* over or under a diagram as a comma ∞-category. When the indexing shape for the diagrams is given by a simplicial set, an equivalent model can be built from Joyal's join and slice constructions. The ∞-categories of cones over or under a diagram feature prominently in the study of the universal properties of limits and colimits in §4.3. There we see that Theorem 3.5.8 specializes to prove Proposition 4.3.1, characterizing a limit of a diagram as a right representation for the ∞-category of cones, while Corollary 3.6.10 specializes in Proposition 4.3.2 to characterize a limit cone as a terminal element in the ∞-category of cones.

Since the proofs of the main results in this chapter appear in Chapter 3 where they are developed in a more general setting, we are able to focus our efforts here on applications. In §4.4 we introduce *pointed ∞-categories*, which have a *zero element* that is both initial and terminal, and show how this may be used to construct the loops ⊢ suspension adjunction. Pointed ∞-categories that admit fiber and cofiber sequences, which define a common family of *exact triangles*, are called *stable*. While exploring the properties of stable ∞-categories, we encounter a number of equivalent characterizations, enumerated in Theorem 4.4.12.

The fibered equivalences that characterize adjunctions, limits, and colimits can be understood as ∞-categorical analogues of Eilenberg and Mac Lane's famous natural equivalences [42]. To express this "naturality," we observe that arrows in the base ∞-categories act covariantly functorially on the fibers of the codomain projection functor and contravariantly functorially on the fibers of the domain projection functor associated to a comma ∞-category. This is the subject of Chapter 5.

4.1 The Universal Property of Adjunctions

An adjunction between an opposing pair of functors can equally be encoded by a "transposing equivalence" between their left and right representations as comma ∞-categories.

PROPOSITION 4.1.1 (adjunction as fibered equivalence). *An opposing pair of functors* $u : A \to B$ *and* $f : B \to A$ *define an adjunction* $f \dashv u$ *if and only if* $\mathrm{Hom}_A(f, A) \simeq_{A \times B} \mathrm{Hom}_B(B, u)$.

This is a special case of Theorem 3.5.8, so no further argument is required, but we proffer a short proof nevertheless to review the results proven in §3.5.

Proof If $f \dashv u$, then its counit $\epsilon : fu \Rightarrow \mathrm{id}_A$ defines an absolute right lifting diagram by Lemma 2.3.7. By Theorem 3.5.8, the functor induced by the left-hand

pasted composite

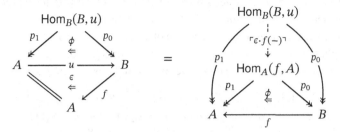

defines a fibered equivalence $\mathrm{Hom}_B(B,u) \simeq_{A \times B} \mathrm{Hom}_A(f,A)$. We interpret this result as asserting that in the presence of an adjunction $f \dashv u$, the right comma cone over u transposes to define the left comma cone over f.[1]

Conversely, from a fibered equivalence $\mathrm{Hom}_B(B,u) \simeq_{A \times B} \mathrm{Hom}_A(f,A)$, Theorem 3.5.8 tells us that one can extract a 2-cell that defines an absolute right lifting diagram

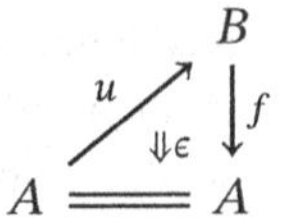

$$
\begin{array}{ccc}
 & & B \\
 & \overset{u}{\nearrow} \Downarrow \epsilon & \downarrow f \\
A & = & A
\end{array}
$$

which by Lemma 2.3.7 then defines the counit of an adjunction $f \dashv u$. □

OBSERVATION 4.1.2 (the transposing equivalence). To justify referring to the induced functor

$$\ulcorner \epsilon \cdot f(-) \urcorner : \mathrm{Hom}_B(B,u) \xrightarrow{\;\approx\;} \mathrm{Hom}_A(f,A)$$

as a **transposing equivalence**, recall that the transpose of a 2-cell $\chi : b \Rightarrow ua$ across the adjunction $f \dashv u$ is computed by the left-hand pasting diagram below:

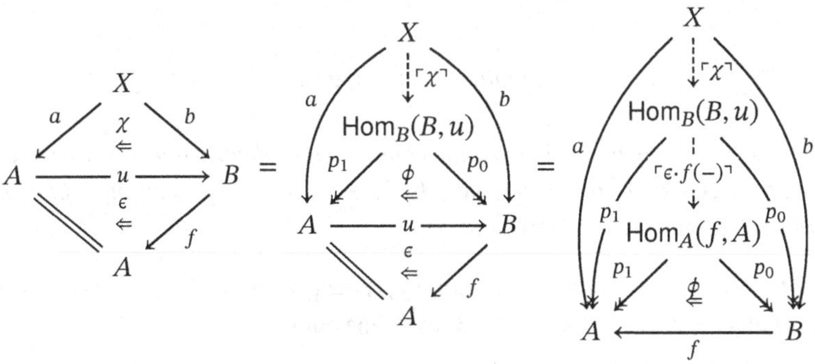

---

[1] If desired, an inverse equivalence can be constructed by applying the dual of Theorem 3.5.8 to the absolute left lifting diagram presented by the unit.

By the weak universal property of the right comma cone over u, the 2-cell χ is represented by the induced functor $\ulcorner\chi\urcorner : X \to \mathrm{Hom}_B(B, u)$, which then composes with the transposing equivalence to define a functor $\ulcorner\epsilon \cdot f(\chi)\urcorner : X \to \mathrm{Hom}_A(f, A)$ that represents the transpose of χ, by the pasting diagram equalities from right to left. This observation also justifies our notation, in which we name the fibered equivalence $\ulcorner\epsilon \cdot f(-)\urcorner$ after the formula for adjoint transposition.

COROLLARY 4.1.3. *An pair of functors* $u : A \to B$ *and* $f : B \to A$ *define an adjunction* $f \dashv u$ *if and only if there is an equivalence* $\mathrm{Hom}_A(fb, a) \simeq_{X \times Y} \mathrm{Hom}_B(b, ua)$ *for any pair of generalized elements* $a : X \to A$ *and* $b : Y \to B$.

Proof When $f \dashv u$, pullback along $a \times b : X \times Y \to A \times B$ defines a cosmological functor that carries the equivalence $\mathrm{Hom}_A(f, A) \simeq_{A \times B} \mathrm{Hom}_B(B, u)$ of Proposition 4.1.1 to an equivalence $\mathrm{Hom}_A(fb, a) \simeq_{X \times Y} \mathrm{Hom}_B(b, ua)$. The converse is proven by the special case where the generalized elements are the identity functors id_A and id_B. $\qquad\Box$

REMARK 4.1.4. In particular, the equivalence of Proposition 4.1.1 pulls back to define an equivalence of internal mapping spaces, introduced in Definition 3.4.9. In Corollary 12.2.15, we see that in an ∞-cosmos of $(\infty, 1)$-categories a natural transformation $\epsilon : fu \Rightarrow \mathrm{id}_A$ defines the counit of an adjunction if and only if the map $\ulcorner\epsilon \cdot f(-)\urcorner : \mathrm{Hom}_B(B, u) \to \mathrm{Hom}_A(f, A)$ defines equivalences of internal mapping spaces $\mathrm{Hom}_B(b, ua) \simeq \mathrm{Hom}_A(fb, a)$ for any pair of elements $a : 1 \to A$ and $b : 1 \to B$.

Comma ∞-categories also provide a vehicle for expressing the universal properties of unit and counit transformations.

PROPOSITION 4.1.5 (the universal property of units and counits). *Consider an adjunction*

$$B \underset{u}{\overset{f}{\underset{\displaystyle\rightleftarrows}{\perp}}} A \qquad \textit{with unit } \eta : \mathrm{id}_B \Rightarrow uf \textit{ and counit } \epsilon : fu \Rightarrow \mathrm{id}_A.$$

Then for each element $a : 1 \to A$, *the component* ϵ_a *defines a terminal element of* $\mathrm{Hom}_A(f, a)$, *and for each element* $b : 1 \to B$, *the component* η_b *defines an initial element of* $\mathrm{Hom}_B(b, u)$.

Proof The fibered equivalence $\mathrm{Hom}_A(f, A) \simeq_{A \times B} \mathrm{Hom}_B(B, u)$ of Proposition 4.1.1 pulls back, by Corollary 4.1.3, to define equivalences

$$\mathrm{Hom}_A(f, a) \simeq_B \mathrm{Hom}_B(B, ua) \qquad \text{and} \qquad \mathrm{Hom}_A(fb, A) \simeq_A \mathrm{Hom}_B(b, u).$$

By Corollary 3.5.10, id_{ua} induces a terminal element in $\mathrm{Hom}_B(B, ua)$ and by

Lemma 2.2.7 its image across the equivalence $\mathrm{Hom}_B(B, ua) \xrightarrow{\sim} \mathrm{Hom}_A(f, a)$ is again a terminal element. By Observation 4.1.2 this element represents the transposed 2-cell: the component of the counit ϵ at the element a.   □

The universal property of unit and counit components captured in Proposition 4.1.5 gives the main idea behind the adjoint functor theorems. In an ∞-cosmos of $(\infty, 1)$-categories, a functor $f : B \to A$ admits a right adjoint just when for each element $a : 1 \to A$, the ∞-category $\mathrm{Hom}_A(f, a)$ admits a terminal element (see Corollary 12.2.7).[2] The image of this terminal element under the domain projection functor $p_0 : \mathrm{Hom}_A(f, a) \twoheadrightarrow B$ defines the element $ua : 1 \to B$ and the comma cone defines the component of the counit at a. The universal property of the counit components is then used to extend the mapping on elements to a functor $u : A \to B$.

An analogous result that is true in a generic ∞-cosmos is obtained by replacing the quantifier "for each element $a : 1 \to A$" with "for each generalized element $a : X \to A$," in which case the meaning of "terminal element" should be enhanced to "terminal element over X" (see Definition 3.6.8). Since every generalized element factors through the universal generalized element, namely the identity functor at A, it suffices to prove:

PROPOSITION 4.1.6. *A functor* $f : B \to A$ *admits a right adjoint if and only if* $\mathrm{Hom}_A(f, A)$ *admits a terminal element over* A. *Dually,* $f : B \to A$ *admits a left adjoint if and only if* $\mathrm{Hom}_A(A, f)$ *admits an initial element over* A.

Proof By Proposition 4.1.1, $f : B \to A$ admits a right adjoint if and only if the comma ∞-category $\mathrm{Hom}_A(f, A)$ is right representable, which by Corollary 3.6.10 is the case just when $\mathrm{Hom}_A(f, A)$ admits a terminal element over A.   □

The same suite of results from §3.5–§3.6 specialize to theorems that encode the universal properties of limits and colimits. Before exploring these, we first construct the ∞-category of cones over or under a diagram.

Exercises

EXERCISE 4.1.i (4.3.13). Specialize Proposition 4.1.1 to the case of adjunctions

$$1 \underset{!}{\overset{i}{\rightleftarrows}} \bot \; A \qquad \text{and} \qquad 1 \underset{t}{\overset{!}{\rightleftarrows}} \bot \; A$$

---

[2] Recall from Example 2.3.11 that a terminal element is a colimit of the identity functor. The technical conditions in Freyd's general adjoint functor theorem and special adjoint functor theorem are deployed to reduce this large colimit to a small colimit and guarantee its existence (see [104, §4.6] for a 1-categorical exposition of these results). Analogous theorems have been proven in the $(\infty, 1)$-categorical context by Nguyen, Raptis, and Schrade [88].

to discover an alternate characterization of initial and terminal elements.

EXERCISE 4.1.ii. For any parallel pair of fully specified adjunctions

$$B \underset{u}{\overset{f}{\rightleftarrows}} A \qquad \text{with unit } \eta : \ \mathrm{id}_B \Rightarrow uf \text{ and counit } \epsilon : \ fu \Rightarrow \mathrm{id}_A, \text{ and}$$

$$B \underset{u'}{\overset{f'}{\rightleftarrows}} A \qquad \text{with unit } \eta' : \ \mathrm{id}_B \Rightarrow u'f' \text{ and counit } \epsilon' : \ f'u' \Rightarrow \mathrm{id}_A.$$

there is a bijection between natural transformations $\alpha : f' \Rightarrow f$ and $\beta : u \Rightarrow u'$ as a special case of the *mates correspondence* (see Definition B.3.3). Argue that the transposing equivalence of Proposition 4.1.1 is natural with respect to precomposing with a 2-cell $\alpha : f' \Rightarrow f$ or postcomposing with its mate $\beta : u \Rightarrow u'$ (see Corollary 3.5.11) by proving that there is a fibered natural isomorphism over $A \times B$ between the functors:

$$
\begin{array}{ccc}
\mathrm{Hom}_A(f, A) & \xrightarrow{\ulcorner \alpha^* \urcorner} & \mathrm{Hom}_A(f', A) \\
{\scriptstyle \ulcorner u(-) \cdot \eta \urcorner} \downarrow \wr & & \wr \downarrow {\scriptstyle \ulcorner u'(-) \cdot \eta' \urcorner} \\
\mathrm{Hom}_B(B, u) & \xrightarrow[\ulcorner \beta_* \urcorner]{} & \mathrm{Hom}_B(B, u')
\end{array}
$$

4.2 ∞-Categories of Cones

The comma ∞-category construction can be used to define the ∞-category of cones over or under a given diagram. Since these ∞-categories feature centrally in the description of the universal properties of limits and colimits, we present a few equivalent models for this construction.

A **cone** over a diagram $d : 1 \to A^J$ with summit $a : 1 \to A$ is a natural transformation $\lambda \Delta a \Rightarrow d$, where $\Delta : A \to A^J$ is the constant diagram functor of Definition 2.3.1. This motivates the following definition.

DEFINITION 4.2.1 (the ∞-category of cones). Let $d : 1 \to A^J$ be a J-shaped diagram in an ∞-category A. The ∞-**category of cones over** d is the comma ∞-category $\mathrm{Hom}_{A^J}(\Delta, d)$ from the constant diagram functor Δ to d, while the

∞-**category of cones under** d is the comma ∞-category $\mathrm{Hom}_{A^J}(d, \Delta)$.

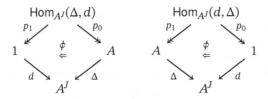

By replacing the diagram leg of the cospans, Definition 4.2.1 can be modified to allow $d : D \to A^J$ to be a family of diagrams. In the universal case, where d is the identity functor $\mathrm{id}_{A^J} : A^J \to A^J$, this defines the ∞-category $\mathrm{Hom}_{A^J}(\Delta, A^J)$ of cones over or under any diagram of shape J.

In the case where the indexing shape J is a simplicial set (as opposed to an ∞-category in a cartesian closed ∞-cosmos), there is another model of the ∞-category of cones over or under a diagram that may be constructed using the simplicial join construction first developed by Ehlers and Porter [40]. The equivalence of models is a consequence of the equivalence between the join operation and the so-called "fat join" introduced by Joyal [63, §9]. As Lemma 4.2.3 reveals, a particular instance of the fat join construction gives the shape of the cones appearing in Definition 4.2.1. We now introduce these notions.

DEFINITION 4.2.2 (fat join). The **fat join** of simplicial sets I and J is the simplicial set constructed by the following pushout:

$$
\begin{array}{ccc}
(I \times J) \amalg (I \times J) & \xrightarrow{\pi_I \amalg \pi_J} & I \amalg J \\
\big\uparrow & & \big\downarrow \\
I \times 2 \times J & \xrightarrow{\hspace{2cm}} & I \diamond J
\end{array}
$$

from which it follows that

$$(I \diamond J)_n := I_n \amalg \left(\coprod_{[n] \to [1]} I_n \times J_n \right) \amalg J_n.$$

Note there is a natural map $I \diamond J \twoheadrightarrow 2$ induced by the projection $\pi : I \times 2 \times J \twoheadrightarrow 2$ so that I is the fiber over 0 and J is the fiber over 1:

$$
\begin{array}{ccc}
I \amalg J & \hookrightarrow & I \diamond J \\
\big\downarrow & \lrcorner & \big\downarrow \\
1 + 1 & \underset{(0,1)}{\hookrightarrow} & 2
\end{array}
$$

The ∞-categories of cones over or under any J-shaped diagram can be redescribed as follows.

LEMMA 4.2.3. *For any simplicial set J and ∞-category A in an ∞-cosmos $\mathcal{K}$, we have natural isomorphisms*

$$\mathrm{Hom}_{A^J}(\Delta, A^J) \cong A^{1 \diamond J} \quad and \quad \mathrm{Hom}_{A^J}(A^J, \Delta) \cong A^{J \diamond 1}.$$

Proof The simplicial cotensor $A^{(-)} : s\mathcal{S}et^{\mathrm{op}} \to \mathcal{K}$ carries the pushout of Definition 4.2.2 to the pullback squares that define the left and right representations of $\Delta : A \to A^J$ as a comma ∞-category:

$$
\begin{array}{ccc}
A^{1 \diamond J} & \longrightarrow & (A^J)^2 \\
\downarrow & \lrcorner & \downarrow {\scriptstyle (p_1, p_0)} \\
A^J \times A & \xrightarrow[\mathrm{id} \times \Delta]{} & A^J \times A^J
\end{array}
\qquad
\begin{array}{ccc}
A^{J \diamond 1} & \longrightarrow & (A^J)^2 \\
\downarrow & \lrcorner & \downarrow {\scriptstyle (p_1, p_0)} \\
A \times A^J & \xrightarrow[\Delta \times \mathrm{id}]{} & A^J \times A^J
\end{array}
\qquad \square
$$

DEFINITION 4.2.4 (join, D.2.6). The **join** of simplicial sets I and J is the simplicial set $I \star J$

$$
\begin{array}{ccc}
I \amalg J & \lhook\joinrel\longrightarrow & I \star J \\
\downarrow & \lrcorner & \downarrow \\
1 + 1 & \xrightarrow[(0,1)]{} & 2
\end{array}
\qquad \text{with} \quad (I \star J)_n := I_n \amalg \Big(\coprod_{0 \le k < n} I_{n-k-1} \times J_k \Big) \amalg J_n
$$

and with the vertices of these n-simplices oriented so that there is a canonical map $I \star J \to 2$ so that I is the fiber over 0 and J is the fiber over 1 (see Definitions D.2.2 and D.2.6 or the original sources [40] and [61, §3] for more details).

The join functor $- \star J : s\mathcal{S}et \to s\mathcal{S}et$ preserves connected colimits but not the initial object or other coproducts, but cocontinuity is achieved by replacing the codomain by the slice category under J: the functor $- \star J : s\mathcal{S}et \to {}^{J/}s\mathcal{S}et$ preserves all colimits (see Lemma D.2.7). Contextualized in this way, the join admits a right adjoint, defined by Joyal's **slice** construction, which carries a simplicial map $f : J \to X$ to a simplicial set traditionally denoted by $X_{/f}$.

PROPOSITION 4.2.5 (join $\dashv$ slice adjunction). *The join functors admit right adjoints defined by the natural bijections:*

$$
s\mathcal{S}et \underset{-/\_}{\overset{I \star -}{\underset{\perp}{\rightleftarrows}}} {}^{I/}s\mathcal{S}et
\qquad
\left\{ \Delta[n] \to {}^{h/}X \right\} :=
\left\{
\begin{array}{c}
{}^{I}\!\!\nearrow \, \searrow^{h} \\[-2pt]
I \star \Delta[n] \longrightarrow X
\end{array}
\right\}
$$

$$
s\mathcal{S}et \underset{-/\_}{\overset{- \star J}{\underset{\perp}{\rightleftarrows}}} {}^{J/}s\mathcal{S}et
\qquad
\left\{ \Delta[n] \to X_{/k} \right\} :=
\left\{
\begin{array}{c}
{}^{J}\!\!\nearrow \, \searrow^{k} \\[-2pt]
\Delta[n] \star J \longrightarrow X
\end{array}
\right\}.
$$

Proof The simplicial set $X_{/k}$ is defined to have n-simplices corresponding to maps $\Delta[n] \star J \to X$ under J, with the right action by the simplicial operators

$[m] \to [n]$ given by precomposition with $\Delta[m] \to \Delta[n]$. Since the join functor $- \star J : s\mathcal{S}et \to {}^{J/}s\mathcal{S}et$ preserves colimits, this extends to a bijection between maps $I \to X_{/k}$ and maps $I \star J \to X$ under J that is natural in I and in $k : J \to X$. □

NOTATION 4.2.6. For any simplicial set J, we write

$$J^{\triangleleft} := \mathbb{1} \star J \qquad \text{and} \qquad J^{\triangleright} := J \star \mathbb{1}$$

and write ⊤ for the **cone vertex** of $J^{\triangleleft}$ and ⊥ for the **cone vertex** of $J^{\triangleright}$ contributed by the terminal simplicial set $\mathbb{1}$. These simplicial sets are equipped with canonical inclusions

$$J^{\triangleleft} \longleftarrow J \longrightarrow J^{\triangleright}$$

As the terminology suggests, the join and fat join constructions define equivalent indexing shapes, in the following sense.

PROPOSITION 4.2.7 (join vs fat join). *For any simplicial sets I and J and any ∞-category A, there is a natural equivalence*

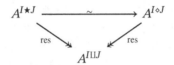

Proof There is a canonical map of simplicial sets

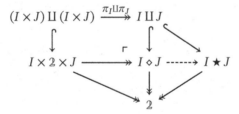

that commutes with the inclusions of the fibers $I \amalg J$ and lies over the projections to 2. An n-simplex in $I \diamond J$ that does not lie in either fiber is given by the data of a triple $(\alpha : [n] \twoheadrightarrow [1], \sigma \in I_n, \tau \in J_n)$. The dashed map carries this simplex to the pair $(\sigma|_{\{0,\dots,k\}} \in I_k, \tau|_{\{k+1,\dots,n\}} \in J_{n-k-1})$ representing an n-simplex of $I \star J$, where $k \in [n]$ is the maximal vertex in $\alpha^{-1}(0)$. Proposition D.6.3, or Lurie's [78, 4.2.1.2], prove that this map induces a natural equivalence $Q^{I \star J} \twoheadrightarrow Q^{I \diamond J}$ of quasi-categories over $Q^J \times Q^I$. Taking Q to be the functor space $\mathsf{Fun}(X, A)$ proves the claimed equivalence for general ∞-categories. □

COROLLARY 4.2.8. *For any simplicial set J and ∞-category A, there are comma squares*

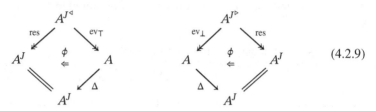

$$(4.2.9)$$

which pull back over a family of diagrams $d : D \to A^J$ *to define equivalent models for the ∞-categories of cones over or under d.*

$$
\begin{array}{ccc}
\mathrm{Hom}_{A^J}(\Delta, d) \simeq A_{/d} & \longrightarrow & A^{J^{\triangleleft}} \\
\downarrow & \lrcorner & \downarrow{\scriptstyle res} \\
D \times A & \xrightarrow[d \times \mathrm{id}]{} & A^J \times A
\end{array}
\qquad
\begin{array}{ccc}
\mathrm{Hom}_{A^J}(d, \Delta) \simeq {}^{d/}A & \longrightarrow & A^{J^{\triangleright}} \\
\downarrow & \lrcorner & \downarrow{\scriptstyle res} \\
A \times D & \xrightarrow[\mathrm{id} \times d]{} & A \times A^J
\end{array}
$$

Proof Proposition 4.2.7 constructs fibered equivalences $A^{\mathbb{1} \diamond J} \simeq_{A^J \times A} A^{J^{\triangleleft}}$ and $A^{J \diamond \mathbb{1}} \simeq_{A \times A^J} A^{J^{\triangleright}}$. By Lemma 4.2.3, $A^{\mathbb{1} \diamond J}$ and $A^{J \diamond \mathbb{1}}$ are comma ∞-categories. Thus, Proposition 3.4.11 implies that the fibered equivalences equip $A^{J^{\triangleleft}}$ and $A^{J^{\triangleright}}$ with comma cones, satisfying the weak universal property of Proposition 3.4.6. The natural transformations in (4.2.9) are represented by the horizontal composites

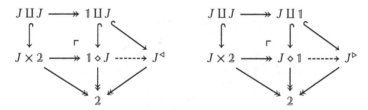

which yield natural transformations upon cotensoring into A:

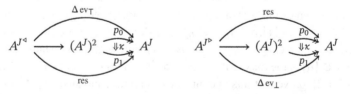

The fibered equivalences pullback to define equivalent models for the ∞-categories of cones over or under a fixed family of diagrams d. $\qquad\square$

WARNING 4.2.10. In the statement of Corollary 4.2.8 and elsewhere it is convenient to borrow Joyal's slice notation for the fibers of the restriction maps over a diagram $d : 1 \to A^J$. This usage is justified by Proposition D.6.4, which proves that $A_{/d} \simeq_A \mathrm{Hom}_{A^J}(\Delta, d)$ and $^{d/}A \simeq_A \mathrm{Hom}_{A^J}(d, \Delta)$ in the ∞-cosmos of quasi-categories. Note, however, that in the ∞-cosmos of quasi-categories the strict fibers are *not* isomorphic to Joyal's slice quasi-categories (see Exercise 4.2.ii) but are merely equivalent to them.

Exercises

EXERCISE 4.2.i. Compute $\Delta[n] \star \Delta[m]$ and $\Delta[n] \diamond \Delta[m]$ and define a section

$$\Delta[n] \star \Delta[m] \to \Delta[n] \diamond \Delta[m]$$

to the map constructed in the proof of Proposition 4.2.7.

EXERCISE 4.2.ii. Compute the fiber of $A^{J^\triangleleft} \twoheadrightarrow A^J$ over $d : 1 \to A^J$ in the ∞-cosmos of quasi-categories and prove that this quasi-category is not isomorphic to $A_{/d}$.

EXERCISE 4.2.iii ([63, 3.5]). The category of simplicial sets, as a category of presheaves, is **locally cartesian closed**, meaning that the pullback functor associated to any map $f : U \to V$ has a right adjoint Π_f called the **dependent product** or **pushforward**.

$$sSet_{/V} \underset{\Pi_f}{\overset{f^*}{\rightleftarrows}} sSet_{/U}$$

Show that the join $I \star J$ can be defined as an object of $sSet_{/2}$ as the dependent product of $! + ! : I + J \to 1 + 1$ along $1 + 1 \hookrightarrow 2$.

4.3 The Universal Property of Limits and Colimits

To describe the universal properties of limits and colimits we return to the general context of Definition 2.3.1, simultaneously considering diagrams valued in an ∞-category that are indexed by either a simplicial set or another ∞-category, in the case where the ambient ∞-cosmos is cartesian closed. As was the case for Proposition 4.1.1, Theorem 3.5.8 specializes immediately to the setting of Definition 2.3.8 to prove:

PROPOSITION 4.3.1 (co/limits represent cones). *A family of diagrams* $d : D \to A^J$ *admits a limit if and only if the ∞-category of cones* $\mathrm{Hom}_{A^J}(\Delta, d)$ *over d is right representable*

$$\mathrm{Hom}_{A^J}(\Delta, d) \simeq_{D \times A} \mathrm{Hom}_A(A, \ell),$$

in which case the representing functor $\ell : D \to A$ *defines the limit functor. Dually, a family of diagrams* $d : D \to A^J$ *admits a colimit if and only if the ∞-category of cones* $\mathrm{Hom}_{A^J}(d, \Delta)$ *under d is left representable*

$$\mathrm{Hom}_{A^J}(d, \Delta) \simeq_{A \times D} \mathrm{Hom}_A(c, A),$$

in which case the representing functor $c : D \to A$ *defines the colimit functor.* □

Corollary 3.6.10 specializes to tell us that such representations can be encoded by terminal or initial elements, a result which is easiest to interpret for a single diagram rather than a family of diagrams.

PROPOSITION 4.3.2 (limits as terminal cones). *A diagram* $d : 1 \to A^J$ *of shape J in an ∞-category A*

   (i) *admits a limit if and only if the ∞-category* $\mathrm{Hom}_{A^J}(\Delta, d)$ *of cones over d admits a terminal element, in which case the terminal element defines a limit cone, and*
   (ii) *admits a colimit if and only if the ∞-category* $\mathrm{Hom}_{A^J}(d, \Delta)$ *of cones under d admits an initial element, in which case the initial element defines the colimit cone.* □

The uniqueness of limit and colimit cones up to isomorphism follows by applying Lemma 2.2.3. Alternatively, this can be proven from the absolute lifting diagram characterization (see Exercise 2.3.vi).

REMARK 4.3.3. Corollary 3.6.10 applies equally to say that a family of diagrams $d : D \to A^J$ admits a limit just when $\mathrm{Hom}_{A^J}(\Delta, d)$ admits a terminal element over D and admits a colimit just when $\mathrm{Hom}_{A^J}(d, \Delta)$ admits an initial element over D.

For aesthetic reasons, we state the following two results for diagrams indexed by simplicial sets so that we may deploy more elegant notation that may be easier to interpret. As Exercise 4.3.ii reveals, there is no mathematical reason to restrict to this special case.[3]

---

[3] Indeed, the proof in fact uses the codomain projection functor $p_1 : \mathrm{Hom}_{A^J}(\Delta, A^J) \twoheadrightarrow A^J$ in place of the equivalent isofibration res $: A^{J^{\triangleleft}} \twoheadrightarrow A^J$, and thus the plainer argument applies equally in the case of diagrams indexed by ∞-categories J in cartesian closed ∞-cosmoi that may or may not have a join operation available for indexing shapes.

PROPOSITION 4.3.4. *An ∞-category A admits a limit of a family of diagrams* $d : D \to A^J$ *indexed by a simplicial set J if and only if there exists an absolute right lifting of* d *through the restriction functor*

$$
\begin{array}{ccc}
& & A^{J^{\triangleleft}} \\
& \nearrow^{\text{rand}} & \downarrow^{\text{res}} \\
& \Downarrow \epsilon & \\
D & \xrightarrow{\;\;d\;\;} & A^J
\end{array}
$$

When these equivalent conditions hold, ε is necessarily an isomorphism and may be chosen to be the identity.

Proof By Definition 2.3.8, a family of diagrams d admits a limit if and only if it admits an absolute right lifting through $\Delta : A \to A^J$. By Proposition 3.6.14, this absolute lifting exists if and only if d admits an absolute right lifting through codomain projection functor $p_1 : \mathrm{Hom}_{A^J}(\Delta, A^J) \twoheadrightarrow A^J$, in which case the natural isomorphism of this latter absolute right lifting diagram is invertible. By Corollary 4.2.8, the restriction functor res : $A^{J^{\triangleleft}} \twoheadrightarrow A^J$ is equivalent to this codomain projection functor, so Exercise 3.5.ii implies that absolute right liftings of d through p_1 are equivalent to absolute right liftings of d through res. If this absolute lifting diagram is inhabited by an invertible 2-cell, the isomorphism lifting property of the isofibration proven in Proposition 1.4.9 can be used to replace the functor ran : $D \to A^{J^{\triangleleft}}$ with an isomorphic functor, yielding a strictly commutative triangle that remains an absolute right lifting diagram by Exercise 2.3.vi. □

Proposition 4.3.4 specializes to give a structured characterization of those ∞-categories that admit all limits or all colimits of a particular shape (see Definition 2.3.2).

COROLLARY 4.3.5. *An ∞-category A admits all limits indexed by a simplicial set J if and only if the restriction functor below-left admits a fibered right adjoint over A^J, and A admits all colimits indexed by a simplicial set J if and only if the restriction functor below-right admits a fibered left adjoint over A^J.*

$$
A^{J^{\triangleleft}} \overset{\text{res}}{\underset{\text{ran}}{\rightleftarrows}} \perp A^J \qquad\qquad A^{J^{\triangleright}} \overset{\text{lan}}{\underset{\text{res}}{\rightleftarrows}} \perp A^J
$$

Proof By Proposition 4.3.4 and Lemma 2.3.7, A admits all J-shaped limits if and only if the functor res : $A^{J^{\triangleleft}} \twoheadrightarrow A^J$ admits a right adjoint right inverse. Since the restriction functor is an isofibration, Lemma 3.6.9 applies to rectify the right adjoint right inverse into a fibered adjunction. □

We now apply the general theory we have developed to particular indexing shapes.

DEFINITION 4.3.6 (tensors and cotensors). Let K be a simplicial set and let $a : 1 \to A$ be an element of an ∞-category A. The **tensor** $K \otimes a$ of a by K is the colimit of the constant K-indexed diagram valued at a, while the **cotensor** a^K of a by K is the limit of the same diagram. Thus, the tensor and cotensor functors can be defined by the absolute lifting diagrams:

$$
\begin{array}{ccc}
& A & \\
K\otimes- \;\nearrow & \Big\downarrow \Delta & \\
\quad\Uparrow\lambda & & \\
A \xrightarrow{\;\;\Delta\;\;} A^K &
\end{array}
\qquad\qquad
\begin{array}{ccc}
& A & \\
(-)^K \;\nearrow & \Big\downarrow \Delta & \\
\quad\Downarrow\rho & & \\
A \xrightarrow{\;\;\Delta\;\;} A^K &
\end{array}
$$

By Theorem 3.5.3, these absolute lifting diagrams define fibered equivalences

$$
\mathrm{Hom}_A(K \otimes -, A) \simeq_{A \times A} \mathrm{Hom}_{A^K}(\Delta, \Delta) \simeq_{A \times A} \mathrm{Hom}_A(A, (-)^K)
$$

which compose to define the fibered equivalence encoding an adjunction between the tensor and cotensor functors:

$$
A \underset{(-)^K}{\overset{K\otimes-}{\rightleftarrows}} \!\!\perp\!\! \; A
$$

By Corollary 4.1.3, the fibered equivalences that express the universal properties of tensors and cotensors pullback over elements $a, x : 1 \to A$ to define equivalences of mapping spaces:

$$
\mathrm{Hom}_A(K \otimes a, x) \simeq \mathrm{Hom}_{A^K}(\Delta a, \Delta x) \cong \mathrm{Hom}_A(a, x)^K \qquad \text{and}
$$
$$
\mathrm{Hom}_A(x, a^K) \simeq \mathrm{Hom}_{A^K}(\Delta x, \Delta a) \cong \mathrm{Hom}_A(x, a)^K.
$$

DEFINITION 4.3.7 (span and cospan). A **span** in an ∞-category A is a diagram indexed by the simplicial set $\ulcorner := \Lambda^0[2]$ formed by gluing two 1-simplices along their domain vertices. Dually, a **cospan** in A is a diagram indexed by the simplicial set $\lrcorner := \Lambda^2[2]$ formed by gluing two 1-simplices along their codomain vertices. Cospans and spans in an ∞-category A may be defined by gluing together a pair of arrows along their common codomains or domains, respectively:

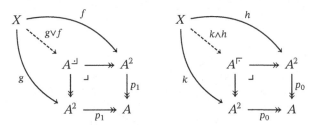

DEFINITION 4.3.8 (pushout and pullback). A **pushout** in an ∞-category A is a colimit indexed by the simplicial set $\ulcorner$, while a **pullback** in an ∞-category A is a limit indexed by the simplicial set $\lrcorner$. Cones over diagrams of shape $\lrcorner$ or cones under diagrams of shape $\ulcorner$ define **commutative squares**, diagrams of shape

$$\square := \ulcorner^{\triangleright} \cong 2 \times 2 \cong \lrcorner^{\triangleleft}.$$

A **pullback square** in an ∞-category A is an element of $A^{\square}$ that defines an absolute right lifting of its underlying cospan:

When A admits all pullbacks, the pullback squares can be characterized as those elements of $A^{\square}$ at which the component of the unit of the adjunction res $\dashv$ ran of Corollary 4.3.5 is an isomorphism (see Exercise 4.3.v). Dually, a **pushout square** in A is an element of $A^{\square}$ that defines an absolute left lifting of its underlying span, i.e., an element at which the component of the counit of the adjunction lan $\dashv$ res is an isomorphism. The notion of a family $X \to A^{\square}$ of pushout or pullback squares is defined analogously.

Pullback squares may be characterized by absolute lifting diagrams, which proves useful for establishing their basic calculus. For this, we make use of the following lemma.

LEMMA 4.3.9. *Consider a family of arrows* $f : X \to A^2$ *representing a natural transformation* $X \underset{a}{\overset{b}{\rightrightarrows}} \Downarrow f \ A$. *Then there is a span*

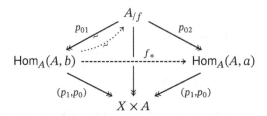

so that any section to the trivial fibration p_{01} composes with p_{02} to define a functor f_* representing postcomposition with f, and every functor in the fibered isomorphism class of f_* arises this way.[4] *Moreover, f represents a natural isomorphism $f : a \cong b$ if and only if the isofibration p_{02} is a trivial fibration.*

Proof By forming the pullbacks of each column, the span of cospans below defines the objects and maps of the span in the statement:

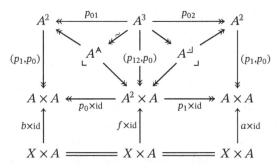

Note that the maps to the pullbacks in the tops squares respectively define a trivial fibration and an isofibration. Thus, by Proposition C.1.12, the induced map $p_{01} : A_{/f} \twoheadrightarrow \mathrm{Hom}_A(A, b)$ is a trivial fibration while the induced map $p_{02} : A_{/f} \twoheadrightarrow \mathrm{Hom}_A(A, a)$ is an isofibration.

In particular, from the pullbacks below-left, we see that sections to the trivial fibration p_{01} correspond to maps $\mathrm{Hom}_A(A, b) \to A^3$ that extend the composable pair of arrows $\phi : \mathrm{Hom}_A(A, b) \to A^2$ and $f p_1 : \mathrm{Hom}_A(A, b) \to A^2$ to a 2-

---

[4] Corollary 3.5.11 defines a bijection between natural transformations $\alpha : b \Rightarrow a$ in the homotopy 2-category and functors $\ulcorner \alpha_* \urcorner : \mathrm{Hom}_A(A, b) \to \mathrm{Hom}_A(A, a)$ up to fibered isomorphism. Here we slightly alter our notation because we are starting from a family of arrows $f : X \to A^2$ rather than from the natural transformation represented by that family.

simplex in $\mathsf{Fun}(\mathsf{Hom}_A(A,b),A)$.

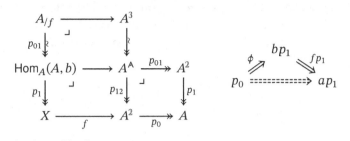

Similarly, the pullback

$$\mathsf{Fun}_{X\times A}(\mathsf{Hom}_A(A,b),\mathsf{Hom}_A(A,a)) \longrightarrow \mathsf{Fun}(\mathsf{Hom}_A(A,b),A)^2$$

$$\downarrow \qquad\qquad \qquad \qquad\qquad \downarrow (p_1,p_0)$$

$$\mathbb{1} \xrightarrow[\ (p_1a,p_0)\ ]{} \mathsf{Fun}(\mathsf{Hom}_A(A,b),A) \times \mathsf{Fun}(\mathsf{Hom}_A(A,b),A)$$

shows that functors $f_* : \mathsf{Hom}_A(A,b) \to \mathsf{Hom}_A(A,a)$ correspond to choices of representatives $fp_1 \cdot \phi : \mathsf{Hom}_A(A,b) \to A^2$ for the composite arrow (compare with the proof of Corollary 3.5.11). Thus, every section defines a map in the correct fibered isomorphism class of functors, and conversely, since each triple of arrows that define a commutative diagram in $\mathsf{hFun}(\mathsf{Hom}_A(A,b),A)$ bound some 2-simplex in $\mathsf{Fun}(\mathsf{Hom}_A(A,b),A)$ (see Lemma 1.1.12) every representing functor arises in this way.

Finally, by Exercise 3.6.ii and Corollary 3.5.11, $f : b \Rightarrow a$ is an isomorphism if and only if the functor $f_* : \mathsf{Hom}_A(A,b) \to \mathsf{Hom}_A(A,a)$ is a fibered equivalence. By the 2-of-3 property, it follows that f is an isomorphism if and only if p_{02} is a trivial fibration. □

LEMMA 4.3.10 (pullbacks as absolute lifting diagrams). *A commutative square in an ∞-category A is a pullback square if and only if the induced natural transformation (id_a, v) is an absolute right lifting diagram*

$$
\begin{array}{ccc}
d & \xrightarrow{\ u\ } & b \\[-2pt]
{\scriptstyle v}\downarrow \ \searrow{\scriptstyle w} & & \downarrow{\scriptstyle f} \\[-2pt]
c & \xrightarrow[\ g\ ]{} & a
\end{array}
\qquad\qquad
\begin{array}{ccc}
 & & \mathsf{Hom}_A(A,b) \\[-2pt]
{\scriptstyle u}\nearrow & \Downarrow{\scriptstyle (\mathrm{id}_a,v)} & \downarrow{\scriptstyle f_*} \\[-2pt]
1 & \xrightarrow[\ g\ ]{} & \mathsf{Hom}_A(A,a)
\end{array}
$$

The statement requires some explanation. A commutative square $s : 1 \to A^{\square}$ defines an element of $\mathsf{Fun}(1,A)^{\square}$, the data of which is given by the four vertices $d,b,c,a : 1 \to A$ and five 1-simplices $u,v,f,g,w : 1 \to A^2$ in the underlying quasi-category $A_0 := \mathsf{Fun}(1,A)$ of A, displayed above-left, together with a pair of unnamed 2-simplices that witness commutativity $fu = w = gv$ in $\mathsf{h}A :=$

hFun$(1, A)$. By Proposition 3.4.7, the composite $f_* u$ is isomorphic to fu. By 2-cell induction, the natural transformation $(\mathrm{id}_a, v) \colon fu \Rightarrow g$ displayed above-right may be constructed by specifying its domain and codomain components, the former of which we take to be $v \colon d \Rightarrow c$ and the latter of which we take to be id_a.

Proof   By Corollary 4.2.8, the fiber of the restriction functor

$$
\begin{array}{ccc}
A^{\square}_{g \vee f} & \longrightarrow & A^{\square} \\
\downarrow & \lrcorner & \downarrow{\scriptstyle \mathrm{res}} \\
1 & \xrightarrow{\;g \vee f\;} & A^{\lrcorner}
\end{array}
$$

is equivalent to the ∞-category of cones over the cospan diagram $g \vee f$. By Proposition 4.3.2, to show that the commutative square defines a pullback diagram is to show that $(v, f, u, g) \colon 1 \to A^{\square}_{g \vee f}$ defines a terminal element. Similarly, by Corollary 3.6.10, the pair $(u, (\mathrm{id}_a, v))$ defines an absolute right lifting diagram if and only if it represents a terminal element in the comma ∞-category $\mathrm{Hom}_{\mathrm{Hom}_A(A,a)}(f_*, g)$. We claim that $A^{\square}_{g \vee f}$ and $\mathrm{Hom}_{\mathrm{Hom}_A(A,a)}(f_*, g)$ are equivalent via maps that identify these elements, which proves the biconditional.

To see this, note that the simplicial square $\square$ can be formed by gluing two 2-simplices along their diagonal edge, giving rise to the pullback below-left:

$$
\begin{array}{ccc}
A^{\square} & \longrightarrow & A^3 \\
\downarrow & \lrcorner & \downarrow{\scriptstyle p_{02}} \\
A^3 & \xrightarrow{\;p_{02}\;} & A^2
\end{array}
\qquad
\begin{array}{ccc}
\mathrm{Hom}_{\mathrm{Hom}_A(A,a)}(f_*, \mathrm{Hom}_A(A,a)) & \longrightarrow & \mathrm{Hom}_A(A,a)^2 \\
\downarrow & \lrcorner & \downarrow{\scriptstyle p_0} \\
\mathrm{Hom}_A(A,b) & \xrightarrow{\;f_*\;} & \mathrm{Hom}_A(A,a)
\end{array}
$$

We argue that maps in the cospan whose pullback defines the comma ∞-category displayed above-right are each equivalent to pullbacks of the functor $p_{02} \colon A^3 \twoheadrightarrow A^2$ to suitable fibers. By Lemma 4.3.9, the map $p_{02} \colon A^3 \twoheadrightarrow A^2$ pulls back to a map equivalent to f_* on the fibers over f and a respectively, so it remains to consider the map p_0.

By applying $(-)^2$ to the pullback diagram that defines $\mathrm{Hom}_A(A, a)$ we obtain a pullback square that factors as a composite of two pullbacks:

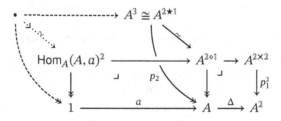

Due to the equivalence $A^{2\star 1} \simeq A^{2\diamond 1}$ of Proposition 4.2.7, the left-hand pullback square shows that $\mathrm{Hom}_A(A, a)^2$ is equivalent to the fiber of $p_2 : A^3 \twoheadrightarrow A$ along $a : 1 \to A$. Modulo this equivalence, the domain projection map $p_0 : \mathrm{Hom}_A(A, a)^2 \twoheadrightarrow \mathrm{Hom}_A(A, a)$ is equivalent to the map induced from

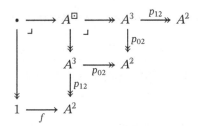

on fibers over $a : 1 \to A$. The codomain projection $p_1 : \mathrm{Hom}_A(A, a)^2 \twoheadrightarrow \mathrm{Hom}_A(A, a)$ is similarly equivalent to the pullback of the fibered projection map $p_{12} : A^3 \twoheadrightarrow A^2$ over $a : 1 \to A$.

Putting this together, the comma ∞-category $\mathrm{Hom}_{\mathrm{Hom}_A(A,a)}(f_*, \mathrm{Hom}_A(A, a))$ is equivalent to the limit of the diagram:

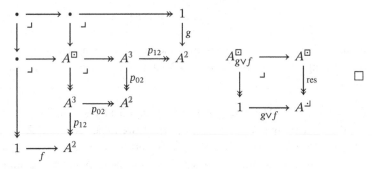

The codomain projection $p_1 : \mathrm{Hom}_{\mathrm{Hom}_A(A,a)}(f_*, \mathrm{Hom}_A(A, a)) \twoheadrightarrow \mathrm{Hom}_A(A, a)$ is the pullback of the top-horizontal composite in the above diagram along the inclusion $\mathrm{Hom}_A(A, a) \to A^2$. So the comma ∞-category $\mathrm{Hom}_{\mathrm{Hom}_A(A,a)}(f_*, g)$ is equivalent to the limit below-left, which rearranges into the pullback below-right that defines the fiber $A^{\square}_{g\vee f}$, proving the claimed equivalence:

$$
\begin{array}{ccccccc}
\bullet & \longrightarrow & \bullet & \longrightarrow & 1 \\
\downarrow & \lrcorner & \downarrow & \lrcorner & \downarrow g \\
\bullet & \longrightarrow & A^{\square} & \twoheadrightarrow & A^3 \xrightarrow{p_{12}} A^2 \\
\downarrow & \lrcorner & \downarrow & \lrcorner & \downarrow p_{02} \\
& & A^3 & \xrightarrow{p_{02}} & A^2 \\
& & \downarrow p_{12} \\
1 & \xrightarrow{f} & A^2
\end{array}
\qquad
\begin{array}{ccc}
A^{\square}_{g\vee f} & \longrightarrow & A^{\square} \\
\downarrow & \lrcorner & \downarrow \text{res} \\
1 & \xrightarrow{g\vee f} & A^{\lrcorner}
\end{array} \qquad \square
$$

There is a nonidentity automorphism of the simplicial set 2×2, which induces a "transposition" automorphism of $A^{\square}$. By symmetry, a commutative square in A is a pullback if and only if its transposed square is a pullback. This gives a dual form of Lemma 4.3.10 with the roles of f and g and of u and v interchanged. As a corollary, we can easily prove that pullback squares compose both "vertically" and "horizontally" and can be cancelled from the "right" and "bottom."

PROPOSITION 4.3.11 (composition and cancelation of pullback squares). *Given a composable pair of commutative squares in A and their composite rectangle defined via the equivalence $A^{3 \times 2} \simeq A^{\square} \underset{A^2}{\times} A^{\square}$*

$$
\begin{array}{ccccc}
p & \xrightarrow{\ x\ } & d & \xrightarrow{\ u\ } & b \\
{\scriptstyle y}\downarrow & {\scriptstyle z}\searrow & {\scriptstyle v}\downarrow & {\scriptstyle w}\searrow & \downarrow{\scriptstyle f} \\
e & \xrightarrow[\ h\ ]{} & c & \xrightarrow[\ g\ ]{} & a
\end{array}
$$

if the right-hand square is a pullback, then the left-hand square is a pullback if and only if the composite rectangle is a pullback.

Proof By Lemma 4.3.10, we are given an absolute right lifting diagram

$$
\begin{array}{ccc}
 & & \mathsf{Hom}_A(A, c) \\
 & {\scriptstyle v}\nearrow & \downarrow{\scriptstyle g_*} \\
 & {\scriptstyle \Downarrow(\mathrm{id}_a, u)} & \\
1 & \xrightarrow[\ f\ ]{} & \mathsf{Hom}_A(A, a)
\end{array}
$$

By Lemma 2.4.1, the composite diagram

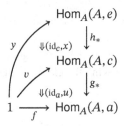

is an absolute right lifting diagram if and only if the top triangle is an absolute right lifting diagram. By Lemma 4.3.10, this is exactly what we wanted to show. □

REMARK 4.3.12. The result of Proposition 4.3.11 also holds for X-indexed families of commutative squares, by which we mean diagrams $X \to A^{\square}$, or equivalently, elements of $\mathsf{Fun}(X, A)^{\square}$. The proof is the same, making use of a generalization of Lemma 4.3.10 which states that an X-indexed commutative square valued in an ∞-category A in an ∞-cosmos $\mathcal{K}$ as below-left is a pullback square if and only if the induced 2-cell (id_a, v) below-right is an absolute right lifting diagram in $\mathcal{K}_{/X}$:

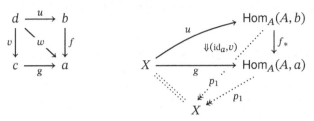

This characterization of X-indexed pullback squares can be proven by re-implementing the construction given in the proof of Lemma 4.3.10, using comma ∞-categories and simplicial limits in the sliced ∞-cosmos $\mathcal{K}_{/X}$, as described in Proposition 1.2.22, in place of the analogous constructions in $\mathcal{K}$. Were it not for the more complicated notation involved, we would have presented this general proof instead of its special case above.

Alternatively, this extension can be deduced from the result we prove here. A diagram $s \colon X \to A^{\boxdot}$ in $\mathcal{K}$ also defines a X-indexed commutative square in the ∞-cosmos $\mathcal{K}_{/X}$ valued in $\pi \colon A \times X \to X$. This takes the form of a functor $(s, \mathrm{id}_X) \colon X \to A^{\boxdot} \times X$ over X. It's easy to verify that a diagram valued in $\pi \colon A \times X \twoheadrightarrow X$ whose component at X is the identity has a limit in $\mathcal{K}_{/X}$ if and only if the A component of the diagram has a limit in $\mathcal{K}$. Since id_X is the terminal object of $\mathcal{K}_{/X}$, this object is the ∞-category $1 \in \mathcal{K}_{/X}$, so Lemma 4.3.10 applies in the ∞-cosmos $\mathcal{K}_{/X}$ to prove the general case of X-indexed families of commutative squares in $\mathcal{K}$.

As discussed in Example 2.3.10, terminal and initial elements are special cases of limits and colimits, respectively, where the diagram shape is empty. For any ∞-category A, the ∞-category $A^{\varnothing} \cong 1$ of empty diagrams in A is terminal. Thus, there is a unique $\varnothing$-indexed diagram in A. It follows immediately from the construction of the comma ∞-categories in Definition 4.2.1, that both of the ∞-categories of cones over or under the unique empty diagram are isomorphic to A. In the case of cones over an empty diagram, the domain-evaluation functor, carrying a cone to its summit, is the identity on A, while in the case of cones under the empty diagram, the codomain-evaluation functor, carrying a cone to its nadir, is the identity on A. The following characterization of terminal elements can be deduced as a special case of Proposition 4.3.1, though we find it easier to argue from Proposition 4.1.1.

PROPOSITION 4.3.13. *An element* $a \colon 1 \to A$ *of an* ∞-*category* A

(i) *defines a terminal element of* A *if and only if the domain projection functor* $p_0 \colon \mathrm{Hom}_A(A, a) \twoheadrightarrow A$ *is a trivial fibration, and*

(ii) defines an initial element of A if and only if the codomain projection functor p_1 : $\mathrm{Hom}_A(a, A) \twoheadrightarrow A$ is a trivial fibration.

Proof   Recall from Definition 2.2.1, that an element is terminal if and only if it is right adjoint to the unique functor

$$1 \underset{a}{\overset{!}{\underset{\longrightarrow}{\overset{\longleftarrow}{\perp}}}} A$$

By Proposition 4.1.1, ! $\dashv a$ if and only if there is an equivalence $\mathrm{Hom}_1(!, 1) \simeq_A \mathrm{Hom}_A(A, t)$. By the defining pullback (3.4.2) for the comma ∞-category, the left representation of ! : $A \to 1$ is A itself, with domain projection functor the identity. So the component of the equivalence $\mathrm{Hom}_A(A, a) \twoheadrightarrow A$ over A must be the domain projection functor p_0 : $\mathrm{Hom}_A(A, a) \twoheadrightarrow A$, and we conclude that a is a terminal element if and only if this isofibration is a trivial fibration. $\square$

DIGRESSION 4.3.14 (terminal elements of a quasi-category). In the ∞-cosmos of quasi-categories, the domain of the isofibration p_0 : $\mathrm{Hom}_A(A, a) \twoheadrightarrow A$ is equivalent over A to the *slice quasi-category* $A_{/a}$, defined in Proposition 4.2.5 (see Corollary D.6.6). Via this equivalence, Proposition 4.3.13 proves that a is terminal if and only if the projection $A_{/a} \twoheadrightarrow A$ is a trivial fibration in the sense of Definition 1.1.25, which transposes to Joyal's original definition of a terminal element of a quasi-category. See Proposition F.1.1 for an expanded discussion.

Exercises

EXERCISE 4.3.i (pointwise limits in functor ∞-categories). Suppose A admits the limit ℓ : $D \to A$ of a family of diagrams d : $D \to A^J$ of shape J. Prove that the diagram ∞-category A^K admits limits of the corresponding family of J-shaped diagrams d^K : $D^K \to (A^J)^K \cong (A^K)^J$ defined "pointwise in K" by the functor ℓ^K : $D^K \to A^K$.[5]

EXERCISE 4.3.ii. State and prove versions of Proposition 4.3.4 and Corollary 4.3.5 that apply to a family of diagrams d : $D \to A^J$ indexed by an ∞-category J in a cartesian closed ∞-cosmos.

EXERCISE 4.3.iii. Let a be an element of an ∞-category A and let K and L be

---

[5] If K is an ∞-category in a cartesian closed ∞-cosmos $\mathcal{K}$ this can be proven directly by arguing in the homotopy 2-category, but another proof applies simultaneously to this case and to the case where K is a simplicial set: use the fact that the cosmological functor $(-)^K$: $\mathcal{K} \to \mathcal{K}$ preserves the equivalence of Proposition 4.3.1.

simplicial sets. Prove that if A has tensors and cotensors then these operations are associative in the sense that the elements

$$K \otimes (L \otimes a) \cong (K \times L) \otimes a \cong L \otimes (K \otimes a) \qquad \text{and} \qquad (a^L)^K \cong a^{K \times L} \cong (a^K)^L$$

are isomorphic.

EXERCISE 4.3.iv. Prove that if A has a terminal element t then for any element a the mapping space $\mathrm{Hom}_A(a, t)$ is **contractible**, i.e., is equivalent to the terminal ∞-category 1.[6]

EXERCISE 4.3.v. Suppose A admits pullbacks and consider a family of commutative squares $d : D \to A^{\square}$. Show that the following are equivalent:

(i) The commutative triangle

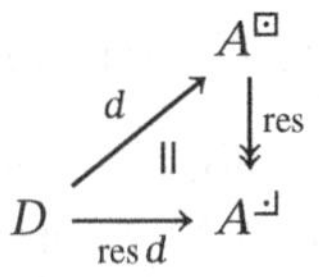

is an absolute right lifting diagram.

(ii) The component of the unit of the adjunction res ⊣ ran

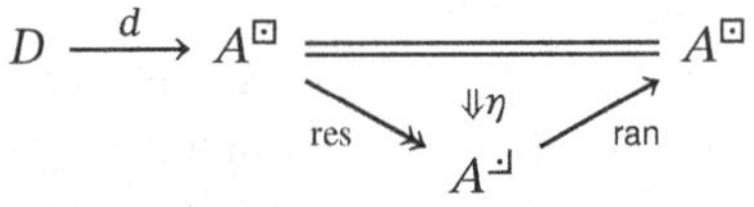

is invertible.

EXERCISE 4.3.vi. Prove that a square in A is a pullback if and only if its "transposed" square, defined by composing with the involution $A^{\square} \cong A^{\square}$ induced from the automorphism of 2×2 that swaps the "off-diagonal" elements, is a pulllback square.

EXERCISE 4.3.vii. Show that any ∞-category that has pullbacks and a terminal element admits binary products.

4.4 Pointed and Stable ∞-Categories

In this section, we study ∞-categories with special exactness properties, admitting certain finite limit and colimit constructions, which coincide.

---

[6] The converse implication holds in ∞-cosmoi of $(\infty, 1)$-categories, as argued in the proof of Proposition F.1.1.

DEFINITION 4.4.1 (pointed ∞-category). An ∞-category A is **pointed** if it admits a **zero element**: an element $* : 1 \to A$ that is both initial and terminal.

Recall Lemma 2.2.2, which enumerates the data required to present an initial or terminal element. To show that an element $* : 1 \to A$ defines a zero element it suffices to define a pair of natural transformations $\rho : *! \Rightarrow \mathrm{id}_A$ and $\xi : \mathrm{id}_A \Rightarrow *!$ so that the components $\rho_* : * \Rightarrow *$ and $\xi_* : * \Rightarrow *$ are isomorphisms in hA. Here ρ is the counit of the adjunction $* \dashv !$ that witnesses the initiality of the zero element and ξ is the unit of the adjunction $! \dashv *$ that witnesses the terminality of the zero element.

The counit ρ is represented by a functor $\ulcorner\rho\urcorner : A \to A^2$, whose domain component is constant at $*$ and whose codomain component is id_A, that we refer to as the **family of points** of A. Dually, the unit ξ is represented by a functor $\ulcorner\xi\urcorner : A \to A^2$, whose domain component is id_A and whose codomain component is constant at $*$, that we refer to as the **family of copoints**.

LEMMA 4.4.2 (pointed ∞-categories of based elements). *If A is an ∞-category with a terminal element $t : 1 \to A$ then the ∞-category $\mathrm{Hom}_A(t, A)$ is pointed, with $\ulcorner\mathrm{id}_t\urcorner : 1 \to \mathrm{Hom}_A(t, A)$ serving as its zero element. Moreover all pointed ∞-categories arise in this manner.*

Proof If A is a pointed ∞-category with zero element $* : 1 \to A$ then by Proposition 4.3.13, $p_1 : \mathrm{Hom}_A(*, A) \twoheadrightarrow A$ defines an equivalence between A and an ∞-category of the form described in the statement. Since the codomain projection functor p_1 carries the element $\ulcorner\mathrm{id}_*\urcorner$ of $\mathrm{Hom}_A(*, A)$ to the zero element of A, Lemma 2.2.7 tells us that $\ulcorner\mathrm{id}_*\urcorner$ must define a zero element of $\mathrm{Hom}_A(*, A)$.

Now suppose only that A has a terminal element $t : 1 \to A$. By Corollary 3.5.10, $\ulcorner\mathrm{id}_t\urcorner$ defines an initial element of $\mathrm{Hom}_A(t, A)$, so it remains only to show that this element is also terminal. By Lemma 2.2.2, our task is to define a natural transformation

$$\mathrm{Hom}_A(t, A) \xLongequal{\quad\quad\quad} \mathrm{Hom}_A(t, A)$$

witnessing the terminality of $\ulcorner\mathrm{id}_t\urcorner$. By Proposition 3.4.6, we may use 2-cell induction to induce η from a pair of natural transformations $(p_1\tau, p_0\tau)$ satisfying a compatibility condition. Here necessarily $p_0\tau = \mathrm{id}_1$, since its codomain ∞-category is terminal, and we define $p_1\tau$ to be ξp_1, where ξ is the unit of the adjunction $! \dashv t$. The compatibility condition of Proposition 3.4.6(ii) follows from the triangle equality relation $\xi_t = \mathrm{id}_t$.

The component $\eta_{\ulcorner \mathrm{id}_t \urcorner}$ is induced from the pair of identity 2-cells $(\xi_t, \mathrm{id}_!)$, so by 2-cell conservativity, $\eta_{\ulcorner \mathrm{id}_t \urcorner}$ is invertible. By Lemma 2.2.2, this is enough to witness the terminality of $\ulcorner \mathrm{id}_t \urcorner$. □

Pointed ∞-categories permit familiar constructions from homotopy theory. By Definition 4.3.7, gluing two copies of the family of points $\ulcorner \rho \urcorner : A \to A^2$ along their codomains defines a family of cospans $\check{\rho} := \ulcorner \rho \urcorner \vee \ulcorner \rho \urcorner : A \to A^{\lrcorner}$. Dually, there is a family of spans $\hat{\xi} := \ulcorner \xi \urcorner \wedge \ulcorner \xi \urcorner : A \to A^{\ulcorner}$ defined by gluing the family of copoints $\ulcorner \xi \urcorner : A \to A^2$ to itself along their domains.

DEFINITION 4.4.3 (loops and suspension). A pointed ∞-category A **admits loops** if it admits a limit of the family of cospans $\check{\rho}$, in which case the limit functor $\Omega : A \to A$ is called the **loops functor**. Dually, a pointed ∞-category A **admits suspensions** if it admits a colimit of the family of spans $\hat{\xi}$, in which case the colimit functor $\Sigma : A \to A$ is called the **suspension functor**.

Importantly, if A admits loops and suspensions, then the loops and suspension functors are adjoint:

PROPOSITION 4.4.4 (the loops-suspension adjunction). *If A is a pointed ∞-category that admits loops and suspensions, then the loops functor is right adjoint to the suspension functor*

$$A \underset{\Omega}{\overset{\Sigma}{\underset{\perp}{\rightleftarrows}}} A$$

The main idea of the proof is easy to describe. If A admits all pullbacks and all pushouts, then Corollary 4.3.5 supplies adjunctions

$$A^{\lrcorner} \underset{\mathrm{ran}}{\overset{\mathrm{res}}{\underset{\perp}{\rightleftarrows}}} A^{\square} \underset{\mathrm{res}}{\overset{\mathrm{lan}}{\underset{\perp}{\rightleftarrows}}} A^{\ulcorner}$$

that are fibered over $A \times A$ upon evaluating at the intermediate vertices of the commutative square. Pulling back along $(*, *) : 1 \to A \times A$, pins these vertices at the zero element. Since the zero element is initial and terminal, the ∞-categories of pullback and pushout diagrams of this form are both equivalent to A and the pulled-back adjoints now coincide with the loops and suspension functors.

The only subtlety in the proof that follows is that we have assumed weaker hypotheses: that A admits only loops and suspensions, but perhaps not all pullbacks and pushouts.

Proof The family of cospans $\check{\rho}$ lands in a subobject $A_*^{\lrcorner}$ of $A^{\lrcorner}$ defined below-left that is comprised of those cospans whose source elements are pinned at the zero element $*$ of A.

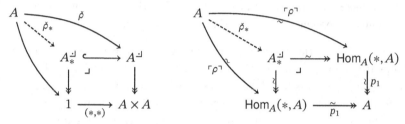

From a second construction of $A_*^{\lrcorner}$ displayed above-right and the characterization of initiality given in Proposition 4.3.13, we conclude from the 2-of-3 property of equivalences first that the family of points $\ulcorner\rho\urcorner : A \to A^2$ restricts to define an equivalence $\ulcorner\rho\urcorner : A \xrightarrow{\sim} \mathrm{Hom}_A(*, A)$ and then that the induced diagram $\check{\rho}_* : A \xrightarrow{\sim} A_*^{\lrcorner}$ is an equivalence. Dually, the family of spans $\check{\xi} : A \to A^{\ulcorner}$ defines an equivalence $\check{\xi}_* : A \xrightarrow{\sim} A_*^{\ulcorner}$ when its codomain is restricted to the subobject of spans whose target elements are pinned at the zero element $*$.

By Proposition 4.3.4, a pointed ∞-category A admits loops or admits suspensions if and only if there exist absolute lifting diagrams as below-left and below-right, respectively

and moreover we may take these natural isomorphisms to be identities. Doing so allows us to define restricted lifts

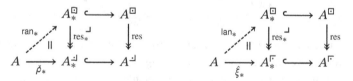

which we argue again define absolute right and left lifting diagrams, respectively. The right- and left-handed arguments are dual, so we focus our attention on the former. By Theorem 3.5.3 the absolute right lifting diagram defines a fibered equivalence $\mathrm{Hom}_{A^\square}(A^\square, \mathrm{ran}) \simeq_{A \times A^\square} \mathrm{Hom}_{A^\lrcorner}(\mathrm{res}, \check{\rho})$, which may be pulled back along the inclusion of the subobject $A_*^\square \hookrightarrow A^\square$ of commutative squares in A whose intermediate vertices are pinned at the zero element to yield a fibered equivalence over $A \times A_*^\square$. We claim that these ∞-categories pull back to ∞-categories that are equivalent to $\mathrm{Hom}_{A^\square}(A_*^\square, \mathrm{ran}_*)$ and $\mathrm{Hom}_{A^\lrcorner}(\mathrm{res}_*, \check{\rho}_*)$, respectively.

To see this, first observe that the universal property of the zero element implies that $\mathrm{Hom}_A(*, *)$ is contractible (see Proposition 4.3.13) and therefore the outer square is equivalent to the pullback

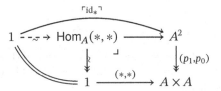

Since the top and bottom faces of the commutative prism are strict pullbacks

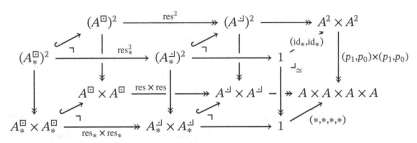

it follows that the left and middle vertical faces are also pullbacks up to equivalence. We use the latter of these and the commutative cube

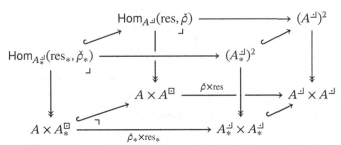

to conclude that the comma ∞-category $\mathrm{Hom}_{A^\lrcorner}(\mathrm{res}, \check{\rho})$ pulls back along the inclusion $A_*^\square \hookrightarrow A^\square$ to an ∞-category that is equivalent to $\mathrm{Hom}_{A^\lrcorner}(\mathrm{res}_*, \check{\rho}_*)$, as claimed. Similarly, from the former pullback up to equivalence, we conclude that $\mathrm{Hom}_{A^\square}(A^\square, \mathrm{ran})$ pulls back to an ∞-category that is equivalent to $\mathrm{Hom}_{A^\square}(A_*^\square, \mathrm{ran}_*)$.

In this way we obtain fibered equivalences

$$\mathrm{Hom}_{A^{\square}}(A^{\square}_*, \mathrm{ran}_*) \simeq_{A \times A^{\square}_*} \mathrm{Hom}_{A^{\lrcorner}_*}(\mathrm{res}_*, \check{\rho}_*) \quad \text{and}$$

$$\mathrm{Hom}_{A^{\square}}(\mathrm{lan}_*, A^{\square}_*) \simeq_{A^{\square}_* \times A} \mathrm{Hom}_{A^{\ulcorner}_*}(\hat{\xi}_*, \mathrm{res}_*).$$

which, by Theorem 3.5.8, encode absolute liftings of $\check{\rho}_*$ and $\hat{\xi}_*$ through the restriction functors:

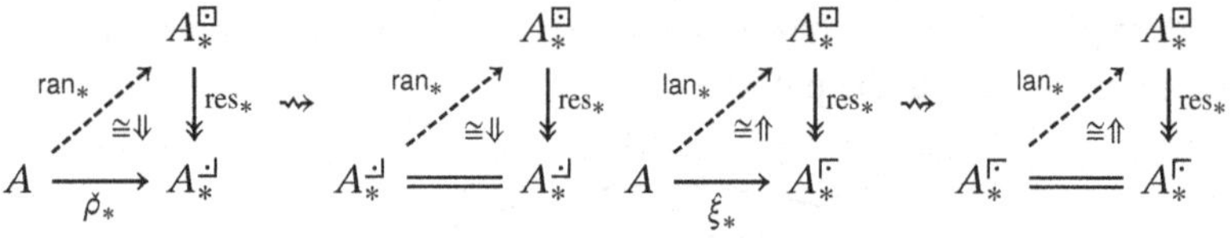

Restricting along the inverse equivalences $A^{\lrcorner}_* \rightsquigarrow A$ and $A^{\ulcorner}_* \rightsquigarrow A$ to $\check{\rho}_*$ and $\hat{\xi}_*$ and pasting with the invertible 2-cell we obtain absolute lifting diagrams whose bottom edge is the identity.

By Lemma 2.3.7, these lifting diagrams define adjunctions:

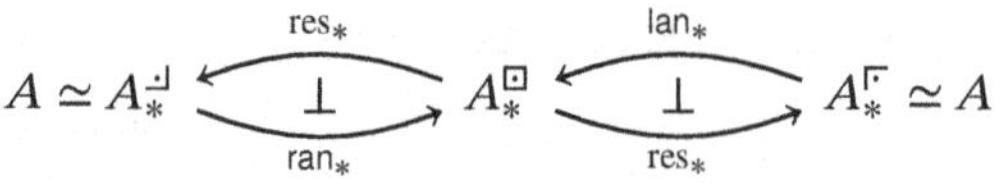

which compose to the desired adjunction $\Sigma \dashv \Omega$. □

DEFINITION 4.4.5 (fiber and cofiber). An arrow $f : 1 \to A^2$ from x to y in a pointed ∞-category A **admits a fiber** if A admits a pullback of the cospan formed by f and the component $\ulcorner \rho_y \urcorner$ of the family of points. The pullback square defined by the absolute right lifting diagram

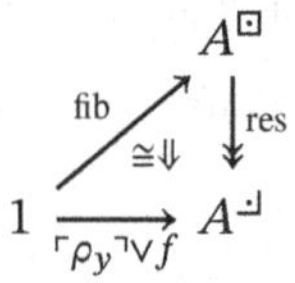

is referred to as the **fiber sequence** for f. Dually, f **admits a cofiber** if A admits a pushout of the span formed by f and the component $\ulcorner \xi_x \urcorner$ of the family of copoints, in which case the pushout square

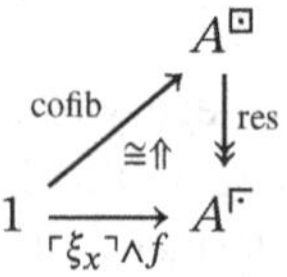

defines the **cofiber sequence** for f.

Fiber and cofiber sequences in A define commutative squares whose lower-left vertex is the zero element $*$.

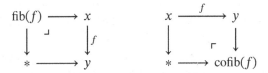

The data of such squares is given by a commutative triangle in A – an element of A^3 – involving a diagonal arrow that we have neglected to draw, together with a **nullhomotopy** of that diagonal edge – a witness that this edge factors through the zero element in hA. A commutative square in A whose lower-left vertex is the zero element is referred to as a **triangle** in A.

We can now state the first of several equivalent characterizations of stable ∞-categories. This notion and the results that follow are due to Lurie first appearing in a preprint [77] later incorporated into the first chapter of *Higher Algebra* [80].

DEFINITION 4.4.6 (stable ∞-category). A **stable ∞-category** is a pointed ∞-category A in which

(i) every morphism admits a fiber and a cofiber: that is, there exist absolute lifting diagrams

$$
\begin{array}{ccc}
 & & A^{\square} \\
 & \nearrow^{\text{fib}} & \downarrow{\text{res}} \\
 & \cong\Downarrow & \\
A^2 & \xrightarrow[\ulcorner\rho_{\text{cod}}\urcorner\wedge\text{id}]{} & A^{\lrcorner}
\end{array}
\qquad
\begin{array}{ccc}
 & & A^{\square} \\
 & \nearrow^{\text{cofib}} & \downarrow{\text{res}} \\
 & \cong\Uparrow & \\
A^2 & \xrightarrow[\ulcorner\xi_{\text{dom}}\urcorner\wedge\text{id}]{} & A^{\ulcorner}
\end{array}
$$

(ii) and a triangle in A defines a fiber sequence if and only if it also defines a cofiber sequence. Such triangles are called **exact triangles**.

As a means of familiarizing ourselves with this definition, we prove:

LEMMA 4.4.7. *Let A be a stable ∞-category and let J be either a simplicial set or another ∞-category in the case where the ambient ∞-cosmos $\mathcal{K}$ is cartesian closed. Then A^J is again a stable ∞-category.*

Proof By Proposition 2.1.7, the cosmological functor $(-)^J : \mathcal{K} \to \mathcal{K}$ preserves the adjunctions

$$
A \overset{!}{\underset{*}{\underrightarrow{\underleftarrow{\quad\bot\quad}}}} 1
\qquad
A \overset{*}{\underset{!}{\underrightarrow{\underleftarrow{\quad\bot\quad}}}} 1
\rightsquigarrow
A^J \overset{!^J}{\underset{*^J}{\underrightarrow{\underleftarrow{\quad\bot\quad}}}} 1^J \cong 1
\qquad
A^J \overset{*^J}{\underset{!^J}{\underrightarrow{\underleftarrow{\quad\bot\quad}}}} 1^J \cong 1
$$

that exhibit the universal properties of the zero element $* : 1 \to A$. Thus, we see that A^J is a pointed ∞-category, whose basepoint is the constant J-shaped functor valued at $*$.

Similarly, by Corollary 3.5.7, the cosmological functor $(-)^J : \mathcal{K} \to \mathcal{K}$ preserves the absolute lifting diagrams of 4.4.6(i) that define fiber and cofiber sequences. Since $\mathrm{res}^J : (A^\square)^J \twoheadrightarrow (A^\lrcorner)^J$ is isomorphic to $\mathrm{res} : (A^J)^\square \twoheadrightarrow (A^J)^\lrcorner$ and similarly for the functor restricting from a square to its underlying span, these absolute lifting diagrams define fiber and cofiber sequences in A^J.

Finally, condition 4.4.6(ii) can be re-expressed as the assertion that the commutative triangle below-left is absolute left lifting and the triangle below-right is absolute right lifting:

which is to say that fiber sequences in A are also cofiber sequences and cofiber sequences in A are also fiber sequences. By applying Corollary 3.5.7 once more, we see that the same exactness property holds in A^J. Thus A^J is stable. $\square$

Stable ∞-categories in fact admit all pushouts and all pullbacks, and such squares coincide. Squares that are both pushouts and pullbacks are called **exact squares**.

PROPOSITION 4.4.8 (pullbacks and pushouts in stable ∞-categories). *A stable ∞-category admits all pushouts and all pullbacks, and moreover, a square is pushout if and only if it is a pullback.*

Proof Given a generic family of cospans $g \vee f : X \to A^\lrcorner$ in A, form the cofiber of f followed by the fiber of the composite map $qg : c \to a \to \mathrm{cofib}\, f$:

$$\begin{array}{ccccc}
\mathrm{fib}(qg) & \dashrightarrow{u} & b & \longrightarrow & * \\
{\scriptstyle v}\big\downarrow & {\scriptstyle \lrcorner} & \big\downarrow{\scriptstyle f} & {\scriptstyle \ulcorner} & \big\downarrow \\
c & \xrightarrow{\;\;g\;\;} & a & \dashrightarrow{q} & \mathrm{cofib}(f)
\end{array} \qquad (4.4.9)$$

By Definition 4.4.6(ii), the cofiber sequence $b \to a \to \mathrm{cofib}(f)$ is also a fiber sequence. By the pullback cancelation result of Proposition 4.3.11, we conclude that $\mathrm{fib}(qg)$ computes the pullback of the cospan $g \vee f$.

To see that this pullback square is also a pushout, form the fiber of the map v:

$$\begin{array}{ccccc}
\mathrm{fib}(v) & \dashrightarrow & \mathrm{fib}(qg) & \xrightarrow{u} & b \\
\big\downarrow & {\scriptstyle \lrcorner} & {\scriptstyle v}\big\downarrow & {\scriptstyle \lrcorner} & \big\downarrow{\scriptstyle f} \\
* & \longrightarrow & c & \xrightarrow{\;\;g\;\;} & a
\end{array}$$

By the pullback composition result of Proposition 4.3.11, $\text{fib}(v)$ is also the fiber of the map f. By Definition 4.4.6(ii), the fiber sequences $\text{fib}(v) \to \text{fib}(qg) \to c$ and $\text{fib}(v) \to b \to a$ are also cofiber sequences. Now by the pushout cancelation result of Proposition 4.3.11, we see that the right-hand pullback square is also a pushout square. A dual argument proves that pushouts exist and coincide with pullbacks. □

DIGRESSION 4.4.10 (on the use of generalized elements to define functors). The first paragraph of the proof just given takes a generic family of cospans and constructs a rectangular diagram (4.4.9), to which Proposition 4.3.11 can be applied (by Remark 4.3.12). By the Yoneda lemma, a construction given as a mapping on generalized elements defines an arrow internally to the ∞-cosmos, in this case taking the form of a functor $A^{\lrcorner} \to A^{3\times 2}$, as we now illustrate by unpacking each of the steps.[7] First, we build, from the generic cospan, the dashed arrow below-left, which forms a diagram that glues this cospan to the cofiber sequence associated to one of its legs:

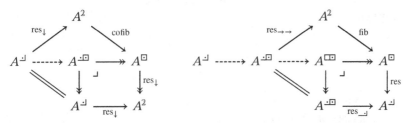

The simplicial set $\llcorner\square \cong \lrcorner \cup_{\downarrow} \square$ does not include the composite 1-simplex from the lower-left vertex to the lower-right vertex but this can be attached by filling an inner horn, resulting in an equivalent ∞-category that we also denote by $A^{\llcorner\square}$. Next we attach the fiber sequence associated to that composite arrow, gluing the exterior rectangle onto the diagram of shape $\llcorner\square$, defining the dashed arrow above-right.

The simplicial set $\square\!\square$ is a subset of the rectangle diagram shape 3×2. In the notation of (4.4.9) what is missing is the map u and the left-hand square, which we induce by the universal property of the fiber sequence $b \to a \to \text{cofib}(f)$, encoded by the absolute right lifting diagram below-left. There is a functor $(g, \text{id}, \text{id}) : \lrcorner \times 2 \to \square\!\square$ inducing the natural transformation γ below-center,

---

[7] Indeed, this functor can be understood as the result of applying the construction to the universal generalized element, which is always given by the identity. The motivation for the conceit of considering a generic cospan $g \vee f : X \to A^{\lrcorner}$ in place of the universal cospan $\text{id} : A^{\lrcorner} \to A^{\lrcorner}$ is to introduce some human-readable notation.

which then factors as below-right:

$$
\begin{array}{ccc}
& A^{\square} & \\
\text{res}_{\square} \nearrow & \Big\| \;\; \Big\downarrow \text{res} & \\
A^{\square\square} & \xrightarrow{\;\;\;\;} & A^{\lrcorner} \\
& \text{res}_{\lrcorner} &
\end{array}
\qquad
\begin{array}{ccc}
A^{\square\square} & \xrightarrow{\;\text{res}_{\square}\;} & A^{\square} \\
\Big\| \quad \Downarrow\gamma & & \Big\downarrow \text{res} \\
A^{\square\square} & \xrightarrow{\;\text{res}_{\lrcorner}\;} & A^{\lrcorner}
\end{array}
\;=\;
\begin{array}{ccc}
A^{\square\square} & \xrightarrow{\;\text{res}_{\square}\;} & A^{\square} \\
\Big\| \quad\Downarrow\upsilon \;\; \nearrow & & \Big\downarrow \text{res} \\
A^{\square\square} & \xrightarrow{\;\text{res}_{\lrcorner}\;} & A^{\lrcorner}
\end{array}
$$

The composite functor

$$
A^{\lrcorner} \longrightarrow A^{\square\square} \xrightarrow{\;\ulcorner\upsilon\urcorner\;} A^{\square\times2} \xrightarrow{\;\text{res}\;} A^{3\times2}
$$

builds the diagram on display in (4.4.9) from a generic cospan.

A stable ∞-category admits loops and suspensions, formed by taking fibers of the arrows in the family of points and cofibers of arrows in the family of copoints, respectively.

Proposition 4.4.11 (loops and suspension in stable ∞-categories). *If A is a stable ∞-category, the loops and suspension functors define inverse adjoint equivalences*

Proof   In the proof of Proposition 4.4.4, the adjunction $\Sigma \dashv \Omega$ is constructed as a composite of adjunctions

$$
A \simeq A_*^{\lrcorner} \underset{\text{ran}_*}{\overset{\text{res}_*}{\rightleftarrows}} A_*^{\square} \underset{\text{res}_*}{\overset{\text{lan}_*}{\rightleftarrows}} A_*^{\ulcorner} \simeq A
$$

that construct fiber and cofiber sequences. By Proposition 2.1.9, the unit and counit of this composite adjunction are given by

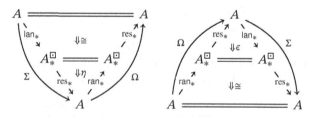

By Definition 4.3.8, the unit of $\text{res}_* \dashv \text{ran}_*$ restricts to an isomorphism on the subobject of pushout squares. In a stable ∞-category, the cofiber sequences in the image of $\text{lan}_* : A \to A_*^{\square}$ are pullback squares, so this tells us that η_{lan_*} is an isomorphism. Dually, the fiber sequences in the image of $\text{ran}_* : A \to A_*^{\square}$

are pushout squares, which tells us that $\epsilon_{\mathrm{ran}_*}$ is an isomorphism. Hence, the unit and counit of $\Sigma \dashv \Omega$ are invertible, so these functors define an adjoint equivalence. □

The results just proven suggest several equivalent characterizations of stable ∞-categories. The equivalence of condition (iii) is due to Groth [49, §3], who works in the closely related setting of stable derivators and also discusses further equivalent conditions not mentioned here. The remaining equivalences are established by Lurie in [77]. The proof that (iv)⇒(i) is an adaptation of a clever argument of Harpaz appearing as [51, 2.4].

THEOREM 4.4.12 (equivalent characterizations of stable ∞-categories). *In a pointed ∞-category A the following are equivalent, and characterize the **stable** ∞-categories:*

- *(i) A admits fibers and cofibers, and fiber and cofiber sequences coincide.*
- *(ii) A admits pullbacks and pushouts, and pullback and pushout squares coincide.*
- *(iii) A admits pullbacks and pushouts, and the pullback functor* $\lim : A^{\lrcorner} \to A$ *preserves pushouts while the pushout functor* $\operatorname{colim} : A^{\ulcorner} \to A$ *preserves pullbacks.*
- *(iv) A admits cofibers and the suspension functor is an equivalence.*
- *(v) A admits fibers and the loops functor is an equivalence.*

Proof Proposition 4.4.8 proves the equivalence (i)⇔(ii), while Proposition 4.4.11 proves (i)⇒(iv) and (i)⇒(v). So it remains to prove (ii)⇔(iii) as well as the converses of these latter implications, which are dual.

Assuming (ii), we may apply Lemma 4.4.7 to see that the diagram ∞-categories $A^{\lrcorner}$ and $A^{\ulcorner}$ are stable. In particular, $\lim : A^{\lrcorner} \to A$ and $\operatorname{colim} : A^{\ulcorner} \to A$ are functors between stable ∞-categories that preserve all limits and all colimits, respectively, by virtue of Theorem 2.4.2. Since pushout and pullback squares coincide, we see that pushouts are preserved by the pullback functor and pullbacks are preserved by the pushout functor, proving (iii).

Now assume (iii). By Exercise 3.6.iv to say that the pushout functor preserves pullbacks is equally to say that the functor $\operatorname{lan} : A^{\ulcorner} \to A^{\Box}$ preserves pullbacks, meaning that the left-hand composite is isomorphic to the right-hand absolute right lifting diagram:

$$
\begin{array}{ccc}
(A^{\ulcorner})^{\Box} \xrightarrow{\operatorname{lan}^{\Box}} (A^{\Box})^{\Box} & & (A^{\Box})^{\Box} \\
{}^{\operatorname{ran}}\nearrow \quad \downarrow^{\operatorname{res}} \qquad \downarrow^{\operatorname{res}} \cong & & {}^{\operatorname{ran}}\nearrow \quad \downarrow^{\operatorname{res}} \\
(A^{\ulcorner})^{\lrcorner} = (A^{\ulcorner})^{\lrcorner} \xrightarrow{\operatorname{lan}^{\lrcorner}} (A^{\Box})^{\lrcorner} & \quad (A^{\ulcorner})^{\lrcorner} \xrightarrow{\operatorname{lan}^{\lrcorner}} (A^{\Box})^{\lrcorner} = (A^{\Box})^{\lrcorner}
\end{array}
$$

To see that any pullback square in A is also a pushout square, consider the diagram $\psi : A^{\lrcorner} \to A^{\ulcorner \times \lrcorner}$ depicted below-left, that expands a cospan $a \to c \leftarrow b$ to a cospan of spans by restricting along an appropriate functor $\ulcorner \times \lrcorner \to \lrcorner$ that sends five elements to the terminal vertex of $\lrcorner$ and two elements apiece to the remaining two vertices.

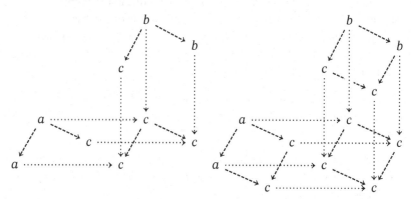

By Exercise 2.3.iv, when we compose with $\mathsf{lan}^{\lrcorner} : (A^{\ulcorner})^{\lrcorner} \to (A^{\square})^{\lrcorner}$ we obtain the diagram above-right in which each of the dashed squares are pushouts. By Exercise 4.3.i, the functor $\mathsf{ran} : (A^{\square})^{\lrcorner} \to (A^{\square})^{\square}$ is naturally isomorphic to the functor $\mathsf{ran}^{\square} : (A^{\lrcorner})^{\square} \to (A^{\square})^{\square}$. Thus, composing with this functor forms the pullbacks of the dotted cospans, which by Exercise 2.3.iv yields the diagram

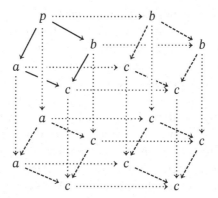

where p is the pullback of the original span $a \to c \leftarrow b$. On account of the natural isomorphism $\mathsf{lan}^{\square} \circ \mathsf{ran} \cong \mathsf{ran} \circ \mathsf{lan}^{\lrcorner}$ this diagram is also produced by first forming the pullbacks of the dotted spans and then taking the pushouts of each cospan. In this way we see that the solid-arrow pullback square above-left is also a pushout. The dual construction completes the proof that (iii)⇒(ii).

It remains to prove (iv)⇒(i). Assuming (iv), the first task is to show that any cofiber sequence $d : D \to A^{\boxdot}$ is also a fiber sequence, which is to say the diagram

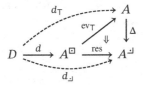

is absolute right lifting, where the natural transformation arises from the functor $\lrcorner \times 2 \to \boxdot$ that defines the canonical cone induced by a commutative square over its underlying cospan. By Theorem 3.5.3, to show that this diagram is absolute right lifting, it suffices to show that the induced map defines a fibered equivalence $\mathrm{Hom}_A(A, d_\top) \simeq_{D \times A} \mathrm{Hom}_{A^\lrcorner}(\Delta, d_\lrcorner)$. Since $\Sigma : A \overset{\sim}{\to} A$ is an equivalence, by Proposition 3.4.5 the maps of cospans

$$
\begin{array}{ccccc}
D & \xrightarrow{d_\top} & A & \!=\!=\! & A \\
\| & & \downdownarrows{\scriptstyle\Sigma} & & \| \\
D & \xrightarrow[\Sigma d_\top]{} & A & \xleftarrow[\Sigma]{} & A
\end{array}
\qquad
\begin{array}{ccccc}
D & \xrightarrow{d_\lrcorner} & A^\lrcorner & \xleftarrow{\Delta} & A \\
\| & & \downdownarrows{\scriptstyle\Sigma^\lrcorner} & & \| \\
D & \xrightarrow[\Sigma d_\lrcorner]{} & A^\lrcorner & \xleftarrow[\Delta\Sigma]{} & A
\end{array}
$$

induce equivalences of comma ∞-categories over $D \times A$ displayed vertically below.

$$
\begin{array}{ccc}
\mathrm{Hom}_A(A, d_\top) & \longrightarrow & \mathrm{Hom}_{A^\lrcorner}(\Delta, d_\lrcorner) \\
\Sigma \downdownarrows & \overset{\simeq}{\diagdown} & \downdownarrows \Sigma^\lrcorner \\
\mathrm{Hom}_A(\Sigma, \Sigma d_\top) & \overset{\simeq}{\longrightarrow} & \mathrm{Hom}_{A^\lrcorner}(\Delta\Sigma, \Sigma d_\lrcorner)
\end{array}
$$

Our task is to define the dashed diagonal morphism in such a way that we may apply Proposition 3.4.11 to argue that the diagram commutes up to fibered isomorphism. By the 2-of-6 property of the equivalences in an ∞-cosmos (see Remark 1.2.21 and Exercise 1.4.iii) it follows that the top horizontal map defines a fibered equivalence witnessing the fact that the cofiber sequence is also a fiber sequence.

To explain the construction of this map, it is helpful to give names to the generalized elements in the cofiber sequence $d : D \to A^{\boxdot}$ as depicted in the square below-left:

$$
\begin{array}{ccc}
a & \longrightarrow b & \longrightarrow * \\
\downarrow & \ulcorner \downarrow & \ulcorner \vdots \\
* & \longrightarrow c & \dashrightarrow \Sigma a
\end{array}
$$

Here $a : D \to A$ represents the generalized element $d_\top$. Since the cofiber sequence $a \to b \to c$ defines a pushout square, when we restrict to its right-hand edge $b \to c$ and form the cofiber, the resulting element is isomorphic to the suspension Σa by Proposition 4.3.11. In particular, we have an absolute left lifting diagram of the following form:

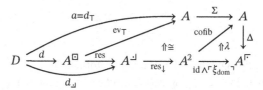

the natural transformation component of which is represented by a functor $\ulcorner \lambda \urcorner$ that we restrict to the ∞-category of cones over $d_{\lrcorner}$ for later use:

$$\mathrm{Hom}_{A^{\lrcorner}}(\Delta, d_{\lrcorner}) \xrightarrow{p_1} D \xrightarrow{\ulcorner \lambda \urcorner} A^{\ulcorner \times 2}$$

$$
\begin{array}{ccccc}
c & \longleftarrow & b & \longrightarrow & * \\
\downarrow & & \downarrow & & \downarrow \\
\Sigma a & = & \Sigma a & = & \Sigma a
\end{array}
$$

By Proposition 4.2.7, the comma ∞-category $\mathrm{Hom}_{A^{\lrcorner}}(\Delta, d_{\lrcorner})$ is equivalent to the pullback

$$
\begin{array}{ccc}
\mathrm{Hom}_{A^{\lrcorner}}(\Delta, d_{\lrcorner}) \simeq A_{/d_{\lrcorner}} & \xrightarrow{\ulcorner \phi \urcorner} & A^{\square} \\
{\scriptstyle (p_1, p_0)} \downarrow & & \downarrow {\scriptstyle (\mathrm{res}, \mathrm{ev}_T)} \\
D \times A & \xrightarrow{d_{\lrcorner} \times A} & A^{\lrcorner} \times A
\end{array}
$$

Here $\ulcorner \phi \urcorner$ defines a square as displayed below-left, where $z : \mathrm{Hom}_{A^{\lrcorner}}(\Delta, d_{\lrcorner}) \to A$ represents the generalized element $p_0 : \mathrm{Hom}_{A^{\lrcorner}}(\Delta, d_{\lrcorner}) \twoheadrightarrow A$ that projects to the summit of a cone over $d_{\lrcorner}$

$$
\begin{array}{ccc}
* & \longleftarrow & z \\
\downarrow & & \downarrow \\
c & \longleftarrow & b
\end{array}
\qquad\qquad
\begin{array}{ccccc}
* & \longleftarrow & z & \longrightarrow & * \\
\downarrow & & \downarrow & & \downarrow \\
c & \longleftarrow & b & \longrightarrow & *
\end{array}
$$

This square may be extended to the map of spans above-right by gluing on the square defined by the functor of co-points:

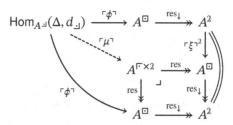

The diagrams $\ulcorner\mu\urcorner$ and $\ulcorner\lambda\urcorner$ glue together to form a diagram $\mathrm{Hom}_{A^{\lrcorner}}(\Delta, d_{\lrcorner}) \to A^{\ulcorner\times 3}$ as below

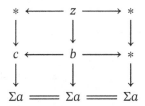

from which we extract a composite map of cospans defining the natural transformation below-left:

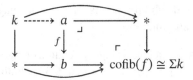

This factors through the absolute left lifting diagram of Definition 4.4.3 defining a natural transformation ζ. By 1-cell induction, ζ defines the sought-for functor $\ulcorner\zeta\urcorner : \mathrm{Hom}_{A^{\lrcorner}}(\Delta, d_{\lrcorner}) \to \mathrm{Hom}_{A}(\Sigma, \Sigma d_{\mathsf{T}})$ over $D \times A$, which completes the proof that the cofiber sequence $a \to b \to c$ is also a fiber sequence.

Now to show that every arrow $f : a \to b$ admits a fiber, start by forming its cofiber. Since Σ is an equivalence, there exists some element k of A so that $\mathrm{cofib}(f) \cong \Sigma k$. By what we have just proven, both of the cofiber sequences $a \to b \to \Sigma k$ and $k \to * \to \Sigma k$ are fiber sequences, and in particular the right-hand square below is both a pushout and a pullback.

$$
\begin{array}{ccc}
k \dashrightarrow a & \longrightarrow & * \\
\downarrow \quad f\downarrow \quad {}^{\lrcorner} & \ulcorner & \downarrow \\
* \longrightarrow b & \longrightarrow & \mathrm{cofib}(f) \cong \Sigma k
\end{array}
$$

Thus, the outer rectangle factors through the right-hand square and this composite rectangle, which is given as a pushout, is also a pullback. Now by pullback cancelation, the left-hand square is a pullback defining the fiber $k \cong \mathrm{fib}(f)$ of $f : a \to b$.

To see that the fiber sequence $k \to a \to b$ is also a cofiber sequence, form the cofiber sequence $k \to a \to c$. Our task is to show that the dashed map from the front face of the cube below-left to the back face is an isomorphism:

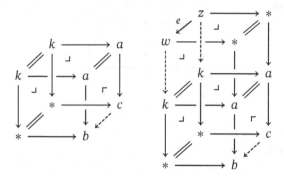

Since Σ is an equivalence, there exists an arrow $e : z \to w$ in A whose image under Σ is isomorphic to the induced map $c \to b$. The cofiber sequences $z \to * \to c$ and $w \to * \to b$ are also fiber sequences, so by pullback cancelation we see that the induced upper front and back squares in the prism above-left are pullbacks. From this we see that $e : z \to w$ is a map between two fibers of $k \to a$. Thus e is an isomorphism and the isomorphism $c \cong \Sigma z \cong \Sigma w \cong b$ reveals that the original fiber sequence $k \to a \to b$ is also a cofiber sequence.   □

REMARK 4.4.13. In fact, stable ∞-categories have all finite limits and finite colimits, meaning limits and colimits indexed by simplicial sets with finitely many nondegenerate simplices. More generally, any ∞-category with pullbacks and a terminal element admits all finite limits, which can be defined by induction over the dimension of the simplicial set (see [78, 4.4.2.4], [112, 6.3.9]). It follows from this construction and condition (iii) of Theorem 4.4.12 that in a stable ∞-category the limit and colimit functors for any finite diagram shape preserve all finite colimits and all finite limits.

DEFINITION 4.4.14. A pointed ∞-category A admits binary **direct sums** when there exists a bifunctor $\oplus : A \times A \to A$ that defines both the binary product and coproduct and so that the legs of the colimit and limit cones

$$
\begin{array}{ccc}
& & A \\
& \overset{\oplus}{\nearrow} & \Big\uparrow \Delta \\
A \times A & \underset{\Uparrow(\iota_1,\iota_2)}{=\!=\!=} & A \times A
\end{array}
\qquad
\begin{array}{ccc}
& & A \\
& \overset{\oplus}{\nearrow} & \Big\uparrow \Delta \\
A \times A & \underset{\Downarrow(\rho_1,\rho_2)}{=\!=\!=} & A \times A
\end{array}
$$

satisfy the following relations involving the zero map (see Exercise 4.4.ii):

$$\rho_1 \circ \iota_1 = \mathrm{id}, \quad \rho_1 \circ \iota_2 = 0, \quad \rho_2 \circ \iota_1 = 0, \quad \text{and} \quad \rho_2 \circ \iota_2 = \mathrm{id}$$

LEMMA 4.4.15. *Stable ∞-categories admit finite direct sums.*

Proof We argue that a stable ∞-category A has binary direct sums and leave it to Exercise 4.4.iii to extend Exercise 2.3.i to show that A has finite direct sums. The direct sum bifunctor can be defined as a pullback over the zero element:

$$\oplus : A \times A \xrightarrow{\ulcorner \xi_{\pi_1} \urcorner \vee \ulcorner \xi_{\pi_2} \urcorner \lrcorner} A^{\lrcorner} \xrightarrow{\lim} A$$

By Exercise 4.3.vii, this construction guarantees that direct sums are products.

To see that the direct sum is also a coproduct, consider the diagram $A \times A \to A^{\boxplus}$ whose image at a generalized element $(a, b) : X \to A \times A$ is depicted below:

$$
\begin{array}{ccccc}
* & \xrightarrow{\ulcorner \rho_a \urcorner} & a & \xrightarrow{\ulcorner \xi_a \urcorner} & * \\
{\scriptstyle \ulcorner \rho_b \urcorner} \downarrow & \lrcorner & {\scriptstyle \iota_a} \downarrow & \lrcorner & \downarrow {\scriptstyle \ulcorner \rho_b \urcorner} \\
b & \xrightarrow{\iota_b} & a \oplus b & \xrightarrow{\pi_b} & b \\
{\scriptstyle \ulcorner \xi_b \urcorner} \downarrow & \lrcorner & {\scriptstyle \pi_a} \downarrow & \lrcorner & \downarrow {\scriptstyle \ulcorner \xi_b \urcorner} \\
* & \xrightarrow{\ulcorner \rho_a \urcorner} & a & \xrightarrow{\ulcorner \xi_a \urcorner} & *
\end{array}
$$

Here the vertical and horizontal composite morphisms are identities, and the right-hand rectangle and lower rectangle are defined by restricting along a projection functor $2 \times 2 \to 2$ and then using the universal property of the lower-right-hand pullback to factor through that square. By Exercise 2.3.iv, each of these rectangles are themselves pullbacks so by Proposition 4.3.11, we see that the upper-right-hand and lower-left-hand squares so constructed are pullbacks as well, and thus so too is the upper-left-hand square, by the same reasoning. Since A is stable, this pullback is a pushout witnessing the fact that the direct sum is also a coproduct.

Note by construction that the composites of the coproduct inclusions and product projections are either identities or nullhomotopic. Thus, these biproducts are direct sums. □

DIGRESSION 4.4.16 (the homotopy category of a stable ∞-category). When an ordinary 1-category has binary direct sums, its hom-sets can be equipped with a canonically defined commutative monoid structure in such a way that composition defines a bilinear map. For a parallel pair of morphisms $f, g : x \to y$, their sum is defined to be the composite

$$f + g := x \xrightarrow{\Delta} x \oplus x \xrightarrow{f \oplus g} y \oplus y \xrightarrow{\nabla} y \qquad (4.4.17)$$

By Lemma 2.3.3, the zero element and finite direct sums in a stable ∞-category descend to define a zero element and finite direct sums on its homotopy category. In fact, the homotopy category hA of a stable ∞-category A is **additive**, meaning that the commutative monoids $hA(x, y)$ defined by (4.4.17) are in fact abelian groups (i.e., each morphism admits an additive inverse) [80, 1.1.2.9].

In fact, Lurie proves in [80, 1.1.2.14] that if A is a stable ∞-category, then its homotopy category hA is triangulated in the sense of Verdier [127]. A **triangulated category** is an additive category hA that admits a self-equivalence $\Sigma : hA \xrightarrow{\sim} hA$ together with specified **distinguished triangles**

$$ x \xrightarrow{f} y \xrightarrow{g} z \xrightarrow{h} \Sigma x $$

satisfying six axioms. From the vantage point of the 1-category, the distinguished triangles are additional data. In particular, there is no canonical way to define the distinguished triangles for a category of diagrams valued in a triangulated category. Lurie's insight is that this structure borne by the homotopy category may be captured by a property of the ∞-category, namely stability. He declares a triple (f, g, h) of morphisms in hA to be a **distinguished triangle** if there exist representing arrows (f, g, h) in A that assemble into a pushout rectangle of the following form:

$$
\begin{array}{ccccc}
x & \xrightarrow{f} & y & \longrightarrow & * \\
\downarrow & & \llcorner \;\downarrow{\scriptstyle g} & & \llcorner\;\downarrow \\
* & \longrightarrow & z & \xrightarrow[h]{} & \Sigma x
\end{array}
$$

Thus, at the level of the ∞-category A, there is an essentially unique way to extend an arrow $f : x \to y$ to a distinguished triangle. In particular, the famous "octahedral axiom" is a consequence of the composition and cancelation property for pushout rectangles of Proposition 4.3.11: given a composable pair of morphisms $k : w \to x$ and $f : x \to y$ in a stable ∞-category A, the diagram of pushout squares

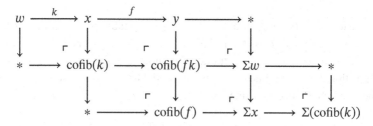

defines a distinguished triangle

$$ \mathrm{cofib}(k) \longrightarrow \mathrm{cofib}(fk) \longrightarrow \mathrm{cofib}(f) \longrightarrow \Sigma(\mathrm{cofib}(k)) $$

compatibly with given distinguished triangles extending k, f, and fk.

Exercises

EXERCISE 4.4.i. Arguing in the homotopy category, show that if an ∞-category A admits an initial element i and a terminal element t, and there exists an arrow $t \to i$, then A is a pointed ∞-category.

EXERCISE 4.4.ii. If A is a pointed ∞-category and $f, g : X \to A$ are functors define a canonical zero map

$$X \underset{g}{\overset{f}{\rightrightarrows}} {\Downarrow 0}\; A$$

that factors through the constant functor at the zero element.

EXERCISE 4.4.iii. Show that any pointed ∞-category that admits binary direct sums admits finite direct sums.

EXERCISE 4.4.iv. A functor $f : A \to B$ between stable ∞-categories is **exact** if it preserves zero elements as well as fiber and cofiber sequences. Show that exact functors also preserve exact squares.

5

Fibrations and Yoneda's Lemma

The aim in this chapter is to describe an ∞-categorical encoding of the contravariant functor represented by an element $b: 1 \to B$ of an ∞-category B, informally defined to send an element x of B to the mapping space $\text{Hom}_B(x, b)$ of Definition 3.4.9. By Proposition 3.4.10, such representable functors take values in discrete ∞-categories, which correspond to "spaces" or "∞-groupoids" in ∞-cosmoi of $(\infty, 1)$-categories.

In contrast with the situation in ordinary 1-category theory, in ∞-category theory it is challenging to explicitly describe the representable functors as functors between ∞-categories. The complexity arises in establishing the ∞-functoriality of this mapping, which must encode homotopy coherently functorial actions of the arrows in B in each dimension, data that proves too elaborate to easily enumerate even in this fundamental special case. What turns out to be easier to describe is the ∞-categorical analogue of the "category of elements" associated to the mapping $x \mapsto \text{Hom}_B(x, b)$ together with its associated projection to B. In fact, we are quite familiar with this ∞-category already: It is the "right represented" comma ∞-category $\text{Hom}_B(B, b)$ equipped with its domain projection functor $p_0 : \text{Hom}_B(B, b) \twoheadrightarrow B$ whose fiber over $x: 1 \to B$ recovers the mapping space $\text{Hom}_B(x, b)$.

It remains to explain the sense in which $p_0 : \text{Hom}_B(B, b) \twoheadrightarrow B$ expresses the contravariantly functorial of arrows in B on its fibers. This functor enjoys a special property that allows one to lift natural transformations valued in B in an essentially unique way to natural transformations valued in $\text{Hom}_B(B, b)$ with specified codomains. A special case of this lifting defines the precomposition functor f^* associated to an arrow $f : x \to y$ in the homotopy category hB:

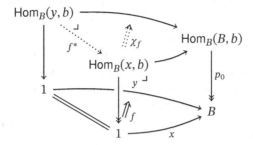

Roughly speaking, an isofibration $p : E \twoheadrightarrow B$ defines a *cartesian fibration* just when the arrows of B act contravariantly functorially on the fibers and a *cocartesian fibration* when the arrows of B act covariantly functorially on the fibers. The functor $p_0 : \mathrm{Hom}_B(B, b) \twoheadrightarrow B$ is an example of a *discrete cartesian fibration*, whose fibers are discrete ∞-categories. When $b : X \to B$ is a generalized element, the domain projection functor $p_0 : \mathrm{Hom}_B(B, b) \twoheadrightarrow B$ remains a cartesian fibration, but loses this discreteness property.

One of the properties that characterizes a cartesian fibration $p : E \twoheadrightarrow B$ is an axiom that says that for any 2-cell with codomain B and specified lift of its target 1-cell, there is a lifted 2-cell with codomain E with that 1-cell as its target. In particular, this lifting property can be applied in the case where the 2-cell in question is a whiskered composite of an arrow in the homotopy category of B as below-left and the lift of the source 1-cell is the canonical inclusion of the fiber over its codomain.

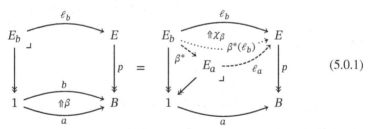

$$(5.0.1)$$

In this case the domain $\beta^*(\ell_a)$ of the lifted cell χ_β displayed above right lies strictly above the codomain of the original 2-cell, and thus factors through the pullback defining its fiber. This defines a functor $\beta^* : E_b \to E_a$, the "action" of the arrow β on the fibers of p. This action is not strict but rather functorial up to isomorphism in a sense explored in Exercise 5.2.ii.[1] The up-to-isomorphism

---

[1] Considerably more is true: There is a contravariant homotopy coherent diagram indexed by the underlying quasi-category of B and valued in the Kan complex enriched category of discrete ∞-categories [111, 6.1.16].

functoriality of these action maps arises from a universal property required of the specified lifted 2-cells, namely that they are *cartesian arrows* in a sense we define in §5.1.

Cartesian fibrations are introduced in §5.2, where the first examples are also established. The main theorem in this section characterizes cartesian fibrations in terms of the presence of adjoints to certain canonically defined functors. The structure-preserving maps between cartesian fibrations are commutative squares called *cartesian functors*, preserving the cartesian natural transformations. In §5.3, we see that these can similarly be characterized relative to the adjunctions constructed in §5.2.

In §5.4 we study the dual cocartesian fibrations, for which there exist lifts of natural transformations with a specified domain functor. By a dual of the construction displayed in (5.0.1), when $p : E \twoheadrightarrow B$ is a cocartesian fibration an arrow $\beta : a \to b$ in hB defines a functor $\beta_* : E_a \to E_b$. An isofibration p that is simultaneously a cartesian fibration and a cocartesian fibration is called a *bifibration*.[2] In this case Proposition 5.4.7 proves that the induced functors $\beta_* \dashv \beta^*$ are adjoints.

In §5.6, we show that when $p : E \twoheadrightarrow B$ is a cartesian fibration in an ∞-cosmos the induced functor $p_* : \mathrm{Fun}(X, E) \twoheadrightarrow \mathrm{Fun}(X, B)$ defines a cartesian fibration of quasi-categories. The converse also holds, under the additional condition that restriction along any $f : Y \to X$ defines a cartesian functor.

The special classes of *discrete cartesian fibrations* and *discrete cocartesian fibrations* are studied in §5.5. This chapter concludes in §5.7 with a version of the Yoneda lemma for the discrete cartesian fibration $p_0 : \mathrm{Hom}_B(B, b) \twoheadrightarrow B$ that is represented by the element $b : 1 \to B$. Its formulation was inspired by a paper of Street "Fibrations and Yoneda's lemma in a 2-category" [118],[3] the debt to which we acknowledge in the title of this section.

5.1 Cartesian Arrows

Before defining the notion of cartesian fibration we describe the weak universal property enjoyed by certain "upstairs" natural transformations. Recall from

Proposition 3.2.6 that any natural transformation $X \underset{e}{\overset{e'}{\rightrightarrows}} \Downarrow\psi \; E$ is represented by a functor $\ulcorner\psi\urcorner : X \to E^2$ so that $p_0\ulcorner\psi\urcorner = e'$ and $p_1\ulcorner\psi\urcorner = e$ and such

---

[2] In [78, §2.4.7], Lurie uses the term "bifibration" to refer to a different class of functors: the *modules* of Chapter 7.

[3] The closest analogue to Street's Yoneda Lemma [118, 16] appears as Theorem 7.4.8.

representations are unique up to fibered isomorphism. The representing X-shaped arrow $\ulcorner\psi\urcorner$ in E defines a 1-arrow (aka a 1-simplex) in the functor space $\mathsf{Fun}(X, E)$ from e' to e.

Definition 5.1.1 (*p*-cartesian arrow). Consider an isofibration $p : E \twoheadrightarrow B$. An X-shaped arrow $\ulcorner\psi\urcorner : X \to E^2$ in E is p-**cartesian** if the dashed map defined by the pullback of the Leibniz cotensor is a trivial fibration:

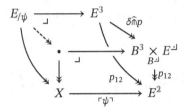

where $\delta : \lrcorner \hookrightarrow 3$ is the inclusion whose image is the cospan $0 \to 2 \leftarrow 1$ in 3 and p_{12} is defined by restricting along the inclusion $i_{12} : 2 \hookrightarrow 3$ with image indicated by the subscript.

A natural transformation $X \underset{e}{\overset{e'}{\rightrightarrows}} \Downarrow\psi \; E$ with codomain E is p-**cartesian** if any representing functor $\ulcorner\psi\urcorner : X \to E^2$ is p-cartesian. By Exercise 3.3.i, this is well-defined. We freely switch between the perspectives presented by a natural transformation $\psi : e' \Rightarrow e$ and a representing arrow $\ulcorner\psi\urcorner : X \to E^2$.[4]

There are various equivalent ways to formulate the definition of p-cartesian arrows.

Lemma 5.1.2. *For any isofibration $p : E \twoheadrightarrow B$, an arrow $\ulcorner\psi\urcorner : X \to E^2$ with codomain $e : X \to E$ is p-cartesian if and only if the dashed square induced by*

---

[4] We also indulge in some streamlined notation, writing $E_{/\psi}$ for the ∞-category defined in Warning 4.2.10 using the notation $E_{/\ulcorner\psi\urcorner}$, and writing $\ulcorner p\psi\urcorner$ and $\ulcorner\psi f\urcorner$ for the composite arrows $p^2\ulcorner\psi\urcorner$ and $\ulcorner\psi\urcorner f$.

the hypercube

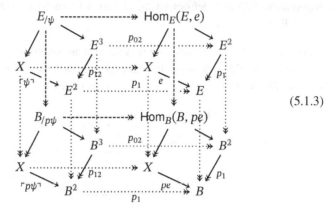

$$(5.1.3)$$

is weakly cartesian, meaning that the induced map is a trivial fibration:

$$E_{/\psi} \xrightarrow{\ \sim\ } B_{/p\psi} \underset{\mathrm{Hom}_B(B,pe)}{\times} \mathrm{Hom}_E(E,e)$$

Proof   In fact the induced map in the dashed square is isomorphic to the induced map of Definition 5.1.1. First note that the solid-arrow squares in the hypercube (5.1.3) are pullbacks, so by the hypercube pullback lemma[5] the pullback in the dashed square is equally the limit of the diagram

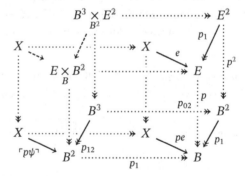

where now the dotted-arrow squares are the pullbacks. Thus, our map is isomorphic to the top dashed map in the cube in which the front and back faces are the

---

[5]   The hypercube pullback lemma observes that the limit of a diagram of shape $\lrcorner \times \lrcorner$ can be computed by first forming the pullbacks in the left factor and then forming the pullback of the resulting cospan, or by first forming the pullbacks in the right factor and then forming the pullback of the resulting cospan.

pullbacks defining its domain and codomain:

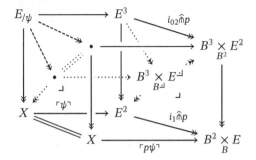

The induced map to the pullback in the right face is $\delta \,\widehat{\pitchfork}\, p : E^3 \twoheadrightarrow B^3 \times_{B^\lrcorner} E^\lrcorner$, as can be verified by constructing another hypercube. By pullback composition and cancellation, this map pulls back to the dashed map in the left face, but since the bottom edge is an identity, this agrees with the top dashed map. Note the diagram displayed in Definition 5.1.1 is embedded as the back prism in this cube. □

The next several lemmas develop various stability properties for the class of p-cartesian transformations defined relative to a fixed isofibration $p : E \twoheadrightarrow B$.

LEMMA 5.1.4 (stability under restriction). *If $\ulcorner \psi \urcorner : X \to E^2$ is p-cartesian then so is its restriction along any functor $f : Y \to X$.*

Proof By Definition 5.1.1, if ψ is p-cartesian we have a trivial fibration

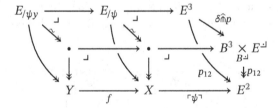

that pulls back to define a trivial fibration which exhibits $\ulcorner \psi f \urcorner$ as a p-cartesian arrow. □

We now demonstrate that the class of p-cartesian transformations is closed under vertical composition and left cancelation in the homotopy 2-category.

LEMMA 5.1.5. *Let $p : E \twoheadrightarrow B$ be an isofibration and consider natural transformations so that $\psi'' = \psi \cdot \psi'$ and so that ψ is p-cartesian. Then ψ' is p-cartesian if and only if ψ'' is p-cartesian.*

Proof By Lemma 1.1.12, for any triple of arrows $\ulcorner\psi''\urcorner, \ulcorner\psi'\urcorner, \ulcorner\psi\urcorner : X \to E^3$ representing natural transformations so that $\psi'' = \psi \cdot \psi'$, there exists a 2-simplex in $\mathsf{Fun}(X, E)$ represented by a functor $\ulcorner\tau\urcorner : X \to E^3$ with boundary given by these specified representative functors:

$$
\begin{array}{ccc}
 & \overset{e'}{} & \\
\psi' \nearrow & \tau & \searrow \psi \\
e'' & \xrightarrow[\psi'']{} & e
\end{array}
$$

By cotensoring the lower-left diagram of categories into E, restricting to the complements of the initial vertices $0 \in \mathfrak{m}$, and pulling back along $\ulcorner\tau\urcorner : X \to E^3$ and its face, we obtain the lower-right diagram of ∞-categories, in which the bottom square is weakly cartesian (see Lemma 4.3.9 for more details about the construction of the outer spans and a proof that the maps p_{01} are trivial fibrations).

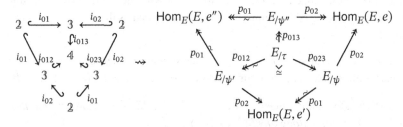

The isofibration $p : E \twoheadrightarrow B$ defines a natural transformation from this diagram onto a similar diagram built for $\ulcorner p\tau\urcorner : X \to B^3$.[6]

The trivial fibration $E_{/\psi} \twoheadrightarrow B_{/p\psi} \times_{\mathsf{Hom}_B(B,pe)} \mathsf{Hom}_E(E, e)$ of Lemma 5.1.2 pulls back to define the displayed dashed trivial fibration that commutes with the induced maps from $E_{/\tau}$.

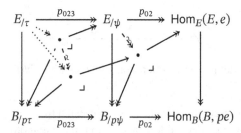

By the 2-of-3 property for equivalences, if either dotted map is a trivial fibration, then both are.

---

[6] A better way to build this diagram is to implement this construction in the ∞-cosmos of isofibrations defined in Proposition 6.1.1, which tells us additionally that the maps to the pullbacks in each naturality square are isofibrations.

These diagrams induce a commutative square displayed in the interior of the cube from the upper dotted map to the map that detects whether ψ' is p-cartesian

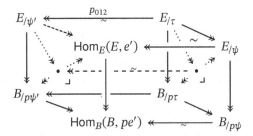

By the 2-of-3 property, if either of these maps is a trivial fibration, so is the other. A similar square defined using $p_{013} : E_{/\tau} \twoheadrightarrow E_{/\psi''}$ demonstrates that the lower dotted map is a trivial fibration if and only if $E_{/\psi''} \twoheadrightarrow B_{/p\psi''} \times_{\mathrm{Hom}_B(B,pe)} \mathrm{Hom}_E(E, e)$ is. Thus, these conditions are equivalent when ψ is p-cartesian. $\square$

The isomorphism stability of the p-cartesian arrows is expressed by the following suite of observations:

LEMMA 5.1.6. *Let* $p : E \twoheadrightarrow B$ *be an isofibration.*

(i) Natural isomorphisms with codomain E define p-cartesian arrows.

(ii) Any p-cartesian lift of a natural isomorphism is a natural isomorphism.

(iii) The class of p-cartesian arrows is closed under pre- and postcomposition with natural isomorphisms.

Proof By Lemma 4.3.9, $\psi : e' \Rightarrow e$ is an isomorphism if and only if the map $p_{02} : E_{/\psi} \twoheadrightarrow \mathrm{Hom}_E(E, e)$ that defines the top horizontal arrow in the weakly cartesian square of Lemma 5.1.2 is a trivial fibration. So if ψ is invertible, then both horizontal arrows of Lemma 5.1.2 are trivial fibrations, and the square is automatically weakly cartesian, proving (i). For (ii), if $p\psi$ is invertible, then the bottom horizontal in this square is a trivial fibration, and thus if ψ is weakly cartesian, $p_{02} : E_{/\psi} \twoheadrightarrow \mathrm{Hom}_E(E, e)$ is a composite of trivial fibrations, proving that ψ is invertible. The final property (iii) follows from (i) and Lemma 5.1.5. $\square$

We get considerable mileage from two more sophisticated characterizations of p-cartesian arrows.

THEOREM 5.1.7. *For an isofibration* $p : E \twoheadrightarrow B$ *and an arrow* $\ulcorner\psi\urcorner : X \to$ E^2 *representing a natural transformation* $X \overset{e'}{\underset{e}{\Longrightarrow}} E$ *the following are equivalent:*

(i) ψ *is p-cartesian.*

(ii) *The commutative triangle defines an absolute right lifting diagram:*

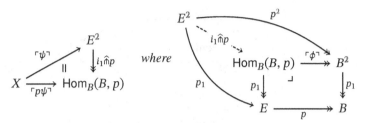

(iii) *There is an absolute right lifting diagram with* $p_1\epsilon = \psi$ *and* $p_0\epsilon = \mathrm{id}_{pe'}$

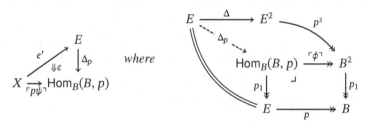

At a high level, the equivalence between the three characterizations of a p-cartesian arrow is easy to explain. By Definition 5.1.1 and Theorem 3.5.3, each of three statements asserts that some map between ∞-categories is an equivalence, and these three maps turn out to be equivalent to each other. But the geometry of this equivalence is quite subtle, as the proof reveals.

Proof   We prove (i)$\Rightarrow$(ii)$\Rightarrow$(iii)$\Rightarrow$(i).

(i)$\Rightarrow$(ii): We use the condition of Definition 5.1.1 to prove that $\ulcorner\psi\urcorner\colon X \to E^2$ defines an absolute right lifting of $\ulcorner p\psi\urcorner$ through $i_1 \hat{\pitchfork} p$. By Theorem 3.5.3, our task is to show that the functor induced by the identity 2-cell defines an equivalence between comma ∞-categories. In this case, the desired map, displayed below-left is a pullback of the Leibniz cotensor of p with the inclusion $\iota\colon \sqcup \hookrightarrow \boxdot$

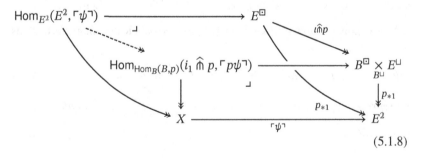

$$(5.1.8)$$

Since the inclusion ι factors as below-left, by Proposition C.2.9(vi), the Leibniz cotensor factors as below-right:

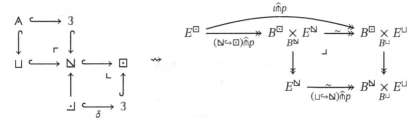

and since $\sqcup \hookrightarrow \boxtimes$ is a pushout of an inner horn inclusion, the second of these maps is a trivial fibration. Since $\boxtimes \hookrightarrow \Box$ is a pushout of δ, $(\boxtimes \hookrightarrow \Box) \,\hat{\pitchfork}\, p$ is a pullback of $\delta \,\hat{\pitchfork}\, p$. From hypothesis (i), we know that the latter map pulls back along $\ulcorner\psi\urcorner \colon X \to E^2$ to a trivial fibration. Thus the former map does as well, and so the dashed map in (5.1.8) is a trivial fibration, proving the claimed absolute lifting diagram in (ii).

(ii)⇒(iii): The constant diagram functor $\Delta \colon E \to E^2$ is defined by restricting along the functor $! \colon 2 \to 1$, and thus is left adjoint right inverse to the domain projection functor p_0:

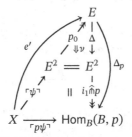

In particular, by Lemma 2.3.7 the counit $\nu \colon \Delta p_0 \Rightarrow$ id defines an absolute right lifting of the identity through Δ. By Lemma 2.4.1, this absolute lifting diagram composes with the absolute lifting diagram of (ii) to define an absolute lifting diagram with the properties required by (iii):

$$
\begin{array}{c}
E \\
\end{array}
$$

$X \xrightarrow[\ulcorner p\psi\urcorner]{} \mathrm{Hom}_B(B, p)$

(iii)⇒(i): By Theorem 3.5.3, an absolute right lifting diagram, such as given in (iii), supplies a fibered equivalence. For a generic absolute right lifting diagram as below-left, the fibered equivalence $\mathrm{Hom}_B(B, r) \twoheadrightarrow_{C \times B} \mathrm{Hom}_A(f, g)$ may be constructed by composing the dashed maps in the diagram below-right, constructing this map as a restriction of the dotted composition map (see Lemma 2.1.14):

$$
\begin{array}{ccc}
& B & \\
\overset{r}{\nearrow} & \downarrow f & \rightsquigarrow \\
C \xrightarrow{\ g\ } A &
\end{array}
\qquad
\begin{array}{ccc}
& & \mathrm{Hom}_A(f,g) \overset{(p_1,p_0)}{\twoheadrightarrow} C \times B \\
& & \downarrow \quad \downarrow g \times f \\
A^3_{\langle \rho, f(-)\rangle} \longrightarrow A^3 \xrightarrow{p_{02}} A^2 \longrightarrow A \times A \\
\uparrow \quad \downarrow & & \uparrow (p_{12}, p_{01}) \\
\mathrm{Hom}_B(B,r) \cong C \underset{C}{\times} \mathrm{Hom}_B(B,r) \xrightarrow[\ulcorner \rho \urcorner \times f^2]{fr} A^2 \underset{A}{\times} A^2
\end{array}
$$

By the 2-of-3 property, the restriction map $\circ : A^3_{\langle \rho, f(-)\rangle} \twoheadrightarrow \mathrm{Hom}_A(f,g)$ is an equivalence.

Applied to the absolute right lifting diagram of (iii), this constructs a trivial fibration

$$
\mathrm{Hom}_B(B,p)^3_{\langle \epsilon, \Delta_p(-)\rangle} \xrightarrow{\ p_{02}\ } \mathrm{Hom}_{\mathrm{Hom}_B(B,p)}(\Delta_p, \ulcorner p\psi \urcorner) \tag{5.1.9}
$$

The domain of (5.1.9) is the limit of the diagram, computed by first pulling back the rows then forming the pullback of the resulting cospan

$$
\begin{array}{ccccc}
B^{2\times 3} & \xrightarrow{\ p_{1*}\ } & B^3 & \xleftarrow{\ p^3\ } & E^3 \\
\downarrow & & \downarrow & & \downarrow \\
B^{2\times 2} \underset{B^2}{\times} B^{2\times 2} & \xrightarrow[p_1]{p_{1*}\times p_{1*}} & B^2 \underset{B}{\times} B^2 & \xleftarrow{\ p^\Lambda\ } & E^2 \underset{E}{\times} E^2 \\
{\scriptstyle \ulcorner \epsilon \urcorner \times \Delta}\uparrow & & {\scriptstyle \ulcorner p\psi \urcorner \times \mathrm{id}}\uparrow & & \uparrow{\scriptstyle \ulcorner \psi \urcorner \times \mathrm{id}} \\
X \underset{B}{\times} B^2 & =\!=\!=\!= & X \underset{B}{\times} B^2 & \xleftarrow{\ p\ } & X \underset{E}{\times} E^2
\end{array}
$$

but by the hypercube pullback lemma it could equally be formed as the pullback of the induced cospan between the pullbacks of the columns:

$$
\mathrm{Hom}_B(B,pe') \underset{B^{2\times A}}{\times} B^{2\times 3} \xrightarrow{\ p_{1*}\ } B_{/p\psi} \xleftarrow{\ p\ } E_{/\psi}
$$

and thus we seek a better understanding of the pullback of the left-hand column.

Since $\epsilon : \Delta_p e' \Rightarrow \ulcorner p\psi \urcorner$ has $p_0\epsilon = \mathrm{id}_{pe'}$, we may choose a representing functor $\ulcorner \epsilon \urcorner$ that factors through the pullback

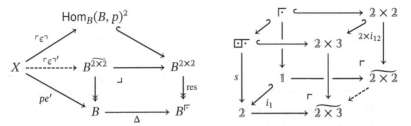

defined by degenerating the initial cospan in the simplicial set 2×2 as indicated by the pushout of simplicial sets in the back face of the cube. Pulling back along $\Delta : B^2 \to B^{2 \times 2}$ implements a similar quotienting in the left-hand square of 2×3, as illustrated in the front face of the cube, where $s : \boxdot \to 2$ sends the left two objects to 0 and the right three objects to 1. Thus, we see that the map (5.1.9) is defined by the pullback of the map q in the following diagram along $\ulcorner \epsilon \urcorner'$.

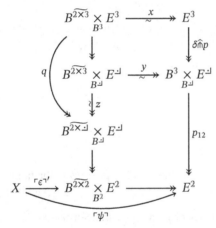

We claim that the maps labeled x, y, and z are all trivial fibrations. It will then follow from the 2-of-3 property that if the pullback of q along $\ulcorner \epsilon \urcorner'$ is an equivalence, as given by the absolute right lifting diagram of (iii), then the pullback of $\delta \widehat{\pitchfork} p$ along $\ulcorner \psi \urcorner$ is an equivalence, proving (i).

The trivial fibrations x and y are both pullbacks of the restriction map $B^{\widetilde{2 \times 3}} \twoheadrightarrow B^3$, which we show is a trivial fibration. The is a sequence of adjunctions between the categories $3 \hookrightarrow \boxed{\nearrow} \hookrightarrow 2 \times 3$ that become simplicial homotopy equivalences between the quotient simplicial sets:

$$\left\{ 0 \to 1 \to 2 \right\} \rightleftarrows_{\perp} \left\{ \begin{array}{c} 0 \to 1 \\ \| \searrow \| \searrow \\ 0 \to 1 \to 2 \end{array} \right\} \rightleftarrows_{\perp} \left\{ \begin{array}{c} 0 \to 1 = 1 \\ \| \searrow \| \searrow \downarrow \\ 0 \to 1 \to 2 \end{array} \right\}$$

Thus, restriction along the composite inclusion defines an equivalence $B^{\widetilde{2\times3}} \twoheadrightarrow B^3$. The map z is a pullback of $B^{\widetilde{2\times3}} \twoheadrightarrow B^{\widetilde{2\times\lrcorner}}$, which is a trivial fibration because the inclusion $\widetilde{2 \times \lrcorner} \hookrightarrow \widetilde{2 \times 3}$ can be filled by "special outer horns" (see Theorem D.5.1 and Corollary D.3.12).

In this way we see that (5.1.9) is equivalent to the map of Definition 5.1.1, and thus we see that ψ is p-cartesian, proving that (iii)$\Rightarrow$(i). $\square$

Observe that the data of the cospan underlying the weakly cartesian square (5.1.3) is determined by a functor $\ulcorner\beta\urcorner : X \to \mathrm{Hom}_B(B, p)$ representing a natural transformation $\beta : b \Rightarrow pe$ whose codomain factors through p along a specified functor $e : X \to E$. Thus it is relevant to ask whether a particular arrow of this form has a p-cartesian lift with codomain e.

DEFINITION 5.1.10 (p-cartesian lifts). Consider an isofibration $p : E \twoheadrightarrow B$. An arrow $\ulcorner\psi\urcorner : X \to E^2$ is said to **lift** an arrow $\ulcorner\beta\urcorner : X \to \mathrm{Hom}_B(B, p)$ if the triangle below-left commutes:

which gives rise to the pasting equality $\beta = p\psi$ in the homotopy 2-category.[7] When $\ulcorner\psi\urcorner$ is p-cartesian, we say it defines a p-**cartesian lift** of $\ulcorner\beta\urcorner$.

The universal property of p-cartesian transformations implies that the natural transformations represented by any two p-cartesian lifts of a common natural transformation $\beta : b \Rightarrow pe$ are fibered isomorphic:

LEMMA 5.1.11 (uniqueness of cartesian lifts). *If the natural transformations*

$$X \underset{e}{\overset{e'}{\rightrightarrows}} \Downarrow\psi \; E \quad \text{and} \quad X \underset{e}{\overset{e''}{\rightrightarrows}} \Downarrow\psi' \; E \quad \text{are p-cartesian lifts of a common natural}$$

transformation $\beta : b \Rightarrow pe$, *then there exists an invertible natural transformation*

$$X \underset{e'}{\overset{e''}{\rightrightarrows}} \cong\Downarrow\zeta \; E \quad \text{so that } \psi' = \psi \cdot \zeta \text{ and } p\zeta = \mathrm{id}_b.$$

---

[7] It makes no essential difference whether the lifting property is phrased in terms of 2-cells in the homotopy 2-category or 1-arrows in the functor spaces of the ∞-cosmos: see Exercise 5.1.i.

Proof If both ψ and ψ' define p-cartesian lifts of a common arrow β, then by Theorem 5.1.7(iii) we have a pair of absolute right liftings (e', ϵ) and (e'', ϵ') of $\ulcorner\beta\urcorner$ through Δ_p with $p_1\epsilon = \psi$, $p_1\epsilon' = \psi'$, and $p_0\epsilon = \mathrm{id}_b = p_0\epsilon'$. By uniqueness of absolute right lifting diagrams, this induces a natural isomorphism

so that $\epsilon' = \epsilon \cdot \Delta_p\zeta$. Whiskering this equation with p_1 we see that $\psi' = \psi \cdot \zeta$, and whiskering with p_0 we see that $\mathrm{id}_b = p\zeta$. $\qquad\square$

Combining these results, we obtain a useful conservativity property:

LEMMA 5.1.12 (cartesian conservativity). *Suppose we have* $X \underset{e}{\overset{e'}{\rightrightarrows}}{\Downarrow\psi}\, E$,

$X \underset{e}{\overset{e''}{\rightrightarrows}}{\Downarrow\psi'}\, E$, *and* $X \underset{e'}{\overset{e''}{\rightrightarrows}}{\Downarrow\zeta}\, E$ *which are natural transformations such that ψ and ψ' are p-cartesian, $\psi' = \psi \cdot \zeta$, and $p\zeta$ is invertible. Then ζ is invertible.*

Proof By Lemmas 5.1.5 and 5.1.6, ζ is a p-cartesian lift of a natural isomorphism and hence must be invertible. $\qquad\square$

The universal property that characterizes the p-cartesian transformations in Theorem 5.1.7 gives rise to induction and conservativity operations at the level of the homotopy 2-category, analogously to those operations considered in Chapter 3.

PROPOSITION 5.1.13 (the weak universal property of a p-cartesian arrow). *Let*

$p : E \twoheadrightarrow B$ *be an isofibration. A p-cartesian arrow* $X \underset{e}{\overset{e'}{\rightrightarrows}}{\Downarrow\psi}\, E$ *has a weak universal property in the homotopy 2-category given by two operations:*

(i) **induction**: *Given any 2-cells* $X \underset{e}{\overset{e''}{\rightrightarrows}}{\Downarrow\tau}\, E$ *and* $X \underset{pe'}{\overset{pe''}{\rightrightarrows}}{\Downarrow\gamma}\, B$ *such* *that* $p\tau = p\psi \cdot \gamma$, *there exists a lift* $X \overset{e''}{\underset{e'}{\rightrightarrows}} \Downarrow\bar{\gamma}\, E$ *of* γ *so that* $\tau = \psi \cdot \bar{\gamma}$.

$$
\begin{array}{ccc}
\begin{array}{c}
e'' \overset{\tau}{\Longrightarrow} e \\
\bar{\gamma} \nwarrow \; \nearrow \psi \\
e'
\end{array} & \in & \mathrm{hFun}(X, E) \\
\Downarrow & & \Big\downarrow {p_*} \\
\begin{array}{c}
pe'' \overset{p\tau}{\Longrightarrow} pe \\
\gamma \nwarrow \; \nearrow p\psi \\
pe'
\end{array} & \in & \mathrm{hFun}(X, B)
\end{array}
$$

(ii) **conservativity:** *Any fibered endomorphism of* ψ *is invertible: if* $\zeta : e' \Rightarrow$ e' *is any natural transformation so that* $\psi \cdot \zeta = \psi$ *and* $p\zeta = \mathrm{id}_{pe'}$ *then* ζ *is invertible.*

Proof The conservativity property (ii) is a special case of the conservativity result observed in Lemma 5.1.12 so it remains to prove (i).

The pair $\tau : e'' \Rightarrow e$ and $\gamma : pe'' \Rightarrow pe'$ satisfy the compatibility condition required by Proposition 3.4.6 to induce a natural transformation σ as below-left satisfying $p_1\sigma = \tau$ and $p_0\sigma = \gamma$. Since ψ is p-cartesian, this 2-cell factors through the absolute right lifting diagram of Theorem 5.1.7(iii)

$$
\begin{array}{ccc}
\begin{array}{c}
 \overset{E}{} \\
e'' \nearrow \; \Big\downarrow \Delta_p \\
\Downarrow\sigma \\
X \underset{\ulcorner p\psi \urcorner}{\longrightarrow} \mathrm{Hom}_B(B, p)
\end{array}
& = &
\begin{array}{c}
e'' \overset{}{\longrightarrow} E \\
\Downarrow\bar{\gamma} \nearrow \; \Big\downarrow \Delta_p \\
e' \; \Downarrow\epsilon \\
X \underset{\ulcorner p\psi \urcorner}{\longrightarrow} \mathrm{Hom}_B(B, p)
\end{array}
\end{array}
$$

Since (e', ϵ) is absolute right lifting, (e'', σ) induces a natural transformation $\bar{\gamma} : e'' \Rightarrow e'$ so that $\sigma = \epsilon \cdot \Delta_p\bar{\gamma}$. Whiskering these equations with p_1 we see that $\tau = \psi \cdot \bar{\gamma}$ and whiskering with p_0 we see that $\gamma = p\bar{\gamma}$. $\square$

The universal properties enumerated by Proposition 5.1.13 are considerably weaker than that expressed by Definition 5.1.1. Indeed they do not express the full conservativity observed in Lemma 5.1.12 nor do they take advantage of the restriction stability of cartesian transformations and absolute right lifting diagrams. Nevertheless, conditions (i) and (ii) suffice to characterize the class of p-cartesian transformations under the condition that p is a *cartesian fibration*, a concept that is introduced in the next section. Even more surprisingly, if (i) and (ii) are enhanced by a restriction stability property, then Proposition 5.2.11 demonstrates that it is possible to define cartesian fibrations entirely from the perspective of the homotopy 2-category, without referencing Definition 5.1.1. Since, however, p-cartesian arrows are of interest in their own right even in cases where p is not itself a cartesian fibration (see Exercise 5.1.ii for instance), we de-emphasize the purely 2-categorical development of the theory of cartesian fibrations and instead refer the reader to [110].

Exercises

EXERCISE 5.1.i. Recall that arrows $\ulcorner\psi\urcorner : X \to E^2$ in an ∞-cosmos correspond to 1-simplices (aka 1-arrows) $\psi : e' \to e$ in the functor space $\mathrm{Fun}(X, E)$, and a parallel pair of 1-arrows represents the same 2-cell $X \overset{e'}{\underset{e}{\rightrightarrows}} E$ if and only if they are homotopic, bounding a 2-arrow in $\mathrm{Fun}(X, E)$ whose 0th or 2nd edge is degenerate (see Exercise 3.2.i).

Show that if $p : E \twoheadrightarrow B$ is an isofibration so that every natural transformation $\beta : b \Rightarrow pe$ admits a lift $\psi : e' \Rightarrow e$ in the homotopy 2-category, then every 1-arrow $\beta : b \to pe$ in $\mathrm{Fun}(X, B)$ admits a 1-arrow lift $\psi : e' \to e$ in the sense of Definition 5.1.10.

EXERCISE 5.1.ii (5.2.10).

(i) Characterize the cartesian arrows for the codomain projection functor $p_1 : A^2 \twoheadrightarrow A$.
(ii) Use your answer to (i) to give an alternate proof of Proposition 4.3.11.

5.2 Cartesian Fibrations

Cartesian fibrations between ∞-categories generalize the *Grothendieck fibrations* between ordinary 1-categories. This notion was first extended to quasi-categories by Joyal [62] and Lurie [78] and to complete Segal spaces by Boavida de Brito [34] and Rasekh [98, 97]. Cartesian fibrations between $(\infty, 1)$-categories have been studied model independently by Mazel–Gee [86] and Ayala–Francis [5].

DEFINITION 5.2.1 (cartesian fibration). An isofibration $p : E \twoheadrightarrow B$ is a **cartesian fibration** if any natural transformation $\beta : b \Rightarrow pe$ as below-center admits a p-cartesian lift $\chi_\beta : \beta^*e \Rightarrow e$ as below-right:

By Exercise 5.1.i, it makes no difference whether we express the lifting property in the homotopy 2-category as displayed above-right, or in terms of the representing arrows as above-left.

REMARK 5.2.2. A guiding moral principle of ∞-category theory is that all ∞-categorical notions should be equivalence invariant, but if E and B are replaced by equivalent ∞-categories E' and B' the equivalent functor $p' : E \to B$ is not necessarily an isofibration. However, by Lemma 1.2.19, p' is equivalent to an isofibration p'' over B', so we could declare p' to be a cartesian fibration just when p'' is an isofibration in the sense of Definition 5.2.1. By Corollary 5.3.1, this definition is now equivalence invariant. For technical reasons, such as Proposition 6.3.14, we prefer to leave Definition 5.2.1 as it is.

Certain stability properties of cartesian fibrations can be proven directly from this definition.

LEMMA 5.2.3. *If $p : E \twoheadrightarrow B$ and $q : B \twoheadrightarrow A$ are cartesian fibrations, then so is $qp : E \twoheadrightarrow A$. Moreover, a natural transformation* $X \underset{e}{\overset{e'}{\rightrightarrows}} {\scriptstyle\Downarrow\psi}\ E$ *is qp-cartesian if and only if ψ is p-cartesian and $p\psi$ is q-cartesian.*

Proof The first claim follows immediately from the second, for the required lifts can be constructed by first taking a q-cartesian lift χ_α and then taking a p-cartesian lift χ_{χ_α} of this lifted cell.

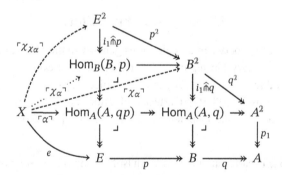

To prove the second claim, first consider a natural transformation $\psi : e' \Rightarrow e$ that is p-cartesian and so that $p\psi$ is q-cartesian. By Lemma 5.1.2, these properties are expressed by the dashed trivial fibrations, the latter of which pulls back to define the dotted trivial fibration:

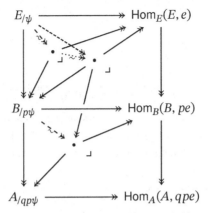

Since trivial fibrations compose, this tells us that $E_{/\psi} \twoheadrightarrow A_{/qp\psi} \times_{\mathrm{Hom}_A(A,qpe)} \mathrm{Hom}_E(E,e)$ is a trivial fibration, and thus ψ is qp-cartesian.

Conversely, if ψ is qp-cartesian, then Lemma 5.1.11 implies it is isomorphic to all other qp-cartesian lifts of $qp\psi$. The construction given above produces a qp-cartesian lift of any 2-cell that is p-cartesian and whose image under p is q-cartesian. By the isomorphism stability of p- and q-cartesian transformations of Lemma 5.1.6, ψ must also have these properties. $\qquad\square$

PROPOSITION 5.2.4 (pullback stability). *In any pullback square*

$$\begin{array}{ccc} F & \overset{h}{\longrightarrow} & E \\ q\downarrow & \lrcorner & \downarrow p \\ A & \underset{k}{\longrightarrow} & B \end{array}$$

if p is a cartesian fibration then q is a cartesian fibration. Moreover, a natural transformation ψ with codomain F is q-cartesian if and only if $h\psi$ is p-cartesian.

Proof The pullback square in the statement induces a pullback square between the Leibniz cotensors of $i_1 \colon \mathbb{1} \hookrightarrow \mathbb{2}$ with q and p. Consider an arrow $\ulcorner\alpha\urcorner \colon X \to \mathrm{Hom}_A(A,q)$ representing a natural transformation $\alpha \colon a \Rightarrow qf$. Since p is a cartesian fibration, the natural transformation $k\alpha \colon ka \Rightarrow kqf = phf$ has a p-cartesian lift $\ulcorner\tau\urcorner$, which induces a lift $\ulcorner\sigma\urcorner \colon X \to F^{\mathbb{2}}$ of $\ulcorner\alpha\urcorner$ by the universal property of the pullback with the property that $\ulcorner h\sigma \urcorner$ is a p-cartesian arrow:

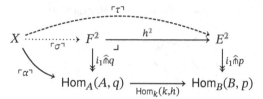

So if we prove the second part of the statement – that a natural transformation $\psi : f' \Rightarrow f$ with codomain F is q-cartesian if and only if $h\psi$ is p-cartesian – then we will have shown that q is itself a cartesian fibration with cartesian cells created by the pullback.

To that end consider the cube:

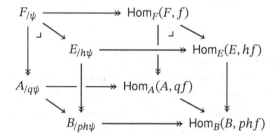

By the hypercube pullback lemma, the left and right faces are strict pullback squares. Hence the map from $E_{/h\psi}$ to the pullback in the front face pulls back to the map from $F_{/\psi}$ to the pullback in the back face. Thus, by Proposition 3.3.4 if $h\psi$ is p-cartesian then ψ is q-cartesian. This completes the proof that q is a cartesian fibration.

Now if ψ is q-cartesian then by Lemma 5.1.11, ψ is isomorphic to the q-cartesian lift σ of $q\psi$ constructed in first part of this proof. Thus $h\psi$ is isomorphic to $h\sigma = \tau$, which is p-cartesian, so by Lemma 5.1.6, $h\psi$ must be p-cartesian as well. $\qquad\square$

In fact, it suffices in Definition 5.2.1 to assume only that the generic transformation whose codomain factors through p admits a p-cartesian lift:

Lemma 5.2.5. *An isofibration* $p : E \twoheadrightarrow B$ *is a cartesian fibration if and only if the right comma cone over* p *displayed below-left admits a* p-*cartesian lift* χ *as* *displayed below-right:*

$$
\begin{array}{ccc}
& E^2 & \mathrm{Hom}_B(B,p) \xrightarrow{p_1} E \\
\ulcorner\chi\urcorner & \Big\downarrow i_1 \widehat{\pitchfork} p \rightsquigarrow & \Uparrow\phi \quad \Big\downarrow p \\
\mathrm{Hom}_B(B,p) = \mathrm{Hom}_B(B,p) & p_0 \twoheadrightarrow B &
\end{array}
\qquad
\begin{array}{c}
\mathrm{Hom}_B(B,p) \overset{p_1}{\underset{r}{\rightrightarrows}} E \\
= \quad \Big\downarrow p \\
B
\end{array}
$$

$$(5.2.6)$$

Proof By Theorem 5.1.7(ii), to say that the right comma cone over p admits a p-cartesian lift χ means that $\ulcorner\chi\urcorner$ defines an absolute right lifting of the identity through $i_1 \widehat{\pitchfork} p$. By Lemma 2.3.6 it follows that the restriction

$$
\begin{array}{ccc}
& & E^2 \\
& \nearrow^{\ulcorner\chi\urcorner} & \Big\downarrow i_1 \widehat{\pitchfork} p \\
X \xrightarrow{\ulcorner\beta\urcorner} \mathrm{Hom}_B(B,p) & = & \mathrm{Hom}_B(B,p)
\end{array}
$$

defines an absolute right lifting for any $\ulcorner\beta\urcorner \colon X \to \mathrm{Hom}_B(B,p)$, and thus by Theorem 5.1.7(ii) any $\beta \colon b \Rightarrow pe$ admits a p-cartesian lift. $\square$

When p is a cartesian fibration, we refer to the universal cartesian arrow $\ulcorner\chi\urcorner \colon \mathrm{Hom}_B(B,p) \twoheadrightarrow E^2$ of (5.2.6) as the **generic p-cartesian lift**.

REMARK 5.2.7 (action of arrows on the fibers of a cartesian fibration). The action of an arrow in the base of a cartesian fibration $p \colon E \twoheadrightarrow B$ on the fibers can be described as follows. Consider a natural transformation β with codomain B and form the fibers of p over its domain and codomain functors

$$
X \underset{b}{\overset{a}{\rightrightarrows}} B \Downarrow\beta
\qquad
\begin{array}{ccc}
E_a \longrightarrow E \longleftarrow E_b \\
p_a \Big\downarrow \quad \lrcorner \quad \Big\downarrow p \quad \llcorner \quad \Big\downarrow p_b \\
X \xrightarrow{a} B \xleftarrow{b} X
\end{array}
$$

The pullback square defining the fiber E_b factors as below-left

$$
\begin{array}{ccc}
E_b \longrightarrow \mathrm{Hom}_B(B,p) \xrightarrow{p_1} E \\
p_b \Big\downarrow \quad \lrcorner \quad q \Big\downarrow \quad \lrcorner \quad \Big\downarrow p \\
X \xrightarrow{\ulcorner\beta\urcorner} B^2 \xrightarrow{p_1} B
\end{array}
\qquad
\begin{array}{ccc}
E_b \longrightarrow \mathrm{Hom}_B(B,p) \xrightarrow{r} E \\
p_b \Big\downarrow \quad \lrcorner \quad q \Big\downarrow \quad \Big\downarrow p \\
X \xrightarrow{\ulcorner\beta\urcorner} B^2 \xrightarrow{p_0} B
\end{array}
$$

and thus the rectangle displayed above-right defines a cone over the pullback defining E_a, inducing a map $\beta^* \colon E_b \to E_a$. The top-horizontal functor in this rectangle recovers the domain component $\beta^*(\ell_b)$ of the p-cartesian lift of β with codomain ℓ_b. This coincides with the description of $\beta^* \colon E_b \to E_a$ given in (5.0.1).

Lemma 5.2.5 extends to give an internal characterization of cartesian fibrations inspired by a similar result of Street [118, 119, 121], which in turn was inspired by previous work of Gray [48] on what he calls a "Chevalley criterion"[8] (see also [131]). As we shall see, the universal property of a cartesian fibration $p : E \twoheadrightarrow B$ can be encoded by the data of a suitable right adjoint to the functor $\Delta_p : E \to \mathrm{Hom}_B(B, p)$ induced from the identity 2-cell or to the functor $i_1 \mathbin{\widehat{\pitchfork}} p : E^2 \to \mathrm{Hom}_B(B, p)$ defined by applying p to the generic arrow with codomain E. In fact, this result follows quite easily by specializing the characterizations of p-cartesian cells of Theorem 5.1.7 to the universal case described in Lemma 5.2.5.

THEOREM 5.2.8 (an internal characterization of cartesian fibrations). *For an isofibration* $p : E \twoheadrightarrow B$ *the following are equivalent:*

(i) $p : E \twoheadrightarrow B$ *defines a cartesian fibration.*

(ii) *The functor* $i_1 \mathbin{\widehat{\pitchfork}} p : E^2 \twoheadrightarrow \mathrm{Hom}_B(B, p)$ *admits a right adjoint right inverse:*[9]

$$
E^2 \underset{\ulcorner \chi \urcorner}{\overset{i_1 \widehat{\pitchfork} p}{\rightleftarrows}} \bot \;\; \mathrm{Hom}_B(B, p)
$$

(iii) *The functor* $\Delta_p : E \to \mathrm{Hom}_B(B, p)$ *admits a right adjoint over* B:[10]

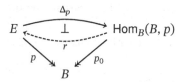

As the proof reveals, the right adjoint of (iii) is the domain component of the generic p-cartesian lift χ of (5.2.6), and χ is recovered as $p_1\varepsilon$, where ε is the counit of the fibered adjunction $\Delta_p \dashv r$. By 1-cell induction, the generic cartesian lift χ can be represented by a functor $\ulcorner \chi \urcorner : \mathrm{Hom}_B(B, p) \to E^2$ and this defines the right adjoint of (ii).

---

[8] Gray attributes [48, 3.11] – the special case of the equivalence of (i)⇔(ii) of Theorem 5.2.8 in the ∞-cosmos $\mathcal{C}at$ – to unpublished notes from a seminar given by Claude Chevalley at Berkeley in 1962.

[9] By Lemma 3.6.9, such an adjunction may be rectified to an adjunction that is fibered over $\mathrm{Hom}_B(B, p)$, which allows us to interpret $\ulcorner \chi \urcorner : \mathrm{Hom}_B(B, p) \to E^2$ as a terminal element in E^2 over $\mathrm{Hom}_B(B, p)$ (see Definition 3.6.8).

[10] By Lemma 3.5.9, $\Delta_p \cong \ulcorner \mathrm{id}_p \urcorner$ is itself right adjoint over E and thus over B to the codomain projection functor $p_1 : \mathrm{Hom}_B(B, p) \twoheadrightarrow E$. Since the counit of the adjunction $p_1 \dashv \Delta_p$ is an isomorphism, it follows formally that the unit of the adjunction $\Delta_p \dashv r$ must also be an isomorphism, whenever the adjunction postulated in (iii) exists (see Lemma B.3.9). Thus r defines a right adjoint left inverse to Δ_p over B.

Proof We prove (i)⇔(ii) and (i)⇔(iii).

(i)⇔(ii): By Theorem 5.1.7(ii) and Lemma 5.2.5, $p : E \twoheadrightarrow B$ is a cartesian fibration if and only if $i_1 \widehat{\pitchfork} p : E^2 \twoheadrightarrow \mathrm{Hom}_B(B, p)$ admits a section that defines an absolute right lifting diagram:

$$
\begin{array}{ccc}
& & E^2 \\
& \overset{\ulcorner\chi\urcorner}{\nearrow} \raisebox{1ex}{$\mathbin{-}\mathbin{-}\mathbin{\rightarrow}$} & \big\downarrow {\scriptstyle i_1 \widehat{\pitchfork} p} \\
\mathrm{Hom}_B(B, p) & \mathrel{=\!=\!=} & \mathrm{Hom}_B(B, p)
\end{array}
$$

By Lemma 2.3.7, such an absolute right lifting defines an adjunction $i_1 \widehat{\pitchfork} p \dashv \ulcorner\chi\urcorner$ with identity counit. Conversely, if $i_1 \widehat{\pitchfork} p \dashv \ulcorner\chi\urcorner$ with invertible counit, then since the left adjoint is an isofibration, this can be rectified into a fibered adjunction in which $\ulcorner\chi\urcorner$ defines a strict section and the counit is the identity. By Lemma 2.3.7, $\ulcorner\chi\urcorner$ then defines an absolute right lifting of the identity functor through $i_1 \widehat{\pitchfork} p$, which proves that p is a cartesian fibration.

(i)⇔(iii): If $p : E \twoheadrightarrow B$ is a cartesian fibration in an ∞-cosmos $\mathcal{K}$, then by Theorem 5.1.7(iii) there is an absolute right lifting diagram

$$
\begin{array}{ccc}
& & E \\
& \overset{r}{\nearrow} & \big\downarrow {\scriptstyle \Delta_p} \\
\mathrm{Hom}_B(B, p) & \underset{\Downarrow\epsilon}{\mathrel{=\!=}} & \mathrm{Hom}_B(B, p)
\end{array}
$$

for which $p_0\epsilon = \mathrm{id}_{p_0}$. By Proposition 3.6.2, we may lift ϵ to a natural transformation in the sliced ∞-cosmos $\mathcal{K}_{/B}$. Applying Lemma 2.3.7 in $\mathcal{K}_{/B}$, we see that r defines a fibered right adjoint to Δ_p. Conversely, if we are given a fibered adjunction over B, we may apply the forgetful 2-functor $\mathfrak{h}(\mathcal{K}_{/B}) \to \mathfrak{h}\mathcal{K}$ to obtain an adjunction[11]

$$
E \underset{r}{\overset{\Delta_p}{\underset{\longleftarrow}{\rightleftarrows}}} \perp \mathrm{Hom}_B(B, p)
$$

and then apply Lemma 2.3.7 to conclude that the counit ϵ defines an absolute right lifting of the identity through Δ_p. Since the counit is unchanged by the process of forgetting that the adjunction is fibered over B, we still have that $p_0\epsilon = \mathrm{id}_{p_0}$, as required. □

Examples of cartesian fibrations are overdue.

---

[11] Recall from Non-Example 1.3.6 that the forgetful functor $\mathcal{K}_{/B} \to \mathcal{K}$ is not cosmological. Nevertheless any simplicial functor between ∞-cosmoi descends to a 2-functor between their homotopy 2-categories, which is all that is needed here.

PROPOSITION 5.2.9 (domain projection fibration). *For any ∞-category A, the domain projection functor $p_0 : A^2 \twoheadrightarrow A$ defines a cartesian fibration. Moreover, a natural transformation ψ with codomain A^2 is p_0-cartesian if and only if $p_1\psi$ is invertible.*

Before giving the proof, we explain the idea. A natural transformation

$$
\begin{array}{ccc}
X & \xrightarrow{\ulcorner\beta\urcorner} & A^2 \\
& \Uparrow\alpha \quad \Big\downarrow p_0 & \\
a & \longrightarrow & A
\end{array}
$$

defines a composable pair of 2-cells $\alpha : a \Rightarrow x$ and $\beta : x \Rightarrow y$ in $\mathsf{hFun}(X, A)$. Composing these we induce a 2-cell $X \underset{\ulcorner\beta\urcorner}{\overset{\ulcorner\beta\circ\alpha\urcorner}{\rightrightarrows}} A^2$ representing the commutative square in $\mathsf{hFun}(X, A)$

$$
\begin{array}{ccc}
a & \xRightarrow{\alpha} & x \\
{\scriptstyle\beta\circ\alpha}\Big\Downarrow & & \Big\Downarrow{\scriptstyle\beta} \\
y & = & y
\end{array}
$$

so that $p_0\psi = \alpha$, as required, and $p_1\psi = \mathrm{id}$.

Proof We use Theorem 5.2.8(ii) and prove that p_0 is cartesian by constructing an appropriate adjoint to the functor

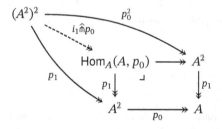 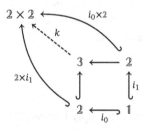

defined by cotensoring with the 1-categories displayed above right.[12]

To construct a right adjoint right inverse to the map $i_1 \hat{\pitchfork} p_0$, it suffices to construct a left adjoint left inverse to the inclusion of 1-categories $k : 3 \hookrightarrow 2 \times 2$ with image $(0, 0) \to (0, 1) \to (1, 1)$. The left adjoint $\ell : 2 \times 2 \to 3$ is a left

---

[12] The cotensor $A^{(-)}$ carries pushouts of simplicial sets to pullbacks of ∞-categories, and the pushout of $2 \cup_1 2$ of simplicial sets is $\mathbb{A} = \Lambda^1[2]$, not $3 = \Delta[2]$. However, on account of the equivalence of ∞-categories $A^3 \simeq A^\mathbb{A}$, no harm comes from making the indicated substitution.

inverse on the image of 3 and sends $(1,0)$ to the terminal element of 3:

$$2 \times 2 \ni \left\{ \begin{array}{ccc} (0,0) & \longrightarrow & (0,1) \\ \downarrow & & \downarrow \\ (1,0) & \longrightarrow & (1,1) \end{array} \right\} \quad \overset{\ell}{\mapsto} \quad \left\{ \begin{array}{ccc} 0 & \longrightarrow & 1 \\ \downarrow & & \downarrow \\ 2 & \!\!=\!\!=\!\! & 2 \end{array} \right\} \in 3$$

Now

$$(A^2)^2 \underset{-\circ\ell}{\overset{i_1\widehat{\mathbb{m}}p_0\simeq-\circ k}{\rightleftarrows}} \bot \quad A^3 \simeq \mathsf{Hom}_A(A, p_0)$$

defines the desired right adjoint right inverse.

The characterization of p_0-cartesian transformations follows from Theorem 5.1.7(ii). For any arrow $\ulcorner\psi\urcorner \colon X \to A^2$ the commutative triangle below-left factors through the absolute lifting diagram defined by the adjunction, and $\ulcorner\psi\urcorner$ is p_0-cartesian if and only if the induced natural transformation ζ is invertible.

$$
\begin{array}{ccc}
 & A^{2\times2} & \\
{}^{\ulcorner\psi\urcorner}\nearrow & {\Big\downarrow}{\scriptstyle -\circ k} & \\
X \underset{\ulcorner p_0\psi\urcorner}{\longrightarrow} A^3 & \| &
\end{array}
\;=\;
\begin{array}{ccc}
 & \overset{\ulcorner\psi\urcorner}{\longrightarrow} & A^{2\times2} \\
\nearrow \;\; \Downarrow\zeta \;\;{\scriptstyle -\circ\ell}\nearrow & \| & {\Big\downarrow}{\scriptstyle -\circ k} \\
X \underset{\ulcorner p_0\psi\urcorner}{\longrightarrow} A^3 & \!\!=\!\!=\!\! & A^3
\end{array}
$$

If ζ is an isomorphism, then $\ulcorner\psi\urcorner$ is isomorphic to an arrow in the image of restriction along ℓ, and thus its codomain component must be invertible. Conversely, if $p_1\psi$ is invertible, to show that ψ is p_0-cartesian it suffices by Theorem 5.1.7(ii) to prove that ζ is an isomorphism. By two applications of 2-cell conservativity, it suffices to show that the four components of ζ indexed by each element of 2×2 are invertible. Since ζ restricts along k to an identity, three of these components are necessarily identities, and the fourth component $X \underset{\ulcorner p_0\psi\urcorner|_e}{\overset{\ulcorner\psi\urcorner}{\underset{\Downarrow\zeta}{\rightrightarrows}}} A^{2\times2} \xrightarrow{P_{(1,0)}} A$

equals $p_1\psi$ in $\mathsf{hFun}(X,A)$. $\qquad\qquad\Box$

For ∞-categories admitting pullbacks (see Definition 4.3.8), the codomain projection functor also defines a cartesian fibration:

PROPOSITION 5.2.10 (codomain projection fibration). *Let A be an ∞-category that admits pullbacks. Then the codomain projection functor $p_1 \colon A^2 \twoheadrightarrow A$ is a cartesian fibration and the p_1-cartesian arrows are the pullback squares.*

Proof Via Theorem 5.2.8(ii), we desire a right adjoint right inverse to the functor $i_1 \;\widehat{\pitchfork}\; p_1$ defined below-left applying $A^{(-)}$ to the diagram of simplicial sets appearing below-right:

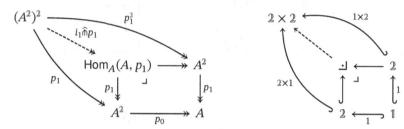

This is provided by Corollary 4.3.5:

$$(A^2)^2 \cong A^{\boxdot} \underset{\mathrm{ran}}{\overset{\mathrm{res}}{\rightleftarrows}} A^{\lrcorner} \cong \mathrm{Hom}_A(A, p_1)$$

By comparing Definition 4.3.8 with Theorem 5.1.7(ii), we see that p_1-cartesian arrows coincide with pullback squares. $\qquad\square$

There is a fourth equivalent definition of cartesian fibrations that we might have added to Theorem 5.2.8, except that it is an "external characterization" of cartesian fibrations, phrased entirely in the setting of the homotopy 2-category, rather than an "internal characterization," that is amenable for proving that cartesian fibrations are preserved by cosmological functors.

PROPOSITION 5.2.11. *An isofibration* $p : E \twoheadrightarrow B$ *is a cartesian fibration if and only if any natural transformation* $\beta : b \Rightarrow pe$ *admits a lift* $\chi_\beta : e' \Rightarrow e$

$$
\begin{array}{ccc}
\begin{array}{c}
X \xrightarrow{\;e\;} E \\
\;\;\Uparrow\beta \;\;\;\; \downarrow p \\
b \;\searrow\;\;\; B
\end{array}
& = &
\begin{array}{c}
X \overset{e}{\underset{e'}{\Longrightarrow}} E \;\; \Uparrow\chi_\beta \\
\;\;\;\; \downarrow p \\
B
\end{array}
\end{array}
$$

so that for any functor $f : Y \to X$ *the 2-cell* $\chi_\beta f$ *satisfies the following properties in the homotopy 2-category:*

   (i) **induction:** *Given any 2-cells* $\tau : e'' \Rightarrow ef$ *and* $\gamma : pe'' \Rightarrow pe'f$ *so that* $p\tau = \beta f \cdot \gamma$, *there exists a lift* $\bar\gamma : e'' \Rightarrow e'f$ *of* γ *so that* $\tau = \chi_\beta f \cdot \bar\gamma$.

   (ii) **conservativity:** *Any fibered endomorphism of* $\chi_\beta f$ *is invertible: for any 2-cell* $\zeta : e'f \Rightarrow e'f$ *so that* $\chi_\beta f \cdot \zeta = \chi_\beta f$ *and* $p\zeta = \mathrm{id}_{pe'f}$ *then* ζ *is invertible.*

Moreover, under these hypothesis, a natural transformation $\psi : e' \Rightarrow e$ *with co-domain* E *is* p-*cartesian if and only if* ψ *satisfies the induction and conservativity conditions in the case* $f = \mathrm{id}$.

To help us stay organized during the proof, we refer to a natural transformation $\psi: e' \Rightarrow e$ satisfying conditions (i) and (ii) as a *weak p-cartesian transformation*. We argue that when p is a cartesian fibration, the class of weak p-cartesian transformations coincides with the class of p-cartesian transformations.

Proof If $p: E \twoheadrightarrow B$ is a cartesian fibration, then any $\beta: b \Rightarrow pe$ admits a p-cartesian lift $\chi_\beta: \beta^* e \Rightarrow e$. By Lemma 5.1.4, the restriction $\chi_\beta f$ along any functor is again a p-cartesian transformation. By Proposition 5.1.13, this $\chi_\beta f$ then satisfies the induction and conservativity properties. In fact Proposition 5.1.13 shows more generally that any p-cartesian transformation is weakly p-cartesian.

Conversely, assume that $p: E \twoheadrightarrow B$ satisfies the hypotheses of the statement. We will use the induction and conservativity properties associated to the lift

of the right comma cone ϕ over p and its restrictions to construct the data of Theorem 5.2.8(iii).

First, we apply the induction property to $\chi \Delta_p: r\Delta_p \Rightarrow \mathrm{id}_E$ to induce a 2-cell $\eta: \mathrm{id}_E \Rightarrow r\Delta_p$ so that $p\eta = \mathrm{id}_p$ and $\chi\Delta_p \cdot \eta = \mathrm{id}$. By construction, $\eta \cdot \chi\Delta_p$ defines a fibered automorphism of $\chi\Delta_p$, so must be invertible by the conservativity property. Thus, the fibered 2-cell η is the inverse isomorphism to $\chi\Delta_p$.

We induce the counit $\epsilon: \Delta_p r \Rightarrow \mathrm{id}$ by 2-cell induction from the pair of 2-cells $p_1 \epsilon = \chi$ and $p_0 \epsilon = \mathrm{id}$. Since $\epsilon\Delta_p$ is induced from a pair of invertible 2-cells, it must be an isomorphism by 2-cell conservativity. The 2-cell $r\epsilon$ can also be seen to be invertible, on account of the naturality of whiskering square associated to the horizontal composite:

Weak p-cartesian cells are stable under isomorphism, so the top right composite is weakly p-cartesian. Since $pr\epsilon = p_0 \epsilon = \mathrm{id}$, we can apply the induction and conservativity properties once more to induce a transformation $\gamma: r \Rightarrow r\Delta_p r$ so that $\gamma \cdot r\epsilon$ and $r\epsilon \cdot \gamma$ are both isomorphisms. Hence $r\epsilon$ is invertible.

By Lemma 2.1.11 this data in an ∞-cosmos $\mathcal{K}$ suffices to define an adjunction in the sliced homotopy 2-category $(\mathfrak{h}\mathcal{K})_{/B}$. By Lemma 3.6.7, this adjunction may then be lifted along the smothering 2-functor $\mathfrak{h}(\mathcal{K}_{/B}) \to (\mathfrak{h}\mathcal{K})_{/B}$ of Proposition 3.6.2 to a genuine fibered adjunction. By Theorem 5.2.8(iii), this proves that p is a cartesian fibration.

Finally, observe that when $p : E \twoheadrightarrow B$ is a cartesian fibration, any $\psi : e' \Rightarrow e$ satisfying (i) and (ii) for $f = \mathrm{id}$ is fibered isomorphic to a p-cartesian lift of $p\psi$ with codomain e. By Lemma 5.1.6, it follows that ψ is then a p-cartesian transformation, proving that when p is a cartesian fibration these classes coincide. $\square$

Exercises

EXERCISE 5.2.i. There is a standard notion of cartesian fibration in a 2-category developed by Street [118] that recovers the Grothendieck fibrations when specialized to the 2-category $\mathcal{C}at$. This is *not* the correct notion of cartesian fibration between ∞-categories as the universal property the usual notion demands of lifted 2-cells is too strict. Compare this definition with 2-categorical definition of Proposition 5.2.11 and consider why the stricter universal property does not hold in the ∞-categorical context, for instance by considering the cartesian fibrations of Proposition 5.2.9.

EXERCISE 5.2.ii. Show that a cartesian fibration $p : E \twoheadrightarrow B$ defines an "incoherent pseudofunctor" $E : \mathsf{h}B^{\mathrm{op}} \rightsquigarrow \mathfrak{h}\mathcal{K}$ given by the data:

- a mapping on objects $b \in \mathsf{h}B \mapsto E_b \in \mathfrak{h}\mathcal{K}$;
- a mapping on 1-cells $\beta : a \to b \in \mathsf{h}B \mapsto \beta^* : E_b \to E_a \in \mathfrak{h}\mathcal{K}$ defined by (5.0.1);
- an invertible 2-cell $E_b \overset{\iota_b \Downarrow \cong}{\underset{\mathrm{id}_b^*}{\rightrightarrows}} E_b \in \mathfrak{h}\mathcal{K}$ for each $b \in \mathsf{h}B$; and
- an invertible 2-cell

$$E_c \overset{\gamma^*}{\underset{(\gamma\circ\beta)^*}{\longrightarrow}} \overset{E_b}{\underset{\alpha_{\beta,\gamma}\Downarrow\cong}{\searrow}} \overset{\beta^*}{\longrightarrow} E_a$$

in $\mathfrak{h}\mathcal{K}$ for each composable pair of $\beta : a \to b$ and $\gamma : b \to c$ of arrows in $\mathsf{h}B$.

The coherence conditions present in the full definition of a **pseudo-functor** (see Definition 10.4.1) are not evident here, but do follow from the extension of this construction to a homotopy coherent diagram indexed by the underlying quasi-category of B [111, 6.1.16].

EXERCISE 5.2.iii (5.5.13). Use either Theorem 5.2.8(iii) or (ii) to prove that for any cospan of functors $C \xrightarrow{g} A \xleftarrow{f} B$ between ∞-categories, the domain projection functor $p_0 : \mathrm{Hom}_A(f,g) \twoheadrightarrow B$ is a cartesian fibration, and moreover, a natural transformation ψ with codomain $\mathrm{Hom}_A(f,g)$ is p_0-cartesian if and only if $p_1\psi$ is invertible.[13]

5.3 Cartesian Functors

We now show that cartesianness is an equivalence invariant property of isofibrations by appealing to Theorem 5.2.8 to study the relationship between the data that witnesses the cartesianness of an isofibration and that data provided by an equivalence.

COROLLARY 5.3.1. *Consider an essentially commutative square between isofibrations whose horizontal functors are equivalences:*

$$
\begin{array}{ccc}
F & \xrightarrow{\ \ h\ \ } & E \\
{\scriptstyle q}\downarrow & \cong & \downarrow{\scriptstyle p} \\
A & \xrightarrow[\ \ k\ \ ]{\sim} & B
\end{array}
$$

Then p is a cartesian fibration if and only if q is a cartesian fibration in which case h preserves and reflects cartesian transformations: ψ is q-cartesian if and only if $h\psi$ is p-cartesian.

Proof By Proposition 3.4.6, an essentially commutative square induces an essentially commutative square

$$
\begin{array}{ccc}
F^2 & \xrightarrow{\quad h^2\quad}_{\sim} & E^2 \\
{\scriptstyle i_1\widehat{\pitchfork}q}\downarrow & \cong & \downarrow{\scriptstyle i_1\widehat{\pitchfork}p} \\
\mathrm{Hom}_A(A,q) & \xrightarrow[\mathrm{Hom}_k(k,h)]{\sim} & \mathrm{Hom}_B(B,p)
\end{array}
$$

---

[13] On account of Proposition 5.2.4 and the pullback square

$$
\begin{array}{ccc}
\mathrm{Hom}_A(f,g) & \longrightarrow & \mathrm{Hom}_A(A,g) \\
{\scriptstyle p_0}\downarrow & \lrcorner & \downarrow{\scriptstyle p_0} \\
B & \xrightarrow[\ \ f\ \ ]{} & A
\end{array}
$$

it suffices to prove that the domain projection functor $p_0 : \mathrm{Hom}_A(A,g) \twoheadrightarrow A$ is a cartesian fibration. There is a sense in which this functor can be understood as a pullback of $p_0 : A^2 \twoheadrightarrow A$, which we explain in the proof of Corollary 5.5.13 where this result appears. The reader opting to reprise the proof of Proposition 5.2.9 might wish to appeal to Proposition 6.3.10 if this construction proves too painful.

whose horizontal functors are the equivalences defined in Proposition 3.4.5. By the equivalence invariance of adjunctions (see Proposition B.3.8) the left-hand vertical functor admits a right adjoint right inverse if and only if the right-hand vertical functor does. By Theorem 5.2.8(ii), it follows that p is cartesian if and only if q is cartesian. By the equivalence invariance of absolute lifting diagrams (see Exercises 2.3.vi and 3.5.ii), it follows similarly from Theorem 5.1.7(ii) that ψ is p-cartesian if and only if $h\psi$ is q-cartesian. $\square$

We have now met a few examples of structure-preserving morphisms between cartesian fibrations.

DEFINITION 5.3.2 (cartesian functor). Let $p : E \twoheadrightarrow B$ and $q : F \twoheadrightarrow A$ be cartesian fibrations. A commutative square defines a **cartesian functor** if its domain component h preserves cartesian transformations: if ψ is q-cartesian then $h\psi$ is p-cartesian.

$$
\begin{array}{ccc}
F & \xrightarrow{\ \ h\ \ } & E \\
{\scriptstyle q}\downarrow & & \downarrow{\scriptstyle p} \\
A & \xrightarrow{\ \ k\ \ } & B
\end{array}
$$

For the purposes of Proposition 6.3.14, we prefer to reserve this terminology for strictly commutative squares, but it can be extended to essentially commutative squares.

EXAMPLE 5.3.3 (pullbacks and equivalences are cartesian). Immediately from Proposition 5.2.4 and Corollary 5.3.1, both pullback squares and commutative squares of equivalences define cartesian functors, which have the special property that the top horizontal functor reflects, as well as preserves, cartesian transformations. These results extend to weakly cartesian squares, since cartesian functors compose and any weakly cartesian square factors as:

$$
\begin{array}{ccccc}
F & \overset{h}{\underset{\sim}{\rightleftarrows}} & P & \longrightarrow & E \\
{\scriptstyle q}\downarrow & & \downarrow{\scriptstyle \lrcorner} & & \downarrow{\scriptstyle p} \\
A & = & A & \xrightarrow{\ \ k\ \ } & B
\end{array}
$$

The internal characterization of cartesian fibrations of Theorem 5.2.8 extends to an internal characterization of cartesian functors.

THEOREM 5.3.4 (an internal characterization of cartesian functors). *For a com-*

mutative square

$$
\begin{array}{ccc}
F & \xrightarrow{\ h\ } & E \\
{\scriptstyle q}\downarrow & & \downarrow{\scriptstyle p} \\
A & \xrightarrow{\ k\ } & B
\end{array}
$$

between cartesian fibrations the following are equivalent:

(i) *The square* (h, k) *defines a cartesian functor from* q *to* p.

(ii) *The mate of the identity in the diagram of functors below-left is an isomorphism:*

$$
\begin{array}{ccc}
F^2 & \xrightarrow{\ h^2\ } & E^2 \\
{\scriptstyle i_1\hat{m}q}\downarrow & {\scriptstyle =\nearrow} & \downarrow{\scriptstyle i_1\hat{m}p} \\
\mathrm{Hom}_A(A, q) & \xrightarrow[\mathrm{Hom}_k(k,h)]{} & \mathrm{Hom}_B(B, p)
\end{array}
\quad\leadsto\quad
\begin{array}{ccc}
F^2 & \xrightarrow{\ h^2\ } & E^2 \\
{\scriptstyle \ulcorner\chi\urcorner}\uparrow & {\scriptstyle \cong\searrow} & \uparrow{\scriptstyle \ulcorner\chi\urcorner} \\
\mathrm{Hom}_A(A, q) & \xrightarrow[\mathrm{Hom}_k(k,h)]{} & \mathrm{Hom}_B(B, p)
\end{array}
$$

(iii) *The mate of the identity in the diagram of functors over* $k \colon A \to B$ *below-left is an isomorphism:*

$$
\begin{array}{ccc}
F & \xrightarrow{\ h\ } & E \\
{\scriptstyle \Delta_q}\downarrow & {\scriptstyle =\nearrow} & \downarrow{\scriptstyle \Delta_p} \\
\mathrm{Hom}_A(A, q) & \xrightarrow[\mathrm{Hom}_k(k,h)]{} & \mathrm{Hom}_B(B, p)
\end{array}
\quad\leadsto\quad
\begin{array}{ccc}
F & \xrightarrow{\ h\ } & E \\
{\scriptstyle r}\uparrow & {\scriptstyle \cong\searrow} & \uparrow{\scriptstyle r} \\
\mathrm{Hom}_A(A, q) & \xrightarrow[\mathrm{Hom}_k(k,h)]{} & \mathrm{Hom}_B(B, p)
\end{array}
$$

The *mates* referenced here generalize the notion of "adjoint" or "transposed" 2-cells (see Definition B.3.3 and §B.3). As noted in Warning B.3.7, the mate of an isomorphism is not necessarily invertible.

Proof We prove (i)⇔(iii) and (i)⇔(ii). To save space, we write $\bar{h}$ as an abbreviation for the functor $\mathrm{Hom}_k(k, h)$.

(i)⇔(ii): The square (h, k) defines a cartesian functor from q to p if and only if it carries the generic q-cartesian lift $\ulcorner\chi\urcorner \colon \mathrm{Hom}_A(A, q) \to F^2$ to a p-cartesian arrow, in other words, if and only if the commutative diagram defines an absolute right lifting:

$$
\begin{array}{ccccc}
 & & F^2 & \xrightarrow{\ h^2\ } & E^2 \\
 & {\scriptstyle \ulcorner\chi\urcorner}\nearrow & \downarrow{\scriptstyle i_1\hat{m}q} & & \downarrow{\scriptstyle i_1\hat{m}p} \\
\mathrm{Hom}_A(A, q) & \!\!=\!\!=\!\! & \mathrm{Hom}_A(A, q) & \xrightarrow[\bar{h}]{} & \mathrm{Hom}_B(B, p)
\end{array}
$$

By Definition B.3.3, this square factors through the absolute right lifting diagram below-right via a whiskered copy $\bar{\eta}h^{2\ulcorner}\chi^\urcorner$ of the counit $\bar{\eta}$ of $i_1 \,\hat{\pitchfork}\, p \dashv {}^\ulcorner\chi^\urcorner$.

$$
\begin{array}{ccccc}
 & F^2 & \xrightarrow{\;h^2\;} & E^2 & =\!=\!= & E^2 \\
{}^\ulcorner\chi^\urcorner \nearrow & \downarrow i_1\hat{\pitchfork}q & & \downarrow i_1\hat{\pitchfork}p & \Downarrow\bar{\eta} \quad {}^\ulcorner\chi^\urcorner \nearrow & \downarrow i_1\hat{\pitchfork}p \\
\mathrm{Hom}_A(A,q) & =\!=\!= & \mathrm{Hom}_A(A,q) & \xrightarrow{\;\bar{h}\;} & \mathrm{Hom}_B(B,p) & =\!=\!= & \mathrm{Hom}_B(B,p)
\end{array}
$$

Thus, by Theorem 5.1.7(ii), $\bar{\eta}h^{2\ulcorner}\chi^\urcorner$ is invertible if and only if $h^{2\ulcorner}\chi^\urcorner$ is p-cartesian.

(i)$\Leftrightarrow$(iii): Since the unit of $\Delta_p \dashv r$ is an isomorphism for the reasons discussed in the statement of Theorem 5.2.8(iii), the mate of the isomorphism on the left-hand side of (iii) is isomorphic to $r\bar{h}\epsilon$, so our task is to show that this natural transformation is invertible if and only if h defines a cartesian functor. This leads us to consider the naturality of whiskering square expressing the horizontal composite of the two counits:

$$
\begin{array}{ccccc}
\mathrm{Hom}_A(A,q) & \overset{r\nearrow{}^{F}\searrow\Delta_q}{\underset{\Downarrow\epsilon}{=\!=\!=}} & \mathrm{Hom}_A(A,q) & \overset{\bar{h}}{\to} \mathrm{Hom}_B(B,p) & \overset{r\nearrow{}^{E}\searrow\Delta_p}{\underset{\Downarrow\epsilon}{=\!=\!=}} \mathrm{Hom}_B(B,p) \\
\downarrow p_1 & & \downarrow p_1 & & \downarrow p_1 \\
F & \xrightarrow{\;h\;} & E & =\!=\!= & E
\end{array}
$$

$$
\begin{array}{ccc}
r\bar{h}\Delta_q r & \xRightarrow{\;r\bar{h}\epsilon\;} & r\bar{h} \\
{\scriptstyle p_1\epsilon\bar{h}\Delta_q r}\Big\|{\scriptstyle\mathbb{R}} & & \Big\|{\scriptstyle p_1\epsilon\bar{h}} \\
p_1\bar{h}\Delta_q r & \xRightarrow[\;p_1\bar{h}\epsilon\;]{} & p_1\bar{h}
\end{array}
$$

Since ϵ is the counit of an adjunction $\Delta_p \dashv r$ with invertible unit, $\epsilon\Delta_p$ is an isomorphism, so $\epsilon\bar{h}\Delta_q = \epsilon\Delta_p h$, is invertible. From the calculation $pr\bar{h}\epsilon = p_0\bar{h}\epsilon = kp_0\epsilon = \mathrm{id}$, we see that the top-horizontal 2-cell lies in a fiber over an identity.

Recall from Theorem 5.2.8(iii) that $p_1\epsilon$ defines the generic p-cartesian lift of Lemma 5.2.5 for the cartesian fibration $p : E \twoheadrightarrow B$. Hence the right-hand vertical 2-cell $p_1\epsilon\bar{h}$ is a p-cartesian lift of $\phi\bar{h}$, where ϕ is the right comma cone over p. By the definition of $\bar{h}$, $\phi\bar{h} = k\phi$, the latter ϕ being the right comma cone over q.

Similarly, the bottom horizontal 2-cell $p_1\bar{h}\epsilon$ is a lift of $k\phi = \phi\bar{h}$. So if h is a cartesian functor, the right-hand vertical and bottom horizontal 2-cells are both p-cartesian lifts of a common 2-cell, and the conservativity property of Lemma 5.1.12 implies that $r\bar{h}\epsilon$ is invertible. Conversely, if $r\bar{h}\epsilon$ is invertible, then $p_1\bar{h}\epsilon = hp_1\epsilon$ is isomorphic to the p-cartesian transformation $p_1\epsilon\bar{h}$ and is consequently p-cartesian. Since every q-cartesian transformation is isomorphic to a restriction of $p_1\epsilon$, this is the case if and only if h is a cartesian functor. $\square$

One of the myriad applications of Theorems 5.2.8 and 5.3.4 is:

COROLLARY 5.3.5. *Cosmological functors preserve cartesian arrows, cartesian fibrations, and cartesian functors.*

Proof By Theorem 5.2.8(ii), an isofibration $p : E \twoheadrightarrow B$ in an ∞-cosmos $\mathcal{K}$ is cartesian if and only if the isofibration $i_1 \mathbin{\widehat{\pitchfork}} p : E^2 \twoheadrightarrow \mathrm{Hom}_B(B, p)$ admits a right adjoint right inverse. A cosmological functor $F : \mathcal{K} \to \mathcal{L}$ preserves the class of isofibrations and the simplicial limits that define this map $i_1 \mathbin{\widehat{\pitchfork}} p$. Since the 2-functor $F : \mathfrak{h}\mathcal{K} \to \mathfrak{h}\mathcal{L}$ associated to a cosmological functor preserves adjunctions and the invertibility of 2-cells, F preserves cartesian fibrations. The preservation of cartesian functors follows similarly from Theorem 5.3.4(ii). By Corollary 3.5.7, cosmological functors preserves absolute lifting diagrams, so the preservation of cartesian arrows follows from Theorem 5.1.7(ii). $\square$

Another family of examples of cartesian functors is given by the following lemma, which can be proven using Theorem 5.3.4.

LEMMA 5.3.6. *A fibered right adjoint functor defines a cartesian functor between cartesian fibrations with a common base:*

$$
\begin{array}{ccc}
F & \xrightarrow[\quad h \quad]{\quad \perp \quad} & E \\
& {}_{q}\searrow \quad {}^{B} \quad \swarrow_{p} &
\end{array}
$$

Proof If ℓ is a fibered left adjoint to h in an ∞-cosmos $\mathcal{K}$, then the cosmological functor defined by pullback $p_1^* : \mathcal{K}_{/B} \to \mathcal{K}_{/B^2}$ carries this fibered adjunction to a fibered adjunction

$$
\mathrm{Hom}_B(B, q) \underset{\mathrm{Hom}_{\mathrm{id}_B}(\mathrm{id}_B, h)}{\overset{\mathrm{Hom}_{\mathrm{id}_B}(\mathrm{id}_B, \ell)}{\rightleftarrows}} \perp \; \mathrm{Hom}_B(B, p)
$$

Similarly, the cosmological functor $(-)^2 : \mathcal{K} \to \mathcal{K}$ carries the adjunction $\ell \dashv h$ to an adjunction $\ell^2 \dashv h^2$.

Now both horizontal functors in the commutative square

$$
\begin{array}{ccc}
F^2 & \xrightarrow{\quad h^2 \quad} & E^2 \\
{\scriptstyle i_1\widehat{\pitchfork}q}\Big\downarrow & {\scriptstyle \cong\,\ell} & \Big\downarrow{\scriptstyle i_1\widehat{\pitchfork}p} \\
\mathrm{Hom}_B(B, q) & \xrightarrow[\mathrm{Hom}_{\mathrm{id}_B}(\mathrm{id}_B, h)]{} & \mathrm{Hom}_B(B, p)
\end{array}
$$

admit left adjoints. A standard result from the calculus of mates tells us that the mate with respect to the vertical adjunctions is an isomorphism if and only if the mate with respect to the horizontal adjunctions is an isomorphism, the latter natural transformation between left adjoints being the transpose of the former natural transformation between their right adjoints (see Exercise B.3.iii). In the present context, the mate with respect to the horizontal adjunctions can be seen to be an isomorphism by 2-cell conservativity for $\mathrm{Hom}_B(B, q)$. $\qquad\square$

Exercises

EXERCISE 5.3.i. Show that the pullback of a cartesian functor h between cartesian fibrations with a common base defines a cartesian functor

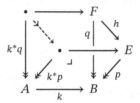

EXERCISE 5.3.ii. Show that cartesian functors compose both vertically and horizontally.

EXERCISE 5.3.iii. Categorify the intuition that cartesian fibrations $p : E \twoheadrightarrow B$ and $q : F \twoheadrightarrow B$ define "contravariant B-indexed functors valued in ∞-categories" by proving that a cartesian functor

$$E \xrightarrow{\ h\ } F$$
$$p \searrow \quad \swarrow q$$
$$B$$

defines a "natural transformation": show that there exists a natural isomorphism in the square of fibers

$$E_b \xrightarrow{\ h\ } F_b$$
$$\beta^* \downarrow \quad \exists \nearrow\cong \quad \downarrow \beta^*$$
$$E_a \xrightarrow{\ h\ } F_a$$

where the action of an arrow β in the homotopy category of B on the fibers is defined by factoring the domain of a p- or q-cartesian lift of β as displayed in (5.0.1).

5.4 Cocartesian Fibrations and Bifibrations

Cocartesian fibrations are dual to cartesian fibrations in the sense that the arrows in the base act covariantly, rather than contravariantly, on the fibers. To make this duality precise, recall from Definition 1.2.25 that for any ∞-cosmos $\mathcal{K}$, there is a **dual ∞-cosmos** $\mathcal{K}^{\mathrm{co}}$ with the same objects but with functor spaces defined by:

$$\mathrm{Fun}_{\mathcal{K}^{\mathrm{co}}}(A, B) := \mathrm{Fun}_{\mathcal{K}}(A, B)^{\mathrm{op}}.$$

The isofibrations, equivalences, and trivial fibrations in $\mathcal{K}^{\mathrm{co}}$ coincide with those of $\mathcal{K}$. Conical limits in $\mathcal{K}^{\mathrm{co}}$ coincide with those in $\mathcal{K}$, while the cotensor of $A \in \mathcal{K}$ with $U \in s\mathcal{S}et$ is defined to be $A^{U^{\mathrm{op}}}$. In particular, the cotensor of an ∞-category with 2 is defined to be $A^{2^{\mathrm{op}}}$, which exchanges the domain and codomain projections from arrow and comma ∞-categories.

With this structure in hand, we can succinctly define a cocartesian fibration in an ∞-cosmos $\mathcal{K}$ to be an isofibration $p : E \twoheadrightarrow B$ that defines a cartesian fibration in the dual ∞-cosmos $\mathcal{K}^{\mathrm{co}}$. Now all of the results proven in §5.1–§5.3 develop the theory of cocartesian arrows, cocartesian fibrations, and the cartesian functors between them. In this section, we tour a few of the highlights before turning our attention to *bifibrations*, isofibrations that define both cartesian and cocartesian fibrations.

DEFINITION 5.4.1 (*p*-cocartesian arrow). Consider an isofibration $p : E \twoheadrightarrow B$. An X-shaped arrow $\ulcorner\psi\urcorner : X \to E^2$ in E is *p*-**cocartesian** if the dashed map defined by the pullback of the Leibniz cotensor is a trivial fibration:

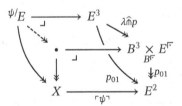

where $\lambda : \ulcorner \hookrightarrow 3$ is the inclusion whose image is the span $2 \leftarrow 0 \to 1$ in 3.

DEFINITION 5.4.2 (cocartesian fibration). An isofibration $p : E \twoheadrightarrow B$ is a **cocart-esian fibration** if any natural transformation $\beta : pe \Rightarrow b$ as below-left admits a p-cocartesian lift $\chi^\beta : e \Rightarrow \beta_* e$ as below-right:

Again it suffices to assume only that the generic arrow whose domain factors through p admits a p-cocartesian lift:

LEMMA 5.4.3. *An isofibration* $p : E \twoheadrightarrow B$ *is a cocartesian fibration if and only if the left comma cone over* p *displayed below-left admits a* p-*cocartesian lift* χ *as displayed below-right:*

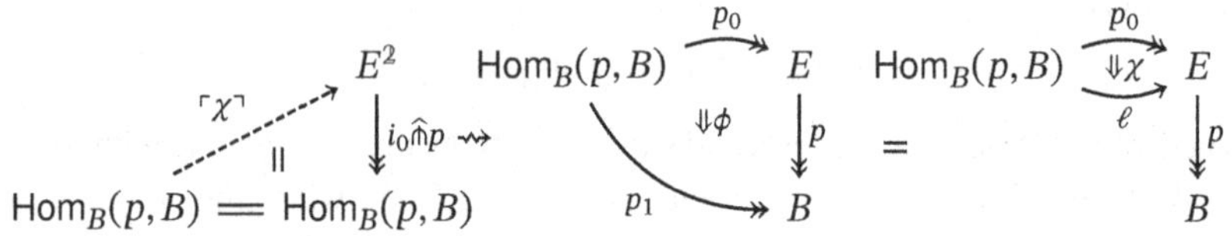

$$
\begin{array}{ccc}
& E^2 & \mathrm{Hom}_B(p,B) \xrightarrow{p_0} E \qquad \mathrm{Hom}_B(p,B) \xrightarrow[\ell]{\quad p_0 \quad \Downarrow\chi} E \\
\ulcorner\chi\urcorner \nearrow & \downarrow i_0 \widehat{\pitchfork} p \rightsquigarrow & \searrow \Downarrow\phi \quad \downarrow p \qquad\qquad = \qquad\qquad \downarrow p \\
\mathrm{Hom}_B(p,B) = \mathrm{Hom}_B(p,B) & & p_1 \searrow B \qquad\qquad\qquad\qquad B
\end{array}
$$

$\square$

Lemma 5.4.3 extends to an internal characterization of cocartesian fibrations. The dual to Theorem 5.2.8 asks for a fibered left adjoint to $\Delta_p : E \to \mathrm{Hom}_B(p,B)$ and a left adjoint right inverse to $i_0 \widehat{\pitchfork} p : E^2 \to \mathrm{Hom}_B(p,B)$ in place of right adjoints (see Exercise 5.4.i).

Propositions 5.2.9 and 5.2.10 dualize to provide the following examples.

PROPOSITION 5.4.4 (codomain projection fibration). *For any* ∞-*category* A:

    (i) *The codomain projection functor* $p_1 : A^2 \twoheadrightarrow A$ *defines a cocartesian fibration. Moreover, a natural transformation* ψ *with codomain* A^2 *is* p_1-*cocartesian if and only if* $p_0\psi$ *is invertible.*

    (ii) *If* A *has pushouts, then the domain projection functor* $p_0 : A^2 \twoheadrightarrow A$ *defines a cocartesian fibration, for which the* p_0-*cocartesian cells are the pushout squares.* $\square$

By this result and its dual, when A has pushouts, the domain projection functor $p_0 : A^2 \twoheadrightarrow A$ is both a cartesian fibration and a cocartesian fibration. Such maps have a special property we now explore.

DEFINITION 5.4.5 (bifibration). An isofibration $p : E \twoheadrightarrow B$ defines a **bifibration** if p is both a cartesian fibration and a cocartesian fibration.

Projections give trivial examples of bifibrations.

EXAMPLE 5.4.6. For any ∞-categories A and B the projection functor $\pi : A \times B \twoheadrightarrow B$ is a bifibration, in which a 2-cell with codomain $A \times B$ is π-cocartesian or π-cartesian if and only if its composite with the projection $\pi : A \times B \twoheadrightarrow A$ is an isomorphism.

PROPOSITION 5.4.7. *Let* $p : E \twoheadrightarrow B$ *be a bifibration. Then any natural transformation* $X \overset{a}{\underset{b}{\Downarrow \beta}} B$ *induces a fibered adjunction between the fibers of* p *over* a *and* b:

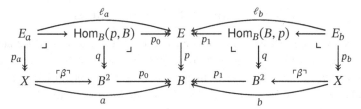

Proof Write $\ulcorner \beta \urcorner : X \to B^2$ for the functor induced by β. Note that the pullbacks defining the fibers over its domain edge a and codomain edge b factor as:

$$
\begin{array}{ccccccccc}
E_a & \overset{\ell_a}{\rightrightarrows} & \mathrm{Hom}_B(p,B) & \underset{p_0}{\twoheadrightarrow} & E & \overset{p_1}{\twoheadleftarrow} & \mathrm{Hom}_B(B,p) & \leftleftarrows & E_b \\
\downarrow p_a & \lrcorner & \downarrow q & \lrcorner & \downarrow p & \llcorner & \downarrow q & \llcorner & \downarrow p_b \\
X & \underset{\ulcorner \beta \urcorner}{\longrightarrow} & B^2 & \underset{p_0}{\twoheadrightarrow} & B & \underset{p_1}{\twoheadleftarrow} & B^2 & \underset{\ulcorner \beta \urcorner}{\longleftarrow} & X \\
& a & & & & & & b &
\end{array}
$$

Theorem 5.2.8(ii) and its dual provide a right adjoint right inverse to the functor $i_1 \widehat{\pitchfork} p : E^2 \twoheadrightarrow \mathrm{Hom}_B(B,p)$ and a left adjoint right inverse to the functor $i_0 \widehat{\pitchfork} p : E^2 \twoheadrightarrow \mathrm{Hom}_B(p,B)$. Composing the former fibered adjunction with $q : \mathrm{Hom}_B(B,p) \twoheadrightarrow B^2$ and the latter fibered adjunction with $q : \mathrm{Hom}_B(p,B) \twoheadrightarrow B^2$ we obtain a composable pair of adjunctions that are fibered over B^2:

$$
\mathrm{Hom}_B(p,B) \underset{i_0 \widehat{\pitchfork} p}{\overset{\ulcorner \chi \urcorner}{\rightleftarrows}} E^2 \underset{\ulcorner \chi \urcorner}{\overset{i_1 \widehat{\pitchfork} p}{\rightleftarrows}} \mathrm{Hom}_B(B,p)
$$

Pulling back the composite adjunction along $\ulcorner \beta \urcorner : X \to B^2$ yields the desired fibered adjunction. $\square$

Note that the construction of the adjoint functors $\beta_* \dashv \beta^*$ given in this proof coincides with the description of the the the action of the 2-cell β on the fibers of p given in Remark 5.2.7.

Exercises

EXERCISE 5.4.i. Formulate the duals to Theorems 5.1.7 and 5.2.8, providing an internal characterization of cocartesian arrows and cocartesian fibrations.

EXERCISE 5.4.ii. Suppose A is an ∞-category with pullbacks. By Propositions 5.2.10 and 5.4.4, the codomain projection functor $p_1 : A^2 \twoheadrightarrow A$ is a bifibration. Describe the action of the left and right adjoints in the adjunction induced from an arrow $\alpha : x \to y$ in hA:

$$1 \underset{y}{\overset{x}{\underset{\Downarrow\alpha}{\rightrightarrows}}} A \quad \rightsquigarrow \quad \mathrm{Hom}_A(A, x) \underset{\alpha^*}{\overset{\alpha_*}{\underset{\longleftarrow}{\overset{\longrightarrow}{\perp}}}} \mathrm{Hom}_A(A, y)$$

5.5 Discrete Cartesian Fibrations

Cartesian and cocartesian fibrations encode families of ∞-categories acted upon contravariantly or covariantly by the base ∞-category. In certain special cases, the ∞-categories arising as fibers of a cartesian or cocartesian fibration are all *discrete* (see Definition 1.2.26). As the analogous functors between 1-categories are called *discrete fibrations* or *discrete opfibrations*, we refer to these maps of ∞-categories as "discrete cartesian fibrations" or "discrete cocartesian fibrations," respectively. Our aim in this section is to study this special class of fibrations.

Before giving the definition, we describe the appropriate sort of "discreteness" required of an isofibration $p : E \twoheadrightarrow B$. Recall from Exercise 1.4.iv that an object E in an ∞-cosmos $\mathcal{K}$ is **discrete** if and only if every natural transformation with codomain E is invertible. Discrete fibrations are isofibrations $p : E \twoheadrightarrow B$ in an ∞-cosmos $\mathcal{K}$ that are discrete when considered as an object of the sliced ∞-cosmos $\mathcal{K}_{/B}$; we call such maps **discrete isofibrations** for short. Using Proposition 3.6.2, there are several equivalent ways to unpack the notion of discrete object in $\mathcal{K}_{/B}$ at the level of the homotopy 2-category h$\mathcal{K}$:

LEMMA 5.5.1. *An isofibration $p : E \twoheadrightarrow B$ is a discrete isofibration if and only if either of the equivalent conditions hold:*

(i) *Any* $X \underset{b}{\overset{a}{\underset{\Downarrow\gamma}{\rightrightarrows}}} E$ *for which $p\gamma$ is an identity is invertible.*

(ii) *Any* $X \underset{b}{\overset{a}{\underset{\Downarrow\gamma}{\rightrightarrows}}} E$ *for which $p\gamma$ is an isomorphism is invertible.*

Proof Exercise 5.5.i. □

Thus, the discrete isofibrations are exactly those isofibrations that define conservative functors in the homotopy 2-category. Neither the domain or codomain of a discrete isofibration need to be discrete ∞-categories (see Exercise 5.5.ii). Instead, the discreteness in the sliced ∞-cosmos is "fiberwise."

LEMMA 5.5.2. *The fibers of a discrete isofibration are discrete ∞-categories.*

Proof Recall from Remark 1.3.3 that discrete ∞-categories are preserved by cosmological functors. In particular, the pullback functor $b^* : \mathcal{K}_{/B} \to \mathcal{K}$ associated to an element $b : 1 \to B$ preserves discrete objects. Hence, the fibers of a discrete isofibration $p : E \twoheadrightarrow B$ are discrete ∞-categories. □

The converse to Lemma 5.5.2 holds in an ∞-cosmos of $(\infty, 1)$-categories: an isofibration $p : E \twoheadrightarrow B$ between $(\infty, 1)$-categories is discrete if and only if its fibers are discrete ∞-categories (see Proposition 12.2.3). Thus, in such ∞-cosmoi, the discrete cartesian fibrations and discrete cocartesian fibrations we presently introduce can be understood as cartesian fibrations with discrete fibers.

DEFINITION 5.5.3 (discrete co/cartesian fibration). An isofibration $p : E \twoheadrightarrow B$ in an ∞-cosmos $\mathcal{K}$ is a **discrete cartesian fibration** if it is a cartesian fibration that is discrete as an object of $\mathcal{K}_{/B}$. Dually, p is a **discrete cocartesian fibration** if it is a cocartesian fibration that is discrete as an object of $\mathcal{K}_{/B}$.

DIGRESSION 5.5.4 (left and right fibrations of quasi-categories). In the ∞-cosmos of quasi-categories, the discrete cocartesian fibrations coincide with Joyal's class of **left fibrations** – those maps that lift against the left horn inclusions – and dually the discrete cartesian fibrations coincide with Joyal's class of **right fibrations** (see Proposition F.4.9). While the terminology of left/right fibrations is more familiar in the ∞-categorical literature, we use the terms "discrete co/cartesian fibrations" to clarify the relationship between these classes of maps and their nondiscrete and 1-categorical analogues.

As in §5.4, the theory of discrete cocartesian fibrations can be obtained by interpreting results about discrete cartesian fibrations in the co-dual ∞-cosmos, so we streamline our exposition by focusing on the class of discrete cartesian fibrations.

LEMMA 5.5.5 (pullback stability). *In any pullback square*

$$
\begin{array}{ccc}
F & \xrightarrow{\;h\;} & E \\
{\scriptstyle q}\downarrow & \lrcorner & \downarrow{\scriptstyle p} \\
A & \xrightarrow{\;k\;} & B
\end{array}
$$

if p is a discrete cartesian fibration then q is a discrete cartesian fibration.

Proof   In light of Proposition 5.2.4 it remains only to verify that q is discrete.
Consider a 2-cell $X \underset{b}{\overset{a}{\rightrightarrows}} F$ with $\Downarrow\gamma$ so that $q\gamma$ is invertible. Then $kq\gamma = ph\gamma$
is invertible and conservativity of p implies that $h\gamma$ is invertible. By 2-cell
conservativity (see Proposition 3.3.1), γ is also invertible.   □

There is a direct 2-categorical characterization of the discrete cartesian fibrations, as those isofibrations $p : E \twoheadrightarrow B$ with the property that every natural
transformation $\beta : b \Rightarrow pe$ has an essentially unique lift with codomain e. This
is closely related to the observation that for a discrete cartesian fibration p,
there is no special class of p-cartesian arrows, unlike the case for the indiscrete
version.

PROPOSITION 5.5.6.

   (i) *If $p : E \twoheadrightarrow B$ is a discrete cartesian fibration, every natural transforma-
       tion with codomain E is p-cartesian.*
   (ii) *An isofibration $p : E \twoheadrightarrow B$ is a discrete cartesian fibration if and only if
        every natural transformation $\beta : b \Rightarrow pe$ has an essentially unique lift
        with codomain e: given $\chi : e' \Rightarrow e$ and $\psi : e'' \Rightarrow e$ so that $p\chi = p\psi = \beta$,
        then there exists an isomorphism $\gamma : e'' \Rightarrow e'$ with $\chi \cdot \gamma = \psi$ and $p\gamma = $ id.*

Proof   By Proposition 5.1.13, if $p : E \twoheadrightarrow B$ is a cartesian fibration, then any
natural transformation ψ with codomain E factors through a p-cartesian lift of
$p\psi$ via a natural transformation γ so that $p\gamma = $ id. When p is discrete, this γ is
an isomorphism, and thus ψ is isomorphic to a p-cartesian transformation, and
thus is itself p-cartesian by Lemma 5.1.6.

From what we have just observed in (i) and the essential uniqueness of p-
cartesian lifts in Lemma 5.1.11, we see that if $p : E \twoheadrightarrow B$ is a discrete cartesian
fibration, then any natural transformation $\beta : b \Rightarrow pe$ has an essentially unique
lift. For the converse, note that any $p : E \twoheadrightarrow B$ satisfying the hypothesis of (ii)
is a discrete isofibration: if $\psi : e' \Rightarrow e$ is so that $p\psi = $ id, then id $: e \Rightarrow e$ is
another lift of $p\psi$ and essential uniqueness provides an inverse isomorphism
$\psi^{-1} : e \Rightarrow e'$.

To complete the proof, it remains to show that any $p : E \twoheadrightarrow B$ satisfying the hypothesis of (ii) is a cartesian fibration. We do this by establishing the 2-categorical characterization of Proposition 5.2.11, showing that any 2-cell $\psi : e' \Rightarrow e$ with codomain E is weakly p-cartesian – and hence, by Proposition 5.2.11 again, p-cartesian in the usual sense.

To that end consider a pair $\tau : e'' \Rightarrow e$ and $\gamma : pe'' \Rightarrow pe'$ so that $p\tau = p\psi \cdot \gamma$. By the hypothesis that every 2-cell admits an essentially unique lift, we can construct a lift $\mu : \bar{e} \Rightarrow e'$ so that $p\mu = \gamma$. Now τ and $\psi \cdot \mu$ are two lifts of $p\tau$ with the same codomain, so there exists an isomorphism $\zeta : e'' \Rightarrow \bar{e}$ with $p\zeta = \mathrm{id}$. The composite $\mu \cdot \zeta$ then defines the desired lift of γ to a cell so that $\tau = \psi \cdot \mu \cdot \zeta$. □

As an immediate consequence of (i):

COROLLARY 5.5.7. *Any commutative square from a cartesian fibration to a discrete cartesian fibration defines a cartesian functor.* □

In analogy with Theorem 5.2.8, there is an internal characterization of discrete cartesian fibrations, which in the discrete case takes a much simpler form.

PROPOSITION 5.5.8 (internal characterization of discrete cartesian fibrations). *An isofibration* $p : E \twoheadrightarrow B$ *is a discrete cartesian fibration if and only if the functor* $i_1 \widehat{\pitchfork} p : E^2 \twoheadrightarrow \mathrm{Hom}_B(B, p)$ *is a trivial fibration.*

Recall from Theorem 5.2.8(ii) that an isofibration $p : E \twoheadrightarrow B$ defines a cartesian fibration if and only if $i_1 \widehat{\pitchfork} p : E^2 \twoheadrightarrow \mathrm{Hom}_B(B, p)$ admits a right adjoint right inverse $\ulcorner \chi \urcorner$. Proposition 5.5.8 asserts that p defines a discrete cartesian fibration if and only if this adjunction defines an adjoint equivalence.

Proof Assume first that $p : E \twoheadrightarrow B$ is a discrete cartesian fibration. By Theorem 5.2.8(ii), $i_1 \widehat{\pitchfork} p : E^2 \rightarrow \mathrm{Hom}_B(B, p)$ then admits a right adjoint $\ulcorner \chi \urcorner$ with invertible counit $\bar{\epsilon} : (i_1 \widehat{\pitchfork} p)\ulcorner \chi \urcorner \cong \mathrm{id}$. We argue that in this case the unit $\bar{\eta} : \mathrm{id} \Rightarrow \ulcorner \chi \urcorner(i_1 \widehat{\pitchfork} p)$ is also invertible, proving that $i_1 \widehat{\pitchfork} p \dashv \ulcorner \chi \urcorner$ defines an adjoint equivalence.

Since the counit of $i_1 \widehat{\pitchfork} p \dashv \ulcorner \chi \urcorner$ is invertible, $(i_1 \widehat{\pitchfork} p)\bar{\eta}$ is an isomorphism. Thus $p_1(i_1 \widehat{\pitchfork} p)\bar{\eta} = p_1\bar{\eta}$ and $p_0(i_1 \widehat{\pitchfork} p)\bar{\eta} = pp_0\bar{\eta}$ are both isomorphisms. By conservativity of the discrete fibration $p : E \twoheadrightarrow B$ proven in Lemma 5.5.1, this implies that $p_0\bar{\eta}$ is invertible and now 2-cell conservativity for E^2 reveals that $\bar{\eta}$ is an isomorphism.

Conversely, if $i_1 \widehat{\pitchfork} p : E^2 \twoheadrightarrow \mathrm{Hom}_B(B, p)$ is a trivial fibration, by Proposition 2.1.12 and Lemma 3.6.9, we may choose a right adjoint right inverse equivalence $\ulcorner \chi \urcorner : \mathrm{Hom}_B(B, p) \twoheadrightarrow E^2$. By composing with $\ulcorner \chi \urcorner$, we see that any arrow $\ulcorner \beta \urcorner : X \rightarrow \mathrm{Hom}_B(B, p)$ has a lift. The unit of this adjoint equivalence is necessarily a fibered isomorphism, so for any arrow $\ulcorner\psi\urcorner\colon X \to E^2$ we have a pasting equality

$$
\begin{array}{c}
E^2 \\
\ulcorner\psi\urcorner \nearrow \quad \downarrow i_1 \widehat{m} p \\
\| \\
X \xrightarrow[\ulcorner p\psi\urcorner]{} \mathrm{Hom}_B(B, p)
\end{array}
\quad = \quad
\begin{array}{c}
E^2 = \!=\!=\!=\!=\!= E^2 \\
\ulcorner\psi\urcorner \nearrow \quad \downarrow i_1 \widehat{m} p \overset{\cong \Downarrow \tilde{\eta}}{\underset{\ulcorner\chi\urcorner}{\;}} \nearrow \downarrow i_1 \widehat{m} p \\
\| \qquad\qquad \| \\
X \xrightarrow[\ulcorner p\psi\urcorner]{} \mathrm{Hom}_B(B, p) = \!=\!= \mathrm{Hom}_B(B, p)
\end{array}
$$

Since the right-hand side is an absolute right lifting diagram, the left-hand side must be as well, and by Theorem 5.1.7(ii) we conclude that every arrow with codomain E is p-cartesian. Now the conservativity property for cartesian arrows of Lemma 5.1.6 applies to all arrows and tells us that $p\colon E \twoheadrightarrow B$ defines a conservative functor, and in particular is discrete. $\qquad\square$

As a consequence of Proposition 5.5.8 it is clear that cosmological functors preserve discrete cartesian fibrations. Using the internal characterization, it is also straightforward to verify that discrete cartesian fibrations compose and cancel on the left:

Lemma 5.5.9. *Suppose $p\colon E \twoheadrightarrow B$ and $q\colon B \twoheadrightarrow A$ are isofibrations and q is a discrete cartesian fibration. Then p is a discrete cartesian fibration if and only if qp is a discrete cartesian fibration.*

Proof The map $E^2 \twoheadrightarrow \mathrm{Hom}_A(A, qp)$ that tests whether $qp\colon E \twoheadrightarrow A$ is a discrete cartesian fibration factors as the map $E^2 \twoheadrightarrow \mathrm{Hom}_B(B, p)$ that tests whether $p\colon E \twoheadrightarrow B$ is a discrete cartesian fibration followed by a pullback of the map $B^2 \twoheadrightarrow \mathrm{Hom}_A(A, q)$ that tests whether $q\colon B \twoheadrightarrow A$ is a discrete cartesian fibration:

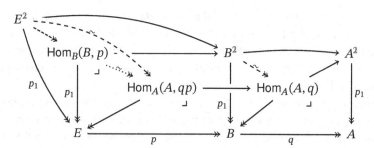

The result now follows from Lemma 1.2.14 and the 2-of-3 property for equivalences. $\qquad\square$

We now turn to examples of discrete cartesian fibrations.

Lemma 5.5.10. *A trivial fibration $p\colon E \overset{\sim}{\twoheadrightarrow} B$ is a discrete bifibration.*

Proof By Lemma 1.2.14, both $i_1 \,\hat{\pitchfork}\, p : E^2 \twoheadrightarrow \mathrm{Hom}_B(B, p)$ and $i_0 \,\hat{\pitchfork}\, p : E^2 \twoheadrightarrow$ $\mathrm{Hom}_B(p, B)$ are trivial fibrations. Now Proposition 5.5.8 and its dual prove that p is a discrete cartesian fibration and also a discrete cocartesian fibration. $\square$

The discussion at the start of this chapter suggests that for any element $b : 1 \to$ B of an ∞-category B, the right representable $p_0 : \mathrm{Hom}_B(B, b) \twoheadrightarrow B$ defines a discrete cartesian fibration. We deduce this in Corollary 5.5.14 as a consequence of a more sophisticated observation concerning the central object of study in Part II. Proposition 5.2.9 proves that for any ∞-category A, the domain projection functor $p_0 : A^2 \twoheadrightarrow A$ defines a cartesian fibration. Unless A is discrete, this functor does not define a discrete cartesian fibration (see Exercise 5.5.ii). However, recall that p_0-cartesian lifts can be constructed to project to identity arrows along $p_1 : A^2 \twoheadrightarrow A$. This suggests that we might productively consider the domain projection functor as a map over A, in which case we have the following result.

PROPOSITION 5.5.11. *For any ∞-category A in an ∞-cosmos $\mathcal{K}$*

$$A^2 \xrightarrow{(p_1, p_0)} A \times A \atop \underset{p_1}{\searrow} \underset{A}{\swarrow} \pi \qquad (5.5.12)$$

defines a discrete cartesian fibration in the sliced ∞-cosmos $\mathcal{K}_{/A}$.

Proof By 2-cell conservativity, (5.5.12) is a discrete object in the sliced ∞-cosmos $(\mathcal{K}_{/A})_{/\pi : A \times A \twoheadrightarrow A} \cong \mathcal{K}_{/A \times A}$. So it remains only to prove that this functor defines a cartesian fibration. We prove this using Theorem 5.2.8(iii). The first step is to compute the right representable comma object for the functor (5.5.12) by interpreting the formula (3.4.2) in the sliced ∞-cosmos $\mathcal{K}_{/A}$. By Proposition 1.2.22, the 2-cotensor of the object $\pi : A \times A \twoheadrightarrow A$ is $\pi : A \times A^2 \twoheadrightarrow A$, and this right representable comma is computed by the left-hand pullback in $\mathcal{K}_{/A}$ below:

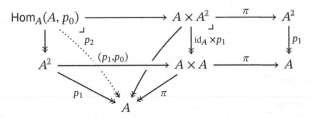

Pasting with the right-hand pullback in $\mathcal{K}$, we recognize that the ∞-category so-constructed coincides with the right representable comma object for the functor $p_0 : A^2 \twoheadrightarrow A$ considered as a map in $\mathcal{K}$. Similarly, the canonical functor

$\Delta_{p_0} : A^2 \to \mathrm{Hom}_A(A, p_0)$ induced by id_{p_0} in $\mathcal{K}$ coincides with the canonical functor $\Delta_{(p_1,p_0)} : A^2 \to \mathrm{Hom}_A(A, p_0)$ over A induced by $\mathrm{id}_{(p_1,p_0)}$ in $\mathcal{K}_{/A}$.

Under the equivalence $\mathrm{Hom}_A(A, p_0) \simeq A^3$ established in the proof of Proposition 5.2.9, the isofibration $p_2 : \mathrm{Hom}_A(A, p_0) \twoheadrightarrow A$ is evaluation at the final element $2 \in 3$ in the composable pair of arrows. Since $p_0 : A^2 \twoheadrightarrow A$ is a cartesian fibration in $\mathcal{K}$, Theorem 5.2.8(iii) tells us that the functor $\Delta_{p_0} : A^2 \to \mathrm{Hom}_A(A, p_0) \simeq A^3$ admits a right adjoint r over A. The proofs of Proposition 5.2.9 and Theorem 5.2.8(ii)$\Rightarrow$(iii) (which can be extracted from the proof of Theorem 5.1.7(ii)$\Rightarrow$(iii)) combine to provide a construction: this adjunction can be defined by cotensoring the composite adjunction of categories below-left into A:

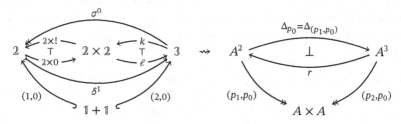

where $\ell \dashv k$ is described in the proof of Proposition 5.2.9. The composite right adjoint is the functor $\sigma^0 : 3 \twoheadrightarrow 2$ that sends 0 and 1 to 0 and 2 to 1, while the composite left adjoint is the functor $\delta^1 : 2 \rightarrowtail 3$ that sends 0 to 0 and 1 to 2. In particular, this adjunction lies in the strict slice 2-category under the inclusion of the "endpoints" of 2 and 3.

It follows that upon cotensoring into A, we obtain a fibered adjunction over $A \times A$, which by Theorem 5.2.8(iii) implies that (5.5.12) is a cartesian fibration in $\mathcal{K}_{/A}$, completing the proof. $\qquad\square$

Combining Propositions 5.2.9 and Proposition 5.5.11, we can now generalize both results to arbitrary comma ∞-categories.

COROLLARY 5.5.13. *For any functors $C \xrightarrow{g} A \xleftarrow{f} B$ between ∞-categories in an ∞-cosmos $\mathcal{K}$:*

    (i) The domain projection functor $p_0 : \mathrm{Hom}_A(f, g) \twoheadrightarrow B$ is a cartesian fibration.[14]

---

[14] Moreover, a natural transformation ψ with codomain $\mathrm{Hom}_A(f, g)$ is p_0-cartesian if and only if $p_1\psi$ is invertible. We defer the proof only because the same argument proves a more general statement (see Lemma 7.4.3).

(ii) The functor

$$\text{Hom}_A(f,g) \xrightarrow{(p_1,p_0)} C \times B$$
$$p_1 \searrow \quad \swarrow \pi$$
$$C$$

defines a discrete cartesian fibration in $\mathcal{K}_{/C}$.

Proof We start with (ii). Since $p_1 : \text{Hom}_A(A,g) \twoheadrightarrow C$ is the pullback of $p_1 : A^2 \twoheadrightarrow A$ along g, we may use the cosmological functor $g^* : \mathcal{K}_{/A} \to \mathcal{K}_{/C}$ to pull back the discrete cartesian fibration of Proposition 5.5.11 to a discrete cartesian fibration in $\mathcal{K}_{/C}$:

$$\text{Hom}_A(A,g) \xrightarrow{(p_1,p_0)} C \times A$$
$$p_1 \searrow \quad \swarrow \pi$$
$$C$$

There is a pullback square in $\mathcal{K}_{/C}$:

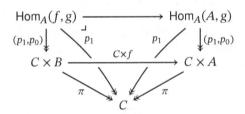

so Lemma 5.5.5 implies that the pullback is also a discrete cartesian fibration.

Using (ii) we can now prove (i). In fact, we show more generally that if

$$E \xrightarrow{(q,p)} C \times B$$
$$q \searrow \quad \swarrow \pi$$
$$C$$

defines a cartesian fibration in $\mathcal{K}_{/C}$ then $p : E \twoheadrightarrow B$ defines a cartesian fibration in $\mathcal{K}$ (see Lemma 7.1.1). By Theorem 5.2.8(iii) applied in $\mathcal{K}_{/C}$, the functor $\Delta_{(q,p)} = \Delta_p$ admits a right adjoint r over $C \times B$:

$$E \underset{r}{\overset{\Delta_{(q,p)}=\Delta_p}{\rightleftarrows}} \perp \text{Hom}_B(B,p)$$
$$(q,p) \searrow \quad \swarrow (qp_1,p_0)$$
$$C \times B$$

Composing with $\pi : C \times B \twoheadrightarrow B$, this fibered adjunction defines an adjunction over B. Applying Theorem 5.2.8(iii) in $\mathcal{K}$ this time allows us to conclude that $p : E \twoheadrightarrow B$ is a cartesian fibration. $\qquad \square$

Note that the domain projection $p_0 : \mathrm{Hom}_A(f, A) \twoheadrightarrow B$ is the pullback of $p_0 : A^2 \twoheadrightarrow A$ along $f : B \to A$, so Propositions 5.2.4 and 5.2.9 imply that this functor is a cartesian fibration, but $p_0 : \mathrm{Hom}_A(A, g) \twoheadrightarrow A$ is not similarly a pullback of $p_0 : A^2 \twoheadrightarrow A$. This is why a more circuitous argument to the result of (i) is needed.

As a corollary, we can finally introduce one of the key examples of discrete cartesian fibrations:

COROLLARY 5.5.14 (domain projection from an element). *For $b : 1 \to B$, the domain projection functor $p_0 : \mathrm{Hom}_B(B, b) \twoheadrightarrow B$ is a discrete cartesian fibration.*

Proof By Corollary 5.5.13, $p_0 : \mathrm{Hom}_B(B, b) \twoheadrightarrow B$ is a cartesian fibration.

Discreteness follows from 2-cell conservativity: if $X \underset{g}{\overset{f}{\rightrightarrows}} \mathrm{Hom}_B(B, b)$

is a natural transformation for which $p_0\gamma$ is an identity, then since $p_1\gamma$ is also an identity – this being a 2-cell whose codomain is the terminal ∞-category – γ must be invertible. $\qquad \square$

Exercises

EXERCISE 5.5.i. Prove Lemma 5.5.1.

EXERCISE 5.5.ii. To explore the discreteness of discrete cartesian fibrations:

(i) Prove that $p_0 : A^2 \twoheadrightarrow A$ is a discrete isofibration if and only if A is a discrete ∞-category.

(ii) By Corollary 5.5.14, for any element $a : 1 \to A$, $p_0 : \mathrm{Hom}_A(A, a) \twoheadrightarrow A$ is a discrete cartesian fibration. Is $\mathrm{Hom}_A(A, a)$ necessarily a discrete ∞-category?

EXERCISE 5.5.iii. Use Theorem 5.3.4 to give an alternate proof of Corollary 5.5.7.

5.6 The Representability of Cartesian Fibrations

In this section we consider the family $p_* : \mathrm{Fun}(X, E) \twoheadrightarrow \mathrm{Fun}(X, B)$ of isofibrations of quasi-categories associated to an isofibration of ∞-categories $p : E \twoheadrightarrow B$

in an ∞-cosmos. Our aim is to show that the notions of cartesian fibration, cartesian functor, and discrete cartesian fibration are each representably defined in various senses:

PROPOSITION 5.6.1. *An isofibration* $p : E \twoheadrightarrow B$ *in an ∞-cosmos* $\mathcal{K}$ *defines a discrete cartesian fibration if and only if for all* $X \in \mathcal{K}$, $p_* : \mathrm{Fun}(X, E) \twoheadrightarrow \mathrm{Fun}(X, B)$ *defines a discrete cartesian fibration of quasi-categories.*

Proof Since equivalences and simplicial limits in an ∞-cosmos are representably defined notions, this follows immediately from the characterization of discrete cartesian fibrations given in Proposition 5.5.8. □

The representable nature of cartesian fibrations is more subtle:

PROPOSITION 5.6.2. *Let* $p : E \twoheadrightarrow B$ *be an isofibration between ∞-categories in an ∞-cosmos* $\mathcal{K}$. *Then* p *is a cartesian fibration if and only if:*

(i) *For all* $X \in \mathcal{K}$, *the isofibration* $p_* : \mathrm{Fun}(X, E) \twoheadrightarrow \mathrm{Fun}(X, B)$ *is a cartesian fibration between quasi-categories.*

(ii) *For all* $f : Y \to X \in \mathcal{K}$, *the square defined by the restriction maps is a cartesian functor:*

$$
\begin{array}{ccc}
\mathrm{Fun}(X, E) & \xrightarrow{\ f^*\ } & \mathrm{Fun}(Y, E) \\
{\scriptstyle p_*}\downarrow & & \downarrow{\scriptstyle p_*} \\
\mathrm{Fun}(X, B) & \xrightarrow[\ f^*\ ]{} & \mathrm{Fun}(Y, B)
\end{array}
$$

Proof If $p : E \twoheadrightarrow B$ is a cartesian fibration, then Theorem 5.2.8(ii) constructs a right adjoint right inverse to $i_1 \widehat{\pitchfork} p : E^2 \twoheadrightarrow \mathrm{Hom}_B(B, p)$. The simplicial bifunctor $\mathrm{Fun} : \mathcal{K}^{\mathrm{op}} \times \mathcal{K} \to \mathcal{QC}at$ defines a 2-functor $\mathrm{Fun} : \mathfrak{h}\mathcal{K}^{\mathrm{op}} \times \mathfrak{h}\mathcal{K} \to \mathfrak{h}\mathcal{QC}at$, which transposes to a Yoneda-type embedding $\mathrm{Fun} : \mathfrak{h}\mathcal{K} \to \mathfrak{h}\mathcal{QC}at^{\mathfrak{h}\mathcal{K}^{\mathrm{op}}}$ from the homotopy 2-category of $\mathcal{K}$ to the 2-category of 2-functors, 2-natural transformations, and modifications (see §B.2). This 2-functor carries the adjunction $i_1 \widehat{\pitchfork} p \dashv \ulcorner\chi\urcorner$ to an adjunction in the 2-category $\mathfrak{h}\mathcal{QC}at^{\mathfrak{h}\mathcal{K}^{\mathrm{op}}}$. This latter adjunction defines, for each $X \in \mathcal{K}$, a right adjoint right inverse adjunction

$$
\begin{array}{ccc}
\mathrm{Fun}(X, E^2) & \underset{\ulcorner\chi_*\urcorner}{\overset{(i_1 \widehat{\pitchfork} p)_*}{\rightleftarrows}} \perp & \mathrm{Fun}(X, \mathrm{Hom}_B(B, p)) \\
\rotatebox{90}{$\approx$} & & \rotatebox{90}{$\approx$} \\
\mathrm{Fun}(X, E)^2 & & \mathrm{Hom}_{\mathrm{Fun}(X,B)}(\mathrm{Fun}(X, B), p_*)
\end{array}
$$

and for each $f : Y \to X$ in $\mathcal{K}$, a strict adjunction morphism, commuting strictly with the left and right adjoints and with the units and counits:[15]

$$
\begin{array}{ccc}
\mathrm{Fun}(X, E^2) & \xrightarrow{\;f^*\;} & \mathrm{Fun}(Y, E^2) \\[4pt]
\ulcorner\chi\urcorner_* \Big(\vdash \!\Big\downarrow\! (i_1\widehat{\pitchfork}p)_* & (i_1\widehat{\pitchfork}p)_* \!\Big\downarrow\! \dashv\Big) \ulcorner\chi\urcorner_* & (5.6.3) \\[4pt]
\mathrm{Fun}(X, \mathrm{Hom}_B(B, p)) & \xrightarrow[f^*]{} & \mathrm{Fun}(Y, \mathrm{Hom}_B(B, p))
\end{array}
$$

By Theorems 5.2.8(ii) and 5.3.4(ii), this demonstrates the two conditions of the statement.

Conversely, if $p : E \twoheadrightarrow B$ satisfies conditions (i) and (ii) then by Theorems 5.2.8(ii) and 5.3.4(ii) there is a commutative square $(i_1 \widehat{\pitchfork} p)_* f^* = f^*(i_1 \widehat{\pitchfork} p)_*$ where both verticals $(i_1 \widehat{\pitchfork} p)_*$ admit right adjoint right inverses $(i_1 \widehat{\pitchfork} p)_* \dashv \bar{r}$ and the mate of the identity $(i_1 \widehat{\pitchfork} p)_* f^* = f^*(i_1 \widehat{\pitchfork} p)_*$ defines an isomorphism $f^* \bar{r} \cong \bar{r} f^*$. Using the right adjoints $\bar{r}$, we extract the internal right adjoint functor $\ulcorner\chi\urcorner : \mathrm{Hom}_B(B, p) \to E^2$ as the image of the identity element

$$
\begin{array}{ccc}
\mathrm{Fun}(\mathrm{Hom}_B(B, p), \mathrm{Hom}_B(B, p)) & \xrightarrow{\;\bar{r}\;} & \mathrm{Fun}(\mathrm{Hom}_B(B, p), E^2) \\[8pt]
\mathrm{id} & \longmapsto & \ulcorner\chi\urcorner
\end{array}
$$

The counit is internalized similarly. The condition on the mates is used to define the unit and verify the triangle equalities equalities that demonstrate that $i_1 \widehat{\pitchfork} p \dashv \ulcorner\chi\urcorner$ (see Proposition B.6.2). Now Theorem 5.2.8(ii) proves that $p : E \twoheadrightarrow B$ is a cartesian fibration. □

An easier argument along the same lines demonstrates:

Corollary 5.6.4. *A commutative square between cartesian fibrations as displayed below-left*

$$
\begin{array}{ccc}
F \xrightarrow{\;h\;} E & & \mathrm{Fun}(X, F) \xrightarrow{\;h_*\;} \mathrm{Fun}(X, E) \\[4pt]
q\Big\downarrow \quad \Big\downarrow p \qquad \rightsquigarrow & & q_*\Big\downarrow \qquad\qquad \Big\downarrow p_* \\[4pt]
A \xrightarrow{\;k\;} B & & \mathrm{Fun}(X, A) \xrightarrow[k_*]{} \mathrm{Fun}(X, B)
\end{array}
$$

defines a cartesian functor in an ∞-cosmos $\mathcal{K}$ if and only if for all $X \in \mathcal{K}$, the square displayed above right defines a cartesian functor between cartesian fibrations of quasi-categories.

Proof Exercise 5.6.i. □

---

[15] In particular, the mate of the identity $(i_1 \widehat{\pitchfork} p)_* f^* = f^*(i_1 \widehat{\pitchfork} p)_*$ is the identity $f^* \ulcorner\chi\urcorner_* = \ulcorner\chi\urcorner_* f^*$.

In particular, if $p : E \twoheadrightarrow B$ is a cartesian fibration, so is $p_* : \mathsf{Fun}(X, E) \twoheadrightarrow \mathsf{Fun}(X, B)$. We now consider the relationship between p-cartesian arrows and p_*-cartesian arrows.

LEMMA 5.6.5. *Consider a cartesian fibration* $p : E \twoheadrightarrow B$ *between* ∞-*categories. A 2-cell as below-left is cartesian for* $p : E \twoheadrightarrow B$ *if and only if the corresponding 2-cell below-right is cartesian for* $p_* : \mathsf{Fun}(X, E) \twoheadrightarrow \mathsf{Fun}(X, B)$.

$$X \underset{e}{\overset{e'}{\Longrightarrow}} {\scriptstyle\Downarrow\psi} E \qquad \leftrightsquigarrow \qquad 1 \underset{e}{\overset{e'}{\Longrightarrow}} {\scriptstyle\Downarrow\psi} \mathsf{Fun}(X, E)$$

The natural transformation on the left defines an arrow in the hom-category $\mathsf{hFun}(X, E)$, while the natural transformation on the right defines an arrow in the hom-category $\mathsf{hFun}(1, \mathsf{Fun}(X, E))$. These hom-categories are isomorphic, justifying our conflating notation for their objects and arrows. There is a similar bijective correspondence between X-shaped arrows $\ulcorner\psi\urcorner : X \to E^2$ in E and 1-arrows $\ulcorner\psi\urcorner : 1 \to \mathsf{Fun}(X, E)^2$ in $\mathsf{Fun}(X, E)$.

Proof Since $p : E \twoheadrightarrow B$ is a cartesian fibration, a natural transformation as above-left is p-cartesian if and only if it satisfies the weak universal properties of Proposition 5.1.13, which are started entirely in reference to the functor $p_* : \mathsf{hFun}(X, E) \twoheadrightarrow \mathsf{hFun}(X, B)$. Similarly, since $p_* : \mathsf{Fun}(X, E) \twoheadrightarrow \mathsf{Fun}(X, B)$ is a cartesian fibration, a natural transformation as above right is p_*-cartesian if and only if it satisfies the weak universal properties of Proposition 5.1.13, started entirely in reference to $p_* : \mathsf{hFun}(1, \mathsf{Fun}(X, E)) \twoheadrightarrow \mathsf{hFun}(1, \mathsf{Fun}(X, B))$. As these functors are isomorphic, the p-cartesian transformations and p_*-cartesian transformations $\psi : e' \Rightarrow e$ coincide. $\qquad\square$

Lemma 5.6.5 characterizes the p_*-cartesian transformations with domain 1. More generally:

LEMMA 5.6.6. *A natural transformation as below left*

$$Q \underset{e}{\overset{e'}{\Longrightarrow}} {\scriptstyle\Downarrow\psi} \mathsf{Fun}(X, E) \qquad 1 \overset{q}{\longrightarrow} Q \underset{e}{\overset{e'}{\Longrightarrow}} {\scriptstyle\Downarrow\psi} \mathsf{Fun}(X, E) \leftrightsquigarrow X \underset{eq}{\overset{e'q}{\Longrightarrow}} {\scriptstyle\Downarrow\psi q} E$$

is p_*-*cartesian if and only if each of its components* ψq *is* p-*cartesian.*

Proof If ψ is p_*-cartesian, then by Lemma 5.1.4 so is its restriction along any element $q : 1 \to Q$. By Lemma 5.6.5 this tells us that ψq defines a p-cartesian transformation.

Conversely, if ψq is a p-cartesian transformation, then Lemma 5.6.5 tells us that ψq is a p_*-cartesian transformation. Now consider the factorization

$\psi = \chi \cdot \zeta$ through p_*-cartesian lift χ of $p_*\psi$. Because the components ψq of ψ are p_*-cartesian, the components ζq of ζ are isomorphisms. By Corollary 1.1.22, an arrow in an exponential quasi-category $\mathrm{Fun}(X, E)^Q$ is an isomorphism if and only if it is a pointwise isomorphism, so this implies that ζ is an isomorphism. By isomorphism stability of cartesian transformations (see Lemma 5.1.6), we thus conclude that ψ is p_*-cartesian.   □

Exercises

EXERCISE 5.6.i. Prove Corollary 5.6.4.

5.7 The Yoneda Lemma

Let $b : 1 \to B$ be an element of an ∞-category B and consider its right representation $\mathrm{Hom}_B(B, b)$ as a comma ∞-category. The codomain projection functor provides no additional information in this case, but the domain projection functor $p_0 : \mathrm{Hom}_B(B, b) \twoheadrightarrow B$ has a special property expressed by Corollary 5.5.14: it defines a discrete cartesian fibration. As the fibers of this map over an element $x : 1 \to B$ are the mapping spaces $\mathrm{Hom}_B(x, b)$ of Definition 3.4.9, we regard $p_0 : \mathrm{Hom}_B(B, b) \twoheadrightarrow B$ as encoding the contravariant functor represented by b.

Our aim in this section is to state and prove the Yoneda lemma in this setting, where contravariant representable functors are encoded as discrete cartesian fibrations. Informally, the Yoneda lemma asserts that "evaluation at the identity defines an equivalence," where the identity element in question is the functor

$$1 \xrightarrow{\ulcorner \mathrm{id}_b \urcorner} \mathrm{Hom}_B(B, b)$$

$$b \searrow \quad \swarrow p_0$$

$$B$$

Technically, $\ulcorner \mathrm{id}_b \urcorner$ does not live in the sliced ∞-cosmos over B because the domain object $b : 1 \to B$ is not an isofibration but nevertheless for any isofibration $p : E \twoheadrightarrow B$, restriction along $\ulcorner \mathrm{id}_b \urcorner$ induces a functor between sliced quasi-categorical functor spaces

$$\mathrm{Fun}_B(\mathrm{Hom}_B(B, b) \xrightarrow{p_0} B, E \xrightarrow{p} B) \xrightarrow{\mathrm{ev}_{\ulcorner \mathrm{id}_b \urcorner}} \mathrm{Fun}_B(1 \xrightarrow{b} B, E \xrightarrow{p} B)$$

Here the codomain is the quasi-category defined by the pullback

$$
\begin{array}{ccc}
\mathrm{Fun}_B(b, p) \cong \mathrm{Fun}(1, E_b) & \longrightarrow & \mathrm{Fun}(1, E) \\
\downarrow & \lrcorner & \downarrow{\scriptstyle p_*} \\
1 & \xrightarrow{\quad b \quad} & \mathrm{Fun}(1, B)
\end{array}
$$

which is isomorphic to the underlying quasi-category of the fiber E_b of $p : E \twoheadrightarrow B$ over b. When $p : E \twoheadrightarrow B$ is a discrete isofibration, $\mathsf{Fun}(1, E_b)$ is a Kan complex and might be referred to more evocatively as the "underlying space" of the fiber E_b.

If a discrete cartesian fibration over B is thought of as a contravariant B-indexed discrete ∞-category-valued functor, then maps of discrete cartesian fibrations over B are "natural transformations," the "naturality in B" arising because the functors are fibered over B. This leads to our first statement of the fibrational Yoneda lemma:

THEOREM 5.7.1 (discrete Yoneda lemma). *If $p : E \twoheadrightarrow B$ is a discrete cartesian fibration, then*

$$\mathsf{Fun}_B(\mathsf{Hom}_B(B, b) \xrightarrow{p_0} B, E \xrightarrow{p} B) \xrightarrow[\sim]{\mathrm{ev}_{\ulcorner \mathrm{id}_b \urcorner}} \mathsf{Fun}(1, E_b)$$

is an equivalence of Kan complexes.

We deduce this result from a "dependent" generalization where the target discrete cartesian fibration has codomain $\mathsf{Hom}_B(B, b)$. In this case, the result provides an equivalence between sections of a discrete cartesian fibration $q : F \twoheadrightarrow \mathsf{Hom}_B(B, b)$ and elements of the fiber over $\ulcorner \mathrm{id}_b \urcorner$. This result is analogous to the "path induction" principle for identity types in homotopy type theory: the inverse equivalence of Theorem 5.7.2 provides a "directed" version of the "transport" operation [106, §9].

THEOREM 5.7.2 (dependent Yoneda lemma). *If $b : 1 \to B$ is an element of an ∞-category B and $q : F \twoheadrightarrow \mathsf{Hom}_B(B, b)$ is a discrete cartesian fibration, then*

$$\mathsf{Fun}_{\mathsf{Hom}_B(B,b)}(\mathsf{Hom}_B(B, b), F) \xrightarrow[\sim]{\mathrm{ev}_{\ulcorner \mathrm{id}_b \urcorner}} \mathsf{Fun}(1, F_{\ulcorner \mathrm{id}_b \urcorner})$$

is an equivalence of Kan complexes.

Theorem 5.7.1 is subsumed by a generalization that allows $p : E \twoheadrightarrow B$ to be any cartesian fibration, not necessarily discrete. In this case, p encodes a contravariant B-indexed ∞-category-valued functor, as does $p_0 : \mathsf{Hom}_B(B, b) \twoheadrightarrow B$. The correct notion of "natural transformation" between two such functors is now given by a cartesian functor over B (see Exercise 5.3.iii). Given a pair of cartesian fibration $q : F \twoheadrightarrow B$ and $p : E \twoheadrightarrow B$, we write

$$\mathsf{Fun}_B^{\mathrm{cart}}(F \xrightarrow{q} B, E \xrightarrow{p} B) \subset \mathsf{Fun}_B(F \xrightarrow{q} B, E \xrightarrow{p} B)$$

for the sub-quasi-category containing all those simplices whose vertices define *cartesian* functors from q to p.[16]

---

[16] For any quasi-category Q and any subset of its vertices, there is a "full" sub-quasi-category containing exactly those vertices and all the simplices of Q that they span.

THEOREM 5.7.3 (Yoneda lemma). *If* $p : E \twoheadrightarrow B$ *is a cartesian fibration, then*

$$\mathsf{Fun}_B^{\mathrm{cart}}(\mathsf{Hom}_B(B, b) \xrightarrow{p_0} B, E \xrightarrow{p} B) \xrightarrow[\sim]{\mathrm{ev}_{\ulcorner\mathrm{id}_b\urcorner}} \mathsf{Fun}(1, E_b)$$

is an equivalence of quasi-categories.

The proofs of these theorems overlap significantly and we develop them in parallel. The basic idea is to use the universal property of $\ulcorner\mathrm{id}_b\urcorner$ as a terminal element of $\mathsf{Hom}_B(B, b)$ in the ∞-cosmos $\mathcal{K}$ (see Corollary 3.5.10) to define a right adjoint to $\mathrm{ev}_{\ulcorner\mathrm{id}_b\urcorner}$ and prove that when $p : E \twoheadrightarrow B$ is discrete or when the domain is restricted to the sub-quasi-category of cartesian functors, this adjunction defines an adjoint equivalence. Note that the functor $\mathrm{ev}_{\ulcorner\mathrm{id}_b\urcorner}$ is the image of the functor $\ulcorner\mathrm{id}_b\urcorner$ under the 2-functor $\mathsf{Fun}_B(-, p) : \mathfrak{h}(\mathcal{K}_{/B})^{\mathrm{op}} \to \mathfrak{h}\mathcal{QC}at$. If the adjunction $! \dashv \ulcorner\mathrm{id}_b\urcorner$ lived in the sliced ∞-cosmos $\mathcal{K}_{/B}$, this would directly construct a right adjoint to $\mathrm{ev}_{\ulcorner\mathrm{id}_b\urcorner}$. The main technical difficulty in following this outline is that the adjunction that witnesses the terminality of $\ulcorner\mathrm{id}_b\urcorner$ does not live in the slice of the homotopy 2-category $\mathfrak{h}\mathcal{K}_{/B}$ but rather in a lax slice $\mathfrak{h}\mathcal{K}_{/\!/B}$ of the homotopy 2-category that we now introduce.

DEFINITION 5.7.4. Consider a 2-category $\mathfrak{h}\mathcal{K}$ and an object $A \in \mathfrak{h}\mathcal{K}$. The **lax slice 2-category** $\mathfrak{h}\mathcal{K}_{/\!/A}$ is the strict 2-category whose

- objects are maps $f : X \to A$ in $\mathfrak{h}\mathcal{K}$ with codomain A;
- 1-cells are diagrams

$$
\begin{array}{ccc}
X & \xrightarrow{\ s\ } & Y \\
& \underset{f}{\searrow} \overset{\sigma}{\underset{\Rightarrow}{}} \swarrow{g} & \\
& A &
\end{array}
\tag{5.7.5}
$$

in $\mathfrak{h}\mathcal{K}$; and

- 2-cells from (s, σ) to (s', σ') are 2-cells $\theta : s \Rightarrow s'$ so that

$$
\begin{array}{ccc}
X \overset{s'}{\underset{\Uparrow\theta}{\underset{s}{\rightrightarrows}}} Y & & X \xrightarrow{\ s'\ } Y \\
\underset{f}{\searrow} \overset{\sigma}{\underset{\Rightarrow}{}} \swarrow{g} & = & \underset{f}{\searrow} \overset{\sigma'}{\underset{\Rightarrow}{}} \swarrow{g} \\
A & & A
\end{array}
\tag{5.7.6}
$$

For instance, the adjunctions that define terminal elements lift to the lax slice 2-category:

LEMMA 5.7.7. *Suppose* $t : 1 \to A$ *defines a terminal element in an ∞-category*

A in an ∞-cosmos $\mathcal{K}$. Then

$$
\begin{array}{c}
1 \xrightarrow{\ t\ } A \\
\searrow_{t} \quad \Uparrow \\
\quad A
\end{array}
\qquad \text{is right adjoint to the unit map} \qquad
\begin{array}{c}
A \xrightarrow{\ !\ } 1 \\
\searrow \ \overset{\eta}{\Rightarrow} \ \swarrow_{t} \\
\quad A
\end{array}
$$

in the lax slice 2-category $\mathfrak{h}\mathcal{K}_{/\!/A}$.

Proof We check that the unit and counit of the adjunction $! \dashv t$ that witnesses the terminality of t lift along the forgetful 2-functor $\mathfrak{h}\mathcal{K}_{/\!/A} \to \mathfrak{h}\mathcal{K}$, which amounts to verifying the condition (5.7.6). The forgetful 2-functor $\mathfrak{h}\mathcal{K}_{/\!/A} \to \mathfrak{h}\mathcal{K}$ is faithful on 1- and 2-cells, so the triangle equalities automatically hold for the lifted cells. These lax compatibility conditions reduce to the pasting equalities

The first of these is trivial, while the second holds by the triangle equality $\eta t = \mathrm{id}_t$. □

Using somewhat nonstandard 2-categorical techniques, we transfer the adjunction of Lemma 5.7.7 to an adjoint equivalence

$$
\mathsf{Fun}_A(A, F) \overset{\mathrm{ev}_t}{\underset{\widetilde{y}}{\overset{\sim}{\rightleftarrows}}} \mathsf{Fun}(1, F_t)
$$

between the Kan complex of sections of a discrete cartesian fibration $q : F \twoheadrightarrow A$ and the underlying space of the fiber F_t over a terminal element $t : 1 \to A$. Because our initial adjunction lives in the lax rather than the strict slice, the construction is somewhat delicate, passing through a pair of auxiliary 2-categories that we now introduce.

DEFINITION 5.7.8. Let $\mathfrak{h}\mathcal{K}$ be the homotopy 2-category of an ∞-cosmos and write $\mathfrak{h}\mathcal{K}^{\lrcorner}$ for the strict 2-category whose

- objects are cospans

$$
A \xrightarrow{\ k\ } B \xleftarrow{\ p\ } E
$$

in which p is a cartesian fibration;

- 1-cells are diagrams of the form

$$
\begin{array}{ccccc}
A' & \xrightarrow{\ k'\ } & B' & \xleftarrow{\ p'\ } & E' \\
{\scriptstyle a}\downarrow & {\scriptstyle \Uparrow\phi} & \downarrow{\scriptstyle b} & & \downarrow{\scriptstyle e} \\
A & \xrightarrow[\ k\ ]{} & B & \xleftarrow[\ p\ ]{} & E
\end{array}
\qquad (5.7.9)
$$

- and whose 2-cells consist of triples $\alpha\colon a \Rightarrow \bar{a}$, $\beta\colon b \Rightarrow \bar{b}$, and $\epsilon\colon e \Rightarrow \bar{e}$ between the verticals of parallel 1-cell diagrams so that $p\epsilon = \beta p'$ and $\bar{\phi}\cdot k\alpha = \beta k'\cdot\phi$.

DEFINITION 5.7.10. Let $\mathfrak{h}\mathcal{K}$ be the homotopy 2-category of an ∞-cosmos and write $\mathfrak{h}\mathcal{K}^{\square}$ for the strict 2-category whose

- objects are pullback squares

$$
\begin{array}{ccc}
F & \xrightarrow{\ h\ } & E \\
{\scriptstyle q}\downarrow & {\scriptstyle \lrcorner} & \downarrow{\scriptstyle p} \\
A & \xrightarrow[\ k\ ]{} & B
\end{array}
$$

whose verticals are cartesian fibrations;

- 1-cells are cubes

$$
(5.7.11)
$$

whose vertical faces commute and in which $\chi\colon h\ell \Rightarrow eh'$ is a p-cartesian lift of $\phi q'$; and

- whose 2-cells are given by quadruples $\alpha\colon a \Rightarrow \bar{a}$, $\beta\colon b \Rightarrow \bar{b}$, $\epsilon\colon e \Rightarrow \bar{e}$, and $\lambda\colon \ell \Rightarrow \bar{\ell}$ in which ϵ and λ are, respectively, lifts of $\beta p'$ and $\alpha q'$ and so that $\phi\cdot k\alpha = \beta k'\cdot\phi$ and $\bar{\chi}\cdot h\lambda = \epsilon h'\cdot\chi$.

These definitions are arranged so that there is an evident forgetful 2-functor $U\colon \mathfrak{h}\mathcal{K}^{\square} \to \mathfrak{h}\mathcal{K}^{\lrcorner}$ that has the strong surjectivity property introduced in Definition 3.6.1.

LEMMA 5.7.12. *The forgetful 2-functor* $U\colon \mathfrak{h}\mathcal{K}^{\square} \to \mathfrak{h}\mathcal{K}^{\lrcorner}$ *is a smothering 2-functor.*

Proof   Proposition 5.2.4 tells us that $\mathfrak{h}\mathcal{K}^{\square} \to \mathfrak{h}\mathcal{K}^{\dashv}$ is surjective on objects. To see that it is full on 1-cells, first form the pullbacks of the cospans in (5.7.9), then define χ to be any p-cartesian lift of $\phi q'$ with codomain eh'. By construction, the domain of χ lies strictly over kaq' and so this functor factors uniquely through the pullback leg h defining the map ℓ of (5.7.11).

To prove that $\mathfrak{h}\mathcal{K}^{\square} \to \mathfrak{h}\mathcal{K}^{\dashv}$ is full on 2-cells, consider a parallel pair of 1-cells in $\mathfrak{h}\mathcal{K}^{\square}$. For one of these we use the notation of (5.7.11) and for the other we denote the diagonal functors by $\bar{a}, \bar{b}, \bar{e}$, and $\bar{\ell}$ and denote the 2-cells by $\bar{\phi}$ and $\bar{\chi}$; the requirement that these 1-cells be parallel implies that the pullback faces are necessarily the same. Now consider a triple $\alpha : a \Rightarrow \bar{a}$, $\beta : b \Rightarrow \bar{b}$, and $\epsilon : e \Rightarrow \bar{e}$ satisfying the conditions of Definition 5.7.8. Our task is to define a fourth 2-cell $\lambda : \ell \Rightarrow \bar{\ell}$ so that $q\lambda = \alpha q'$ and $\bar{\chi} \cdot h\lambda = \epsilon h' \cdot \chi$.

To achieve this, we first define a 2-cell $\gamma : h\ell \Rightarrow h\bar{\ell}$ using the induction property of the p-cartesian cell $\bar{\chi} : h\bar{\ell} \Rightarrow \bar{e}h'$ applied to the composite 2-cell $\epsilon h' \cdot \chi : h\ell \Rightarrow \bar{e}h'$ and the factorization $p\epsilon h' \cdot p\chi = \bar{\phi}q' \cdot k\alpha q'$. By construction $p\gamma = k\alpha q'$ so the pair $\alpha q'$ and γ induces a 2-cell $\lambda : \ell \Rightarrow \bar{\ell}$ so that $q\lambda = \alpha q'$ and $h\lambda = \gamma$. The quadruple $(\alpha, \beta, \epsilon, \lambda)$ now defines the required 2-cell in $\mathfrak{h}\mathcal{K}^{\square}$.

Finally, for 2-cell conservativity, suppose α, β, and ϵ as above are isomorphisms. By the conservativity property for pullbacks described in Proposition 3.3.1, to show that λ is an isomorphism, it suffices to prove that $q\lambda = \alpha q'$ is, which we know already, and that $h\lambda = \gamma$ is invertible. But γ was constructed as a factorization $\epsilon h' \cdot \chi = \bar{\chi} \cdot \gamma$ with $p\gamma = k\alpha q'$. Since ϵ is an isomorphism, $\epsilon h' \cdot \chi$ is p-cartesian, so Lemma 5.1.12 proves that γ is an isomorphism. $\square$

We cannot directly define a pullback 2-functor $\mathfrak{h}\mathcal{K}^{\dashv} \to \mathfrak{h}\mathcal{K}$ in the homotopy 2-category because the 2-categorical universal property of pullbacks in $\mathfrak{h}\mathcal{K}$ is weak and not strict (see Proposition 3.3.1). Instead, the zigzag of 2-functors

$$\mathfrak{h}\mathcal{K}^{\dashv} \xleftarrow{\ U\ } \mathfrak{h}\mathcal{K}^{\square} \xrightarrow{\ \mathrm{ev_T}\ } \mathfrak{h}\mathcal{K},$$

in which the backwards map is a smothering 2-functor and the forwards map evaluates at the pullback vertex, defines a reasonable replacement.

PROPOSITION 5.7.13. *Let $t : 1 \to A$ define a terminal element of A and let $q : F \twoheadrightarrow A$ be a cartesian fibration. Then evaluation at t admits a right adjoint*

$$\mathsf{Fun}_A(A, F) \underset{y}{\overset{\mathrm{ev}_t}{\underset{\perp}{\rightleftarrows}}} \mathsf{Fun}(1, F_t)$$

that defines an adjoint equivalence of Kan complexes in the case where q is a discrete cartesian fibration.

Proof The desired adjunction is obtained by transferring the adjunction of Lemma 5.7.7 through the sequence of 2-functors

$$\mathfrak{h}\mathcal{QC}at^{\square} \xrightarrow{\text{ev}_\top} \mathfrak{h}\mathcal{QC}at$$
$$\downarrow U$$
$$\mathfrak{h}\mathcal{K}^{\text{op}}_{/\!/A} \xrightarrow{-\vee q_*} \mathfrak{h}\mathcal{QC}at^{\lrcorner}$$

using Lemma 3.6.7 to lift along the middle smothering 2-functor.

For a fixed cartesian fibration $q : F \twoheadrightarrow A$ in an ∞-cosmos $\mathcal{K}$, there is a 2-functor $- \vee q_* : \mathfrak{h}\mathcal{K}^{\text{op}}_{/\!/A} \to \mathfrak{h}\mathcal{QC}at^{\lrcorner}$ that carries a 1-cell (5.7.5) to

$$
\begin{array}{ccccc}
1 & \xrightarrow{\;g\;} & \text{Fun}(Y,\Lambda) & \xleftarrow{\;q_*\;} & \text{Fun}(Y,F) \\
\| & \quad\Uparrow\sigma & \downarrow{s^*} & & \downarrow{s^*} \\
1 & \xrightarrow{\;f\;} & \text{Fun}(X,A) & \xleftarrow{\;q_*\;} & \text{Fun}(X,F)
\end{array}
$$

and a 2-cell $\theta : s \Rightarrow s'$ to the 2-cell that acts via pre-whiskering with θ in its two nonidentity components. By Corollary 5.3.5, the functors q_* are cartesian fibration of quasi-categories.

We now apply the 2-functor $- \vee q_* : \mathfrak{h}\mathcal{K}^{\text{op}}_{/\!/A} \to \mathfrak{h}\mathcal{QC}at^{\lrcorner}$ to the adjunction of Lemma 5.7.7 to obtain an adjunction in $\mathfrak{h}\mathcal{QC}at^{\lrcorner}$ and then use the smothering 2-functor of Lemma 5.7.12 and Lemma 3.6.7 to lift this to an adjunction in $\mathfrak{h}\mathcal{QC}at^{\square}$. As elaborated in Exercise 3.6.iii, the lifted adjunction in $\mathfrak{h}\mathcal{QC}at^{\square}$ can be constructed using any lifts of the objects, 1-cells, and either the unit or counit of the adjunction in $\mathfrak{h}\mathcal{QC}at^{\lrcorner}$.

In particular, we may take the left and right adjoints of the lifted adjunction in $\mathfrak{h}\mathcal{QC}at^{\square}$ to be any lifts of the images in $\mathfrak{h}\mathcal{QC}at^{\lrcorner}$ of the right and left adjoints of the adjunction $! \dashv t$ in $\mathfrak{h}\mathcal{K}_{/\!/A}$:

$$
\begin{array}{ccc}
\text{Fun}_A(A,F) & \longrightarrow & \text{Fun}(A,F) \\
& \quad {}^{\lrcorner}\text{ev}_t \nearrow & \qquad\qquad {}_{t^*}\searrow \\
\downarrow & = \quad \text{Fun}(1,F_t) \xrightarrow{\quad} & \Big| q_* \quad \text{Fun}(1,F) \\
& \qquad\qquad {}^{\lrcorner} & \quad \downarrow \\
1 \xRightarrow{\;\;\text{id}_A\;\;} & \Big| \longrightarrow \text{Fun}(A,A) & \qquad {}_{t^*}\searrow \;\; \Big| q_* \\
& \qquad = \quad \downarrow & \\
& 1 \xrightarrow{\quad t \quad} & \text{Fun}(1,A)
\end{array}
$$

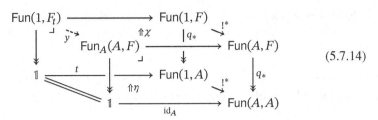

$$(5.7.14)$$

which shows that the left adjoint is the desired functor. Since the counit of $! \dashv t$ is an identity, the counit of the lifted adjunction may also be taken to be an identity. Finally, we compose with the forgetful 2-functor $\mathfrak{h}\mathcal{Q}\mathcal{C}at^{\square} \to \mathfrak{h}\mathcal{Q}\mathcal{C}at$ that evaluates at the pullback vertex to project our adjunction in $\mathfrak{h}\mathcal{Q}\mathcal{C}at^{\square}$ to the desired adjunction in $\mathfrak{h}\mathcal{Q}\mathcal{C}at$.

When $q : F \twoheadrightarrow A$ is a discrete isofibration, both $\mathrm{Fun}(1, F_t)$ and $\mathrm{Fun}_A(A, F)$ are Kan complexes. Any adjunction between Kan complexes is automatically an adjoint equivalence, since it follows from Corollary 1.1.22 that any natural transformation whose codomain is a Kan complex is a natural isomorphism. $\qquad\square$

A special case of Proposition 5.7.13 proves the dependent Yoneda lemma.

Proof of Theorem 5.7.2   Recall from Corollary 3.5.10 that for any element $b : 1 \to B$, its identity arrow $\ulcorner\mathrm{id}_b\urcorner : 1 \to \mathrm{Hom}_B(B, b)$ defines a terminal element. So the dependent Yoneda lemma follows immediately as a special case of Proposition 5.7.13. $\qquad\square$

Using Theorem 5.7.2, we now prove the discrete Yoneda lemma.

Proof of Theorem 5.7.1   Let $p : E \twoheadrightarrow B$ be a discrete cartesian fibration and consider an element $b : 1 \to B$. By Lemma 5.5.5, the pullback

$$
\begin{array}{ccc}
F & \longrightarrow & E \\
{\scriptstyle q}\downarrow \quad {\scriptstyle \ulcorner} & & \downarrow{\scriptstyle p} \\
\mathrm{Hom}_B(B, b) & \xrightarrow{\;p_0\;} & B
\end{array}
\qquad(5.7.15)
$$

defines a discrete cartesian fibration over $\mathrm{Hom}_B(B, b)$. By pullback composition, the fibers $F_{\ulcorner\mathrm{id}_b\urcorner} \cong E_b$ are isomorphic and similarly the space of sections of q is isomorphic to the functor space $\mathrm{Fun}_B(p_0, p)$. So in this context, the equivalence of Theorem 5.7.2 specializes to the desired equivalence of Kan complexes:

$$
\begin{array}{ccc}
\mathrm{Fun}_{\mathrm{Hom}_B(B,b)}(\mathrm{Hom}_B(B, b), F) & \xrightarrow[\sim]{\mathrm{ev}_{\ulcorner\mathrm{id}_b\urcorner}} & \mathrm{Fun}(1, F_{\ulcorner\mathrm{id}_b\urcorner}) \\
\wr\| & & \|\wr \\
\mathrm{Fun}_B(\mathrm{Hom}_B(B, b) \xrightarrow{p_0} B, E \xrightarrow{p} B) & \xrightarrow[\mathrm{ev}_{\ulcorner\mathrm{id}_b\urcorner}]{\sim} & \mathrm{Fun}(1, E_b)
\end{array}
\qquad\square
$$

Specializing to the case of two right representable discrete cartesian fibrations, we conclude that the Kan complex of natural transformations is equivalent to the underlying space of the corresponding mapping space.

COROLLARY 5.7.16. *For any elements* $x, y : 1 \to A$ *in an* ∞-*category* A, *evaluation at the identity of* x *induces an equivalence of Kan complexes*

$$\mathsf{Fun}_A(\mathsf{Hom}_A(A, x), \mathsf{Hom}_A(A, y)) \xrightarrow[\sim]{\mathsf{ev}^{\ulcorner}\mathsf{id}_x{}^{\urcorner}} \mathsf{Fun}(1, \mathsf{Hom}_A(x, y)) \qquad \square$$

It remains to prove the general case of Theorem 5.7.3. When $p : E \twoheadrightarrow B$ is a cartesian fibration the pullback (5.7.15) defines a cartesian fibration $q : F \twoheadrightarrow \mathsf{Hom}_B(B, b)$ and Proposition 5.7.13 provides an adjunction

$$
\begin{array}{ccc}
\mathsf{Fun}_{\mathsf{Hom}_B(B,b)}(\mathsf{Hom}_B(B,b), F) & \underset{y}{\overset{\mathsf{ev}^{\ulcorner}\mathsf{id}_b{}^{\urcorner}}{\underset{\perp}{\rightleftarrows}}} & \mathsf{Fun}(1, F_{\ulcorner\mathsf{id}_b{}^{\urcorner}}) \\[4pt]
\wr\| & & \| \wr \qquad\qquad (5.7.17) \\[4pt]
\mathsf{Fun}_B(\mathsf{Hom}_B(B,b) \xrightarrow{p_0} B, E \xrightarrow{p} B) & \underset{y}{\overset{\mathsf{ev}^{\ulcorner}\mathsf{id}_b{}^{\urcorner}}{\underset{\perp}{\rightleftarrows}}} & \mathsf{Fun}(1, E_b)
\end{array}
$$

The next step is to observe that the right adjoint lands in the sub-quasi-category of cartesian functors from p_0 to p.

LEMMA 5.7.18. *For any cartesian fibration* $p : E \twoheadrightarrow B$ *and element* $b : 1 \to B$, *for each vertex* e *in* $\mathsf{Fun}(1, E_b) \cong \mathsf{Fun}_B(b, p)$ *below-left*

$$\mathsf{Fun}_B(b, p) \xrightarrow{\quad y \quad} \mathsf{Fun}_B(p_0, p)$$

$$
\begin{array}{ccc}
1 \xrightarrow{\ e\ } E & & \mathsf{Hom}_B(B, b) \xrightarrow{\ ye\ } E \\
{}_{b}\searrow \ \ {}^{\Swarrow}{}_{p} & \mapsto & {}_{p_0}\searrow \ \ {}^{\Swarrow}{}_{p} \\
B & & B
\end{array}
$$

the functor ye *in* $\mathsf{Fun}_B(p_0, p)$ *above-right defines a cartesian functor from* p_0 *to* p.

Proof In the proof of Proposition 5.7.13, the right adjoint is defined by the diagram (5.7.14) as a factorization of the domain component of the q_*-cartesian lift of the unit η of the adjunction witnessing the terminal element. Here q is a pullback (5.7.15) of the cartesian fibration p, and so by Proposition 5.2.4, y can equally be described as a factorization of the domain component of the p_*-cartesian lift of $p_0\eta$, which equals the right comma cone $\phi : p_0 \Rightarrow b!$. In summary, the functor y is defined by:

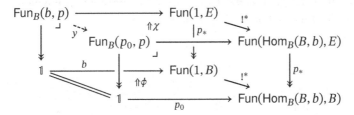

Thus, by Lemma 5.6.5, we see that ye is the domain component of a p-cartesian lift χe of the natural transformation $\phi : p_0 \Rightarrow pe!$

Since $p_0 : \mathrm{Hom}_B(B, b) \twoheadrightarrow B$ is discrete, every natural transformation ψ with codomain $\mathrm{Hom}_B(B, b)$ is p_0-cartesian, so to prove that ye defines a cartesian functor, we must show that $ye\psi$ is p-cartesian. To that end, consider the horizontal composite:

By naturality of whiskering, $\chi g \cdot ye\psi = e!\psi \cdot \chi f$, and since 1 is the terminal ∞-category, $e!\psi$ is an identity. Thus, by left cancelation of p-cartesian transformations (see Lemma 5.1.5), $ye\psi$ is p-cartesian. $\qquad\square$

To complete the proof of Theorem 5.7.3, it remains to argue that this restricted adjunction defines an adjoint equivalence.

Proof of Theorem 5.7.3 By Lemma 5.7.18, the adjunction (5.7.17) restricts to define an adjunction

Since the counit of the original adjunction $! \dashv \ulcorner \mathrm{id}_b \urcorner$ is an isomorphism and smothering 2-functors are conservative on 2-cells, the counit of the adjunction of Proposition 5.7.13 and hence also of the restricted adjunction is an isomorphism. As in the proof of Theorem 5.7.1, we prove that $\mathrm{ev}_{\ulcorner\mathrm{id}_b\urcorner}$ is an equivalence by demonstrating that the unit of the restricted adjunction is also invertible. By Corollary 1.1.22, it suffices to verify this elementwise, proving that the component of the unit indexed by a cartesian functor

$$\mathrm{Hom}_B(B, b) \xrightarrow{\;h\;} E$$
$$p_0 \searrow \quad \swarrow p$$
$$B$$

is an isomorphism.

Unpacking the proof of Proposition 5.7.13, the unit $\hat{\eta}$ of $\mathrm{ev}_{\ulcorner \mathrm{id}_b \urcorner} \dashv y$ is defined to be a factorization

$$\mathrm{Fun}_B(p_0, p) \lhook\joinrel\longrightarrow \mathrm{Fun}(\mathrm{Hom}_B(B, b), E) \xrightarrow{\ulcorner \mathrm{id}_b \urcorner^*} \mathrm{Fun}(1, E)$$

$$\Uparrow \mathrm{Fun}(\eta, E) \qquad \downarrow !^*$$

$$// \qquad\qquad\qquad \mathrm{Fun}(\mathrm{Hom}_B(B, b), E)$$

$$\mathrm{Fun}_B(p_0, p) \xrightarrow{\mathrm{ev}_{\ulcorner \mathrm{id}_b \urcorner}} \mathrm{Fun}_B(b, p) \lhook\joinrel\longrightarrow \mathrm{Fun}(1, E)$$

$$\Uparrow \hat{\eta} \qquad \downarrow y \qquad \Uparrow \chi \qquad \downarrow !^*$$

$$\mathrm{Fun}_B(p_0, p) \lhook\joinrel\longrightarrow \mathrm{Fun}(\mathrm{Hom}_B(B, b), E)$$

of the restriction of $\mathrm{Fun}(\eta, E)$ through the p_*-cartesian lift χ, where η is the unit of the adjunction $! \dashv \ulcorner \mathrm{id}_b \urcorner$. The component of the restricted 2-cell $\mathrm{Fun}(\eta, E)$ at the cartesian functor h is $h\eta$. Since $p_0 : \mathrm{Hom}_B(B, b) \twoheadrightarrow B$ is a discrete cartesian fibration, any 2-cell, such as η, which has codomain $\mathrm{Hom}_B(B, b)$ is p_0-cartesian, and since h is a cartesian functor, we then see that $h\eta$ is p-cartesian.

By Lemma 5.6.6, the components of the p_*-cartesian cell χ define p-cartesian natural transformations in the ambient ∞-cosmos. As $\hat{\eta}$ is a natural transformation with codomain $\mathrm{Fun}_B(p_0, p)$ its components project along p to the identity. In this way, we see that $\hat{\eta}h$ is a factorization of the p-cartesian transformation $h\eta$ through a p-cartesian lift of ϕ over an identity, and Lemma 5.1.12 proves that $\hat{\eta}h$ is an isomorphism, as desired.   □

In ∞-cosmology we have access to the following trick: any result, such as Theorem 5.7.3, that is proven in a generic ∞-cosmos can then be applied to a sliced ∞-cosmos. This can often be used to extend a result about elements of an ∞-category to generalized elements of that ∞-category (see Remark 4.3.12 for instance). By this technique, Theorem 5.7.3 implies the following generalization, replacing the element $b : 1 \to B$ by a generalized element $b : X \to B$.

CОROLLARY 5.7.19 (generalized Yoneda lemma). *For any cartesian fibration* $p : E \twoheadrightarrow B$ *and functor* $b : X \to B$, *restricting along the canonical induced functor* $\ulcorner\mathrm{id}_b\urcorner$ *defines an equivalence of quasi-categories:*

$$X \xrightarrow{\ulcorner\mathrm{id}_b\urcorner} \mathrm{Hom}_B(B,b) \qquad \mathrm{Fun}_B^{\mathrm{cart}}(\mathrm{Hom}_B(B,b) \xrightarrow{p_0} B, E \xrightarrow{p} B)$$

$$b \searrow \quad \swarrow p_0 \qquad \rightsquigarrow \qquad \downarrow \mathrm{ev}_{\ulcorner\mathrm{id}_b\urcorner}$$

$$B \qquad\qquad \mathrm{Fun}_B(X \xrightarrow{b} B, E \xrightarrow{p} B)$$

Corollary 5.7.19 can be interpreted as defining a "left biadjoint" to the inclusion of the subcategory of cartesian fibrations and cartesian functors, reflecting an arbitrary functor $b : X \to B$ into a cartesian fibration $p_0 : \mathrm{Hom}_B(B,b) \twoheadrightarrow B$.

Proof Given the stated data in an ∞-cosmos $\mathcal{K}$, Theorem 5.7.3 applies in $\mathcal{K}_{/X}$ to the cartesian fibration $p \times X : E \times X \twoheadrightarrow B \times X$ and the element $(b,X) : X \to B \times X$ to define an equivalence

$$\mathrm{Fun}_{B \times X}^{\mathrm{cart}}((p_1, p_0), p \times X) \xrightarrow[\sim]{\mathrm{ev}_{\ulcorner(b,X)\urcorner}} \mathrm{Fun}_{B \times X}((b,X), p \times X)$$

$$\wr\| \qquad\qquad\qquad\qquad\qquad \wr\|$$

$$\mathrm{Fun}_B^{\mathrm{cart}}(\mathrm{Hom}_B(B,b) \xrightarrow{p_0} B, E \xrightarrow{p} B) \xrightarrow[\sim]{\mathrm{ev}_{\ulcorner\mathrm{id}_b\urcorner}} \mathrm{Fun}_B(X \xrightarrow{b} B, E \xrightarrow{p} B)$$

which transposes under the simplicial adjunction

$$\mathcal{K}_{/X} \underset{-\times X}{\overset{U}{\rightleftarrows}} \perp \ \mathcal{K}$$

to the equivalence of the statement. $\square$

Despite its name, the generalized Yoneda lemma is not the most general form of the Yoneda lemma we require. A "two-sided" version of this result appears in Theorem 7.3.2 and is proven with the same trick, by applying Corollary 5.7.19 in a more exotic ∞-cosmos: namely, of cocartesian fibrations with a fixed base. For this reason, we invite the reader to accompany us at an interlude where we derive further examples of ∞-cosmoi.

Exercises

EXERCISE 5.7.i. Given an element $f : 1 \to \mathrm{Hom}_A(x,y)$ in the mapping space between a pair of elements in an ∞-category A, use the explicit description of the inverse equivalence to the map of Corollary 5.7.16 to construct a map

$$\mathrm{Hom}_A(A,x) \xdashrightarrow{f_*} \mathrm{Hom}_A(A,y)$$

$$p_0 \searrow \quad \swarrow p_0$$

$$A$$

which represents the "natural transformation" defined by postcomposing with f.[17]

---

[17] Hint: this construction is a special case of the construction given in the first half of the proof of Lemma 5.7.18.

AN INTERLUDE ON ∞-COSMOLOGY

6

Exotic ∞-Cosmoi

Morally an ∞-cosmos can be described as "an $(\infty, 2)$-category with $(\infty, 2)$-categorical limits," but the precise axiomatization given in Definition 1.2.1 employs a particularly strict interpretation of this phrase, taking advantage of the strictness that is available in so many examples to simplify proofs. We define ∞-cosmoi to be quasi-categorically enriched categories – selecting the strictest model of $(\infty, 2)$-categories in common use (see [79, 0.0.3-4]) – and we construct $(\infty, 2)$-categorical limits by certain homotopically well-behaved simplicially enriched limits.

Our aim in this chapter is to develop further examples of ∞-cosmoi, which immediately allows us to apply all of the theorems proven in Part I in more exotic contexts, where the ∞-categories of an ∞-cosmos should not be thought of as "$(\infty, 1)$-categories in some model." Some of these examples, such as the ∞-cosmos of isofibrations introduced in §6.1, can be established easily by directly verifying the axioms of Definition 1.2.1.

A larger family of examples, appearing in §6.3, arise as subcategories of previously defined ∞-cosmoi. For instance, in Proposition 6.3.14 we prove that the ∞-cosmos of isofibrations $\mathcal{K}^{\downarrow}$ in any ∞-cosmos $\mathcal{K}$ has a sub ∞-cosmos $\mathcal{C}art(\mathcal{K})$ of cartesian fibrations and cartesian functors between them. This result, and many others of a similar flavor, follows from a common paradigm appearing as Proposition 6.3.3, which states that a *replete* subcategory (see Definition 6.3.1) of an ∞-cosmos inherits an ∞-cosmos structure, provided that it is closed under *flexible weighted limits*.

This leads us to the second main theme of this chapter: an elaboration of the $(\infty, 2)$-categorical limits present in any ∞-cosmos. In §6.2, we discover that the *cosmological limit notions* enumerated in axiom 1.2.1(i) generate a much larger class of simplicially enriched limits that exist in any ∞-cosmos, which are precisely those simplicially enriched limits that deserve to be called

"$(\infty, 2)$-categorical limits." We dub these simplicially enriched limits as *flexible weighted limits*, borrowing a term from 2-category theory (see Digression 6.2.7).

To explain the intuition, we make use of the formalism of weighted limits from enriched category theory (see §A.6). An ordinary limit defines a universal cone over a given diagram, each cone leg being an arrow whose source is the limit object. A weighted limit similarly defines a universal cone over a given diagram, but now that cone might take a more exotic shape. In the simplicially enriched context, each cone leg may take the shape of an arbitrary simplicial set, with the cone-commutativity conditions specified by a simplicial-set-valued simplicial functor, referred to as the *weight*. For example, an ordinary cone over a cospan defines a commutative square, but the cones for a different choice of weight are squares inhabited by a 1-simplex. In this way, the comma ∞-categories of Definition 3.4.1 arise as weighted limits of cospans (see Example A.6.14).

Intuitively, the *flexible* weighted limits are those whose weights define cone shapes that do not impose any strict commutativity conditions: Pullbacks are not flexible weighted limits, while comma objects are. Flexible weighted limits are invariant under pointwise equivalence between diagrams, while general weighted limits need not be. These are the senses in which flexible weighted limits correspond to $(\infty, 2)$-categorical limit notions.

After establishing the homotopical properties of flexible weighted limits, we also see that the cosmological limits notions, such as pullbacks of isofibrations, are really flexible weighted limits in disguise. The requirement that certain arrows in the diagram are isofibrations means that strictly commuting cones correspond to pseudo-commutative cones, providing the required flexibility.

This chapter closes by illustrating a few sample applications of the general ∞-cosmology developed here. More applications follow in Part II, where we use the ∞-cosmoi constructed here to develop the theory of two-sided fibrations and modules.

6.1 The ∞-Cosmos of Isofibrations

Our first example of an "exotic ∞-cosmos" is a special case of a more general result that is left as Exercise 6.1.iii. The walking arrow category 2 is an *inverse Reedy category*, where the domain of the nonidentity arrow is assigned degree one and the codomain is assigned degree zero. This Reedy structure motivates the definitions in the **∞-cosmos of isofibrations** that we now introduce:

PROPOSITION 6.1.1 (∞-cosmoi of isofibrations). *For any ∞-cosmos $\mathcal{K}$ there is an ∞-cosmos $\mathcal{K}^{\pitchfork}$ whose*

(i) objects are isofibrations $p : E \twoheadrightarrow B$ in $\mathcal{K}$,

(ii) functor spaces, say from $q : F \twoheadrightarrow A$ to $p : E \twoheadrightarrow B$, are defined by pullback

$$
\begin{array}{ccc}
\mathrm{Fun}(F \overset{q}{\twoheadrightarrow} A, E \overset{p}{\twoheadrightarrow} B) & \longrightarrow & \mathrm{Fun}(F, E) \\
\downarrow & \lrcorner & \downarrow{\scriptstyle p_*} \\
\mathrm{Fun}(A, B) & \underset{q^*}{\longrightarrow} & \mathrm{Fun}(F, B),
\end{array}
$$

(iii) isofibrations from q to p are commutative squares

$$
\begin{array}{ccc}
F & \overset{g}{\longrightarrow} & E \\
{\scriptstyle q}\downarrow & \lrcorner \quad & \downarrow{\scriptstyle p} \\
A & \underset{f}{\longrightarrow} & B
\end{array}
$$

in which the horizontals and the induced map from the initial vertex to the pullback of the cospan are isofibrations in $\mathcal{K}$,

(iv) limits are defined pointwise in $\mathcal{K}$,

(v) and in which a map

$$
\begin{array}{ccc}
F & \overset{g}{\underset{\sim}{\longrightarrow}} & E \\
{\scriptstyle q}\downarrow & & \downarrow{\scriptstyle p} \\
A & \underset{\sim}{\underset{f}{\longrightarrow}} & B
\end{array}
$$

is an equivalence in the ∞-cosmos $\mathcal{K}^{\downarrow}$ if and only if g and f are equivalences in $\mathcal{K}$.

Relative to these definitions, the domain, codomain, and identity functors

$$
\mathcal{K}^{\downarrow} \underset{\mathrm{cod}}{\overset{\mathrm{dom}}{\underset{\longleftarrow}{\overset{\longrightarrow}{\longleftarrow \mathrm{id} \longrightarrow}}}} \mathcal{K}
$$

are all cosmological.

Proof   The diagram category $\mathcal{K}^{\downarrow}$ inherits its simplicially enriched limits, defined pointwise, from $\mathcal{K}$. The functor spaces described in (ii) are the usual ones for an enriched category of diagrams. This verifies 1.2.1(i). Note that the definitions of functor spaces, limits, and isofibrations make each of the domain, codomain, and identity functors cosmological.

For axiom 1.2.1(ii) note that the product and simplicial cotensor functors carry pointwise isofibrations to isofibrations. The pullback of an isofibration as in (iii) along a commutative square from an isofibration r to p may be formed in $\mathcal{K}$. Our task is to show that the induced map t is an isofibration and also that the square from t to r is an isofibration in the sense of (iii):

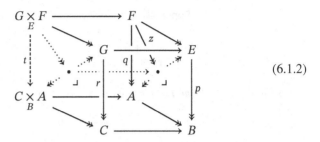

$$(6.1.2)$$

The map t factors as a pullback of z followed by a pullback of r as displayed above, and is thus an isofibration, as claimed. This observation also verifies that the square from t to r defines an isofibration. A similar argument verifies the Leibniz stability of the isofibrations and that the limit of a tower of isofibrations is an isofibration. This proves that $\mathcal{K}^{\pitchfork}$ defines an ∞-cosmos in such a way so that the domain, codomain, and identity functors are cosmological.

Finally, by Proposition 3.3.4, a pair of equivalences as in (v) induces an equivalence between the functor spaces defined in (ii). The converse, that an equivalence in $\mathcal{K}^{\pitchfork}$ defines a pair of equivalences in $\mathcal{K}$, follows from Lemma 1.3.2 and the fact that domain and codomain projection functors are cosmological. □

In close analogy with Proposition 3.6.2, we have a smothering 2-functor that relates the homotopy 2-category of $\mathcal{K}^{\pitchfork}$ to the 2-category of isofibrations, commutative squares, and parallel natural transformations in the homotopy 2-category of $\mathcal{K}$.

LEMMA 6.1.3. *There is an identity on objects and 1-cells smothering 2-functor* $\mathfrak{h}(\mathcal{K}^{\pitchfork}) \to (\mathfrak{h}\mathcal{K})^{\pitchfork}$ *whose codomain is the 2-category whose*

- *objects are isofibrations in $\mathcal{K}$,*
- *1-cells are commutative squares between such, and*
- *2-cells are pairs of 2-cells in $\mathfrak{h}\mathcal{K}$*

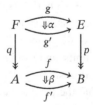

so that $p\alpha = \beta q$.

Proof Exercise 6.1.i. □

Similarly, any ∞-cosmos admits an ∞-**cosmos of trivial fibrations**, which can be defined as a full subcategory of the ∞-cosmos of isofibrations. The following general result from abstract ∞-cosmology explains how it inherits its ∞-cosmos structure.

LEMMA 6.1.4. *Let $\mathcal{K}$ be an ∞-cosmos and let $\mathcal{L} \subset \mathcal{K}$ be a full subcategory. Then $\mathcal{L}$ inherits an ∞-cosmos structure from $\mathcal{K}$ created from the inclusion $\mathcal{L} \hookrightarrow \mathcal{K}$ if and only if $\mathcal{L}$ is closed in $\mathcal{K}$ under the cosmological limit notions.*

In practice, the full subcategories we consider have the property that any object of $\mathcal{K}$ that is equivalent to an object in $\mathcal{L}$ in fact lies in $\mathcal{L}$ – in other words, these subcategories are replete in the sense of Definition 6.3.1. In the proof below, we tacitly assume that $\mathcal{L}$ is at least closed under isomorphism so that if a limit of a diagram lies in $\mathcal{L}$ then all of the limits of that diagram do, but this assumption is only used for linguistic convenience and is inessential.

Proof As a full subcategory, $\mathcal{L}$ inherits its quasi-categorical enrichment from $\mathcal{K}$, and we define a map to be an isofibration in $\mathcal{L}$ if and only if it is an isofibration in $\mathcal{K}$. Note this definition makes axiom 1.2.1(ii) follow immediately once that we have shown that the limits required by 1.2.1(i) coincide with the corresponding limits in $\mathcal{K}$. But this is exactly what is asserted by the hypothesis that $\mathcal{L}$ is closed in $\mathcal{K}$ under the cosmological limit notions.

We have shown that a closed full subcategory inherits an ∞-cosmos structure defined in such a way that the inclusion $\mathcal{L} \hookrightarrow \mathcal{K}$ is a cosmological functor that reflects isofibrations and cosmological limits. Clearly, this inclusion reflects representably defined equivalences, since $\mathcal{L}$ is a full subcategory of $\mathcal{K}$. But by Lemma 1.3.2 the cosmological functor $\mathcal{L} \hookrightarrow \mathcal{K}$ preserves them as well, which tells us that equivalences in $\mathcal{L}$ are created from $\mathcal{K}$ along with the isofibrations and cosmological limits. □

With this result in hand, further ∞-cosmoi are easy to establish.

PROPOSITION 6.1.5 (∞-cosmoi of trivial fibrations). *Let $\mathcal{K}$ be an ∞-cosmos.*

(i) *For any ∞-category B in $\mathcal{K}$, the full subcategory $\mathcal{K}^{\simeq}_{/B} \hookrightarrow \mathcal{K}_{/B}$ spanned by the trivial fibrations with codomain B defines an ∞-cosmos, with limits, isofibrations, equivalences, and trivial fibrations created by the inclusion.*

(ii) *The full subcategory $\mathcal{K}^{\downarrow} \hookrightarrow \mathcal{K}^{\downarrow}$ spanned by the trivial fibrations defines an ∞-cosmos, with limits, isofibrations, equivalences, and trivial fibrations created by the inclusion.*

Proof The details, which are similar to Propositions 1.2.22 and 6.1.1, are left to Exercise 6.1.ii.   □

Note that the sliced ∞-cosmoi $\mathcal{K}^{\simeq}_{/B}$ of trivial fibrations are weakly contractible in the sense that the unique functor $\mathcal{K}^{\simeq}_{/B} \twoheadrightarrow \mathbb{1}$ to the terminal ∞-cosmos is a cosmological biequivalence. In particular, each functor in $\mathcal{K}^{\simeq}_{/B}$ is an equivalence, so this ∞-cosmos is more a curiosity than a structure of substantial interest.

As promised in §1.2, Lemma 6.1.4 allows us to generalize Proposition 1.2.12 to show that the discrete ∞-categories in any ∞-cosmos form an ∞-cosmos:

PROPOSITION 6.1.6 (∞-cosmoi of discrete ∞-categories). *The full subcategory* $\mathcal{K}^{\simeq} \hookrightarrow \mathcal{K}$ *spanned by the discrete ∞-categories in any ∞-cosmos inherits an* ∞*-cosmos structure created from the inclusion.*

Proof Recall from Lemma 1.2.27 that an ∞-category E in an ∞-cosmos $\mathcal{K}$ is discrete if and only if the map $E^{\mathbb{I}} \twoheadrightarrow E^2$ is a trivial fibration. This says that the full subcategory $\mathcal{K}^{\mathbb{I}} \hookrightarrow \mathcal{K}$ of discrete ∞-categories is defined by the pullback

$$
\begin{array}{ccc}
\mathcal{K}^{\simeq} & \longrightarrow & \mathcal{K}^{\downarrow} \\
\downarrow & \lrcorner & \downarrow \\
\mathcal{K} & \xrightarrow{\ I\ } & \mathcal{K}^{\downarrow}
\end{array}
$$

along the cosmological functor I that sends an ∞-category E to the isofibration $E^{\mathbb{I}} \twoheadrightarrow E^2$. By Lemma 6.1.4, to show that $\mathcal{K}^{\simeq}$ admits an ∞-cosmos structure inherited from $\mathcal{K}$, we need only show that the discrete ∞-categories are closed in $\mathcal{K}$ under the limit constructions of 1.2.1(i). For the simplicial cotensors, for instance, this follows easily from the defining universal property and the fact that the Kan complexes form an exponential ideal in the category of simplicial sets. A common argument can be given for each of the conical limits; for the sake of concreteness, consider a tower of isofibrations between discrete ∞-categories E_n and form the limit in $\mathcal{K}$

$$
E := \lim \left(\cdots \longrightarrow E_n \longrightarrow E_{n-1} \longrightarrow \cdots \longrightarrow E_1 \longrightarrow E_0 \right).
$$

The cosmological functor I carries this to a limit diagram in the ∞-cosmos $\mathcal{K}^{\downarrow}$

$$
\begin{array}{c}
E^{\mathbb{I}} \\
\downarrow \\
E^2
\end{array}
:= \lim
\left(
\begin{array}{ccccccccc}
\cdots & \longrightarrow & E_n^{\mathbb{I}} & \longrightarrow & E_{n-1}^{\mathbb{I}} & \longrightarrow & \cdots & \longrightarrow & E_1^{\mathbb{I}} & \longrightarrow & E_0^{\mathbb{I}} \\
 & & \downarrow{\scriptstyle\wr} & & \downarrow{\scriptstyle\wr} & & & & \downarrow{\scriptstyle\wr} & & \downarrow{\scriptstyle\wr} \\
\cdots & \longrightarrow & E_n^2 & \longrightarrow & E_{n-1}^2 & \longrightarrow & \cdots & \longrightarrow & E_1^2 & \longrightarrow & E_0^2
\end{array}
\right).
$$

Since each E_n is a discrete ∞-category, each of the objects in this diagram is a trivial fibration. Hence, by Proposition 6.1.5, the limit $E^{\parallel} \twoheadrightarrow E^2$ is a trivial fibration as well. This proves that E is discrete. $\qquad\qquad\qquad\qquad\qquad\square$

The proof of Proposition 6.1.1 reveals that it is tedious to manually verify the limit axiom in the construction of new ∞-cosmoi; indeed, even in that relatively basic example, we omitted some details. In the following sections, we develop machinery that allows us to attack this problem more systematically.

Exercises

EXERCISE 6.1.i. Prove Lemma 6.1.3.

EXERCISE 6.1.ii. Prove Proposition 6.1.5 using Lemma 6.1.4.

EXERCISE 6.1.iii. Let $\mathcal{K}$ be an ∞-cosmos and let $\mathcal{J}$ be an inverse category (see Definition C.1.16). Guided by Proposition C.1.23, prove that there is an ∞-cosmos $\mathcal{K}^{\mathcal{J}}$ whose:

- objects are the *fibrant diagrams* of Definition C.1.19,
- isofibrations are the *fibrant natural transformations* of Definition C.1.19,
- functor spaces are the simplicial hom-spaces of Definition A.3.8,
- and in which the simplicial limits and equivalences are defined pointwise in $\mathcal{K}$.

Use Proposition C.1.21 to demonstrate that limits of fibrant $\mathcal{J}$-indexed diagrams exist in $\mathcal{K}$ and moreover that the functor $\lim : \mathcal{K}^{\mathcal{J}} \to \mathcal{K}$ is cosmological.

6.2 Flexible Weighted Limits

Our aim in this section is to introduce a special class of simplicial-set-valued weights whose associated weighted limit notions are homotopically well-behaved. Borrowing a term from 2-category theory, we refer to these weights as *flexible*. All of the cosmological limit notions can be understood as flexible weighted limits. In fact, we prove that ∞-cosmoi admit all flexible weighted limits because these can be built out of the axiomatized cosmological limits. In §6.3, we make use of this observation to more efficiently verify the limit axiom for newly constructed ∞-cosmoi.

Roughly speaking, the flexible weighted limits are the simplicially enriched limits for which the simplices appearing in a cone are freely attached, relative to the diagram shape. Ordinary "conical" cones involve commutative triangles formed by two of the cone legs and one arrow in the diagrams. This commutativity is not permitted in the cone shapes proscribed by a flexible weight; instead a diagram of 0-arrows might commute up to a higher cell. If the notion of weighted limits in enriched category theory is unfamiliar, we suggest reading §A.4–§A.6 before proceeding.

DEFINITION 6.2.1 (flexible weights as projective cell complexes). For a simplicial category $\mathcal{A}$, consider the category $sSet^{\mathcal{A}}$ of simplicial functors, called **weights**, and simplicial natural transformations.

- For any $n \geq 0$ and object a of $\mathcal{A}$, a simplicial natural transformation of the form

$$\partial\Delta[n] \times \mathcal{A}(a, -) \hookrightarrow \Delta[n] \times \mathcal{A}(a, -)$$

  is called a **projective n-cell at a**.
- A simplicial natural transformation $\alpha : V \hookrightarrow W$ that can be expressed as a countable composite of pushouts of coproducts of projective cells is called a **projective cell complex**.
- A weight W is **flexible** just when $\varnothing \hookrightarrow W$ is a projective cell complex.

A **flexible weighted limit** of a diagram $F : \mathcal{A} \to \mathcal{K}$ valued in an ∞-cosmos is a weighted limit, in the sense of §A.6, whose weight $W : \mathcal{A} \to sSet$ is flexible.

REMARK 6.2.2. Since any monomorphism of simplicial sets $U \hookrightarrow V$ can be decomposed as a countable composite of pushouts of coproducts of boundary inclusions $\partial\Delta[n] \hookrightarrow \Delta[n]$, the class of projective cell complexes may also be described as the class of maps in $sSet^{\mathcal{A}}$ that can be expressed as a countable composite of pushouts of coproducts of monomorphisms of the form $U \times \mathcal{A}(a, -) \hookrightarrow V \times \mathcal{A}(a, -)$ for some $U \hookrightarrow V \in sSet$ and $a \in \mathcal{A}$.

EXAMPLE 6.2.3 (cotensors are flexible). Recall from §A.4 that the **cotensor** of an ∞-category A in an ∞-cosmos $\mathcal{K}$ with a simplicial set U is the object $A^U \in \mathcal{K}$ characterized by the simplicial natural isomorphism

$$\mathsf{Fun}(X, A^U) \cong \mathsf{Fun}(X, A)^U.$$

This object can be regarded as the limit of the diagram $A : \mathbb{1} \to \mathcal{K}$ weighted by $U : \mathbb{1} \to sSet$. This weight is defined by a single generalized projective cell of shape $\varnothing \hookrightarrow U$ at the unique object of $\mathbb{1}$.

Recall from §A.5 that the **conical limit** of a diagram $F : \mathcal{A} \to \mathcal{K}$ is the limit weighted by the terminal weight $* : \mathcal{A} \to sSet$.

EXAMPLE 6.2.4 (products are flexible). Conical products also define flexible weighted limits, built by attaching one projective 0-cell at each object in the indexing set.

NON-EXAMPLE 6.2.5. Conical limits indexed by any 1-category that contains nonidentity arrows are *not* flexible because the legs of a conical cone are required to define a strictly commutative triangle over each 0-arrow in the diagram. The specifications for a flexible weight allow us to freely attach n-arrows of any dimension but do not provide a mechanism for demanding strict commutativity of any diagram of n-arrows – only commutativity up to the presence of a higher cell.

EXAMPLE 6.2.6 (commas are flexible). In an ∞-cosmos $\mathcal{K}$, the limit of the diagram $\lrcorner \to \mathcal{K}$ given by the cospan

$$ C \xrightarrow{\ g\ } A \xleftarrow{\ f\ } B $$

weighted by the diagram $\lrcorner \to s\mathcal{S}et$ given by the cospan

$$ \mathbb{1} \overset{1}{\hookrightarrow} 2 \xleftarrow{\ 0\ } \mathbb{1} $$

is the comma ∞-category $\mathrm{Hom}_A(f, g)$ (see Example A.6.14). Since this weight can be built by attaching two projective 0-cells at the corners of the cospan followed by a projective 1-cell at the terminal object of the cospan, comma objects are flexible weighted limits.

DIGRESSION 6.2.7 (on flexible limits in 2-category theory). Simplicial limits weighted by flexible weights should be thought of as analogous to **flexible 2-limits**, i.e., category enriched limits built out of products, inserters, equifiers, and retracts (splittings of idempotents) [18]. Because we define flexible weights as countable composites of pushouts of coproducts – and not retracts thereof – the flexible weighted limits of Definition 6.2.1 are more exactly analogous to the **PIE limits**, built from just products, inserters, and equifiers. The PIE limits also include iso-inserters, descent objects, comma objects, and Eilenberg–Moore objects, as well as all pseudo, lax, and oplax limits. Many important 2-categories, such as the 2-category of accessible categories and accessible functors, fail to admit all 2-categorical limits, but do admit all PIE limits [83].

The weights for flexible 2-limits indexed by a 2-category $\mathcal{A}$ are the cofibrant objects in a model structure on the diagram 2-category $\mathcal{C}at^{\mathcal{A}}$ that is enriched over the folk model structure on $\mathcal{C}at$; the PIE weights are exactly the *cellular* cofibrant objects (see Definition C.2.4). Correspondingly, the projective cell complexes of Definition 6.2.1 are exactly the cellular cofibrations in the projective model structure on $s\mathcal{S}et^{\mathcal{A}}$.

This suggests that "PIE limits" would be a more precise name for the flexible weighted limits of Definition 6.2.1, and we do not disagree. With apologies to the Australian 2-category theory diaspora, we cannot resist adopting the more evocative term.

Our interest in flexible weights stems from their homotopical properties, which we now explore. In a $\mathcal{V}$-model category $\mathcal{M}$, the fibrant objects are closed under weighted limits whose weights are projective cofibrant (see Corollary C.3.15). For instance, the fibrant objects in a $\mathcal{C}at$-enriched model structure are closed under flexible weighted limits [72, 5.4] in the sense of [18]. As explained in §E.1, ∞-cosmoi are very closely related to categories of fibrant objects associated to a model category that is enriched over the Joyal model structure on simplicial sets. Thus, we may adapt the proofs of results from enriched model category theory to obtain the following:

PROPOSITION 6.2.8 (flexible weights are homotopical). *Let $W \colon \mathcal{A} \to sSet$ be a flexible weight and let $\mathcal{K}$ be an ∞-cosmos.*

(i) *The weighted limit $\lim_W F$ of any diagram $F \colon \mathcal{A} \to \mathcal{K}$ may be expressed as a countable inverse limit of pullbacks of products of isofibrations*

$$Fa^{\Delta[n]} \longrightarrow Fa^{\partial\Delta[n]} \qquad\qquad (6.2.9)$$

one for each projective n-cell at a in the given projective cell complex presentation of W. Hence, ∞-cosmoi admit all flexible weighted limits and cosmological functors preserve them.

(ii) *If $V \hookrightarrow W \in sSet^{\mathcal{A}}$ is a projective cell complex between flexible weights, then for any diagram $F \colon \mathcal{A} \to \mathcal{K}$, the induced map between weighted limits is an isofibration:*

$$\lim_W F \longrightarrow \lim_V F$$

(iii) *If $\alpha \colon F \Rightarrow G$ is a simplicial natural transformation between a pair of diagrams $F, G \colon \mathcal{A} \to \mathcal{K}$ whose components $\alpha_a \colon Fa \Rightarrow Ga$ are isofibration, trivial fibrations, equivalences, then the induced map*

$$\lim_W F \xrightarrow{\ \alpha\ } \lim_W G$$

is an isofibration, trivial fibration, or equivalence, respectively.

Proof To begin, observe that the axioms of Definition A.6.1 imply that the limit of a diagram F weighted by the weight $U \times \mathcal{A}(a, -)$, for $U \in sSet$ and $a \in \mathcal{A}$, is the cotensor Fa^U. Thus, the map of weighted limits induced by the projective n-cell at a is the isofibration (6.2.9). By definition, any flexible weight is built as a countable composite of pushouts of coproducts of these projective cells and the weighted limit functor $\lim_- F$ carries each of these conical colimits to the corresponding limit notion. So it follows that $\lim_W F$ may be expressed as a countable inverse limit of pullbacks of products of the maps (6.2.9). This proves (i).

The same argument proves (ii). By definition, a relative cell complex $V \hookrightarrow W$ is built as a countable composite of pushouts of coproducts of these projective cells and the weighted limit functor $\lim_- F$ carries each of these conical colimits to the corresponding limit notion. So it follows that $\lim_W F$ is the limit of a countable tower of isofibrations whose base is $\lim_V F$, where each of these isofibrations is the pullback of products of the maps (6.2.9) appearing in the projective cell complex decomposition of $V \hookrightarrow W$. As products, pullbacks, and limits of towers of isofibrations are isofibrations, (ii) follows.

For (iii), suppose first that α is a componentwise isofibration. Then the simplicial natural transformation $\alpha : F \Rightarrow G$ defines a simplicial functor $A : \mathcal{A} \to \mathcal{K}^{\twoheadrightarrow}$ valued in the ∞-cosmos of isofibrations of Proposition 6.1.1. By (i), this diagram admits a W-weighted limit $\lim_W A$, which is then necessarily an isofibration. Since the domain and codomain functors $\mathrm{dom}, \mathrm{cod} : \mathcal{K} \to \mathcal{K}$ are cosmological, it is clear that the isofibration $\lim_W A$ coincides with the induced map on weighted limits $\alpha : \lim_W F \twoheadrightarrow \lim_W G$.

If α is a componentwise trivial fibration, then the above diagram and thus its W-weighted limit lies in the sub ∞-cosmos $\mathcal{K}^{\twoheadrightarrow} \hookrightarrow \mathcal{K}^{\twoheadrightarrow}$ of trivial fibrations established in Proposition 6.1.5. By the analogous argument, the induced map $\alpha : \lim_W F \twoheadrightarrow \lim_W G$ is a trivial fibration in this case. The final statement for equivalences now follows from the first two statements by Ken Brown's Lemma C.1.10. $\square$

Proposition 6.2.8(i) proves that the flexible weighted limit of any diagram in an ∞-cosmos can be constructed out of the *cosmological limits*, i.e., the limits of diagrams of isofibrations axiomatized in 1.2.1(i). Over a series of lemmas, we describe a converse of sorts, constructing each of the cosmological limits as a flexible weighted limit. It follows that any quasi-categorically enriched category equipped with a class of representably defined isofibrations that possesses flexible weighted limits admits all of the simplicial limits of 1.2.1(i).

To start, simplicial cotensors are flexible weighted limits, as discussed in Example 6.2.3. This leaves only the conical limits: products, pullbacks of isofibrations, and inverse limits of towers of isofibrations. Example 6.2.4 notes that the weights for products are flexible. However, for the reasons discussed in 6.2.5, the weights for conical pullbacks or limits of towers of isofibrations are not flexible because the definition of a cone over either diagram shape imposes composition relations on 0-arrows.

Our strategy is to modify the weights for pullbacks and for limits of countable towers so that each composition equation involved in defining cones over such diagrams is replaced by the insertion of an invertible arrow one dimension up, where we must also take care to define this invertibility without specifying any equations between arrows in the next dimension. We have a device for specifying just this sort of isomorphism: recall the homotopy coherent isomorphism $\mathbb{I}$ from Exercise 1.1.v(i). A diagram $\mathbb{I} \to \mathsf{Fun}(A,B)$ specifies a homotopy coherent isomorphism between a pair of 0-arrows f and g from A to B, given by:

- a pair of 1-arrows $\alpha : f \to g$ and $\beta : g \to f$,
- a pair of 2-arrows

$$
\begin{array}{cc}
\overset{g}{\underset{\alpha \nearrow \ \Phi \ \searrow \beta}{}} & \overset{f}{\underset{\beta \nearrow \ \Psi \ \searrow \alpha}{}} \\
f =\!\!=\!\!= f & g =\!\!=\!\!= g \,,
\end{array}
$$

- a pair of 3-arrows whose outer faces are Φ and Ψ and whose inner faces are degenerate,
- a pair of 4-simplices whose outer faces are these 3-simplices and whose inner faces are degenerate, and so on.

We now introduce the weight for pullback diagrams whose cone shapes are given by squares inhabited by a homotopy coherent isomorphism.

DEFINITION 6.2.10 (iso-commas). The **iso-comma object** $C \overset{.}{\times}_A B$ of a cospan

$$
C \overset{g}{\longrightarrow} A \overset{f}{\longleftarrow} B
$$

in a simplicially enriched and cotensored category is the limit weighted by a weight $W_{\overset{.}{\times}} : \lrcorner \to s\mathcal{S}et$ defined by the cospan

$$
\mathbb{1} \overset{1}{\hookrightarrow} \mathbb{I} \overset{0}{\longleftarrow} \mathbb{1}
$$

Under the simplification of Remark A.6.11, the formula for the weighted limit reduces to the equalizer of the pair of maps

$$
\begin{array}{ccc}
 & \overset{\pi}{\longrightarrow} A^{\mathbb{I}} \overset{(q_1,q_0)}{\searrow} & \\
C \times A^{\mathbb{I}} \times B & & A \times A \\
 & \underset{\pi}{\longrightarrow} C \times B \underset{g \times f}{\nearrow} &
\end{array}
$$

where the maps $(q_1, q_0) : A^{\mathbb{I}} \to A \times A$ are defined by restricting along the endpoint inclusion $\mathbb{1} + \mathbb{1} = \partial\mathbb{I} \hookrightarrow \mathbb{I}$. In an ∞-cosmos, this map is an isofibration and the equalizer defining the iso-comma object is computed by the pullback

$$
\begin{array}{ccc}
C \overset{\times}{\underset{A}{}} B & \longrightarrow & A^{\mathbb{I}} \\
{\scriptstyle (q_1, q_0)} \downarrow & \lrcorner & \downarrow {\scriptstyle (q_1, q_0)} \\
C \times B & \underset{g \times f}{\longrightarrow} & A \times A
\end{array}
\tag{6.2.11}
$$

LEMMA 6.2.12. *Iso-comma objects are flexible weighted limits and in particular exist in any ∞-cosmos.*

Proof Writing $b \to a \leftarrow c$ for the objects in the cospan category $\lrcorner$, the weight $W_{\times}$ is constructed by the pushout

$$
\begin{array}{ccc}
\partial\mathbb{I} \times \lrcorner(a, -) & \longrightarrow & \lrcorner(b, -) \sqcup \lrcorner(c, -) \\
\uparrow & & \uparrow \\
\mathbb{I} \times \lrcorner(a, -) & \underset{\ulcorner}{\longrightarrow} & W_{\times}
\end{array}
$$

where the attaching map picks out the two arrows in the cospan. As a projective cell complex, $W_{\times}$ is built from a projective 0-cell at b, a projective 0-cell at c, and two projective k-cells at a for each $k > 0$, corresponding to the nondegenerate simplices of $\mathbb{I}$. As described by Remark 6.2.2, these may be attached all at once. In this way, we see that $W_{\times}$ is a flexible weight, so Proposition 6.2.8(i) tells us that iso-comma objects exist in any ∞-cosmos, a fact that is also evident from the pullback (6.2.11). $\qquad\square$

REMARK 6.2.13. In the homotopy 2-category of an ∞-cosmos, there is a canonical invertible 2-cell defining the **iso-comma cone**:

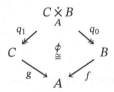

that has a weak universal property analogous to that of the comma cone presented in Proposition 3.4.6 (see [110, §3]). The proof makes use of the fact that $A^{\mathbb{I}}$ is the weak $\mathbb{I}$-cotensor in the homotopy 2-category [108, 3.3.13].

Our notation for iso-commas is deliberately similar to the usual notation for pullbacks. In an ∞-cosmos, iso-commas can be used to compute "homotopy pullbacks" of diagrams in which neither map is an isofibration. When at least one map of the cospan is an isofibration, the iso-comma is equivalent to the conical pullback.

LEMMA 6.2.14 (iso-commas and pullbacks). *For any cospan in an ∞-cosmos involving at least one isofibration, the pullback and the iso-comma are equivalent. More precisely, given a pullback square as below-left and an iso-comma square as below-right*

$$
\begin{array}{ccc}
P & \xrightarrow{b} & B \\
{\scriptstyle c}\downarrow & \lrcorner & \downarrow{\scriptstyle f} \\
C & \xrightarrow{g} & A
\end{array}
\qquad\qquad
\begin{array}{ccc}
C\underset{A}{\overset{\downarrow}{\times}}B & \xrightarrow{q_0} & B \\
{\scriptstyle q_1}\downarrow & \phi_{\cong} & \downarrow{\scriptstyle f} \\
C & \xrightarrow{g} & A
\end{array}
$$

$P \simeq C \underset{A}{\overset{\downarrow}{\times}} B$ *over C and up to isomorphism over B.*

Proof Applying Lemma 1.2.19 to the functor $b : P \to B$, we can replace the span $(c, b) : P \to C \times B$ by a span $(cq, p) : Pb \twoheadrightarrow C \times B$ whose legs are both isofibrations that is related via an equivalence $s : P \overset{\sim}{\to} Pb$ that lies over C on the nose and over B up to isomorphism. We claim that under the hypothesis that f is an isofibration, this new span is equivalent to the iso-comma span.

To see this, note that the factorization constructed in (1.2.20) is in fact defined using an iso-comma, constructed via the pullback in the top square of the diagram below-left. Since the map b is itself defined by a pullback, the bottom square of the diagram below-left is also a pullback, defining the left-hand pullback rectangle:

$$
\begin{array}{ccc}
Pb & \longrightarrow & B^{\mathbb{I}} \\
{\scriptstyle (q,p)}\downarrow\ \lrcorner & & \downarrow{\scriptstyle (q_1,q_0)} \\
P \times B & \xrightarrow{b\times B} & B \times B \\
{\scriptstyle c\times B}\downarrow\ \lrcorner & & \downarrow{\scriptstyle f\times B} \\
C \times B & \xrightarrow{g\times B} & A \times B
\end{array}
\qquad
\begin{array}{ccccc}
C\underset{A}{\overset{\downarrow}{\times}}B & \longrightarrow & A\underset{A}{\overset{\downarrow}{\times}}B & \twoheadrightarrow & A^{\mathbb{I}} \\
{\scriptstyle (q_1,q_0)}\downarrow\ \lrcorner & & {\scriptstyle (q_1,q_0)}\downarrow\ \lrcorner & & \downarrow{\scriptstyle (q_1,q_0)} \\
C \times B & \xrightarrow{g\times B} & A \times B & \xrightarrow{A\times f} & A \times A
\end{array}
$$

Now the iso-comma is constructed by a similar pullback rectangle, displayed above-right. And because f is an isofibration, Lemma 1.2.14 tells us that the Leibniz tensor $i_0 \,\widehat{\pitchfork}\, f : B^{\mathbb{I}} \twoheadrightarrow A \underset{A}{\overset{\downarrow}{\times}} B$ of $i_0 : \mathbb{1} \hookrightarrow \mathbb{I}$ with $f : B \twoheadrightarrow A$ is a trivial fibration. This equivalence commutes with the projections to $A \times B$ and hence the maps $(cq, p) : Pb \twoheadrightarrow C \times B$ and $(q_1, q_0) : C \underset{A}{\overset{\downarrow}{\times}} B \twoheadrightarrow C \times B$, defined as pullbacks of an equivalent pair of isofibrations along $g \times B$, are equivalent as claimed. □

We now introduce a flexible weight for diagrams given by a countable tower of 0-arrows whose cone shapes have a homotopy coherent isomorphism in the triangle over each generating arrow.

DEFINITION 6.2.15 (iso-towers). Recall the category ω whose objects are natural numbers and whose morphisms are freely generated by maps $\iota_{n,n+1} : n \to n+1$ for each n. The **iso-tower** of a ω^{op}-shaped diagram

$$F := \qquad \cdots \xrightarrow{f_{n+2,n+1}} F_{n+1} \xrightarrow{f_{n+1,n}} F_n \xrightarrow{f_{n,n-1}} \cdots \xrightarrow{f_{2,1}} F_1 \xrightarrow{f_{1,0}} F_0$$

in a simplicially enriched and cotensored category is the limit weighted by the diagram $W^{\leftarrow} : \omega^{\mathrm{op}} \to sSet$ defined by the pushout in $sSet^{\omega^{\mathrm{op}}}$.

$$
\begin{array}{ccc}
\coprod_{n\in\omega} \partial\mathbb{I} \times \omega(-,n) & \xrightarrow{(\mathrm{id}_n,\iota_{n,n+1})} & \coprod_{m\in\omega} \omega(-,m) \\
\downarrow & & \downarrow \\
\coprod_{n\in\omega} \mathbb{I} \times \omega(-,n) & \longrightarrow & W^{\leftarrow}
\end{array}
\qquad (6.2.16)
$$

By Definition A.6.1(ii), in an ∞-cosmos the iso-tower of F is constructed by the pullback

$$
\begin{array}{ccc}
\lim_{W^{\leftarrow}} F & \xrightarrow{\phi} & \prod_{n\in\omega} F_n^{\mathbb{I}} \\
\rho\downarrow & & \downarrow \prod(q_1,q_0) \\
\prod_{m\in\omega} F_m & \xrightarrow{(f_{n+1,n},\mathrm{id}_{F_n})} & \prod_{n\in\omega} F_n \times F_n
\end{array}
\qquad (6.2.17)
$$

The limit cone is generated by a 0-arrow $\rho_n : \lim_{W^{\leftarrow}} F \to F_n$ for each $n \in \omega$ together with a homotopy coherent isomorphism ϕ_n in each triangle over a generating arrow $F_{n+1} \to F_n$ in the ω^{op}-indexed diagram.

LEMMA 6.2.18. *Iso-towers are flexible weighted limits and in particular exist in any ∞-cosmos.*

Proof The weight $W^{\leftarrow}$ is a projective cell complex built by attaching one projective 0-cell at each $n \in \omega$ – forming the coproduct appearing in the upper right-hand corner of (6.2.16) – and then by attaching a projective k-cell at each $n \in \omega$ for each nondegenerate k-simplex of $\mathbb{I}$ with $k > 0$. Rather than attach each projective k-cell for fixed $n \in \omega$ in sequence, by Remark 6.2.2 these can all be attached at once by taking a single pushout of the "generalized projective cell at n" defined by the map $\partial\mathbb{I} \times \omega(-,n) \hookrightarrow \mathbb{I} \times \omega(-,n)$. These are the maps appearing as the left-hand vertical of (6.2.16). Now Proposition 6.2.8(i) or the formula (6.2.17) make it clear that such objects exist in any ∞-cosmos. $\qquad\square$

As is the case for iso-commas and pullbacks, iso-towers give a way to compute inverse limits of diagrams of arbitrary maps. When those maps are isofibrations, the iso-tower is equivalent to the conical limit.

LEMMA 6.2.19 (iso-towers and inverse limits). *The inverse limit of a countable tower of isofibrations in an ∞-cosmos is equivalent to the iso-pullback of that tower.*

Proof We will rearrange the limit (6.2.17) to construct the iso-tower $\lim_{W\leftarrow} F$ in an ∞-cosmos $\mathcal{K}$ as an inverse limit of a countable tower of isofibrations $P: \omega^{\mathrm{op}} \to \mathcal{K}$ that is pointwise equivalent to the tower of isofibrations $F: \omega^{\mathrm{op}} \to \mathcal{K}$.

$$
\begin{array}{ccccccccccc}
\lim P & \cong & \cdots \xrightarrow{p_{n+2,n+1}} & P_{n+1} & \xrightarrow{p_{n+1,n}} & P_n & \xrightarrow{p_{n,n-1}} & \cdots & \xrightarrow{p_{2,1}} & P_1 & \xrightarrow{p_{1,0}} & P_0 \\
\Vert & & & \downarrow e_n & & \downarrow e_n & & & & \downarrow e_1 & & \downarrow e_0 \\
\lim F & \cong & \cdots \xrightarrow{f_{n+2,n+1}} & F_{n+1} & \xrightarrow{f_{n+1,n}} & F_n & \xrightarrow{f_{n,n-1}} & \cdots & \xrightarrow{f_{2,1}} & F_1 & \xrightarrow{f_{1,0}} & F_0
\end{array}
$$

$$(6.2.20)$$

The equivalence invariance of the inverse limit of a diagram of isofibrations implies that the limits $\lim_{W\leftarrow} F \cong \lim P$ and $\lim F$ are equivalent as claimed.

The ∞-categories P_n are defined as conical limits of truncated versions of the diagram (6.2.17). To start define $P_0 := F_0$ and e_0 to be the identity, then define P_1, $p_{1,0}$, and e_1 via the pullback

$$
\begin{array}{ccc}
 & \overset{p_{1,0}}{\overgroup{\hspace{4cm}}} & \\
P_1 \twoheadrightarrow F_0^{\mathbb{I}} & \xrightarrow[q_0]{\sim} & F_0 \\
e_1 \downarrow \quad \lrcorner & & \downarrow q_1 \\
F_1 \xrightarrow[f_{1,0}]{} & & F_0
\end{array}
$$

Note that $P_1 \cong F_1 \underset{F_0}{\times} F_0$ computes the iso-comma objects of the cospan given by id_{F_0} and $f_{1,0}$.

Now define P_2, $p_{2,1}$, and e_2 using the composite pullback

$$
\begin{array}{ccccccc}
 & \overset{p_{2,1}}{\overgroup{\hspace{4cm}}} & & & & & \\
P_2 \twoheadrightarrow \bullet & \xrightarrow{\sim} & P_1 & \longrightarrow & F_0^{\mathbb{I}} & \xrightarrow[q_0]{\sim} & F_0 \\
\lrcorner \downarrow & \lrcorner \downarrow & e_1 \downarrow & \lrcorner & \downarrow q_1 & & \\
\bullet \longrightarrow & F_1^{\mathbb{I}} \xrightarrow[q_0]{\sim} & F_1 & \xrightarrow[f_{1,0}]{} & F_0 & & \\
\downarrow & \lrcorner \downarrow q_1 & & & & & \\
F_2 \xrightarrow[f_{2,1}]{} & F_1 & & & & &
\end{array}
$$

Continuing inductively, P_n, $p_{n,n-1}$, and e_n are defined by appending the diagram

$$
\begin{array}{ccc}
F_{n-1}^{\parallel} & \xrightarrow[q_0]{\sim} & F_{n-1} \\
{\scriptstyle q_1}\downarrow & & \\
F_n & \xrightarrow[f_{n,n-1}]{} & F_{n-1}
\end{array}
$$

to the limit cone defining P_{n-1} and taking the limit of this composite diagram. In the limit, the composite diagram is an unraveling of (6.2.17). Hence $\lim_{W\leftarrow} F \cong \lim P$.

It remains to use the maps $e_n : P_n \twoheadrightarrow F_n$ to compare $\lim P$ with $\lim F$. There is one small problem with the construction just given: it defines a diagram (6.2.20) in which each square commutes up to isomorphism – the isomorphism encoded by the map $P_n \to F_{n-1}^{\parallel}$ – not on the nose. But because the maps $f_{n+1,n}$ are isofibrations this is not a problem. The isomorphism inhabiting the square $e_0 p_{1,0} \cong f_{1,0} e_1$ can be lifted along $f_{1,0}$ to define a new map $e_1' : P_1 \twoheadrightarrow F_1$ isomorphic to e_1. By Exercise 1.4.iii this e_1' is then also an equivalence (though no longer necessarily a trivial fibration), so we replace e_1 with e_1', and then continue inductively to lift away the isomorphisms in the square $e_1' p_{2,1} \cong f_{2,1} e_2$.

Since inverse limits of towers of isofibrations are equivalence invariant by Proposition C.1.15, it follows that $\lim P \simeq \lim F$. By construction $\lim P \cong \lim_{W\leftarrow} F$, so it follows that $\lim_{W\leftarrow} F \simeq \lim F$, which is what we wanted to show. $\qquad\square$

Exercises

EXERCISE 6.2.i ([112, 2.2.2]). Show that ∞-cosmoi admit **wide pullbacks**: limits of finite or countable diagrams of the following form

$$
\cdots \; A_n \quad\quad A_{n-1} \quad\quad \cdots \quad\quad A_1 \quad\quad A_0
$$
$$
{}^{p_{n-1}}\searrow \; {}^{f_{n-1}}\swarrow \; {}_{p_{n-2}}\searrow \; {}^{f_{n-2}}\swarrow \; {}_{p_1}\searrow \; {}^{f_1}\swarrow \; {}_{p_0}\searrow \; {}^{f_0}\swarrow
$$
$$
\cdots \; B_{n-1} \quad\quad B_{n-2} \quad \cdots \quad B_1 \quad\quad B_0
$$

and that their construction is invariant under pointwise equivalence between diagrams.

6.3 Cosmologically Embedded ∞-Cosmoi

In this section, we generalize Lemma 6.1.4 – which was used to show that the subcategory of discrete ∞-categories inherits an ∞-cosmos structure – to subcategories of ∞-cosmoi that are not full. As we shall discover, there are many interesting examples.

For any ∞-cosmos $\mathcal{K}$ and any subcategory of its underlying 1-category – that is for any subset of its objects and subcategory of its 0-arrows – one can form a quasi-categorically enriched subcategory $\mathcal{L} \subset \mathcal{K}$ that contains exactly those objects and 0-arrows and all higher dimensional arrows that they span. We call such subcategories **full on positive-dimensional arrows**; note the functor spaces of $\mathcal{L}$ are quasi-categories because all inner horn inclusions are bijective on vertices. We take particular interest in subcategories that satisfy a further "repleteness" condition.

DEFINITION 6.3.1. Let $\mathcal{K}$ be an ∞-cosmos. A subcategory $\mathcal{L} \subset \mathcal{K}$ is **replete** in $\mathcal{K}$ if it is full on positive-dimensional arrows and moreover:

  (i) Every ∞-category in $\mathcal{K}$ that is equivalent to an object in $\mathcal{L}$ lies in $\mathcal{L}$.
  (ii) Any equivalence in $\mathcal{K}$ between objects in $\mathcal{L}$ lies in $\mathcal{L}$.
  (iii) Any 0-arrow in $\mathcal{K}$ that is isomorphic in $\mathcal{K}$ to an 0-arrow in $\mathcal{L}$ lies in $\mathcal{L}$.

The inclusion of a replete subcategory of an ∞-cosmos is both a monomorphism and also an isofibration of $(∞, 2)$-categories, in a sense explored in Exercise 6.3.i.

LEMMA 6.3.2. *Suppose $\mathcal{L} \subset \mathcal{K}$ is a replete subcategory of an ∞-cosmos. Then any map $p : E \to B$ in $\mathcal{L}$ that defines an isofibration in $\mathcal{K}$ is a representably defined isofibration in $\mathcal{L}$: that is for all $X \in \mathcal{L}$, $p_* : \mathrm{Fun}_{\mathcal{L}}(X, E) \twoheadrightarrow \mathrm{Fun}_{\mathcal{L}}(X, B)$ is an isofibration of quasi-categories.*

Proof Since $\mathcal{K}$ is an ∞-cosmos, the isofibration axiom 1.2.1(ii) requires that $p_* : \mathrm{Fun}_{\mathcal{K}}(X, E) \twoheadrightarrow \mathrm{Fun}_{\mathcal{K}}(X, B)$ is an isofibration of quasi-categories. Because the inner horn inclusions are bijective on vertices and $\mathrm{Fun}_{\mathcal{L}}(X, E) \hookrightarrow \mathrm{Fun}_{\mathcal{K}}(X, E)$ is full on positive-dimensional arrows, it follows immediately that the restricted map $p_* : \mathrm{Fun}_{\mathcal{L}}(X, E) \twoheadrightarrow \mathrm{Fun}_{\mathcal{L}}(X, B)$ lifts against the inner horn inclusions. Thus it remains only to solve lifting problems of the form displayed below-left

$$
\begin{array}{ccccc}
\mathbb{1} & \xrightarrow{\ e\ } & \mathrm{Fun}_{\mathcal{L}}(X, E) & \hookrightarrow & \mathrm{Fun}_{\mathcal{K}}(X, E) \\
\Big\downarrow & \nearrow & \Big\downarrow{\scriptstyle p_*} & & \Big\downarrow{\scriptstyle p_*} \\
\mathbb{1} & \xrightarrow[\ \beta\ ]{} & \mathrm{Fun}_{\mathcal{L}}(X, B) & \hookrightarrow & \mathrm{Fun}_{\mathcal{K}}(X, B)
\end{array}
$$

The lifting problem defines a 0-arrow $e : X \to E$ in $\mathcal{L}$ and a homotopy coherent isomorphism $\beta : b \cong pe$ in $\mathcal{L}$. Its solution in $\mathcal{K}$ defines a 0-arrow $e' : X \to E$ in $\mathcal{K}$ so that $pe' = b$ together with a homotopy coherent isomorphism $e \cong e'$

in $\mathcal{K}$. By fullness on positive-dimensional arrows, to show that this lift factors through the inclusion $\mathrm{Fun}_{\mathcal{L}}(X, E) \hookrightarrow \mathrm{Fun}_{\mathcal{K}}(X, E)$, we need only argue that the map e' lies in $\mathcal{L}$, but this is the case by condition (iii) of Definition 6.3.1. □

The following result describes a condition under which a replete subcategory $\mathcal{L} \subset \mathcal{K}$ inherits an ∞-cosmos structure created from $\mathcal{K}$.

PROPOSITION 6.3.3. *Suppose $\mathcal{L} \subset \mathcal{K}$ is a replete subcategory of an ∞-cosmos. If $\mathcal{L}$ is closed under flexible weighted limits in $\mathcal{K}$, then $\mathcal{L}$ defines an ∞-cosmos with isofibrations, equivalences, trivial fibrations and simplicial limits created by the inclusion $\mathcal{L} \hookrightarrow \mathcal{K}$, which then defines a cosmological functor.*

When these conditions hold, we refer to $\mathcal{L}$ as a **cosmologically embedded ∞-cosmos** of $\mathcal{K}$ and $\mathcal{L} \hookrightarrow \mathcal{K}$ as a **cosmological embedding**. The notation reflects both the embedding and isofibration-like properties of replete subcategory inclusions.

Proof To say that a replete subcategory $\mathcal{L} \hookrightarrow \mathcal{K}$ is closed under flexible weighted limits means that for any diagram in $\mathcal{L}$ and any limit cone in $\mathcal{K}$ over that diagram, then the limit cone lies in $\mathcal{L}$ and satisfies the appropriate simplicially enriched universal property of Definition A.6.5 in the subcategory $\mathcal{L}$. We must verify that each of the limits of axiom 1.2.1(i) exist in $\mathcal{L}$. Immediately, $\mathcal{L}$ has a terminal object, products, and simplicial cotensors, since all of these are flexible weighted limits, with each of these limits inherited from $\mathcal{K}$. By Lemmas 6.2.12 and 6.2.18, $\mathcal{L}$ also admits the construction of iso-comma objects and of iso-towers, again formed in $\mathcal{K}$.

Define the class of isofibrations in $\mathcal{L}$ to be those maps in $\mathcal{L}$ that define isofibrations in $\mathcal{K}$. By Lemmas 6.2.14 and 6.2.19, pullbacks and limits of towers of isofibrations are equivalent in $\mathcal{K}$ to the iso-commas and iso-towers formed over the same diagrams. Since these latter limit cones lie in $\mathcal{L}$ by hypothesis, so do the equivalent former cones by repleteness of $\mathcal{L}$ in $\mathcal{K}$.

There is a little more still to verify: namely that pullbacks and limits of towers of isofibrations satisfy the simplicially enriched universal property as conical limits in $\mathcal{L}$. In the case of a pullback diagram

$$
\begin{array}{ccc}
P & \xrightarrow{\ h\ } & B \\
{\scriptstyle k}\Big\downarrow & \lrcorner & \Big\downarrow{\scriptstyle f} \\
C & \xrightarrow[\ g\ ]{} & A
\end{array}
$$

in $\mathcal{L}$ we must show that for each $X \in \mathcal{L}$, the functor space $\mathrm{Fun}_{\mathcal{L}}(X, P)$ is isomorphic to the pullback $\mathrm{Fun}_{\mathcal{L}}(X, C) \times_{\mathrm{Fun}_{\mathcal{L}}(X, A)} \mathrm{Fun}_{\mathcal{L}}(X, B)$ of functor spaces.

We have such an isomorphism for functor spaces in $\mathcal{K}$ and on account of the commutative diagram

$$\begin{array}{ccc}
\text{Fun}_{\mathcal{L}}(X,P) & \dashrightarrow & \text{Fun}_{\mathcal{L}}(X,C) \underset{\text{Fun}_{\mathcal{L}}(X,A)}{\times} \text{Fun}_{\mathcal{L}}(X,B) \\
\big\uparrow & & \big\uparrow \\
\text{Fun}_{\mathcal{K}}(X,P) & \xrightarrow{\ \cong\ } & \text{Fun}_{\mathcal{K}}(X,C) \underset{\text{Fun}_{\mathcal{K}}(X,A)}{\times} \text{Fun}_{\mathcal{K}}(X,B)
\end{array}$$

and fullness on positive-dimensional arrows, we need only verify surjectivity of the dotted map on 0-arrows. So consider a cone $(b : X \to B, c : X \to C)$ over the pullback diagram in $\mathcal{L}$. By the universal property of the iso-comma $C \overset{}{\underset{A}{\times}} B$, there exists a factorization $y : X \to C \overset{}{\underset{A}{\times}} B$ in $\mathcal{L}$. Composing with the equivalence $C \overset{}{\underset{A}{\times}} B \simeq P$, this map is equivalent to the factorization $z : X \to P$ of the cone (b, c) through the limit cone (h, k) in $\mathcal{K}$ that exists on account of the strict universal property of the pullback in there. By repleteness, the isomorphism between z and the composite of y with the equivalence suffices to show that z lies in $\mathcal{L}$. Hence, the functor spaces in $\mathcal{L}$ are isomorphic. A similar argument invoking Lemma 6.2.19 proves that inverse limits of towers of isofibrations define conical limits in $\mathcal{L}$. This completes the proof of the limit axiom 1.2.1(i).

Since the isofibrations in $\mathcal{L}$ are a subset of the isofibrations in $\mathcal{K}$ and the limit constructions in both contexts coincide, most of the closure properties of 1.2.1(ii) are inherited from the closure properties in $\mathcal{K}$. The one exception is the requirement that the isofibrations in $\mathcal{L}$ define isofibrations of quasi-categories representably, which was proven for any replete subcategory in Lemma 6.3.2. This proves that $\mathcal{L}$ defines an ∞-cosmos.

Finally, we argue that the equivalences in $\mathcal{L}$ are created from the equivalences of $\mathcal{K}$, which will imply that the trivial fibrations in $\mathcal{L}$ coincide with those of $\mathcal{K}$ as well. Condition (ii) of Definition 6.3.1 implies that for any arrow in $\mathcal{L}$ that defines an equivalence in $\mathcal{K}$, its equivalence inverse and witnessing homotopy coherent isomorphisms of Lemma 1.2.15 lie in $\mathcal{L}$. Because we have already shown that $\mathcal{L}$ admits cotensors with $\mathbb{I}$ preserved by the inclusion $\mathcal{L} \hookrightarrow \mathcal{K}$, Lemma 1.2.15 implies that this data defines an equivalence in $\mathcal{L}$. Conversely, any equivalence in $\mathcal{L}$ extends to the data of (1.2.16) and since $\mathcal{L} \hookrightarrow \mathcal{K}$ preserves $\mathbb{I}$-cotensors, this data defines an equivalence in $\mathcal{K}$. Thus, by construction, the ∞-cosmos structure of $\mathcal{L}$ is preserved and reflected by the inclusion $\mathcal{L} \hookrightarrow \mathcal{K}$ as claimed. $\square$

In practice, the repleteness condition of Definition 6.3.1 is satisfied by any subcategory of objects and 0-arrows that is characterized by some ∞-categorical property, so the main task in verifying that a subcategory defines an ∞-cosmos is verifying the closure under flexible weighted limits. In our first example, which acts as a sort of base case, we must do this by hand. In subsequent examples, we deploy various techniques to bootstrap the flexible weighted limit closure from previously-established facts.

PROPOSITION 6.3.4. *For any ∞-cosmos $\mathcal{K}$, let $\mathcal{K}_T$ denote the quasi-categorically enriched category whose*

(i) *objects are ∞-categories in $\mathcal{K}$ that possess a terminal element and*
(ii) *functor spaces $\mathrm{Fun}_T(A, B) \subset \mathrm{Fun}(A, B)$ are the sub-quasi-categories whose 0-arrows preserve terminal elements and containing all n-arrows these functors span.*

Then the inclusion $\mathcal{K}_T \hookrightarrow \mathcal{K}$ creates an ∞-cosmos structure on $\mathcal{K}_T$ from $\mathcal{K}$, and moreover for each object of $\mathcal{K}_T$ defined as a flexible weighted limit of some diagram in $\mathcal{K}_T$, its terminal element is created by the 0-arrow legs of the limit cone.

Proof We apply Proposition 6.3.3. Lemma 2.2.7 and Proposition 2.1.10 verify the repleteness conditions of Definition 6.3.1, so it remains only to prove closure under flexible weighted limits. We do so by induction over the tower of isofibrations constructed in Proposition 6.2.8(i), which expresses a flexible weighted limit $\lim_W F$ as the inverse limit of a tower of isofibrations

$$\lim_W F \twoheadrightarrow \cdots \twoheadrightarrow \lim_{W_{k+1}} F \twoheadrightarrow \lim_{W_k} F \twoheadrightarrow \cdots \twoheadrightarrow \lim_{W_0} F \twoheadrightarrow 1$$

each of which is a pullback of products of maps of the form (6.2.9) indexed by the projective cells of the flexible weight W. We argue inductively that each ∞-category in this tower possesses a terminal element that is preserved and jointly created by the legs of the limit cone.

For the base case, if $(A_i)_{i \in I}$ is a family of ∞-categories possessing terminal elements $t_i : 1 \to A_i$, then the product of the adjunctions $! \dashv t_i$ defines an adjunction

$$1 \cong \prod_{i \in I} 1 \underset{(t_i)_{i \in I} \cong \prod_i t_i}{\overset{!}{\rightleftarrows}} \prod_{i \in I} A_i$$

exhibiting $(t_i)_{i \in I}$ as a terminal element of $\prod_{i \in I} A_i$. By construction, this terminal element is jointly created by the legs of the product cone. In particular, the product projection functors preserve this terminal element and the map into the product ∞-category $\prod_{i \in I} A_i$ induced by any family of terminal element preserving functors $(f : X \to A_i)_{i \in I}$ preserves terminal elements. This verifies that the subcategory $\mathcal{K}_T$ is closed under products.

Similarly, if A is an ∞-category with a terminal element $t : 1 \to A$, then by Proposition 2.1.7(iii), the constant diagram at t, which we also denote by t, defines a terminal element in the cotensor A^U by any simplicial set U:

$$1 \cong 1^U \underset{t := t^U}{\overset{!}{\rightleftarrows}} \prod_{i \in I} A^U$$

It remains to argue that terminal elements in A^U are jointly created by the 0-arrow components of the limit cone, namely by evaluation on each of the vertices of the cotensoring simplicial set. To that end, suppose $s : 1 \to A^U$ has the property that

$$s_u := 1 \xrightarrow{\ s\ } A^U \xrightarrow{\ \mathrm{ev}_u\ } A$$

is terminal in A for each vertex u of U. By terminality of $t : 1 \to A^U$ there is a natural transformation $\alpha : s \Rightarrow t$, and since both s_u and t define terminal elements of A, the component of α at each u defines an isomorphism $\alpha_u : s_u \cong t$ in A. By Corollary 1.1.22, it follows that α is an isomorphism, which tells us that s is also a terminal element of A^U.

For the inductive step consider a pullback diagram

$$
\begin{array}{ccc}
\lim_{W_{k+1}} F & \longrightarrow & A^{\Delta[n]} \\
\downarrow & \lrcorner & \downarrow \\
\lim_{W_k} F & \underset{\ell}{\longrightarrow} & A^{\partial \Delta[n]}
\end{array}
$$

that arises from the attaching map for a projective n-cell. The inductive hypothesis tells us that $\lim_{W_k} F$ admits a terminal element t_k and for each vertex of $i \in \partial\Delta[n]$, the corresponding component $\ell_i : \lim_{W_k} F \to A$ of the limit cone preserves it. Since F is a diagram valued in $\mathcal{K}_T$ and A is an ∞-category in its image, we know that A must possess a terminal element $t : 1 \to A$. Thus the constant diagram $t : 1 \to A^{\partial\Delta[n]}$ defines a terminal element. By terminality, there is a natural transformation $\alpha : \ell(t_k) \Rightarrow t$ whose components at each $i \in \partial\Delta[n]$ are isomorphisms in A. By Corollary 1.1.22, it follows that α is also an isomorphism, which tells us that $\ell(t_k)$ is also a terminal element of $A^{\partial\Delta[n]}$. Thus, we see that ℓ preserves terminal elements. The proof is now completed by the following pair of lemmas.   □

LEMMA 6.3.5. *Consider a pullback diagram*

$$
\begin{array}{ccc}
P & \xrightarrow{\ h\ } & B \\
k \downarrow & \lrcorner & \downarrow f \\
C & \underset{g}{\longrightarrow} & A
\end{array}
$$

*in which the ∞-categories A, B, and C possess a terminal element and the
functors f and g preserve them. Then P possesses a terminal element that is
created by the legs of the pullback cone h and k.*

Proof If $b : 1 \to B$ and $c : 1 \to C$ are terminal, then uniqueness of terminal
elements implies that $f(b) \cong g(c) \in A$. Using the fact that f is an isofibration,
there is a lift $b' \cong b$ of this isomorphism along f that then defines another
terminal element of B. The pair (c, b') now induces an element t of P that we
claim is terminal.

To see this, we apply Proposition 4.3.13, which proves that t is a terminal
element of P if and only if the domain projection functor $p_0 : \mathrm{Hom}_P(P, t) \twoheadrightarrow P$
is a trivial fibration. By construction of t, we know that the domain projection
functors for the elements $ht = b'$, $kt = c$, and $fht = gkt$ are all trivial fibrations
and moreover, by the hypercube pullback lemma, the top and bottom faces of
the cube

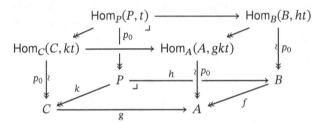

are pullbacks. By Proposition 3.3.4, the fact that the three maps between the
cospans are equivalences implies that the map between their pullbacks is also
an equivalence, as required.   □

LEMMA 6.3.6. *Consider a tower of isofibrations*

$$A := \lim \left(\quad \cdots \longtwoheadrightarrow A_n \xrightarrow{\;q_n\;} A_{n-1} \longtwoheadrightarrow \cdots \longtwoheadrightarrow A_1 \xrightarrow{\;q_1\;} A_0 \right)$$

*in which each ∞-category has a terminal element each fibration preserves
terminal elements. Then the limit of the tower has a terminal element created by
the legs of the limit cone.*

Proof The hypothesis provides terminal elements $s_n : 1 \to A_n$ so that $q_n s_n \cong
s_{n-1}$. By lifting away these isomorphisms inductively starting from the bottom,
we can choose a different family of terminal elements $t_n : 1 \to A_n$ so that
$q_n t_n = t_{n-1}$. This cone then induces an element $t : 1 \to A$ in the limit A that
we claim is terminal. This follows from Proposition 4.3.13 and the observation
that $p_0 : \mathrm{Hom}_A(A, t) \twoheadrightarrow A$ is the limit in the ∞-cosmos of isofibrations of a
tower whose objects are the trivial fibrations $p_0 : \mathrm{Hom}_{A_n}(A_n, t_n) \twoheadrightarrow A_n$. By
Proposition 6.1.5, $p_0 : \mathrm{Hom}_A(A, t) \twoheadrightarrow A$ is then a trivial fibration as well.   □

Applying the result of Proposition 6.3.4 to $\mathcal{K}^{co}$ constructs an ∞-cosmos $\mathcal{K}_\perp$ whose objects are ∞-categories in $\mathcal{K}$ that possess an initial object and whose 0-arrows are initial-element-preserving functors. It is straightforward to verify that the inclusion $\mathcal{K}_\perp \hookrightarrow \mathcal{K}$ is a cosmological embedding.[1] Combining these results, we get an ∞-cosmos for the pointed ∞-categories of Definition 4.4.1, those that possess a zero element.

PROPOSITION 6.3.7. *For any ∞-cosmos $\mathcal{K}$, let $\mathcal{K}_*$ denote the quasi-categorically enriched category of pointed ∞-categories – ∞-categories that possess a zero element – and functors that preserve them. Then the inclusion $\mathcal{K}_* \hookrightarrow \mathcal{K}$ is a cosmological embedding, creating an ∞-cosmos structure on $\mathcal{K}_*$ from $\mathcal{K}$, and moreover for each object of $\mathcal{K}_*$ defined as a flexible weighted limit of some diagram in $\mathcal{K}_*$, its zero element is created by the 0-arrow legs of the limit cone.*

Proof This result follows directly from Proposition 6.3.4 and its dual, since the ∞-cosmos $\mathcal{K}_*$ of ∞-categories in $\mathcal{K}$ possessing a zero object is isomorphic to $(\mathcal{K}_\top)_\perp \cong (\mathcal{K}_\perp)_\top$, the insight being that an initial element $a : 1 \to A$ in $\mathcal{K}_\top$ is encoded by a terminal-element preserving functor, which says exactly that a is a zero element in A. □

Applying the result of Proposition 6.3.4 or its dual to the ∞-cosmos $\mathcal{K}_{/B}$ of isofibrations over $B \in \mathcal{K}$, we obtain two new ∞-cosmoi of interest.

COROLLARY 6.3.8. *For any ∞-category B in an ∞-cosmos $\mathcal{K}$, the sliced ∞-cosmos $\mathcal{K}_{/B}$ admits cosmologically embedded ∞-cosmoi*

$$\mathcal{R}ari(\mathcal{K})_{/B} \hookrightarrow \mathcal{K}_{/B} \hookleftarrow \mathcal{L}ari(\mathcal{K})_{/B}$$

whose

- *objects are isofibrations over B admitting a right adjoint right inverse or left adjoint right inverse, respectively, and*
- *0-arrows are functors over B that commute with the respective right or left adjoints up to fibered isomorphism over B.*

Proof By Lemma 3.6.9, these ∞-cosmoi are defined by

$$\mathcal{R}ari(\mathcal{K})_{/B} := (\mathcal{K}_{/B})_\top \quad \text{and} \quad \mathcal{L}ari(\mathcal{K})_{/B} := (\mathcal{K}_{/B})_\perp. \qquad □$$

Leveraging Corollary 6.3.8, we can establish similar cosmological embeddings

$$\mathcal{R}ari(\mathcal{K}) \hookrightarrow \mathcal{K}^\downarrow \hookleftarrow \mathcal{L}ari(\mathcal{K})$$

---

[1] Other duals of the ∞-cosmoi constructed here are established similarly (see Exercise 6.3.iii).

defining ∞-cosmoi of right adjoint right inverse or left adjoint right inverse adjunctions with varying bases. To apply Proposition 6.3.3, we must check closure under flexible weighted limits. We argue separately for cotensors, which are easy, and for the conical limits, which are harder – as the reader who attempted to solve Exercise 5.2.iii using Theorem 5.2.8(ii) may have already discovered. To treat all of the conical limits at once, we make use of a general 1-categorical result that constructs limits in the total space $\mathcal{E}$ of a Grothendieck fibration $P : \mathcal{E} \to \mathcal{B}$ out of limits in the base $\mathcal{B}$ and limits in the fibers $\mathcal{E}_B$.

LEMMA 6.3.9. *Let* $P : \mathcal{E} \to \mathcal{B}$ *be a Grothendieck fibration between 1-categories. Suppose that* $\mathcal{J}$ *is a small category, that* $D : \mathcal{J} \to \mathcal{E}$ *is a diagram, and that*

  (i) the diagram $PD : \mathcal{J} \to \mathcal{B}$ *has a limit* L *in* $\mathcal{B}$ *with limit cone* $\lambda : \Delta L \Rightarrow PD$,

  (ii) the diagram $\lambda^* D : \mathcal{J} \to \mathcal{E}_L$ *constructed by lifting the cone* λ *to a cartesian natural transformation* $\chi : \lambda^* D \Rightarrow D$

$$
\begin{array}{ccc}
\mathcal{J} \xrightarrow{\ D\ } \mathcal{E} & & \mathcal{J} \overset{D}{\underset{\lambda^* D}{\overset{\Uparrow \chi}{\rightrightarrows}}} \mathcal{E} \\
{\scriptstyle \Uparrow \lambda} \ \ \downarrow {\scriptstyle P} & = & \qquad\quad \downarrow {\scriptstyle P} \\
\Delta L \searrow\ \mathcal{B} & & \mathcal{B}
\end{array}
$$

  has a limit M *in the fibre* $\mathcal{E}_L$ *with limit cone* $\mu : \Delta M \Rightarrow \lambda^* D$, *and*

  (iii) the limit cone $\mu : \Delta M \Rightarrow \lambda^* D$ *is preserved by the reindexing functor* $u^* : \mathcal{E}_L \to \mathcal{E}_B$ *associated with any arrow* $u : B \to L$ *in* $\mathcal{B}$.

Then the composite cone $\Delta M \xRightarrow{\ \mu\ } \lambda^* D \xRightarrow{\ \chi\ } D$ *displays* M *as a limit of the diagram* D *in* $\mathcal{E}$.

Proof Any arrow $f : E \to E'$ in the domain of a Grothendieck fibration $P : \mathcal{E} \to \mathcal{B}$ factors uniquely up to isomorphism through a "vertical" arrow in the fiber $\mathcal{E}_{PE}$ followed by a "horizontal" cartesian lift of Pf with codomain E'.

Given a cone $\alpha : \Delta E \Rightarrow D$ with summit $E \in \mathcal{E}$ over D, by (i) its image $P\alpha : \Delta PE \Rightarrow PD$ factors uniquely through the limit cone $\lambda : \Delta L \Rightarrow D$ via a map $b : PE \to L \in \mathcal{B}$. By the universal property of the cartesian lift χ of λ constructed in (ii), it follows that α factors uniquely through χ via a natural transformation $\beta : \Delta E \Rightarrow \lambda^* D$ so that $P\beta = \Delta b$. This cone factors uniquely up to isomorphism via "vertical" natural transformation $\gamma : \Delta E \Rightarrow b^* \lambda^* D$ followed by a "horizontal" cartesian lift of b. By (iii), the limit cone $\mu : \Delta M \Rightarrow \lambda^* D$ in $\mathcal{E}_L$ pulls back along b to a limit cone in $\mathcal{E}_{PE}$ through which the pullback of β factors via a map $k : E \to b^* M$. Thus, β itself factors uniquely through μ via the composite of this map $k : E \to b^* M$ with the cartesian arrow $b^* M \to M$ lifting $b : PE \to L$. $\square$

For example, the codomain projection functor cod : $\mathcal{K}^{\downarrow} \to \mathcal{K}$ is a Grothendieck fibration of underlying 1-categories, with cod-cartesian lifts defined by pullback squares. Since right adjoint right inverses to isofibrations can be constructed as fibered adjunctions by Lemma 3.6.9, and fibered adjunctions are stable under pullback by Lemma 3.6.6, this Grothendieck fibration restricts to define a Grothendieck fibration cod : $\mathcal{R}ari(\mathcal{K}) \to \mathcal{K}$, which allows us to appeal to Lemma 6.3.9 in the proof of the following result.

PROPOSITION 6.3.10. *For any ∞-cosmos $\mathcal{K}$, the ∞-cosmos of isofibrations admits cosmologically embedded ∞-cosmoi*

$$\mathcal{R}ari(\mathcal{K}) \overset{}{\hookrightarrow\!\!\!\twoheadrightarrow} \mathcal{K}^{\downarrow} \overset{}{\twoheadleftarrow\!\!\!\longrightarrow} \mathcal{L}ari(\mathcal{K})$$

whose

- *objects are isofibrations admitting a right adjoint right inverse or left adjoint right inverse, respectively, and*
- *0-arrows are commutative squares between the right or left adjoints, respectively, whose mates are isomorphisms.*

We refer to a commutative square between right adjoints whose mate is an isomorphism as an **exact square**.

Proof By Proposition B.3.8 and Exercise B.4.i, the quasi-categorically enriched subcategory $\mathcal{R}ari(\mathcal{K})$ is replete in $\mathcal{K}^{\downarrow}$, so by Proposition 6.3.3 we need only check that $\mathcal{R}ari(\mathcal{K}) \hookrightarrow \mathcal{K}^{\downarrow}$ is closed under flexible weighted limits. We argue separately for cotensors and for the conical limits.

If $p : E \twoheadrightarrow B$ is an isofibration admitting a right adjoint right inverse in $\mathcal{K}$ and U is a simplicial set, then the cosmological functor $(-)^U : \mathcal{K} \to \mathcal{K}$ carries this data to a right adjoint right inverse to $p^U : E^U \twoheadrightarrow B^U$, which proves that the simplicial cotensor in $\mathcal{K}^{\downarrow}$ of an object in $\mathcal{R}ari(\mathcal{K})$ lies in $\mathcal{R}ari(\mathcal{K})$. To conclude that $\mathcal{R}ari(\mathcal{K})$ is closed in $\mathcal{K}^{\downarrow}$ under simplicial cotensors, we must also verify that p^U has the enriched universal property of the simplicial cotensor in $\mathcal{R}ari(\mathcal{K})$: that is, we must show that the natural isomorphism $\mathsf{Fun}(q, p^U) \cong \mathsf{Fun}(q, p)^U$ of functor spaces in $\mathcal{K}^{\downarrow}$ restricts to define a natural isomorphism of functor spaces in $\mathcal{R}ari(\mathcal{K})$. Since the inclusion $\mathcal{R}ari(\mathcal{K}) \hookrightarrow \mathcal{K}^{\downarrow}$ is full on 0-arrows, this amounts to verifying that certain 0-arrows in $\mathcal{K}^{\downarrow}$ are exact squares.

The weighted limit cone for the cotensor p^U in $\mathcal{K}^{\downarrow}$ is given by the canonical map of simplicial sets $U \to \mathsf{Fun}(p^U, p)$ defined on each vertex $u : \mathbb{1} \to U$ by the commutative square

$$
\begin{array}{ccc}
E^U & \xrightarrow{\ u^*\ } & E \\
{\scriptstyle p^U}\downarrow & & \downarrow{\scriptstyle p} \\
B^U & \xrightarrow[\ u^*\ ]{} & B
\end{array}
\tag{6.3.11}
$$

The maps u^* define the components of a simplicial natural transformation from $(-)^U$ to the identity functor and thus the mate of this commutative square is an identity, so the limit cone for the U-cotensor lies in $\mathcal{R}ari(\mathcal{K})$. Finally, to verify the universal property of the cotensor in $\mathcal{R}ari(\mathcal{K})$, we must show that any commutative square whose domain q is an isofibration admitting a right adjoint right inverse

$$
\begin{array}{ccc}
F & \longrightarrow & E^U \\
{\scriptstyle q}\downarrow & & \downarrow{\scriptstyle p^U} \\
A & \longrightarrow & B^U
\end{array}
$$

and which composes with each of the squares (6.3.11) to an exact square is itself exact. To see this, observe that the mate of the identity defines a natural transformation that is represented by a 1-arrow in $\mathsf{Fun}(A, E^U) \cong \mathsf{Fun}(A, E)^U$ and the hypothesis says that the components of this 1-arrow are invertible for each vertex of U. Corollary 1.1.22 then tells us that this 1-arrow is invertible as required.

A similar and slightly easier argument proves that $\mathcal{R}ari(\mathcal{K})$ is closed in $\mathcal{K}^{\downarrow}$ under products. It remains only to show that this subcategory is closed under the remaining conical limits. As argued above cod : $\mathcal{R}ari(\mathcal{K}) \to \mathcal{K}$ is a Grothendieck fibration on underlying 1-categories whose fibers are the ∞-cosmoi $\mathcal{R}ari(\mathcal{K})_{/B}$. By Lemma 6.3.9, 1-categorical limit cones in $\mathcal{R}ari(\mathcal{K})$ $\subset \mathcal{K}^{\downarrow}$ can be calculated as composites of cartesian lifts of limit cones in the base with limit cones of fiberwise diagrams: in this case, the recipe is to pull back along the limit cone formed by the codomains and then form the limit in the sliced ∞-cosmos over the limit object for the base diagram. By Corollary 6.3.8, these fiberwise limits in the sliced ∞-cosmos $\mathcal{K}_{/B}$ of diagrams in $\mathcal{R}ari(\mathcal{K})_{/B}$ lie in $\mathcal{R}ari(\mathcal{K})_{/B} \hookrightarrow \mathcal{R}ari(\mathcal{K})$. Moreover, these 1-categorical limits are preserved by the simplicial cotensor, which by Proposition A.5.5 implies that their universal property enriches to define conical limits. In this way we see that $\mathcal{R}ari(\mathcal{K}) \hookrightarrow \mathcal{K}^{\downarrow}$ is closed under flexible weighted limits and thus defines a cosmological embedding, as claimed. □

Now that we have established many examples of ∞-cosmoi and cosmologically embedded replete subcategories, we can make use of the following result:

PROPOSITION 6.3.12. *Suppose* $F \colon \mathcal{K}' \to \mathcal{K}$ *is a cosmological functor and* $\mathcal{L} \subset \mathcal{K}$ *is a replete subcategory that defines a cosmologically embedded ∞-cosmos. Then the simplicially enriched category defined by the pullback*

$$
\begin{array}{ccc}
\mathcal{L}' & \longrightarrow & \mathcal{L} \\
\big\downarrow & \lrcorner & \big\downarrow \\
\mathcal{K}' & \xrightarrow{\ F\ } & \mathcal{K}
\end{array}
$$

is replete and defines a cosmologically embedded sub ∞-cosmos of $\mathcal{K}'$, and the restricted functor $F \colon \mathcal{L}' \to \mathcal{L}$ is cosmological.

Proof The objects and n-arrows of the pullback $\mathcal{L}'$ are defined to be the objects and n-arrows of $\mathcal{K}'$ whose image under F lies in the simplicial subcategory $\mathcal{L}$. In particular, since the inclusion $\mathcal{L} \hookrightarrow \mathcal{K}$ is full on positive-dimensional arrows, the inclusion $\mathcal{L}' \hookrightarrow \mathcal{K}'$ is as well, and in particular $\mathcal{L}'$ is quasi-categorically enriched. Since the cosmological functor F preserves equivalences and $\mathcal{L} \hookrightarrow \mathcal{K}$ is replete, the quasi-categorically enriched subcategory $\mathcal{L}'$ is replete in $\mathcal{K}'$, satisfying the three conditions of Definition 6.3.1.

To prove that the replete subcategory $\mathcal{L}' \hookrightarrow \mathcal{K}'$ admits an ∞-cosmos structure created by the inclusion, it remains to argue that $\mathcal{L}'$ is closed under flexible weighted limits in $\mathcal{K}'$. Consider a diagram $D \colon A \to \mathcal{L}'$ and a flexible weight and form its flexible weighted limit in $\mathcal{L}$ and in $\mathcal{K}'$. The functors to $\mathcal{K}$ carry these to a pair of equivalent cones over the same diagram and since the inclusion $\mathcal{L} \hookrightarrow \mathcal{K}$ is replete, there exists a limit cone over D in $\mathcal{L}$ whose image in $\mathcal{K}$ is equal to the image of the limit cone in $\mathcal{K}'$ under the functor F. Now the universal property of the pullback allows us to lift this cone to $\mathcal{L}$, and a similar argument demonstrates that the lifted cone is a flexible weighted limit cone over the original diagram. Note by construction that the flexible weighted limit in $\mathcal{L}'$ is preserved by the functors to both $\mathcal{K}'$ and to $\mathcal{L}$.

Proposition 6.3.3 applies to conclude that $\mathcal{L}' \hookrightarrow \mathcal{K}'$ is a cosmological embedding. Since the isofibrations in $\mathcal{L}'$ are created in $\mathcal{K}'$ and preserved by F, and $\mathcal{L} \hookrightarrow \mathcal{K}$ is a cosmological embedding, it follows that the restricted functor $F \colon \mathcal{L}' \to \mathcal{L}$ preserves isofibrations. Since $F \colon \mathcal{L}' \to \mathcal{L}$ preserves flexible weighted limits, we see that this functor is cosmological. $\square$

This result allows us to construct further ∞-cosmoi of interest.

PROPOSITION 6.3.13. *For any ∞-cosmos $\mathcal{K}$ and simplicial set J, there cosmologically embedded ∞-cosmoi*

$$
\mathcal{K}_{\top,J} \hookrightarrow \mathcal{K} \hookleftarrow \mathcal{K}_{\bot,J}
$$

whose

- *objects are ∞-categories in $\mathcal{K}$ that admit all limits of shape J or all colimits of shape J, respectively,*
- *0-arrows are the functors that preserve them*

Moreover for each object of $\mathcal{K}_{\top,J}$ or $\mathcal{K}_{\perp,J}$ defined as a flexible weighted limit of some diagram in that ∞-cosmos, its J-shaped limits or colimits are created by the 0-arrow legs of the limit or colimit cones, respectively.

Proof We prove this in the case of ∞-categories admitting J-shaped colimits, the other case being dual. For any fixed simplicial set J, there is a cosmological functor $F_J : \mathcal{K} \to \mathcal{K}^\downarrow$ defined on objects by mapping an ∞-category A to the isofibration $A^{J^\triangleright} \twoheadrightarrow A^J$ in the notation of 4.2.6 and a functor $f : A \to B$ to the commutative square

$$\begin{array}{ccc} A^{J^\triangleright} & \xrightarrow{f^{J^\triangleright}} & B^{J^\triangleright} \\ \downarrow & & \downarrow \\ A^J & \xrightarrow{f^J} & B^J \end{array}$$

By Corollary 4.3.5, A admits colimits of shape J if and only if this isofibration admits a left adjoint right inverse, and $f : A \to B$ preserves these colimits if and only if the square displayed above is exact. In summary, the quasi-categorically enriched subcategory $\mathcal{K}_{\perp,J}$ is defined by the pullback

$$\begin{array}{ccc} \mathcal{K}_{\perp,J} & \longrightarrow & \mathcal{L}ari(\mathcal{K}) \\ \downarrow & \lrcorner & \downarrow \\ \mathcal{K} & \xrightarrow{F_J} & \mathcal{K}^\downarrow \end{array}$$

By Proposition 6.3.10, $\mathcal{L}ari(\mathcal{K}) \hookrightarrow \mathcal{K}^\downarrow$ is a cosmological embedding, so Proposition 6.3.12 proves that the inclusion $\mathcal{K}_{\perp,J} \hookrightarrow \mathcal{K}$ creates an ∞-cosmos structure, as claimed.

In particular, the closure of the subcategory $\mathcal{K}_{\perp,J}$ under flexible weighted limits in $\mathcal{K}$ implies that J-shaped colimits in an ∞-category defined as a flexible weighted limit are created by the 0-arrow legs of the limit cone, as we explain. Certainly the colimits in an ∞-category in $\mathcal{K}_{\perp,J}$, formed as a weighted limit of a diagram of ∞-categories in $\mathcal{K}_{\perp,J}$, are preserved by the 0-arrow legs of the weighted limit cone, since the 0-arrows in $\mathcal{K}_{\perp,J}$ are J-shaped-colimit-preserving functors. And since the J-colimit completeness of an ∞-category that is defined as the flexible weighted limit in $\mathcal{K}$ can be deduced whenever that diagram lies in the sub ∞-cosmos $\mathcal{K}_{\perp,J}$, these J-colimits are also created. □

PROPOSITION 6.3.14. *For any ∞-cosmos $\mathcal{K}$, the ∞-cosmos of isofibrations admits cosmologically embedded ∞-cosmoi*

$$\mathcal{C}art(\mathcal{K}) \longhookrightarrow\!\!\!\!\!\twoheadrightarrow \mathcal{K}^{\downarrow} \longleftarrow\!\!\!\!\! \longleftarrow co\mathcal{C}art(\mathcal{K})$$

whose objects are cartesian or cocartesian fibrations, respectively, and whose 0-arrows are cartesian functors. Similarly, for any ∞-category B in an ∞-cosmos $\mathcal{K}$, the sliced ∞-cosmos $\mathcal{K}_{/B}$ admits cosmologically embedded ∞-cosmoi

$$\mathcal{C}art(\mathcal{K})_{/B} \longhookrightarrow\!\!\!\!\!\twoheadrightarrow \mathcal{K}_{/B} \longleftarrow\!\!\!\!\! \longleftarrow co\mathcal{C}art(\mathcal{K})_{/B}$$

whose objects are cartesian or cocartesian fibrations over B, respectively, and whose 0-arrows are cartesian functors, with the ∞-cosmos structures created by the inclusions.

Proof By Theorems 5.2.8 and 5.3.4, the quasi-categorically enriched category $\mathcal{C}art(\mathcal{K})$ is defined by the pullback

$$
\begin{array}{ccc}
\mathcal{C}art(\mathcal{K}) & \longrightarrow & \mathcal{R}ari(\mathcal{K}) \\
\downarrow & \lrcorner & \downarrow \\
\mathcal{K}^{\downarrow} & \xrightarrow{\ K\ } & \mathcal{K}^{\downarrow}
\end{array}
$$

along the simplicial functor K that sends an isofibration $p : E \twoheadrightarrow B$ to the isofibration $i_1 \,\widehat{\pitchfork}\, p : E^2 \twoheadrightarrow \mathrm{Hom}_B(B, p)$. To see this, recall from Proposition 6.3.10 that the 0-arrows in the functor spaces of $\mathcal{R}ari(\mathcal{K})$ are commutative squares between isofibrations admitting a right adjoint right inverse so that the mate of the identity 2-cell induces an isomorphism in the corresponding square involving the right adjoints. By Theorem 5.3.4, this condition pulls back along the functor K to tell us that 0-arrows in $\mathcal{C}art(\mathcal{K})$ are commutative squares between cocartesian fibrations that define cartesian functors in the sense of Definition 5.3.2.

The simplicial functor K is constructed out of weighted limits and thus preserves all weighted limits. In addition, it preserves the isofibrations of Proposition 6.1.1: if $r : F \twoheadrightarrow P$ is the isofibration from the initial vertex to the pullback of the cospan displayed in 6.1.1(iii), then the corresponding map under the image of the functor K is the Leibniz cotensor $i_1 \,\widehat{\pitchfork}\, r : F^2 \twoheadrightarrow \mathrm{Hom}_P(P, r)$. Thus, K is a cosmological functor. By Proposition 6.3.12, the inclusion $\mathcal{C}art(\mathcal{K}) \hookrightarrow \mathcal{K}^{\downarrow}$ creates an ∞-cosmos structure.

We have a similar pullback defining the quasi-categorically enriched category of cartesian fibrations with a fixed base

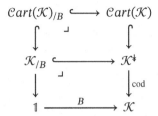

but we cannot appeal to Proposition 6.3.12 because the inclusion $\mathcal{K}_{/B} \hookrightarrow \mathcal{K}^{\downarrow}$ is neither cosmological nor replete. Instead we must appeal to Proposition 6.3.3. By Corollary 5.3.1 and Theorem 5.3.4, $\mathcal{C}art(\mathcal{K})_{/B} \hookrightarrow \mathcal{K}_{/B}$ is a replete subcategory. The connected conical limits in $\mathcal{C}art(\mathcal{K})_{/B}$ and $\mathcal{K}_{/B}$ are created by the inclusions into $\mathcal{C}art(\mathcal{K})$ and into $\mathcal{K}^{\downarrow}$, so these are created by the inclusion $\mathcal{C}art(\mathcal{K})_{/B} \hookrightarrow \mathcal{K}_{/B}$. It remains only to argue directly that product and simplicial cotensors are created by this inclusion.

By Proposition 1.2.22, the product of isofibrations $p_i : E_i \twoheadrightarrow B$ is formed by the pullback

$$\begin{array}{ccc} \times^B_i E_i & \longrightarrow & \prod_i E_i \\ \scriptstyle\times^B_i p_i \downarrow & \lrcorner & \downarrow \scriptstyle\prod_i p_i \\ B & \xrightarrow{\ \Delta\ } & \prod_i B \end{array}$$

If each p_i is a cartesian fibration, then since $\mathcal{C}art(\mathcal{K})$ is an ∞-cosmos, so is $\prod_i p_i$, and the legs of the limit cone are cartesian functors. By Proposition 5.2.4 it follows that their product in $\mathcal{K}_{/B}$ is a cartesian fibration and the legs of the limit cone, defined by composing the pullback square with the product projections, are cartesian functors. Since cartesian transformations are created by pullbacks, it follows easily that a square with codomain $\times^B_i p_i$ is a cartesian functor if and only if each of the composite squares with codomain p_i are cartesian functors. This proves that products are created by the inclusion $\mathcal{C}art(\mathcal{K})_{/B} \hookrightarrow \mathcal{K}_{/B}$

Similarly, by Proposition 1.2.22, the cotensor of an isofibration $p : E \twoheadrightarrow B$ with a simplicial set U is formed by the pullback

$$\begin{array}{ccc} U \pitchfork_B p & \longrightarrow & E^U \\ \downarrow & \lrcorner & \downarrow \scriptstyle p^U \\ B & \xrightarrow{\ \Delta\ } & B^U \end{array}$$

and hence is a cartesian fibration if p is. A similar argument shows that this object has the universal property required of the simplicial cotensor in $\mathcal{C}art(\mathcal{K})_{/B}$, completing the proof. $\qquad\square$

The ∞-cosmoi of cartesian fibrations provide a fertile setting to explore the parametrized ∞-category theory developed by Shah [115].

PROPOSITION 6.3.15. *For any ∞-cosmos $\mathcal{K}$, there exist cosmologically embedded ∞-cosmoi*

$$\mathcal{D}isc\mathcal{C}art(\mathcal{K}) \xhookrightarrow{\qquad\qquad} \mathcal{K}^{\downarrow} \xleftarrow{\qquad\qquad} \mathcal{D}iscco\mathcal{C}art(\mathcal{K})$$
$$\mathcal{C}art(\mathcal{K}) \qquad\qquad co\mathcal{C}art(\mathcal{K})$$

whose objects are discrete cartesian or discrete cocartesian fibrations. Similarly, for any ∞-category B in an ∞-cosmos $\mathcal{K}$, there exist cosmologically embedded ∞-cosmoi

$$\mathcal{D}isc\mathcal{C}art(\mathcal{K})_{/B} \xhookrightarrow{\qquad\qquad} \mathcal{K}_{/B} \xleftarrow{\qquad\qquad} \mathcal{D}iscco\mathcal{C}art(\mathcal{K})_{/B}$$
$$\mathcal{C}art(\mathcal{K})_{/B} \qquad\qquad co\mathcal{C}art(\mathcal{K})_{/B}$$

whose objects are discrete cartesian or discrete cocartesian fibrations over B, respectively.

Proof By Exercise 6.3.iv, the ∞-cosmoi of discrete fibrations with a fixed base arise as ∞-cosmoi of discrete objects:

$$\mathcal{D}isc\mathcal{C}art(\mathcal{K})_{/B} \cong (\mathcal{C}art(\mathcal{K})_{/B})^{\simeq} \text{ and } \mathcal{D}iscco\mathcal{C}art(\mathcal{K})_{/B} \simeq (co\mathcal{C}art(\mathcal{K})_{/B})^{\simeq}$$

and hence their existence is guaranteed by Proposition 6.1.6.

The ∞-cosmoi of discrete fibrations with varying bases arise as pullbacks

$$
\begin{array}{ccccc}
\mathcal{D}isc\mathcal{C}art(\mathcal{K}) & \longrightarrow & \mathcal{K}^{\downarrow} & \longleftarrow & \mathcal{D}iscco\mathcal{C}art(\mathcal{K}) \\
\downarrow & \lrcorner & \downarrow & \llcorner & \uparrow \\
\mathcal{C}art(\mathcal{K}) & \xrightarrow{\ K\ } & \mathcal{K}^{\downarrow} & \xleftarrow{\ K\ } & co\mathcal{C}art(\mathcal{K})
\end{array}
$$

along restrictions of the cosmological functor K defined in the proof of Proposition 6.3.14, and hence exist by Proposition 6.3.12. $\qquad\square$

PROPOSITION 6.3.16. *For any ∞-cosmos $\mathcal{K}$, the replete subcategory $\mathcal{S}tab(\mathcal{K})$ of stable ∞-categories and exact functors in $\mathcal{K}$ is an ∞-cosmos and the inclusion $\mathcal{S}tab(\mathcal{K}) \hookrightarrow \mathcal{K}$ is a cosmological embedding.*

Recall from Exercise 4.4.iv that in this context that a functor between stable ∞-categories is called **exact** just when it preserves zero elements and the so-called exact squares, that define both pushouts and pullbacks.

Proof Simplifying the notation of Proposition 6.3.13, write $\mathcal{K}_\lrcorner$ and $\mathcal{K}_\sqsubset$ for the ∞-cosmoi of ∞-categories in $\mathcal{K}$ that admit pullbacks and pushouts, respectively. We argue that the ∞-cosmoi $(\mathcal{K}_\lrcorner)_\sqsubset \cong (\mathcal{K}_\sqsubset)_\lrcorner$ are isomorphic as replete subcategories of $\mathcal{K}$. An object of $(\mathcal{K}_\lrcorner)_\sqsubset$ is an ∞-category A in $\mathcal{K}_\lrcorner$ that admits an adjunction

$$A^\square \underset{\text{res}}{\overset{\text{lan}}{\rotatebox{0}{$\perp$}}} A^\sqsubset$$

in the ∞-cosmos $\mathcal{K}_\lrcorner$, meaning that the functor $\text{lan}: A^\sqsubset \to A^\square$ preserves the pullbacks that both of these ∞-categories admit. Since A itself admits pullbacks, the pullbacks in $A^\sqsubset$ and $A^\square$ may be defined pointwise, via the adjunctions

$$(A^\sqsubset)^\square \underset{\text{ran}^\sqsubset}{\overset{\text{res}}{\rotatebox{0}{$\perp$}}} (A^\sqsubset)^\lrcorner \quad \text{and} \quad (A^\square)^\square \underset{\text{ran}^\square}{\overset{\text{res}}{\rotatebox{0}{$\perp$}}} (A^\square)^\lrcorner$$

and so are automatically preserved by the restriction functor. Thus the condition that this adjunction lies in $\mathcal{K}_\sqsubset$ amounts to the condition that the canonical natural transformation

$$
\begin{array}{ccc}
(A^\sqsubset)^\square & \xrightarrow{\text{lan}^\square} & (A^\square)^\square \\
\text{ran}^\lrcorner \uparrow & \searrow & \uparrow \text{ran}^\square \\
(A^\sqsubset)^\lrcorner & \xrightarrow[\text{lan}^\lrcorner]{} & (A^\square)^\lrcorner
\end{array}
$$

is an isomorphism.

Dually, an object of $(\mathcal{K}_\sqsubset)_\lrcorner$ is an ∞-category A that admits pushouts and also admits an adjunction

$$A^\square \underset{\text{ran}}{\overset{\text{res}}{\rotatebox{0}{$\perp$}}} A^\lrcorner$$

in the ∞-cosmos $\mathcal{K}_\sqsubset$, meaning that the pullback functor $\text{ran}: A^\lrcorner \to A^\square$ preserves the pointwise defined pushouts: i.e., that the canonical natural transformation

$$
\begin{array}{ccc}
(A^\lrcorner)^\square & \xrightarrow{\text{ran}^\square} & (A^\square)^\square \\
\text{lan}^\lrcorner \uparrow & \nwarrow & \uparrow \text{lan}^\square \\
(A^\lrcorner)^\sqsubset & \xrightarrow[\text{ran}^\sqsubset]{} & (A^\square)^\sqsubset
\end{array}
$$

is an isomorphism. Thus, we see that the objects of $(\mathcal{K}_\lrcorner)_\sqsubset$ and $(\mathcal{K}_\sqsubset)_\lrcorner$ coincide. Similarly, the 0-arrows in each of $(\mathcal{K}_\lrcorner)_\sqsubset$ and $(\mathcal{K}_\sqsubset)_\lrcorner$ are functors that preserve both pullbacks and pushouts. This proves that $(\mathcal{K}_\lrcorner)_\ulcorner$ and $(\mathcal{K}_\ulcorner)_\lrcorner$ define isomorphic subcategories of $\mathcal{K}$, as claimed. A similar simpler observation justifies the assertion $(\mathcal{K}_\top)_\bot \cong (\mathcal{K}_\bot)_\top$ made in the proof of Proposition 6.3.7.

By Theorem 4.4.12(iii), an ∞-category is stable just when it is pointed, admits pullbacks and pushouts, and the pullback functor preserves pushouts, and the pushout functor preserves pullbacks. By the proof of Theorem 4.4.12(iii) ⇒ (ii) these conditions imply that pullback and pushout squares coincide. In this way we see that $\mathcal{S}tab(\mathcal{K})$ is the intersection

$$
\begin{array}{ccc}
\mathcal{S}tab(\mathcal{K}) & \lhook\joinrel\longrightarrow\joinrel\rightarrow & (\mathcal{K}_\lrcorner)_\ulcorner \cong (\mathcal{K}_\ulcorner)_\lrcorner \\
\big\downarrow\!\!\!\upharpoonright & \lrcorner & \big\downarrow\!\!\!\upharpoonright \\
(\mathcal{K}_\top)_\bot \cong (\mathcal{K}_\bot)_\top & \lhook\joinrel\longrightarrow\joinrel\rightarrow & \mathcal{K}
\end{array}
$$

of cosmologically embedded ∞-cosmoi and thus is itself a cosmologically embedded ∞-cosmos by Proposition 6.3.12.   □

The results in this section can be used to prove technical results in the ∞-categorical literature (see Exercise 6.3.v). We close with a few illustrations of results along these lines.

LEMMA 6.3.17. *Consider a pullback diagram of ∞-categories, whose vertical functors are isofibrations.*

$$
\begin{array}{ccc}
F & \xrightarrow{\ h\ } & E \\
q\big\downarrow & \lrcorner & \big\downarrow p \\
A & \xrightarrow[\ k\ ]{} & B
\end{array}
$$

Then q creates and h preserves any class of limits or colimits that k preserves and p creates.

Proof The hypotheses ensure that the underlying cospan of the pullback lies in the ∞-cosmos of Proposition 6.3.13. Since this is a cosmologically embedded ∞-cosmos, it follows that the pullback of ∞-categories lies in there as well, which tells us that F admits and q and h preserve any limits or colimits that are present in E, A, and B and preserved by p and k.

It remains only to argue that such (co)limits are created by $q : F \twoheadrightarrow A$. Suppose $d : D \to F^J$ is a family of diagrams and $\eta : d \Rightarrow \Delta c$ is a cone under this diagram so that $q\eta$ is a colimit cone for qd in A. Then $k q\eta = p h\eta$ is a colimit cone for $kqd = phd$ in B. Since $p : E \twoheadrightarrow B$ is assumed to create colimits, this tells us that $h\eta$ is a colimit cone for hd in E. Now the second statement of Proposition 6.3.13, which tells us that (co)limits in F are jointly created by the functors q and h, allows us to conclude that η is a colimit cone in F as claimed.   □

Our final application reproves a result [80, 1.4.2.24] relevant to the theory of stable ∞-categories.

LEMMA 6.3.18. *Suppose A is a pointed ∞-category admitting pullbacks. Then the homotopy limit of the tower of loops functors defines a stable ∞-category.*

$$\mathrm{Sp}A := \lim\left(\;\cdots\; \longrightarrow A \xrightarrow{\;\Omega\;} A \xrightarrow{\;\Omega\;} A \xrightarrow{\;\Omega\;} A \;\right)$$

Proof Since the loops functors are not isofibrations, the homotopy limit of the tower of loops functors is calculated by first replacing each map by an isofibration in the ambient ∞-cosmos $\mathcal{K}$ and then forming the inverse limit. Alternatively, by Lemmas 6.2.18 and 6.2.19, $\mathrm{Sp}A$ is defined by the iso-tower:

$$
\begin{array}{ccc}
\mathrm{Sp}A & \xrightarrow{\;\phi\;} & \prod_{n\in\omega} A^{\mathbb{I}} \\[4pt]
\rho \downarrow \quad {\lrcorner} & & \downarrow \Pi(q_1,q_0) \\[4pt]
\prod_{m\in\omega} A & \xrightarrow[(\Omega,\mathrm{id})]{} & \prod_{n\in\omega} A \times A
\end{array}
\qquad (6.3.19)
$$

which defines $\mathrm{Sp}A$ to be the universal ∞-category equipped with a cone

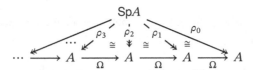

in which each triangle is inhabited by a homotopy coherent isomorphism.

Since A is pointed and admits pullbacks, it lies in the ∞-cosmos $(\mathcal{K}_*)_{\lrcorner} \hookrightarrow \mathcal{K}$. Pullback functors automatically preserve zero elements, so $(\mathcal{K}_*)_{\lrcorner} \cong \mathcal{K}_* \cap \mathcal{K}_{\lrcorner}$. For the same reason, the loops functor preserves zero elements and pullbacks, so the loops functor $\Omega : A \to A$ also lies in $(\mathcal{K}_*)_{\lrcorner}$. Since this ∞-cosmos is cosmologically embedded in $\mathcal{K}$, it follows that the iso-tower $\mathrm{Sp}A$ is again a pointed ∞-category that admits pullbacks, and each of the functors $\rho_n : \mathrm{Sp}A \twoheadrightarrow A$ preserve them.

Indeed, by Propositions 6.3.7 and 6.3.13 the zero element and products are created by the legs ρ_n of the limit cone, which tells us that its loops functor can be chosen to commute with these projections:

$$
\begin{array}{ccc}
\mathrm{Sp}A & \overset{\Omega}{\dashrightarrow} & \mathrm{Sp}A \\
\rho \downarrow & & \downarrow \rho \\
\displaystyle\prod_{n\in\omega} A & \overset{\prod\Omega}{\longrightarrow} & \displaystyle\prod_{n\in\omega} A
\end{array}
$$

By the universal property of $\mathrm{Sp}A$ as a flexible weighted limit we can induce another endofunctor $\Sigma^* : \mathrm{Sp}A \to \mathrm{Sp}A$ from the data of the weighted cone

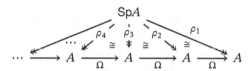

obtained by restricting along the inclusion $\Sigma : n \mapsto n + 1 : \omega^{\mathrm{op}} \to \omega^{\mathrm{op}}$ to shift the data in the weighted limit cone. To analyze this functor more formally, consider the factorization of the natural isomorphism $\sigma : W^{\leftarrow} \cong W^{\leftarrow}\Sigma$ through the left Kan extension:

$$
\begin{array}{ccc}
\omega^{\mathrm{op}} \xrightarrow{\ W^{\leftarrow}\ } s\mathcal{S}et & & \omega^{\mathrm{op}} \xrightarrow{\ W^{\leftarrow}\ } s\mathcal{S}et \\
{}_{\Sigma}\searrow \ {}_{\cong\Downarrow\sigma}\ \nearrow_{W^{\leftarrow}} & = & {}_{\Sigma}\searrow\ {}_{\cong\Downarrow\eta}\ {}^{\mathrm{lan}_{\Sigma}W^{\leftarrow}} \diagup_{\exists!\Downarrow\alpha}\ \nearrow_{W^{\leftarrow}} \\
\omega^{\mathrm{op}} & & \omega^{\mathrm{op}}
\end{array}
$$

In this way, the map Σ^* fits into a commutative diagram of induced maps on weighted limits

$$
\begin{array}{ccc}
\lim_{W^{\leftarrow}} F \xrightarrow{\ \Sigma^*\ } \lim_{W^{\leftarrow}\Sigma} F\Sigma \xrightarrow[\approx]{\ \sigma^*\ } \lim_{W^{\leftarrow}} F\Sigma \\
\alpha^* \downarrow \qquad\qquad \wr\downarrow \alpha\Sigma^* \qquad\qquad \| \\
\lim_{\mathrm{lan}_{\Sigma}W^{\leftarrow}} F \xrightarrow[\approx]{\ \Sigma^*\ } \lim_{\mathrm{lan}_{\Sigma}W^{\leftarrow}\Sigma} F\Sigma \xrightarrow[\approx]{\ \eta^*\ } \lim_{W^{\leftarrow}} F\Sigma
\end{array}
$$

Note the bottom horizontal composite exhibits the isomorphism of Lemma A.6.19, a general result of enriched category theory that says that the weighted limit of a restricted diagram agrees with the limit weighted by the left Kan extension of the weight (see [68, 4.63]). From this result and the isomorphisms $\sigma : W^{\leftarrow} \cong W^{\leftarrow}\Sigma$ and $F \cong F\Sigma$ we see that all of the objects in this diagram are isomorphic to $\mathrm{Sp}A$.

Since Σ is fully faithful, η is an isomorphism, and hence so is $\alpha\Sigma$, so the map $\Sigma^* : \mathrm{Sp}A \to \mathrm{Sp}A$ of interest is isomorphic to the map induced on weighted limits from the inclusion $\alpha : \mathrm{lan}_{\Sigma}W^{\leftarrow} \hookrightarrow W^{\leftarrow}$. This map is a projective cell complex between flexible weights, defined by attaching a single generalized projective cell $\mathbb{1} \times \omega(-,0) \hookrightarrow \mathbb{1} \times \omega(-,0)$ at 0. Put another way, it is an iso-morphism on all but one component; via the isomorphism $W^{\leftarrow} \cong \mathrm{lan}_{\Sigma} W^{\leftarrow} \Sigma$, its zeroth component is $\alpha_0 : W_1^{\leftarrow} \hookrightarrow W_0^{\leftarrow}$. Thus, we see that α defines a pointwise Joyal weak equivalence between projective cell complexes. It follows from the pullback diagram

$$
\begin{array}{ccc}
\lim_{W \leftarrow} F & \longrightarrow & \lim_{\mathbb{1} \times \omega(-,0)} F \cong A^{\mathbb{1}} \\
\alpha^* \downarrow \wr & \lrcorner & \downarrow \wr \\
\lim_{\mathrm{lan}_{\Sigma} W \leftarrow} F & \longrightarrow & \lim_{\mathbb{1} \times \omega(-,0)} F \cong A
\end{array}
$$

that the induced map on weighted limits $\Sigma^* : \mathrm{Sp}A \to \mathrm{Sp}A$ is an equivalence.

By construction the composite $\Omega\Sigma^* : \mathrm{Sp}A \to \mathrm{Sp}A$ is naturally isomorphic to the identity functor, where this isomorphism can be induced by the weak universal property of the pullback (6.3.19) from the isomorphisms $\Omega\rho_{n+1} \cong \rho_n$ of the weighted limit cone. Thus, by 2-of-3 and stability of equivalences under natural isomorphism, Ω is an equivalence.

We have shown that $\mathrm{Sp}A$ is a pointed ∞-category with pullbacks for which the loops functor is an equivalence. By Theorem 4.4.12(v), $\mathrm{Sp}A$ is a stable ∞-category as claimed. □

Exercises

EXERCISE 6.3.i. Let $\mathcal{L} \hookrightarrow \mathcal{K}$ be a replete subcategory of an ∞-cosmos. Show that for all $A, B \in \mathcal{L}$, the map

$$
\mathrm{Fun}_{\mathcal{L}}(A, B) \overset{\subset}{\longrightarrow} \mathrm{Fun}_{\mathcal{K}}(A, B)
$$

is both a monomorphism and an isofibration between quasi-categories. This latter property may be summarized by saying that the simplicial functor $\mathcal{L} \hookrightarrow \mathcal{K}$ is a *local isofibration*.[2]

EXERCISE 6.3.ii. For any ∞-cosmos $\mathcal{K}$, compare the cosmologically embedded ∞-cosmoi $\mathcal{K}_{\perp} \cap \mathcal{K}_{\top}$ and $\mathcal{K}_* := (\mathcal{K}_{\top})_{\perp} \cong (\mathcal{K}_{\perp})_{\top}$ appearing in the discussion surrounding Proposition 6.3.7.

EXERCISE 6.3.iii. Prove that $\mathcal{L}ari(\mathcal{K}) \cong \mathcal{R}ari(\mathcal{K}^{\mathrm{co}})^{\mathrm{co}}$.

EXERCISE 6.3.iv. For a cosmologically embedded ∞-cosmos $\mathcal{L} \hookrightarrow \mathcal{K}$, show that $A \in \mathcal{L}$ is discrete if and only if A is discrete as an object of $\mathcal{K}$.

---

[2] The inclusions of replete subcategories are also *global isofibrations* in a sense appropriate for (∞, 2)-category theory, namely the full data of an equivalence in $\mathcal{K}$ involving one object in the subcategory $\mathcal{L}$ can be lifted to $\mathcal{L}$.

EXERCISE 6.3.V ([46, 4.1.5]). Given a diagram where the vertical maps are cocartesian fibrations and the squares define cartesian functors

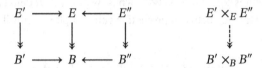

verify that the induced functor between the pullbacks is a cocartesian fibration and the projections define cartesian functors. Show also that if the vertical maps are discrete cocartesian fibrations so is this induced functor.

PART TWO

THE CALCULUS OF MODULES

By convention we refer to an object A in an ∞-cosmos as an "∞-category." The ∞-cosmos provides access to its elements $z : 1 \to A$ and to its mapping spaces $\mathrm{Hom}_A(x, y)$, defined as bifibers

$$
\begin{array}{ccc}
\mathrm{Hom}_A(x, y) & \longrightarrow & A^2 \\
\downarrow & \quad\lrcorner & \downarrow {\scriptstyle (p_1, p_0)} \\
1 & \xrightarrow{\ (y, x)\ } & A \times A
\end{array}
$$

over a pair of elements. Proposition 3.4.10 proves that mapping spaces are discrete ∞-categories, which, in an ∞-cosmos of $(\infty, 1)$-categories, are the ∞-groupoids in the ∞-cosmos. These structures justify part of the intuition that $(\infty, 1)$-categories are categories weakly enriched over ∞-groupoids.

These observations suggest that the arrow ∞-category is of particular importance. Propositions 3.2.5 and 3.2.6 describe its weak 2-categorical universal property: natural transformations

$$
X \underset{y}{\overset{x}{\rightrightarrows}} {\Downarrow \beta}\ A
$$

correspond to functors $\ulcorner \beta \urcorner : X \to A^2$ with $p_0 \ulcorner \beta \urcorner = x$ and $p_1 \ulcorner \beta \urcorner = y$ up to fibered isomorphism over $A \times A$. By Proposition 3.2.10 this weak universal property characterizes the arrow ∞-category up to fibered equivalence over $A \times A$. However, it does not capture the additional fact that natural transformations from X to A can be composed vertically defining commuting contravariant and covariant actions on the domains and codomains of the natural transformation $\beta : x \Rightarrow y$.

In Chapter 5, we discovered one way to express the actions, proving in Propositions 5.2.9 and Proposition 5.4.4 that the domain projection functor $p_0 : A^2 \twoheadrightarrow A$ and the codomain projection functor $p_1 : A^2 \twoheadrightarrow A$, respectively, define a *cartesian fibration* and a *cocartesian fibration*. Particular p_0-cartesian lifts and

p_1-cocartesian lifts can be used to define pre- and postcomposition functors:

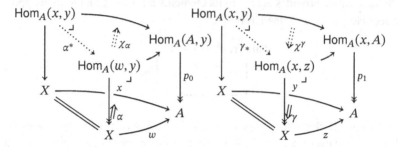

with universal properties that imply that $\alpha^*\beta^* \cong (\beta\alpha)^*$ and $\gamma_*\beta_* \cong (\gamma\beta)_*$.

To see that the post- and precomposition actions commute with each other, defining an essentially commutative square

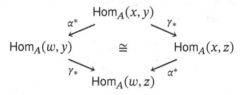

of functors over X, necessitates considering the codomain and domain projection functors as a span $A \xleftarrow{p_1} A^2 \xrightarrow{p_0} A$, rather than individually. This is the aim of Chapter 7, which introduces *two-sided fibrations*, given by a span whose left leg is a cocartesian fibration and right leg is a cartesian fibration, with commuting "fiberwise" actions.

The span $A \xleftarrow{p_1} A^2 \xrightarrow{p_0} A$ is two-sided fibration that is discrete as an object in the sliced ∞-cosmos over $A \times A$. In particular, this implies that the fibers over a pair of elements are discrete ∞-categories. In terminology introduced in §7.4, this discreteness means that E defines a *module* from A to B.

In Chapter 8, we develop the calculus of modules, which closely resembles the calculus of (bi)modules between unital rings. In Chapter 9, we deploy this calculus to further develop the formal category theory of ∞-categories, specifically by defining and developing pointwise right and left Kan extensions, which are notably missing from Part I. We note also that several results in previous chapters – Theorem 3.5.3, Corollary 3.5.6, Proposition 4.1.1, Proposition 4.3.1 – encode ∞-categorical notions as equivalences of modules. Accordingly, modules form the cornerstone of our proof of the model independence of $(\infty, 1)$-category theory in Part III.

7

Two-Sided Fibrations and Modules

In Chapter 5 we studied those isofibrations for which the arrows in the base act covariantly or contravariantly on the fibers. A prototypical example of a so-called *cocartesian fibration* is the codomain projection functor $p_1 : A^2 \twoheadrightarrow A$ associated to the arrow ∞-category, where the fiberwise action is by postcomposition. Dually, the domain projection functor $p_0 : A^2 \twoheadrightarrow A$ is a *cartesian fibration*, with fiberwise action by precomposition. In this section, we refine and sharpen our understanding of these fibration structures by considering the codomain and domain projection functors as a span $A \xleftarrow{p_1} A^2 \xrightarrow{p_0} A$ rather than separately. A few observations come quickly to mind. First, p_1-cocartesian lifts can be chosen to define natural transformations with codomain A^2 whose domain components are identities; that is, p_1-cocartesian lifts can be chosen to lie "over $p_0 : A^2 \twoheadrightarrow A$." Dually, p_0-cartesian lifts can be chosen to lie over p_1 (see Proposition 5.5.11).

Second, the action on fibers $\alpha^* : \mathrm{Hom}_A(x, A) \to \mathrm{Hom}_A(w, A)$ by precomposition with a natural transformation $\alpha : w \Rightarrow x$ with codomain A, commutes with the action $\gamma_* : \mathrm{Hom}_A(A, y) \to \mathrm{Hom}_A(A, z)$ by postcomposition with $\gamma : y \Rightarrow z$ in a suitable sense. These first two properties are summarized by saying that the span associated to the arrow ∞-category defines a *two-sided fibration* from A to A.

Finally, the bifibers of the isofibration $(p_1, p_0) : A^2 \twoheadrightarrow A \times A$ define discrete ∞-categories, namely the mapping spaces of A. Indeed, A^2 is itself discrete when considered as an object in the sliced ∞-cosmos over $A \times A$. This additional property means that $A \xleftarrow{p_1} A^2 \xrightarrow{p_0} A$ defines a *discrete two-sided fibration* from A to A, which we abbreviate by saying that A^2 defines a *module* from A to A.

Recall from Proposition 6.3.14 that for any ∞-category B in an ∞-cosmos $\mathcal{K}$, the quasi-categorically enriched categories $co\mathcal{C}art(\mathcal{K})_{/B}$ and $\mathcal{C}art(\mathcal{K})_{/B}$ define ∞-cosmoi, inheriting their limits, isofibrations, and equivalences from their cosmological embedding into $\mathcal{K}_{/B}$. In this chapter, we introduce another

279

cosmologically embedded ∞-cosmos $_{A\backslash}\mathcal{F}ib(\mathcal{K})_{/B} \hookrightarrow \mathcal{K}_{/A \times B}$ whose objects are *two-sided fibrations* from A to B. Several equivalent definitions of this notion are given in §7.1. Iterating Proposition 6.3.14 reveals that $_{A\backslash}\mathcal{F}ib(\mathcal{K})_{/B}$ is again an ∞-cosmos, which we study in §7.2.

A two-sided fibration from B to 1 is simply a cocartesian fibration over B, while a two-sided fibration from 1 to B is a cartesian fibration over B, so results about two-sided fibrations simultaneously generalize these one-sided notions. Certain closure properties of two-sided fibrations developed in §7.2, such as closure under pullback, are familiar from Chapter 5, while others, such as closure under span composition, are specific to the two-sided context.

In §7.3, we introduce two-sided representables and prove a two-sided version of the Yoneda lemma, generalizing Theorem 5.7.3 and Corollary 5.7.19. This is the formulation of the Yoneda lemma that proves the most useful going forward.

The main reason for our interest in two-sided fibrations is the fact that the discrete objects in $_{A\backslash}\mathcal{F}ib(\mathcal{K})_{/B}$ are precisely the *modules*[1] from A to B, which we define and study in §7.4. The calculus of modules, developed in Chapter 8, is the main site of the formal category theory of ∞-categories, which is the subject of Chapter 9.

7.1 Two-Sided Fibrations

By factoring, any span in an ∞-cosmos $\mathcal{K}$ from A to B may be replaced up to equivalence by a **two-sided isofibration**, a span $A \xleftarrow{q} E \xrightarrow{p} B$ for which the functor $(q, p) \colon E \twoheadrightarrow A \times B$ is an isofibration. Two-sided isofibrations from A to B are the objects of the ∞-cosmos $\mathcal{K}_{/A \times B}$. In this section, we describe what it means for a two-sided isofibration to be *cocartesian on the left* or *cartesian on the right*, and then introduce *two-sided fibrations*, which integrate these notions.

To motivate these concepts, consider the span $A \xleftarrow{p_1} A^2 \xrightarrow{p_0} A$ associated to the arrow ∞-category of A in an ∞-cosmos $\mathcal{K}$. In Proposition 5.5.11, we observed that

$$A^2 \xrightarrow{(p_1,p_0)} A \times A$$
$$p_1 \searrow \quad \swarrow \pi$$
$$A$$

defines a cartesian fibration in the sliced ∞-cosmos $\mathcal{K}_{/A}$ – in fact a discrete cartesian fibration, though we postpone consideration of discreteness until we

---

[1] In the 1- and ∞-categorical literature, the names "distributor" [10], "profunctor" [132], and "correspondence" [5, 78] are all used as synonyms for "module."

introduce *modules* in §7.4. The proof invokes Theorem 5.2.8(iii), which characterizes cartesian fibrations via a fibered adjunction over the codomain. In this instance, the fibered adjunction lies in the ∞-cosmos $(\mathcal{K}_{/A})_{/\pi\colon A\times A\twoheadrightarrow A}\cong\mathcal{K}_{/A\times A}$ and is equivalent to the canonical adjunction

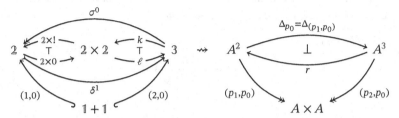

arising from an adjunction between the indexing categories. In terminology we presently introduce this is what it means to say that the two-sided isofibration $A\xleftarrow{p_1}A^2\xrightarrow{p_0}A$ is *cartesian on the right*.

When we postcompose with the projection $\pi\colon A\times A\twoheadrightarrow A$ onto the right factor, we are left with an adjunction

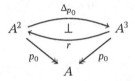

that witnesses the fact established in Proposition 5.2.9 that $p_0\colon A^2\twoheadrightarrow A$ is a cartesian fibration. So part of what it means for a two-sided isofibration to be cartesian on the right is that its right leg is a cartesian fibration. To say that this adjunction lies over $A\times A$ not merely over A amounts to the assertion that p_0-cartesian lifts can be chosen to lie in fibers of $p_1\colon A^2\twoheadrightarrow A$, whiskering along the left leg to identity arrows. It follows that every p_0-cartesian transformation whiskers with p_1 to an isomorphism.

These two properties can be expressed in another way: by demanding that the functor

$$A^2\xrightarrow{(p_1,p_0)}A\times A$$
$$\underset{A}{\overset{p_0}{\searrow}\ \nearrow\pi}$$

lies in the ∞-cosmos $\mathcal{C}art(\mathcal{K})_{/A}\subset\mathcal{K}_{/A}$. This asserts first that $p_0\colon A^2\twoheadrightarrow A$ is a cartesian fibration and second that (p_1,p_0) defines a cartesian functor from this cartesian fibration to the cartesian fibration $\pi\colon A\times A\twoheadrightarrow A$ of Example 5.4.6. As observed there, π-cartesian cells are those natural transformations that whisker with the other projection to isomorphisms. Thus to say that (p_1,p_0) is a cartesian functor is exactly to say that p_1 carries p_0-cartesian transformations to isomorphisms.

We now formalize these intuitions:

LEMMA 7.1.1 (cartesian on the right). *For a two-sided isofibration $A \xleftarrow{q} E \xtwoheadrightarrow{p} B$ in an ∞-cosmos $\mathcal{K}$, the following are equivalent:*

(i) *The functor*

$$E \xrightarrow{(q,p)} A \times B$$
$$q \searrow \quad \swarrow \pi$$
$$A$$

is a cartesian fibration in the slice ∞-cosmos $\mathcal{K}_{/A}$.

(ii) *The functor*

$$E \xrightarrow{(q,p)} A \times B$$
$$p \searrow \quad \swarrow \pi$$
$$B$$

in $\mathcal{K}_{/B}$ lies in the sub ∞-cosmos $\mathcal{C}art(\mathcal{K})_{/B}$.

(iii) *The functor induced by id_p admits a right adjoint in $\mathcal{K}_{/A \times B}$.*

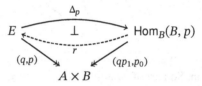

(iv) *The isofibration $p : E \twoheadrightarrow B$ is a cartesian fibration in $\mathcal{K}$ and for every*

p-cartesian transformation $X \underset{e}{\overset{e'}{\rightrightarrows}} \Downarrow\psi \; E$, the whiskered composite

$X \underset{e}{\overset{e'}{\rightrightarrows}} \Downarrow\psi \; E \xrightarrow{q} A$ is an isomorphism.

*A two-sided isofibration $A \xleftarrow{q} E \xtwoheadrightarrow{p} B$ is **cartesian on the right** when these equivalent conditions are satisfied*

Proof   The equivalence (i)⇔(iii) is exactly the interpretation of the equivalence of Theorem 5.2.8(i)⇔(iii) applied to the isofibration $(q, p) : E \twoheadrightarrow A \times B$ in the ∞-cosmos $\mathcal{K}_{/A}$. This latter result asserts that the isofibration $(q, p) : E \twoheadrightarrow A \times B$ is a cartesian fibration in $\mathcal{K}_{/A}$ if and only if the functor induced by $\mathrm{id}_{(q,p)}$ from E to the right representation of the functor $(q, p) : E \twoheadrightarrow A \times B$ admits a right adjoint over the codomain $\pi : A \times B \twoheadrightarrow A$; since $(\mathcal{K}_{/A})_{/\pi : A\times B\twoheadrightarrow A} \cong \mathcal{K}_{/A\times B}$ this is the same as asserting this adjunction over $A \times B$.

The only subtlety in interpreting Theorem 5.2.8 in $\mathcal{K}_{/A}$ has to do with the correct interpretation of the left representable comma ∞-category in $\mathcal{K}_{/A}$ for the functor $(q, p) : E \twoheadrightarrow A \times B$. This comma ∞-category is constructed as a subobject of the 2-cotensor of the object $\pi : A \times B \to A$ in $\mathcal{K}_{/A}$, which Proposition 1.2.22 tells us is formed as the left-hand vertical of the pullback diagram

$$
\begin{array}{ccc}
A \times B^2 & \longrightarrow & (A \times B)^2 \\
{\scriptstyle \pi}\downarrow & \lrcorner & \downarrow{\scriptstyle \pi^2} \\
A & \xrightarrow{\;\;\Delta\;\;} & A^2
\end{array}
$$

By (3.4.2) the comma ∞-category is constructed by the pullback in $\mathcal{K}_{/A}$

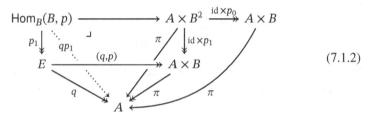

(7.1.2)

which is created by the forgetful functor $\mathcal{K}_{/A} \to \mathcal{K}$, and its domain projection functor is the top composite $(qp_1, p_0) : \mathrm{Hom}_B(B, p) \twoheadrightarrow A \times B$. Now we see that the interpretation of Theorem 5.2.8(i)$\Leftrightarrow$(iii) in $\mathcal{K}_{/A}$ is exactly the equivalence (i)$\Leftrightarrow$(iii).

It remains to demonstrate the equivalence with (ii) and (iv). Assuming (iii) and composing with $\pi : A\times B \twoheadrightarrow B$ yields a fibered adjunction that demonstrates that p is a cartesian fibration. The counit of both fibered adjunctions is the same, and by Theorem 5.2.8(iii) the composite

$$
\mathrm{Hom}_B(B, p) \; \underset{\Downarrow\epsilon}{\overset{\overset{r}{\longrightarrow} E \overset{\Delta_p}{\longrightarrow}}{\rightrightarrows}} \; \mathrm{Hom}_B(B, p) \; -p_1 \twoheadrightarrow E
$$

is the generic p-cartesian cell. Since ϵ is fibered over $A \times B$, when we postcompose with q we get an identity, which tells us that $q : E \twoheadrightarrow A$ carries p-cartesian cells to isomorphisms. This proves that (iii) implies (iv).

By Example 5.4.6, the cartesian cells for the cartesian fibration $\pi : A \times B \twoheadrightarrow B$ are precisely those 2-cells whose component with codomain A is an isomorphism, so (iv) says exactly that $(q, p) : E \twoheadrightarrow A \times B$ defines a cartesian functor from p to π. Thus (iv) implies (ii).

To complete the argument, we show that (ii) implies (i). Assume (ii) and consider a 2-cell in $\mathcal{K}_{/A}$

$$
\begin{array}{ccc}
X & \xrightarrow{\ e\ } & E \\
& \underset{(a,b)}{\overset{\Uparrow(\mathrm{id},\beta)}{\searrow}} & \downarrow {\scriptstyle(q,p)} \\
& & A \times B
\end{array}
$$

Because p is a cartesian fibration, $\beta \colon b \Rightarrow pe$ has a p-cartesian lift $\chi \colon e' \Rightarrow e$, and since (q,p) is a cartesian functor, the whiskered 2-cell $q\chi \colon qe' \Rightarrow a$ is an isomorphism. Because $(q,p) \colon E \twoheadrightarrow A \times B$ is an isofibration, we may lift the 2-cell $(q\chi^{-1}, \mathrm{id}) \colon (a,b) \Rightarrow e'$ to an invertible 2-cell $\gamma \colon e'' \Rightarrow e'$ with $p\gamma = \mathrm{id}_b$. The composite $\chi \cdot \gamma \colon e'' \Rightarrow e$ is a lift of (id, β) along (q,p) over A, which is easily verified to satisfy the weak universal properties that characterize a (q,p)-cartesian lift of (id, β) in $\mathcal{K}_{/A}$ as established by Proposition 5.2.11. This proves (i). □

Combining Lemma 7.1.1 and its dual, which defines the class of two-sided isofibrations that are **cocartesian on the left**:

COROLLARY 7.1.3. *A two-sided isofibration $A \xleftarrow{q} E \xrightarrow{p} B$ in an ∞-cosmos $\mathcal{K}$ is cocartesian on the left and cartesian on the right if the following equivalent conditions are satisfied:*

(i) *The functor*

$$
\begin{array}{ccc}
E & \xrightarrow{\ (q,p)\ } & A \times B \\
& {\scriptstyle q}\searrow \quad \swarrow{\scriptstyle \pi} & \\
& A &
\end{array}
$$

lies in $\mathrm{co}\mathcal{C}\mathrm{art}(\mathcal{K})_{/A}$ and defines a cartesian fibration in $\mathcal{K}_{/A}$.

(ii) *The functor*

$$
\begin{array}{ccc}
E & \xrightarrow{\ (q,p)\ } & A \times B \\
& {\scriptstyle p}\searrow \quad \swarrow{\scriptstyle \pi} & \\
& B &
\end{array}
$$

lies in $\mathcal{C}\mathrm{art}(\mathcal{K})_{/B}$ and defines a cocartesian fibration in $\mathcal{K}_{/B}$. □

A **two-sided fibration** is a span that is cocartesian on the left, cartesian on the right, and satisfies a further compatibility condition that can be stated in a number of equivalent ways, which boil down to the assertion that the processes of taking q-cocartesian and p-cartesian lifts commute:

THEOREM 7.1.4. *For a two-sided isofibration $A \xleftarrow{q} E \xrightarrow{p} B$ in an ∞-cosmos $\mathcal{K}$, the following are equivalent:*

(i) *The functor*

$$E \xrightarrow{(q,p)} A \times B$$

with q and π to A

defines a cartesian fibration in $\mathrm{co}\mathcal{C}\mathrm{art}(\mathcal{K})_{/A}$.

(ii) *The functor*

$$E \xrightarrow{(q,p)} A \times B$$

with p and π to B

defines a cocartesian fibration in $\mathcal{C}\mathrm{art}(\mathcal{K})_{/B}$.

(iii) *The canonical functors admit the displayed adjoints in* $\mathcal{K}_{/A \times B}$

$$
\begin{array}{ccc}
E & \underset{\underset{\Delta_p}{\overset{r}{\rightleftarrows}}}{\top} & \mathrm{Hom}_B(B,p) \\
\ell \dashv \Delta_q \big\uparrow & & \big\uparrow (\Delta_q p_1,\mathrm{id}) \dashv \ell \\
\mathrm{Hom}_A(q,A) & \underset{r}{\overset{(\mathrm{id},\Delta_p p_0)}{\rightleftarrows}} \underset{1}{\top} & \mathrm{Hom}_A(q,A) \underset{E}{\times} \mathrm{Hom}_B(B,p)
\end{array}
$$

and the mate of the identity 2-cell in this displayed commutative square defines an isomorphism $\ell r \cong r \ell$.

(iv) *The two-sided isofibration* $A \xleftarrow{q} E \xrightarrow{p} B$ *is cocartesian on the left, cartesian on the right, and satisfies a further condition: Given any pair of natural transformations as below-left together with a q-cocartesian lift* $\psi : e \Rightarrow \alpha_* e$ *of* α *over B and a p-cartesian lift* $\chi : \beta^* e \Rightarrow e$ *of* β *over A as below-right:*

$$
\begin{array}{ccc}
& X & \\
a \swarrow \quad {\scriptstyle\alpha \Leftarrow} e \, {\scriptstyle\Leftarrow \beta} \quad \searrow b & & \\
A \xLeftarrow{q} E \xrightarrow{p} B & \rightsquigarrow & \alpha_* e \; \underset{\Leftarrow \, \downarrow \, \Leftarrow}{\overset{\psi \; e \; \chi}{\cdots}} \; \beta^* e \\
& & A \xLeftarrow{q} E \xrightarrow{p} B
\end{array}
$$

then there is an isomorphism $\alpha_* \beta^* e \cong \beta^* \alpha_* e$ *over* $A \times B$ *commuting with the q-cocartesian lift of* $\alpha \cdot q\chi$ *over B and the p-cartesian lift of* $p\psi \cdot \beta$ *over A.*

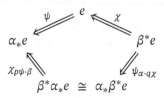

A two-sided isofibration $A \xleftarrow{q} E \xrightarrow{p} B$ defines a **two-sided fibration** from A to B when these equivalent conditions are satisfied.

Proof Our strategy is to show that condition (i) is equivalent to (iii), an equationally witnessed condition in the slice ∞-cosmos $\mathcal{K}_{/A \times B}$. A dual argument shows that condition (ii) is equivalent to (iii). We then unpack this condition to prove its equivalence with (iv).

If (i) holds, then $A \xleftarrow{q} E \xrightarrow{p} B$ also satisfies condition (i) of Corollary 7.1.3, since the property of being a cartesian fibration is preserved by the cosmological functor $co\mathcal{C}art(\mathcal{K})_{/A} \hookrightarrow \mathcal{K}_{/A}$. So $A \xleftarrow{q} E \xrightarrow{p} B$ is in particular cocartesian on the left and cartesian on the right and is thus equipped with adjunctions

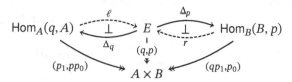

By Lemma 3.6.6, we may pull back the left-hand adjunction along $A \times p_1 : A \times B^2 \twoheadrightarrow A \times B$ and then compose with $A \times p_0 : A \times B^2 \twoheadrightarrow A \times B$ to obtain the right-hand fibered adjunction below, and also pull back the right-hand adjunction along $p_0 \times B : A^2 \times B \twoheadrightarrow A \times B$ and then compose with $p_1 \times B : A^2 \times B \twoheadrightarrow A \times B$ to obtain the left-hand fibered adjunction below:

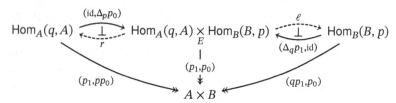

It remains to account for the assumption in (i) that (q, p) defines a cartesian fibration *in* $co\mathcal{C}art(\mathcal{K})_{/A}$ rather than merely in $\mathcal{K}_{/A}$. By Theorem 5.2.8(iii) this additional condition is equivalent to the assertion that the right adjoint

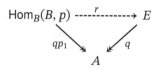

defines a cartesian functor from the cocartesian fibration p_1 to the cocartesian fibration q. By Theorem 5.3.4(iii) this is equivalent to the assertion that the mate of the isomorphism in the solid-arrow square

$$
\begin{array}{ccc}
\mathrm{Hom}_B(B, p) & \xrightarrow{\quad r \quad} & E \\
\ell \uparrow \dashv \ (\Delta_q p_1, \mathrm{id}) & & \Delta_q \ \vdash \ \uparrow \ell \\
\mathrm{Hom}_A(q, A) \underset{E}{\times} \mathrm{Hom}_B(B, p) & \underset{r}{\longrightarrow} & \mathrm{Hom}_A(q, A)
\end{array}
$$

defines an isomorphism.[2] This proves that (i) implies (iii).

Conversely assuming (iii), by Lemma 7.1.1 and Corollary 7.1.3 we conclude from two of the adjunctions that the functor

$$
\begin{array}{ccc}
E & \xrightarrow{\ (q,p)\ } & A \times B \\
& {}_{q}\searrow \quad \swarrow_{\pi} & \\
& A &
\end{array}
$$

lies in $co\mathcal{C}art(\mathcal{K})_{/A}$ and defines a cartesian fibration in $\mathcal{K}_{/A}$. By Theorem 5.2.8(iii) it defines a cartesian fibration in $co\mathcal{C}art(\mathcal{K})_{/A}$ if and only if the right adjoint r defines a cartesian functor from $qp_1 : \mathrm{Hom}_B(B, p) \twoheadrightarrow A$ to $q : E \twoheadrightarrow A$; recall that the inclusion $co\mathcal{C}art(\mathcal{K})_{/A} \hookrightarrow \mathcal{K}_{/A}$ is full on positive dimensional arrows so there is no comparable condition on the unit and counit. By Theorem 5.3.4(iii) this follows from the invertibility of the mate $\ell r \cong r\ell$. In this way we see that (iii) implies (i).

Thus, we have shown that condition (i) is equivalent to (iii) positing the existence of adjunctions in $\mathcal{K}_{/A \times B}$ so that all of the mates of the solid-arrow diagram are isomorphisms. Dualizing this argument, we see that (iii) is equivalent to condition (ii).

Finally, (iii) and (iv) are equivalent since the existence of the left adjoints in (iii) is equivalent to the span being cocartesian on the left, the existence of the right adjoints is equivalent to being cartesian on the right, and the compatibility condition for the cartesian and cocartesian lifts is the meaning of the isomorphism $\ell r \cong r\ell$. □

The internal characterization of two-sided fibrations has a familiar consequence:

COROLLARY 7.1.5. *Any two-sided isofibration* $(a, b) : X \twoheadrightarrow A \times B$ *that is equivalent over* $A \times B$ *to a two-sided fibration* $(q, p) : E \twoheadrightarrow A \times B$ *is a two-sided fibration.*

---

[2] The isomorphism in the solid-arrow square is the inverse of the mate that witnesses the fact that Δ_q defines a cartesian functor from p to p_0, which is part of what it means for (q, p) to be cartesian on the right.

Proof The assertion of Theorem 7.1.4(iii) is invariant under fibered equivalence.   □

Theorem 7.1.4 may be used to establish an important family of examples involving the ordinal categories from Definition 1.1.4.

PROPOSITION 7.1.6. *For any ∞-category A and any $n \geq 2$, the two-sided isofibration $A \xleftarrow{p_{n-1}} A^n \xrightarrow{p_0} A$ defines a two-sided fibration.*

This result is a generalization of Proposition 5.5.11 and its dual and the proof uses similar ideas.

Proof We use Theorem 7.1.4(iii). The right representable comma ∞-category associated to the evaluation at the initial object $p_0 : A^n \twoheadrightarrow A$ is constructed by the pullback

$$
\begin{array}{ccc}
A^{2 \vee n} & \longrightarrow & A^2 \\
\downarrow & \lrcorner & \downarrow p_1 \\
A^n & \xrightarrow{\quad p_0 \quad} & A
\end{array}
$$

which is equivalent to $(p_n, p_0) : A^{n+1} \twoheadrightarrow A \times A$ over the endpoint evaluation maps. The canonical functor that tests whether $(p_{n-1}, p_0) : A^n \twoheadrightarrow A \times A$ is cartesian on the right is given by restriction along the epimorphism $\sigma^0 : n+1 \to n$ that sends the objects $0, 1 \in n+1$ to $0 \in n$. This functor admits a left adjoint under the endpoint inclusions

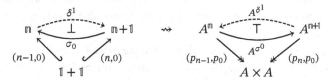

which provides the desired fibered right adjoint displayed above left.

A dual argument shows that $(p_{n-1}, p_0) : A^n \twoheadrightarrow A \times A$ is cocartesian on the left. The final condition asks that the mate of the commutative square defined by the degeneracy maps

$$
\begin{array}{ccc}
n & \xleftarrow{\;\perp\;\delta^1\;} & n+1 \\
{\scriptstyle\delta^n}\uparrow \vdash \; {\scriptstyle\sigma^n} & \xrightarrow{\;\sigma^0\;} & {\scriptstyle\sigma^{n+1}}\; \dashv \downarrow {\scriptstyle\delta^{n+1}} \\
n+1 & \xrightarrow[\;\delta^1\;]{\;\top\;\sigma^0\;} & n+2
\end{array}
$$

is an isomorphism, encoding one of the familiar simplicial identities. The square in Theorem 7.1.4(iii) is obtained by applying $A^{(-)}$. □

We defer further discussion of the closure properties of the class of two-sided fibrations to the next section, where we deploy cosmological arguments to streamline their proofs. These observations allow us to further enlarge our family of examples.

To prepare for that work, we consider the structure-preserving morphisms between two-sided fibrations, proving a relative analogue of Theorem 7.1.4:

PROPOSITION 7.1.7. *For map of spans between a pair of two-sided fibrations from A to B in an ∞-cosmos* $\mathcal{K}$

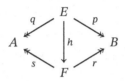

the following are equivalent:

   (i) *h defines a cartesian functor between cartesian fibrations in* $co\mathcal{C}art(\mathcal{K})_{/A}$.
   (ii) *h defines a cartesian functor between cocartesian fibrations in* $\mathcal{C}art(\mathcal{K})_{/B}$.
  (iii) *The mates of the canonical isomorphisms*

$$
\begin{array}{ccc}
E \xrightarrow{\;\;h\;\;} F & \qquad & E \xrightarrow{\;\;h\;\;} F \\
{\scriptstyle \ell}\big\uparrow \quad \cong\!\!\diagdown \quad \big\uparrow{\scriptstyle \ell} & & {\scriptstyle r}\big\uparrow \quad \cong\!\!\diagdown \quad \big\uparrow{\scriptstyle r} \\
\mathrm{Hom}_A(q,A) \xrightarrow[\mathrm{Hom_{id}}(h,\mathrm{id})]{} \mathrm{Hom}_A(s,A) & & \mathrm{Hom}_B(B,p) \xrightarrow[\mathrm{Hom_{id}}(\mathrm{id},h)]{} \mathrm{Hom}_B(B,r)
\end{array}
$$

    define isomorphisms in $\mathcal{K}_{/A\times B}$.
  (iv) *h defines a cartesian functor between the cocartesian fibrations q and s and a cartesian functor between the cartesian fibrations p and r in* $\mathcal{K}$.

The map of spans defines a **cartesian functor** *between two-sided fibrations from A to B when these equivalent conditions are satisfied.*

Proof By Theorem 5.3.4(iii), condition (iv) is equivalent to demanding that the two mates on display in (iii) are isomorphisms in $\mathcal{K}_{/A}$ and $\mathcal{K}_{/B}$, respectively. But since the spans are two-sided fibrations from A to B and h is a map over $A \times B$, both natural transformations lie in $\mathcal{K}_{/A\times B}$. By Proposition 3.6.2, if these natural transformations admit inverses in $\mathcal{K}_{/A}$ and $\mathcal{K}_{/B}$, then these inverse transformations lift to an inverse isomorphism in $\mathcal{K}_{/A\times B}$. This proves that (iv) is equivalent to (iii).

We next prove that (i) is equivalent to (iii). The proof of the equivalence of (ii) and (iii) is dual. To assert, as in (i) that the map

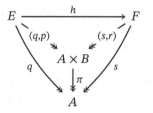

defines a cartesian functor between cartesian fibrations in $co\mathcal{C}art(\mathcal{K})_{/A}$ means first that the mate

$$
\begin{array}{ccc}
E & \xrightarrow{\ h\ } & F \\
\Delta_q \downarrow & \cong \nearrow & \downarrow \Delta_s \\
\mathrm{Hom}_A(q,A) & \xrightarrow[\mathrm{Hom}_{\mathrm{id}}(h,\mathrm{id})]{} & \mathrm{Hom}_A(s,A)
\end{array}
\quad \rightsquigarrow \quad
\begin{array}{ccc}
E & \xrightarrow{\ h\ } & F \\
\ell \uparrow & \cong \nwarrow & \uparrow \ell \\
\mathrm{Hom}_A(q,A) & \xrightarrow[\mathrm{Hom}_{\mathrm{id}}(h,\mathrm{id})]{} & \mathrm{Hom}_A(s,A)
\end{array}
$$

defines an isomorphism in $\mathcal{K}_{/A}$, because h defines a cartesian functor between the cocartesian fibrations q and s, over A and second that the mate

$$
\begin{array}{ccc}
E & \xrightarrow{\ h\ } & F \\
\Delta_p \downarrow & \cong \nearrow & \downarrow \Delta_r \\
\mathrm{Hom}_B(B,p) & \xrightarrow[\mathrm{Hom}_{\mathrm{id}}(\mathrm{id},h)]{} & \mathrm{Hom}_B(B,r)
\end{array}
\quad \rightsquigarrow \quad
\begin{array}{ccc}
E & \xrightarrow{\ h\ } & F \\
r \uparrow & \cong \nwarrow & \uparrow r \\
\mathrm{Hom}_B(B,p) & \xrightarrow[\mathrm{Hom}_{\mathrm{id}}(\mathrm{id},h)]{} & \mathrm{Hom}_B(B,r)
\end{array}
$$

is an isomorphism in $(co\mathcal{C}art(\mathcal{K})_{/A})_{/\pi}$. Since the spans are two-sided fibrations and h is a map over $A \times B$, the first pair of mates lies in $\mathcal{K}_{/A \times B}$. And since the inclusion $(co\mathcal{C}art(\mathcal{K})_{/A})_{/\pi} \hookrightarrow \mathcal{K}_{/A \times B}$ is full on positive-dimensional arrows, to ask that the second pair of mates are isomorphisms in $(co\mathcal{C}art(\mathcal{K})_{/A})_{/\pi}$ is equivalent to asking the same condition in $\mathcal{K}_{/A \times B}$. This proves the equivalence with (iii). □

It follows from the internal characterizations of Theorem 7.1.4 and Proposition 7.1.7.

COROLLARY 7.1.8. *Any cosmological functor preserves two-sided fibrations and cartesian functors between them.*

Proof By Proposition 1.3.4(vi) any cosmological functor $F\colon \mathcal{K} \to \mathcal{L}$ induces a cosmological functor $F\colon \mathcal{K}_{/A \times B} \to \mathcal{L}_{/FA \times FB}$ for any pair of ∞-categories A and B in $\mathcal{K}$. This functor preserves the fibered adjunctions and invertible mates of Theorem 7.1.4(iii). Thus $F\colon \mathcal{K} \to \mathcal{L}$ preserves two-sided fibrations.

Similarly, any cosmological functor preserves the invertible mates of Proposition 7.1.7(iii). □

Exercises

EXERCISE 7.1.i. Prove that the product projection span $A \xleftarrow{\pi} A \times B \xrightarrow{\pi} B$ defines a two-sided fibration for any ∞-categories A and B.

EXERCISE 7.1.ii. Prove that

(i) A two-sided isofibration $1 \xleftarrow{!} E \xrightarrow{p} B$ defines a two-sided fibration from 1 to B if and only if $p : E \twoheadrightarrow B$ is a cartesian fibration.

(ii) A map of spans as below defines a cartesian functor of two-sided fibrations if and only if $h : E \to F$ defines a cartesian functor from p to q.

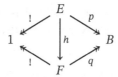

EXERCISE 7.1.iii. Suppose $A \xleftarrow{q} E \xrightarrow{p} B$ and $A \xleftarrow{s} F \xrightarrow{r} B$ are two-sided fibrations and $\alpha : h \cong h'$ is a fibered natural isomorphism

$$A \xleftarrow{q} E \xrightarrow{p} B \qquad h' \left(\underset{\cong}{\overset{\alpha}{\Downarrow}}\right) h \qquad A \xleftarrow{s} F \xrightarrow{r} B$$

so that $r\alpha = \mathrm{id}_p$ and $s\alpha = \mathrm{id}_q$; in particular, h and h' each define maps of spans. Show that h is a cartesian functor if and only if h' is a cartesian functor.

7.2 The ∞-Cosmos of Two-Sided Fibrations

The first pair of equivalent conditions of Theorem 7.1.4 and Proposition 7.1.7 provide two equivalent ways to define the ∞-cosmos of two-sided fibrations, using the ∞-cosmoi of Proposition 6.3.14.

DEFINITION 7.2.1 (the ∞-cosmos of two-sided fibrations). By Theorem 7.1.4

and Proposition 7.1.7, for any ∞-categories A and B in an ∞-cosmos $\mathcal{K}$ the pair of quasi-categorically enriched subcategories

$$\mathcal{C}art(co\mathcal{C}art(\mathcal{K})_{/A})_{/\pi\,:\,A\times B \twoheadrightarrow A} \quad \text{and} \quad co\mathcal{C}art(\mathcal{C}art(\mathcal{K})_{/B})_{/\pi\,:\,A\times B \twoheadrightarrow B}$$

of $\mathcal{K}_{/A\times B}$ coincide. We write

$$_{A\backslash}\mathcal{F}ib(\mathcal{K})_{/B} \subset \mathcal{K}_{/A\times B}$$

for this common subcategory and refer to it as the ∞-**cosmos of two-sided fibrations** from A to B, employing terminology that will be justified momentarily. By definition its functor space from $A \xleftarrow{q} E \xrightarrow{p} B$ to $A \xleftarrow{s} F \xrightarrow{r} B$

$$\mathsf{Fun}_{{}_{A\backslash}\mathcal{F}ib(\mathcal{K})_{/B}}(E, F) := \mathsf{Fun}^{\mathrm{cart}}_{A\times B}(E, F) \subset \mathsf{Fun}_{A\times B}(E, F)$$

is the quasi-category of maps of spans that define cartesian functors from E to F (see Proposition 7.1.7).

Since a two-sided isofibration from A to B is simply an isofibration over $A \times B$, we may refer to $\mathcal{K}_{/A\times B}$ as the ∞-**cosmos of two-sided isofibrations** from A to B.

PROPOSITION 7.2.2. *For any ∞-categories A and B in an ∞-cosmos $\mathcal{K}$, the ∞-cosmos of two-sided fibrations is cosmologically embedded in the ∞-cosmos of isofibrations ${}_{A\backslash}\mathcal{F}ib(\mathcal{K})_{/B} \hookrightarrow \mathcal{K}_{/A\times B}$, with the inclusion creating an ∞-cosmos structure.*

Proof   This inclusion factors as

$$\begin{array}{ccc} {}_{A\backslash}\mathcal{F}ib(\mathcal{K})_{/B} & & \mathcal{K}_{/A\times B} \\ \rotatebox{90}{=} & & \rotatebox{90}{=} \\ co\mathcal{C}art(\mathcal{C}art(\mathcal{K})_{/B})_{/A\times B \twoheadrightarrow B} \hookrightarrow (\mathcal{C}art(\mathcal{K})_{/B})_{/A\times B \twoheadrightarrow B} & \hookrightarrow & (\mathcal{K}_{/B})_{/A\times B \twoheadrightarrow B} \end{array}$$

and by Proposition 6.3.14, both inclusions are cosmological embeddings.   □

OBSERVATION 7.2.3 (two-sided fibrations generalize co/cartesian fibrations).   By Exercise 7.1.ii, a two-sided fibration from B to 1 is a cocartesian fibration over B, while a two-sided fibration from 1 to B is a cartesian fibration over B. Indeed, as ∞-cosmoi

$$co\mathcal{C}art(\mathcal{K})_{/B} \cong {}_{B\backslash}Fib(\mathcal{K})_{/1} \quad \text{and} \quad \mathcal{C}art(\mathcal{K})_{/B} \cong {}_{1\backslash}Fib(\mathcal{K})_{/B}.$$

In this sense, statements about two-sided fibrations simultaneously generalize statements about cartesian and cocartesian fibrations.

We now turn our attention to the promised closure properties:

PROPOSITION 7.2.4. *For any pair of functors* $a : A' \to A$ *and* $b : B' \to B$, *the cosmological pullback functor restricts to define a cosmological functor*

$$
\begin{array}{ccc}
{}_{A\backslash}\mathcal{F}ib(\mathcal{K})_{/B} & \lhook\joinrel\longrightarrow & \mathcal{K}_{/A\times B} \\
{\scriptstyle (a,b)^*}\Big\downarrow & & \Big\downarrow{\scriptstyle (a,b)^*} \\
{}_{A'\backslash}\mathcal{F}ib(\mathcal{K})_{/B'} & \lhook\joinrel\longrightarrow & \mathcal{K}_{/A'\times B'}
\end{array}
$$

In particular, the pullback of a two-sided fibration is again a two-sided fibration.

Proof By factoring the functor (a, b) as $A' \times B' \xrightarrow{\mathrm{id}\times b} A' \times B \xrightarrow{a\times\mathrm{id}} A \times B$ we see that it suffices to consider pullback along one side at a time. Proposition 5.2.4 and Exercise 5.3.i prove that pullback along $b : B' \to B$ preserves cartesian fibrations and cartesian functors, defining a restricted functor

$$
\begin{array}{ccc}
\mathcal{C}art(\mathcal{K})_{/B} & \lhook\joinrel\longrightarrow & \mathcal{K}_{/B} \\
{\scriptstyle b^*}\Big\downarrow & & \Big\downarrow{\scriptstyle b^*} \\
\mathcal{C}art(\mathcal{K})_{/B'} & \lhook\joinrel\longrightarrow & \mathcal{K}_{/B'}
\end{array}
$$

Since limits and isofibrations in $\mathcal{C}art(\mathcal{K})_{/B}$ are created in $\mathcal{K}_{/B}$, this restricted functor is cosmological (see Lemma 10.1.1). Applying this result to the map

$$
A \times B' \xrightarrow{\mathrm{id}\times b} A \times B
$$
$$
{\scriptstyle \pi}\searrow \quad\; A \;\quad \swarrow{\scriptstyle \pi}
$$

in the ∞-cosmos $co\mathcal{C}art(\mathcal{K})_{/A}$, we conclude that pullback restricts to define a cosmological functor

$$
\begin{array}{ccccc}
\mathcal{C}art(co\mathcal{C}art(\mathcal{K})_{/A})_{/\pi:\,A\times B\twoheadrightarrow A} & \lhook\joinrel\twoheadrightarrow & (co\mathcal{C}art(\mathcal{K})_{/A})_{/\pi:\,A\times B\twoheadrightarrow A} & \lhook\joinrel\twoheadrightarrow & \mathcal{K}_{/A\times B} \\
{\scriptstyle (\mathrm{id}\times b)^*}\Big\downarrow & & {\scriptstyle (\mathrm{id}\times b)^*}\Big\downarrow & & \Big\downarrow{\scriptstyle (\mathrm{id}\times b)^*} \\
\mathcal{C}art(co\mathcal{C}art(\mathcal{K})_{/A})_{/\pi:\,A\times B'\twoheadrightarrow A} & \lhook\joinrel\twoheadrightarrow & (co\mathcal{C}art(\mathcal{K})_{/A})_{/\pi:\,A\times B'\twoheadrightarrow A} & \lhook\joinrel\twoheadrightarrow & \mathcal{K}_{/A\times B'}
\end{array}
$$

$\square$

LEMMA 7.2.5. *If* $A \xleftarrow{q} E \xrightarrow{p} B$ *is a two-sided fibration from* A *to* B, $v : A \twoheadrightarrow C$ *is a cocartesian fibration and* $u : B \twoheadrightarrow D$ *is a cartesian fibration, then the composite span*

$$
C \xleftarrow{\;v\;} A \xleftarrow{\;q\;} E \xrightarrow{\;p\;} B \xrightarrow{\;u\;} D
$$

defines a two-sided fibration from C *to* D. *Moreover, a cartesian functor* h

between two-sided fibrations from A to B induces a cartesian functor between two-sided fibrations from C to D:

$$
\begin{array}{ccccccc}
 & & & E & & & \\
 & & {\scriptstyle q}\nearrow & & \searrow{\scriptstyle p} & & \\
C & \xleftarrow{\;v\;} & A & \Big\downarrow h & & B & \xrightarrow{\;u\;} D \\
 & & {\scriptstyle s}\searrow & & \nearrow{\scriptstyle r} & & \\
 & & & F & & &
\end{array}
$$

This whiskering composition does not define a cosmological functor from $_{A\backslash}\mathcal{F}ib(\mathcal{K})_{/B}$ to $_{C\backslash}\mathcal{F}ib(\mathcal{K})_{/D}$ since the functor defined by composition with an isofibration $(v \times u)_* : \mathcal{K}_{/A \times B} \to \mathcal{K}_{/C \times D}$ does not preserve flexible weighted limits (see Non-Example 1.3.6).

Proof By Theorem 7.1.4, it suffices to consider composition on one side at a time, say with a cocartesian fibration $v : A \twoheadrightarrow C$. Working in the ∞-cosmos $\mathcal{C}art(\mathcal{K})_{/B}$, we are given cocartesian fibrations

$$
\begin{array}{ccc}
E \xrightarrow{\;(q,p)\;} A \times B & \qquad & A \times B \xrightarrow{\;v \times \mathrm{id}\;} C \times B \\
{\scriptstyle p}\searrow \quad \swarrow{\scriptstyle \pi} & & {\scriptstyle \pi}\searrow \quad \swarrow{\scriptstyle \pi} \\
\quad B & & \quad B
\end{array}
$$

These compose to define a cocartesian fibration

$$
\begin{array}{c}
E \xrightarrow{\;(vq,p)\;} C \times B \\
{\scriptstyle p}\searrow \quad \swarrow{\scriptstyle \pi} \\
\quad B
\end{array}
$$

and hence a two-sided fibration from C to B, as desired.

By Lemma 5.2.3, the vq-cocartesian cells are the q-cocartesian lifts of the v-cocartesian cells. If h is a cartesian functor from q to s, then these are clearly preserved, proving that h also defines a cartesian functor from vq to vs. By Proposition 7.1.7(iv) this proves that h defines a cartesian functor between the whiskered two-sided fibrations. □

Proposition 7.2.4 and Lemma 7.2.5 combine to prove that two-sided fibrations can be composed "horizontally."

PROPOSITION 7.2.6. *The pullback of a two-sided fibration from A to B along a two-sided fibration from B to C defines a two-sided fibration from A to C as displayed.*

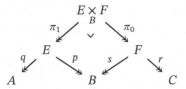

Proof   The composite two-sided fibration is constructed in two stages, first by pulling back

$$
\begin{array}{ccc}
E \underset{B}{\times} F & \longrightarrow & F \\
{\scriptstyle(\pi_1, r\pi_0)}\downarrow & \lrcorner & \downarrow{\scriptstyle(s,r)} \\
E \times C & \underset{p \times C}{\longrightarrow} & B \times C
\end{array}
$$

and then by composing the left leg with the cocartesian fibration $q : E \twoheadrightarrow A$. By Proposition 7.2.4 and Lemma 7.2.5, the result is a two-sided fibration from A to C. Alternatively, the composite can be constructed by pulling back along $A \times s$ and composing with the cartesian fibration $r : F \twoheadrightarrow C$, resulting in another two-sided fibration from A to C that is canonically isomorphic to the first. $\square$

With these results in hand, we may add to our library of examples:

EXAMPLE 7.2.7. If $p : E \twoheadrightarrow B$ is a cartesian fibration and $q : F \twoheadrightarrow A$ is a cocartesian fibration, then the span formed by composing with the product-projections

$$
A \overset{q}{\twoheadleftarrow} F \overset{\pi}{\twoheadleftarrow} F \times E \overset{\pi}{\twoheadrightarrow} E \overset{p}{\twoheadrightarrow} B
$$

defines a two-sided fibration from A to B by Exercise 7.1.i.

EXAMPLE 7.2.8. By Proposition 7.1.6 and Proposition 7.2.4, a general comma span

$$
C \overset{p_1}{\twoheadleftarrow} \mathrm{Hom}_A(f, g) \overset{p_0}{\twoheadrightarrow} B
$$

is a two-sided fibration, as a pullback of $A \overset{p_1}{\twoheadleftarrow} A^2 \overset{p_0}{\twoheadrightarrow} A$.

EXAMPLE 7.2.9. By Proposition 7.2.6 and Example 7.2.8, horizontal composites of comma spans are also two-sided fibrations. For instance, the two-sided fibrations $A \overset{p_n}{\twoheadleftarrow} A^{\mathbb{n}+1} \overset{p_0}{\twoheadrightarrow} A$ are n-ary horizontal composites of the arrow span $A \overset{p_1}{\twoheadleftarrow} A^2 \overset{p_0}{\twoheadrightarrow} A$.

Other notable two-sided fibrations include horizontal composites of comma spans, one instance of which features prominently in the next section.

Exercises

Exercise 7.2.i. Consider a diagram in which h and k define cartesian functors between two-sided fibrations from A to B and from B to C, respectively:

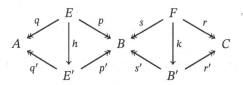

Prove that h and k pull back to define a cartesian functor $(h,k) : E \underset{B}{\times} F \to E' \underset{B}{\times} F'$ between two-sided fibrations from A to C.

7.3 The Two-Sided Yoneda Lemma

In this section, we prove a two-sided version of the Yoneda lemma using the following notion of representable two-sided fibration:

Definition 7.3.1. For any span of generalized elements $A \overset{a}{\leftarrow} X \overset{b}{\to} B$, the span

$$\mathrm{Hom}_A(a,A) \underset{X}{\times} \mathrm{Hom}_B(B,b)$$

with π_1 to $\mathrm{Hom}_A(a,A)$ and π_0 to $\mathrm{Hom}_B(B,b)$, p_1, p_0 to A, p_0, p_1 to X, and p_0 to B

defines a two-sided fibration from A to B that we refer to as the **two-sided fibration represented by a and b**. As is the case for one-sided representables, there is a canonical generalized element $(\ulcorner\mathrm{id}_a\urcorner, \ulcorner\mathrm{id}_b\urcorner) : X \to \mathrm{Hom}_A(a,A) \underset{X}{\times} \mathrm{Hom}_B(B,b)$ in the fiber over $(a,b) : X \to A \times B$.

The terminology of Definition 7.3.1 is justified by the Yoneda lemma for two-sided fibrations, which we prove using the Yoneda lemma for generalized elements of Corollary 5.7.19.

Theorem 7.3.2 (Yoneda lemma). *For any span $A \overset{a}{\leftarrow} X \overset{b}{\to} B$ and any two-sided fibration $A \overset{q}{\twoheadleftarrow} E \overset{p}{\twoheadrightarrow} B$, restriction along the element $(\ulcorner\mathrm{id}_a\urcorner, \ulcorner\mathrm{id}_b\urcorner) : X \to \mathrm{Hom}_A(a,A) \underset{X}{\times} \mathrm{Hom}_B(B,b)$ defines an equivalence of quasi-categories*

$$\mathrm{Fun}^{\mathrm{cart}}_{A\times B}\left(\begin{matrix} \mathrm{Hom}_A(a,A) \underset{X}{\times} \mathrm{Hom}_B(B,b) & & E \\ & | & | \\ (p_1\pi_1, p_0\pi_0) & , & (q,p) \\ & \downarrow & \downarrow \\ & A\times B & A\times B \end{matrix}\right) \overset{\mathrm{ev}(\ulcorner\mathrm{id}_a\urcorner, \ulcorner\mathrm{id}_b\urcorner)}{\underset{\sim}{\longrightarrow}} \mathrm{Fun}_{A\times B}\left(\begin{matrix} X & E \\ | & | \\ (a,b) & , & (q,p) \\ \downarrow & \downarrow \\ A\times B & A\times B \end{matrix}\right)$$

Proof We prove this result using the perspective of Theorem 7.1.4(i), considering the target two-sided fibration $A \xleftarrow{q} E \xrightarrow{p} B$ in an ∞-cosmos $\mathcal{K}$ as a cartesian fibration over $\pi : A \times B \twoheadrightarrow A$ in the ∞-cosmos $co\mathcal{C}art(\mathcal{K})_{/A \times B}$. We deduce this result by applying Corollary 5.7.19 twice: first for cartesian fibrations over $\pi : A \times B \twoheadrightarrow A$ in $co\mathcal{C}art(\mathcal{K})_{/A}$ and then for cocartesian fibrations over A in $\mathcal{K}$.

To prepare for the first application of Yoneda, we need to produce a generalized element with codomain $\pi : A \times B \twoheadrightarrow A$ in $co\mathcal{C}art(\mathcal{K})_{/A}$ from the map

$$X \xrightarrow{(a,b)} A \times B \qquad a \searrow \quad \swarrow \pi \qquad A \tag{7.3.3}$$

over A. Corollary 5.7.19 supplies an equivalence

$$\mathsf{Fun}_A^{\mathrm{cart}}(\mathsf{Hom}_A(a,A) \xrightarrow{p_1} A, A \times B \xrightarrow{\pi} A) \xrightarrow{\mathrm{ev}_{\ulcorner \mathrm{id}_a \urcorner}} \mathsf{Fun}_A(X \xrightarrow{a} A, A \times B \xrightarrow{\pi} A)$$

which tells us that the reflection of $a : X \to A$ into $co\mathcal{C}art(\mathcal{K})_{/A}$ is given by $p_1 : \mathsf{Hom}_A(a,A) \twoheadrightarrow A$. So we take our generalized element to be the cartesian functor between cocartesian fibrations over A

$$\mathsf{Hom}_A(a,A) \xrightarrow{(p_1,bp_0)} A \times B \qquad p_1 \searrow \quad \swarrow \pi \qquad A \tag{7.3.4}$$

which restricts along $\ulcorner \mathrm{id}_a \urcorner : X \to \mathsf{Hom}_A(a,A)$ to (7.3.3).

By Corollary 5.7.19 once more, the reflection of the object (7.3.4) from $(co\mathcal{C}art(\mathcal{K})_{/A})_{/\pi : A \times B \twoheadrightarrow A}$ into ${}_A \backslash \mathcal{F}ib(\mathcal{K})_{/B}$ is given by the codomain projection from its right comma representation, formed in $co\mathcal{C}art(\mathcal{K})_{/A}$. This recovers the two-sided fibration

$$\mathsf{Hom}_A(a,A) \underset{X}{\times} \mathsf{Hom}_B(B,b) \xrightarrow{(p_1\pi_1, p_0\pi_0)} A \times B \qquad p_1\pi_1 \searrow \quad \swarrow \pi \qquad A$$

Thus, that result supplies an equivalence

$$\mathsf{Fun}_{A \times B}^{\mathrm{cart}} \left(\begin{array}{c} \mathsf{Hom}_A(a,A) \underset{X}{\times} \mathsf{Hom}_B(B,b) \\ \downarrow \\ A \times B \end{array} , \begin{array}{c} E \\ \downarrow \\ A \times B \end{array} \right) \xrightarrow[\sim]{\mathrm{ev}_{(\mathrm{id}, \ulcorner \mathrm{id}_{bp_0} \urcorner)}} \mathsf{Fun}_{A \times B}^{\mathrm{cart}} \left(\begin{array}{c} \mathsf{Hom}_A(a,A) \\ (p_1, bp_0) \downarrow \\ A \times B \end{array} , \begin{array}{c} E \\ \downarrow \\ A \times B \end{array} \right)$$

where the right-hand functor space is defined by the pullback

$$\mathrm{Fun}^{\mathrm{cart}}_{A\times B}(\mathrm{Hom}_A(a,A),E) \longrightarrow \mathrm{Fun}^{\mathrm{cart}}_A(\mathrm{Hom}_A(a,A) \overset{p_1}{\twoheadrightarrow} A, E \overset{q}{\twoheadrightarrow} A)$$

$$\downarrow \qquad\qquad \lrcorner \qquad\qquad\qquad \downarrow (q,p)_*$$

$$1 \overset{(p_1,bp_0)}{\longrightarrow} \mathrm{Fun}^{\mathrm{cart}}_A(\mathrm{Hom}_A(a,A) \overset{p_1}{\twoheadrightarrow} A, A\times B \overset{\pi}{\twoheadrightarrow} A)$$

of functor spaces in $co\mathcal{C}art(\mathcal{K})_{/A}$. Corollary 5.7.19 again supplies natural equivalences

$$\mathrm{Fun}^{\mathrm{cart}}_A(\mathrm{Hom}_A(a,A) \overset{p_1}{\twoheadrightarrow} A, E \overset{q}{\twoheadrightarrow} A) \overset{\mathrm{ev}_{\ulcorner\mathrm{id}_a\urcorner}}{\underset{\sim}{\longrightarrow}} \mathrm{Fun}_A(X,E)$$

$$(q,p)_* \downarrow \qquad\qquad\qquad\qquad\qquad \downarrow (q,p)_*$$

$$\mathrm{Fun}^{\mathrm{cart}}_A(\mathrm{Hom}_A(a,A) \overset{p_1}{\twoheadrightarrow} A, A\times B \overset{\pi}{\twoheadrightarrow} A) \overset{\mathrm{ev}_{\ulcorner\mathrm{id}_a\urcorner}}{\underset{\sim}{\longrightarrow}} \mathrm{Fun}_A(X,A\times B)$$

which induce an equivalence between the fiber of the left-hand isofibration over (p_1, bp_0) and the fiber of the right-hand isofibration over $(p_1, bp_0) \cdot \ulcorner\mathrm{id}_a\urcorner = (a,b)$. To compute this fiber, observe that the right-hand isofibration is itself a pullback

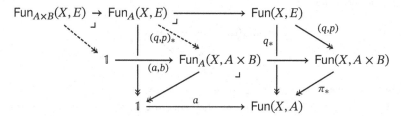

The composite equivalence is the map defined by precomposition with the canonical generalized element completing the proof.

$$\mathrm{Fun}^{\mathrm{cart}}_{A\times B}(\mathrm{Hom}_A(a,A) \underset{X}{\times} \mathrm{Hom}_B(B,b),E) \overset{\mathrm{ev}_{(\mathrm{id},\ulcorner\mathrm{id}_{bp_0}\urcorner)}}{\underset{\sim}{\longrightarrow}} \mathrm{Fun}^{\mathrm{cart}}_{A\times B}(\mathrm{Hom}_A(a,A),E)$$

$$\overset{\mathrm{ev}_{(\ulcorner\mathrm{id}_a\urcorner,\ulcorner\mathrm{id}_b\urcorner)}}{\underset{\sim}{\searrow}} \qquad\qquad \wr \downarrow \mathrm{ev}_{\mathrm{id}_a}$$

$$\mathrm{Fun}_{A\times B}(X,E)$$

$\square$

As with Corollary 5.7.19, we interpret Theorem 7.3.2 as defining a left biadjoint to the inclusion of the subcategory of two-sided fibrations, sending a span $A \overset{a}{\leftarrow} X \overset{b}{\rightarrow} B$ to the two-sided isofibration constructed in Definition 7.3.1. In the case of spans represented by a single nonidentity functor a "one-sided" version of Theorem 7.3.2, which is much more simply established, may be preferred:

PROPOSITION 7.3.5 (one-sided Yoneda for two-sided fibrations). *For any functor* $f : A \to B$ *and two-sided isofibration* $A \xleftarrow{q} E \xrightarrow{p} B$, *restriction along* $\ulcorner\mathrm{id}_f\urcorner : A \to \mathrm{Hom}_B(B, f)$ *induces an equivalence of quasi-categories*

$$
\mathrm{Fun}^{\mathrm{cart}}_{A \times B}
\begin{pmatrix}
\mathrm{Hom}_B(B, f) & E \\
\scriptstyle(p_1, p_0) \downarrow & \downarrow \scriptstyle(q,p) \\
A \times B & A \times B
\end{pmatrix}
\xrightarrow[\sim]{\mathrm{ev}_{\ulcorner\mathrm{id}_f\urcorner}}
\mathrm{Fun}_{A \times B}
\begin{pmatrix}
A & E \\
\scriptstyle(\mathrm{id}_A, f) \downarrow & \downarrow \scriptstyle(q,p) \\
A \times B & A \times B
\end{pmatrix}.
$$

Proof This follows by applying the Yoneda lemma of Theorem 5.7.3 to the element $(\mathrm{id}, f) : A \to A \times B$ and the cartesian fibration $(q, p) : E \twoheadrightarrow A \times B$ in the ∞-cosmos $co\mathcal{C}art(\mathcal{K})_{/A}$. □

Exercises

EXERCISE 7.3.i. State and prove the other one-sided Yoneda lemma for two-sided fibrations, establishing an equivalence of functor spaces induced by restricting along the functor $\ulcorner\mathrm{id}_f\urcorner : B \to \mathrm{Hom}_A(f, A)$.

$$
\mathrm{Fun}^{\mathrm{cart}}_{A \times B}
\begin{pmatrix}
\mathrm{Hom}_A(f, A) & E \\
\scriptstyle(p_1, p_0) \downarrow & \downarrow \scriptstyle(q,p) \\
A \times B & A \times B
\end{pmatrix}
\xrightarrow{\sim}
\mathrm{Fun}_{A \times B}
\begin{pmatrix}
B & E \\
\scriptstyle(f, \mathrm{id}) \downarrow & \downarrow \scriptstyle(q,p) \\
A \times B & A \times B
\end{pmatrix}
$$

7.4 Modules as Discrete Two-Sided Fibrations

We now turn our attention to the special properties of certain two-sided fibrations, such as comma spans, which are discrete as objects in a sliced ∞-cosmos.

DEFINITION 7.4.1. A **module** from A to B in an ∞-cosmos $\mathcal{K}$ is a two-sided fibration $A \xleftarrow{q} E \xrightarrow{p} B$ that is a discrete object in $_{A\backslash}\mathcal{F}ib(\mathcal{K})_{/B}$.

An object in the cosmologically embedded ∞-cosmos $_{A\backslash}\mathcal{F}ib(\mathcal{K})_{/B} \hookrightarrow \mathcal{K}_{/A \times B}$ is discrete in there if and only if it is discrete as an object of $\mathcal{K}_{/A \times B}$ (see Exercise 6.3.iv). Our work in this chapter enables us to give a direct characterization of modules:

PROPOSITION 7.4.2. *A two-sided isofibration* $A \xleftarrow{q} E \xrightarrow{p} B$ *defines a module from A to B if and only if it is*

(i) *cocartesian on the left,*
(ii) *cartesian on the right,*

(iii) *and discrete as an object of* $\mathcal{K}_{/A \times B}$.

Proof   Comparing Definition 7.4.1 with the statement, we see that we need only argue that these three properties suffice to prove that E is a two-sided fibration from A to B. Theorem 7.1.4(iv) tells us that a two-sided fibration E from A to B in $\mathcal{K}$ is a two-sided isofibration that is cocartesian on the left, cartesian on the right, and for which a certain natural transformation with codomain E in $\mathcal{K}_{/A \times B}$ is an isomorphism. When E is a discrete object in $\mathcal{K}_{/A \times B}$ all natural transformations with codomain E are invertible, so properties (i), (ii), and (iii) suffice.   □

The following properties of modules are easily deduced from our results about two-sided fibrations.

LEMMA 7.4.3. *If* $A \xleftarrow{q} E \xrightarrow{p} B$ *defines a module from* A *to* B *in an* ∞-*cosmos* $\mathcal{K}$, *then*

(i) *The functors displayed below define a discrete cocartesian fibration and a discrete cartesian fibration, respectively, in the* ∞-*cosmoi* $\mathcal{C}art(\mathcal{K})_{/B}$ *and* $co\mathcal{C}art(\mathcal{K})_{/A}$.

$$E \xrightarrow{(q,p)} A \times B \qquad\qquad E \xrightarrow{(q,p)} A \times B$$
$$\;\;\;\;{}^{p}\searrow\;\;\nwarrow{}^{\pi} \qquad\qquad\qquad\;\;\;\;{}^{q}\searrow\;\;\nwarrow{}^{\pi}$$
$$\qquad B \qquad\qquad\qquad\qquad\qquad\quad A$$

(ii) *The functors* $q : E \twoheadrightarrow A$ *and* $p : E \twoheadrightarrow B$ *respectively define a cocartesian fibration and a cartesian fibration in* $\mathcal{K}$.

(iii) *For any natural transformation* ψ *with codomain* E, ψ *is* p-*cartesian if and only if* $q\psi$ *is invertible, and* ψ *is* q-*cocartesian if and only if* $p\psi$ *is invertible.*

(iv) *In particular, any natural transformation that is fibered over* $A \times B$ *is both* p- *and* q-*cocartesian and any map of spans from a two-sided fibration*

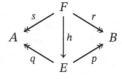

to a module defines a cartesian functor of two-sided fibrations, and also a cartesian functor from s *to* q *and from* r *to* p.

Proof   By Lemma 7.1.1, conditions (ii) and (iii) of Proposition 7.4.2 combine to tell us that the two-sided isofibration $(q, p) : E \twoheadrightarrow A \times B$ defines a discrete cartesian fibration in $\mathcal{K}_{/A}$. Since modules are two-sided fibrations and discreteness is reflected by cosmological embeddings, in fact $(q, p) \colon E \twoheadrightarrow A \times B$ defines a discrete cartesian fibration in $co\mathcal{C}art(\mathcal{K})_{/A}$, proving (i).

Statement (ii) holds for any two-sided fibration (see Lemma 7.1.1(iv)). The point of reasserting it here is that it is to emphasize that the legs of a module are not necessarily discrete fibrations themselves (see Exercise 7.4.i).

One direction of statement (iii) is proven as Lemma 7.1.1(iv). For the converse, consider a natural transformation ψ with codomain E and suppose $q\psi$ is invertible. Composing with a lift of $q\psi^{-1}$ along the isofibration $q \colon E \twoheadrightarrow A$, we see that ψ is isomorphic to a natural transformation ψ' with $q\psi'$ an identity. By isomorphism stability of p-cartesian transformations, it suffices to prove that ψ' is p-cartesian. By Proposition 3.6.2, ψ' can be lifted along the smothering 2-functor $\mathfrak{h}(\mathcal{K}_{/A}) \to (\mathfrak{h}\mathcal{K})_{/A}$ to a natural transformation in the sliced ∞-cosmos $\mathcal{K}_{/A}$ with codomain $q \colon E \twoheadrightarrow A$; as we typically perform such liftings without comment, we retain the notation ψ' for the fibered natural transformation in $\mathcal{K}_{/A}$. Since $(q, p) \colon E \twoheadrightarrow A \times B$ is a discrete cartesian fibration in $\mathcal{K}_{/A}$, every such natural transformation is (q, p)-cartesian. By Theorem 5.2.8(iii), it follows that ψ' is isomorphic to a whiskered copy of the counit of the fibered adjunction $\Delta_p \dashv r$ over $A \times B$. As the same data, regarded this time as a fibered adjunction over B, witnesses the fact that $p \colon E \twoheadrightarrow B$ is a cartesian fibration, Theorem 5.2.8(iii) also tells us that ψ' and thus ψ is p-cartesian, as claimed.

Statement (iv) follows from (iii) and Proposition 7.1.7(iv). $\qquad\square$

As the modules are exactly the discrete objects in the ∞-cosmos of two-sided fibrations:

COROLLARY 7.4.4. *For any ∞-categories A and B in an ∞-cosmos $\mathcal{K}$, there is a cosmologically embedded ∞-cosmos*

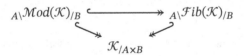

defined as a full subcategory of either the ∞-cosmos of two-sided fibrations from A to B or of the ∞-cosmos of isofibrations from A to B.

Proof Proposition 6.1.6 proves that $_A\backslash\mathcal{M}od(\mathcal{K})_{/B} \hookrightarrow {_A\backslash\mathcal{F}ib(\mathcal{K})_{/B}}$ is a cosmologically embedded sub ∞-cosmos in the ∞-cosmos of Proposition 7.2.2, while Lemma 7.4.3(iv) proves that the inclusion $_A\backslash\mathcal{M}od(\mathcal{K})_{/B} \hookrightarrow \mathcal{K}_{/A \times B}$ is also full. $\qquad\square$

An important property of modules is that they are stable under pullback:

PROPOSITION 7.4.5. *If $A \twoheadleftarrow E \twoheadrightarrow B$ is a module from A to B and $a : A' \to A$ and $b : B' \to B$ are any pair of functors, then the pullback defines a module $A' \twoheadleftarrow E(b, a) \twoheadrightarrow B'$ from A' to B'.*

$$
\begin{array}{ccc}
E(b, a) & \longrightarrow & E \\
\downarrow & \lrcorner & \downarrow \\
A' \times B' & \xrightarrow{a \times b} & A \times B
\end{array}
$$

Proof The cosmological functor $(a, b)^* : {}_{A\backslash}\mathcal{F}ib(\mathcal{K})_{/B} \longrightarrow {}_{A'\backslash}\mathcal{F}ib(\mathcal{K})_{/B'}$ of Proposition 7.2.4, like all cosmological functors, preserves discrete objects. □

Applying Proposition 7.4.5 to a pair of elements $a : 1 \to A$ and $b : 1 \to B$, we see that a module from A to B is a two-sided fibration whose fibers $E(b, a)$ are discrete ∞-categories. The converse does not generally hold, as being discrete as an object of the sliced ∞-cosmos $\mathcal{K}_{/A \times B}$ is a stronger condition than merely having discrete fibers. However, when $\mathcal{K}$ is an ∞-cosmos of $(\infty, 1)$-categories, Proposition 12.2.3 proves that discreteness in a slice is implied by fiberwise discreteness.

The prototypical examples of modules are given by the arrow and comma ∞-category constructions.

PROPOSITION 7.4.6 (comma ∞-categories are modules).

  (i) *For any ∞-category A, the arrow ∞-category A^2 defines a module from A to A.*

  (ii) *For any cospan $C \xrightarrow{g} A \xleftarrow{f} B$, the comma ∞-category $\mathrm{Hom}_A(f, g)$ defines a module from C to B.*

As just remarked, the fact that $A^2 \twoheadrightarrow A \times A$ is discrete is related to but stronger than the fact, proven in Proposition 3.4.10, that the mapping space between any pair of elements of A is a discrete ∞-category.

Proof By Proposition 7.4.5, the second statement follows from the first, but it is no harder to prove both statements at once from our suite of established results. Proposition 7.1.6 and Example 7.2.8 prove that arrow and comma spans define two-sided fibrations, so it remains only to verify the discreteness conditions. By Lemma 5.5.1, discreteness of $A^2 \twoheadrightarrow A \times A$ in the sliced ∞-cosmos over $A \times A$ and $\mathrm{Hom}_A(f, g) \twoheadrightarrow C \times B$ in the sliced ∞-cosmos over $C \times B$ are immediate consequences of 2-cell conservativity of Proposition 3.2.5 and Proposition 3.4.6: for the latter, if τ is any 2-cell with codomain $\mathrm{Hom}_A(f, g)$ so that $p_1\tau$ and $p_0\tau$ are invertible, then τ is itself invertible. □

Two special cases of these comma modules, those studied in §3.5, deserve a special name:

DEFINITION 7.4.7. To any functor $f : A \to B$ between ∞-categories

(i) the module $\mathrm{Hom}_B(B, f)$ from A to B is **right** or **covariantly represented** by f, while
(ii) the module $\mathrm{Hom}_B(f, B)$ from B to A is **left** or **contravariantly represented** by f.

More generally, a module is **covariantly** or **contravariantly represented** by f if it is fibered equivalent to the left or right represented modules.

As in §5.7, the Yoneda lemma for two-sided fibrations simplifies when mapping into a module on account of the observation in Lemma 7.4.3(iv) that any map of spans from a two-sided fibration to a module defines a cartesian functor.

THEOREM 7.4.8 (Yoneda for modules). *For any span $A \xleftarrow{a} X \xrightarrow{b} B$ and any module $A \xleftarrow{q} E \xrightarrow{p} B$, restriction along $(\ulcorner id_a \urcorner, \ulcorner id_b \urcorner) : X \to \mathrm{Hom}_A(a, A) \underset{X}{\times} \mathrm{Hom}_B(B, b)$ defines an equivalence of Kan complexes*

$$\mathrm{Fun}_{A \times B} \begin{pmatrix} \mathrm{Hom}_A(a,A) \underset{X}{\times} \mathrm{Hom}_B(B,b) & E \\ \ \ (p_1\pi_1, p_0\pi_0) \!\!\downarrow & , \ (q,p) \!\!\downarrow \\ A \times B & A \times B \end{pmatrix} \xrightarrow[\sim]{\mathrm{ev}_{(\ulcorner id_a \urcorner, \ulcorner id_b \urcorner)}} \mathrm{Fun}_{A \times B} \begin{pmatrix} X & E \\ | & | \\ (a,b) & , \ (q,p) \\ \downarrow & \downarrow \\ A \times B & A \times B \end{pmatrix}$$

and for any functors $f : A \to B$ or $g : B \to A$, restriction along $\ulcorner id_f \urcorner : A \to \mathrm{Hom}_B(B, f)$ or $\ulcorner id_g \urcorner : B \to \mathrm{Hom}_A(g, A)$ define equivalences of Kan complexes

$$\mathrm{Fun}_{A \times B} \begin{pmatrix} \mathrm{Hom}_B(B,f) & E \\ (p_1,p_0)\!\!\downarrow & , \ \downarrow\!\!(q,p) \\ A \times B & A \times B \end{pmatrix} \xrightarrow[\sim]{\mathrm{ev}_{\ulcorner id_f \urcorner}} \mathrm{Fun}_{A \times B} \begin{pmatrix} A & E \\ (\mathrm{id},f)\!\!\downarrow & , \ \downarrow\!\!(q,p) \\ A \times B & A \times B \end{pmatrix}$$

$$\mathrm{Fun}_{A \times B} \begin{pmatrix} \mathrm{Hom}_A(g,A) & E \\ (p_1,p_0)\!\!\downarrow & , \ \downarrow\!\!(q,p) \\ A \times B & A \times B \end{pmatrix} \xrightarrow[\sim]{\mathrm{ev}_{\ulcorner id_g \urcorner}} \mathrm{Fun}_{A \times B} \begin{pmatrix} B & E \\ (g,\mathrm{id})\!\!\downarrow & , \ \downarrow\!\!(q,p) \\ A \times B & A \times B \end{pmatrix}. \ \square$$

We have not yet explained the curious terminology – why we refer to discrete two-sided fibrations as "modules." This moniker is intended to suggest a deep structural analogy between the calculus of modules and the calculus of the more familiar algebraic structure, a subject to which we now turn.

Exercises

EXERCISE 7.4.i. Demonstrate by means of an example that if $A \xleftarrow{q} E \xrightarrow{p} B$ defines a module from A to B then it is not necessarily the case that $p : E \twoheadrightarrow B$ is a discrete cartesian fibration or $q : E \twoheadrightarrow A$ is a discrete cocartesian fibration.

EXERCISE 7.4.ii.

(i) Explain why the two-sided fibration $(p_{n-1}, p_0) : A^{\mathbb{m}} \twoheadrightarrow A \times A$ of Proposition 7.1.6 does not define a module for $n > 2$.

(ii) Conclude that the horizontal composite of modules, as defined in Proposition 7.2.6, is not necessarily a module.

EXERCISE 7.4.iii. State and prove an analogue of Exercise 7.1.ii characterizing modules from 1 to B and modules from B to 1, and resolve any apparent conflicts between this statement and Exercise 7.4.i.

8

The Calculus of Modules

The calculus of modules between ∞-categories bears a strong resemblance to the calculus of (bi)modules between unital rings. Here ∞-categories take the place of rings, with functors between ∞-categories playing the role of ring homomorphisms, which we display vertically in the following table. A module E from A to B, like the two-sided fibrations considered in Chapter 7, is an ∞-category on which A "acts on the left" and B "acts on the right" and these actions commute; this is analogous to the notation of A–B bimodule in ring theory and explains our choice of terminology. A module $A \xleftarrow{q} E \xrightarrow{p} B$ will now be depicted as $A \xrightarrow{E} B$ whenever explicit names for the legs of the constituent span are not needed.

unital rings	A	∞-categories
ring homomorphisms	$\begin{matrix} A' \\ f\downarrow \\ A \end{matrix}$	∞-functors
bimodules between rings	$A \xrightarrow{E} B$	modules between ∞-categories
module maps	$\begin{matrix} A' \xrightarrow{E'} B' \\ f\downarrow \quad \Downarrow\alpha \quad \downarrow g \\ A \xrightarrow[E]{} B \end{matrix}$	module maps

Finally, there is a notion of module map whose boundary is a square comprised of two modules and two functors as displayed. In ring theory, a module map with this boundary is given by an A'–B' module homomorphism $E' \to E(g, f)$, whose codomain is the A'–B' bimodule defined by restricting the scalar multiplication in the A–B module E along the ring homomorphisms f and g. A similar idea is encoded by Definition 8.1.4.

The analogy extends deeper than this: unital rings, ring homomorphisms, bimodules, and module maps define a *proarrow equipment*, in the sense of Wood [132].[1] Our main result in this chapter is Theorem 8.2.6, which asserts that ∞-categories, functors, modules, and module maps in any ∞-cosmos define a *virtual equipment*, in the sense of Cruttwell and Shulman [32].

As a first step, in §8.1 we introduce the double category of two-sided isofibrations, which restricts to define a *virtual double category* of modules. A *double category* is a sort of 2-dimensional category with objects; two varieties of 1-morphisms, the "horizontal" and the "vertical"; and 2-dimensional cells fitting into "squares" whose boundaries consist of horizontal and vertical 1-morphisms with compatible domains and codomains (see Definition B.1.9). A motivating example from abstract algebra is the double category of modules: objects are rings, vertical morphisms are ring homomorphisms, horizontal morphisms are bimodules, and whose squares are bimodule homomorphisms. Vertical composition is by composing homomorphisms, a strictly associative and unital operation, while horizontal composition is by tensor product of modules, which is associative and unital up to coherent natural isomorphism. In the literature, this sort of structure is sometimes called a *pseudo double category* – morphisms and squares compose strictly in the "vertical" direction but only up to isomorphism in the "horizontal" direction – but we refer to this simply as a "double category" as it is the only variety considered in the main text.[2]

Our aim in §8.1 is to describe a similar structure whose objects and vertical morphisms are the ∞-categories and functors in any fixed ∞-cosmos, whose horizontal morphisms are modules, and whose squares are module maps. If the horizontal morphisms are replaced by the larger class of two-sided fibrations or the still larger class of two-sided isofibrations, then these morphisms assemble into a double category with the horizontal composition operation defined by Proposition 7.2.6 (see Proposition 8.1.6). However, this horizontal composition operation does not preserve modules: the arrow ∞-category A^2 defines a module from A to A whose horizontal composite with itself is equivalent to the two-sided fibration $(p_2, p_0) : A^3 \twoheadrightarrow A \times A$ of Proposition 7.1.6, which is not discrete in the sliced ∞-cosmos over $A \times A$. To define a genuine "tensor product for modules" operation requires a two-stage construction: first forming the pullback that defines a composite two-sided fibration as in Proposition 7.2.6, and then reflecting this into a two-sided *discrete* fibration by means of some sort of

---

[1] This can be seen as a special case of the prototypical equipment comprised of $\mathcal{V}$-categories, $\mathcal{V}$-functors, $\mathcal{V}$-modules, and $\mathcal{V}$-natural transformations between them, for any closed symmetric monoidal category $\mathcal{V}$. The equipment for rings is obtained from the case where $\mathcal{V}$ is the category of abelian groups by restricting to abelian group enriched categories with a single object.

[2] Strict double categories make a brief appearance in Theorem B.3.6 to express the functoriality of the mates correspondence.

"homotopy coinverter" construction. As colimits that are not within the purview of the axioms of an ∞-cosmos, this presents somewhat of an obstacle.[3]

Rather than leave the comfort of our axiomatic framework in pursuit of a double category of modules, we instead describe the structure that naturally arises within the axiomatization, which turns out to be familiar to category theorists and robust enough for our desired applications. We first demonstrate that ∞-categories, functors, modules, and module maps assemble into a *virtual double category*, a weaker structure than a double category in which cells are permitted to have a multi horizontal source, as a "virtual" replacement for horizontal composition of modules. Virtual double categories relate to double categories as multicategories relate to monoidal categories; indeed virtual double categories are the "generalized multicategories" defined relative to the free category monad on directed graphs [32, 76].

A *virtual equipment* is a virtual double category satisfying two additional axioms. One of these provides a "restriction of scalars" operation, allowing a horizontal morphism $A \xrightarrow{E} B$ to be pulled back along a pair of vertical morphisms $f : A' \to A$ and $g : B' \to B$. The other condition requires each object to have a horizontal "unit" morphism, satisfying a suitable universal property that serves as a substitute, in the absence of a composition operation, for the unital composition rules. Once the definition of a virtual equipment is given in §8.2, these axioms follow easily from the results of Chapter 7. The final two sections are devoted to exploring the consequences of this structure, which are put to full use in the development of the formal category theory of ∞-categories in Chapter 9. In §8.3, we explain how certain horizontal composites of modules can be recognized in the virtual equipment, even if the general construction of the tensor product of an arbitrary composable pair of modules is not known. The final §8.4 collects together many special properties of the modules $A \xrightarrow{\mathrm{Hom}_B(B,k)} B$ and $B \xrightarrow{\mathrm{Hom}_B(k,B)} A$ represented by a functor $k : A \to B$ of ∞-categories, revisiting some of the properties first established in §3.5.

8.1 The Double Category of Two-Sided Isofibrations

Recall from Corollary 7.4.4 that the ∞-cosmos of modules may be defined as a full subcategory of either the ∞-cosmos of two-sided fibrations or the ∞-cosmos of two-sided isofibrations. This presents us with two options for constructing the virtual double category of modules, which may be realized as a

---

[3] The ∞-cosmoi that one encounters in practice in fact admit all flexible weighted homotopy colimits – including homotopy coinverters in particular – as they tend to be *accessible* in a sense defined by Bourke, Lack, and Vokřínek [22] (see Digression E.1.8).

full subcategory of the double category of two-sided fibrations or of the double category of two-sided isofibrations. We adopt the latter tactic as it is marginally simpler, but the reader is invited to opt for the former route instead (see Exercise 8.1.iv).

Our first task is to define the 2-dimensional morphisms in the double categories that we will introduce.

DEFINITION 8.1.1. A **fibered map of two-sided isofibrations** from $A \xleftarrow{q} E \xtwoheadrightarrow{p} B$ to $A \xleftarrow{s} F \xtwoheadrightarrow{r} B$ is a fibered isomorphism class of strictly commuting functors

$$
\begin{array}{ccc}
 & E & \\
{}^{q}\swarrow & \downarrow & \searrow^{p} \\
A & \big\downarrow a & B \\
{}_{s}\nwarrow & \downarrow & \nearrow_{r} \\
 & F & \\
\end{array}
$$

where two such functors a and a' are considered equivalent if there exists a natural isomorphism $\gamma : a \cong a'$ so that $s\gamma = \mathrm{id}_q$ and $r\gamma = \mathrm{id}_p$.

When $A \xleftarrow{q} E \xtwoheadrightarrow{p} B$ and $A \xleftarrow{s} F \xtwoheadrightarrow{r} B$ are modules, we refer to a fibered map of two-sided isofibrations between them as **fibered map of modules** from E to F.

OBSERVATION 8.1.2 (the 1-categories of modules and fibered maps). The 1-category of two-sided isofibrations from A to B in an ∞-cosmos $\mathcal{K}$ and fibered maps may be obtained as a quotient of the quasi-categorically enriched category $\mathcal{K}_{/A \times B}$, or of its homotopy 2-category $\mathfrak{h}(\mathcal{K}_{/A \times B})$, or of the slice homotopy 2-category $\mathfrak{h}\mathcal{K}_{/A \times B}$. The quotient 1-category has the same collection of objects and has isomorphism classes of functors as its morphisms.

By Lemma 7.4.3, the 1-category of modules from A to B and fibered maps is a full subcategory, or alternatively may be regarded as a quotient of the Kan complex enriched category $_{A\backslash}\mathcal{M}od(\mathcal{K})_{/B}$, or of its homotopy 2-category $\mathfrak{h}(_{A\backslash}\mathcal{M}od(\mathcal{K})_{/B})$ in the same manner.

The 1-categories of Observation 8.1.2 are of interest because they precisely capture the correct notion of equivalence between two-sided isofibrations or modules first introduced in Definition 3.2.7.

LEMMA 8.1.3. *In an ∞-cosmos $\mathcal{K}$:*

(i) *A pair of two-sided isofibrations are equivalent in $\mathcal{K}_{/A \times B}$ if and only if they are isomorphic in the 1-category of two-sided isofibrations from A to B.*

(ii) *A pair of modules are equivalent over A times B if and only if they are isomorphic in the 1-category of modules from A to B.* □

Each of the definitions just presented admits a common generalization, which defines the 2-dimensional maps inhabiting squares.

DEFINITION 8.1.4 (maps in squares). A **map of modules** or a **map of two-sided isofibrations** from $A' \xleftarrow{q'} E' \xrightarrow{p'} B'$ to $A \xleftarrow{q} E \xrightarrow{p} B$ over $f : A' \to A$ and $g : B' \to B$, depicted as

$$
\begin{array}{ccc}
A' & \xrightarrow{E'} & B' \\
f\downarrow & \Downarrow\alpha & \downarrow g \\
A & \xrightarrow[E]{} & B
\end{array}
$$

is a fibered isomorphism class of strictly commuting functors a

$$
\begin{array}{ccccc}
A' & \xleftarrow{q'} & E' & \xrightarrow{p'} & B' \\
f\downarrow & & \downarrow a & & \downarrow g \\
A & \xleftarrow[q]{} & E & \xrightarrow[p]{} & B
\end{array}
$$

where two such functors a and a' are considered equivalent if there exists a natural isomorphism $\gamma : a \cong a'$ so that $q\gamma = \mathrm{id}_{aq'}$ and $p\gamma = \mathrm{id}_{bp'}$.

OBSERVATION 8.1.5. In the case of modules or two-sided isofibrations, the functor space $\mathrm{Fun}_{f \times g}(E', E)$ of maps from E' to E over $f \times g$ is defined by the pullback

$$
\begin{array}{ccc}
\mathrm{Fun}_{f \times g}(E', E) & \longrightarrow & \mathrm{Fun}(E', E) \\
\downarrow & \lrcorner & \downarrow (q,p) \\
\mathbb{1} & \xrightarrow[(fq', gp')]{} & \mathrm{Fun}(E', A \times B)
\end{array}
$$

As in Observation 8.1.2, module maps are defined to be isomorphism classes of objects in this functor space.

We now introduce the double category of two-sided isofibrations. This structure can be viewed either as a collection of data present in the homotopy 2-category $\mathfrak{h}\mathcal{K}$ of an ∞-cosmos or as a quotient of quasi-categorically enriched structures described in Exercise 8.1.i. In some sense the latter point of view is more natural, since its horizontal composition is characterized by a strict universal property and no isomorphism classes of maps are required, but the 2-categorical approach is more familiar and provides a convenient setting within which to develop the formal category theory of ∞-categories.

PROPOSITION 8.1.6 (the double category of two-sided isofibrations). *The homotopy 2-category of an ∞-cosmos $\mathcal{K}$ supports a (non-unital) **double category of two-sided isofibrations** whose:*

- *objects are ∞-categories,*
- *vertical arrows are functors,*
- *horizontal arrows $A \xrightarrow{E} B$ are two-sided isofibrations $A \xleftarrow{q} E \xrightarrow{p} B$, and*
- *cells with boundary*

$$
\begin{array}{ccc}
A' & \xrightarrow{E'} & B' \\
f\downarrow & \Downarrow\alpha & \downarrow g \\
A & \xrightarrow{E} & B
\end{array}
$$

are maps of two-sided isofibrations over $f \times g$, or equivalently, are isomorphism classes of objects in the quasi-category $\mathrm{Fun}_{f\times g}(E', E)$.

Proof Vertical composition of arrows and cells is by composition in $\mathcal{K}$. The composition of horizontal arrows is defined in Proposition 7.2.6, while the horizontal composition of cells is defined in Exercise 7.2.i. By simplicial functoriality of pullback and composition in $\mathcal{K}$, both constructions are associative up to canonical natural isomorphism.   □

The double category just introduced does not contain horizontal unit arrows. For technical reasons, we find it most convenient to leave them out.

REMARK 8.1.7 (why the horizontal unit is missing). The unit for the span composition operation defined in Proposition 7.2.6 is the identity span, which is not typically a two-sided isofibration. Rather than formally adjoin horizontal units to the "double category" of Proposition 8.1.6, we find it less confusing to leave them out because when we restrict to the structure of greatest interest – the virtual equipment of modules – we will see that the arrow ∞-category plays the role of the horizontal unit for composition in a sense to be described in Proposition 8.2.4, even though it does not define a horizontal composition unit for the span composition operation.

By Exercise 7.4.ii, modules do not form double category under span composition, which leads us to search for another categorical structure to axiomatize their behavior. A virtual double category is a multicategorical analogue of a double category appropriate for settings when horizontal composition and units may or may not be defined. This notion has been studied by Burroni [24] and Leinster [76] under various names, though we adopt the terminology and notation of Cruttwell and Shulman [32].

DEFINITION 8.1.8 (virtual double category). A **virtual double category** consists of

- a category of **objects** and **vertical arrows**
- for any pair of objects A, B, a collection of **horizontal arrows** $A \nrightarrow B$
- **cells**, with boundary depicted as follows

$$
\begin{array}{ccccccc}
A_0 & \xrightarrow{E_1} & A_1 & \xrightarrow{E_2} & \cdots & \xrightarrow{E_n} & A_n \\
f \downarrow & & & \Downarrow\alpha & & & \downarrow g \\
B_0 & & & \xrightarrow{\quad\quad F \quad\quad} & & & B_n
\end{array}
\tag{8.1.9}
$$

including, in the case $A_0 = A_n$, those whose horizontal source has length zero

- a **composite cell** as below-right, for any configuration as below-left

$$
\begin{array}{ccccccc}
A_0 & \overset{E_{11},\ldots,E_{1k_1}}{\cdots\!\cdots\!\rightarrow} & A_1 & \overset{E_{21},\ldots,E_{2k_2}}{\cdots\!\cdots\!\rightarrow} & \cdots & \overset{E_{n1},\ldots,E_{nk_n}}{\cdots\!\cdots\!\rightarrow} & A_n \\
f_0 \downarrow & \Downarrow & \downarrow f_1 & \Downarrow & \downarrow\cdots & \Downarrow & \downarrow f_n \\
B_0 & \xrightarrow{F_1} & B_1 & \xrightarrow{F_2} & \cdots & \xrightarrow{F_n} & B_n \\
g \downarrow & & & \Downarrow & & & \downarrow h \\
C_0 & & & \xrightarrow{\quad G \quad} & & & C_n
\end{array}
\quad =:
\begin{array}{ccc}
A_0 & \overset{E_{11}\; E_{12},\ldots,E_{nk_n-1}E_{nk_n}}{\xrightarrow{\;\cdots\!\cdots\!\rightarrow\;}} & A_n \\
gf_0 \downarrow & \Downarrow & \downarrow hf_n \\
C_0 & \xrightarrow{\quad G \quad} & C_n
\end{array}
$$

$$\tag{8.1.10}$$

- an **identity cell** for every horizontal arrow

$$
\begin{array}{ccc}
A & \xrightarrow{E} & B \\
\| & \Downarrow \mathrm{id}_E & \| \\
A & \xrightarrow[E]{} & B
\end{array}
$$

so that composition of cells is associative and unital in the usual multicategorical sense.

LEMMA 8.1.11. *There exists a virtual double category of two-sided isofibrations whose:*

- *objects are ∞-categories,*
- *vertical arrows are functors,*
- *horizontal arrows are two-sided isofibrations, and*

- *n-ary cells (8.1.9) are maps of two-sided isofibrations over $f \times g$*

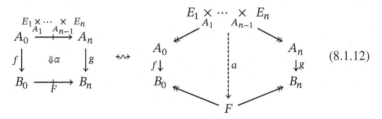

$$(8.1.12)$$

whose single vertical source is the $(n-1)$-fold span composite of the sequence of spans comprising the vertical source in (8.1.9).

Proof The required composition laws can be defined by embedding the virtual double categories into the double categories of Proposition 8.1.6. For instance, the composition of a configuration as on the left of (8.1.10) can be defined by horizontally composing the k top cells in the double category of isofibrations, and then vertically composing the result with the bottom cell. The result is a $k_1 + \cdots + k_n$-ary cell in the virtual double category of the correct form.

It remains only to define the nullary cells[4] which have an empty sequence as their vertical domain

$$
\begin{array}{ccc}
A \;=\!=\; A \\
f\big\downarrow \quad \Downarrow\alpha \quad \big\downarrow g \\
B \xrightarrow{\;\;F\;\;} C
\end{array}
\qquad \leftrightsquigarrow \qquad
\begin{array}{c}
A \\
\overset{f}{\swarrow} \; \big\downarrow a \; \overset{g}{\searrow} \\
B \qquad\qquad C \\
\underset{s}{\nwarrow} \; \big\downarrow \; \overset{r}{\nearrow} \\
F
\end{array}
$$

which we interpret as a 0-fold pullback, this being the identity span from A to A. So the nullary cells displayed above-left are fibered isomorphism classes of maps above-right where a and a' lie in the same equivalence class if there exists a natural isomorphism $\gamma : a \cong a'$ so that $s\gamma = \mathrm{id}_f$ and $r\gamma = \mathrm{id}_g$. □

REMARK 8.1.13. For instance, the map

$$
\begin{array}{c}
A \\
/\!\!/ \; \big\downarrow \Delta \; \backslash\!\backslash \\
A \qquad\qquad A \\
\underset{q_1}{\nwarrow} \; \big\downarrow \; \overset{q_0}{\nearrow} \\
A^{\shortparallel}
\end{array}
\qquad \leftrightsquigarrow \qquad
\begin{array}{c}
A \;=\!=\; A \\
\big\| \quad \Downarrow\delta \quad \big\| \\
A \xrightarrow{\;A^{\shortparallel}\;} A
\end{array}
$$

defines a nullary morphism with codomain $A \xrightarrow{A^{\shortparallel}} A$ in the virtual double category of two-sided isofibrations. Note, however, that despite the fact that $\Delta : A \rightsquigarrow A^{\shortparallel}$

---

[4] Note the use of the symbol $A \;=\!=\; A$ to denote the nullary source of a nullary cell, adopted so such cells fit naturally into pasting diagrams of cells in a virtual double category.

defines an equivalence in the ambient ∞-cosmos, this cell does not define an isomorphism in the virtual double category of any kind. This nullary morphism endows $A \xrightarrow{A^{\parallel}} A$ with the structure of a unit in a sense suitable to virtual double categories (see Exercise 8.2.iii). It is for this sort of reason that we left out the identity horizontal arrows in Proposition 8.1.6.

Our main example of interest is a full sub virtual double category defined by restricting the class of horizontal arrows and taking all cells between them. Since the only operations given in the structure of a virtual double category are vertical sources and targets, vertical identities, and vertical composition, it is clear that this substructure is closed under all of these operations, and thus inherits the structure of a virtual double category.

COROLLARY 8.1.14. *For any ∞-cosmos $\mathcal{K}$, there is a virtual double category of modules $\mathrm{Mod}(\mathcal{K})$ defined as a full subcategory of the virtual double categories of isofibrations whose*

- *objects are ∞-categories,*
- *vertical arrows are functors,*
- *horizontal arrows $A \xrightarrow{E} B$ are modules E from A to B,*
- *n-ary cells are fibered isomorphism classes of maps of two-sided isofibrations over $f \times g$:*

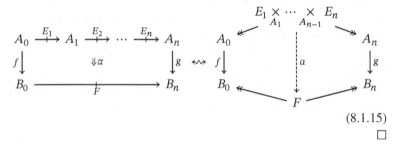

$$(8.1.15)$$

□

The *n*-**ary module maps** of (8.1.15) can be thought of as special cases of the *n*-ary cells of Lemma 8.1.11 where $E_1, \ldots, E_n, F$ are all required to be modules: the single horizontal source in the diagram (8.1.12) is the two-sided fibration defined by the $(n-1)$-fold pullback of the sequence of modules comprising the horizontal source in the left-hand diagram. We refer to the finite sequence of modules occurring as the horizontal domain of an *n*-ary module map as a **compatible sequence of modules**, which just means that their horizontal sources and targets are compatible in the evident way.

A hint at the relevance of this notion of *n*-ary module map is given by the following special case.

LEMMA 8.1.16. *There is a bijection between n-ary module maps whose codo-main module is a comma module as displayed below-left and natural transfor-mations in the homotopy 2-category whose boundary is displayed above-right.*

$$
\begin{array}{ccccc}
A_0 & \xrightarrow{E_1} & A_1 \xrightarrow{E_2} \cdots \xrightarrow{E_n} & A_n \\
f\downarrow & & \Downarrow & \downarrow g \\
B & \xrightarrow[\mathrm{Hom}_D(h,k)]{} & & C
\end{array}
\quad \leftrightsquigarrow \quad
\begin{array}{ccc}
 & E_1 \times \cdots \times E_n \\
 & {}_{A_1} \qquad {}_{A_{n-1}} \\
A_0 \swarrow & & \searrow A_n \\
f\downarrow & \Leftarrow & \downarrow g \\
B & & C \\
 \searrow_k & D & \swarrow_h
\end{array}
$$

Proof   Combine Definition 8.1.4 with Proposition 3.4.7.   □

For any pair of objects A and B in the virtual double category of modules, there is a vertical 1-category of modules from A to B and module maps over a pair of identity functors, which coincides with the 1-category of modules from A to B introduced in Observation 8.1.2. In this context, Lemma 8.1.3 may be restated as follows:

LEMMA 8.1.17. *A parallel pair of modules $A \xrightarrow{E} B$ and $A \xrightarrow{F} B$ are isomorphic as objects of vertical 1-category of modules in the virtual double category of modules if and only if $E \simeq_{A \times B} F$.*   □

For consistency with the rest of the text, we write $E \simeq F$ or $E \simeq_{A \times B} F$ whenever the modules $A \xrightarrow{E} B$ and $A \xrightarrow{F} B$ are isomorphic as objects of the vertical 1-category of modules from A to B. For instance, Proposition 4.1.1 proves that a functor $f : B \to A$ is left adjoint to a functor $u : A \to B$ if and only if $\mathrm{Hom}_A(f, A) \simeq_{A \times B} \mathrm{Hom}_B(B, u)$, that is, if and only if the modules $B \xrightarrow{\mathrm{Hom}_A(f,A)} A$ and $A \xrightarrow{\mathrm{Hom}_B(B,u)} B$ are isomorphic as objects of the vertical 1-category of modules from A to B.

Exercises

EXERCISE 8.1.i.   By Exercise 6.1.iii, for any ∞-cosmos $\mathcal{K}$, there is an ∞-cosmos $\mathcal{K}^\times$ whose objects are two-sided isofibrations between an arbitrary pair of ∞-categories, these being exactly the "fibrant diagrams" indexed by the span, considered as an inverse category. Use this ∞-cosmos together with the endpoint evaluation functors $\mathcal{K}^\times \to \mathcal{K}$ to give a second description of the double cate-gory of two-sided isofibrations as as quotient of a nonunital internal category defined at the level of ∞-cosmoi, cosmological functors, and simplicial natural isomorphisms.[5]

EXERCISE 8.1.ii. Prove that any double category defines a virtual double category.[6]

EXERCISE 8.1.iii. Use the three statements of Theorem 7.4.8 to describe three bijections between cells with various boundary shapes in the virtual double category of modules.

EXERCISE 8.1.iv. Let $A \xleftarrow{q} E \xrightarrow{p} B$ and $A \xleftarrow{s} F \xrightarrow{r} B$ be two-sided fibrations. A **map of two-sided fibrations** from $A \xleftarrow{q} E \xrightarrow{p} B$ to $A \xleftarrow{s} F \xrightarrow{r} B$ is a fibered isomorphism class of strictly commuting functors a so that the left square defines a cartesian functor between the cocartesian fibrations and the right square defines a cartesian functor between the cartesian fibrations (see Proposition 7.1.7).

$$
\begin{array}{ccc}
A' & \xleftarrow{q'} E' \xrightarrow{p'} & B' \\
f\downarrow & \quad\downarrow a & \quad\downarrow g \\
A & \xleftarrow{q} E \xrightarrow{p} & B
\end{array}
$$

Define a double category of two-sided fibrations as a nonfull subcategory of the double category of two-sided isofibrations of Proposition 8.1.6.

8.2 The Virtual Equipment of Modules

The virtual double category of modules $\mathbb{M}od(\mathcal{K})$ in an ∞-cosmos $\mathcal{K}$ has two special properties that characterize a *virtual equipment*. Before stating the definition, we explore each of these in turn.

PROPOSITION 8.2.1 (restriction). *Any diagram in $\mathbb{M}od(\mathcal{K})$ as below-left can be completed to a **cartesian cell** as below-right*

$$
\begin{array}{ccc}
A' & & B' \\
a\downarrow & & \downarrow b \\
A & \xrightarrow[E]{} & B
\end{array}
\quad\rightsquigarrow\quad
\begin{array}{ccc}
A' & \xrightarrow{E(b,a)} & B' \\
a\downarrow & \Downarrow\rho & \downarrow b \\
A & \xrightarrow[E]{} & B
\end{array}
$$

---

[5] We consider it to be an open problem whether it is useful to take a "double quasi-categorical" on the nonunital double category of two-sided isofibrations or the virtual double category of modules.

[6] If the double category lacks horizontal identity morphisms, the corresponding virtual double category may lack nullary morphisms – unless these can be defined in some other way as we did in the proof of Lemma 8.1.11.

characterized by the universal property that any cell as displayed below-left factors uniquely through ρ as below-right:

$$X_0 \xrightarrow{E_1} X_1 \xrightarrow{E_2} \cdots \xrightarrow{E_n} X_n$$

$$X_0 \xrightarrow{E_1} X_1 \xrightarrow{E_2} \cdots \xrightarrow{E_n} X_n \qquad\qquad f \downarrow \qquad\qquad \exists! \Downarrow \qquad\qquad \downarrow g$$

$$af \downarrow \qquad\qquad \Downarrow \qquad\qquad \downarrow bg \quad =: \quad A' \xrightarrow{\quad E(b,a) \quad} B'$$

$$A \xrightarrow{\qquad\qquad E \qquad\qquad} B \qquad\qquad a \downarrow \qquad\qquad \Downarrow \rho \qquad\qquad \downarrow b$$

$$A \xrightarrow{\qquad\qquad E \qquad\qquad} B$$

Proof The horizontal source of the cartesian cell is defined by restricting the module $A \xrightarrow{E} B$ along the functors a and b:

$$
\begin{array}{ccc}
E(b,a) & \xrightarrow{\ \rho\ } & E \\
\downarrow & \lrcorner & \downarrow \\
A' \times B' & \xrightarrow{a \times b} & A \times B
\end{array}
\qquad\qquad (8.2.2)
$$

By Proposition 7.4.5, this left-hand isofibration defines a module from A' to B', while by Definition 8.1.4 the top horizontal functor represents a module map inhabiting the desired square. By Observation 8.1.5, the simplicial pullback in $\mathcal{K}$ induces an equivalence[7] of functor spaces:

$$\mathrm{Fun}_{f \times g}(E_1 \underset{A_1}{\times} \cdots \underset{A_{n-1}}{\times} E_n, E(b,a)) \xrightarrow[\sim]{\rho \circ -} \mathrm{Fun}_{af \times bg}(E_1 \underset{A_1}{\times} \cdots \underset{A_{n-1}}{\times} E_n, E)$$

which descends to a bijection on isomorphism classes of objects. This defines the unique factorization of cells as displayed above-left through the cartesian restriction cell ρ. $\qquad\square$

We refer to the module $A' \xrightarrow{E(b,a)} B'$ as the **restriction** of $A \xrightarrow{E} B$ along the functors a and b, because the pullback (8.2.2) is analogous to the restriction of scalars of a bimodule along a pair of ring homomorphisms.

EXAMPLE 8.2.3. The module $C \xrightarrow{\mathrm{Hom}_A(f,g)} B$ is the restriction of the module encoded by the arrow ∞-category $A \xleftarrow{p_1} A^2 \xrightarrow{p_0} A$ along $g: C \to A$ and $f: B \to A$. To make this restriction relationship more transparent, we typically

---

[7] If the pullbacks are defined strictly, then in fact pullback induces an *isomorphism* of functor spaces, but even if $E(b,a)$ is replaced by an equivalent module, the functor spaces are still equivalent, which is enough to induce a bijection on isomorphism classes of objects.

write $A \xrightarrow{\mathrm{Hom}_A} A$ when regarding the arrow ∞-category as a module.

$$
\begin{array}{ccc}
C & \xrightarrow{\mathrm{Hom}_A(f,g)} & B \\
g\downarrow & \Downarrow\rho & \downarrow f \\
A & \xrightarrow[\mathrm{Hom}_A]{} & A
\end{array}
$$

Since the common notation for hom bifunctors places the contravariant variable on the left and the covariant variable on the right, we have adopted a similar notation convention for restrictions in Proposition 8.2.1.

PROPOSITION 8.2.4 (units). *Any object A in* $\mathbb{Mod}(\mathcal{K})$ *is equipped with a nullary* **cocartesian cell**

$$
\begin{array}{ccc}
A & \xLongequal{\quad} & A \\
\| & \Downarrow\iota & \| \\
A & \xrightarrow[\mathrm{Hom}_A]{} & A
\end{array}
$$

characterized by the universal property that any cell in $\mathbb{Mod}(\mathcal{K})$ *whose horizontal source includes the object A factors uniquely through* ι *as below:*

$$
\begin{array}{ccccccc}
X & \xrightarrow{E_1} \cdots \xrightarrow{E_n} & A & \xrightarrow{F_1} \cdots \xrightarrow{F_m} & Y \\
f\downarrow & & \Downarrow & & \downarrow g & =: \\
B & & \xrightarrow[G]{} & & C
\end{array}
$$

$$
\begin{array}{ccccccccc}
X & \xrightarrow{E_1} \cdots \xrightarrow{E_n} & A & \xLongequal{\quad} & A & \xrightarrow{F_1} \cdots \xrightarrow{F_m} & Y \\
\| & \Downarrow\mathrm{id}_{E_1} \|\cdots\| \Downarrow\mathrm{id}_{E_n} & \| & \Downarrow\iota & \| & \Downarrow\mathrm{id}_{F_1} \|\cdots\| \Downarrow\mathrm{id}_{F_m} & \| \\
X & \xrightarrow{E_1} \cdots \xrightarrow{E_n} & A & \xrightarrow{\mathrm{Hom}_A} & A & \xrightarrow{F_1} \cdots \xrightarrow{F_m} & Y \\
f\downarrow & & & \Downarrow\exists! & & & \downarrow g \\
B & & & \xrightarrow[G]{} & & & C
\end{array}
$$

Proof The nullary cell is represented by the map of spans induced by the identity 2-cell at the identity 1-cell at A.

Elsewhere this functor is denoted by $\ulcorner\mathrm{id}_{\mathrm{id}_A}\urcorner\colon A \to A^2$; recall from Example

8.2.3 that we write $A \xrightarrow{\mathsf{Hom}_A} A$ for the module encoded by the arrow ∞-category construction to help us recognize its restrictions.

In the case where both of the sequences E_i and F_j are empty, the one-sided version of the Yoneda lemma for modules given by Theorem 7.4.8 tells us that restriction along this map induces an equivalence of functor spaces

$$\mathsf{Fun}_{A \times A}(\mathsf{Hom}_A, G(g, f)) \xrightarrow{\sim} \mathsf{Fun}_{A \times A}(A, G(g, f)).$$

Taking isomorphism classes of objects gives the bijection of the statement.

In the case where one or both of the sequences are nonempty, we may form their horizontal composite two-sided fibrations $X \xleftarrow{q} E \xrightarrow{p} A$ and $A \xleftarrow{s} F \xrightarrow{r} Y$ and then form either the horizontal composite $E \underset{A}{\times} \mathsf{Hom}_A$ or the horizontal composite $\mathsf{Hom}_A \underset{A}{\times} F$ of the composable triple below:

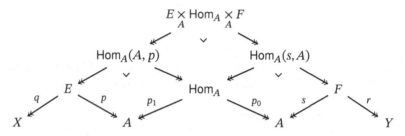

The horizontal composite of the left pair of modules is the two-sided fibration $(qp_1, p_0) \colon \mathsf{Hom}_A(A, p) \twoheadrightarrow X \times A$, while the composite of the right pair is the two-sided fibration $(p_1, rp_0) \colon \mathsf{Hom}_A(s, A) \twoheadrightarrow A \times Y$. By Theorem 7.1.4, these two-sided fibrations give rise to fibered adjunctions:

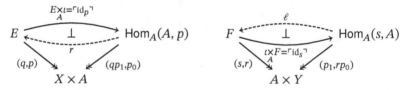

By inspection, the solid-arrow adjoints $\ulcorner\mathrm{id}_p\urcorner$ and $\ulcorner\mathrm{id}_s\urcorner$ can be constructed by pulling back the map $\iota \colon A \to \mathsf{Hom}_A$.

We only require one of these adjunctions, so without loss of generality we use the former. This fibered adjunction pulls back along s and pushes forward along r to define a fibered adjunction

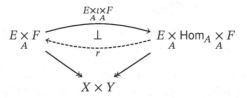

between two-sided fibrations. Upon mapping into the discrete object $G(g, f) \in \mathcal{K}_{/X \times Y}$, this adjunction becomes an adjoint equivalence. In particular, restriction along ι induces an equivalence of Kan complexes

$$\mathrm{Fun}_{X \times Y}(E \underset{A}{\times} \mathrm{Hom}_A \underset{A}{\times} F, G(g, f)) \xrightarrow[\approx]{-\circ(E \underset{A}{\times} \iota \underset{A}{\times} F)} \mathrm{Fun}_{X \times Y}(E \underset{A}{\times} F, G(g, f)),$$

and once again taking isomorphism classes of objects gives the bijection of the statement. □

Propositions 8.2.1 and 8.2.4 imply that the virtual double category of modules is a *virtual equipment* in the sense introduced by Cruttwell and Shulman [32, §7].

DEFINITION 8.2.5. A **virtual equipment** is a virtual double category so that

(i) For any horizontal arrow $A \xrightarrow{E} B$ and pair of vertical arrows $a : A' \to A$ and $b : B' \to B$, there exists a horizontal arrow $B' \xrightarrow{E(b,a)} A'$ and unary cartesian cell ρ satisfying the universal property of Proposition 8.2.1.

(ii) Every object A admits a **unit** horizontal arrow $A \xrightarrow{\mathrm{Hom}_A} A$ equipped with a nullary cocartesian cell ι satisfying the universal property of Proposition 8.2.4.

Thus, Propositions 8.2.1 and 8.2.4 combine to prove:

THEOREM 8.2.6. *The virtual double category* $\mathbb{M}\mathrm{od}(\mathcal{K})$ *of modules in an ∞-cosmos $\mathcal{K}$ is a virtual equipment.* □

By abstract nonsense, the relatively simple axioms (i) and (ii) established in Theorem 8.2.6 establish a robust "calculus of modules." In an effort to familiarize the reader with this little-known categorical structure and expedite the proofs of the formal category theory of ∞-categories in Chapter 9, we devote the remainder of this chapter to proving a plethora of results that actually follow formally from this axiomatization: namely, Lemmas 8.3.10 and 8.3.15, Proposition 8.4.1, Theorem 8.4.4, Corollary 8.4.6, Proposition 8.4.7, Corollary 8.4.8, Corollary 8.4.9, and the bijection of Proposition 8.4.11 between unary cells in the virtual equipment of modules.[8] One additional formal result is left as Exercise 8.4.v for the reader.

NOTATION 8.2.7. The following notational conventions streamline certain virtual equipment diagrams.

- We adopt the convention that an unlabeled unary cell whose vertical boundaries are identities and whose horizontal sources and targets agree is an identity cell.

$$
\begin{array}{ccc}
A \xrightarrow{E} B & & A \xrightarrow{E} B \\
\Big\| \quad\quad \Big\| & := & \Big\| \quad {\Downarrow \mathrm{id}_E} \quad \Big\| \\
A \xrightarrow[E]{} B & & A \xrightarrow[E]{} B
\end{array}
$$

- Cells whose vertical boundary functors are identities and therefore whose source and target spans lie between the same pair of ∞-categories

$$
\mu : E_1 \mathbin{\times} \cdots \mathbin{\times} E_n \Rightarrow E \quad := \quad
\begin{array}{ccc}
A_0 \xrightarrow{E_1} A_1 \xrightarrow{E_2} \cdots \xrightarrow{E_n} A_n \\
\Big\| \quad\quad\quad {\Downarrow \mu} \quad\quad\quad \Big\| \\
A_0 \xrightarrow[E]{\hspace{4cm}} A_n
\end{array}
$$

may be displayed in line using the notation $\mu : E_1 \mathbin{\times} \cdots \mathbin{\times} E_n \Rightarrow E$, an expression which implicitly asserts that the modules appearing in the domain define a compatible sequence, with the symbol "$\times$" meant to suggest the pullback appearing as the horizontal domain of (8.1.9) rather than a product. In the virtual equipment of modules, cells of the form $\mu : E_1 \mathbin{\times} \cdots \mathbin{\times} E_n \Rightarrow E$ correspond to *fibered* maps, in the sense introduced in Definition 8.1.1.

- Given a compatible sequence of modules $E_1 \mathbin{\times} \cdots \mathbin{\times} E_n$ from A_0 to A_n we write $A_0 \twoheadleftarrow \vec{E} \twoheadrightarrow A_n$ for the composite two-sided fibration and similarly abbreviate an n-ary module map (8.1.9) with this sequence as its source as below-left:

$$
\begin{array}{ccc}
A_0 \xdashrightarrow{\vec{E}} A_n & & A_0 \xrightarrow{E_1} A_1 \xrightarrow{E_2} \cdots \xrightarrow{E_n} A_n \\
{\scriptstyle f}\Big\downarrow \quad {\Downarrow \alpha} \quad \Big\downarrow{\scriptstyle g} & := & {\scriptstyle f}\Big\downarrow \quad\quad {\Downarrow \alpha} \quad\quad \Big\downarrow{\scriptstyle g} \\
B_0 \xrightarrow[F]{} B_n & & B_0 \xrightarrow[F]{\hspace{4cm}} B_n
\end{array}
$$

---

[8] A point of minor distinction is that we observe here that the composites referenced in many of these statements are "strong," a notion that has no meaning in a generic virtual equipment.

Exercises

EXERCISE 8.2.i. Given a diagram of functors $E \xrightarrow{k} C \xrightarrow{g} A \xleftarrow{f} B \xleftarrow{h} D$, compute the restriction of the module $C \xrightarrow{\mathrm{Hom}_A(f,g)} B$ along the functors k and h.

EXERCISE 8.2.ii. Prove that unital rings, ring homomorphisms, bimodules, and bimodule maps also define a virtual equipment.

EXERCISE 8.2.iii. Prove that the virtual double category of two-sided isofibrations forms a virtual equipment with the units cells of Remark 8.1.13.

8.3 Composition of Modules

In a virtual equipment, there is no assumption that a generic pair $A \xrightarrow{E} B$ and $B \xrightarrow{F} C$ of horizontal arrows admits a composite but there is a mechanism that recognizes a particular horizontal composite $A \xrightarrow{G} C$ when it happens to exist. When G is the horizontal composite of E and F, we write $E \otimes F \simeq G$ or $\mu \colon E \otimes F \simeq G$ to reinforce the intuition provided by the analogy with bimodules.

In the virtual equipment of modules, there are two possible meanings we can ascribe to a "horizontal composite" of modules. In the first of these, a composition relation $\mu \colon E \otimes F \simeq G$ is witnessed by a fibered module map $\mu \colon E \times F \Rightarrow G$ that defines a cocartesian cell in a sense analogous to the universal property stated in Proposition 8.2.4 (see Definition 8.3.1). This definition of composition can be given in any virtual double category and is well-behaved in any virtual equipment, such as the virtual equipment of modules. Composites of n-ary sequences of composable modules are defined analogously.

In the virtual equipment of modules, all of our formally defined composites satisfy a stronger universal property expressed at the level of the ∞-cosmos. Recall that a fibered map $\mu \colon E \times F \Rightarrow G$ between modules $A \xrightarrow{E} B$, $B \xrightarrow{F} C$, and $A \xrightarrow{G} C$ corresponds to a fibered isomorphism class of maps $\ulcorner\mu\urcorner \colon E \times_B F \to G$ of spans over $A \times C$. Part of what it means to say that μ witnesses a *strong composite* in Definition 8.3.5 is the requirement that its representing functor induces an equivalence of Kan complexes $- \circ \ulcorner\mu\urcorner \colon \mathrm{Fun}_{A \times C}(G, H) \twoheadrightarrow \mathrm{Fun}_{A \times C}(E \times_B F, H)$ for all modules $A \xrightarrow{H} C$. When this is the case, the module G may be understood as the "reflection" of the two-sided isofibration $E \times_B F$ into the subcategory of modules; note however that G and $E \times_B F$ are not necessarily equivalent over $A \times C$. A special case of Theorem 7.3.2, reappearing as Proposition 8.3.11, is one instance of a strong composite.

As this terminology suggests, strong composites are necessarily composites.

Since the weaker universal property of being a composite characterizes the composite module up to fibered equivalence, if we are working in an ∞-cosmos where strong composites are guaranteed to exist, then the 2-categorical universal property we introduce now suffices to detect them.[9]

DEFINITION 8.3.1. A compatible sequence of modules

$$A \xrightarrow{E_1} A_1, A_1 \xrightarrow{E_2} A_2, \ldots, A_{n-1} \xrightarrow{E_n} B$$

admits a **composite** if there exists a module $A \xrightarrow{E} B$ and a cocartesian cell

$$\mu : E_1 \times \cdots \times E_n \Rightarrow E$$

characterized by the universal property that any cell of the form

$$
\begin{array}{ccccccccc}
X & \xrightarrow{F_1} & \cdots & \xrightarrow{F_m} & A & \xrightarrow{E_1} & \cdots & \xrightarrow{E_n} & B & \xrightarrow{G_1} & \cdots & \xrightarrow{G_k} & Y \\
f\downarrow & & & & & & \Downarrow & & & & & & \downarrow g \\
C & & & & & & \xrightarrow[H]{} & & & & & & D
\end{array}
$$

factors uniquely through μ as follows:

$$
\begin{array}{ccccccccccccc}
X & \xrightarrow{F_1} & \cdots & \xrightarrow{F_m} & A & \xrightarrow{E_1} & \cdots & \xrightarrow{E_n} & B & \xrightarrow{G_1} & \cdots & \xrightarrow{G_k} & Y \\
\| & & \|\cdots\| & & \| & & \Downarrow\mu & & \| & & \|\cdots\| & & \| \\
X & \xrightarrow{F_1} & \cdots & \xrightarrow{F_m} & A & & \xrightarrow{E} & & B & \xrightarrow{G_1} & \cdots & \xrightarrow{G_k} & Y \\
f\downarrow & & & & & & \Downarrow\exists! & & & & & & \downarrow g \\
C & & & & & & \xrightarrow[H]{} & & & & & & D
\end{array}
$$

We have already seen one instance of this definition. Proposition 8.2.4 proves that units define nullary composites $\iota : \varnothing \Rightarrow \mathsf{Hom}_A$ in the virtual equipment of modules.

OBSERVATION 8.3.2 (uniqueness of composites). Immediately from this universal property, composites are unique up to vertical isomorphism in the virtual equipment of modules. Recall from Lemma 8.1.17 that parallel modules are vertically isomorphic in the virtual equipment if and only if they are equivalent in the usual fibered sense.

A composite $\mu : E_1 \times \cdots \times E_n \Rightarrow E$ can be used to reduce the domain of a cell by replacing any occurrence of the compatible sequence $E_1 \times \cdots \times E_n$ from A to B by a single module E. Particularly in the case of binary composites,

---

[9] This is analogous to the relationship between the universal properties that define weak p-cartesian transformations and p-cartesian transformations considered in Proposition 5.2.11.

we write $E_1 \otimes E_2$ to denote the **composite** of the modules E_1 and E_2, appearing as the codomain of the cocartesian cell $E_1 \times E_2 \Rightarrow E_1 \otimes E_2$. Note that the universal property of Definition 8.3.1 demands more than simply the condition that $\mu : E_1 \times E_2 \Rightarrow E_1 \otimes E_2$ is universal among binary cells whose source is $E_1 \times E_2$. As observed by Hermida, that weaker universal property is not enough to prove that the tensor product of modules is associative (see [53, §8.1] and Exercise 8.3.ii). By contrast, the universal property of Definition 8.3.1 allows us to use an expression like $E_1 \otimes E_2 \otimes E_3$, without parentheses, for ternary and higher composites, since composition is "associative" in the following sense:

LEMMA 8.3.3. *Suppose the cells* $\mu_i : E_{i1} \times \cdots \times E_{ik_i} \Rightarrow E_i$ *exhibit composites for* $i = 1, \ldots, n$.

(i) *If* $\mu : E_1 \times \cdots \times E_n \Rightarrow E$ *exhibits a composite then the composite cell*

$$E_{11} \times \cdots \times E_{nk_n} \xRightarrow{\mu_1 \times \cdots \times \mu_n} E_1 \times \cdots \times E_n \xRightarrow{\mu} E$$

exhibits E *as a composite of the sequence* $E_{11} \times \cdots \times E_{nk_n}$.

(ii) *If the composite cell*

$$E_{11} \times \cdots \times E_{nk_n} \xRightarrow{\mu_1 \times \cdots \times \mu_n} E_1 \times \cdots \times E_n \xRightarrow{\mu} E$$

exhibits E *as a composite of the sequence* $E_{11} \times \cdots \times E_{nk_n}$, *then* $\mu : E_1 \times \cdots \times E_n \Rightarrow E$ *exhibits* E *as a composite of the sequence* $E_1 \times \cdots \times E_n$.

Proof For (i), the required bijection factors as a composite of $n + 1$ bijections induced by the maps $\mu_1, \ldots, \mu_n$ and μ. For (ii), the required bijection induced by μ composes with the bijections supplied by the maps $\mu_1, \ldots, \mu_n$ to a bijection and is thus itself a bijection. $\square$

REMARK 8.3.4. On account of the universal property of restrictions established in Proposition 8.2.1, to prove that a cell $\mu : E_1 \times \cdots \times E_n \Rightarrow E$ exhibits a composite in a virtual equipment, it suffices to prove the factorization property of Definition 8.3.1 in the case where the vertical functors are identities, i.e., when all of the cells are "fibered."

DEFINITION 8.3.5. A compatible sequence of modules

$$A \xrightarrow{E_1} A_1, A_1 \xrightarrow{E_2} A_2, \ldots, A_{n-1} \xrightarrow{E_n} B$$

admits a **strong composite** if there exists a module $A \xrightarrow{E} B$ and a fibered cell

$$\mu : E_1 \times \cdots \times E_n \Rightarrow E$$

represented by a functor $\ulcorner \mu \urcorner : E_1 \times \cdots \times E_n \to E$ over $A \times B$ that induces an equivalence of Kan complexes

$$\mathsf{Fun}_{X \times Y}(\vec{F} \times E \times \vec{G}, H) \xrightarrow[\sim]{-\circ(\vec{F} \times \ulcorner \mu \urcorner \times \vec{G})} \mathsf{Fun}_{X \times Y}(\vec{F} \times E_1 \times \cdots \times E_n \times \vec{G}, H)$$

for all compatible sequences of modules $F_1, \ldots, F_m$ from X to A and $G_1, \ldots, G_k$ from B to Y and all modules $X \xrightarrow{H} Y$.

LEMMA 8.3.6 (strong composites are composites). *If $\mu : E_1 \times \cdots \times E_n \Rightarrow E$ is a strong composite then μ exhibits E as the composite $E_1 \otimes \cdots \otimes E_n$.*

Proof Given a compatible sequence of modules $F_1, \ldots, F_m$ from X to A and a compatible sequence of modules $G_1, \ldots, G_k$ from B to Y, form the composite two-sided fibrations $X \xleftarrow{q} \vec{F} \xrightarrow{p} A$ and $B \xleftarrow{s} \vec{G} \xrightarrow{r} Y$. By Remark 8.3.4, it suffices to check the universal property of the composite cell for fibered maps whose codomain is a module $X \xrightarrow{H} Y$. The hypothesis that μ is a strong composite provides an equivalence of Kan complexes

$$\mathsf{Fun}_{X \times Y}(\vec{F} \underset{A}{\times} E \underset{B}{\times} \vec{G}, H) \xrightarrow[\sim]{-\circ(\vec{F} \underset{A}{\times} \ulcorner \mu \urcorner \underset{B}{\times} \vec{G})} \mathsf{Fun}_{A \times B}(\vec{F} \underset{A}{\times} E_1 \underset{A_1}{\times} \cdots \underset{A_{n-1}}{\times} E_n \underset{B}{\times} \vec{G}, H)$$

An equivalence of Kan complexes defines a bijection on path components, with each path component corresponding to an isomorphism class of functors by Observation 8.1.2. □

At first blush, the definition of "strong composite" appears unreasonably strong. Examples will arise from the same structure used in the proof of Proposition 8.2.4 to demonstrate that units define nullary composites in the virtual equipment of modules.

LEMMA 8.3.7. *Consider an n-ary module morphism $\mu : E_1 \times \cdots \times E_n \Rightarrow E$ in the virtual equipment of modules whose codomain is a module from A to B. If any representing map of spans*

$$E_1 \times \cdots \times E_n \xrightarrow{\ulcorner \mu \urcorner} E$$
$$\searrow \qquad \swarrow$$
$$A \times B$$

admits a fibered adjoint over $A \times B$, then μ exhibits E as the strong composite $E_1 \otimes \cdots \otimes E_n$.

Proof To verify the universal property of Definition 8.3.5, consider a compatible sequence of modules $F_1, \ldots, F_m$ from X to A and a compatible sequence of modules $G_1, \ldots, G_k$ from B to Y, and form the horizontal composite two-sided fibrations $X \xleftarrow{q} F \xrightarrow{p} A$ and $B \xleftarrow{s} G \xrightarrow{r} Y$. The fibered adjunction of the statement pulls back along $p \times s : F \times G \twoheadrightarrow A \times B$ and pushes forward along $q \times r : F \times G \to X \times Y$ to a fibered adjoint to

$$
F \times E_1 \underset{A_1}{\times} \cdots \underset{A_{n-1}}{\times} E_n \underset{B}{\times} G \xrightarrow{F \times \ulcorner\mu\urcorner \times G \atop A \qquad B} F \underset{A}{\times} E \underset{B}{\times} G
$$
$$
\searrow \qquad \swarrow
$$
$$
X \times Y
$$

Via Remark 8.3.4, it suffices to verify the universal property of the composite for modules $X \xrightarrow{H} Y$. Since modules are discrete, applying $\mathsf{Fun}_{X \times Y}(-, H)$ transforms this fibered adjunction into an adjoint equivalence

$$
\mathsf{Fun}_{X \times Y}(F \underset{A}{\times} E \underset{B}{\times} G, H) \xrightarrow[\sim]{-\circ (F \times \ulcorner\mu\urcorner \times G) \atop A \qquad B} \mathsf{Fun}_{X \times Y}(F \times E_1 \underset{A_1}{\times} \cdots \underset{A_{n-1}}{\times} E_n \underset{B}{\times} G, H).
$$

establishing the universal property that characterizes strong composites. $\square$

Strong composites satisfy an associativity property analogous to Lemma 8.3.3 with a similar proof:

LEMMA 8.3.8. *Suppose the cells $\mu_i : E_{i1} \times \cdots \times E_{ik_i} \Rightarrow E_i$ exhibit strong composites for $i = 1, \ldots, n$.*

(i) *If $\mu : E_1 \times \cdots \times E_n \Rightarrow E$ exhibits a strong composite then the composite cell*

$$
E_{11} \times \cdots \times E_{nk_n} \xRightarrow{\mu_1 \times \cdots \times \mu_n} E_1 \times \cdots \times E_n \xRightarrow{\mu} E
$$

exhibits E as a strong composite of the sequence $E_{11} \times \cdots \times E_{nk_n}$.

(ii) *If the composite cell*

$$
E_{11} \times \cdots \times E_{nk_n} \xRightarrow{\mu_1 \times \cdots \times \mu_n} E_1 \times \cdots \times E_n \xRightarrow{\mu} E
$$

exhibits E as a strong composite of the sequence $E_{11} \times \cdots \times E_{nk_n}$, then $\mu : E_1 \times \cdots \times E_n \Rightarrow E$ exhibits E as a strong composite of the sequence $E_1 \times \cdots \times E_n$.

Proof Exercise 8.3.iii. $\square$

We now turn to examples. The trivial instances of composites are easily verified:

LEMMA 8.3.9. *In the virtual equipment of modules:*

(i) *The units ι: $\varnothing \Rightarrow \mathsf{Hom}_A$ of Proposition 8.2.4 define nullary strong composites.*

(ii) *A unary cell μ: $E \Rightarrow F$ is a composite if and only if it is an isomorphism in the vertical category of modules from A to B and fibered module maps. When this is the case, μ: $E \Rightarrow F$ is a strong composite.*

Proof   Exercise 8.3.iv.   □

As one might hope, the unit modules $A \xrightarrow{\mathsf{Hom}_A} A$ are units for the composition notion introduced in Definition 8.3.1 in the following sense: for any module $A \xrightarrow{E} B$, there is a strong composite relation $\mathsf{Hom}_A \otimes E \otimes \mathsf{Hom}_B \simeq E$.

LEMMA 8.3.10 (composites with units). *For any module $A \xrightarrow{E} B$ the unique cell $\circ$: $\mathsf{Hom}_A \times E \times \mathsf{Hom}_B \Rightarrow E$ defined using the universal property of the unit cell*

$$
\begin{array}{ccccccc}
A & = & A & \xrightarrow{E} & B & = & B \\
\| & \Downarrow\iota & \| & & \| & \Downarrow\iota & \| \\
A & \xrightarrow{\mathsf{Hom}_A} & A & \xrightarrow{E} & B & \xrightarrow{\mathsf{Hom}_B} & B \\
\| & & & \Downarrow\circ & & & \| \\
A & & & \xrightarrow{\hspace{2cm}E\hspace{2cm}} & & & B
\end{array}
\quad :=\quad
\begin{array}{ccc}
A & \xrightarrow{E} & B \\
\| & & \| \\
A & \xrightarrow{E} & B
\end{array}
$$

displays E as the strong composite $\mathsf{Hom}_A \otimes E \otimes \mathsf{Hom}_B$.

Proof   The result is immediate from Lemma 8.3.8(ii) and Lemma 8.3.9.   □

PROPOSITION 8.3.11.   *Let $A \xrightarrow{E} B$ be a module encoded by the span $A \xleftarrow{q} E \xrightarrow{p} B$. Then the binary module map represented by composite left and right adjoints of Theorem 7.1.4(iii)*

$$
\begin{array}{ccc}
A \xrightarrow{\mathsf{Hom}_A(q,A)} E \xrightarrow{\mathsf{Hom}_B(B,p)} B \\
\| \qquad \Downarrow\mu \qquad \| \\
A \xrightarrow{\hspace{3cm}E\hspace{3cm}} B
\end{array}
$$

exhibits a strong composite $\mathsf{Hom}_A(q,A) \otimes \mathsf{Hom}_B(B, p) \simeq E$ expressing the module as the composite of the modules representing its legs.

Proof   The result is immediate from Theorem 7.1.4(iii) and Lemma 8.3.7 applied twice.[10]   □

---

[10] Of course the composite of a left and a right adjoint is not an adjoint but here we are effectively composing adjoint equivalences in which case the direction does not matter.

REMARK 8.3.12. Nothing in the proof of Proposition 8.3.11 requires that the span $A \xleftarrow{q} E \xrightarrow{p} B$ is actually a module, rather than a mere two-sided fibration, except for the interpretation that μ is a binary cell in the virtual equipment of modules. For any two-sided fibration $A \xleftarrow{q} E \xrightarrow{p} B$, it is still the case that restriction along the map of spans $\ulcorner\mu\urcorner$ defines a bijection between maps of two-sided isofibrations whose source includes the span E and whose codomain is a module and maps whose source instead includes $\mathrm{Hom}_A(q, A) \times \mathrm{Hom}_B(B, p)$. In particular, for any compatible sequence of modules $E_1 \times \cdots \times E_n$ from A to B whose span composite defines the two-sided fibration $A \xleftarrow{q} \vec{E} \xrightarrow{p} B$, there is a bijection between module maps in the virtual equipment of modules:

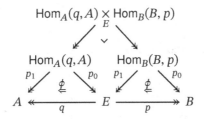

REMARK 8.3.13. The composite map $\mu : \mathrm{Hom}_A(q, A) \times \mathrm{Hom}_B(B, p) \Rightarrow E$ in Proposition 8.3.11 and Remark 8.3.12 can be described in another way. The comma cones for $\mathrm{Hom}_A(q, A)$ and $\mathrm{Hom}_B(B, p)$ define a pair of natural transformations to which the premises of Theorem 7.1.4(iv) apply.

$$\mathrm{Hom}_A(q, A) \underset{E}{\times} \mathrm{Hom}_B(B, p)$$

$$\mathrm{Hom}_A(q, A) \qquad \mathrm{Hom}_B(B, p)$$

$$A \xleftarrow{\quad q \quad} E \xrightarrow{\quad p \quad} B$$

The conclusion of that result asserts that there is a well-defined fibered isomorphism class of functors $\ulcorner\mu\urcorner : \mathrm{Hom}_A(q, A) \times_E \mathrm{Hom}_B(B, p) \to E$ defined by taking the q-cocartesian lift of the left comma cone, composing with the right comma cone, and then taking the codomain of a p-cartesian lift of this composite cell – or by first taking the p-cartesian lift, composing, and then taking the domain of a q-cartesian lift of this composite – this being the functor $\ell r \simeq r\ell$ in the notation of Theorem 7.1.4(iii). In the case where $A \xrightarrow{E} B$ is itself a comma module, the resulting fibered isomorphism class of functors $\ulcorner\mu\urcorner$ is the one that classifies the 2-cell defined by pasting the displayed composite with the comma cone under E.

Any virtual double category has a vertical identity cell for each horizontal arrow $A \xrightarrow{E} B$ whose vertical boundary arrows are identities. In a virtual equipment, we also have a horizontal unary **unit cell** for each vertical arrow $f : A \to B$ whose horizontal boundary is given by the unit modules $A \xrightarrow{\mathrm{Hom}_A} A$ and $B \xrightarrow{\mathrm{Hom}_B} B$ of Proposition 8.2.4.

DEFINITION 8.3.14. Using the unit modules in $\mathbb{M}\mathrm{od}(\mathcal{K})$, for any functor $f : A \to B$ we may define a unary **unit cell** as displayed below-left by appealing to the universal property of the nullary unit cell $\iota : \varnothing \Rightarrow \mathrm{Hom}_A$ for A in the equation below-right:

$$
\begin{array}{ccc}
A \xrightarrow{\mathrm{Hom}_A} A \\
f\downarrow \;\;\Downarrow\mathrm{Hom}_f\;\; \downarrow f \\
B \xrightarrow[\mathrm{Hom}_B]{} B
\end{array}
\qquad \rightsquigarrow \qquad
\begin{array}{c}
A =\!\!=\!\!= A \\
\| \quad \Downarrow\iota \quad \| \\
A \xrightarrow{\mathrm{Hom}_A} A \\
f\downarrow \;\Downarrow\mathrm{Hom}_f\; \downarrow f \\
B \xrightarrow[\mathrm{Hom}_B]{} B
\end{array}
\;\; := \;\;
\begin{array}{c}
A =\!\!=\!\!= A \\
f\downarrow \qquad \downarrow f \\
B =\!\!=\!\!= B \\
\| \quad \Downarrow\iota \quad \| \\
B \xrightarrow[\mathrm{Hom}_B]{} B
\end{array}
$$

In the characterization of Lemma 8.1.16, both sides of the pasting equality defining the unary unit cell correspond to the identity 2-cell $A \underset{f}{\overset{f}{\rightrightarrows}} \Downarrow\mathrm{id}_f\; B$.

As one might hope, the unit cells are units for the vertical composition of cells in the virtual equipment of modules.

LEMMA 8.3.15 (composition with unit cells). *Any cell α as below-right equals the pasted composite below-left:*

$$
\begin{array}{c}
A_0 =\!\!=\!\!= A_0 \cdots\!\overset{\vec{E}}{\to} A_n =\!\!=\!\!= A_n \\
\| \;\;\Downarrow\iota\;\; \| \qquad \| \;\;\Downarrow\iota\;\; \| \\
A_0 \xrightarrow{\mathrm{Hom}_{A_0}} A_0 \cdots\!\overset{\vec{E}}{\to} A_n \xrightarrow{\mathrm{Hom}_{A_n}} A_n \\
f\downarrow \;\Downarrow\mathrm{Hom}_f\; \downarrow f \;\;\Downarrow\alpha\; g\downarrow \;\Downarrow\mathrm{Hom}_g\; \downarrow g \\
B \xrightarrow[\mathrm{Hom}_B]{} B \xrightarrow[E]{} C \xrightarrow[\mathrm{Hom}_C]{} C \\
\| \qquad\qquad \Downarrow\circ \qquad\qquad \| \\
B \xrightarrow[\qquad\qquad E \qquad\qquad]{} C
\end{array}
\qquad = \qquad
\begin{array}{c}
A_0 \cdots\!\overset{\vec{E}}{\to} A_n \\
f\downarrow \;\;\Downarrow\alpha\;\; \downarrow g \\
B \xrightarrow[E]{} C
\end{array}
$$

Proof By Definition 8.3.14 and the laws for composition with identity cells in a virtual double category stated in Definition 8.1.8, the left-hand composite of the statement equals the left-hand composite cell displayed below and the right-hand side of the statement equals the right-hand composite cell displayed below:

$$
\begin{array}{ccccc}
A_0 & \xlongequal{\quad} & A_0 & \xdashrightarrow{\;\vec{E}\;} & A_n & \xlongequal{\quad} & A_n \\
f\downarrow & & \downarrow f & \Downarrow\alpha & \downarrow g & & \downarrow g \\
B & \xlongequal{\quad} & B & \xrightarrow{\;E\;} & C & \xlongequal{\quad} & C \\
\| & \Downarrow\iota & \| & & \| & \Downarrow\iota & \| \\
B & \xrightarrow{\;\mathrm{Hom}_B\;} & B & \xrightarrow{\;E\;} & C & \xrightarrow{\;\mathrm{Hom}_C\;} & C \\
\| & & & \Downarrow\circ & & & \| \\
B & & & \xrightarrow{\qquad E\qquad} & & & C
\end{array}
\qquad = \qquad
\begin{array}{ccc}
A_0 & \xdashrightarrow{\;\vec{E}\;} & A_n \\
f\downarrow & \Downarrow\alpha & \downarrow g \\
B & \xrightarrow{\;E\;} & C \\
\| & & \| \\
B & \xrightarrow{\;E\;} & C
\end{array}
$$

By Lemma 8.3.10, the left-hand side equals the right-hand side. $\qquad\square$

DEFINITION 8.3.16 (horizontal composition of cells). If given a horizontally compatible sequence of unary cells

$$
\begin{array}{ccccccc}
A_0 & \xrightarrow{\;E_1\;} & A_1 & \xrightarrow{\;E_2\;} & \cdots & \xrightarrow{\;E_n\;} & A_n \\
f_0\downarrow & \Downarrow\alpha_1 & f_1\downarrow & \Downarrow\alpha_2 & \Downarrow\cdots & \Downarrow\alpha_n & \downarrow f_n \\
B_0 & \xrightarrow{\;F_1\;} & B_1 & \xrightarrow{\;F_2\;} & \cdots & \xrightarrow{\;F_n\;} & B_n
\end{array}
$$

for which the compatible sequences $E_1 \times \cdots \times E_n$ and $F_1 \times \cdots \times F_n$ both admit composites, then there exists a **horizontal composite unary cell** $\alpha_1 * \cdots * \alpha_n$ that is uniquely determined up to the specification of the composites $\circ : E_1 \times \cdots \times E_n \Rightarrow E$ and $F_1 \times \cdots \times F_n \Rightarrow F$ by the pasting equality:

$$
\begin{array}{ccccccc}
A_0 & \xrightarrow{\;E_1\;} & A_1 & \xrightarrow{\;E_2\;} & \cdots & \xrightarrow{\;E_n\;} & A_n \\
\| & & \Downarrow\circ & & & & \| \\
A_0 & & \xrightarrow{\qquad E\qquad} & & & & A_n \\
f_0\downarrow & & \Downarrow\alpha_1*\cdots*\alpha_n & & & & \downarrow f_n \\
B_0 & & \xrightarrow{\qquad F\qquad} & & & & B_n
\end{array}
\;:=\;
\begin{array}{ccccccc}
A_0 & \xrightarrow{\;E_1\;} & A_1 & \xrightarrow{\;E_2\;} & \cdots & \xrightarrow{\;E_n\;} & A_n \\
f_0\downarrow & \Downarrow\alpha_1 & f_1\downarrow & \Downarrow\alpha_2 & \Downarrow\cdots & \Downarrow\alpha_n & \downarrow f_n \\
B_0 & \xrightarrow{\;F_1\;} & B_1 & \xrightarrow{\;F_2\;} & \cdots & \xrightarrow{\;F_n\;} & B_n \\
\| & & & \Downarrow\circ & & & \| \\
B_0 & & & \xrightarrow{\qquad F\qquad} & & & B_n
\end{array}
$$

(8.3.17)

By an argument very similar to the proof of Lemma 8.3.15 using the composites of Lemma 8.3.10, the horizontal composite $\mathrm{Hom}_f * \alpha * \mathrm{Hom}_g$ of a unary cell α with the unit cells Hom_f and Hom_g at its vertical boundary functors recovers α (see Exercise 8.3.v). Using the horizontal composition of unary cells of Definition 8.3.16, we can understand $\mathrm{Mod}(\mathcal{K})$ to contain various "vertical" and "horizontal" bicategories.

PROPOSITION 8.3.18 (the vertical 2-category in the virtual equipment). *Any virtual equipment contains a **vertical 2-category** whose objects are the objects of the virtual equipment, whose arrows are the vertical arrows, and whose 2-cells are those unary cells*

$$
\begin{array}{ccc}
A & \xrightarrow{\mathrm{Hom}_A} & A \\
g\downarrow & \Downarrow\alpha & \downarrow f \\
B & \xrightarrow[\mathrm{Hom}_B]{} & B
\end{array}
$$

whose horizontal boundary arrows are given by the unit modules.

Proof   To prove that these structures define a 2-category we must – adopting the standard terminology from Definition B.1.1 – define "horizontal" composition of 2-cells (composing along a boundary 0-cell) and "vertical" composition of 2-cells (composing along a boundary 1-cell). The "horizontal" composition in the 2-category is defined via the vertical composition in the virtual double category described in Definition 8.1.8. The "vertical" composition in the 2-category is defined by Definition 8.3.16. To see that this yields a 2-category and not a bicategory note that any bicategory in which the composition of 1-cells is strictly associative and unital is a 2-category; in this case, the 1-cells are the vertical arrows of the virtual double category, which do indeed compose strictly.   □

In Proposition 8.4.11 we prove that the vertical 2-category in the virtual equipment $\mathbb{Mod}(\mathcal{K})$ is isomorphic to the homotopy 2-category $\mathfrak{h}\mathcal{K}$.

REMARK 8.3.19 (horizontal bicategories in the virtual equipment). Via Definition 8.3.16, a virtual equipment can also be understood to contain various "horizontal" bicategories, defined by taking the 1-cells to be composable modules and the 2-cells to be unary module maps whose vertical boundary functors are identities. Particular horizontal bicategories of interest are described in Definition 8.4.12.

Exercises

EXERCISE 8.3.i. Extending Exercise 8.1.ii, prove that in a virtual double category arising from an actual double category, every compatible sequence of horizontal arrows admits a composite in the sense of Definition 8.3.1.

EXERCISE 8.3.ii ([53, 8.5],[76, 3.3.4]).

  (i) Suppose that for every compatible sequence $E_1 \times \cdots \times E_n$ in a virtual equipment there exists a cell $\mu : E_1 \times \cdots \times E_n \Rightarrow E$ that is *weakly cocartesian* meaning that every n-ary cell $E_1 \times \cdots \times E_n \Rightarrow H$ factors uniquely through μ via a unary cell $E \Rightarrow H$. If moreover the weakly cocartesian cells are closed under composition in the virtual double category, show that every weakly cocartesian cell is in fact cocartesian, satisfying the universal property of Definition 8.3.1, and conclude that the virtual equipment then admits all composites.

(ii) Suppose $\mathcal{K}$ is an ∞-cosmos in which every two-sided fibration $A \xleftarrow{q} E \xrightarrow{p} B$ can be reflected into a module, meaning that that there exists a map of spans with $\bar{E}$ a module so that for every module $A \xrightarrow{H} B$, μ induces an equivalence of Kan complexes

$$
\begin{array}{c}
\begin{array}{ccc}
 & E & \\
{}^{q}\swarrow & \downarrow \mu & \searrow^{p} \\
A & & B \\
{}_{\bar{q}}\nwarrow & \downarrow & \nearrow_{\bar{p}} \\
 & \bar{E} &
\end{array}
\quad \rightsquigarrow \quad
\mathsf{Fun}_{A \times B}(\bar{E}, H) \xrightarrow[\sim]{-\circ\mu} \mathsf{Fun}_{A \times B}(E, H).
\end{array}
$$

Conclude that the virtual equipment $\mathsf{Mod}(\mathcal{K})$ has all composites.

EXERCISE 8.3.iii. Prove Lemma 8.3.8.

EXERCISE 8.3.iv. Prove Lemma 8.3.9.

EXERCISE 8.3.v. For any unary cell as displayed show that the horizontal composite $\mathsf{Hom}_f * \alpha * \mathsf{Hom}_g$ equals α.

$$
\begin{array}{ccc}
A & \xrightarrow{E} & B \\
f \downarrow & \Downarrow \alpha & \downarrow g \\
C & \xrightarrow{F} & D
\end{array}
$$

8.4 Representable Modules

Any vertical arrow $f : A \to B$ in a virtual equipment has a pair of associated horizontal arrows $B \xrightarrow{\mathsf{Hom}_B(f,B)} A$ and $A \xrightarrow{\mathsf{Hom}_B(B,f)} B$, defined as restrictions of the horizontal unit arrows, that have universal properties similar to *companions* and *conjoints* in an ordinary double category [47]. In the virtual equipment of modules, these are sensibly referred to as the *left* and *right representations* of a functor as a module, using the familiar terminology and notation because these coincide exactly with the left and right representables introduced in Definitions 3.5.1 and 7.4.7.

PROPOSITION 8.4.1 (companion and conjoint relations for representables). *To any functor $f : A \to B$ in the virtual equipment of modules, there exist canonical restriction cells displayed below-left and application cells displayed below-right*

$$
\begin{array}{cc}
\begin{array}{ccc}
B & \xrightarrow{\mathrm{Hom}_B(f,B)} & A \\
\Big\| & \Downarrow\rho & \Big\downarrow f \\
B & \xrightarrow[\mathrm{Hom}_B]{} & B
\end{array}
&
\begin{array}{ccc}
A & \xrightarrow{\mathrm{Hom}_B(B,f)} & B \\
f\Big\downarrow & \Downarrow\rho & \Big\| \\
B & \xrightarrow[\mathrm{Hom}_B]{} & B
\end{array}
\\[3em]
\begin{array}{ccc}
A & \xrightarrow{\mathrm{Hom}_A} & A \\
f\Big\downarrow & \Downarrow\kappa & \Big\| \\
B & \xrightarrow[\mathrm{Hom}_B(f,B)]{} & A
\end{array}
&
\begin{array}{ccc}
A & \xrightarrow{\mathrm{Hom}_A} & A \\
\Big\| & \Downarrow\kappa & \Big\downarrow f \\
A & \xrightarrow[\mathrm{Hom}_B(B,f)]{} & B
\end{array}
\end{array}
$$

defining unary maps between the unit modules $A \xrightarrow{\mathrm{Hom}_A} A$ and $B \xrightarrow{\mathrm{Hom}_B} B$ and the left and right representable modules $B \xrightarrow{\mathrm{Hom}_B(f,B)} A$ and $A \xrightarrow{\mathrm{Hom}_B(B,f)} B$. These satisfy the identities:

$$
\begin{array}{ccccc}
\begin{array}{ccc}
A & \xrightarrow{\mathrm{Hom}_A} & A \\
f\Big\downarrow & \Downarrow\kappa & \Big\| \\
B & \xrightarrow{\mathrm{Hom}_B(f,B)} & A \\
\Big\| & \Downarrow\rho & \Big\downarrow f \\
B & \xrightarrow[\mathrm{Hom}_B]{} & B
\end{array}
& = &
\begin{array}{ccc}
A & \xrightarrow{\mathrm{Hom}_A} & A \\
f\Big\downarrow & \Downarrow\mathrm{Hom}_f & \Big\downarrow f \\
B & \xrightarrow[\mathrm{Hom}_B]{} & B
\end{array}
& = &
\begin{array}{ccc}
A & \xrightarrow{\mathrm{Hom}_A} & A \\
\Big\| & \Downarrow\kappa & \Big\downarrow f \\
A & \xrightarrow{\mathrm{Hom}_B(B,f)} & B \\
f\Big\downarrow & \Downarrow\rho & \Big\| \\
B & \xrightarrow[\mathrm{Hom}_B]{} & B
\end{array}
\end{array} \tag{8.4.2}
$$

and

$$
\begin{array}{ccc}
\begin{array}{ccccc}
B & \xrightarrow{\mathrm{Hom}_B(f,B)} & A & \xrightarrow{\mathrm{Hom}_A} & A \\
\Big\| & \Downarrow\rho & f\Big\downarrow & \Downarrow\kappa & \Big\| \\
B & \xrightarrow{\mathrm{Hom}_B} & B & \xrightarrow{\mathrm{Hom}_B(f,B)} & A \\
\Big\| & & \Downarrow\circ & & \Big\| \\
B & & \xrightarrow[\mathrm{Hom}_B(f,B)]{} & & A
\end{array}
& = &
\begin{array}{ccc}
B & \xrightarrow{\mathrm{Hom}_B(f,B)} A \xrightarrow{\mathrm{Hom}_A} & A \\
\Big\| & \Downarrow\circ & \Big\| \\
B & \xrightarrow[\mathrm{Hom}_B(f,B)]{} & A
\end{array}
\\[4em]
\begin{array}{ccc}
A & \xrightarrow{\mathrm{Hom}_A} A \xrightarrow{\mathrm{Hom}_B(B,f)} & B \\
\Big\| & \Downarrow\circ & \Big\| \\
A & \xrightarrow[\mathrm{Hom}_B(B,f)]{} & B
\end{array}
& = &
\begin{array}{ccccc}
A & \xrightarrow{\mathrm{Hom}_A} & A & \xrightarrow{\mathrm{Hom}_B(B,f)} & B \\
\Big\| & \Downarrow\kappa & f\Big\downarrow & \Downarrow\rho & \Big\| \\
A & \xrightarrow{\mathrm{Hom}_B(B,f)} & B & \xrightarrow{\mathrm{Hom}_B} & A \\
\Big\| & & \Downarrow\circ & & \Big\| \\
A & & \xrightarrow[\mathrm{Hom}_B(B,f)]{} & & B
\end{array}
\end{array} \tag{8.4.3}
$$

Proof The unary module maps κ are defined by the equations (8.4.2) by appealing to the universal property of the restriction cells in Proposition 8.2.1. The relations (8.4.3) could also be verified directly from the axioms of Theorem 8.2.6 via Propositions 8.2.4 and Lemmas 8.3.10 and 8.3.15. For sake of variety, we appeal to Lemma 8.1.16 to characterize each of the cells in the virtual equipment as natural transformations in the homotopy 2-category.

We prove this for the right representables. By Remark 8.3.13, the binary module morphism on the left-hand side of the rightmost equality of (8.4.3) represents the natural transformation below-left, while the right-hand composite is below-right:

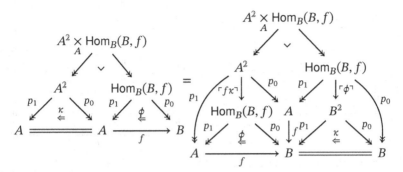

By the definition of the induced functors $\phi^\ulcorner f\kappa^\urcorner = f\kappa$ and $\phi = \kappa^\ulcorner \phi^\urcorner$. Thus, the left-hand side equals the right-hand side. $\qquad\square$

The companions and conjoints of Proposition 8.4.1 can be deployed to "bend" vertical functors into horizontal modules, producing bijections between cells whose boundaries involve these arrows. The following result is often called the "spider lemma," in reference to a graphical calculus that can be used to illustrate virtual equipment cells and their composites [87].

THEOREM 8.4.4 (spider lemma). *In the virtual equipment of modules there are natural bijections between cells of the following four forms implemented by composing with the canonical cells κ and ρ of Proposition 8.4.1 and with the composition and nullary cells associated to the units.*

$$
\begin{array}{ccc}
A & \xrightarrow{\;\vec{E}\;} & B \\
{\scriptstyle f}\big\downarrow & \Downarrow & \big\downarrow{\scriptstyle g} \\
C & \xrightarrow[F]{} & D
\end{array}
$$

$$
\begin{array}{ccccc}
C \xrightarrow{\mathrm{Hom}_C(f,C)} A & \xrightarrow{\;\vec{E}\;} & B & \qquad & A \xrightarrow{\;\vec{E}\;} B \xrightarrow{\mathrm{Hom}_D(D,g)} D \\
\big\| \qquad\qquad \Downarrow & & \big\downarrow{\scriptstyle g} & \qquad & {\scriptstyle f}\big\downarrow \qquad \Downarrow \qquad\quad \big\| \\
C \xrightarrow[\qquad F \qquad]{} & & D & \qquad & C \xrightarrow[\qquad F \qquad]{} D
\end{array}
$$

$$
\begin{array}{c}
C \xrightarrow{\mathrm{Hom}_C(f,C)} A \xrightarrow{\;\vec{E}\;} B \xrightarrow{\mathrm{Hom}_D(D,g)} D \\
\big\| \qquad\qquad \Downarrow \qquad\qquad \big\| \\
C \xrightarrow[\qquad\qquad F \qquad\qquad]{} D
\end{array}
$$

Proof The composite bijection displayed vertically in the statement carries cells α and β to the cells displayed below-left and below-right, respectively:

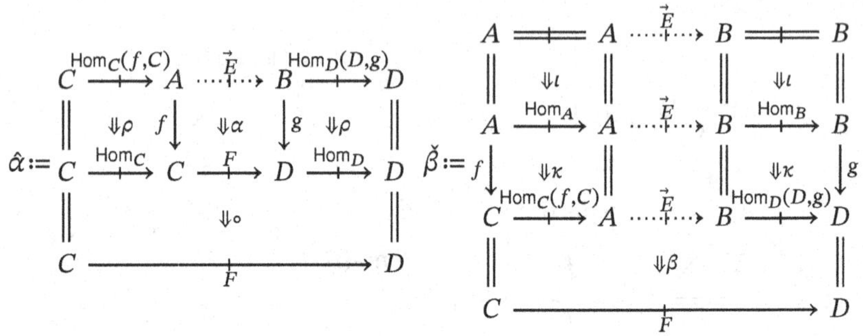

$$
\hat{\alpha} :=
\begin{array}{c}
C \xrightarrow{\mathrm{Hom}_C(f,C)} A \dashrightarrow^{\vec{E}} B \xrightarrow{\mathrm{Hom}_D(D,g)} D \\
\Vert \quad \Downarrow\rho \quad f\downarrow \quad \Downarrow\alpha \quad \downarrow g \quad \Downarrow\rho \quad \Vert \\
C \xrightarrow{\mathrm{Hom}_C} C \xrightarrow{F} D \xrightarrow{\mathrm{Hom}_D} D \\
\Vert \quad\quad \Downarrow\circ \quad\quad \Vert \\
C \xrightarrow{\qquad F \qquad} D
\end{array}
$$

$$
\check{\beta} :=
\begin{array}{c}
A = A \dashrightarrow^{\vec{E}} B = B \\
\Vert \quad \Downarrow\iota \quad \Vert \quad\quad \Vert \quad \Downarrow\iota \quad \Vert \\
A \xrightarrow{\mathrm{Hom}_A} A \dashrightarrow^{\vec{E}} B \xrightarrow{\mathrm{Hom}_B} B \\
f\downarrow \quad \Downarrow\kappa \quad \Vert \quad\quad \Vert \quad \Downarrow\kappa \quad \downarrow g \\
C \xrightarrow{\mathrm{Hom}_C(f,C)} A \dashrightarrow^{\vec{E}} B \xrightarrow{\mathrm{Hom}_D(D,g)} D \\
\Vert \quad\quad \Downarrow\beta \quad\quad \Vert \\
C \xrightarrow{\qquad F \qquad} D
\end{array}
$$

By Proposition 8.4.1 and Lemma 8.3.15

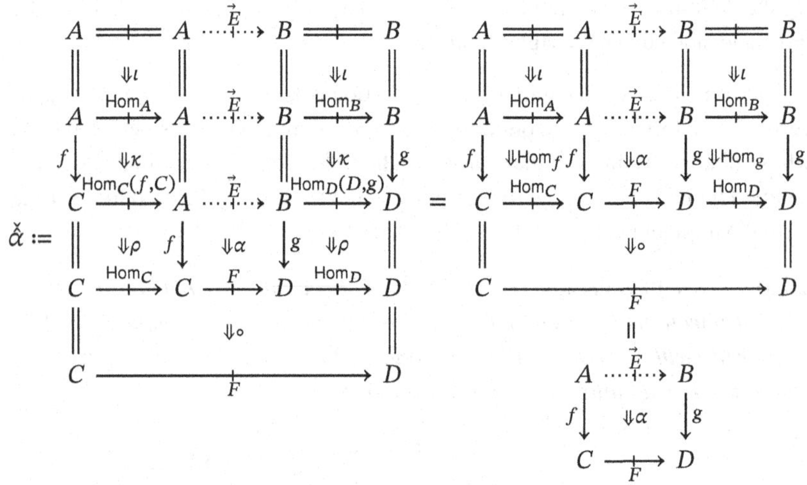

$$
\check{\hat{\alpha}} :=
\begin{array}{c}
A = A \dashrightarrow^{\vec{E}} B = B \\
\Vert \ \Downarrow\iota \ \Vert \quad\quad \Vert \ \Downarrow\iota \ \Vert \\
A \xrightarrow{\mathrm{Hom}_A} A \dashrightarrow^{\vec{E}} B \xrightarrow{\mathrm{Hom}_B} B \\
f\downarrow \ \Downarrow\kappa \ \Vert \quad\quad \Vert \ \Downarrow\kappa \ \downarrow g \\
C \xrightarrow{\mathrm{Hom}_C(f,C)} A \dashrightarrow^{\vec{E}} B \xrightarrow{\mathrm{Hom}_D(D,g)} D \\
\Vert \ \Downarrow\rho \ f\downarrow \ \Downarrow\alpha \ \downarrow g \ \Downarrow\rho \ \Vert \\
C \xrightarrow{\mathrm{Hom}_C} C \xrightarrow{F} D \xrightarrow{\mathrm{Hom}_D} D \\
\Vert \quad \Downarrow\circ \quad \Vert \\
C \xrightarrow{\qquad F \qquad} D
\end{array}
\ = \
\begin{array}{c}
A = A \dashrightarrow^{\vec{E}} B = B \\
\Vert \ \Downarrow\iota \ \Vert \quad\quad \Vert \ \Downarrow\iota \ \Vert \\
A \xrightarrow{\mathrm{Hom}_A} A \dashrightarrow^{\vec{E}} B \xrightarrow{\mathrm{Hom}_B} B \\
f\downarrow \ \Downarrow\mathrm{Hom}_f f\downarrow \ \Downarrow\alpha \ \downarrow g \ \Downarrow\mathrm{Hom}_g \ \downarrow g \\
C \xrightarrow{\mathrm{Hom}_C} C \xrightarrow{F} D \xrightarrow{\mathrm{Hom}_D} D \\
\Vert \quad \Downarrow\circ \quad \Vert \\
C \xrightarrow{\qquad F \qquad} D \\
\Vert \\
A \dashrightarrow^{\vec{E}} B \\
f\downarrow \ \Downarrow\alpha \ \downarrow g \\
C \xrightarrow{F} D
\end{array}
$$

The other composite is:

$$\overset{\circ}{\beta} := \qquad (8.4.5)$$

By Lemma 8.3.10, we have

since both composites equal β. By the universal property of the unit cells in Proposition 8.2.4, the bottom two rows of these diagrams are equal, so we may substitute the bottom two rows of the right-hand diagram for the bottom two rows of (8.4.5) to obtain:

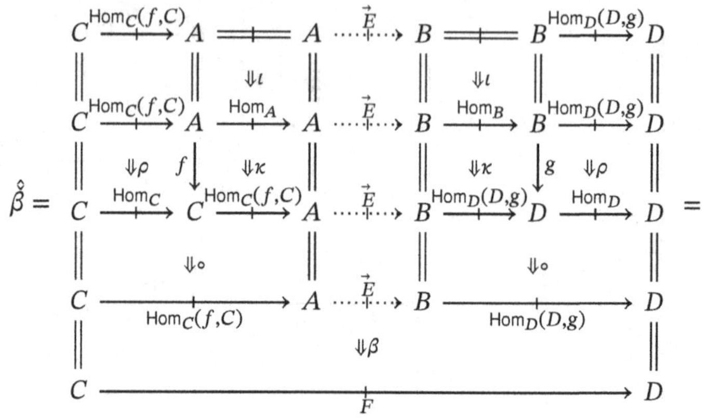

$$\mathring{\beta} =$$

By Proposition 8.4.1 and Lemma 8.3.10 this reduces to β.

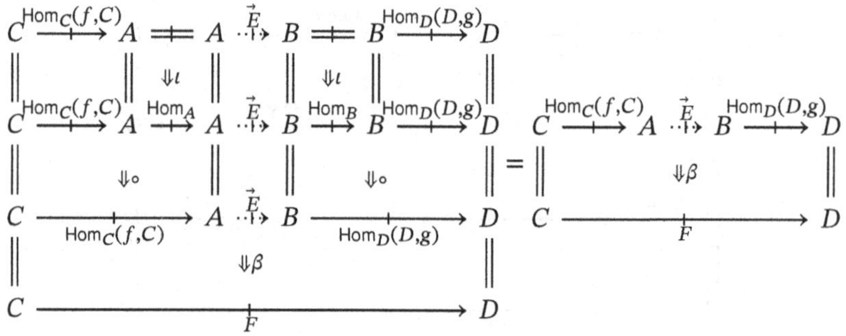

By vertically bisecting this construction, one obtains the one-sided bijections of the statement. □

We frequently apply this result in the following form:

COROLLARY 8.4.6. *Given a compatible sequence of modules* $E_1 \times \cdots \times E_n$ *from A to B, a module F from C to D, and functors* $f : A \to C$ *and* $g : B \to D$, *there is a bijection between fibered module maps* $\mathrm{Hom}_C(f, C) \times E_1 \times \cdots \times E_n \times \mathrm{Hom}_D(D, g) \Rightarrow F$ *and fibered module maps* $E_1 \times \cdots \times E_n \Rightarrow F(g, f)$, *i.e., between cells*

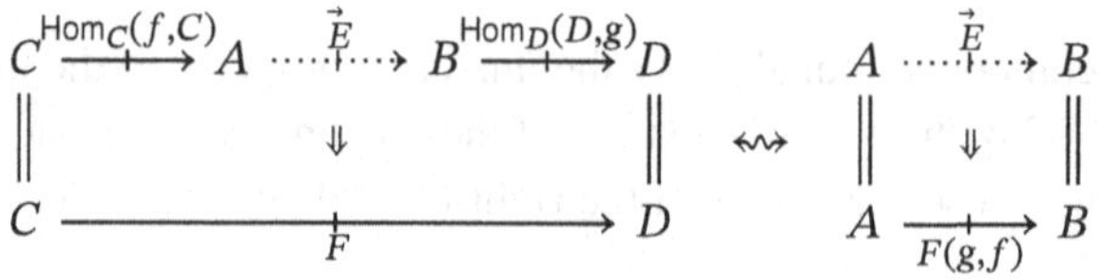

Proof　Combine Theorem 8.4.4 with Proposition 8.2.1.　□

PROPOSITION 8.4.7. *For any module $A \xrightarrow{E} B$ and pair of functors $a \colon X \to A$ and $b \colon Y \to B$, the strong composite $\mathsf{Hom}_A(A, a) \otimes E \otimes \mathsf{Hom}_B(b, B)$ exists and is given by the restriction $E(b, a)$, with the ternary strong composite map $\mu \colon \mathsf{Hom}_A(A, a) \times E \times \mathsf{Hom}_B(b, B) \Rightarrow E(b, a)$ defined by the universal property of the restriction by the pasting diagram:*

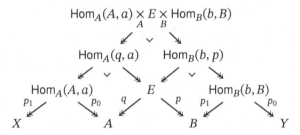

Proof The horizontal composite two-sided fibration of the compatible sequence is

$$
\begin{array}{c}
\mathsf{Hom}_A(A, a) \underset{A}{\times} E \underset{B}{\times} \mathsf{Hom}_B(b, B) \\
\swarrow \quad \vee \quad \searrow \\
\mathsf{Hom}_A(q, a) \qquad \mathsf{Hom}_B(b, p) \\
\swarrow \quad \vee \quad \searrow \quad \swarrow \quad \vee \quad \searrow \\
\mathsf{Hom}_A(A, a) \qquad E \qquad \mathsf{Hom}_B(b, B) \\
{}_{p_1}\swarrow \quad \searrow{}_{p_0} \quad {}_{q}\swarrow \quad \searrow{}_{p} \quad {}_{p_1}\swarrow \quad \searrow{}_{p_0} \\
X \qquad\qquad A \qquad\qquad B \qquad\qquad Y
\end{array}
$$

from which we see that the binary composite cell

$$
\begin{array}{c}
\mathsf{Hom}_A(q, A) \underset{E}{\times} \mathsf{Hom}_B(B, p) \longrightarrow E \\
\searrow \qquad \swarrow {}_{(q,p)} \\
A \times B
\end{array}
$$

of Proposition 8.3.11 pulls back along $a \times b \colon X \times Y \to A \times B$ to define the map μ. By Lemma 8.3.7, we conclude that $\mu \colon \mathsf{Hom}_A(A, a) \times E \times \mathsf{Hom}_B(b, B) \Rightarrow E(b, a)$ is a strong composite. □

When one of the three modules appearing in the source of the ternary composite of Proposition 8.4.7 is a unit module, we can use Lemma 8.3.8 and the nullary composite cell associated with the unit to reduce this ternary composite to a binary composite (see Exercise 8.4.iii). This argument proves the following pair of corollaries.

COROLLARY 8.4.8. *For any cospan of functors $C \xrightarrow{g} A \xleftarrow{f} B$, the comma module $\mathsf{Hom}_A(f, g)$ factors as a strong composite*

$$
\mathsf{Hom}_A(A, g) \otimes \mathsf{Hom}_A(f, A) \simeq \mathsf{Hom}_A(f, g). \qquad \square
$$

A similar reduction (see Exercise 8.2.i) proves that right representable modules can always be composed with each other and dually left representable modules can always be composed with each other:

COROLLARY 8.4.9. *Any composable pair of functors* $A \xrightarrow{f} B \xrightarrow{g} C$ *defines a strongly composable pair of right represented modules and a strongly composable pair of left represented modules*

$$A \xrightarrow{\mathrm{Hom}_B(B,f)} B \xrightarrow{\mathrm{Hom}_C(C,g)} C \qquad C \xrightarrow{\mathrm{Hom}_C(g,C)} B \xrightarrow{\mathrm{Hom}_B(f,B)} A$$

and moreover the strong composites are again represented:

$$\mathrm{Hom}_B(B,f) \otimes \mathrm{Hom}_C(C,g) \simeq \mathrm{Hom}_C(C,gf) \quad and$$
$$\mathrm{Hom}_C(g,C) \otimes \mathrm{Hom}_B(f,B) \simeq \mathrm{Hom}_C(gf,C). \qquad \square$$

This result combines with Proposition 8.3.11 to prove a generalization of Theorem 3.5.12, which allows us to detect representable modules.

PROPOSITION 8.4.10. *Let* $A \xleftarrow{q} E \xrightarrow{p} B$ *encode a module.*

(i) *The module* $A \xrightarrow{E} B$ *is right representable just when its left leg* $q : E \twoheadrightarrow A$ *has a right adjoint* $r : A \to E$ *in which case* $E \simeq \mathrm{Hom}_B(B, pr)$.

(ii) *The module* $A \xrightarrow{E} B$ *is left representable just when its right leg* $p : E \twoheadrightarrow B$ *has a left adjoint* $\ell : B \to E$ *in which case* $E \simeq \mathrm{Hom}_A(q\ell, A)$.

Proof By Lemma 3.5.9 and Theorem 3.5.12, the claimed adjoints and fibered equivalences exist for representable modules, so it remains only to prove the converse. By Proposition 8.3.11, any module E can be expressed as a composite $\mathrm{Hom}_A(q, A) \otimes \mathrm{Hom}_B(B, p) \simeq E$ of the left representation of its left leg followed by the right representation of its right leg. If $q \dashv r$, then by Proposition 4.1.1 $\mathrm{Hom}_A(q, A) \simeq \mathrm{Hom}_E(E, r)$ as modules from A to E, so by Lemma 8.3.3, $\mathrm{Hom}_E(E, r) \otimes \mathrm{Hom}_B(B, p) \simeq E$. By Corollary 8.4.9, $\mathrm{Hom}_E(E, r) \otimes \mathrm{Hom}_B(B, p) \simeq \mathrm{Hom}_B(B, pr)$ so by Lemma 8.3.3 again, $\mathrm{Hom}_B(B, pr) \simeq E$. $\square$

Finally, we revisit the "cheap" version of the Yoneda lemma presented in Corollary 3.5.11, which encodes natural transformations in the homotopy 2-category as maps of represented modules.

PROPOSITION 8.4.11. *For any parallel pair of functors there are natural bijections between 2-cells in the homotopy 2-category*

$$A \underset{g}{\overset{f}{\rightrightarrows}}{\Downarrow\alpha} B$$

and cells in the virtual equipment of modules:

$$
\begin{array}{ccccc}
A \xrightarrow{\operatorname{Hom}_B(B,f)} B & \qquad & A \xrightarrow{\operatorname{Hom}_A} A & \qquad & B \xrightarrow{\operatorname{Hom}_B(g,B)} A \\
\| \quad \Downarrow\alpha_* \quad \| & \rightsquigarrow \quad g\Big\downarrow \quad \Downarrow\bar{\alpha} \quad \Big\downarrow f \quad \rightsquigarrow & \| \quad \Downarrow\alpha^* \quad \| \\
A \xrightarrow[\operatorname{Hom}_B(B,g)]{} B & & B \xrightarrow[\operatorname{Hom}_B]{} B & & B \xrightarrow[\operatorname{Hom}_B(f,B)]{} A
\end{array}
$$

Proof   The bijections in $\mathsf{Mod}(\mathcal{K})$ can then be derived from Theorem 8.4.4 and Corollary 8.4.6. The bijection between 2-cells in the homotopy 2-category and unary module maps between left or right representables is simply a restatement of Corollary 3.5.11.   □

DEFINITION 8.4.12 (the covariant and contravariant embeddings). Proposition 8.4.11 defines the action on 2-cells of two identity-on-objects locally fully faithful homomorphisms

$$
\mathfrak{h}\mathcal{K} \lhook\joinrel\longrightarrow \mathsf{Mod}(\mathcal{K}) \qquad\qquad \mathfrak{h}\mathcal{K}^{\mathrm{coop}} \lhook\joinrel\longrightarrow \mathsf{Mod}(\mathcal{K})
$$

$$
A \xrightarrow{f} B \longmapsto A \xrightarrow{\operatorname{Hom}_B(B,f)} B \qquad\qquad A \xrightarrow{f} B \longmapsto B \xrightarrow{\operatorname{Hom}_B(f,B)} A
$$

that embed the homotopy 2-category fully faithfully into the sub "bicategory" of $\mathsf{Mod}(\mathcal{K})$ containing only those unary cells whose vertical boundaries are identities.

This substructure of $\mathsf{Mod}(\mathcal{K})$ is not quite a bicategory because not all horizontally composable modules can be composed, but if we restrict only to the right representable modules or only to the left representable modules, then by Corollary 8.4.9 the composites do exist and the embeddings define genuine bicategorical homomorphisms: given $A \xrightarrow{f} B \xrightarrow{g} C$ we have $\operatorname{Hom}_B(B,f) \otimes \operatorname{Hom}_C(C,g) \simeq \operatorname{Hom}_C(C,gf)$ and $\operatorname{Hom}_C(g,C) \otimes \operatorname{Hom}_B(f,B) \simeq \operatorname{Hom}_C(gf,C)$. We refer to these as the **covariant** and **contravariant** embeddings, respectively.

REMARK 8.4.13. In addition to the covariant and contravariant embeddings, there is a third locally fully faithful embedding of the homotopy 2-category $\mathfrak{h}\mathcal{K}$ into $\mathsf{Mod}(\mathcal{K})$ that is identity on objects, sends $f : A \to B$ to the corresponding vertical 1-cell, and uses the third bijection of Proposition 8.4.11 to define the action on 2-cells. Since $\operatorname{Hom}_A \otimes \operatorname{Hom}_A \simeq \operatorname{Hom}_A$, the unary cells in this image of this embedding can be composed horizontally as well as vertically, and this embedding is functorial in both directions: vertical composites of natural transformations in $\mathfrak{h}\mathcal{K}$ coincide with horizontal composites of unary cells and horizontal composites of natural transformations in $\mathfrak{h}\mathcal{K}$ coincide with vertical composites of unary cells. The image is precisely the vertical 2-category of Proposition 8.3.18. We make much greater use of the covariant and contravariant embeddings of Definition 8.4.12 however.

Exercises

EXERCISE 8.4.i. Prove Proposition 8.4.1 in any virtual equipment, without appealing to Lemma 8.1.16.

EXERCISE 8.4.ii. State and prove a generalization of Theorem 8.4.4 in which the bottom $n + 2$-ary cell appearing in the statement is not fibered, but has vertical boundary defined by an arbitrary pair of nonidentity functors.

EXERCISE 8.4.iii. Prove a binary version of Proposition 8.4.7: that for any module $A \xrightarrow{E} B$ and any functors $a : X \to A$ and $b : Y \to B$ the strong composites $\mathrm{Hom}_A(A, a) \otimes E$ and $E \otimes \mathrm{Hom}_B(b, B)$ exist and are given by the restrictions $E(1, a)$ and $E(b, 1)$, respectively.

EXERCISE 8.4.iv. Prove the nonstrong version of Proposition 8.4.7 in any virtual equipment, without appealing to Lemma 8.3.7.

EXERCISE 8.4.v. For any functor $f : A \to B$, define a canonical unary cell $\eta : \mathrm{Hom}_A \Rightarrow \mathrm{Hom}_B(f, f) \simeq \mathrm{Hom}_B(B, f) \otimes \mathrm{Hom}_B(f, B)$ and a binary cell $\epsilon : \mathrm{Hom}_B(f, B) \times \mathrm{Hom}_B(B, f) \Rightarrow \mathrm{Hom}_B$. Use this data to demonstrate that the modules $A \xrightarrow{\mathrm{Hom}_B(B, f)} B$ and $B \xrightarrow{\mathrm{Hom}_B(f, B)} A$ are "adjoint" in a suitable sense.

9

Formal ∞-Category Theory in a Virtual Equipment

Mac Lane famously asserted that "all concepts are Kan extensions" [81, §X.7], at least in category theory. Right and left extensions of a functor $f : A \to C$ along a functor $k : A \to B$ can be defined internally to any 2-category (see Definition 9.1.1) – at this level of generality the eponym "Kan" is typically dropped. However, in the homotopy 2-category of an ∞-cosmos, the universal property defining left and right extensions is too weak, and indeed the correct universal property is associated to the stronger notion of a *pointwise extension*, for which the values of a right or left extension at an element of B can be computed as limits or colimits indexed by the appropriate comma ∞-category (see Proposition 9.4.9 for a precise statement). Indeed, Kelly later amended Mac Lane's assertion, arguing that the *pointwise Kan extensions*, which he calls simply "Kan extensions" are the important ones, writing "Our present choice of nomenclature is based on our failure to find a single instance where a [nonpointwise] Kan extension plays any mathematical role whatsoever" [68, §4].

Using the calculus of modules, we can now add the theory of pointwise Kan extensions of functors between ∞-categories to the basic ∞-category theory developed in Part I. In fact, we give multiple definitions of pointwise extension. One is fundamentally 2-categorical: a pointwise extension is an ordinary 2-categorical extension in the homotopy 2-category that is stable under pasting with comma squares. Another definition is that a natural transformation defines a pointwise right extension if and only if its image under the covariant embedding into the virtual equipment of modules defines a right extension there. Theorem 9.3.3 proves that these two notions coincide.

In §9.1, we introduce right liftings and right extensions in the virtual equipment of modules and establish their elementary properties. Before turning our attention to pointwise extensions, we first introduce *exact squares* in §9.2, a collection of squares in the homotopy 2-category that includes comma squares, which are used to characterize the pointwise extensions internally to the homotopy 2-category. Pointwise extensions are introduced in a variety of equivalent ways in §9.3 and deployed in §9.4 to develop a few aspects of the formal theory of ∞-categories.

The explorations of §9.4 begin with a discussion of adjunctions and fully faithful functors, revisiting well-trodden ground. The vital applications of pointwise right and left extensions arrive in the later part of that section, where we define limits and colimits of ∞-category indexed diagrams in general ∞-cosmoi, that are not necessarily cartesian closed. In §9.5, we introduce a more general class of limits and colimits for diagrams between ∞-categories that are *weighted* by an arbitrary module and quickly establish the key components of the calculus of weighted limits and colimits. These results can be understood as revealing that many aspects of ∞-category are automatically enriched over discrete ∞-categories.

There is one question we do not address in this chapter: namely criteria that guarantee the existence of pointwise left and right extensions. We return to this topic in §12.3 where we take advantage of the results about pointwise defined universal properties established in Chapter 12 to prove the expected converse to Proposition 9.4.9 in an ∞-cosmos of $(\infty, 1)$-categories, reducing the question of the existence of pointwise left or right extensions to the existence of certain colimits or limits.

9.1 Liftings and Extensions of Modules

In this section we introduce and study liftings and extensions in the virtual equipment of modules. To motivate Definition 9.1.2, we briefly recall the standard 2-categorical notion:

DEFINITION 9.1.1. A **right extension** of a 1-cell $f : A \to C$ along a 1-cell $k : A \to B$ is given by a pair $(r : B \to C, \nu : rk \Rightarrow f)$ as below-left

so that any similar pair as above-center factors uniquely through ν as above right.

Dually, a **left extension** of a 1-cell $f : A \to C$ along a 1-cell $k : A \to B$ is given by a pair $(\ell : B \to C, \lambda : f \Rightarrow \ell k)$ as below-left

$$
\begin{array}{ccc}
A \xrightarrow{f} C & A \xrightarrow{f} C & A \xrightarrow{f} C \\
k \downarrow \; {\scriptstyle \Downarrow \lambda} \; \nearrow^{\ell} & k \downarrow \; {\scriptstyle \Downarrow \delta} \nearrow & = \; k \downarrow \nearrow^{\ell}_{\Downarrow \exists !} \nearrow \\
B & B \; \; g & B \xrightarrow{\quad} g
\end{array}
$$

so that any similar pair as above-center factors uniquely through λ as above right.

The op-duals of Definition 9.1.1 define **right** and **left lifting** diagrams in any 2-category. Analogous notions of right extension and right lifting can be defined for horizontal arrows in a virtual double category, where in the presence of the restrictions of a virtual equipment it suffices to consider "fibered" cells, whose vertical boundary arrows are identities. We specialize our language to the virtual equipment of modules, as this is the case of interest:

DEFINITION 9.1.2. A **right extension** of a module $A \overset{F}{\nrightarrow} C$ along a module $A \overset{K}{\nrightarrow} B$ consists of a pair given by a module $B \overset{R}{\nrightarrow} C$ together with a binary cell

$$
\begin{array}{ccc}
A & \xrightarrow{K} B \xrightarrow{R} & C \\
\| & {\scriptstyle \Downarrow \nu} & \| \\
A & \xrightarrow{F} & C
\end{array}
$$

with the property that every $n + 1$-ary cell of the form displayed below-left factors uniquely through $\nu : K \times R \Rightarrow F$ as below-right:

$$
\begin{array}{ccc}
A \xrightarrow{K} B \xrightarrow{E_1} \cdots \xrightarrow{E_n} C & & A \xrightarrow{K} B \xrightarrow{E_1} \cdots \xrightarrow{E_n} C \\
\| \quad {\scriptstyle \Downarrow} \quad \| & = & \| \quad \| \quad {\scriptstyle \Downarrow \exists !} \quad \| \\
A \xrightarrow{F} C & & A \xrightarrow{K} B \xrightarrow{R} C \\
& & \| \quad {\scriptstyle \Downarrow \nu} \quad \| \\
& & A \xrightarrow{F} C
\end{array}
$$

Dually, a **right lifting** of $A \overset{F}{\nrightarrow} C$ through $B \overset{H}{\nrightarrow} C$ consists of a pair given by a module $A \overset{L}{\nrightarrow} B$ together with a binary cell

$$
\begin{array}{ccc}
A & \xrightarrow{L} B \xrightarrow{H} & C \\
\| & {\scriptstyle \Downarrow \lambda} & \| \\
A & \xrightarrow{F} & C
\end{array}
$$

with the property that every $n + 1$-ary cell of the form displayed below-left factors uniquely through $\lambda : L \times H \Rightarrow F$ as below-right:

$$
\begin{array}{ccc}
A \xrightarrow{E_1} \cdots \xrightarrow{E_n} B \xrightarrow{H} C \\
\| \quad\quad \Downarrow \quad\quad \| \\
A \xrightarrow{F} C
\end{array}
\;=\;
\begin{array}{ccc}
A \xrightarrow{E_1} \cdots \xrightarrow{E_n} B \xrightarrow{H} C \\
\| \quad \Downarrow \exists! \quad \| \quad \| \\
A \xrightarrow{L} B \xrightarrow{H} C \\
\| \quad\quad \Downarrow \lambda \quad\quad \| \\
A \xrightarrow{F} C
\end{array}
$$

Because of the asymmetry in Definition 8.1.8, there is no corresponding notion of *left* extension or *left* lifting. It follows easily from these definitions that right extensions or right liftings are unique up to vertical isomorphism in $\mathsf{Mod}(\mathcal{K})$ (see Exercise 9.1.i).

LEMMA 9.1.3. *For any functor* $f : A \to B$, *there is a binary cell*

$$
\begin{array}{ccc}
B \xrightarrow{\mathrm{Hom}_B(f,B)} A \xrightarrow{\mathrm{Hom}_B(B,f)} B \\
\| \quad\quad \Downarrow \epsilon \quad\quad \| \\
B \xrightarrow{\mathrm{Hom}_B} B
\end{array}
$$

that defines both a right extension of $B \xrightarrow{\mathrm{Hom}_B} B$ *through* $B \xrightarrow{\mathrm{Hom}_B(f,B)} A$ *and a right lifting of* $B \xrightarrow{\mathrm{Hom}_B} B$ *through* $A \xrightarrow{\mathrm{Hom}_B(B,f)} B$.

Proof The binary cell $\epsilon : \mathrm{Hom}_B(f,B) \times \mathrm{Hom}_B(B,f) \Rightarrow \mathrm{Hom}_B$, which also appears as a counit of sorts in Exercise 8.4.v, corresponds to the unary unit cell Hom_f under the bijections of Theorem 8.4.4 and Proposition 8.2.4. The verification of the universal properties is left as Exercise 9.1.ii. □

The result of Lemma 9.1.3 is a special case of a more general family of examples:

LEMMA 9.1.4. *For any module* $A \xrightarrow{E} B$ *and any pair of functors* $g : C \to A$ *and* $f : D \to B$ *the canonical cells*

$$
\begin{array}{ccc}
A \xrightarrow{\mathrm{Hom}_A(g,A)} C \xrightarrow{E(1,g)} B \\
\| \quad\quad \Downarrow \rho \quad\quad \| \\
A \xrightarrow{E} B
\end{array}
\qquad
\begin{array}{ccc}
A \xrightarrow{E(f,1)} D \xrightarrow{\mathrm{Hom}_B(B,f)} B \\
\| \quad\quad \Downarrow \rho \quad\quad \| \\
A \xrightarrow{E} B
\end{array}
$$

exhibit $E(1,g)$ *as the right extension of* E *through* $\mathrm{Hom}_A(g,A)$ *and exhibit* $E(f,1)$ *as the right lifting of* E *through* $\mathrm{Hom}_B(B,f)$.

Proof The canonical cells of the statement are defined by applying the bijection of Theorem 8.4.4 to the restriction cells of Proposition 8.2.1, which is to say that in the case of the right extension diagram above-left this cell is obtained as the composite

$$
\begin{array}{ccccc}
A & \xrightarrow{\mathrm{Hom}_A(g,A)} & C & \xrightarrow{E(1,g)} & B \\[2pt]
\| & \Downarrow\rho & \downarrow g & \Downarrow\rho & \| \\[2pt]
A & \xrightarrow[\mathrm{Hom}_A]{} & A & \xrightarrow[E]{} & B \\[2pt]
\| & & \Downarrow\circ & & \| \\[2pt]
A & & \xrightarrow[E]{} & & B
\end{array}
$$

The universal property of the right extension is provided by Corollary 8.4.6. $\square$

In particular:

COROLLARY 9.1.5. *Any module $A \xrightarrow{E} B$ is the right extension of itself along $A \xrightarrow{\mathrm{Hom}_A} A$ as well as the right lifting of itself through $B \xrightarrow{\mathrm{Hom}_B} B$.* $\square$

Right extensions and right liftings can be understood as right adjoints to horizontal composition with a module on the left or on the right, respectively (see Theorem 12.3.6). This leads to the following "associativity" result, which we formulate for right extensions, leaving the dual result for right liftings to the reader.

PROPOSITION 9.1.6. *Suppose $A \xrightarrow{K} B$, $B \xrightarrow{H} C$, and $A \xrightarrow{F} D$ are modules so that the composite $K \otimes H$ and right extension $B \xrightarrow{R_K F} D$ modules exist. Then the right extension of $R_K F$ along H exists if and only if the right extension of F along $K \otimes H$ exists, in which case $R_H(R_K F) \simeq_{C \times D} R_{K \otimes H} F$.*

Proof The universal property of the binary composite cell $\mu : K \times H \Rightarrow K \otimes H$ and right extension cell $\nu : K \times R_K F \Rightarrow F$ provide a bijection between cells involving an arbitrary compatible sequence of modules $R_1 \times \cdots \times R_n$ from C to D (see Notation 8.2.7):

$$
\begin{array}{ccccc}
A \xrightarrow{K} B \xrightarrow{H} C \xdashrightarrow{\vec{R}} D & & & A \xrightarrow{K} B \xrightarrow{H} C \xdashrightarrow{\vec{R}} D \\[2pt]
\| \quad \Downarrow\mu \quad \| \quad \| & & & \| \quad \quad \| \quad \Downarrow\exists! \quad \| \\[2pt]
A \xrightarrow[K \otimes H]{} C \xdashrightarrow{\vec{R}} D & \leftrightsquigarrow & & A \xrightarrow{K} B \xrightarrow{R_K F} D \\[2pt]
\| \quad \Downarrow\exists! \quad \| & & & \| \quad \quad \Downarrow\nu \quad \| \\[2pt]
A \xrightarrow[F]{} D & & & A \xrightarrow[F]{} D
\end{array}
$$

In the case of an empty sequence of modules, this bijection encodes an adjointness between fibered module maps $K \otimes H \Rightarrow F$ and fibered module maps $H \Rightarrow R_K F$.

If the right extension $R_{K \otimes H} F$ exists we take this for R and use the binary cell on the lower-left to induce the binary cell on the upper-right, which can be shown to exhibit $R_{K \otimes H} F$ as the right extension $R_H(R_K F)$ of $R_K F$ along H. Conversely, if the right extension $R_H(R_K F)$ exists we use the binary cell on the upper-right to induce the binary cell on the lower-left, which can be shown to exhibit $R_H(R_K F)$ as the right extension $R_{K \otimes H} F$ of F along $K \otimes H$. This transference of universal properties is straightforward, following again from the bijection just exhibited.

Finally, if separately the right extensions $R_H(R_K F)$ and $R_{K \otimes H} F$ are known to exist, then the argument given above combined with the uniqueness of right extensions combines to show that $R_{K \otimes H} F \simeq R_H(R_K F)$ as modules from C to D (see Exercise 9.1.i).   □

We now explain how Definition 9.1.2 relates to Definition 9.1.1 via the covariant and contravariant embeddings of Definition 8.4.12.

LEMMA 9.1.7. *If*

$$
\begin{array}{ccc}
A \xrightarrow{\mathrm{Hom}_B(B,k)} B \xrightarrow{\mathrm{Hom}_C(C,r)} C \\
\Big\| \qquad \Downarrow\nu_* \qquad \Big\| \\
A \xrightarrow[\mathrm{Hom}_C(C,f)]{} C
\end{array}
$$

defines a right extension in the virtual equipment of modules, then $\nu : rk \Rightarrow f$ *defines a right extension in the homotopy 2-category. Dually if*

$$
\begin{array}{ccc}
A \xrightarrow{\mathrm{Hom}_A(\ell,A)} B \xrightarrow{\mathrm{Hom}_B(h,B)} C \\
\Big\| \qquad \Downarrow\lambda^* \qquad \Big\| \\
A \xrightarrow[\mathrm{Hom}_A(g,A)]{} C
\end{array}
$$

defines a right lifting in the virtual equipment of modules, then $\lambda : g \Rightarrow \ell h$ *is a left extension in the homotopy 2-category.*

Proof   By Corollary 8.4.9 binary cells $\nu_* : \mathrm{Hom}_B(B,k) \times \mathrm{Hom}_C(C,r) \Rightarrow \mathrm{Hom}_C(C,f)$ correspond to unary cells $\nu_* : \mathrm{Hom}_C(C,rk) \Rightarrow \mathrm{Hom}_C(C,f)$, and by Proposition 8.4.11, these correspond to natural transformations $\nu : rk \Rightarrow f$ in the homotopy 2-category. Under this correspondence the universal property of Definition 9.1.2 clearly subsumes that of Definition 9.1.1 by restricting to right represented modules. The left extension case is similar, via the contravariant embedding of Definition 8.4.12.   □

A sharper characterization of the right extension diagrams of modules in the image of the covariant embedding appears in Theorem 9.3.3, but we can characterize the right lifting diagrams of modules in the image of the covariant embedding now. Recall the notion of absolute right lifting diagram introduced in Definition 2.3.5.

PROPOSITION 9.1.8. *A natural transformation in the homotopy 2-category of an ∞-cosmos as below-left defines an absolute right lifting diagram if and only if the corresponding binary cell displayed below-right defines a right lifting in the virtual equipment of modules:*

$$
\begin{array}{ccc}
& B & \\
{}^{r}\nearrow & {\Downarrow\rho}\; \Big\downarrow f & \\
C \xrightarrow{\ g\ } A &
\end{array}
\qquad \leftrightsquigarrow \qquad
\begin{array}{ccc}
C \xrightarrow{\mathrm{Hom}_B(B,r)} B \xrightarrow{\mathrm{Hom}_A(A,f)} A \\
\| \qquad\quad \Downarrow\rho_* \qquad\quad \| \\
C \xrightarrow[\mathrm{Hom}_A(A,g)]{} A
\end{array}
$$

Dually, a 2-cell in the homotopy 2-category of an ∞-cosmos as below-left defines an absolute left lifting diagram if and only if the corresponding binary cell displayed below-right defines a right extension in the virtual equipment of modules:

$$
\begin{array}{ccc}
& B & \\
{}^{\ell}\nearrow & {\Uparrow\lambda}\; \Big\downarrow f & \\
C \xrightarrow{\ g\ } A &
\end{array}
\qquad \leftrightsquigarrow \qquad
\begin{array}{ccc}
A \xrightarrow{\mathrm{Hom}_A(f,A)} B \xrightarrow{\mathrm{Hom}_B(\ell,B)} A \\
\| \qquad\quad \Downarrow\lambda^* \qquad\quad \| \\
A \xrightarrow[\mathrm{Hom}_A(g,A)]{} C
\end{array}
$$

Proof By Proposition 8.4.11, natural transformations in the homotopy 2-category of an ∞-cosmos correspond bijectively to unary squares in the virtual equipment of modules of various forms. By this result and Corollaries 8.4.6 and 8.4.9, there are canonical bijections:

$$
\begin{array}{ccc}
& X & \\
{}^{c}\swarrow & {}^{\chi}_{\Leftarrow} & \searrow^{b} \\
C & & B \\
{}_{g}\searrow & {}_{f}\swarrow & \\
& A &
\end{array}
\ \leftrightsquigarrow\
\begin{array}{c}
X \xrightarrow{\mathrm{Hom}_A(A,fb)} A \\
\| \quad \Downarrow\chi_* \quad \| \\
X \xrightarrow[\mathrm{Hom}_A(A,gc)]{} A
\end{array}
\ \leftrightsquigarrow\
\begin{array}{c}
C \xrightarrow{\mathrm{Hom}_C(c,C)} X \xrightarrow{\mathrm{Hom}_B(B,b)} B \xrightarrow{\mathrm{Hom}_A(A,f)} A \\
\| \qquad\qquad \Downarrow\chi \qquad\qquad \| \\
C \xrightarrow[\mathrm{Hom}_A(A,g)]{} A
\end{array}
$$

$$\tag{9.1.9}$$

If the binary cell $\rho_* : \mathrm{Hom}_B(B,r) \times \mathrm{Hom}_A(A,f) \Rightarrow \mathrm{Hom}_A(A,g)$ defines a right lifting diagram in the virtual equipment of modules, then there is a unique

factorization

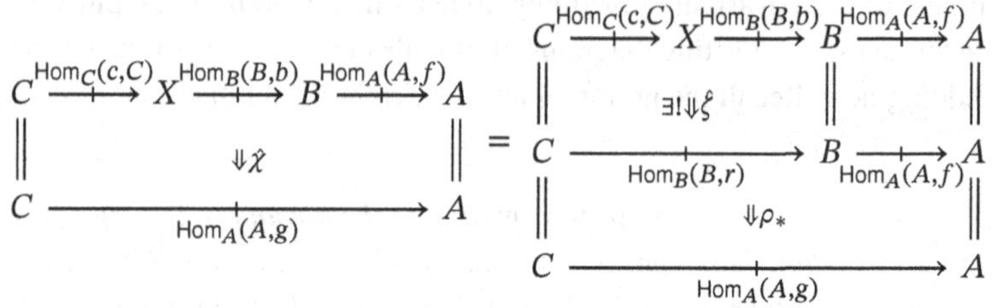

$$
\begin{array}{ccc}
C \xrightarrow{\mathrm{Hom}_C(c,C)} X \xrightarrow{\mathrm{Hom}_B(B,b)} B \xrightarrow{\mathrm{Hom}_A(A,f)} A \\
\| \qquad\qquad \Downarrow\hat{\chi} \qquad\qquad \| \qquad\qquad \| \\
C \xrightarrow[\mathrm{Hom}_A(A,g)]{} A
\end{array}
\quad = \quad
\begin{array}{ccc}
C \xrightarrow{\mathrm{Hom}_C(c,C)} X \xrightarrow{\mathrm{Hom}_B(B,b)} B \xrightarrow{\mathrm{Hom}_A(A,f)} A \\
\| \quad \exists!\Downarrow\hat{\zeta} \qquad \| \qquad\qquad \| \\
C \xrightarrow[\mathrm{Hom}_B(B,r)]{} B \xrightarrow[\mathrm{Hom}_A(A,f)]{} A \\
\| \qquad\qquad \Downarrow\rho_* \qquad\qquad \| \\
C \xrightarrow[\mathrm{Hom}_A(A,g)]{} A
\end{array}
$$

Reversing the canonical bijection (9.1.9), this defines the desired unique factorization in the homotopy 2-category:

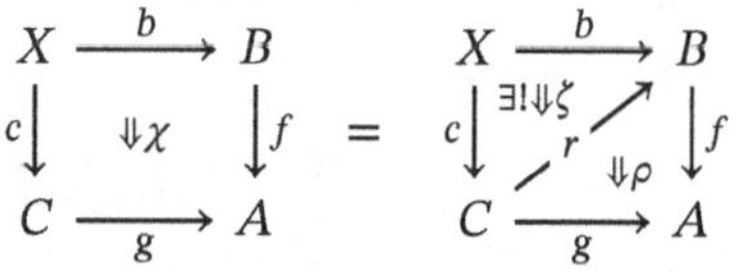

$$
\begin{array}{ccc}
X & \xrightarrow{b} & B \\
c\downarrow & \Downarrow\chi & \downarrow f \\
C & \xrightarrow[g]{} & A
\end{array}
\quad = \quad
\begin{array}{ccc}
X & \xrightarrow{b} & B \\
c\downarrow & {\scriptstyle\exists!\Downarrow\zeta}\;\nearrow^{r}\;\Downarrow\rho & \downarrow f \\
C & \xrightarrow[g]{} & A
\end{array}
$$

Thus if $\rho_* :\ \mathrm{Hom}_B(B,r) \times \mathrm{Hom}_A(A,f) \Rightarrow \mathrm{Hom}_A(A,g)$ is a right lifting, then $\rho :\ fr \Rightarrow g$ is an absolute right lifting.

Conversely, suppose $\rho :\ fr \Rightarrow g$ is an absolute right lifting and consider a cell in the virtual equipment of modules of the following form:

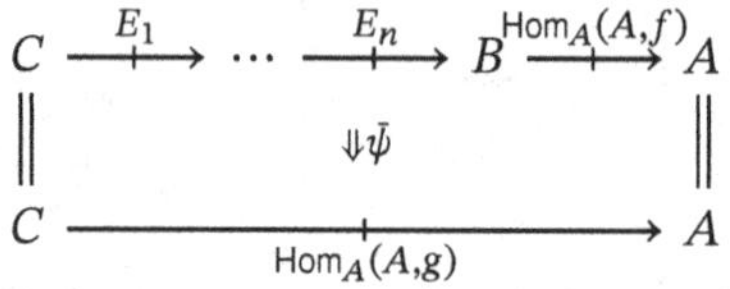

$$
\begin{array}{ccc}
C \xrightarrow{E_1} \cdots \xrightarrow{E_n} B \xrightarrow{\mathrm{Hom}_A(A,f)} A \\
\| \qquad\qquad \Downarrow\psi \qquad\qquad \| \\
C \xrightarrow[\mathrm{Hom}_A(A,g)]{} A
\end{array}
$$

Let $C \xleftarrow{q} \vec{E} \xrightarrow{p} B$ denote the composite two-sided fibration $E_1 \times \cdots \times E_n$. By Remark 8.3.12, module maps $\bar{\psi} :\ E_1 \times \cdots \times E_n \times \mathrm{Hom}_A(A,f) \Rightarrow \mathrm{Hom}_A(A,g)$ correspond to module maps $\hat{\psi} :\ \mathrm{Hom}_C(q,C) \times \mathrm{Hom}_B(B,p) \times \mathrm{Hom}_A(A,f) \Rightarrow \mathrm{Hom}_A(A,g)$, as displayed below-left. As argued in (9.1.9), these stand in canonical bijection with natural transformations as below-center:

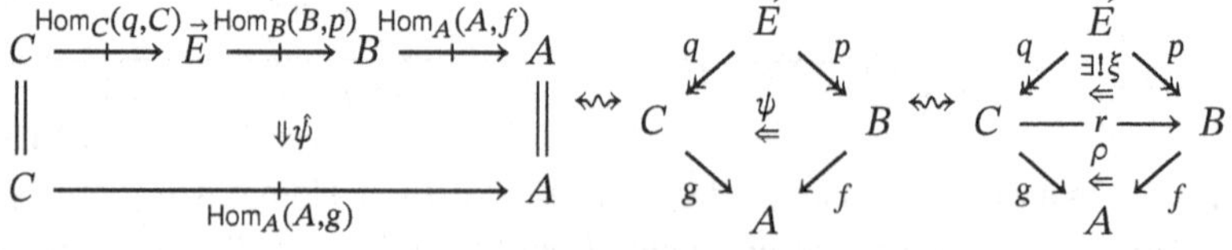

$$
\begin{array}{ccc}
C \xrightarrow{\mathrm{Hom}_C(q,C)} E \xrightarrow{\mathrm{Hom}_B(B,p)} B \xrightarrow{\mathrm{Hom}_A(A,f)} A \\
\| \qquad\qquad \Downarrow\hat{\psi} \qquad\qquad \| \\
C \xrightarrow[\mathrm{Hom}_A(A,g)]{} A
\end{array}
\qquad \leftrightsquigarrow \qquad
\begin{array}{c}
\vec{E} \\
q\swarrow \quad \searrow p \\
C \quad \Downarrow\psi \quad B \\
g\searrow \quad \swarrow f \\
A
\end{array}
\quad \leftrightsquigarrow \quad
\begin{array}{c}
\vec{E} \\
q\swarrow \; {\scriptstyle\exists!\xi} \; \searrow p \\
C \xrightarrow{r} B \\
g\searrow \; {\scriptstyle\Downarrow\rho} \; \swarrow f \\
A
\end{array}
$$

Since $\rho :\ fr \Rightarrow g$ is assumed to be an absolute right lifting, ψ factors uniquely through ρ to define a corresponding 2-cell $\xi :\ p \Rightarrow rq$ as above-right. Applying (9.1.9) again, this constructs a unique factorization in the virtual equipment of modules

$$
\begin{array}{c}
C \xrightarrow{\mathrm{Hom}_C(q,C)} E \xrightarrow{\mathrm{Hom}_B(B,p)} B \xrightarrow{\mathrm{Hom}_A(A,f)} A \\
\| \qquad \Downarrow\tilde\psi \qquad \| \\
C \xrightarrow{\qquad \mathrm{Hom}_A(A,g) \qquad} A
\end{array}
\;=\;
\begin{array}{c}
C \xrightarrow{\mathrm{Hom}_C(q,C)} E \xrightarrow{\mathrm{Hom}_B(B,p)} B \xrightarrow{\mathrm{Hom}_A(A,f)} A \\
\| \qquad \exists!\Downarrow\xi \qquad \| \qquad \| \\
C \xrightarrow{\mathrm{Hom}_B(B,r)} B \xrightarrow{\mathrm{Hom}_A(A,f)} A \\
\| \qquad \Downarrow\rho_* \qquad \| \\
C \xrightarrow{\mathrm{Hom}_A(A,g)} A
\end{array}
$$

By Remark 8.3.12, this defines a bijection

$$
\begin{array}{c}
C \xrightarrow{E_1} \cdots \xrightarrow{E_n} B \xrightarrow{\mathrm{Hom}_A(A,f)} A \\
\| \qquad \Downarrow\tilde\psi \qquad \| \\
C \xrightarrow{\qquad \mathrm{Hom}_A(A,g) \qquad} A
\end{array}
\;=\;
\begin{array}{c}
C \xrightarrow{E_1} \cdots \xrightarrow{E_n} B \xrightarrow{\mathrm{Hom}_A(A,f)} A \\
\| \qquad \exists!\Downarrow\xi \qquad \| \qquad \| \\
C \xrightarrow{\mathrm{Hom}_B(B,r)} B \xrightarrow{\mathrm{Hom}_A(A,f)} A \\
\| \qquad \Downarrow\rho_* \qquad \| \\
C \xrightarrow{\mathrm{Hom}_A(A,g)} A
\end{array}
$$

Thus if $\rho : fr \Rightarrow g$ defines an absolute right lifting, then $\rho_* : \mathrm{Hom}_B(B,r) \times \mathrm{Hom}_A(A, f) \Rightarrow \mathrm{Hom}_A(A, g)$ defines a right lifting. □

In Theorem 9.3.3 we discover that right extensions of modules in the image of the covariant embedding are precisely characterized by the sought-for *pointwise* right extensions in the homotopy 2-category; dually pointwise left extensions correspond to right liftings of modules in the image of the contravariant embedding. In the next section, we build toward the 2-categorical definition of this notion.

Exercises

EXERCISE 9.1.i. Suppose $B \xrightarrow{R} C$ and $B \xrightarrow{S} C$ both define right extensions of a module $A \xrightarrow{K} C$ along a module $A \xrightarrow{F} B$ in the sense of Definition 9.1.2. Prove that $R \simeq_{B \times C} S$.

EXERCISE 9.1.ii. Complete the proof of Lemma 9.1.3 without appealing to Lemma 9.1.4.

EXERCISE 9.1.iii. A virtual equipment is **closed** if the right extensions and right liftings of Definition 9.1.2 always exist. In a closed virtual equipment, prove that a cell $\mu : E_1 \times \cdots \times E_n \Rightarrow E$ is a composite if and only if restriction along μ defines a bijection between fibered unary cells $E \Rightarrow F$ and fibered cells $E_1 \times \cdots \times E_n \Rightarrow F$ for all modules F that are parallel to E.

EXERCISE 9.1.iv. Dualize Proposition 9.1.6 to characterize right liftings through a composite of two modules.

EXERCISE 9.1.v. Establish the dual of Proposition 9.1.8, by verifying that absolute left lifting diagrams $\lambda : g \Rightarrow f\ell$ correspond to right extension diagrams $\lambda^* : \mathrm{Hom}_A(f, A) \times \mathrm{Hom}_B(\ell, B) \Rightarrow \mathrm{Hom}_A(g, A)$.

9.2 Exact Squares

To motivate the main definition of this section, let us try to guess the universal property of a pointwise right extension in an ∞-cosmos by considering a special case that we already understand. If the ambient ∞-cosmos is cartesian closed, then the pointwise right extension of a diagram $f : A \to C$ along a functor $k : A \to B$ is intended to define the value of a right adjoint, which may or may not exist in toto, to the restriction functor $\mathrm{res}_k : C^B \to C^A$ at the element $f : 1 \to C^A$. In the case of extensions along a functor $! : A \to 1$, the restriction functor is the constant diagram functor $\Delta : C \to C^A$ considered in Definition 2.3.2, and so via Definition 2.3.8 we can understand the pointwise right extension as computing the limit of f. The following lemma describes the transposed form of this universal property.

LEMMA 9.2.1. *In a cartesian closed ∞-cosmos, the triangle below-left is an absolute right lifting diagram – defining the limit element and limit cone of f – if and only if the transposed triangle below-center has the property that for any ∞-category X, the composite diagram below-right is a right extension diagram.*

$$
\begin{array}{ccc}
 & C & \\
\lim f \nearrow \Big\downarrow \Delta & & \\
\Downarrow \epsilon & & \\
1 \xrightarrow{\;f\;} C^A & &
\end{array}
\quad\longleftrightarrow\quad
\begin{array}{c}
A \xrightarrow{\;f\;} C \\
! \Big\downarrow \ \Uparrow\epsilon \nearrow \\
\nearrow \lim f \\
1
\end{array}
\qquad
\begin{array}{c}
X \times A \xrightarrow{\;\pi\;} A \xrightarrow{\;f\;} C \\
\pi \Big\downarrow \quad\lrcorner\quad ! \Big\downarrow \ \Uparrow\epsilon \nearrow \\
\nearrow \lim f \\
X \xrightarrow{\;!\;} 1
\end{array}
$$

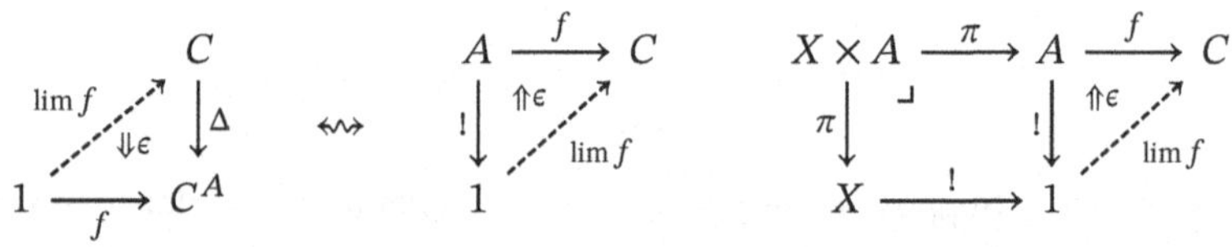

Proof A factorization of a cone with summit X through the absolute right lifting of f along the constant diagram functor

$$
\begin{array}{ccc}
X \xrightarrow{\;c\;} C \\
! \Big\downarrow \ \Downarrow\chi \ \Big\downarrow \Delta \\
1 \xrightarrow{\;f\;} C^A
\end{array}
\quad = \quad
\begin{array}{ccc}
X \xrightarrow{\;c\;} C \\
! \Big\downarrow \ \exists! \Downarrow\zeta \nearrow \Big\downarrow \Delta \\
\lim f \ \Downarrow\epsilon \\
1 \xrightarrow{\;f\;} C^A
\end{array}
$$

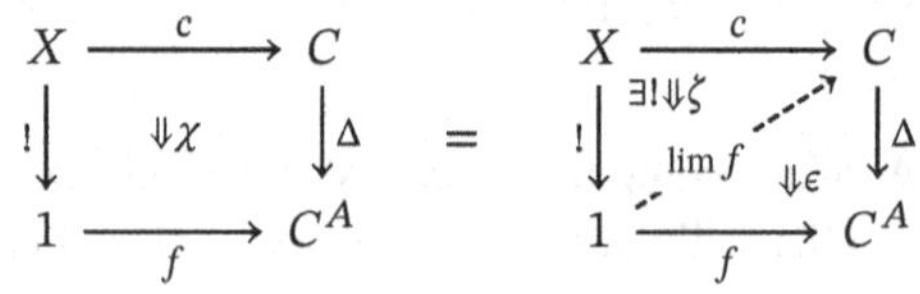

transposes to a factorization as below:

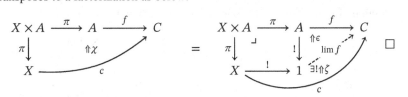

Lemma 9.2.1 reveals that to define the limit of $f : A \to C$ in an ∞-cosmos that is not necessarily cartesian closed, it is not enough to form the right extension of $! : A \to 1$. In terminology introduced in Definition 9.3.1, we must ask in addition that the right extension diagram is *stable under pasting* with squares of the form:

$$
\begin{array}{ccc}
X \times A & \xrightarrow{\pi} & A \\
\pi \downarrow & \lrcorner & \downarrow ! \\
X & \xrightarrow{!} & 1
\end{array}
$$

How might we characterize such squares? First, they are pullbacks each of whose legs is a bifibration. Second, they are comma squares, where the comma cone is an identity 2-cell best regarded as pointing in a direction compatible with ν in the statement of Lemma 9.2.1. By Lemmas 9.2.6 and 9.2.7, we shall see that both of these are instances of *exact squares*, which we now introduce.

By Proposition 8.4.11, natural transformations in the homotopy 2-category of an ∞-cosmos correspond bijectively to unary squares in the virtual equipment of modules of various forms, and in particular, that result, Theorem 8.4.4, Proposition 8.2.4, and Proposition 8.2.1 defines a canonical bijection:

$$
\left(
\begin{array}{ccc}
 & D & \\
k \swarrow & & \searrow h \\
C & \underset{\Leftarrow}{\overset{\alpha}{}} & B \\
g \searrow & & \swarrow f \\
 & A &
\end{array}
\right)
\rightsquigarrow
\begin{array}{ccc}
D & \xrightarrow{\mathrm{Hom}_D} & D \\
gk \downarrow & \Downarrow \tilde{\alpha} & \downarrow fh \\
A & \xrightarrow{\mathrm{Hom}_A} & A
\end{array}
\rightsquigarrow
\begin{array}{ccc}
C & \xrightarrow{\mathrm{Hom}_C(k,C)} D \xrightarrow{\mathrm{Hom}_B(B,h)} & B \\
\| & \Downarrow \hat{\alpha} & \| \\
C & \xrightarrow{\mathrm{Hom}_A(f,g)} & B
\end{array}
$$

DEFINITION 9.2.2 (exact square). A square in the homotopy 2-category of an ∞-cosmos

$$
\begin{array}{ccc}
 & D & \\
k \swarrow & & \searrow h \\
C & \underset{\Leftarrow}{\overset{\alpha}{}} & B \\
g \searrow & & \swarrow f \\
 & A &
\end{array}
$$

is **exact**[1] if and only if the corresponding cell below-left, which under the bijection of Lemma 8.1.16 encodes the below-right pasted composite

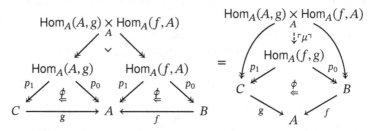

$$(9.2.3)$$

displays $\mathrm{Hom}_A(f, g)$ as the composite $\mathrm{Hom}_C(k, C) \otimes \mathrm{Hom}_B(B, h)$ as defined in Definition 8.3.1.

When the boundary square is clear from context, for economy of language we may write that "$\alpha : fh \Rightarrow gk$ is an exact square" but note that the meaning of the exactness condition is changed if the positions of any of the four boundary arrows of the square inhabited by the 2-cell α are shifted.[2]

REMARK 9.2.4 (exactness as a Beck–Chevalley condition). By Corollary 8.4.8, the map of spans defined by 1-cell induction

encodes a binary cell $\mu : \mathrm{Hom}_A(A, g) \times \mathrm{Hom}_A(f, A) \Rightarrow \mathrm{Hom}_A(f, g)$ which is a composite. Exactness says that α induces an isomorphism

$$\hat{\alpha} : \mathrm{Hom}_C(k, C) \otimes \mathrm{Hom}_B(B, h) \simeq \mathrm{Hom}_A(A, g) \otimes \mathrm{Hom}_A(f, A)$$

of modules from C to B.

---

[1] Unfortunately the terminology "exact square" is used in a variety of different settings. We hope that the context makes it clear that the present notion has nothing to do with squares that are both pushouts and pullbacks in a stable ∞-category. Exercise 9.4.ii reveals a connection between this notion of exact square and commuting squares between adjoint functors whose mates are isomorphisms (see Proposition 6.3.10).

[2] For instance, compare the statements of Lemmas 9.2.8 and 9.4.4.

DIGRESSION 9.2.5 (strong exactness). Recall from §8.3 that many of the formally defined composite cells that are found in any virtual equipment are in fact strong composites in the virtual equipment of modules associated to an ∞-cosmos. Based on this experience, one might expect that many of the exact squares we encounter in the virtual equipment of modules are in fact **strongly exact**, meaning that the cell $\hat{\alpha} : \mathrm{Hom}_C(k, C) \times \mathrm{Hom}_B(B, h) \Rightarrow \mathrm{Hom}_A(f, g)$ is a strong composite, and indeed this is the case. In fact, in an ∞-cosmos of $(\infty, 1)$-categories, all composites exist and are strong: the tensor product of modules can be defined by a fiberwise coinverter construction that is stable under pullback. Since the composite of any compatible sequence of modules is equivalent to the strong composite, in such ∞-cosmoi, all composites, and hence all exact squares, automatically satisfy the stronger universal property.

The primary role played by exact squares here is in developing the notion of pointwise left and right extensions, and for this the weaker 2-categorical universal property suffices. We note in passing that our formally defined exact squares are strong, but typically only appeal to the weaker exactness property in proofs. There is one instance when a stronger notion of exactness may be preferable, which we discuss in Remark 9.4.12.

The remainder of this section is devoted to examples of exact squares.

LEMMA 9.2.6 (comma squares are exact). *For any cospan* $C \xrightarrow{g} A \xleftarrow{f} B$, *the comma cone defines a strongly exact square:*

$$
\begin{array}{ccc}
 & \mathrm{Hom}_A(f, g) & \\
{}^{p_1}\swarrow & & \searrow^{p_0} \\
C & \overset{\phi}{\Leftarrow} & B \\
{}_{g}\searrow & & \swarrow_{f} \\
 & A &
\end{array}
$$

Proof By Proposition 8.3.11, the module $C \xrightarrow{\mathrm{Hom}_A(f,g)} B$ is the strong composite of the left representation of its left leg followed by the right representation of its right leg. By Remark 8.3.13, the binary composition cell $\mathrm{Hom}_C(p_1, C) \times \mathrm{Hom}_B(B, p_0) \Rightarrow \mathrm{Hom}_A(f, g)$ corresponds under the bijection of Lemma 8.1.16

to the pasted composite

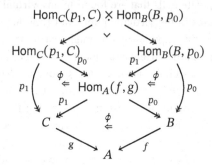

This recovers the cell $\hat{\phi}$ defined by (9.2.3) that tests the exactness of comma square $\phi : f p_0 \Rightarrow g p_1$.   □

LEMMA 9.2.7. *If $g : C \to A$ is a cartesian fibration or $f : B \to A$ is a cocartesian fibration, then the pullback square is strongly exact:*

$$
\begin{array}{ccc}
 & P & \\
\pi_1 \swarrow & \vee & \searrow \pi_0 \\
C & & B \\
g \searrow & & \swarrow f \\
 & A &
\end{array}
$$

Proof The two statements are dual though the positions of the cocartesian and cartesian fibrations cannot be interchanged, as the proof reveals. If $f : B \twoheadrightarrow A$ is a cocartesian fibration, observe that the functor $\Delta : P \to \mathrm{Hom}_A(f, g)$ induced by the identity 2-cell $f \pi_0 = g \pi_1$ is a pullback of the functor $\Delta_f : B \to \mathrm{Hom}_A(f, A)$ induced by the identity 2-cell id_f.

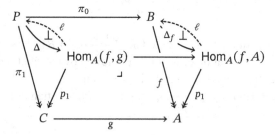

Since f is a cocartesian fibration, Theorem 5.2.8(iii) tells us that $\Delta_f : B \to \mathrm{Hom}_A(f, A)$ has a fibered left adjoint over A. This fibered adjunction pulls back along $g : C \to A$ to define a fibered left adjoint to $\Delta : P \to \mathrm{Hom}_A(f, g)$.

Since $\pi_0 = p_0 \Delta$, Corollary 8.4.9 implies that the canonical cell

$$\mathrm{Hom}_{\mathrm{Hom}_A(f,g)}(\mathrm{Hom}_A(f,g), \Delta) \times \mathrm{Hom}_B(B, p_0) \Rightarrow \mathrm{Hom}_B(B, \pi_0)$$

is a strong composite. Since $\ell \dashv \Delta$, there is an equivalence $\mathrm{Hom}_P(\ell, P) \simeq \mathrm{Hom}_{\mathrm{Hom}_A(f,g)}(\mathrm{Hom}_A(f,g), \Delta)$ by Proposition 4.1.1. And since $p_1 = \pi_1 \ell$, by Corollary 8.4.9 the canonical cell

$$\mathrm{Hom}_C(\pi_1, C) \times \mathrm{Hom}_P(\ell, P) \Rightarrow \mathrm{Hom}_C(p_1, C)$$

is a strong composite. Composing these bijections, we see that cells with domain $\mathrm{Hom}_C(\pi_1, C) \times \mathrm{Hom}_B(B, \pi_0)$ correspond bijectively to cells with domain $\mathrm{Hom}_C(p_1, C) \times \mathrm{Hom}_B(B, p_0)$.

Since $\ell \dashv \Delta$ is a fibered adjunction, the transpose of the natural transformation $\phi \colon f p_0 \Rightarrow g p_1 = g \pi_1 \ell$ along $\ell \dashv \Delta$ equals $\phi \Delta = \mathrm{id} \colon f \pi_0 = f p_0 i \Rightarrow g p_1 i = g \pi_1$. This tells us that the cells

$$\widehat{\mathrm{id}} \colon \mathrm{Hom}_C(\pi_1, C) \times \mathrm{Hom}_B(B, \pi_0) \Rightarrow \mathrm{Hom}_A(f, g) \qquad \text{and}$$

$$\widehat{\phi} \colon \mathrm{Hom}_C(p_1, C) \times \mathrm{Hom}_B(B, p_0) \Rightarrow \mathrm{Hom}_A(f, g)$$

correspond under the bijection just described. Since Lemma 9.2.6 proves that $\widehat{\phi}$ is a strong composite, by Lemma 8.3.8 so is $\widehat{\mathrm{id}}$. □

For later use, we note some trivial examples of exact squares:

LEMMA 9.2.8. *For any pair of functors* $k \colon A \to B$ *and* $h \colon C \to D$ *the pullback square is strongly exact:*

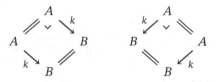

In particular, the identity cells define strongly exact squares:

Proof Exercise 9.2.ii. □

Finally, the exact squares just established can be composed to yield further exact squares:

LEMMA 9.2.9 (composites of exact squares). *Given a diagram of squares in the*

homotopy 2-category

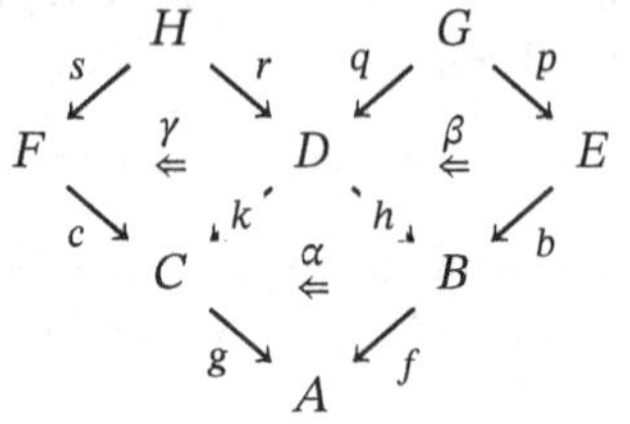

if α : $fh \Rightarrow gk$, β : $bp \Rightarrow hq$, and γ : $kr \Rightarrow cs$ are all exact squares then so are the composite rectangles $\alpha q \cdot f\beta$ and $g\gamma \cdot \alpha r$, and these are strongly exact if the constituent squares are. Consequently, arbitrary "double categorical" composites of (strongly) exact squares define (strongly) exact squares.

Proof   The two cases are co-duals, so it suffices to prove that the rectangle $\alpha q \cdot f\beta$: $g(kq) \Rightarrow (fb)p$ is exact. The corresponding cell $\widehat{\alpha q \cdot f\beta}$ displayed below-left factors as below-right through the composite $\hat{\beta}$

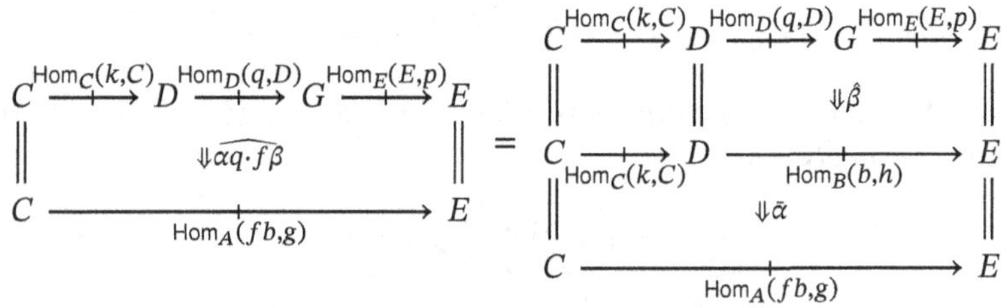

and by comparing the formulae of (9.2.3) with Lemma 8.1.16, we see that the cell $\bar{\alpha}$ satisfies the pasting equality:

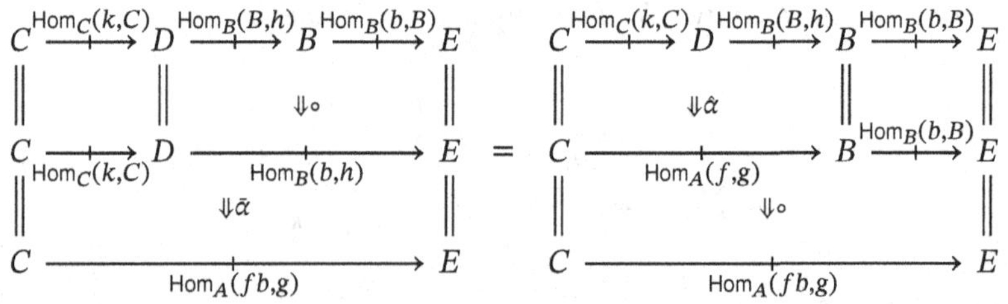

Both canonical cells named $\circ$ are composites, as is $\hat{\alpha}$ by exactness of α, so by Lemma 8.3.3 or 8.3.8, $\bar{\alpha}$ is a composite as well. By exactness of β and Lemma 8.3.3 or 8.3.8 again it now follows that $\widehat{\alpha q \cdot f\beta}$ is also a composite, proving exactness of the rectangle $\alpha q \cdot f\beta$: $g(kq) \Rightarrow (fb)p$.   □

Exercises

EXERCISE 9.2.i. For any exact square

show that the square obtained by composing with any isomorphisms $\beta : f' \cong f$, $\gamma : g \cong g'$, $\delta : h' \cong h$, and $\epsilon : k \cong k'$ is also exact.

EXERCISE 9.2.ii. Prove Lemma 9.2.8.

9.3 Pointwise Right and Left Kan Extensions

In this section, we give four definitions of pointwise right Kan extensions for functors between ∞-categories and prove they are equivalent. Our proof reveals that the general 2-categorical notion of right extensions (see Definition 9.1.1) is too weak on its own. The dual theory of pointwise left Kan extensions is previewed by Lemma 9.1.7 (see Exercise 9.3.i).

DEFINITION 9.3.1 (stability of extensions under pasting). A right extension diagram $\nu : rk \Rightarrow f$ in a 2-category is said to be **stable under pasting with a square** α if the pasted composite

$$(9.3.2)$$

defines a right extension $rb : E \to C$ of $fg : D \to C$ along $h : D \to E$.

THEOREM 9.3.3 (pointwise right extensions). *For a diagram*

in the homotopy 2-category of an ∞-cosmos $\mathcal{K}$ the following are equivalent:

(i) $\nu : rk \Rightarrow f$ *defines a right extension in $\mathfrak{h}\mathcal{K}$ that is stable under pasting with exact squares.*

(ii) $\nu : rk \Rightarrow f$ *defines a right extension in* $\mathfrak{h}\mathcal{K}$ *that is stable under pasting with comma squares.*

(iii) $\nu_* : \mathrm{Hom}_B(B,k) \times \mathrm{Hom}_C(C,r) \Rightarrow \mathrm{Hom}_C(C,f)$ *defines a right extension in* $\mathbb{M}\mathrm{od}(\mathcal{K})$.

(iv) *For any exact square*

$(\nu g \cdot r\alpha)_* : \mathrm{Hom}_E(E,h) \times \mathrm{Hom}_C(C,rb) \Rightarrow \mathrm{Hom}_C(C,fg)$ *defines a right extension in* $\mathbb{M}\mathrm{od}(\mathcal{K})$.

When these conditions hold, we say $r : B \to C$ *defines a **pointwise right extension** of* $f : A \to C$ *along* $k : A \to B$.

Proof   Lemma 9.2.6 proves (i)⇒(ii).

To show (ii)⇒(iii), suppose $\nu : rk \Rightarrow f$ defines a right extension in $\mathfrak{h}\mathcal{K}$ that is stable under pasting with comma squares and consider a cell in $\mathbb{M}\mathrm{od}(\mathcal{K})$:

$$
\begin{array}{ccccccc}
A & \xrightarrow{\mathrm{Hom}_B(B,k)} & B & \xrightarrow{E_1} & \cdots & \xrightarrow{E_n} & C \\
\| & & & \Downarrow\bar{\beta} & & & \| \\
A & & \xrightarrow[\mathrm{Hom}_C(C,f)]{} & & & & C
\end{array}
$$

Let $B \xleftarrow{q} \vec{E} \xrightarrow{p} C$ denote the composite two-sided fibration $E_1 \times \cdots \times E_n$. By Remark 8.3.12, module maps $\beta : \mathrm{Hom}_B(B,k) \times E_1 \times \cdots \times E_n \Rightarrow \mathrm{Hom}_C(C,f)$ correspond to module maps $\hat{\beta} : \mathrm{Hom}_B(B,k) \times \mathrm{Hom}_B(q,B) \times \mathrm{Hom}_C(C,p) \Rightarrow \mathrm{Hom}_C(C,f)$. By Corollary 8.4.6 and Proposition 8.4.7, such module maps stand in bijection with module maps $\check{\beta} : \mathrm{Hom}_B(q,k) \Rightarrow \mathrm{Hom}_C(p,f)$. By Lemma 8.1.16, these module maps correspond bijectively to 2-cells

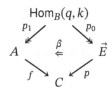

in the homotopy 2-category. By the hypothesis (ii),

$$
\begin{array}{ccccc}
\mathrm{Hom}_B(q,k) & \xrightarrow{\;p_1\;} & A & \xrightarrow{\;f\;} & C \\
{\scriptstyle p_0}\downarrow & {\scriptstyle \nearrow\!\phi} & {\scriptstyle k}\downarrow{\scriptstyle\Uparrow\nu} & \nearrow{\scriptstyle r} & \\
\vec{E} & \xrightarrow[q]{} & B & &
\end{array}
$$

defines a right extension in $\mathfrak{h}\mathcal{K}$, so β factors uniquely through this pasted composite via a natural transformation $\gamma \colon p \Rightarrow rq$. By Proposition 8.4.11, this defines a cell $\gamma_* \colon \mathrm{Hom}_C(C,p) \Rightarrow \mathrm{Hom}_C(C,rq)$, which by Corollary 8.4.6 gives rise to a canonical cell $\hat{\gamma} \colon \mathrm{Hom}_B(q,B) \times \mathrm{Hom}_C(C,p) \Rightarrow \mathrm{Hom}_C(C,r)$. By Remark 8.3.12 again, this produces the desired unique factorization

$$
\begin{array}{c}
\begin{array}{ccccccc}
A & \xrightarrow{\mathrm{Hom}_B(B,k)} & B & \xrightarrow{E_1} & \cdots & \xrightarrow{E_n} & C \\
\| & & \Downarrow\bar{\beta} & & & & \| \\
A & \xrightarrow[\mathrm{Hom}_C(C,f)]{} & & & & & C
\end{array}
\end{array}
\;=\;
\begin{array}{c}
\begin{array}{ccccccc}
A & \xrightarrow{\mathrm{Hom}_B(B,k)} & B & \xrightarrow{E_1} & \cdots & \xrightarrow{E_n} & C \\
\| & & \| & & \exists!\Downarrow\hat{\gamma} & & \| \\
A & \xrightarrow{\mathrm{Hom}_B(B,k)} & B & \xrightarrow{\mathrm{Hom}_C(C,r)} & & & C \\
\| & & & \Downarrow\nu_* & & & \| \\
A & \xrightarrow[\mathrm{Hom}_C(C,f)]{} & & & & & C
\end{array}
\end{array}
$$

To show (iii)$\Rightarrow$(iv), consider a diagram (9.3.2) in which $\alpha \colon bh \Rightarrow kg$ is exact and $\nu_* \colon \mathrm{Hom}_B(B,k) \times \mathrm{Hom}_C(C,r) \Rightarrow \mathrm{Hom}_C(C,f)$ defines a right extension diagram in $\mathbb{M}\mathrm{od}(\mathcal{K})$. Now by Corollary 8.4.6, a cell

$$
\bar{\beta} \colon \mathrm{Hom}_E(E,h) \times E_1 \times \cdots \times E_n \Rightarrow \mathrm{Hom}_C(C,fg)
$$

corresponds to a cell

$$
\hat{\beta} \colon \mathrm{Hom}_A(g,A) \times \mathrm{Hom}_E(E,h) \times E_1 \times \cdots \times E_n \Rightarrow \mathrm{Hom}_C(C,f).
$$

By exactness of α, this corresponds to a cell

$$
\hat{\beta} \colon \mathrm{Hom}_B(b,k) \times E_1 \times \cdots \times E_n \Rightarrow \mathrm{Hom}_C(C,f),
$$

or equivalently, upon restricting along the composite map $\circ \colon \mathrm{Hom}_B(B,k) \times \mathrm{Hom}_B(b,B) \Rightarrow \mathrm{Hom}_B(b,k)$ to a cell

$$
\hat{\beta} \colon \mathrm{Hom}_B(B,k) \times \mathrm{Hom}_B(b,B) \times E_1 \times \cdots \times E_n \Rightarrow \mathrm{Hom}_C(C,f).
$$

Since $B \xrightarrow{\mathrm{Hom}_C(C,r)} C$ is the right extension of $A \xrightarrow{\mathrm{Hom}_C(C,f)} C$ along $A \xrightarrow{\mathrm{Hom}_B(B,k)} B$, this corresponds to a cell

$$
\hat{\gamma} \colon \mathrm{Hom}_B(b,B) \times E_1 \times \cdots \times E_n \Rightarrow \mathrm{Hom}_C(C,r),
$$

which transposes via Corollary 8.4.6 to a cell

$$\bar{\gamma}\colon E_1 \times \cdots \times E_n \Rightarrow \mathsf{Hom}_C(C, rb),$$

which gives the factorization required to prove that the module $E \xrightarrow{\ \mathsf{Hom}_C(C,rb)\ } C$ is the right extension of $D \xrightarrow{\ \mathsf{Hom}_C(C,fg)\ } C$ along $D \xrightarrow{\ \mathsf{Hom}_E(E,h)\ } E$. A slightly more delicate argument is required to see that this bijection is implemented by composing with the map $(\nu g \cdot r\alpha)_* \colon \mathsf{Hom}_E(E, h) \times \mathsf{Hom}_C(C, rb) \Rightarrow \mathsf{Hom}_C(C, fg)$ corresponding to the pasted composite (9.3.2), but for this it suffices by the Yoneda lemma to start with the identity cell $\mathrm{id}_{\mathsf{Hom}_C(C,rb)}$ and trace back up through the bijection just described. Using Lemma 8.1.16 this is straightforward.

Finally Lemma 9.1.7 and the trivial example of Lemma 9.2.8 prove that (iv)⇒(i).   □

Using the various characterizations of Theorem 9.3.3, we can establish some basic stability properties of pointwise right extensions.

COROLLARY 9.3.4. *The pasted composite (9.3.2) of a pointwise right extension with an exact square is a pointwise right extension.*

Proof Lemma 9.2.9, the pasted composite of two exact squares remains an exact square, so by Theorem 9.3.3(i), the pasted composite of a pointwise right extension remains stable under pasting with exact squares.   □

COROLLARY 9.3.5. *Consider a diagram of functors and natural transformations*

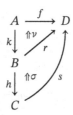

where $(r, \nu\colon rk \Rightarrow f)$ *is a pointwise right extension of* f *along* k. *Then* $(s, \sigma\colon sh \Rightarrow r)$ *is a pointwise right extension of* r *along* h *if and only if* $(s, \nu \cdot \sigma k\colon shk \Rightarrow f)$ *is a pointwise right extension of* f *along* hk.

Proof By Theorem, 9.3.3, a natural transformation in the homotopy 2-category is a pointwise right extension if and only if its covariant embedding into the virtual equipment of modules is a right extension. The binary module maps $\sigma_*\colon \mathsf{Hom}_C(C, h) \times \mathsf{Hom}_D(D, s) \Rightarrow \mathsf{Hom}_D(D, r)$ and $(\nu \cdot \sigma k)_*\colon \mathsf{Hom}_C(hk) \times$

$\mathrm{Hom}_D(D,s) \Rightarrow \mathrm{Hom}_D(D,f)$ correspond under the bijection of Proposition 9.1.6:

$$
\begin{array}{ccc}
A \xrightarrow{\mathrm{Hom}_B(B,k)} B \xrightarrow{\mathrm{Hom}_C(C,k)} C \xrightarrow{\mathrm{Hom}_D(D,s)} D \\[2pt]
\Vert \quad \Downarrow\mu \quad \Vert \quad \Vert \\[2pt]
A \xrightarrow[\mathrm{Hom}_C(C,hk)]{} C \xrightarrow[\mathrm{Hom}_D(D,s)]{} D \\[2pt]
\Vert \quad \Downarrow(\nu\cdot\sigma k)_* \quad \Vert \\[2pt]
A \xrightarrow[\mathrm{Hom}_D(D,f)]{} D
\end{array}
\quad\leftrightsquigarrow\quad
\begin{array}{ccc}
A \xrightarrow{\mathrm{Hom}_B(B,k)} B \xrightarrow{\mathrm{Hom}_C(C,h)} C \xrightarrow{\mathrm{Hom}_D(D,s)} D \\[2pt]
\Vert \quad \Vert \quad \Downarrow\sigma_* \quad \Vert \\[2pt]
A \xrightarrow[\mathrm{Hom}_B(B,k)]{} B \xrightarrow[\mathrm{Hom}_D(D,r)]{} D \\[2pt]
\Vert \quad \Downarrow\nu_* \quad \Vert \\[2pt]
A \xrightarrow[\mathrm{Hom}_D(D,f)]{} D
\end{array}
$$

Thus, by that result, σ_* is a right extension if and only if $(\nu \cdot \sigma k)_*$ is a right extension, as claimed. $\qquad\square$

Exercises

EXERCISE 9.3.i. Using Lemma 9.1.7 as a hint, state and prove a dual version of Theorem 9.3.3 defining pointwise left extensions in the homotopy 2-category of an ∞-cosmos.

9.4 Formal Category Theory in a Virtual Equipment

One reason for our interest in the virtual equipment of modules is that it captures their calculus within an ∞-cosmos. A stronger justification is provided by the theorems about ∞-category theory that can be proven within a virtual equipment. In this section, we revisit adjunctions and fully faithful functors from a module-theoretic point of view, before turning our attention to limits and colimits of functors between ∞-categories.

PROPOSITION 9.4.1. *For any pair of functors* $f \colon B \to A$ *and* $u \colon A \to B$ *and natural transformation* $\epsilon \colon fu \Rightarrow \mathrm{id}_A$ *the following are equivalent:*

(i) *The natural transformation* ϵ *is the counit of an adjunction* $f \dashv u$.
(ii) *The unary cell representing the functor* $\ulcorner\epsilon \cdot f(-)\urcorner \colon \mathrm{Hom}_B(B,u) \to \mathrm{Hom}_A(f,A)$ *defines a vertical isomorphism in the virtual equipment of modules:*

$$
\begin{array}{ccc}
A & \xrightarrow{\mathrm{Hom}_B(B,u)} & B \\[2pt]
\Vert & \Downarrow & \Vert \\[2pt]
A & \xrightarrow[\mathrm{Hom}_A(f,A)]{} & B
\end{array}
$$

(iii) *The counit defines an exact square:*

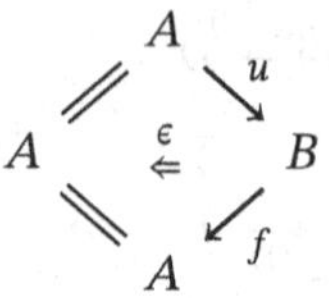

$$
\begin{array}{ccc}
 & A & \\
\nearrow\!\!\!\nearrow & \overset{\epsilon}{\Leftarrow} & \searrow^{u} \\
A & & B \\
\searrow\!\!\!\searrow & & \nearrow_{f} \\
 & A &
\end{array}
$$

(iv) *The counit defines a pointwise right extension diagram that is* **absolute**, *preserved by any functor* $h : A \to C$.

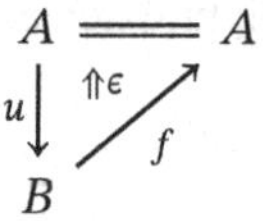

$$
\begin{array}{ccc}
A & =\!\!=\!\!= & A \\
{\scriptstyle u}\downarrow & \overset{\Uparrow\epsilon}{\nearrow} & \\
B & & f
\end{array}
\qquad (9.4.2)
$$

Proof The equivalence (i)⇔(ii) is proven by Proposition 4.1.1, via Lemma 8.1.17.

To say that $\epsilon : fu \Rightarrow \mathrm{id}_A$ is exact is to say that the binary cell below-left is a composite:

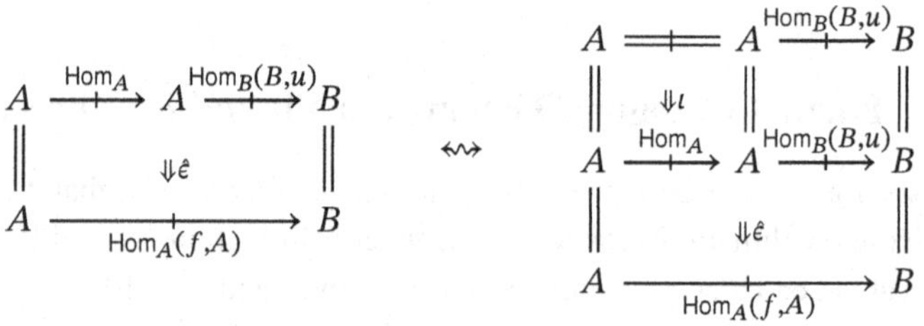

$$
\begin{array}{ccc}
A \xrightarrow{\mathrm{Hom}_A} A \xrightarrow{\mathrm{Hom}_B(B,u)} B \\
\| \quad\quad \Downarrow\hat{\epsilon} \quad\quad \| \\
A \xrightarrow[\mathrm{Hom}_A(f,A)]{} B
\end{array}
\quad\rightsquigarrow\quad
\begin{array}{ccc}
A =\!\!=\!\!= A \xrightarrow{\mathrm{Hom}_B(B,u)} B \\
\| \quad \Downarrow\iota \quad \| \quad\quad \| \\
A \xrightarrow{\mathrm{Hom}_A} A \xrightarrow{\mathrm{Hom}_B(B,u)} B \\
\| \quad\quad \Downarrow\hat{\epsilon} \quad\quad \| \\
A \xrightarrow[\mathrm{Hom}_A(f,A)]{} B
\end{array}
$$

By Lemmas 8.3.3 and 8.3.9(i), this is the case if and only if the unary cell above-right is a composite, and by Lemma 8.3.9(ii) this is exactly the assertion made in (ii). Thus (ii)⇔(iii).

To prove that (ii)⇒(iv), we must show that the binary cell displayed below-left defines a right extension diagram in the virtual equipment of modules.

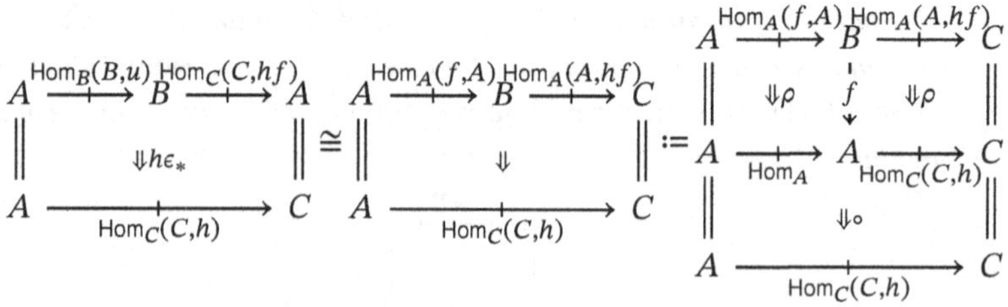

$$
\begin{array}{ccc}
A \xrightarrow{\mathrm{Hom}_B(B,u)} B \xrightarrow{\mathrm{Hom}_C(C,hf)} A \\
\| \quad\quad \Downarrow h\epsilon_* \quad\quad \| \\
A \xrightarrow[\mathrm{Hom}_C(C,h)]{} C
\end{array}
\quad
\begin{array}{ccc}
A \xrightarrow{\mathrm{Hom}_A(f,A)} B \xrightarrow{\mathrm{Hom}_A(A,hf)} C \\
\| \cong \| \quad\quad \Downarrow \quad\quad \| \\
A \xrightarrow[\mathrm{Hom}_C(C,h)]{} C
\end{array}
\quad :=\quad
\begin{array}{ccc}
A \xrightarrow{\mathrm{Hom}_A(f,A)} B \xrightarrow{\mathrm{Hom}_A(A,hf)} C \\
\| \quad \Downarrow\rho \quad {\scriptstyle f}\!\downarrow \quad \Downarrow\rho \quad \| \\
A \xrightarrow{\mathrm{Hom}_A} A \xrightarrow{\mathrm{Hom}_C(C,h)} C \\
\| \quad\quad \Downarrow\circ \quad\quad \| \\
A \xrightarrow[\mathrm{Hom}_C(C,h)]{} C
\end{array}
$$

By (ii), this binary cell is isomorphic to a binary cell of the form displayed above-center. Indeed, Lemma 8.1.16 and Remark 8.3.13 can be used to check that the cell defined by the formula above-right composes with the vertical isomorphism $\ulcorner \epsilon \cdot f(-) \urcorner : \mathrm{Hom}_B(B, u) \simeq \mathrm{Hom}_A(f, A)$ to yield the binary cell above-left. This is why we used the notation "ϵ" for the binary cell in Lemma 9.1.3. Thus, by Lemma 9.1.4, these binary cells are right extensions in the virtual equipment of modules, so we see that (ii)$\Rightarrow$(iv).

The proof that (iv)$\Rightarrow$(i) can be given entirely in the homotopy 2-category and does not require the adjective "pointwise." Indeed, this is the op-dual of the part of Lemma 2.3.7 left as Exercise 2.3.iii. $\qquad\square$

It follows that right extensions along right adjoints and left extensions along left adjoints are easy to calculate:

COROLLARY 9.4.3. *The pointwise right extension of any functor along a right adjoint* $u \colon A \to B$ *is given by its restriction along the left adjoint* $f \colon B \to A$, *while the pointwise left extension of any functor along a left adjoint is given by its restriction along the right adjoint:*

$$
\begin{array}{cc}
\begin{array}{ccc}
A & \xrightarrow{\ h\ } & C \\
\scriptstyle u \Big\downarrow & \Uparrow he \nearrow & \\
B & \underset{\mathrm{ran}_u h \cong hf}{\cdots\nearrow} &
\end{array}
&
\begin{array}{ccc}
B & \xrightarrow{\ g\ } & C \\
\scriptstyle f \Big\downarrow & \Downarrow g\eta \nearrow & \\
A & \underset{\mathrm{lan}_f g \cong gu}{\cdots\nearrow} &
\end{array}
\end{array}
$$

Proof To say in Proposition 9.4.1 that the counit ϵ of an adjunction $f \dashv u$ is an absolute pointwise right extension means that for any functor $h \colon A \to C$ the diagram

$$
\begin{array}{ccccc}
A & == & A & \xrightarrow{\ h\ } & C \\
\scriptstyle u \Big\downarrow & \Uparrow \epsilon \nearrow & & & \\
B & & & &
\end{array}
$$

is a pointwise right extension. In particular, $hf \cong \mathrm{ran}_u h$. $\qquad\square$

Corollary 3.5.6 describes a number of equivalent characterizations of fully faithful functors between ∞-categories, including one characterization that was implicitly module-theoretic: a functor $k \colon A \to B$ is fully faithful if and only if $\ulcorner \mathrm{id}_k \urcorner \colon A^2 \twoheadrightarrow \mathrm{Hom}_B(k, k)$ is a fibered equivalence over $A \times A$. In the virtual equipment of modules, this condition can be rephrased in a number of ways:

LEMMA 9.4.4. *A functor* $k \colon A \to B$ *is **fully faithful** when any of the following equivalent conditions hold:*

(i) The square is exact:

$$
\begin{array}{ccc}
 & A & \\
 {}^{/\!/}\nearrow & & \nwarrow^{\backslash\!\backslash} \\
A & & A \\
{}_{k}\searrow & & \swarrow_{k} \\
 & B &
\end{array}
$$

(ii) The module map $\ulcorner \text{id}_k \urcorner$ defines a vertical isomorphism in the virtual equipment of modules:

$$
\begin{array}{ccc}
A & \xrightarrow{\;\text{Hom}_A\;} & A \\[2pt]
\parallel & \Downarrow & \parallel \\[2pt]
A & \xrightarrow[\text{Hom}_B(k,k)]{} & A
\end{array}
$$

Proof   After unpacking the definitions, the equivalence follows from the nullary and unary composites of Lemma 8.3.9 and the composition relation $\text{Hom}_A \otimes \text{Hom}_A \simeq \text{Hom}_A$ of Lemma 8.3.10.   □

Famously, pointwise left and right extensions along fully faithful functors are genuine extensions, in the sense that the universal natural transformation is an isomorphism.

PROPOSITION 9.4.5 (extensions along fully faithful functors). *If $k : A \to B$ is fully faithful then for any pointwise right extension the natural transformation $\nu : rk \Rightarrow f$ is an isomorphism.*

$$
\begin{array}{ccc}
A & \xrightarrow{\;f\;} & C \\
{\scriptstyle k}\downarrow & \overset{\Uparrow \nu}{\nearrow} & \\
B & & {\scriptstyle r}
\end{array}
$$

Proof   Pasting the pointwise right extension in the statement with the exact square of Lemma 9.4.4 yields a pointwise right extension diagram, whose universal property in the homotopy 2-category can be used to construct an inverse isomorphism to ν and prove it defines a two-sided inverse:

$$
\begin{array}{ccc}
A \xrightarrow{\;f\;} C & A \xrightarrow{\;f\;} C & A \xrightarrow{\;f\;} C \\
\parallel \overset{\Uparrow \nu}{\nearrow}{\scriptstyle rk} & \parallel \overset{\Uparrow \text{id}_f}{\nearrow}{\scriptstyle f} \quad = & \parallel \overset{\Uparrow \nu}{\nearrow}{\scriptstyle rk}{\scriptstyle \Uparrow \exists!} \\
A & A & A \quad {\scriptstyle f}
\end{array}
\qquad □
$$

PROPOSITION 9.4.6. *A right adjoint $u : A \to B$ is fully faithful if and only if the counit $\epsilon : fu \Rightarrow \text{id}_A$ is an isomorphism.*

Proof   If $f \dashv u$ with counit $\epsilon : fu \Rightarrow \text{id}_A$, then by Proposition 9.4.1, the counit $\epsilon : fu \Rightarrow \text{id}_A$ is exact. If u is fully faithful, then by Lemma 9.2.9 the composite rectangle below is also exact.

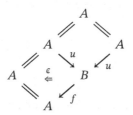

Unpacking Definition 9.2.2, this is to say, by Lemmas 8.3.10 and 8.3.9, that that the image of ϵ under the contravariant embedding of Definition 8.4.12 induces an equivalence of modules $\epsilon^* : \mathrm{Hom}_A \simeq \mathrm{Hom}_A(fu, A)$. By Proposition 8.4.11, this embedding is fully faithful, so it follows that $\epsilon : fu \Rightarrow \mathrm{id}_A$ is an isomorphism.

Conversely, assume $f \dashv u$ with invertible counit $\epsilon : fu \cong \mathrm{id}_A$. The fibered equivalences of Exercise 3.6.ii and Corollary 4.1.3

$$\mathrm{Hom}_A \xrightarrow[\sim]{\epsilon^*} \mathrm{Hom}_A(fu, A) \xrightarrow[\sim]{\ulcorner u(-)\cdot\eta\urcorner} \mathrm{Hom}_B(u, u)$$

compose to define the module map $\ulcorner \mathrm{id}_u \urcorner : \mathrm{Hom}_A \Rightarrow \mathrm{Hom}_B(u, u)$. Thus, by Lemma 9.4.4, u is fully faithful. $\qquad\square$

We now turn to limits and colimits of diagrams indexed by ∞-categories. Recall from Definition 1.2.23 that an ∞-cosmos $\mathcal{K}$ is **cartesian closed** if for any $J \in \mathcal{K}$ there is a cosmological functor $(-)^J : \mathcal{K} \to \mathcal{K}$ equipped with a simplicial natural isomorphism

$$\mathrm{Fun}(X \times J, A) \cong \mathrm{Fun}(X, A^J).$$

It follows that elements $f : 1 \to A^J$ in the ∞-category of J-shaped diagrams in A correspond bijectively to functors $f : J \to A$ between ∞-categories. In such contexts, Definition 2.3.8 defines limits and colimits of an ∞-category indexed diagram $f : J \to A$ to be absolute right and left liftings, respectively:

$$
\begin{array}{cc}
\begin{array}{ccc}
 & A & \\
{\scriptstyle \lim f}\nearrow & \Downarrow\epsilon & \downarrow{\scriptstyle \Delta} \\
1 & \xrightarrow{\;\;f\;\;} & A^J
\end{array}
&
\begin{array}{ccc}
 & A & \\
{\scriptstyle \mathrm{colim}\, f}\nearrow & \Uparrow\eta & \downarrow{\scriptstyle \Delta} \\
1 & \xrightarrow{\;\;f\;\;} & A^J
\end{array}
\end{array}
$$

We now argue that such limits and colimits can also be expressed as pointwise Kan extensions, extending the previously developed theory to ∞-cosmoi that are not necessarily cartesian closed.

DEFINITION 9.4.7. A **limit** of a diagram of ∞-categories $f : J \to A$ is given by a pointwise right extension as below-left, while a **colimit** of a diagram of ∞-categories $f : J \to A$ is given by a pointwise left extension as below-right.

As in Definition 2.3.8, we refer to the natural transformations ϵ and η as the **limit cone** and **colimit cone**, respectively. Our first task is to reconcile these definitions:

PROPOSITION 9.4.8. *In a cartesian closed ∞-cosmos any limit as encoded by the absolute right lifting diagram below-right transposes to define a pointwise right extension diagram as below-left:*

Conversely, any pointwise right extension diagram of this form transposes to define a limit in A.

Proof By Lemma 9.2.1, the universal property of an absolute right lifting diagram as below-left transposes to the universal property of a right extension diagram that is stable under pasting with pullback squares:

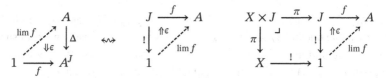

Since comma squares over cospans whose codomain is the terminal ∞-category reduce to pullbacks, we conclude from Theorem 9.3.3(ii) that $\epsilon : \Delta \lim f \Rightarrow f$ is an absolute right lifting diagram if and only if $\epsilon : \lim f! \Rightarrow f$ is a pointwise right extension.   □

In classical category theory, there is a well-known formula that calculates the values of a pointwise right Kan extension "pointwise" as limits. A dual colimit formula calculates pointwise left Kan extensions. These can now be recovered, essentially as tautologies.

PROPOSITION 9.4.9. *For any pointwise right extension as below-left and element* $b: 1 \to B$, *the element* $rb: 1 \to C$ *is the limit of the diagram below-right:*

$$
\begin{array}{ccc}
A & \xrightarrow{f} & C \\
{\scriptstyle k}\downarrow & {\scriptstyle \Uparrow\nu}\nearrow & \\
B & {\scriptstyle r} &
\end{array}
\qquad\qquad
\mathrm{Hom}_B(b,k) \xrightarrow{p_1} A \xrightarrow{f} C
$$

while for any pointwise left extension as below-left and element $b: 1 \to B$, *the element* $\ell b: 1 \to C$ *is the colimit of the diagram below-right:*

$$
\begin{array}{ccc}
A & \xrightarrow{f} & C \\
{\scriptstyle k}\downarrow & {\scriptstyle \Downarrow\lambda}\nearrow & \\
B & {\scriptstyle \ell} &
\end{array}
\qquad\qquad
\mathrm{Hom}_B(k,b) \xrightarrow{p_0} A \xrightarrow{f} C
$$

Proof By Theorem 9.3.3(ii), the composite

$$
\begin{array}{ccccc}
\mathrm{Hom}_B(b,k) & \xrightarrow{p_1} & A & \xrightarrow{f} & C \\
{\scriptstyle !}\downarrow & {\scriptstyle \nearrow\phi} & {\scriptstyle k}\downarrow & {\scriptstyle \Uparrow\nu}\nearrow & \\
1 & \xrightarrow{b} & B & {\scriptstyle r} &
\end{array}
$$

is a pointwise right extension. By Definition 9.4.7 this can be interpreted as saying that $rb: 1 \to B$ defines the limit of the restriction of the diagram $f: A \to C$ along $p_1: \mathrm{Hom}_B(b,k) \twoheadrightarrow A$. $\square$

In Definition 2.2.1, initial and terminal elements in A are defined respectively as left or right adjoints to the unique functor $!: A \to 1$

$$
1 \underset{t}{\overset{i}{\rightleftarrows}} A
$$

or equivalently, by Example 2.3.10, as colimits or limits for the empty diagram in $A^{\varnothing} \cong 1$. As observed in Example 2.3.11, an initial element may also be characterized as a *limit* and a terminal element may be characterized as a *colimit* of the identity diagram $\mathrm{id}_A: A \to A$, as we now show for ∞-cosmoi that are not necessarily cartesian closed:

PROPOSITION 9.4.10. *For an* ∞-*category A:*

(i) *An element* $t: 1 \to A$ *is terminal if and only if it defines a colimit for the identity functor* $\mathrm{id}_A: A \to A$ *in which case the unit for the adjunction* $! \dashv t$ *defines the colimit cone.*

    (ii) An element $i : 1 \to A$ is initial if and only if it defines a limit for the identity functor $\mathrm{id}_A : A \to A$ in which case the counit for $i \dashv !$ defines the limit cone.

Proof We prove (ii). By Definition 2.2.1, an element $i : 1 \to A$ is initial if and only if it defines a right adjoint to the unique functor $! : A \to 1$. If this adjunction exists, then by Proposition 9.4.1, the counit defines a pointwise right extension diagram

$$
\begin{array}{ccc}
A & = = & A \\
{\scriptstyle !}\downarrow & \raise2pt{\Uparrow\epsilon}\nearrow & \\
1 & & {\scriptstyle i}
\end{array}
$$

which by Definition 9.4.7 expresses i as the limit of the diagram $\mathrm{id}_A : A \to A$.

    Conversely, we must show that a pointwise right extension diagram $(i, \epsilon : i! \Rightarrow \mathrm{id}_A)$ gives rise to an adjunction $i \dashv !$ with counit ϵ. By Lemma 2.2.2, it suffices to show that $\epsilon i = \mathrm{id}_i$. By naturality of whiskering (see Lemma B.1.3), the horizontal composite of ϵ with itself gives rise to an equation $i! \epsilon \cdot \epsilon = \epsilon \cdot \epsilon i!$. Since the ∞-category 1 is 2-terminal, $i! \epsilon = \mathrm{id}_{i!}$, so this reduces to the pasting equation:

$$
\begin{array}{ccc}
A = = A \\
{\scriptstyle !}\downarrow \; \raise2pt{\Uparrow\epsilon}\nearrow \\
1 \qquad {\scriptstyle i}
\end{array}
\quad = \quad
\begin{array}{ccc}
A = = A \\
{\scriptstyle !}\downarrow \, \raise2pt{\Uparrow\epsilon}{\scriptstyle i}\;{\scriptstyle \Uparrow\epsilon i}\nearrow \\
1 \longrightarrow {\scriptstyle i}
\end{array}
$$

By the uniqueness statement in the universal property of right extensions, it follows that $\epsilon i = \mathrm{id}_i$ as desired. □

    Recall from Definition 2.4.5 that a functor $k : I \to J$ is **final** if and only if for any ∞-category A, the square

$$
\begin{array}{ccc}
A & = = & A \\
{\scriptstyle \Delta}\downarrow & & \downarrow{\scriptstyle \Delta} \\
A^J & \xrightarrow[A^k]{} & A^I
\end{array}
$$

preserves and reflects all absolute left lifting diagrams. Dually a functor $k : I \to J$ is **initial** if this square preserves and reflects all absolute right lifting diagrams. We can now give a more concise formulation of these notions.

DEFINITION 9.4.11. A functor $k : I \to J$ is **final** if and only if the square below-left is exact and **initial** if and only if the square below-right is exact.

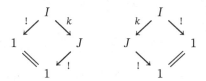

Note that the functor $!: J \to 1$ is represented on the right and on the left by the modules $J \xrightarrow{J} 1$ and $1 \xrightarrow{J} J$. So we see that $k: I \to J$ is **final** if and only if the map of spans below-left is a composite in the virtual equipment of modules

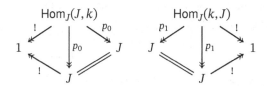

while $k: I \to J$ is **initial** if and only if the map of spans above-right is a composite.

REMARK 9.4.12. As discussed in Digression 9.2.5, exact squares in an ∞-cosmos of $(\infty, 1)$-categories are automatically strongly exact, in which case the universal property satisfied by final functors can be enriched. Since modules from 1 to J reduce to discrete cartesian fibrations over J, in such contexts finality of $k: I \to J$ implies that for every discrete cartesian fibration $p: E \twoheadrightarrow J$, the map

$$\operatorname{Fun}_J(J, E) \xrightarrow[\underset{-\circ k}{\sim}]{\overset{-\circ p_0}{\sim}} \operatorname{Fun}_J(\operatorname{Hom}_J(J, k), E) \xrightarrow{-\circ \ulcorner id_k \urcorner} \operatorname{Fun}_J(I, E) \qquad (9.4.13)$$

is an equivalence of Kan complexes, the first equivalence by strong exactness and the second by the Yoneda lemma of Corollary 5.7.19. In the ∞-cosmos of quasi-categories, the condition (9.4.13) characterizes final functors between quasi-categories, a definition that Lurie attributes to Joyal [78, 4.1.1.1].

As just argued, in an ∞-cosmos of $(\infty, 1)$-categories, our notion of final functor implies this characterization, and we now argue the converse holds as well. Corollary 12.3.8 proves that the ∞-cosmoi of $(\infty, 1)$-categories are strongly closed, meaning that right extensions and right liftings of modules always exist and satisfy a universal property expressed as an equivalence of Kan complexes, not merely a bijection of cells in the virtual equipment. It follows, by the strong analog of Exercise 9.1.iii, that (9.4.13) defines an equivalence for all discrete cartesian fibrations over J if and only if the square of Definition 9.4.11

is strongly exact. Thus, in the ∞-cosmos of quasi-categories, our notion of final functor is equivalent to the standard definition.

We now show that limits are preserved and reflected by reindexing along initial functors.

PROPOSITION 9.4.14. *If $k : I \to J$ is initial and $f : J \to A$ is any diagram, then a limit of f also defines a limit of $fk : I \to A$ and conversely if the limit of $fk : I \to A$ exists, then it also defines a limit of $f : J \to A$.*

Proof By Definition 9.4.7, a limit of f defines a pointwise right extension as below left, which by Corollary 9.3.4 and Definition 9.4.11 gives rise to another pointwise right extension below-right.

$$
\begin{array}{ccc}
J & \xrightarrow{\ f\ } & A \\
{\scriptstyle !}\downarrow & \Uparrow\nu \nearrow_{\ell} & \\
1 & &
\end{array}
\qquad
\begin{array}{ccccc}
I & \xrightarrow{\ k\ } & J & \xrightarrow{\ f\ } & A \\
{\scriptstyle !}\downarrow & & {\scriptstyle !}\downarrow & \Uparrow\nu \nearrow_{\ell} & \\
1 & = & 1 & &
\end{array}
$$

By Definition 9.4.7 again, this tells us that ℓ is the limit of fk.

For the converse, suppose we are given a pointwise right extension diagram

$$
\begin{array}{ccc}
I & \xrightarrow{\ fk\ } & A \\
{\scriptstyle !}\downarrow & \Uparrow\mu \nearrow_{\ell} & \\
1 & &
\end{array}
$$

By Theorem 9.3.3(iii), this means that for any compatible sequence of modules $E_1 \times \cdots \times E_n$ from 1 to A, composing with $\mu : \mathrm{Hom}_1(1,!) \times \mathrm{Hom}_A(A,\ell) \Rightarrow \mathrm{Hom}_A(A,fk)$ defines a bijection between cells $E_1 \times \cdots \times E_n \Rightarrow \mathrm{Hom}_A(A,\ell)$ and the cells below-left:

$$
\begin{array}{ccc}
I \xrightarrow{\mathrm{Hom}_1(1,!)} 1 \dashrightarrow^{\vec{E}} A & & J \xrightarrow{\mathrm{Hom}_J(k,J)} I \xrightarrow{\mathrm{Hom}_1(1,!)} 1 \dashrightarrow^{\vec{E}} A \\
\parallel \quad \Downarrow \quad \parallel & \leftrightsquigarrow & \parallel \quad\quad \Downarrow \quad\quad \parallel \\
I \xrightarrow[\mathrm{Hom}_A(A,fk)]{} A & & J \xrightarrow[\mathrm{Hom}_A(A,f)]{} A
\end{array}
$$

By Corollary 8.4.6, the cells above-left stand in bijection with the cells above-right.

To say that $k : I \to J$ is initial means, by Definition 9.4.11, that the map $\mathrm{Hom}_J(k,J) \times \mathrm{Hom}_1(1,!) \Rightarrow \mathrm{Hom}_1(1,!)$ of modules from J to 1 induced by the identity is a composite. Hence, the cells above-right stand in bijection with cells of the form:

$$
\begin{array}{ccccc}
J & \xrightarrow{\mathrm{Hom}_1(1,!)} & 1 & \xdashrightarrow{\vec{E}} & A \\
\| & & \Downarrow & & \| \\
J & \xrightarrow[\mathrm{Hom}_A(A,f)]{} & & & A
\end{array}
$$

The composite bijection asserts that the cell $\nu : \mathrm{Hom}_1(1,!) \times \mathrm{Hom}_A(A,\ell) \Rightarrow \mathrm{Hom}_A(A,f)$ corresponding to the module id $: \mathrm{Hom}_A(A,\ell) \Rightarrow \mathrm{Hom}_A(A,\ell)$ displays $1 \xrightarrow{\mathrm{Hom}_A(A,\ell)} A$ as the right extension of $J \xrightarrow{\mathrm{Hom}_A(A,f)} A$ along $J \xrightarrow{\mathrm{Hom}_1(1,!)} 1$. Unpacking the bijections that defined ν, we see that $\nu k = \mu$. Thus $\nu : \ell! \Rightarrow f$ is a pointwise right extension and by Proposition 9.4.8 we conclude that $\ell : 1 \to A$ defines a limit of $f : J \to A$ as claimed. □

It remains to extend Theorem 2.4.2 – that right adjoints preserve limits and left adjoints preserve colimits – to diagrams indexed by ∞-categories in general ∞-cosmoi. In Theorem 9.5.7, we prove a generalization of this result that covers limits and colimits weighted by a module, in a sense we now introduce.

Exercises

EXERCISE 9.4.i. Show that the unit of an adjunction $f \dashv u$ defines an exact square:

EXERCISE 9.4.ii. Show that any square α involving parallel adjunctions $f \dashv u$ and $k \dashv r$ as below-left whose mate defines an isomorphism is exact.

EXERCISE 9.4.iii. Show that if $k : I \to J$ is a right adjoint then the square is exact, reproving Proposition 2.4.6.

9.5 Weighted Limits and Colimits in ∞-Categories

In this section we generalize the limits and colimits of diagrams between ∞-categories encoded by Definition 9.4.7 to limits and colimits weighted by a module. The notions of weighted limit and colimit subsume Definition 9.4.7 (see Exercise 9.5.i) and are quite natural from the point of view of the virtual equipment of modules. This can be understood as the ∞-categorical version of the theory introduced in §A.6 in the context of enriched categories, though from the outset we allow our weights to be profunctors, as in Exercise A.6.iv.

DEFINITION 9.5.1. Given a module $A \xrightarrow{W} B$ and a functor $f : A \to C$, a functor $\lim_W f : B \to C$ defines the W-**weighted limit** of f if it covariantly represents the right extension of $\mathrm{Hom}_C(C, f)$ along W. Dually, given $A \xrightarrow{W} B$ and a functor $g : B \to C$, a functor $\mathrm{colim}_W g : A \to C$ defines the W-**weighed colimit** of g if it contravariantly represents the right lifting of $\mathrm{Hom}_C(g, C)$ through W.

$$
\begin{array}{ccc}
A \xrightarrow{\;W\;} B \xdashrightarrow{\mathrm{Hom}_C(C,\lim_W f)} C & \qquad & C \xdashrightarrow{\mathrm{Hom}_C(\mathrm{colim}_W g,C)} A \xrightarrow{\;W\;} B \\
\| \quad \Downarrow\lambda \quad \| & & \| \quad \Downarrow\gamma \quad \| \\
A \xrightarrow[\mathrm{Hom}_C(C,f)]{} C & & C \xrightarrow[\mathrm{Hom}_C(g,C)]{} B
\end{array}
$$

By comparing Definition 9.5.1 with Theorem 9.3.3(iii) we see that pointwise right and left extensions can be understood as special cases of weighted limits and colimits (see also Example A.6.15).

LEMMA 9.5.2. *The pointwise right extension of $f : A \to C$ along $k : A \to B$ is the limit of f weighted by $A \xrightarrow{\mathrm{Hom}_B(B,k)} B$, while the pointwise left extension of f along k is the colimit of f weighted by $B \xrightarrow{\mathrm{Hom}_B(k,B)} A$.*

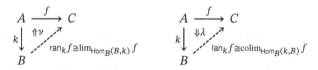

Proof By Theorem 9.3.3(iii) a pointwise right extension defines a right extension of modules while a pointwise left extension defines a right lifting of modules:

$$A \xrightarrow{\operatorname{Hom}_B(B,k)} B \dashrightarrow{\operatorname{Hom}_C(C,\operatorname{ran}_k f)} C \qquad C \dashrightarrow{\operatorname{Hom}_C(\operatorname{lan}_k f,C)} B \xrightarrow{\operatorname{Hom}_B(k,B)} A$$
$$\|\qquad\qquad \Downarrow \nu_* \qquad\qquad \| \qquad\qquad \| \qquad\qquad \Downarrow \lambda^* \qquad\qquad \|$$
$$A \xrightarrow[\operatorname{Hom}_C(C,f)]{} C \qquad\qquad\qquad C \xrightarrow[\operatorname{Hom}_C(f,C)]{} A$$

By Definition 9.5.1, thus $\operatorname{ran}_k f \cong \lim_{\operatorname{Hom}_B(B,k)} f$ and $\operatorname{lan}_k f \cong \operatorname{colim}_{\operatorname{Hom}_B(k,B)} f$ as claimed. $\qquad\square$

In enriched category theory, weighted limits or colimits with representable weights are computed by evaluating at the representing object (see Definition A.6.1). By analogy:

LEMMA 9.5.3. *For any functor $f : A \to C$ and generalized element $a : X \to A$, the restriction $fa : X \to C$ is the limit of f weighted by $A \xrightarrow{\operatorname{Hom}_A(a,A)} X$ and also the colimit of f weighted by $X \xrightarrow{\operatorname{Hom}_A(A,a)} A$.*

Proof By Lemma 9.1.4, the right extension of the module $A \xrightarrow{\operatorname{Hom}_C(C,f)} C$ along $A \xrightarrow{\operatorname{Hom}_A(a,A)} X$ is given by $X \xrightarrow{\operatorname{Hom}_C(C,fa)} C$, while the right lifting of the module $C \xrightarrow{\operatorname{Hom}_C(f,C)} A$ through $X \xrightarrow{\operatorname{Hom}_A(A,a)} A$ is given by $C \xrightarrow{\operatorname{Hom}_C(fa,C)} X$.

$$A \xrightarrow{\operatorname{Hom}_A(a,A)} X \dashrightarrow{\operatorname{Hom}_C(C,fa)} C \qquad A \dashrightarrow{\operatorname{Hom}_C(fa,C)} X \xrightarrow{\operatorname{Hom}_A(A,a)} A$$
$$\|\qquad \Downarrow \rho \qquad \| \qquad\qquad \| \qquad \Downarrow \rho \qquad \|$$
$$A \xrightarrow[\operatorname{Hom}_C(C,f)]{} C \qquad\qquad C \xrightarrow[\operatorname{Hom}_C(f,C)]{} A$$

In the terminology introduced in Definition 9.5.1, this says that the composite functor $fa : X \to C$ is the limit of f weighted by the contravariant module representing a and also the colimit of f weighted by the covariant module representing a. $\qquad\square$

By Proposition 9.1.6 and its dual we have immediately:

PROPOSITION 9.5.4. *For any functors $f : A \to E$ and $g : C \to E$ and weights $A \xrightarrow{W} B$ and $B \xrightarrow{V} C$, the functors $\lim_{W \otimes V} f$ and $\lim_V(\lim_W f)$ are isomorphic and the functors $\operatorname{colim}_{W \otimes V} g$ and $\operatorname{colim}_W(\operatorname{colim}_V g)$ are isomorphic whenever the composite weight and these weighted limits and colimits exist.* $\qquad\square$

We derive various formulae for computing weighted limits and colimits from this result. The first of these determines the values of a weighted limit or colimit functor at a generalized element of its domain.

LEMMA 9.5.5. *If $f : A \to C$ admits a limit $\lim_W f : B \to C$ weighted by $A \xrightarrow{W} B$ then for any generalized element $b : X \to B$, the functor $\lim_W f \circ b : X \to C$ is isomorphic to the limit of f weighted by $A \xrightarrow{W(b,1)} X$. Dually, if $g : B \to C$ admits a colimit $\operatorname{colim}_W g : A \to C$ weighted by W then for any generalized element $a : Y \to A$, the functor $\operatorname{colim}_W g \circ A : Y \to C$ is isomorphic to the colimit of g weighted by $Y \xrightarrow{W(1,a)} B$.*

Proof Recall from Proposition 8.4.7 that $W(b,1) \cong W \otimes \operatorname{Hom}_B(b, B)$. Thus, by Proposition 9.5.4, the $W(b,1)$-weighted limit of f, if it exists, is the functor that right represents the right extension of the module $\operatorname{Hom}_C(C, \lim_W f)$ along the module $\operatorname{Hom}_B(b, B)$. By Lemma 9.5.3, the right extension of the module $\operatorname{Hom}_C(C, \lim_W f)$ along $\operatorname{Hom}_B(b, B)$ is the module $\operatorname{Hom}_C(C, \lim_W f \circ h)$. By Definition 9.5.1, it follows that $\lim_W f \circ b : X \to C$ defines the $W(b,1)$-weighted limit of $f : A \to C$, as claimed. $\qquad\square$

A second useful computational result reduces general weighted limits or weighted colimits to right or left extensions in the homotopy 2-category.

LEMMA 9.5.6. *Consider a weight $A \xrightarrow{W} B$ encoded by the two-sided isofibration $A \xleftarrow{q} W \xrightarrow{p} B$.*

(i) *For any functor $f : A \to C$ the weighted limit $\lim_W f : B \to C$ is given by the pointwise right extension*

$$
\begin{array}{ccc}
W & \xrightarrow{\; q \;} A & \xrightarrow{\; f \;} C \\
& \searrow\!\!{}_{p} \quad \Uparrow \quad \nearrow & \\
& B & \operatorname{ran}_p(fq) \cong \lim_W f
\end{array}
$$

(ii) *For any functor $g : B \to C$ the weighted colimit $\operatorname{colim}_W g : A \to C$ is given by the pointwise left extension*

$$
\begin{array}{ccc}
W & \xrightarrow{\; p \;} B & \xrightarrow{\; g \;} C \\
& \searrow\!\!{}_{q} \quad \Downarrow \quad \nearrow & \\
& A & \operatorname{lan}_q(fp) \cong \operatorname{colim}_W f
\end{array}
$$

In particular, limits weighted by modules from A to 1 and colimits weighted by modules from 1 to B reduce to ordinary limits and colimits.

Proof Recall from Proposition 8.3.11 that any module $A \xleftarrow{q} W \xrightarrow{p} B$ factors as a composite $\operatorname{Hom}_A(q, A) \otimes \operatorname{Hom}_B(B, p) \simeq W$. Thus by Proposition 9.5.4, Lemma 9.5.5, and Lemma 9.5.2

$$
\lim_W f \cong \lim_{\operatorname{Hom}_B(B,p)} (\lim_{\operatorname{Hom}_A(q,A)} f) \cong \lim_{\operatorname{Hom}_B(B,p)} (fq) \cong \operatorname{ran}_p(fq). \qquad \square
$$

In a cartesian closed ∞-cosmos, Theorem 2.4.2 proves right adjoints preserve limits and left adjoints preserve colimits of ∞-category indexed diagrams. We can now extend this result to weighted limits and colimits, while dropping the hypothesis that the ambient ∞-cosmos is cartesian closed.

THEOREM 9.5.7. *Right adjoints preserve weighted limits and left adjoints preserve weighted colimits.*

Proof Consider a weight $J \xrightarrow{W} K$, a diagram $d : J \to A$, and an adjunction with right adjoint $u : A \to B$ and left adjoint $f : B \to A$. The weighted limit $\lim_W d : K \to A$ defines a right extension of modules

$$
\begin{array}{ccc}
J & \xrightarrow{\ W\ } K \xdashrightarrow{\ \mathrm{Hom}_A(A,\lim_W d)\ } & A \\
\| & \Downarrow\lambda & \| \\
J & \xrightarrow[\mathrm{Hom}_A(A,d)]{} & A
\end{array}
$$

Our task is to demonstrate that the cell $u\lambda$ defined by the unique factorization of the cell below-left through the composite below-right

$$
\begin{array}{ccc}
J \xtwoheadrightarrow{W} K \xrightarrow{\mathrm{Hom}_A(A,\lim_W d)} A \xrightarrow{\mathrm{Hom}_B(B,u)} B \\
\| \quad\ \Downarrow\lambda \quad\ \| \quad\ \| \\
J \xrightarrow[\mathrm{Hom}_A(A,d)]{} A \xrightarrow[\mathrm{Hom}_B(B,u)]{} B \\
\| \quad\quad\ \Downarrow\circ \quad\quad\ \| \\
J \xrightarrow[\mathrm{Hom}_B(B,ud)]{} B
\end{array}
\ =\
\begin{array}{ccc}
J \xtwoheadrightarrow{W} K \xrightarrow{\mathrm{Hom}_A(A,\lim_W d)} A \xrightarrow{\mathrm{Hom}_B(B,u)} B \\
\| \quad\ \| \quad\ \Downarrow\circ \quad\ \| \\
J \xtwoheadrightarrow{W} K \xrightarrow[\mathrm{Hom}_B(B,u\,\lim_W d)]{} B \\
\| \quad\quad\ \exists!\Downarrow u\lambda \quad\quad\ \| \\
J \xrightarrow[\mathrm{Hom}_B(B,ud)]{} B
\end{array}
$$

is again a right extension of modules. To that end consider a cell

$$
\begin{array}{ccc}
J & \xrightarrow{\ W\ } K \cdots\overset{\vec{E}}{\cdots\!\!\rightarrow} & B \\
\| & \Downarrow & \| \\
J & \xrightarrow[\mathrm{Hom}_B(B,ud)]{} & B
\end{array}
$$

By Corollary 4.1.3, there is an equivalence $\mathrm{Hom}_B(B, ud) \simeq \mathrm{Hom}_A(f, d)$ and hence by Proposition 8.2.1 there is a restriction cell

$$
\begin{array}{ccc}
J & \xrightarrow{\ \mathrm{Hom}_B(B,ud)\ } & B \\
\| & \Downarrow\rho & \downarrow f \\
J & \xrightarrow[\mathrm{Hom}_A(A,d)]{} & A
\end{array}
$$

so that composition with ρ induces a bijection between the cells $W \times \vec{E} \Rightarrow$

$\operatorname{Hom}_B(B, ud)$ and cells of the form displayed below-left:

$$
\begin{array}{ccc}
J \xrightarrow{W} K \cdots\overset{\vec{E}}{\cdots}\!\!> B & & J \xrightarrow{W} K \cdots\overset{\vec{E}}{\cdots}\!\!> B \xrightarrow{\operatorname{Hom}_A(A,f)} A \\[2pt]
\| \quad\; \Downarrow \quad\;\; \Big\downarrow f \;\;\leftsquigarrow\!\!\rightsquigarrow & & \| \qquad\quad\; \Downarrow \qquad\qquad\quad \| \\[2pt]
J \xrightarrow[\operatorname{Hom}_A(A,d)]{} A & & J \xrightarrow[\operatorname{Hom}_A(A,d)]{} A
\end{array}
$$

By Theorem 8.4.4 these cells are in natural bijection with cells of the form displayed above-right. Since $\operatorname{Hom}_A(A, \lim_W d)$ is a right extension of $\operatorname{Hom}_A(A, d)$ along W, such cells are in bijection with cells of the form below-left

$$
\begin{array}{ccc}
K \cdots\overset{\vec{E}}{\cdots}\!\!> B \xrightarrow{\operatorname{Hom}_A(A,f)} A & & K \cdots\overset{\vec{E}}{\cdots}\!\!> B \\[2pt]
\| \qquad\quad \Downarrow \qquad\qquad \| \;\;\leftsquigarrow\!\!\rightsquigarrow & & \| \quad\; \Downarrow \quad\; \Big\downarrow f \\[2pt]
K \xrightarrow[\operatorname{Hom}_A(A,\lim_W d)]{} A & & K \xrightarrow[\operatorname{Hom}_A(A,\lim_W d)]{} A
\end{array}
$$

and by Theorem 8.4.4 again these cells are in natural bijection with cells of the form displayed above-right. By Proposition 8.2.1, such cells stand in bijection with the cells defined by factoring through the restriction $\operatorname{Hom}_A(f, \lim_W d)$ of $\operatorname{Hom}_A(A, \lim_W d)$ along $f : B \to A$. By Corollary 4.1.3, there is an equivalence of modules $\operatorname{Hom}_A(f, \lim_W d) \simeq \operatorname{Hom}_B(B, u \lim_W d)$ so via this restriction the cells are in bijection to cells of the form

$$
\begin{array}{c}
K \cdots\overset{\vec{E}}{\cdots}\!\!> B \\[2pt]
\| \quad\; \Downarrow \quad\;\; \| \\[2pt]
K \xrightarrow[\operatorname{Hom}_B(B,u \lim_W d)]{} B
\end{array}
$$

When we implement this bijection starting from the identity cell at the module $\operatorname{Hom}_B(B, u \lim_W d)$ and reverse this composite bijection we obtain the cell $u\lambda : W \times \operatorname{Hom}_B(B, u\lim_W d) \Rightarrow \operatorname{Hom}_B(B, ud)$, so this proves that $u\lambda$ displays $\operatorname{Hom}_B(B, u \lim_W d)$ as the right extension of $\operatorname{Hom}_B(B, ud)$ along W. Thus $u \lim_W d$ is the W-weighted limit of ud as claimed. $\qquad\square$

Theorem 9.5.7 should be compared with Proposition A.6.20, which observes that the right adjoints of enriched category theory preserve weighted limits, as defined in §A.6. In ∞-cosmoi of $(\infty, 1)$-categories, we can think of the results of this section as developing an analogous theory of weighted limits and colimits for categories weakly enriched over ∞-groupoids. In particular, adjunctions between $(\infty, 1)$-categories are automatically enriched over ∞-groupoids, as alluded to in Remark 4.1.4, so they should be expected to preserve limits weighted by modules, whose fibers are ∞-groupoids.

Exercises

EXERCISE 9.5.i. Determine weights $W_{\lim}$ and W_{colim} so that the $W_{\lim}$-weighted limit of a diagram f is the ordinary limit of f, as defined in 9.4.7, and the W_{colim}-weighted colimit is the ordinary colimit.

EXERCISE 9.5.ii. Define the **tensor** and **cotensor** of an element $a : 1 \to A$ of an ∞-category by a discrete ∞-category S as a weighted colimit and weighted limit, respectively, and compare this construction with Definition 4.3.6.

PART THREE

MODEL INDEPENDENCE

The rapidly proliferating literature on $(\infty, 1)$-categories begs the following question: if a theorem has been proven for one model of $(\infty, 1)$-categories, does it apply to them all?

To discuss this concern, it is useful to distinguish between "synthetically proven" theorems and "analytically proven" theorems about $(\infty, 1)$-categories. Synthetically proven theorems include the myriad results found in Parts I and II of this text about the objects in an arbitrary ∞-cosmos. While these results can of course be specialized to any particular ∞-cosmos of $(\infty, 1)$-categories, their statements and proofs are entirely agnostic to any particular model that may ultimately be used. Such results are "model independent" in a strong sense.

The situation is more delicate for analytically proven theorems, whose statements and proofs might take advantage of the features of a particular model. For instance, Barwick, Glasman, and Nardin prove that for any cartesian fibration $p: E \twoheadrightarrow B$ between quasi-categories, there is a cocartesian fibration $p^\vee: E^\vee \twoheadrightarrow B^{\mathrm{op}}$ so that the fibers $E_b \simeq E_b^\vee$ are equivalent [7].[3] This result would be useful to have in all models.

A minor obstacle is presented by the statement, which references the opposite $(\infty, 1)$-category. While an opposite-category involution is not axiomatized as part of the structure of an ∞-cosmos, in practice a construction along these lines is easy to give or can be transferred from another model (see §12.1). Once opposite $(\infty, 1)$-categories are understood, a larger challenge is presented by the proof, which appeals to the *twisted arrow quasi-category* construction [28, §5.6]. Certainly one could transfer that particular quasi-category to another model, but the analytic aspects of the proof, involving explicit horns and simplices in that quasi-category, are no longer so easy to express.

The larger question is not just about the formal theory of $(\infty, 1)$-categories but also concerns concrete examples. The prototypical $(\infty, 1)$-category is the $(\infty, 1)$-category of spaces, but how can it be defined? One strategy is give a clever characterization of its universal property so that it can be characterized in any model: for instance, the $(\infty, 1)$-category of spaces is freely generated by the point under colimits [78, 5.1.5.6]. But even so, it is necessary to prove that there exists an $(\infty, 1)$-category with this property, which involves an explicit construction in a particular model.

To address these sorts of questions, we begin in Chapter 10 with the formal study of the best behaved class of change-of-model functors, namely the *cosmological biequivalences* (see Definition 1.3.8). Not all ∞-cosmoi are biequivalent, but there typically exist biequivalences connecting ∞-cosmoi whose objects are

---

[3] In fact, they prove that both fibrations classify the same contravariant B-indexed functor $b \mapsto E_b$ valued in the $(\infty, 1)$-category of $(\infty, 1)$-categories, but to simplify this discussion, we focus on the first part of the statement.

infinite-dimensional categories of the same type, supplied by the experts who have developed the various models (see Appendix E). In Proposition 10.3.6, we prove that cosmological biequivalences preserve, reflect, and create a lengthy list of ∞-categorical structures. Cosmological biequivalences also induce various bijections: between ∞-categories up to equivalence, between ∞-functors up to isomorphism modulo these equivalences, between modules up to fibered equivalence, and between natural transformations with specified boundary. These results can be applied to transfer an explicit adjoint to a given functor or a colimit of a given diagram between models, as we illustrate in Chapter 11. More systematically, any biequivalence of ∞-cosmoi induces a biequivalence between their virtual equipments of modules.

We consider this result, recorded as Theorem 11.1.6, as the basis for the model independence of ∞-category, given how much of the theory of ∞-categories can be expressed in a virtual equipment leveraging the various embeddings of the homotopy 2-category. For instance, it subsumes the model invariance results recorded in Proposition 10.3.6. More profoundly, the virtual equipment of modules forms the basis for the formal language for model independent ∞-category theory that is introduced in §11.3. The study of the model independence of ∞-category theory concludes with Corollary 11.3.10, which proves that formulae written in this formal language are invariant under biequivalence of ∞-cosmoi precisely because such a biequivalence induces a biequivalence between the virtual equipments of modules.

The key takeaway is that the conclusions of both synthetically and analytically proven theorems about ∞-categories can be transferred to biequivalent ∞-cosmoi. In Chapter 12, we specialize to ∞-cosmoi of (∞, 1)-categories (see Example 1.2.24 and Definition 1.3.10) to illustrate applications of this transfer principle to (∞, 1)-category theory: introducing opposite ∞-categories and their ∞-groupoid cores, establishing the pointwise nature of universal properties, and proving an existence theorem for pointwise right and left Kan extensions.

10

Change-of-Model Functors

In this chapter, we study a certain class of cosmological functors between ∞-cosmoi that do not merely preserve ∞-categorical structure but also reflect and create it. We refer to these functors as *cosmological biequivalences* because the 2-functors they induce between homotopy 2-categories are *biequivalences*: surjective on objects up to equivalence and defining a local equivalence of hom-categories. Informally, we refer to cosmological biequivalences as "change-of-model functors." For example, the four ∞-cosmoi of $(\infty, 1)$-categories introduced in Example 1.2.24 are connected by the following biequivalences briefly described in Example 1.3.9 and revisited in §E.2:

$$
\begin{array}{ccc}
\mathcal{C}SS & \xrightarrow{\quad \text{disc} \quad} & \mathcal{S}egal \\
\end{array}
$$

$$(10.0.1)$$

In §10.1, we collect together a number of results about cosmological functors that are scattered throughout the text. In §10.2, we reintroduce the special class of biequivalences and discuss general examples. In particular, we discover that the ∞-cosmology of Chapter 6 is biequivalence invariant: for instance, a cosmological biequivalence $\mathcal{K} \rightsquigarrow \mathcal{L}$ induces a cosmological biequivalence $\mathcal{C}art(\mathcal{K}) \rightsquigarrow \mathcal{C}art(\mathcal{L})$.

Since our biequivalences between ∞-cosmoi are required to be cosmological functors, resembling enriched right Quillen adjoints, the relation "admits a biequivalence to" is not symmetric. Thus, when we say that two ∞-cosmoi "are biequivalent" we mean that there exists a finite zigzag of biequivalences connecting them, in other words, that they lie in the same equivalence class under the symmetric transitive closure of the relation defined by the presence of a cosmological biequivalence. In particular, under this definition, an ∞-*cosmos of* $(\infty, 1)$-*categories* is an ∞-cosmos that is connected by a finite zigzag of cosmological biequivalences to any of the ∞-cosmoi in (10.0.1). In this special case, a simpler characterization is established in Proposition 10.2.1, which proves that an ∞-cosmos $\mathcal{K}$ is an ∞-cosmos of $(\infty, 1)$-categories just when the underlying quasi-category functor $(-)_0 := \mathrm{Fun}(1, -) \colon \mathcal{K} \rightsquigarrow \mathcal{QC}at$ is a cosmological biequivalence – retroactively justifying Definition 1.3.10.

In §10.3, we establish the basic 2-categorical properties of biequivalences, which provides an essential ingredient in the proof of the model independence results in Chapter 11. Finally, in §10.4, we explore the properties of formally defined "inverses" to cosmological biequivalences. These are not guaranteed to be cosmological functors, nor even simplicial functors in the customary strict sense. The situation is analogous to 2-category theory: a 2-functor that defines a biequivalence admits an inverse biequivalence but this may only be a *pseudofunctor*, a notion recalled in Definition 10.4.1. Accordingly, the inverse to a cosmological biequivalence defines a "quasi-categorically enriched pseudofunctor" or *quasi-pseudofunctor* for short that is not cosmological but does define a biequivalence and so that the composite endofunctors are *quasi-pseudonaturally equivalent* to the identity functors on each ∞-cosmos. These structures reappear in Chapter 12 as tools to transport analytically defined structures between biequivalent ∞-cosmoi.

10.1 Cosmological Functors Revisited

Recall from Definition 1.3.1 that a **cosmological functor** is a simplicial functor $F \colon \mathcal{K} \to \mathcal{L}$ between ∞-cosmoi that preserves

- the specified classes of isofibrations and
- all of the cosmological limits.

Lemma 1.3.2 demonstrates that cosmological functors also preserve the equivalences and the trivial fibrations. By Proposition 6.2.8(i), cosmological functors also preserve all flexible weighted limits.

Examples of cosmological functors abound – for instance, see Proposition 1.3.4 and the cosmological embeddings of replete sub ∞-cosmoi constructed in §6.3. There are also cosmological functors connecting the ∞-cosmoi of fibrant diagrams indexed by an inverse category (see Exercise 6.1.iii), such as the domain, codomain, and identity functors

$$\mathcal{K}^{\downarrow} \underset{\substack{\longleftarrow \\ \text{cod}}}{\overset{\substack{\text{dom} \\ \longrightarrow \\ \longleftarrow \text{id} \longrightarrow}}{}} \mathcal{K}$$

which are shown to be cosmological in Proposition 6.1.1.

Cosmological functors frequently restrict to define further cosmological functors between parallel replete sub ∞-cosmoi:

LEMMA 10.1.1. *Suppose that* $\mathcal{K}' \hookrightarrow \mathcal{K}$ *and* $\mathcal{L}' \hookrightarrow \mathcal{L}$ *are cosmological embeddings of replete sub ∞-cosmoi. If* $F: \mathcal{K} \to \mathcal{L}$ *is a cosmological functor that carries objects and 0-arrows of* $\mathcal{K}'$ *to objects and 0-arrows of* $\mathcal{L}'$ *then the restricted functor is cosmological:*

$$\begin{array}{ccc} \mathcal{K}' & \overset{F}{\dashrightarrow} & \mathcal{L}' \\ \Big\downarrow & & \Big\downarrow \\ \mathcal{K} & \underset{F}{\longrightarrow} & \mathcal{L} \end{array}$$

Proof Recall that the repleteness of Definition 6.3.1 includes the requirement that the inclusions $\mathcal{K}' \hookrightarrow \mathcal{K}$ and $\mathcal{L}' \hookrightarrow \mathcal{L}$ are full on positive dimensional arrows. So if the simplicial functor F carries objects and 0-arrows of $\mathcal{K}'$ to objects and 0-arrows of $\mathcal{L}'$ then it restricts to define a simplicial functor $F': \mathcal{K}' \to \mathcal{L}'$. As cosmological embeddings create isofibrations and the cosmological limit notions, the restricted functor is cosmological. □

For example, by Proposition 5.2.4 and Exercise 5.3.i, pullback $f^*: \mathcal{K}_{/B} \to \mathcal{K}_{/A}$ preserves cartesian fibrations and cartesian functors. Thus, pullback along any functor $f: A \to B$ in $\mathcal{K}$ restricts to define a cosmological functor (see Proposition 7.2.4):

$$\begin{array}{ccc} \mathcal{C}art(\mathcal{K})_{/B} & \overset{f^*}{\dashrightarrow} & \mathcal{C}art(\mathcal{K})_{/A} \\ \Big\downarrow & & \Big\downarrow \\ \mathcal{K}_{/B} & \underset{f^*}{\longrightarrow} & \mathcal{K}_{/A} \end{array}$$

Our aim in this section is to show that cosmological functors preserve all of the ∞-categorical structures we have introduced – with a single notable exception discussed in Warning 10.1.5. In many cases this is not evident from the original 2-categorical definitions (e.g., of cartesian fibrations in Definition 5.2.1) but can be deduced quite easily from the accompanying "internal characterization" of each categorical notion (such as given in Theorem 5.2.8(ii)).

Importantly:

PROPOSITION 10.1.2. *Cosmological functors preserve comma ∞-categories: if* $F\colon \mathcal{K} \to \mathcal{L}$ *is a cosmological functor and the diagram below-left is a comma cone in* $\mathcal{K}$, *then the diagram below-right is a comma cone in* $\mathcal{L}$.

$$
\begin{array}{ccc}
 & E & \\
{}^{e_1}\swarrow & & \searrow{}^{e_0} \\
C & \overset{\epsilon}{\Leftarrow} & B \\
{}_g\searrow & & \swarrow{}_f \\
 & A &
\end{array}
\quad\overset{F}{\rightsquigarrow}\quad
\begin{array}{ccc}
 & FE & \\
{}^{Fe_1}\swarrow & & \searrow{}^{Fe_0} \\
FC & \overset{F\epsilon}{\Leftarrow} & FB \\
{}_{Fg}\searrow & & \swarrow{}_{Ff} \\
 & FA &
\end{array}
\qquad (10.1.3)
$$

Proof The simplicial pullback (3.4.2) that constructs the comma cone is preserved by any cosmological functor. By Proposition 3.4.11, any comma cone as above left arises from a fibered equivalence $\ulcorner\epsilon\urcorner\colon E \simeq_{C\times B} \mathrm{Hom}_A(f,g)$ where $\epsilon = \phi\ulcorner\epsilon\urcorner$, and any fibered equivalence of this form defines a comma cone. Since F defines a cosmological functor $F\colon \mathcal{K}_{/C\times B} \to \mathcal{L}_{/FC\times FB}$, $F\ulcorner\epsilon\urcorner\colon FE \simeq_{FC\times FB} F(\mathrm{Hom}_A(f,g)) \cong \mathrm{Hom}_{FA}(Ff,Fg)$, and we conclude that the right-hand data again defines a comma cone. □

Using Proposition 10.1.2, we can quickly establish the following preservation properties of cosmological functors. For ease of reference, this list includes the preservation properties established elsewhere.

PROPOSITION 10.1.4. *Cosmological functors preserve:*

   (i) *Equivalences between ∞-categories.*
  (ii) *Invertible natural transformations and mates.*
 (iii) *Adjunctions between ∞-categories, including right adjoint right inverse adjunctions and left adjoint right inverse adjunctions.*
 (iv) *Fibered adjunctions and equivalences.*
  (v) *Isofibrations, trivial fibrations and discrete ∞-categories.*
 (vi) *Flexible weighted limits.*
(vii) *Comma spans and comma cones.*
(viii) *Absolute right and left lifting diagrams.*
 (ix) *Limits or colimits of diagrams indexed by a simplicial set and co/limit-preserving functors.*
  (x) *Stable ∞-categories and exact functors.*
 (xi) *Cartesian and cocartesian fibrations and cartesian functors between them.*
(xii) *Discrete cartesian fibrations and discrete cocartesian fibrations.*
(xiii) *Two-sided fibrations and cartesian functors between them.*
(xiv) *Modules and represented modules.*

Proof By Lemma 1.4.4, a cosmological functor induces a 2-functor between homotopy 2-categories, and an arbitrary 2-functor preserves the structures of (i), (ii), and (iii) (see e.g., Lemma 2.1.3). As cosmological functors induce cosmological functors between sliced ∞-cosmoi, (iv) can be understood as a special case of (i) and (iii).

The preservation of trivial fibrations is established in Lemma 1.3.2 and the preservation of discrete ∞-categories follows (see Remark 1.3.3). Proposition 6.2.8(i) proves that cosmological functors preserve all flexible weighted limits as stated in (vi). Proposition 10.1.2 proves that cosmological functors preserve comma spans in the ∞-cosmos and comma cones in the homotopy 2-category as stated in (vii).

The preservation property (viii), first observed in Corollary 3.5.7, follows from Theorem 3.5.3, which characterizes absolute lifting diagrams as fibered equivalences of comma ∞-categories, Proposition 10.1.2, which says that cosmological functors preserve commas, and the fact that cosmological functors preserve equivalences. Now (ix) follows from this by Definition 2.3.8 and the fact that cosmological functors preserve simplicial tensors, with the statement about co/limit preserving functors following from (ii). By Theorem 4.4.12(iii), (x) can be understood as a special case of (ix).

The preservation properties (xi) and (xii) also follow from Proposition 10.1.2 and the fact that cosmological functors preserve right or left adjoint right inverse adjunctions, mates, and trivial fibrations via the characterizations of Theorem 5.2.8(ii), Theorem 5.3.4(ii), and Proposition 5.5.8. More details are given in the proof of Corollary 5.3.5, which also observes that cartesian natural transformations are preserved by cosmological functors.

Directly from the internal characterization of Theorem 7.1.4 and the preservation of adjunctions and invertible natural transformations, cartesian functors preserve two-sided fibrations and cartesian functors between them, as observed in Corollary 7.1.8. This establishes (xiii). By Proposition 1.3.4(vi), a cosmological functor induces a direct image cosmological functor between sliced ∞-cosmoi, which then preserves discrete objects by (v). Thus, modules are also preserved. By specializing Proposition 10.1.2 to cospans involving identities, it becomes clear that left and right representable commas are preserved. Since a module is representable if and only if it is fibered equivalent to one of these, representable modules are preserved as well, completing the proof of (xiv). □

WARNING 10.1.5. Conspicuously missing from the list of ∞-categorical structures that are preserved by cosmological functors are composites in the virtual equipment of modules and then a variety of further structures that were defined in reference to this notion: exact squares, pointwise right and left extensions, and (weighted) limits and colimits of ∞-category indexed diagrams.

There are a number of factors that contribute to the failure of composites and strong composites to be preserved by cosmological functors in general. One immediate issue is presented by the universal properties given in Definitions 8.3.1 and 8.3.5, which each involve universal quantifiers. As cosmological functors need not be essentially surjective, a universal quantifier in the domain need not scope over the codomain.

In other contexts, universal quantifiers are done away with by means of an "internal characterization" of the ∞-categorical notion, but no such characterization is given in this case. At issue is the fact that the composite module should be understood as defined using a "pullback stable fiberwise coinverter," reflecting the two-sided fibration formed by the composite of the spans into discrete two-sided fibrations. We have not presented such a construction because our axiomatic notion of ∞-cosmos does not include colimits, but even if it did – and Digression E.1.8 reveals that homotopy colimits of ∞-categories are often present in examples – cosmological functors, being "right-adjoint like" would not preserve them.

That said, any strong composites that are obtained by applying Lemma 8.3.7 are preserved by cosmological functors, since fibered adjunctions are preserved. This includes all of the formally defined composites established in §8.3 and §8.4. Similar remarks apply to exact squares. Because generic composites are not preserved by cosmological functors, generic exact squares need not be preserved either. But the formally defined exact squares established in §9.2 are preserved by cosmological functors. For instance, Lemma 9.2.6 proves that comma squares are exact squares in any ∞-cosmos. Since cosmological functors preserve comma square, this class of exact squares is then preserved.

As we shall discover, cosmological biequivalences *do* preserve all of these ∞-categorical notions, as well as reflect and create them. It is to this subject that we now turn.

Exercises

EXERCISE 10.1.i. Argue that any cosmological functor $F : \mathcal{K} \to \mathcal{L}$ induces a cosmological functor $F : {}_{A\backslash}\mathcal{F}ib(\mathcal{K})_{/B} \to {}_{FA\backslash}\mathcal{F}ib(\mathcal{L})_{/FB}$ for each pair of ∞-categories A and B in $\mathcal{K}$.

EXERCISE 10.1.ii. Exercise 6.3.iv shows that cosmological embeddings reflect discrete ∞-categories in addition to preserving them. What other ∞-categorical properties are reflected or created by cosmological embeddings?

10.2 Cosmological Biequivalences

Special cosmological functors, the biequivalences, reflect and create the ∞-category theory developed in this text as well as preserve it. Recall from Definition 1.3.8 that a **cosmological biequivalence** is a cosmological functor $F \colon \mathcal{K} \to \mathcal{L}$ that is

- surjective on objects up to equivalence: for all $C \in \mathcal{L}$ there exists $A \in \mathcal{K}$ so that $FA \simeq C$; and
- a local equivalence of quasi-categories: for every pair $A, B \in \mathcal{K}$, the map

$$\mathsf{Fun}(A, B) \xrightarrow{\ \sim\ } \mathsf{Fun}(FA, FB)$$

is an equivalence of quasi-categories.

In this section, we pursue further examples of cosmological biequivalences. In the next section, we explore their role as change-of-model functors.

For example, the functors (10.0.1) are all biequivalences (see Example 1.3.9 and §E.2). Except for the functor $\natural \colon \mathcal{QCat} \to 1\text{-}\mathcal{Comp}$, each arises from a simplicially enriched right Quillen equivalence between model categories enriched over the Joyal model structure with all objects cofibrant. Corollary E.1.2 demonstrates that any functor of this form encodes a cosmological biequivalence.

Two ∞-cosmoi are **biequivalent** if there exists a finite zigzag of biequivalences connecting them. Recall that any ∞-cosmos has an **underlying quasi-category functor**

$$\mathcal{K} \xrightarrow{\ (-)_0 := \mathsf{Fun}(1, -)\ } \mathcal{QCat}$$

defined by mapping out of the terminal ∞-category. We now show that the underlying quasi-category functor of any ∞-cosmos that is biequivalent to $\mathcal{QCat}$ is a cosmological biequivalence. This justifies the characterization of an **∞-cosmos of $(\infty, 1)$-categories** given in Definition 1.3.10.

PROPOSITION 10.2.1. *If an ∞-cosmos $\mathcal{K}$ is biequivalent to $\mathcal{QCat}$, then the underlying quasi-category functor $(-)_0 \colon \mathcal{K} \twoheadrightarrow \mathcal{QCat}$ is a cosmological biequivalence.*

Proof We will show that in the presence of cosmological biequivalences

$$\mathcal{K} \xleftarrow{\ G\ }_{\sim} \mathcal{L} \xrightarrow{\ F\ }_{\sim} \mathcal{QCat}$$

the underlying quasi-category functor $(-)_0 \colon \mathcal{K} \to \mathcal{QCat}$ is a cosmological biequivalence. Note this formulation includes the special cases where one of the functors F or G is an identity. By induction, the same conclusion holds for any ∞-cosmos connected by a finite zigzag of biequivalences to $\mathcal{QCat}$.

As F is a biequivalence, for each quasi-category Q, there exists an ∞-category $B \in \mathcal{L}$ so that $FB \simeq Q$. Because F and G are both local equivalences preserving the terminal ∞-category 1 – for which we adopt the same notation in each of $\mathcal{K}, \mathcal{L}$, and $\mathcal{QC}at$ – there is then a zigzag of equivalences of quasi-categories

$$(GB)_0 = \mathsf{Fun}(1, GB) \xleftarrow{\;\sim\;} \mathsf{Fun}(1, B) \xrightarrow{\;\sim\;} \mathsf{Fun}(1, FB) \cong FB \simeq Q.$$

This proves that there exists an ∞-category $GB \in \mathcal{K}$ whose underlying quasi-category is equivalent to Q.

To show that the underlying quasi-category functor $(-)_0 : \mathcal{K} \to \mathcal{QC}at$ is a local equivalence, consider a pair of ∞-categories $A, B \in \mathcal{K}$. By essential surjectivity of G, there exist ∞-categories $X, Y \in \mathcal{L}$ so that $GX \simeq A$ and $B \simeq GY$. By pre- and postcomposing with these equivalences, Corollary 1.4.8 implies that $\mathsf{Fun}(A, B) \to \mathsf{Fun}(A_0, B_0)$ is equivalent to $\mathsf{Fun}(GX, GY) \to \mathsf{Fun}((GX)_0, (GY)_0)$, so it suffices to prove that the latter map is an equivalence of quasi-categories.

By simplicial functoriality (see Definition A.2.6), the actions on functor spaces of F and G commutes with the composition map

$$\mathsf{Fun}(X, Y) \times \mathsf{Fun}(1, X) \xrightarrow{\;\circ\;} \mathsf{Fun}(1, Y)$$

which transposes to define the action on functor spaces of the underlying quasi-category functor. Thus, there is a commutative diagram whose vertical maps are equivalences

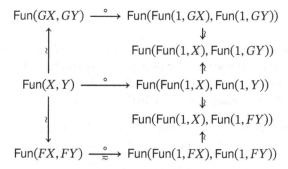

Any quasi-category is isomorphic to its underlying quasi-category, so the bottom horizontal map is an isomorphism. By the 2-of-3 property, it follows that the top horizontal map is an equivalence, which is what we wanted to show. □

Recall from Proposition 1.3.4(vi) that a cosmological functor $F : \mathcal{K} \to \mathcal{L}$ induces a cosmological functor $F : \mathcal{K}_{/B} \to \mathcal{L}_{/FB}$ for any $B \in \mathcal{K}$.

Proposition 10.2.2. *If $F : \mathcal{K} \rightsquigarrow \mathcal{L}$ is a cosmological biequivalence, then*

for any B ∈ 𝒦 the induced functor F : 𝒦$_{/B}$ ⇸ ℒ$_{/FB}$ is also a cosmological biequivalence.

Proof We first argue that the F defines a local equivalence of functor spaces, as defined in Proposition 1.2.22(ii). Given a pair of isofibration $p : E ↠ B$ and $p' : E' ↠ B$ in 𝒦, the induced map on fibered functor spaces is defined by the pullback

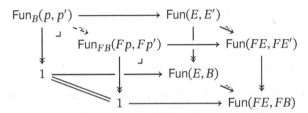

As the maps between the cospans in 𝒬𝒞at are equivalences, by Proposition 3.3.4 so is the induced map between the pullbacks.

For surjectivity up to equivalence, consider an isofibration $q : L ↠ FB$ in ℒ. As F is surjective on objects up to equivalence, there exists some $A ∈ 𝒦$ together with an equivalence $i : FA ⇸ L ∈ ℒ$. As F defines a local equivalence of mapping quasi-categories, there is moreover a functor $f : A → B$ in ℒ so that $Ff : FA → FB$ is naturally isomorphic to qi (see Exercise 10.2.i). The map f need not be an isofibration, but Lemma 1.2.19 allows us to factor f as $A ⇸ K \xrightarrow{p} B$. Choosing an equivalence inverse $j : K ⇸ A$, this data defines a diagram in h̲ℒ that commutes up to isomorphism:

$$FK \xrightarrow{\ Fj\ } FA \xrightarrow{\ i\ } L$$
$$Fp \searrow \ \overset{≅}{\underset{FB}{\underset{\downarrow}{Ff}}} \ \overset{≅}{\nwarrow} q$$

Proposition 1.4.9 tells us that isofibrations in ∞-cosmoi define isofibrations in the homotopy 2-category. In particular, we may lift the displayed isomorphism along the isofibration $q : L ↠ FB$ to define a commutative triangle:

$$FK \xrightarrow{\ Fj\ } FA \xrightarrow{\ i\ } L \qquad FK \xrightarrow{\ i·Fj\ } L$$
$$Fp \searrow \ \overset{≅}{\underset{FB}{\underset{\downarrow}{Ff}}} \ \overset{≅}{\nwarrow} q \quad = \quad Fp \searrow \ \overset{≅}{\underset{e}{}} \ \nwarrow q$$

By Exercise 1.4.iii, since e is isomorphic to an equivalence $i · Fj$, it must also define an equivalence. Thus, by Proposition 1.2.22(vii), we have shown that the isofibration $p : K \twoheadrightarrow B$ maps under F to an isofibration that is equivalent to our chosen $q : L \twoheadrightarrow FB$. □

A similar argument proves that a cosmological biequivalence induces a biequivalence between the corresponding ∞-cosmoi of isofibrations of Proposition 6.1.1.

PROPOSITION 10.2.3. *If* $F : \mathcal{K} \overset{\sim}{\twoheadrightarrow} \mathcal{L}$ *is a cosmological biequivalence then the induced functor* $F : \mathcal{K}^{\Downarrow} \overset{\sim}{\twoheadrightarrow} \mathcal{L}^{\Downarrow}$ *is a biequivalence.*

Proof Exercise 10.2.ii. □

We establish another family of biequivalences of sliced ∞-cosmoi:

PROPOSITION 10.2.4. *If* $f : A \overset{\sim}{\twoheadrightarrow} B$ *is an equivalence in* $\mathcal{K}$, *then the pullback functor* $f^* : \mathcal{K}_{/B} \overset{\sim}{\twoheadrightarrow} \mathcal{K}_{/A}$ *is a cosmological biequivalence.*

Proof To see that $f^* : \mathcal{K}_{/B} \to \mathcal{K}_{/A}$ is essentially surjective, consider an object $r : D \twoheadrightarrow A$ and use Lemma 1.2.19 to factor the composite map $fr : D \to B$ as an equivalence followed by an isofibration, and pull the result back along f.

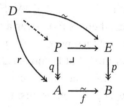

By Proposition 3.3.3, the pullback of f is an equivalence, so by the 2-of-3 property, r is equivalent to the isofibration $q : P \twoheadrightarrow A$, which is in the image of $f^* : \mathcal{K}_{/B} \to \mathcal{K}_{/A}$.

To show that this simplicial functor is a local equivalence, consider a pair of isofibrations $p : E \twoheadrightarrow B$ and $q : F \twoheadrightarrow B$. We will show that the quasi-category of functors over B is equivalent to the quasi-category of functors over A between their pullbacks

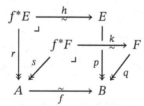

To define the comparison map $\mathrm{Fun}_B(E, F) \to \mathrm{Fun}_A(f^*E, f^*F)$ consider the

following commutative prism

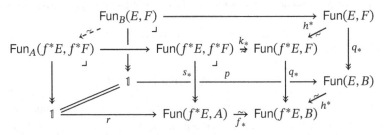

The front-right square is a pullback by the simplicial universal property of f^*F, while the front-left square and back face are the pullbacks that define $\mathrm{Fun}_A(f^*E, f^*F)$ and $\mathrm{Fun}_B(E, F)$. The universal property of the composite front pullback rectangle induces the map $\mathrm{Fun}_B(E, F) \to \mathrm{Fun}_A(f^*E, f^*F)$. As this functor is the pullback of the equivalences h^* of Corollary 1.4.8, by Proposition 3.3.4 the induced map defines an equivalence of quasi-categories. $\qquad\square$

Further induced cosmological biequivalences arise by application of the following lemma, which revisits the setting of Proposition 6.3.3, a result used to construct cosmologically embedded ∞-cosmoi. We leave its applications to the exercises and return to this topic in Corollary 10.3.7.

LEMMA 10.2.5. *Consider a pullback diagram of ∞-cosmoi and cosmological functors in which* $F\colon \mathcal{K} \twoheadrightarrow \mathcal{L}$ *is a cosmological biequivalence and* $\mathcal{L}'\hookrightarrow\mathcal{L}$ *and* $\mathcal{K}'\hookrightarrow\mathcal{K}$ *are cosmological embeddings.*[1]

$$
\begin{array}{ccc}
\mathcal{K}' & \xrightarrow{\ F\ } & \mathcal{L}' \\
\downarrow & \lrcorner & \downarrow \\
\mathcal{K} & \xrightarrow[\ \sim\ ]{F} & \mathcal{L}
\end{array}
$$

Then $F\colon \mathcal{K}' \twoheadrightarrow \mathcal{L}'$ *is a cosmological biequivalence.*

Proof The essential point is the repleteness of the cosmological embedding $\mathcal{L}'\hookrightarrow\mathcal{L}$ enumerated in Definition 6.3.1 and explored in Exercise 6.3.i. To see that $F\colon \mathcal{K}' \to \mathcal{L}'$ is essentially surjective, consider an ∞-category $L \in \mathcal{L}'$. By essential surjectivity of the cosmological biequivalence $F\colon \mathcal{K} \twoheadrightarrow \mathcal{L}$ there exists an ∞-category $K \in \mathcal{K}$ so that $FK \simeq L$ in $\mathcal{L}$. By repleteness of $\mathcal{L}'\hookrightarrow\mathcal{L}$, the equivalence $FK \simeq L$ also lies in $\mathcal{L}'$ and hence K lies in the pullback $\mathcal{K}'$ and maps via F to an object equivalent to L in $\mathcal{L}'$.

---

[1] If the ∞-cosmos structure on $\mathcal{K}'$ is created by this pullback, then $\mathcal{K}'\hookrightarrow\mathcal{K}$ is automatically a cosmological embedding and the functor $\mathcal{K}' \to \mathcal{L}'$ is automatically cosmological (see Proposition 6.3.3).

For the local equivalence, the induced mappings on functor spaces associated to a pair of ∞-categories A and B in $\mathcal{K}'$ form a pullback of quasi-categories

$$
\begin{array}{ccc}
\mathrm{Fun}_{\mathcal{K}'}(A,B) & \xrightarrow{\ F\ } & \mathrm{Fun}_{\mathcal{L}'}(FA,FB) \\
\Big\uparrow & \lrcorner & \Big\uparrow \\
\mathrm{Fun}_{\mathcal{K}}(A,B) & \xrightarrow[\ \widetilde{\ }\ ]{F} & \mathrm{Fun}_{\mathcal{L}}(FA,FB)
\end{array}
$$

By Exercise 6.3.i, the vertical functors are isofibrations between quasi-categories. Thus, by Proposition 3.3.3, the local equivalence of $F \colon \mathcal{K} \to \mathcal{L}$ pulls back to define a local equivalence for $F \colon \mathcal{K}' \to \mathcal{L}'$. So we conclude that the restricted functor remains a cosmological biequivalence.   □

Exercises

EXERCISE 10.2.i (10.3.1). Demonstrate that an equivalence of quasi-categories $f \colon K \xrightarrow{\sim} L$ induces a bijection on isomorphism classes of vertices, i.e., a bijection on isomorphism classes of objects in the homotopy categories of K and L. Conclude that for any cosmological biequivalence $F \colon \mathcal{K} \xrightarrow{\sim} \mathcal{L}$ and any pair of ∞-categories $A, B \in \mathcal{K}$

(i) For any functor $f' \colon FA \to FB$ there exists a functor $f \colon A \to B$ and a natural isomorphism $Ff \cong f'$ in $\mathcal{L}$.

(ii) If $f, g \colon A \to B$ are functors so that $Ff \cong Fg$ in $\mathcal{L}$ then $f \cong g$ in $\mathcal{K}$.

EXERCISE 10.2.ii. Prove Proposition 10.2.3.

EXERCISE 10.2.iii. If $F \colon \mathcal{K} \xrightarrow{\sim} \mathcal{L}$ is a biequivalence and $A \in \mathcal{K}$ and $B \in \mathcal{L}$ are so that $FA \simeq B$ prove that the slice ∞-cosmoi $\mathcal{K}_{/A}$ and $\mathcal{L}_{/B}$ are biequivalent.

EXERCISE 10.2.iv (10.3.7). Explore the applicability of Lemma 10.2.5 to the task of verifying that the ∞-cosmoi constructed in §6.3 are biequivalence invariant.

10.3 Cosmological Biequivalences as Change-of-Model Functors

We refer to biequivalences between ∞-cosmoi as *change-of-model functors*. In this section, we enumerate their basic properties. First, we observe that cosmological biequivalences descend to biequivalences between homotopy 2-categories, hence the name:

PROPOSITION 10.3.1. *A cosmological biequivalence* $F \colon \mathcal{K} \xrightarrow{\sim} \mathcal{L}$ *induces a biequivalence* $F \colon \mathfrak{h}\mathcal{K} \xrightarrow{\sim} \mathfrak{h}\mathcal{L}$ *of homotopy 2-categories: i.e., the 2-functor is*

(i) *surjective on objects up to equivalence and*
(ii) *defines a local equivalence of categories* $\mathrm{hFun}(A, B) \xrightarrow{\sim} \mathrm{hFun}(FA, FB)$ *for all $A, B \in \mathcal{K}$.*

Note that the local equivalences of (ii) necessarily have the properties enumerated in Exercise 10.2.i.

Proof By Theorem 1.4.7, ∞-cosmos-level equivalences coincide with 2-categorical equivalences, proving (i), and by Lemma 1.2.15 the homotopy category functor $\mathrm{h} \colon \mathcal{QC}at \to \mathcal{C}at$ carries equivalences of quasi-categories to equivalences of categories, proving (ii). $\qquad\square$

In particular, it follows that the homotopy category of an ∞-category is invariant under change of model (see Exercise 10.3.i). More generally, any biequivalence between 2-categories induces a variety of local and global bijections, as enumerated below:

COROLLARY 10.3.2. *Consider any cosmological biequivalence $F \colon \mathcal{K} \xrightarrow{\sim} \mathcal{L}$.*

(i) *The biequivalence F preserves, reflects, and creates equivalences between ∞-categories and defines a bijection between equivalence classes of objects.*
(ii) *The biequivalence F induces local bijections between natuarl isomorphism classes of functors extending the bijections of (i): choosing any pairs of objects $A, B \in \mathcal{K}$ and $A', B' \in \mathcal{L}$ and equivalences $a \colon A' \xrightarrow{\sim} FA$ and $b \colon FB \xrightarrow{\sim} B'$, the map*

$$\mathrm{hFun}(A, B) \xrightarrow{\sim} \mathrm{hFun}(FA, FB) \xrightarrow{\sim} \mathrm{hFun}(A', B') \qquad (10.3.3)$$

defines a bijection between isomorphism classes of functors $A \to B$ in $\mathcal{K}$ and isomorphism classes of functors $A' \to B'$ in $\mathcal{L}$.
(iii) *The biequivalence F defines local bijections between natural transformations with specified boundary extending the bijections of (i) and (ii): choosing any pairs of objects $A, B \in \mathcal{K}$ and $A', B' \in \mathcal{L}$, specified equivalences $a \colon A' \simeq FA$ and $b \colon FB \simeq B'$, functors $f, g \colon A \to B$ and $f', g' \colon A' \to B'$, and natural isomorphisms*

$$
\begin{array}{ccc}
FA & \xrightarrow{Ff} & FB \\
a \uparrow \wr & \cong\alpha & \wr \downarrow b \\
A' & \xrightarrow{f'} & B'
\end{array}
\qquad
\begin{array}{ccc}
FA & \xrightarrow{Fg} & FB \\
a \uparrow \wr & \cong\beta & \wr \downarrow b \\
A' & \xrightarrow{g'} & B'
\end{array}
$$

the map (10.3.3) induces a bijection between natural transformations $f \Rightarrow g$ in $\mathcal{K}$ and natural transformations $f' \Rightarrow g'$ in $\mathcal{L}$.

Proof   Lemma 1.3.12 proves that cosmological biequivalences preserve, reflect, and create equivalences via a fundamentally 2-categorical argument that the reader is invited to revisit. This, together with essential surjectivity of cosmological biequivalences implies that such functors induce a bijection on equivalence classes of objects. This proves (i).

For (ii), by Corollary 1.4.8, the chosen equivalences $FA \simeq A'$ and $FB \simeq B'$ induce an equivalence of quasi-categories

$$\mathsf{Fun}(A, B) \overset{\sim}{\longrightarrow} \mathsf{Fun}(FA, FB) \overset{\sim}{\longrightarrow} \mathsf{Fun}(A', B')$$

which descends to the equivalence of homotopy categories (10.3.3). Since equivalences of quasi-categories induce bijections between isomorphism classes of vertices, this yields in particular a bijection between isomorphism classes of functors.

For (iii), the equivalence (10.3.3) is full and faithful, inducing a bijection between natural transformations $f \Rightarrow g$ and $b \cdot Ff \cdot a \Rightarrow b \cdot Fg \cdot a$. This bijection can be transported along any chosen isomorphisms α and β to yield a bijection between natural transformations $f \Rightarrow g$ in $\mathsf{hFun}(A, B)$ in $\mathcal{K}$ and natural transformations $f' \Rightarrow g'$ in $\mathsf{hFun}(A', B')$ in $\mathcal{L}$.   □

As an application of Corollary 10.3.2, we now fulfill a promise made in §2.3, establishing an equivalence between the internal hom B^A between ∞-categories A and B in an ∞-cosmos of $(\infty, 1)$-categories and the simplicial cotensor B^{A_0} of B with the underlying quasi-category of A.

OBSERVATION 10.3.4.   Even if an ∞-cosmos of $(\infty, 1)$-categories $\mathcal{K}$ is not cartesian closed, its homotopy 2-category $\mathfrak{h}\mathcal{K}$ is cartesian closed in the bicategorical sense, replacing the natural isomorphisms of Proposition 1.4.5(ii) with natural equivalences. On account of the biequivalence $(-)_0 : \mathcal{K} \twoheadrightarrow \mathcal{QC}at$ of Proposition 10.2.1, we define $B^A \in \mathcal{K}$ to be any ∞-category whose underlying quasi-category is equivalent to $B_0^{A_0}$. By composing equivalences

$$
\begin{array}{ccc}
\mathsf{Fun}(X, B^A) & \overset{\sim}{\longrightarrow} & \mathsf{Fun}(X_0, B_0^{A_0}) \\
{\scriptstyle \wr}\downarrow & \cong & \downarrow{\scriptstyle \wr} \\
\mathsf{Fun}(X \times A, B) & \overset{\sim}{\longrightarrow} & \mathsf{Fun}(X_0 \times A_0, B_0)
\end{array}
$$

we see that $\mathsf{Fun}(X, B^A) \simeq \mathsf{Fun}(X \times A, B)$ for any X. In the terminology of Definition 10.4.13, the map $\mathsf{Fun}(X, B^A) \twoheadrightarrow \mathsf{Fun}(X \times A, B)$ is a *quasi-pseudonatural equivalence*. Note that if $\mathcal{K}$ is cartesian closed, this universal property demonstrates that the weak exponentials are equivalent to the strictly defined ones.

For this reason, the statement of Proposition 10.3.5 does not require that the ambient ∞-cosmos is cartesian closed; the exponentials can be inferred to exist a posteriori.

Proposition 10.3.5. *In an ∞-cosmos of* $(\infty, 1)$-*categories, for any ∞-categories A and B, the exponential B^A is equivalent to the cotensor B^{A_0} of B with the underlying quasi-category of A.*

Proof By Corollary 10.3.2, cosmological biequivalences reflect equivalences of ∞-categories. Thus, to prove $B^A \simeq B^{A_0}$, it suffices by Proposition 10.2.1 and Corollary 10.3.2 to prove that B^A and B^{A_0} have equivalent underlying quasi-categories. The defining universal properties of the exponential and cotensor provide equivalences, which compose with the local equivalence of the underlying quasi-category functor to provide the desired equivalence:

$$\mathrm{Fun}(1, B^A) \simeq \mathrm{Fun}(A, B) \overset{\sim}{\longrightarrow} \mathrm{Fun}(A_0, B_0) \cong \mathrm{Fun}(1, B^{A_0}) \qquad \square$$

We now prove that biequivalences reflect and create, as well as preserve, the ∞-categorical structures considered in Proposition 10.1.4.

Proposition 10.3.6. *A cosmological biequivalence $F \colon \mathcal{K} \overset{\sim}{\to} \mathcal{L}$*

   (i) *Preserves, reflects, and creates equivalences.*
  (ii) *Preserves and reflects the invertibility of natural transformations and creates natural isomorphisms between given functors.*
 (iii) *Preserves, reflects, and creates adjunctions between ∞-categories, including right adjoint right inverse adjunctions and left adjoint right inverse adjunctions.*
  (iv) *Preserves, reflects, and creates fibered adjunctions and equivalences.*
   (v) *Preserves and reflects discreteness.*
  (vi) *Preserves and reflects comma ∞-categories: a cell defines a comma cone in $\mathcal{K}$ if and only if its image is a comma cone in $\mathcal{L}$.*
 (vii) *Preserves, reflects, and creates absolute right and left lifting diagrams over a given cospan.*
(viii) *Preserves and reflects limits or colimits of diagrams indexed by a simplicial set and creates the property of an ∞-category admitting a limit or colimit of a given diagram.*
  (ix) *Preserves and reflects the stability of an ∞-category and the exactness of functors between such.*
   (x) *Preserves and reflects cartesian and cocartesian fibrations and cartesian functors between them.*
  (xi) *Preserves and reflects discrete cartesian fibrations and discrete cocartesian fibrations.*

(xii) *Preserves and reflects two-sided fibrations and cartesian functors between them.*

(xiii) *Preserves and reflects modules and represented modules and induces a bijection on equivalence classes of modules between a fixed pair of ∞-categories.*

Proof The preservation results are proven in Proposition 10.1.4 under the weaker hypothesis that F is a mere cosmological functor. So it remains only to address reflection of properties and creation of ∞-categorical structures.

Properties (i), (ii), and (iii) hold for any biequivalence between 2-categories, such as $F \colon \mathfrak{h}\mathcal{K} \overset{\sim}{\to} \mathfrak{h}\mathcal{L}$. The aspects that have not already been discussed are left to Exercise 10.3.ii as a useful exercise to familiarize oneself with the 2-categorical notion of biequivalence. By Proposition 10.2.2, (iv) is a special case of (i) and (iii).

Property (v) follows from (ii): if FE is discrete, then the image under F of any 2-cell in $\mathcal{K}$ with codomain E is invertible, which implies that that 2-cell is invertible in E.

Since both $\mathcal{K}$ and $\mathcal{L}$ admit comma ∞-categories and Proposition 3.4.11 shows that comma spans are characterized by a fibered equivalence class of two-sided isofibrations, (vi) follows from (iv).

The reflection properties of (vii) and (viii) follow from Theorem 3.5.3 and the creation properties follow from Theorem 3.5.12 and (vi). Then (ix) can be argued from any of the equivalent characterizations in Theorem 4.4.12 using (ii) and (viii).

Proposition (x) follows from (iii) and (ii) via Theorem 5.2.8, and (xi) follows by applying (i) to the morphism considered in Proposition 5.5.8. Property (xii) follows similarly from Theorem 7.1.4(iii) and (iii) and (ii). Preservation and reflection of modules now follows from this and (v) and the bijection between equivalence classes follows from (iv).[2] The representability statement of (xiii) combines (iv) with (vi), as elaborated upon in Proposition 11.1.5. □

COROLLARY 10.3.7. *If $F \colon \mathcal{K} \overset{\sim}{\to} \mathcal{L}$ is a cosmological biequivalence then the following induced cosmological functors are all biequivalences:*

(i) $F \colon \mathcal{D}isc(\mathcal{K}) \overset{\sim}{\to} \mathcal{D}isc(\mathcal{L})$

(ii) $F \colon \mathcal{K}_{\top,J} \overset{\sim}{\to} \mathcal{L}_{\top,J}$ *and* $F \colon \mathcal{K}_{\perp,J} \overset{\sim}{\to} \mathcal{L}_{\perp,J}$

(iii) $F \colon \mathcal{R}ari(\mathcal{K})_{/B} \overset{\sim}{\to} \mathcal{R}ari(\mathcal{L})_{/FB}$ *and* $F \colon \mathcal{L}ari(\mathcal{K})_{/B} \overset{\sim}{\to} \mathcal{L}ari(\mathcal{L})_{/FB}$

(iv) $F \colon \mathcal{R}ari(\mathcal{K}) \overset{\sim}{\to} \mathcal{R}ari(\mathcal{L})$ *and* $F \colon \mathcal{L}ari(\mathcal{K}) \overset{\sim}{\to} \mathcal{L}ari(\mathcal{L})$

(v) $F \colon \mathcal{C}art(\mathcal{K})_{/B} \overset{\sim}{\to} \mathcal{C}art(\mathcal{L})_{/FB}$ *and* $F \colon co\mathcal{C}art(\mathcal{K})_{/B} \overset{\sim}{\to} co\mathcal{C}art(\mathcal{L})_{/FB}$

(vi) $F \colon \mathcal{C}art(\mathcal{K}) \overset{\sim}{\to} \mathcal{C}art(\mathcal{L})$ *and* $F \colon co\mathcal{C}art(\mathcal{K}) \overset{\sim}{\to} co\mathcal{C}art(\mathcal{L})$

---

[2] A more precise statement appears as Proposition 11.1.4.

(vii) $F\colon \mathcal{D}isc\mathcal{C}art(\mathcal{K}) \righttwoheadrightarrow \mathcal{D}isc\mathcal{C}art(\mathcal{L})$ *and* $F\colon \mathcal{D}iscco\mathcal{C}art(\mathcal{K}) \righttwoheadrightarrow \mathcal{D}isc$ *co$\mathcal{C}$art$(\mathcal{L})$*

(viii) $F\colon \mathcal{S}tab(\mathcal{K}) \righttwoheadrightarrow \mathcal{S}tab(\mathcal{L})$

(ix) $F\colon {}_{A\backslash}\mathcal{F}ib(\mathcal{K})_{/B} \righttwoheadrightarrow {}_{FA\backslash}\mathcal{F}ib(\mathcal{L})_{/FB}$

(x) $F\colon {}_{A\backslash}\mathcal{M}od(\mathcal{K})_{/B} \righttwoheadrightarrow {}_{FA\backslash}\mathcal{M}od(\mathcal{L})_{/FB}$

Proof In each case we start with a cosmological biequivalence – for instance $\mathcal{K}_{/B} \righttwoheadrightarrow \mathcal{L}_{/FB}$ or $\mathcal{K}^{\downarrow} \righttwoheadrightarrow \mathcal{L}^{\downarrow}$ – and must show that the restricted cosmological functor of Lemma 10.1.1 is again a biequivalence between the cosmologically embedded ∞-cosmoi. Each of the arguments is similar; for concreteness' sake, we prove (viii). Proposition 10.3.6 proves that the property that characterizes the objects and 0-arrows of the sub ∞-cosmos is preserved and reflected by any biequivalence. Thus, the diagram of cosmological functors is a pullback:

$$
\begin{array}{ccc}
\mathcal{S}tab(\mathcal{K}) & \xrightarrow{\ F\ } & \mathcal{S}tab(\mathcal{L}) \\
\Big\uparrow & \lrcorner & \Big\uparrow \\
\mathcal{K} & \xrightarrow[\ \sim\ ]{F} & \mathcal{L}
\end{array}
$$

so Lemma 10.2.5 allows us to conclude that the induced functor $F\colon \mathcal{S}tab(\mathcal{K}) \righttwoheadrightarrow \mathcal{S}tab(\mathcal{L})$ is a cosmological biequivalence. $\qquad\square$

Warning 10.1.5 mentioned some ∞-categorical properties that are not necessarily preserved by cosmological functors. Importantly, these notions are preserved, reflected, and created by cosmological biequivalences. Lemma 11.1.7 proves this in the case of composites and exact squares, two notions which are situated in the virtual equipment of modules, but we address the case of pointwise right and left extensions now.

PROPOSITION 10.3.8. *A cosmological biequivalence* $F\colon \mathcal{K} \righttwoheadrightarrow \mathcal{L}$ *preserves, reflects, and creates pointwise left and right extensions:*

(i) A diagram in $\mathcal{K}$ of the form

$$
\begin{array}{ccc}
A & \xrightarrow{\ f\ } & C \\
{\scriptstyle k}\Big\downarrow & {\scriptstyle\Uparrow\nu}\ \nearrow & \\
B & {\scriptstyle r} &
\end{array}
\qquad or \qquad
\begin{array}{ccc}
A & \xrightarrow{\ f\ } & C \\
{\scriptstyle k}\Big\downarrow & {\scriptstyle\Downarrow\lambda}\ \nearrow & \\
B & {\scriptstyle \ell} &
\end{array}
$$

defines a pointwise right or left extension in $\mathcal{K}$, respectively, if and only if its image under F defines, respectively, a pointwise right or left extension in $\mathcal{L}$.

(ii) If $f\colon A \to C$ and $k\colon A \to B$ are functors in $\mathcal{K}$ so that $Ff\colon FA \to FC$ admits a pointwise right or left extension along $Fk\colon FA \to FB$ in $\mathcal{L}$,

> then f admits a pointwise right or left extension, respectively, along k in
> $\mathcal{K}$, and its image under F is isomorphic to the corresponding data in $\mathcal{L}$.

To explain the idea of the proof, we first show that cosmological biequivalences preserve and reflect right extensions; they create them as well, but this can be proven just as easily in the pointwise case, which is discussed subsequently.

Suppose first that $\nu : rk \Rightarrow f$ is a right extension in $\mathcal{K}$. To prove that its image defines a right extension in $\mathcal{L}$, we must show that for all $d : FB \to FC$ the map

$$\mathsf{hFun}(FB, FC)(d, Fr) \xrightarrow{F\nu\circ-} \mathsf{hFun}(FA, FC)(d \cdot Fk, Ff)$$

$$d \xRightarrow{\gamma} Fr \longmapsto d \cdot Fk \xRightarrow{\gamma Fk} Fr \cdot Fk \xRightarrow{F\nu} Ff$$

defines a bijection between sets of 2-cells. By Corollary 10.3.2(ii), there exists a functor $c : B \to C$ in $\mathcal{K}$ together with a natural isomorphism $\delta : d \cong Fc$. By Corollary 10.3.2(iii), application of F and composition with δ defines a bijection

$$\mathsf{hFun}(B, C)(c, r) \xrightarrow{F} \mathsf{hFun}(FB, FC)(Fc, Fr) \xrightarrow{-\circ\delta} \mathsf{hFun}(FB, FC)(d, Fr)$$

$$c \xRightarrow{\alpha} r \longmapsto Fc \xRightarrow{F\alpha} Fr \longmapsto d \xRightarrow{\delta} Fc \xRightarrow{F\alpha} Fr$$

There is a similar bijection defined from the invertible 2-cell $\delta Fk : d \cdot Fk \cong Fc \cdot Fk$. Since $\nu : rk \Rightarrow f$ is a right extension, composition with ν induces its own family of bijections. From the commutative square of functions

$$\begin{array}{ccc}
\mathsf{hFun}(B, C)(c, r) & \xrightarrow[\approx]{\nu\circ-} & \mathsf{hFun}(A, C)(c \cdot k, f) \\
{\scriptstyle F-\circ\delta}\downarrow{\scriptstyle \wr} & & {\scriptstyle \wr}\downarrow{\scriptstyle F-\circ\delta Fk} \\
\mathsf{hFun}(FB, FC)(d, Fr) & \xrightarrow{F\nu\circ-} & \mathsf{hFun}(FA, FC)(d \cdot Fk, Ff)
\end{array}$$

three of which are known to be bijections, we see that $F\nu : Fr \cdot Fk \Rightarrow Ff$ is a right extension in $\mathcal{L}$, as claimed.

Now suppose $F\nu : Fr \cdot Fk \Rightarrow Ff$ is a right extension in $\mathcal{L}$. To see that $\nu : rk \Rightarrow f$ is a right extension in $\mathcal{K}$, note that for any functor $c : B \to C$, there is a commutative square of functions

$$\begin{array}{ccc}
\mathsf{hFun}(B, C)(c, r) & \xrightarrow{\nu\circ-} & \mathsf{hFun}(A, C)(c \cdot k, f) \\
{\scriptstyle F}\downarrow{\scriptstyle \wr} & & {\scriptstyle \wr}\downarrow{\scriptstyle F} \\
\mathsf{hFun}(FB, FC)(Fc, Fr) & \xrightarrow[\approx]{F\nu\circ-} & \mathsf{hFun}(FA, FC)(Fc \cdot Fk, Ff)
\end{array}$$

three of which are known to be bijections. This proves that $\nu : rk \Rightarrow f$ has the universal property of a right extension.

The argument for pointwise right extensions is essentially the same, where we additionally make use of the fact that comma ∞-categories are preserved by cosmological functors and invariant under equivalence, in the sense of Exercise 3.4.iv.

Proof Throughout we use the 2-categorical definition of a pointwise right extension appearing in Theorem 9.3.3(ii). The idea is to take advantage of the local bijections provided by Corollary 10.3.2.

Assume that $\nu : rk \Rightarrow f$ is a pointwise right extension in $\mathcal{K}$. We must show that for every $h : L \to FB$ in $\mathcal{L}$ the diagram

$$
\begin{array}{ccccc}
\mathrm{Hom}_{FB}(h, Fk) & \xrightarrow{\;p_1\;} & FA & \xrightarrow{\;Ff\;} & FC \\
\scriptstyle p_0 \downarrow & \quad \Uparrow \phi & Fk \downarrow \;\; \stackrel{\Uparrow F\nu}{} & \nearrow & \\
L & \xrightarrow[\;\;h\;\;]{} & FB & & Fr
\end{array}
$$

is a right extension diagram in $\mathcal{L}$, meaning that for all $d : L \Rightarrow FC$ the map

$$
\mathrm{hFun}(L, FC)(d, Fr \cdot h) \xrightarrow{\;F\nu p_1 \circ Fr\phi\circ -\;} \mathrm{hFun}(\mathrm{Hom}_{FB}(h, Fk))(dp_0, Ff \cdot p_1)
$$

$$
d \overset{\zeta}{\Rightarrow} Fr\cdot h \longmapsto dp_0 \overset{\zeta p_0}{\Rightarrow} Fr\cdot hp_0 \overset{Fr\phi}{\Rightarrow} Fr\cdot Fk\cdot p_1 \overset{F\nu p_1}{\Rightarrow} Ff\cdot p_1
$$

defines a bijection between sets of 2-cells. By Corollary 10.3.2(i) there exists an ∞-category $K \in \mathcal{K}$ and an equivalence $e : L \overset{\sim}{\to} FK$, and by Corollary 10.3.2(ii) there exist functors $b : K \to B$ and $c : K \to C$ and natural isomorphisms $\beta : Fb \cdot e \cong h$ and $\delta : d \cong Fc \cdot e$. By Corollary 10.3.2(iii), application of F and composition with e, β, and δ defines a bijection

$$
\mathrm{hFun}(K, C)(c, rb) \xrightarrow{\;\;\cong\;\;} \mathrm{Fun}(L, FC)(d, Fr \cdot h)
$$

$$
c \overset{\gamma}{\Rightarrow} rb \longmapsto d \overset{\delta}{\Rightarrow} Fc \cdot e \overset{F\gamma e}{\Rightarrow} Fr \cdot Fb \cdot e \overset{Fr\beta}{\Rightarrow} Fr \cdot h
$$

Since $\nu : rk \Rightarrow f$ is a pointwise right extension in $\mathcal{K}$, the diagram

$$
\begin{array}{ccccc}
\mathrm{Hom}_{B}(b, k) & \xrightarrow{\;p_1\;} & A & \xrightarrow{\;f\;} & C \\
\scriptstyle p_0 \downarrow & \quad \Uparrow \phi & k \downarrow \;\; \stackrel{\Uparrow \nu}{} & \nearrow & \\
K & \xrightarrow[\;\;b\;\;]{} & B & & r
\end{array}
\qquad (10.3.9)
$$

is a (pointwise) right extension as well, defining a bijection

$$
\mathrm{hFun}(K, C)(c, rb) \xrightarrow{\;\;\cong\;\;} \mathrm{hFun}(\mathrm{Hom}_{B}(b, k), C)(cp_0, fp_1)
$$

$$
c \overset{\gamma}{\Rightarrow} rb \longmapsto cp_0 \overset{\gamma p_0}{\Rightarrow} rbp_0 \overset{r\phi}{\Rightarrow} rkp_1 \overset{\nu p_1}{\Rightarrow} fp_1
$$

By Exercise 3.4.iv, there is an equivalence $y : \mathrm{Hom}_{FB}(h, Fk) \overset{\sim}{\to} \mathrm{Hom}_{FB}(Fb, Fk)$

$\cong F\mathrm{Hom}_B(b,k)$ over $e \times \mathrm{id} : L \times FA \twoheadrightarrow FK \times FA$. By Corollary 10.3.2(iii), application of F, composition with y, β, and δ defines the right-hand bijection in the commutative square of functions

$$
\begin{array}{ccc}
\mathrm{hFun}(K,C)(c,rb) & \xrightarrow{\ \cong\ } & \mathrm{hFun}(\mathrm{Hom}_B(b,k),C)(cp_0,fp_1) \\
\cong\downarrow & & \downarrow{\cong} \\
\mathrm{hFun}(L,FC)(d,Fr\cdot h) & \longrightarrow & \mathrm{hFun}(\mathrm{Hom}_{FB}(h,Fk),FC)(dp_0,Ff\cdot p_1)
\end{array}
$$

and thus our pasted diagram in $\mathcal{L}$ is a right extension diagram, as desired.

To see that pointwise right extensions are reflected we must show that (10.3.9) is a right extension diagram in $\mathcal{K}$ under the hypothesis that $Fv : Fr \cdot Fk \Rightarrow Ff$ is a pointwise right extension diagram in $\mathcal{L}$. For any functor $c : K \to C$, we have a commutative square of functions:

$$
\begin{array}{ccc}
\mathrm{hFun}(K,C)(c,rb) & \longrightarrow & \mathrm{hFun}(\mathrm{Hom}_B(b,k),C)(cp_0,fp_1) \\
F\downarrow{\cong} & & {\cong}\downarrow F \\
\mathrm{hFun}(FK,FC)(Fc,Fr\cdot Fb) & \xrightarrow{\ \cong\ } & \mathrm{hFun}(\mathrm{Hom}_{FB}(Fb,Fk),FC)(Fc\cdot p_0,Ff\cdot p_1)
\end{array}
$$

Since comma squares are preserved by cosmological functors, the bottom map is a bijection, as our the vertical functions, defined by application of F. Thus, $v : r \cdot k \Rightarrow f$ satisfies the universal property of a pointwise right extension in $\mathcal{K}$.

Finally, to see that pointwise right extension diagrams are created by cosmological biequivalences suppose we are given a pointwise right extension diagram

$$
\begin{array}{ccc}
FA & \xrightarrow{\ Ff\ } & FC \\
Fk\downarrow & {\Uparrow\sigma}\nearrow & \\
FB & \raise6pt{\hbox{$\scriptstyle s$}} &
\end{array}
$$

in $\mathcal{L}$. By Corollary 10.3.2, there exists a functor $r : B \to C$ together with an isomorphism $\alpha : s \cong Fr$ and a natural transformation $v : rk \Rightarrow f$ mapping to σ under the bijection

$$
\mathrm{hFun}(A,C)(rk,f) \xrightarrow{\ F(-)\circ\alpha Fk\ } \mathrm{hFun}(FA,FC)(s\cdot Fk,Ff)
$$
$$
rk \overset{v}{\Rightarrow} f \longmapsto s\cdot Fk \overset{\sigma}{\Rightarrow} Ff
$$

Since Fv is isomorphic to a pointwise right extension diagram in $\mathcal{L}$ it is a pointwise right extension diagram in $\mathcal{L}$, and since pointwise right extension diagrams are reflected by cosmological biequivalences, v exhibits r as a pointwise right extension of f along k as claimed. $\qquad\square$

Immediately from Definition 9.4.7 and Lemma 9.5.6:

COROLLARY 10.3.10. *Cosmological biequivalences preserve, reflect, and create (weighted) limits and colimits of ∞-category indexed diagrams.* □

Exercises

EXERCISE 10.3.i. Let $F : \mathcal{K} \rightsquigarrow \mathcal{L}$ be a cosmological biequivalence. Show, that F induces an equivalence of homotopy categories $hA \rightsquigarrow hFA$ for any ∞-category $A \in \mathcal{K}$ (see Definition 1.4.11).

EXERCISE 10.3.ii. Consider a 2-functor $F : \mathcal{C} \to \mathcal{D}$ that defines a biequivalence as in Proposition 10.3.1. Prove that:

(i) A 2-cell $A \underset{g}{\overset{f}{\underset{\Downarrow\alpha}{\rightrightarrows}}} B$ in $\mathcal{C}$ is invertible if and only if $F\alpha$ is invertible in $\mathcal{D}$.

(ii) A 1-cell $u : A \to B$ admits a left adjoint in $\mathcal{C}$ if and only if $Fu : FA \to FB$ admits a left adjoint in $\mathcal{D}$, in which case F preserves the adjunction.

EXERCISE 10.3.iii. Prove that cosmological biequivalences between cartesian closed ∞-cosmoi preserve exponential objects up to equivalence.

10.4 Inverse Cosmological Biequivalences

Two ∞-cosmoi are **biequivalent** just when they are connected by a finite zigzag of cosmological biequivalences. In this section, we establish a few useful properties of the "composite" of such a zigzag, the analysis of which immediately reduces to the base case: describing the inverse $G : \mathcal{L} \rightsquigarrow \mathcal{K}$ to a cosmological biequivalence $F : \mathcal{K} \rightsquigarrow \mathcal{L}$. The definitions introduced here describe the ∞-categorical structures that transfer to biequivalent ∞-cosmoi, such as the weak exponentials discussed in Observation 10.3.4. The reader might consider skipping this section for now and referring back to it with the applications of Chapter 12 in mind.

To explain what to expect at the level of $(\infty, 2)$-categories, consider the analogous 2-categorical case. By Proposition 10.3.1, a cosmological biequivalence $F : \mathcal{K} \rightsquigarrow \mathcal{L}$ induces a biequivalence $F : \mathfrak{h}\mathcal{K} \rightsquigarrow \mathfrak{h}\mathcal{L}$ of homotopy 2-categories, this being a 2-functor that is:

- surjective on objects up to equivalence and

- defines a local equivalence of categories $\mathsf{hFun}(A, B) \xrightarrow{\sim} \mathsf{hFun}(FA, FB)$ for all $A, B \in \mathfrak{h}\mathcal{K}$.

From these properties we may attempt to define an inverse biequivalence G as follows:

- For each $C \in \mathfrak{h}\mathcal{L}$, we choose an $A \in \mathfrak{h}\mathcal{K}$ together with a specified equivalence $\epsilon_C : FA \simeq C$ and define $GC := A$.
- For each pair $C, D \in \mathfrak{h}\mathcal{L}$, we define the action of G on hom-categories to be the composite

$$G_{C,D} := \mathsf{hFun}(C, D) \xrightarrow[\sim]{(-\circ\epsilon_C, \epsilon_D^{-1}\circ-)} \mathsf{hFun}(FGC, FGD) \xrightarrow[\sim]{F_{GC,GD}^{-1}} \mathsf{hFun}(GC, GD)$$

of the equivalence defined by pre- and postcomposing with the maps of the specified equivalences $\epsilon_C : FGC \simeq C$ and $\epsilon_D : FGD \simeq D$ together with an inverse of the equivalence defined by the action of F.

These choices are suitably unique: the action of G on objects is well-defined up to equivalence and the action of G on hom-categories is well-defined up to natural isomorphism. However, these mappings cannot in general be chosen to define a 2-functor (see Definition B.2.1 and [71, 3.1]): for instance, while the triangle on the top commutes on the nose – expressing the unit axiom for the 2-functor F – the composite triangle on the bottom only commutes up to isomorphism:

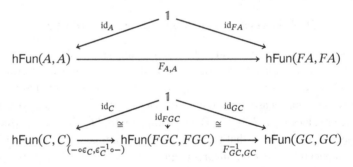

Instead, the mapping $G : \mathfrak{h}\mathcal{L} \leadsto \mathfrak{h}\mathcal{K}$ defines a pseudofunctor between the homotopy 2-categories, a notion we now recall.

DEFINITION 10.4.1. A **pseudofunctor** $G : \mathcal{C} \leadsto \mathcal{D}$ between 2-categories $\mathcal{C}$ and $\mathcal{D}$ is given by:

- a mapping on objects $\mathcal{C} \ni x \mapsto Gx \in \mathcal{D}$;
- a mapping on hom-categories $G_{x,y} : \mathcal{C}(x, y) \to \mathcal{D}(Gx, Gy)$ for each $x, y \in \mathcal{C}$;

- an invertible 2-cell for each $x \in \mathcal{C}$

$$
\begin{array}{ccc}
 & \overset{\mathbb{1}}{\underset{\iota_x \Downarrow \cong}{\nearrow \quad \searrow}} & \\
\mathrm{id}_x \swarrow & & \searrow \mathrm{id}_{Gx} \\
\mathcal{C}(x,x) & \xrightarrow[G_{x,x}]{} & \mathcal{D}(Gx, Gx)
\end{array}
$$

defining an isomorphism $\iota_x : \mathrm{id}_{Gx} \cong G \, \mathrm{id}_x$ in $\mathcal{D}(Gx, Gx)$; and

- an natural isomorphism for each triple of objects $x, y, z \in \mathcal{C}$

$$
\begin{array}{ccc}
\mathcal{C}(y,z) \times \mathcal{C}(x,y) & \xrightarrow{G \times G} & \mathcal{D}(Gy, Gz) \times \mathcal{D}(Gx, Gy) \\
\circ \downarrow & \quad \alpha^{x,y,z} \Downarrow \cong & \downarrow \circ \\
\mathcal{C}(x,z) & \xrightarrow[G]{} & \mathcal{D}(Gx, Gz)
\end{array}
$$

satisfying three coherence conditions encoded by the pasting equalities:

$$
\begin{array}{ccc}
\mathcal{C}(y,z) \times \mathcal{C}(x,y) \times \mathcal{C}(w,x) & \xrightarrow{G \times G \times G} & \mathcal{D}(Gy, Gz) \times \mathcal{D}(Gx, Gy) \times \mathcal{D}(Gw, Gx) \\
\circ \downarrow & \alpha^{x,y,z} \times \mathrm{id} \Downarrow \cong & \downarrow \circ \\
\mathcal{C}(x,z) \times \mathcal{C}(w,x) & \xrightarrow{\quad G \times G \quad} & \mathcal{D}(Gx, Gz) \times \mathcal{D}(Gw, Gx) \\
\circ \downarrow & \alpha^{w,x,z} \Downarrow \cong & \downarrow \circ \\
\mathcal{C}(w,z) & \xrightarrow[G]{} & \mathcal{D}(Gw, Gz)
\end{array}
$$

$$\parallel$$

$$
\begin{array}{ccc}
\mathcal{C}(y,z) \times \mathcal{C}(x,y) \times \mathcal{C}(w,x) & \xrightarrow{G \times G \times G} & \mathcal{D}(Gy, Gz) \times \mathcal{D}(Gx, Gy) \times \mathcal{D}(Gw, Gx) \\
\circ \downarrow & \mathrm{id} \times \alpha^{w,x,y} \Downarrow \cong & \downarrow \circ \\
\mathcal{C}(y,z) \times \mathcal{C}(w,y) & \xrightarrow{\quad G \times G \quad} & \mathcal{D}(Gy, Gz) \times \mathcal{D}(Gw, Gy) \\
\circ \downarrow & \alpha^{w,y,z} \Downarrow \cong & \downarrow \circ \\
\mathcal{C}(w,z) & \xrightarrow[G]{} & \mathcal{D}(Gw, Gz)
\end{array}
$$

$$\begin{array}{ccc}
\mathcal{C}(x,y) & \xrightarrow{\quad G \quad} & \mathcal{D}(Gx, Gy) \\
{\scriptstyle \mathrm{id}\times\mathrm{id}_x}\Big\downarrow & {\scriptstyle \mathrm{id}\times\iota_x\Downarrow\cong} & \Big\downarrow{\scriptstyle \mathrm{id}\times\mathrm{id}_{Gx}} \\
\mathcal{C}(x,y)\times\mathcal{C}(x,x) & \xrightarrow{G\times G} & \mathcal{D}(Gx,Gy)\times\mathcal{D}(Gx,Gx) \\
{\scriptstyle\circ}\Big\downarrow & {\scriptstyle \alpha^{x,x,y}\Downarrow\cong} & \Big\downarrow{\scriptstyle\circ} \\
\mathcal{C}(x,y) & \xrightarrow[\quad G \quad]{} & \mathcal{D}(Gx, Gy)
\end{array}$$

$$\|$$

$$\begin{array}{ccc}
\mathcal{C}(x,y) & \xrightarrow{\quad G \quad} & \mathcal{D}(Gx, Gy) \\
{\scriptstyle \mathrm{id}_y\times\mathrm{id}}\Big\downarrow & {\scriptstyle \iota_y\times\mathrm{id}\Downarrow\cong} & \Big\downarrow{\scriptstyle \mathrm{id}_{Gy}\times\mathrm{id}} \\
\mathcal{C}(y,y)\times\mathcal{C}(x,y) & \xrightarrow{G\times G} & \mathcal{D}(Gy,Gy)\times\mathcal{D}(Gx,Gy) \\
{\scriptstyle\circ}\Big\downarrow & {\scriptstyle \alpha^{x,y,y}\Downarrow\cong} & \Big\downarrow{\scriptstyle\circ} \\
\mathcal{C}(x,y) & \xrightarrow[\quad G \quad]{} & \mathcal{D}(Gx, Gy)
\end{array}$$

where both of these latter composites equal the unit 2-cell $\mathrm{id}_{G_{x,y}}$.

A **2-functor** is a pseudofunctor in which the unit and composition cells ι and α are identities. The notion of a 2-natural transformation (see Definition B.2.2) between 2-functors similarly generalizes to a pseudonatural transformation between pseudofunctors.

DEFINITION 10.4.2. For any 2-categories $\mathcal{C}$ and $\mathcal{D}$ and parallel pseudofunctors $F, G \colon \mathcal{C} \rightsquigarrow \mathcal{D}$, a **pseudonatural transformation** $\phi \colon F \Rrightarrow G$ is given by:

- a 1-cell $\phi_x \colon Fx \to Gx \in \mathcal{D}$ for every object $x \in \mathcal{C}$ and
- an invertible 2-cell in $\mathcal{D}$ for each 1-cell $f \colon x \to y \in \mathcal{C}$

$$\begin{array}{ccc}
Fx & \xrightarrow{\quad Ff \quad} & Fy \\
{\scriptstyle\phi_x}\Big\downarrow & {\scriptstyle\phi_f\Downarrow\cong} & \Big\downarrow{\scriptstyle\phi_y} \\
Gx & \xrightarrow[\quad Gf \quad]{} & Gy
\end{array}$$

so that this data

- is natural, in the sense that for each 2-cell $x \underset{g}{\overset{f}{\rightrightarrows}}{\scriptstyle\Downarrow\gamma}\, y$ in $\mathcal{C}$ the pasted composites are equal

$$
\begin{array}{ccc}
Fx \xrightarrow[\Downarrow F\gamma]{Ff} Fy & & Fx \xrightarrow{Ff} Fy \\
\phi_x \downarrow \;\; {}^{Fg}_{\phi_g \Downarrow \cong} \;\; \downarrow \phi_y & = & \phi_x \downarrow \;\; {}^{\phi_f \Downarrow \cong}_{Gf} \;\; \downarrow \phi_y \quad \text{and}\\
Gx \xrightarrow[Gg]{} Gy & & Gx \xrightarrow[\Downarrow G\gamma]{Gg} Gy
\end{array}
$$

- respects the composition and unit constraints specified by the pseudofunctors

$$
\begin{array}{ccc}
\begin{array}{c}
Fx \xrightarrow[\;\; {}^{\alpha^{x,y,z}_{f,k} \Downarrow \cong}]{Ff \nearrow {}^{Fy} \searrow Fk} Fz \\
\phi_x \downarrow \quad {}^{F(kf)}_{\phi_{kf} \Downarrow \cong} \quad \downarrow \phi_z \\
Gx \xrightarrow[G(kf)]{} Gz
\end{array}
& = &
\begin{array}{c}
Fx \xrightarrow[\phi_f \Downarrow \cong]{Ff \nearrow {}^{Fy} \searrow Fk} Fz \\
\phi_x \downarrow \;\; {}^{Gf \searrow}_{} \;{}^{Gy}_{\alpha^{x,y,z}_{f,k} \Downarrow \cong}\; {}^{\searrow Gk}_{\phi_k \Downarrow \cong} \downarrow \phi_z \\
Gx \xrightarrow[G(kf)]{} Gz
\end{array}
\end{array}
$$

$$
\text{and} \quad
\begin{array}{ccc}
Fx \xrightarrow[\iota_x \Downarrow \cong]{\mathrm{id}_{Fx}} Fx & & Fx \xrightarrow{\mathrm{id}_{Fx}} Fx \\
\phi_x \downarrow \;\; {}^{F\,\mathrm{id}_x}_{\phi_{\mathrm{id}_x} \Downarrow \cong} \;\; \downarrow \phi_x & = & \phi_x \downarrow \;\; {}^{\mathrm{id}_{Gx}}_{} \;\; \downarrow \phi_x \\
Gx \xrightarrow[G\,\mathrm{id}_x]{} Gx & & Gx \xrightarrow[\iota_y \Downarrow \cong]{G\,\mathrm{id}_x} Gx
\end{array}
$$

One context where pseudofunctors emerge are as inverses to 2-functors that define biequivalences. The pseudofunctors that arise in this manner are themselves **biequivalences**: surjective on objects up to equivalence and defining local equivalences on hom-categories. These functors are inverses in the sense that there exist **pseudonatural equivalences** between the composites and the identities, these being pseudonatural transformations that are componentwise equivalences (see Exercise 10.4.ii for an alternate characterization). Collectively, this data defines an equivalence of 2-categories in a sense appropriate to bicategory theory:

PROPOSITION 10.4.3. *If* $F \colon \mathcal{C} \to \mathcal{D}$ *is a 2-functor between 2-categories* $\mathcal{C}$ *and* $\mathcal{D}$ *and a biequivalence then there exists a pseudofunctor* $G \colon \mathcal{D} \rightsquigarrow \mathcal{C}$ *that is also a biequivalence and is a pseudoinverse to* F *in the sense that there exist pseudonatural equivalences* $\mathrm{id}_\mathcal{C} \rightsquigarrow GF$ *and* $FG \rightsquigarrow \mathrm{id}_\mathcal{D}$.

Proof Exercise 10.4.v. □

Proposition 10.4.3 describes a classical result in bicategory theory that Johnson and Yau refer to as "the bicategorical Whitehead theorem" [59, 7.4.1], so we feel content to leave its proof to the exercises (see also Proposition 10.4.16).

Recall from Definition 1.2.1 that an ∞-cosmos is, among other things, a category enriched in the cartesian closed category of quasi-categories, while a cosmological functor is, among other things, an enriched functor between quasi-categorically enriched categories (see §A.2). To define a quasi-categorically enriched pseudofunctor, we need to extend the 1-category of quasi-categories and functors to a 2-category, so that we may use its 2-cells to encode the unit and composition coherences. Fortunately, we have such a 2-category at our disposal: the homotopy 2-category of quasi-categories $\mathfrak{h}\mathcal{QC}at$. By Proposition 1.4.5, this 2-category of quasi-categories, functors between them, and natural transformations between these is cartesian closed, so we have well-behaved cartesian product and transposition operations on these 2-cells.

The extra dimension in the 2-category $\mathfrak{h}\mathcal{QC}at$ enables us to define quasi-categorically enriched pseudofunctors as follows:

DEFINITION 10.4.4. For quasi-categorically enriched categories $\mathcal{K}$ and $\mathcal{L}$, a **quasi-categorically enriched pseudofunctor** $G \colon \mathcal{K} \rightsquigarrow \mathcal{L}$ – a **quasi-pseudofunctor** for short – is given by:

- a mapping on objects $\mathcal{K} \ni x \mapsto Gx \in \mathcal{L}$;
- a functor of hom quasi-categories $G_{x,y} \colon \mathcal{K}(x,y) \to \mathcal{L}(Gx, Gy)$ for each $x, y \in \mathcal{K}$;
- an invertible 2-cell in the homotopy 2-category of quasi-categories for each $x \in \mathcal{K}$

$$
\begin{array}{ccc}
 & \overset{\mathbb{1}}{} & \\
\mathrm{id}_x \swarrow & \iota_x \Downarrow \cong & \searrow \mathrm{id}_{Gx} \\
\mathcal{K}(x,x) & \xrightarrow[G_{x,x}]{} & \mathcal{L}(Gx, Gx)
\end{array}
$$

- an invertible 2-cell in the homotopy 2-category of quasi-categories for each triple of objects $x, y, z \in \mathcal{K}$

$$
\begin{array}{ccc}
\mathcal{K}(y,z) \times \mathcal{K}(x,y) & \xrightarrow{G \times G} & \mathcal{L}(Gy, Gz) \times \mathcal{L}(Gx, Gy) \\
\circ \downarrow & \alpha^{x,y,z} \Downarrow \cong & \downarrow \circ \\
\mathcal{K}(x,z) & \xrightarrow[G]{} & \mathcal{L}(Gx, Gz)
\end{array}
$$

satisfying three coherence conditions encoded by the pasting equalities:

$$\begin{array}{ccc}
\mathcal{K}(y,z) \times \mathcal{K}(x,y) \times \mathcal{K}(w,x) & \xrightarrow{G \times G \times G} & \mathcal{L}(Gy,Gz) \times \mathcal{L}(Gx,Gy) \times \mathcal{L}(Gw,Gx) \\
\circ \downarrow & \alpha^{x,y,z} \times \mathrm{id} \Downarrow \cong & \downarrow \circ \\
\mathcal{K}(x,z) \times \mathcal{K}(w,x) & \xrightarrow{\quad G \times G \quad} & \mathcal{L}(Gx,Gz) \times \mathcal{L}(Gw,Gx) \\
\circ \downarrow & \alpha^{w,x,z} \Downarrow \cong & \downarrow \circ \\
\mathcal{K}(w,z) & \xrightarrow{\qquad\qquad G \qquad\qquad} & \mathcal{L}(Gw,Gz)
\end{array}$$

$$\parallel$$

$$\begin{array}{ccc}
\mathcal{K}(y,z) \times \mathcal{K}(x,y) \times \mathcal{K}(w,x) & \xrightarrow{G \times G \times G} & \mathcal{L}(Gy,Gz) \times \mathcal{L}(Gx,Gy) \times \mathcal{L}(Gw,Gx) \\
\circ \downarrow & \mathrm{id} \times \alpha^{w,x,y} \Downarrow \cong & \downarrow \circ \\
\mathcal{K}(y,z) \times \mathcal{K}(w,y) & \xrightarrow{\quad G \times G \quad} & \mathcal{L}(Gy,Gz) \times \mathcal{L}(Gw,Gy) \\
\circ \downarrow & \alpha^{w,y,z} \Downarrow \cong & \downarrow \circ \\
\mathcal{K}(w,z) & \xrightarrow{\qquad\qquad G \qquad\qquad} & \mathcal{L}(Gw,Gz)
\end{array}$$

$$(10.4.5)$$

$$\begin{array}{ccc}
\mathcal{K}(x,y) & \xrightarrow{\quad G \quad} & \mathcal{L}(Gx,Gy) \\
\mathrm{id} \times \mathrm{id}_x \downarrow & \mathrm{id} \times \iota_x \Downarrow \cong & \downarrow \mathrm{id} \times \mathrm{id}_{Gx} \\
\mathcal{K}(x,y) \times \mathcal{K}(x,x) & \xrightarrow{G \times G} & \mathcal{L}(Gx,Gy) \times \mathcal{L}(Gx,Gx) \quad = \\
\circ \downarrow & \alpha^{x,x,y} \Downarrow \cong & \downarrow \circ \\
\mathcal{K}(x,y) & \xrightarrow{\quad G \quad} & \mathcal{L}(Gx,Gy)
\end{array}$$

$$\begin{array}{ccc}
\mathcal{K}(x,y) & \xrightarrow{\quad G \quad} & \mathcal{L}(Gx,Gy) \\
\mathrm{id}_y \times \mathrm{id} \downarrow & \iota_y \times \mathrm{id} \Downarrow \cong & \downarrow \mathrm{id}_{Gy} \times \mathrm{id} \\
\mathcal{K}(y,y) \times \mathcal{K}(x,y) & \xrightarrow{G \times G} & \mathcal{L}(Gy,Gy) \times \mathcal{L}(Gx,Gy) \\
\circ \downarrow & \alpha^{x,y,y} \Downarrow \cong & \downarrow \circ \\
\mathcal{K}(x,y) & \xrightarrow{\quad G \quad} & \mathcal{L}(Gx,Gy)
\end{array}$$

where both of these latter composites equal the unit 2-cell $\mathrm{id}_{G_{x,y}}$.

REMARK 10.4.6. To emphasize the analogy between Definitions 10.4.1 and 10.4.4, we write $\mathbb{1}$ for the terminal quasi-category, which is also the nerve of the terminal category; elsewhere we write 1 for the terminal ∞-category in a generic ∞-cosmos. For any pair of objects $a, b \in \mathcal{L}$ in a quasi-categorically enriched category,

$$\mathsf{Fun}(1, \mathcal{L}(a, b)) \cong \mathcal{L}(a, b),$$

and so $\mathsf{hFun}(1, \mathcal{L}(a, b)) \cong \mathsf{h}\mathcal{L}(a, b)$. Thus 2-cells in the homotopy 2-category of quasi-categories $\mathfrak{h}\mathcal{QC}at$ with domain 1 and codomain $\mathcal{L}(a, b)$ correspond to 2-cells in the homotopy 2-category of $\mathcal{L}$ – defined exactly as in Definition 1.4.1 – from a to b.

In particular, the data of the invertible 2-cell ι_x is no more and no less than

an invertible 2-cell $\quad Gx \underset{G\mathrm{id}_x}{\overset{\mathrm{id}_{Gx}}{\underset{\iota_x \Downarrow \cong}{\rightrightarrows}}} Gx \quad$ in the homotopy 2-category of $\mathcal{L}$.

The notion of "quasi-pseudofunctor" introduced in Definition 10.4.4 should be regarded as a 2-categorical truncation of an $(\infty, 2)$-categorical notion in the following sense:

DIGRESSION 10.4.7 (quasi-pseudofunctors as $(\infty, 2)$-functors). A quasi-categorically enriched functor between quasi-categorically enriched categories can be understood as a "functor of $(\infty, 2)$-categories." Indeed, the category of simplicially enriched categories has a model structure that presents the $(\infty, 1)$-category of $(\infty, 2)$-categories, in which the fibrant objects are exactly the quasi-categorically enriched categories [79, 0.0.4]. However, the fibrant objects in this category are not necessarily cofibrant. The cofibrant objects are the *simplicial computads* first defined by Dwyer and Kan [37, 4.5]. So to model a generic $(\infty, 2)$-categorical functor from $\mathcal{K}$ to $\mathcal{L}$ as a quasi-categorically enriched functor, one must first replace $\mathcal{K}$ by a weakly equivalent simplicial computad.

This explains why the inverse to a cosmologically biequivalence is not necessarily a strict simplicial functor but something weaker. The 2-cell coherences that enumerate the data of a quasi-categorically enriched pseudofunctor can be understood as a truncation of the higher coherences of a functor of $(\infty, 2)$-categories, much like the homotopy 2-category of quasi-categories is a truncation of the quasi-categorically enriched category of quasi-categories. Since a theme of this text is that much of ∞-category theory can be developed in the truncated homotopy 2-category rather than the full $(\infty, 2)$-category, we decline to enumerate the higher coherences of an inverse to a cosmological biequivalence as part of Definition 10.4.4.

DEFINITION 10.4.8. A quasi-pseudofunctor $G \colon \mathcal{K} \rightsquigarrow \mathcal{L}$ whose codomain $\mathcal{L}$ is an ∞-cosmos is a **biequivalence** when it is:

(i) surjective on objects up to equivalence: if for all $a \in \mathcal{L}$ there exists $x \in \mathcal{K}$ so that $Fx \simeq a$; and

(ii) a local equivalence of quasi-categories: if for every pair $x, y \in \mathcal{K}$, the map

$$\mathcal{K}(x, y) \xrightarrow[\sim]{G_{x,y}} \mathcal{L}(Gx, Gy)$$

is an equivalence of quasi-categories.

REMARK 10.4.9. We find it convenient to assume that $\mathcal{L}$ is an ∞-cosmos in Definition 10.4.8 because that provides us access to the various characterizations of the equivalences in $\mathcal{L}$ given by Theorem 1.4.7. In what follows we ask that an equivalence $a \simeq b$ in $\mathcal{L}$

- defines an equivalence in the homotopy 2-category of $\mathcal{L}$ and
- induces an equivalence of quasi-categories $\mathcal{L}(x, a) \twoheadrightarrow \mathcal{L}(x, b)$ in the homotopy 2-category of quasi-categories that is 2-natural in x.

The latter of these properties implies the former, so if we required a notion of quasi-pseudofunctorial biequivalence between general quasi-categorically enriched categories, we could use this notion of equivalence in Definition 10.4.8. But we make no use of the concept outside of the context provided by ∞-cosmoi and so prefer the simpler terminology. Note that we permit the domain $\mathcal{K}$ to be merely quasi-categorically enriched.

Similarly:

DEFINITION 10.4.10. For quasi-categorically enriched categories $\mathcal{K}$ and $\mathcal{L}$ and quasi-pseudofunctors $F, G : \mathcal{K} \rightsquigarrow \mathcal{L}$, a **quasi-categorically enriched pseudo-natural transformation** – a **quasi-pseudonatural transformation** for short – $\phi : F \Rrightarrow G$ is given by:

- a 0-arrow $\phi_x : Fx \to Gx \in \mathcal{L}$ for every object $x \in \mathcal{K}$ and
- an invertible 2-cell in the homotopy 2-category of quasi-categories, for each pair of objects $x, y \in \mathcal{K}$

$$\begin{array}{ccc} \mathcal{K}(x, y) & \xrightarrow{F_{x,y}} & \mathcal{L}(Fx, Fy) \\ {\scriptstyle G_{x,y}} \downarrow & {\scriptstyle \phi^{x,y} \Downarrow \cong} & \downarrow {\scriptstyle \phi_y \circ -} \\ \mathcal{L}(Gx, Gy) & \xrightarrow[-\circ \phi_x]{} & \mathcal{L}(Fx, Gy) \end{array}$$

so that this data respects the composition and unit constraints specified by the quasi-pseudofunctors, as expressed by the following two pasting diagrams

$$
\begin{array}{ccc}
\mathcal{K}(y,z)\times\mathcal{K}(x,y) & \xrightarrow{\;F\times F\;} & \mathcal{L}(Fy,Fz)\times\mathcal{L}(Fx,Fy) \\
\end{array}
$$

the diagram with vertices:
$\mathcal{K}(y,z)\times\mathcal{K}(x,y) \xrightarrow{F\times F} \mathcal{L}(Fy,Fz)\times\mathcal{L}(Fx,Fy)$, $\alpha^{x,y,z}\!\Downarrow\cong$, $\circ$, $\mathcal{K}(x,z) \xrightarrow{\quad F\quad} \mathcal{L}(Fx,Fz)$, $G\times G$, $\alpha^{x,y,z}\!\Downarrow\cong$, $\mathcal{L}(Gy,Gz)\times\mathcal{L}(Gx,Gy)$, G, $\phi^{x,z}\!\Downarrow\cong$, $\phi_z\circ{-}$, $\circ$, $\|$, $\mathcal{L}(Gx,Gz) \xrightarrow{-\circ\phi_x} \mathcal{L}(Fx,Gz)$

$$
\begin{array}{ccc}
\mathcal{K}(y,z)\times\mathcal{K}(x,y) & \xrightarrow{\;F\times F\;} & \mathcal{L}(Fy,Fz)\times\mathcal{L}(Fx,Fy) \\
\end{array}
$$

the second diagram with vertices:
$G\times F$, $\phi^{y,z}\times\mathrm{id}\Downarrow\cong$, $(\phi_z\circ{-})\times\mathrm{id}$, $G\times G$, $\mathcal{L}(Gy,Gz)\times\mathcal{L}(Fx,Fy) \xrightarrow{(-\circ\phi_y)\times\mathrm{id}} \mathcal{L}(Fy,Gz)\times\mathcal{L}(Fx,Fy)$, $\mathrm{id}\times\phi^{x,y}\Downarrow\cong$, $\mathcal{L}(Gy,Gz)\times\mathcal{L}(Gx,Gy)$, $\mathrm{id}\times(\phi_y\circ{-})$, $=$, $\circ$, $\mathrm{id}\times(-\circ\phi_x)$, $\mathcal{L}(Gy,Gz)\times\mathcal{L}(Fx,Gy) \xrightarrow{\quad\circ\quad} \mathcal{L}(Fx,Gz)$

and

$$
\begin{array}{ccc}
 & \mathbb{1} & \\
\mathrm{id}_x\swarrow & \iota_x\Downarrow\cong & \searrow\mathrm{id}_{Fx} \\
\mathcal{K}(x,x)\xrightarrow[F_{x,x}]{} & & \mathcal{L}(Fx,Fx) \\
G_{x,x}\downarrow & \phi^{x,y}\Downarrow\cong & \downarrow\phi_x\circ{-} \\
\mathcal{L}(Gx,Gx)\xrightarrow[-\circ\phi_x]{} & & \mathcal{L}(Fx,Gx)
\end{array}
\;=\;
\begin{array}{ccc}
 & \mathbb{1} & \\
\mathrm{id}_x\swarrow & \iota_x\Downarrow\cong & \searrow\mathrm{id}_{Gx} \\
\mathcal{K}(x,x)\xrightarrow[G_{x,x}]{} & & \mathcal{L}(Gx,Gx) \\
G_{x,x}\downarrow & & \downarrow-\circ\phi_x \\
\mathcal{L}(Gx,Gx)\xrightarrow[-\circ\phi_x]{} & & \mathcal{L}(Fx,Gx)
\end{array}
$$

REMARK 10.4.11. There is an analogy between Definitions 10.4.2 and 10.4.10 that parallels the analogy between Definitions 10.4.1 and 10.4.4 that is concealed by our presentation. The naturality requirement for the invertible 2-cell components of a pseudonatural transformation $\phi\colon F\Rrightarrow G$ between pseudofunctors $F,G\colon\mathcal{C}\rightsquigarrow\mathcal{D}$ tells us that they assemble into a natural isomorphism between the functors

$$
\begin{array}{ccc}
\mathcal{C}(x,y) & \xrightarrow{\;F_{x,y}\;} & \mathcal{D}(Fx,Fy) \\
G_{x,y}\downarrow & \phi^{x,y}\Downarrow\cong & \downarrow\phi_y\circ{-} \\
\mathcal{D}(Gx,Gy) & \xrightarrow[-\circ\phi_x]{} & \mathcal{D}(Fx,Gy)
\end{array}
$$

for every $x,y\in\mathcal{C}$. The coherence conditions that express the compatibility of these natural isomorphisms with the composition and unit constraints can then be expressed in the form appearing in Definition 10.4.10, but since two natural isomorphisms are equal just when their components are equal, we can reduce these coherence conditions to the two remaining pasting identities of Definition 10.4.2 expressed in terms of the components of ϕ.

The data of a natural isomorphism between functors between quasi-categories is more intricate than the data of a natural isomorphism between functors between categories, which is why the coherence conditions in Definition 10.4.10 must be expressed in a more universal way. However, the coherence conditions of Definition 10.4.2 can be extracted from these as follows.

Recall the 0-arrows in $\mathcal{K}(x, y)$ correspond to functors $f : \mathbb{1} \to \mathcal{K}(x, y)$. By Remark 10.4.6, the restriction $\phi^{x,y} f$ defines the component

$$
\begin{array}{ccc}
Fx & \xrightarrow{Ff} & Fy \\
\phi_x \downarrow & \phi_f \Downarrow \cong & \downarrow \phi_y \\
Gx & \xrightarrow[Gf]{} & Gy
\end{array}
$$

of an invertible 2-cell in the homotopy 2-category of $\mathfrak{h}\mathcal{L}$. This 2-cell is automatically natural, in the sense that for each 2-cell $x \underset{g}{\overset{f}{\Downarrow\gamma}} y$ in the homotopy 2-category of $\mathcal{K}$ the pasted composites

$$
\begin{array}{ccc}
Fx & \overset{Ff}{\underset{Fg}{\Downarrow F\gamma}} & Fy \\
\phi_x \downarrow & \phi_g \Downarrow \cong & \downarrow \phi_y \\
Gx & \underset{Gg}{\longrightarrow} & Gy
\end{array}
\quad = \quad
\begin{array}{ccc}
Fx & \overset{Ff}{\longrightarrow} & Fy \\
\phi_x \downarrow & \begin{array}{c}\phi_f \Downarrow \cong \\ Gf\end{array} & \downarrow \phi_y \\
Gx & \underset{Gg}{\overset{\Downarrow G\gamma}{\longrightarrow}} & Gy
\end{array}
$$

are equal in the homotopy 2-category of $\mathcal{L}$. This follows by naturality of whiskering (see Lemma B.1.3), since both pasted composites are represented by the horizontal composite

$$
\mathbb{1} \underset{g}{\overset{f}{\Downarrow\gamma}} \mathcal{K}(x, y) \underset{(-\circ\phi_x)\circ G_{x,y}}{\overset{(\phi_y\circ-)\circ F_{x,y}}{\phi^{x,y}\Downarrow\cong}} \mathcal{L}(Fx, Gy)
$$

in the homotopy 2-category of quasi-categories.

Similarly, the composition and unit diagrams of Definition 10.4.10 imply that the corresponding diagrams displayed in Definition 10.4.2 commute in the homotopy 2-category of $\mathcal{L}$ (see Exercise 10.4.iii).

For any quasi-categorically enriched category $\mathcal{K}$, the hom bifunctor $\mathcal{K}(-,-) : \mathcal{K}^{\mathrm{op}} \times \mathcal{K} \to \mathcal{QC}at$ is a quasi-categorically enriched functor.

LEMMA 10.4.12. *The action on homs of a quasi-pseudofunctor $F \colon \mathcal{K} \rightsquigarrow \mathcal{L}$ between quasi-categorically enriched categories defines a quasi-pseudonatural transformation*

$$F_{-,-} \colon \mathcal{K}(-,-) \Rrightarrow \mathcal{L}(F-, F-)$$

between quasi-pseudonatural functors $\mathcal{K}^{\mathrm{op}} \times \mathcal{K} \rightsquigarrow \mathcal{QC}at$.

Proof The 2-cell component of the quasi-pseudonatural transformation associated to a pair of objects (x, y) and (w, z) in $\mathcal{K}^{\mathrm{op}} \times \mathcal{K}$ is given by

$$
\begin{array}{ccc}
\mathcal{K}(y, z) \times \mathcal{K}(w, x) & \xrightarrow{\ \circ\ } & \mathcal{K}(w, z)^{\mathcal{K}(x, y)} \\
F \downarrow & \alpha^{w, x, y, z} \Downarrow \cong & \downarrow F_{w, z} \circ - \\
\mathcal{L}(Fw, Fz)^{\mathcal{L}(Fx, Fy)} & \xrightarrow[- \circ F_{x, y}]{} & \mathcal{L}(Fw, Fz)^{\mathcal{K}(x, y)}
\end{array}
$$

where $\alpha^{w, x, y, z}$ is a transpose of the common 2-cell defined by (10.4.5). We leave the verification of the composition and unit axioms to Exercise 10.4.iv. □

DEFINITION 10.4.13. A quasi-pseudonatural transformation $\phi \colon F \Rrightarrow G$ between quasi-pseudofunctors $F, G \colon \mathcal{K} \rightsquigarrow \mathcal{L}$ whose codomain is an ∞-cosmos is a **quasi-pseudonatural equivalence** if each of its components $\phi_x \colon Fx \to Gx$ defines an equivalence in the homotopy 2-category of $\mathcal{L}$.

For the reasons noted in Remark 10.4.9, it is convenient to assume that $\mathcal{L}$ is an ∞-cosmos so we need not be more explicit about the appropriate notion of equivalence in the target category. Our interest in the class of quasi-pseudonatural equivalences stems from the following result, which can be understood as a version of the bicategorical Yoneda lemma in the context of quasi-categorically enriched categories, quasi-pseudofunctors, and quasi-pseudonatural transformations.

LEMMA 10.4.14. *If there exists a quasi-pseudonatural equivalence*

$$\phi \colon \mathcal{K}(-, a) \Rrightarrow \mathcal{K}(-, b)$$

between the simplicial functors $\mathcal{K}^{\mathrm{op}} \to \mathcal{QC}at$ represented by a pair of objects a, b in an ∞-cosmos $\mathcal{K}$, then a and b are equivalent in $\mathcal{K}$.

Proof We will show that the 0-arrow $y := \phi_a(\mathrm{id}_a) \colon a \to b$ is an equivalence in $\mathcal{K}$. First observe that for every $x \in \mathcal{K}$ the component $\phi_x \colon \mathcal{K}(x, a) \twoheadrightarrow \mathcal{K}(x, b)$ – the top-right composite in the following diagram – is isomorphic in the homotopy 2-category of quasi-categories to postcomposition with y – the lower-left composite in the diagram:

$$\begin{array}{ccccc}
\mathcal{K}(x,a) & \xrightarrow{\circ} & \mathcal{K}(x,a)^{\mathcal{K}(a,a)} & \xrightarrow{\mathrm{ev}_{\mathrm{id}_a}} & \mathcal{K}(x,a) \\
\downarrow{\scriptstyle\circ} & \phi^{x,a}\Downarrow\cong & \downarrow{\scriptstyle\phi_x\circ-} & & \downarrow{\scriptstyle\phi_x} \\
\mathcal{K}(x,b)^{\mathcal{K}(a,b)} & \xrightarrow[-\circ\phi_a]{} & \mathcal{K}(x,b)^{\mathcal{K}(a,a)} & \xrightarrow{\mathrm{ev}_{\mathrm{id}_a}} & \mathcal{K}(x,b) \\
& \searrow^{\mathrm{ev}_y} & \mathcal{K}(x,b) & = & \mathcal{K}(x,b)
\end{array}
\quad =: \quad \mathcal{K}(x,a) \underset{y\circ-}{\overset{\phi_x}{\underset{\Downarrow\cong}{\rightrightarrows}}} \mathcal{K}(x,b)$$

In particular, by Exercise 1.4.iii, the map $y \circ - : \mathcal{K}(x,a) \twoheadrightarrow \mathcal{K}(x,b)$ is an equivalence for any $x \in \mathcal{K}$, which means that $y : a \twoheadrightarrow b$ is an equivalence in the ∞-cosmos $\mathcal{K}$. □

Quasi-pseudonatural equivalences may be constructed as adjoint equivalence inverses of simplicial natural transformations that define componentwise equivalences.

Lemma 10.4.15. *Consider a simplicial natural transformation* $\mathcal{K} \underset{G}{\overset{F}{\underset{\Downarrow\phi}{\rightrightarrows}}} \mathcal{L}$ *between quasi-categorically enriched functors between ∞-cosmoi in which the 0-arrow components $\phi_x : Fx \twoheadrightarrow Gx$ all define equivalences in $\mathcal{L}$. Then any choice of adjoint equivalence inverses $\psi_x : Gx \twoheadrightarrow Fx$ assemble into the components of a quasi-pseudonatural transformation $\psi : G \rightsquigarrow F$ that is a quasi-pseudonatural equivalence.*

Proof The components of the quasi-pseudonatural transformation ψ are defined by the adjoint equivalence inverse arrows $\psi_x : Gx \twoheadrightarrow Fx$ and by the pasted composite natural transformation

$$\psi^{x,y} := \quad \begin{array}{ccccc}
\mathcal{K}(x,y) & \xrightarrow{G_{x,y}} & \mathcal{L}(Gx,Gy) & & \\
\downarrow{\scriptstyle F_{x,y}} & & \downarrow{\scriptstyle -\circ\phi_x} & \eta_x\cong & \\
\mathcal{L}(Fx,Fy) & \xrightarrow{\phi_y\circ-} & \mathcal{L}(Fx,Gy) & \xrightarrow[-\circ\psi_x]{} & \mathcal{L}(Gx,Gy) \\
& \epsilon_y\cong & \downarrow{\scriptstyle\psi_y\circ-} & & \downarrow{\scriptstyle\psi_y\circ-} \\
& & \mathcal{L}(Fx,Fy) & \xrightarrow[-\circ\psi_x]{} & \mathcal{L}(Gx,Fy)
\end{array}$$

involving the unit and counit isomorphisms of the adjoint equivalence.

Since F and G are simplicial functors, the unit condition simplifies to ask only that the component of this pasted natural transformation at the identity arrow $\mathrm{id}_x : \mathbb{1} \to \mathcal{K}(x,x)$ is an identity 2-cell id_{ψ_x}. This component is the pasted composite

$$Gx \overline{\qquad\qquad} Gx$$

which is indeed the identity, since the specified data defines an adjoint equivalence.

Similarly, to verify the composition axiom, we must show that the composite

$\mathcal{K}(y,z) \times \mathcal{K}(x,y)$

equals

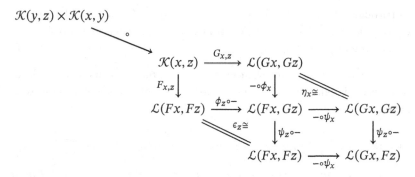

The pre- and postcomposition maps appearing in this diagram are 2-natural (see Definition B.2.2), so for instance the whiskered composite of ϵ_y and $- \circ \psi_x$ can be formed in either order. Using this commutativity property repeatedly in the second pasting diagram and applying the triangle identity $\phi_y \epsilon_y \cdot \eta_y \phi_y = \mathrm{id}_{\phi_y}$, the second pasting diagram reduces to the first one. $\qquad\square$

We now show that any biequivalence $F\colon \mathcal{K} \twoheadrightarrow \mathcal{L}$ between ∞-cosmoi admits a quasi-pseudofunctorial "inverse" $G\colon \mathcal{L} \rightsquigarrow \mathcal{K}$ equipped with quasi-pseudonatural equivalences $\eta\colon \mathrm{id}_{\mathcal{K}} \overset{\approx}{\Rightarrow} GF$ and $\epsilon\colon FG \overset{\approx}{\Rightarrow} \mathrm{id}_{\mathcal{L}}$.

PROPOSITION 10.4.16. *If* $F\colon \mathcal{K} \to \mathcal{L}$ *is a quasi-categorically enriched functor between ∞-cosmoi and a biequivalence then there exists a quasi-pseudofunctor* $G\colon \mathcal{L} \rightsquigarrow \mathcal{K}$ *that is also a biequivalence. Moreover G is a quasi-pseudoinverse to F in the sense that there exist quasi-pseudonatural equivalences* $\mathrm{id}_{\mathcal{K}} \overset{\approx}{\Rightarrow} GF$ *and* $FG \overset{\approx}{\Rightarrow} \mathrm{id}_{\mathcal{L}}$.

Proof To coherently define an inverse to a biequivalence $F\colon \mathcal{K} \twoheadrightarrow \mathcal{L}$, we "fully specify" its data, choosing:

(β) fully specified adjoint equivalences $\epsilon_a\colon Fx_a \simeq a$ for each $a \in \mathcal{L}$ and
(γ) fully specified inverse adjoint equivalences of quasi-categories

$$\mathcal{K}(x,y) \xrightarrow[\approx]{F_{x,y}} \mathcal{L}(Fx, Fy)$$

for each pair $x, y \in \mathcal{K}$ whose inverse is quasi-pseudonatural in x and y.

In (γ), we apply to the simplicial natural transformation $F_{-,-}\colon \mathcal{K}(-,-) \to \mathcal{L}(F-, F-)$ to observe that the pointwise adjoint equivalences to these maps assemble into a quasi-pseudonatural transformation, which is also a pointwise equivalence.

Now, to define $G\colon \mathcal{L} \rightsquigarrow \mathcal{K}$, use ($\beta$) to specify for each $a \in \mathcal{L}$ an object $Ga := x_a \in \mathcal{K}$ together with an equivalence $\epsilon_a\colon FGa \simeq a$ in $\mathcal{L}$. This defines the mapping of G on objects and the 0-arrow components of the quasi-pseudonatural transformation ϵ. To define the action of G on functor spaces, use this data and (γ) to define

$$G_{a,b} := \mathcal{L}(a,b) \xrightarrow[\approx]{(-\circ\epsilon_a, \epsilon_b^{-1}\circ-)} \mathcal{L}(FGa, FGb) \xrightarrow[\approx]{F^{-1}_{Ga,Gb}} \mathcal{K}(Ga, Gb)$$

For each $a \in \mathcal{L}$ define $\iota_a\colon \mathrm{id}_{Ga} \cong G_{a,a}\,\mathrm{id}_a$ to be the composite

of the isomorphism $\beta_a\colon \epsilon_a^{-1} \circ \epsilon_a \cong \mathrm{id}_{FGa}$ in the homotopy 2-category of $\mathcal{L}$ with the component of the isomorphism $\gamma\colon F^{-1}_{Ga,Ga} \circ F_{Ga,Ga} \cong \mathrm{id}_{\mathcal{K}(Ga,Ga)}$ at

id_{Ga}. For each $a, b, c \in \mathcal{L}$, define $\alpha^{a,b,c}$ to be the composite

$$
\begin{array}{ccc}
\mathcal{L}(b,c) \times \mathcal{L}(a,b) & \xrightarrow{\;\circ\;} & \mathcal{L}(a,c) \\
{\scriptstyle (-\circ\epsilon_b,\epsilon_c^{-1}\circ-)\times(-\circ\epsilon_a,\epsilon_b^{-1}\circ-)}\Big\downarrow & \beta\cong & \Big\downarrow{\scriptstyle (-\circ\epsilon_a,\epsilon_c^{-1}\circ-)} \\
\mathcal{L}(FGb,FGc) \times \mathcal{L}(FGa,FGb) & \xrightarrow{\;\circ\;} & \mathcal{L}(FGa,FGc) \\
{\scriptstyle F_{Gb,Gc}^{-1}\times F_{Ga,Gb}^{-1}}\Big\downarrow & \gamma\cong & \Big\downarrow{\scriptstyle F_{Ga,Gc}^{-1}} \\
\mathcal{K}(Gb,Gc) \times \mathcal{K}(Ga,Gb) & \xrightarrow{\;\circ\;} & \mathcal{K}(Ga,Gc)
\end{array}
$$

of the canonical natural transformations built from the data of (β) and (γ).

We next verify that these choices make G into a quasi-pseudofunctor. For the unit condition, we must verify that the composite

$$
\begin{array}{ccccc}
\mathcal{L}(a,b) & \xrightarrow{\mathrm{id}_b \times \mathrm{id}} & \mathcal{L}(b,b) \times \mathcal{L}(a,b) & \xrightarrow{\;\circ\;} & \mathcal{L}(a,b) \\
{\scriptstyle (-\circ\epsilon_a,\epsilon_b^{-1}\circ-)}\downarrow & \beta\cong & {\scriptstyle (-\circ\epsilon_b,\epsilon_b^{-1}\circ-)\times(-\circ\epsilon_a,\epsilon_b^{-1}\circ-)}\downarrow & \beta\cong & \downarrow{\scriptstyle (-\circ\epsilon_a,\epsilon_b^{-1}\circ-)} \\
\mathcal{L}(FGa,FGb) & \xrightarrow{\mathrm{id}_{FGb}\times\mathrm{id}} & \mathcal{L}(FGb,FGb) \times \mathcal{L}(FGa,FGb) & \xrightarrow{\;\circ\;} & \mathcal{L}(FGa,FGb) \\
{\scriptstyle F_{Ga,Gb}^{-1}}\Big\downarrow & \gamma\cong & {\scriptstyle F_{Gb,Gb}^{-1}\times F_{Ga,Gb}^{-1}}\downarrow & \gamma\cong & \Big\downarrow{\scriptstyle F_{Ga,Gb}^{-1}} \\
\mathcal{K}(Ga,Gb) & \xrightarrow{\mathrm{id}_{Gb}\times\mathrm{id}} & \mathcal{K}(Gb,Gb) \times \mathcal{K}(Ga,Gb) & \xrightarrow{\;\circ\;} & \mathcal{K}(Ga,Gb)
\end{array}
$$

is the identity; in fact, each pair of vertical composites is the identity. On the left-hand side, this is on account of one of the triangle equality relations for the adjoint equivalence ϵ_b. On the right-hand side, this is a consequence of quasi-pseudonaturality of the pair $F_{Ga,Gb}^{-1}$ and γ established in Lemma 10.4.15. The right unit constraint and associativity conditions are similar. This completes the proof that $G \colon \mathcal{L} \rightsquigarrow \mathcal{K}$ defines a quasi-pseudofunctor.

By construction, the quasi-pseudofunctor G is a local equivalence: its action on homs is defined by composing an equivalence with a map induced by pre- and postcomposing with equivalences in the ∞-cosmos $\mathcal{L}$, which is then an equivalence by Corollary 1.4.8. We use this local equivalence to argue that for each $x \in \mathcal{K}$, there is an equivalence $\eta_x \colon x \rightsquigarrow GFx$, proving essential surjectivity of G. This component is defined by applying the specified inverse adjoint equivalence $F_{x,Gfx}^{-1} \colon \mathcal{L}(Fx,FGFx) \rightsquigarrow \mathcal{K}(x,GFx)$ of (γ) to the inverse of the specified adjoint equivalence $\epsilon_{Fx}^{-1} \colon Fx \rightarrow FGFx$ of (β). Since F is a cosmological biequivalence, which carries the map η_x to an equivalence in $\mathcal{L}$, η_x is itself an equivalence in $\mathcal{K}$. Thus, the quasi-pseudofunctor G is an biequivalence.

It remains only to check quasi-pseudonaturality of η and ϵ. For the latter, we define the component natural isomorphism by the pasting diagram

$$\epsilon^{a,b} := \begin{array}{c} \mathcal{L}(a,b) \xrightarrow{-\circ\epsilon_a} \mathcal{L}(FGa,b) \xrightarrow{\epsilon_b^{-1}\circ-} \mathcal{L}(FGa,FGb) \xrightarrow{F_{Ga,Gb}^{-1}} \mathcal{K}(Ga,Gb) \end{array}$$

For the former, using the definition $\eta_x := F^{-1}\epsilon_{Fx}^{-1}$ and the quasi-pseudonaturality of $F_{-,-}^{-1}$, we have a pasting diagram

$$\eta^{x,y} := \begin{array}{c} \mathcal{K}(x,y) \xrightarrow{F_{x,y}} \mathcal{L}(Fx,Fy) \xrightarrow{(-\circ\epsilon_{Fx},\epsilon_{Fy}^{-1}\circ-)} \mathcal{L}(FGFx,FGFy) \xrightarrow{F_{GFx,GFy}^{-1}} \mathcal{K}(GFx,GFy) \end{array}$$

which defines component natural isomorphism. We leave the verification that these natural transformations satisfy the unit and composition coherence conditions to define quasi-pseudonatural equivalences $\eta : \mathrm{id}_{\mathcal{K}} \rightsquigarrow GF$ and $\epsilon : FG \rightsquigarrow \mathrm{id}_{\mathcal{L}}$ to the reader. □

It follows direction from the definitions that composites of quasi-pseudofunctors are quasi-pseudofunctors and composites of biequivalences are biequivalences. Hence:

COROLLARY 10.4.17. *Any zigzag of cosmological biequivalences composes to define a quasi-pseudofunctor $\mathcal{K} \rightsquigarrow \mathcal{L}$ between ∞-cosmoi that is also a biequivalence.* □

Moreover, the preservation and reflection properties of cosmological biequivalences established in Proposition 10.3.6 extend to their quasi-pseudofunctorial inverses, as the reader is invited to explore in Exercise 10.4.vi.

Exercises

EXERCISE 10.4.i. For a fixed pair of 2-categories $\mathcal{C}$ and $\mathcal{D}$, show that the collection of pseudofunctors $\mathcal{C} \rightsquigarrow \mathcal{D}$, pseudonatural transformations between them, and modifications (see Definition B.2.3) between these assemble into a 2-category.

EXERCISE 10.4.ii ([59, 6.2.16]). For a pseudonatural transformation $\phi : F \Rrightarrow G$ between pseudofunctors $F, G : \mathcal{C} \rightsquigarrow \mathcal{D}$ between 2-categories $\mathcal{C}$ and $\mathcal{D}$, show that the following are equivalent.

- Each 1-cell component $\phi_x : Fx \xrightarrow{\sim} Gx$ is an equivalence in $\mathcal{D}$.
- The 1-cell ϕ defines an equivalence in the 2-category described in Exercise 10.4.i.

EXERCISE 10.4.iii. Let $\phi : F \Rrightarrow G$ be a quasi-pseudonatural transformation between quasi-pseudofunctors $F, G : \mathcal{K} \rightsquigarrow \mathcal{L}$. For any pair of 0-arrows $f : x \rightarrow y$ and $k : y \rightarrow z$ in the $\mathcal{K}$, verify the following pasting equalities in the homotopy 2-category of $\mathcal{L}$.

and

EXERCISE 10.4.iv. Finish the proof of Lemma 10.4.12.

EXERCISE 10.4.v. Derive a proof of Proposition 10.4.3 from the proof of Proposition 10.4.16, modified according to Remark 10.4.9.

EXERCISE 10.4.vi. Develop a heuristic argument that explains why any ∞-categorical property or structure that is preserved and reflected by a cosmological biequivalence is also preserved and reflected by its inverse quasi-pseudofunctor.

11

Model Independence

Our aim in this chapter is to prove that the theory of ∞-categories is invariant under change of model. Part of the meaning of this result is established in Chapter 10, where we prove that the ∞-categorical notions developed in this text are preserved, reflected, and created, by cosmological biequivalences, which provide our change-of-model functors. But our aim here is to establish something stronger: that statements about ∞-categories that may have been proven by "analytic" techniques particular to a single ∞-cosmos are also model independent, provided that the statements are expressible in a suitable equivalence invariant language. Chapter 12 illustrates applications of this transfer principle.

The problem with our naïve explorations of the model independence of ∞-category theory – such as the results enumerated in Proposition 10.3.6 – is they are ad hoc and tiresome. In §11.1, we pursue a more systematic result. We review the construction of the virtual equipment of modules associated to an ∞-cosmos and explain why it describes a suitable context for proving the model independence of the fundamental ∞-categorical notions. We then prove in Theorem 11.1.6 that a cosmological biequivalence induces a biequivalence of virtual equipments, and revisit a few of our ad hoc model independence statements from the vantage point of this result.

Informally, Theorem 11.1.6 means that any ∞-categorical notion that can be encoded as a property of the virtual equipment of modules is model independent: a biequivalence of virtual equipments gives a mechanism by which results proven with one model of ∞-categories can be transferred to another model. But how can we be sure when a statement about ∞-categories that has been expressed in the language of virtual equipments is in fact invariant under a biequivalence of virtual equipments?

Building on past work of Blanc [19], Freyd [43], Preller [92], and Cartmell [25], Makkai has been in pursuit of a higher categorical foundation of mathematics. In particular, his *First-Order Logic with Dependent Sorts* (FOLDS) [82] provides a formal language for writing mathematical sentences about finite-dimensional higher categorical structures. In §11.2, we sketch the key ideas behind Makkai's FOLDS framework and explain how his work specializes to show that any statement written in the "language of 2-categories" is invariant under a biequivalence between 2-categories.

Since ∞-categories live as objects in an infinite-dimensional category, it is not immediately apparent that Makkai's theory is applicable to the situation at hand. However, a key theme of this text is that a significant chunk of ∞-category theory can be developed in a truncated finite-dimensional framework: namely the virtual equipment of modules. In §11.3, we adapt Makkai's framework to prove that any statement that can be expressed in the "language of virtual equipments," as we define it in a precise sense, gives rise to a model independent statement about ∞-categories.

11.1 A Biequivalence of Virtual Equipments

In this section, we show that a cosmological biequivalence $F \colon \mathcal{K} \xrightarrow{\approx} \mathcal{L}$ induces a *biequivalence of virtual equipments* $F \colon \mathrm{Mod}(\mathcal{K}) \xrightarrow{\approx} \mathrm{Mod}(\mathcal{L})$. We then explain the interpretation of this result: that the category theory of ∞-categories is preserved, reflected, and created by any "change-of-model" functor of this form.

The claim that "the theory of ∞-categories is model independent" should certainly encompass assertions like:

- A functor between ∞-categories admits a left adjoint in a particular model if and only if it admits a left adjoint in every model.
- An ∞-category-valued diagram has a limit in one model if and only if it has a limit in every model.

In our setting, change-of-model functors such as (10.0.1) define cosmological biequivalences, or zigzags thereof. Thus, along these lines, an ad hoc approach to proving the model independence of the basic category theory of ∞-categories is developed in §10.3, where we observe that ad hoc translations of ∞-categorical data and properties between biequivalent ∞-cosmoi can be given relatively mechanically and in excruciating detail, if desired.

A more systematic approach to model independence makes use of most comprehensive framework for the formal category theory of ∞-categories, namely the virtual equipment of modules. We briefly review its essential features. Recall from Chapter 8 that the **virtual double category of modules** $\mathrm{Mod}(\mathcal{K})$ in an ∞-cosmos $\mathcal{K}$ consists of:

- a category of **objects** and **vertical arrows**, here the ∞-categories and ∞-functors, drawn vertically
- for any pair of objects A, B, a collection of **horizontal arrows** $A \xrightarrow{E} B$, here the modules $(q, p) : E \twoheadrightarrow A \times B$ from A to B.
- **cells**, with boundary depicted as follows

$$
\begin{array}{ccc}
A & \overset{\vec{E}}{\dashrightarrow} & B \\
{\scriptstyle f}\downarrow & {\scriptstyle \Downarrow\alpha} & \downarrow{\scriptstyle g} \\
C & \underset{F}{\longrightarrow} & D
\end{array}
$$

where $\vec{E}$ abbreviates a compatible sequence of modules $E_1 \times \cdots \times E_n$ from A to B, which may be empty in the case where $A = B$. Here, a cell with the displayed boundary is an isomorphism class of objects in the functor space $\mathrm{Fun}_{f \times g}(\vec{E}, F)$ of maps from the two-sided fibration $A \twoheadleftarrow \vec{E} \twoheadrightarrow B$ to $C \twoheadleftarrow F \twoheadrightarrow D$ over $f \times g$.

- a **composite cell**, for any configuration

$$
\begin{array}{ccccccc}
A_0 & \overset{\vec{E}_1}{\dashrightarrow} & A_1 & \overset{\vec{E}_2}{\dashrightarrow} & \cdots & \overset{\vec{E}_n}{\dashrightarrow} & A_n \\
{\scriptstyle f_0}\downarrow & {\scriptstyle \Downarrow\alpha_1} & \downarrow{\scriptstyle f_1} \ {\scriptstyle \Downarrow\alpha_2} & \downarrow{\scriptstyle \cdots} & & {\scriptstyle \Downarrow\alpha_n} & \downarrow{\scriptstyle f_n} \\
B_0 & \overset{F_1}{\longrightarrow} & B_1 & \overset{F_2}{\longrightarrow} & \cdots & \overset{F_n}{\longrightarrow} & B_n \\
{\scriptstyle g}\downarrow & & & {\scriptstyle \Downarrow\beta} & & & \downarrow{\scriptstyle h} \\
C_0 & & & \underset{G}{\longrightarrow} & & & C_n
\end{array}
$$

defined by pulling back and then composing fibered isomorphism classes of maps of spans.

- an **identity cell** for every horizontal arrow

$$
\begin{array}{ccc}
A & \overset{E}{\longrightarrow} & B \\
\| & {\scriptstyle \Downarrow\mathrm{id}_E} & \| \\
A & \underset{E}{\longrightarrow} & B
\end{array}
$$

so that composition of cells is strictly associative and unital in the usual multicategorical sense.

LEMMA 11.1.1. *A cosmological functor induces a functor of virtual double categories, preserving all of the structure.*

Proof In Corollary 8.1.14, the categorical structures in the virtual double category of modules are inherited from the double category of two-sided isofibrations. We can understand the action of a cosmological functor $F \colon \mathcal{K} \to \mathcal{L}$ on the virtual double category of modules by taking a similar route.

As outlined in Exercise 8.1.i, the nonunital double category of two-sided isofibrations can be understood as a quotient of the "nonunital internal category" (see Definition B.1.8) of ∞-cosmoi and cosmological functors

$$\mathcal{K}^{\times} \underset{\mathcal{K}}{\times} \mathcal{K}^{\times} \xrightarrow{\;\circ\;} \mathcal{K}^{\times} \underset{\text{r-cod}}{\overset{\text{l-cod}}{\rightrightarrows}} \mathcal{K}$$

where $\mathcal{K}^{\times}$ is the ∞-cosmos of two-sided isofibrations and l-cod, r-cod $\colon \mathcal{K}^{\times} \to \mathcal{K}$ refer to the left and right codomain functors that map a two-sided isofibration $A \xleftarrow{q} E \xrightarrow{p} B$ to the base ∞-categories A and B of the span. Note these functors combine to define a Grothendieck fibration cod $\colon \mathcal{K}^{\times} \to \mathcal{K} \times \mathcal{K}$. The pullback

$$
\begin{array}{ccc}
\mathcal{K}^{\times} \underset{\mathcal{K}}{\times} \mathcal{K}^{\times} & \longrightarrow & \mathcal{K}^{\times} \\
\downarrow & \lrcorner & \downarrow{\scriptstyle \text{l-cod}} \\
\mathcal{K}^{\times} & \underset{\text{r-cod}}{\longrightarrow} & \mathcal{K}
\end{array}
$$

is an ∞-cosmos (see Exercise 6.1.iii), namely the ∞-cosmos of horizontally composable pairs of two-sided isofibrations $A \xleftarrow{q} E \xrightarrow{p} B \xleftarrow{r} F \xrightarrow{s} C$. The composition functor $\circ \colon \mathcal{K}^{\times} \times_{\mathcal{K}} \mathcal{K}^{\times} \to \mathcal{K}^{\times}$, which sends a horizontal composable pair of two-sided isofibrations to a chosen composite span, is associative only up to simplicial natural isomorphism, which accounts for the "pseudo-ness" in the horizontal composition in the double category of two-sided isofibrations. The (nonunital pseudo) double category of two-sided isofibrations can be understood as the quotient of this structure obtained by replacing the ∞-cosmos $\mathcal{K}$ by its underlying category and replacing the fibers of cod $\colon \mathcal{K}^{\times} \to \mathcal{K} \times \mathcal{K}$ by the quotient 1-categories with the same objects but with hom-sets defined to be the isomorphism classes of vertices in the sliced functor spaces.

Now any cosmological functor $F \colon \mathcal{K} \to \mathcal{L}$ induces cosmological functors

$$
\begin{array}{ccc}
\mathcal{K}^{\times} \underset{\mathcal{K}}{\times} \mathcal{K}^{\times} & \xrightarrow{\;\circ\;} \mathcal{K}^{\times} & \overset{\text{l-cod}}{\underset{\text{r-cod}}{\rightrightarrows}} \mathcal{K} \\
\downarrow{\scriptstyle F} \quad \cong \quad & \downarrow{\scriptstyle F} & \downarrow{\scriptstyle F} \\
\mathcal{L}^{\times} \underset{\mathcal{L}}{\times} \mathcal{L}^{\times} & \xrightarrow{\;\circ\;} \mathcal{L}^{\times} & \overset{\text{l-cod}}{\underset{\text{r-cod}}{\rightrightarrows}} \mathcal{L}
\end{array}
$$

that commute strictly with the codomain functors and up to isomorphism with the composition functor. Thus, a cosmological functor induces a double functor between the double categories of two-sided isofibrations that strictly preserves vertical composition of functors and squares and preserves horizontal composition of spans and squares up to coherent natural isomorphism.

It follows that a cosmological functor $F \colon \mathcal{K} \to \mathcal{L}$ induces a functor of virtual double categories of modules $F \colon \mathsf{Mod}(\mathcal{K}) \to \mathsf{Mod}(\mathcal{L})$ that preserves all of the structure strictly, since the multicategorical composition of module maps is derived from the strictly defined vertical composition of squares in the double category of two-sided isofibrations. $\square$

Of crucial importance to its utility as a setting for formal category theory is the fact that the virtual double category of modules is a **virtual equipment**, which means that it satisfies the two further properties:

(i) For any module and pair of functors as displayed on the left, there exists a **restriction** module and **cartesian** cell as displayed on the right

$$
\begin{array}{ccc}
A' & & B' \\
a\downarrow & & \downarrow b \\
A & \xrightarrow{\ \ E\ \ } & B
\end{array}
\quad\rightsquigarrow\quad
\begin{array}{ccc}
A' & \xrightarrow{E(b,a)} & B' \\
a\downarrow & \Downarrow\rho & \downarrow b \\
A & \xrightarrow{\ \ E\ \ } & B
\end{array}
$$

characterized by the universal property that any cell as displayed below-left factors uniquely through ρ as below-right:

$$
\begin{array}{ccc}
X & \cdots\overset{\vec{E}}{\cdots}\!\!\to & Y \\
af\downarrow & \Downarrow & \downarrow bg \\
A & \xrightarrow{\ \ E\ \ } & B
\end{array}
\quad=\quad
\begin{array}{ccc}
X & \cdots\overset{\vec{E}}{\cdots}\!\!\to & Y \\
f\downarrow & \exists!\Downarrow & \downarrow g \\
A' & \xrightarrow{E(b,a)} & B' \\
a\downarrow & \Downarrow\rho & \downarrow b \\
A & \xrightarrow{\ \ E\ \ } & B
\end{array}
$$

The restriction module is defined by pulling back a module $A \overset{E}{\nrightarrow} B$ along functors $a \colon A' \to A$ and $b \colon B' \to B$. The simplicial pullback defining $E(b,a)$ induces an equivalence of functor spaces

$$
\mathsf{Fun}_{af\times bg}(E_1 \times \cdots \times E_n, E) \simeq \mathsf{Fun}_{f\times g}(E_1 \times \cdots \times E_n, E(b,a)),
$$

which gives rise to the universal property (see Proposition 8.2.1).

(ii) Every object A admits a **unit** module equipped with a nullary **cocartesian** cell

$$
\begin{array}{ccc}
A & =\!\!=\!\!= & A \\
\| & \Downarrow\iota & \| \\
A & \xrightarrow[\mathsf{Hom}_A]{} & A
\end{array}
$$

satisfying the universal property that any cell in the virtual double category of modules whose horizontal source includes the object A, as displayed on the left

$$
\begin{array}{ccc}
X \dashrightarrow^{\vec{E}} A \dashrightarrow^{\vec{F}} Y \\
f \downarrow \quad\quad \Downarrow \quad\quad \downarrow g \\
B \xrightarrow{\quad G \quad} C
\end{array}
\quad = \quad
\begin{array}{c}
X \dashrightarrow^{\vec{E}} A =\!\!=\!\!= A \dashrightarrow^{\vec{F}} Y \\
\| \quad \Downarrow\text{id} \quad \| \quad \Downarrow\iota \quad \| \quad \Downarrow\text{id} \quad \| \\
X \dashrightarrow^{\vec{E}} A \xrightarrow{\text{Hom}_A} A \dashrightarrow^{\vec{F}} Y \\
f \downarrow \quad\quad \Downarrow \exists! \quad\quad \downarrow g \\
B \xrightarrow{\quad G \quad} C
\end{array}
$$

factors uniquely through ι as displayed on the right. The unit module is the arrow ∞-category, given the notation $A \xrightarrow{\text{Hom}_A} A$ when considered as a module from A to A. The universal property follows from the Yoneda lemma (see Proposition 8.2.4).

Lemma 11.1.1 extends to the virtual equipments of modules:

PROPOSITION 11.1.2. *A cosmological functor* $F \colon \mathcal{K} \to \mathcal{L}$ *induces a functor* $F \colon \mathsf{Mod}(\mathcal{K}) \to \mathsf{Mod}(\mathcal{L})$ *of virtual equipments, preserving all of the structure.*

Proof In light of Lemma 11.1.1, it remains to consider unit and restriction modules and cells. It follows immediately from the constructions given in the proofs of Propositions 8.2.4 and 8.2.1 that these are preserved by cosmological functors:

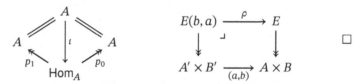

The functors and natural transformations in the homotopy 2-category $\mathfrak{h}\mathcal{K}$ embed into the virtual equipment $\mathsf{Mod}(\mathcal{K})$ in three ways. A functor $f \colon A \to B$ is represented as a vertical arrow and also by the pair of horizontal modules $A \xrightarrow{\text{Hom}_B(B,f)} B$ and $B \xrightarrow{\text{Hom}_B(f,B)} A$, which behave like adjoints is a sense suitable to a virtual double category: the module $A \xrightarrow{\text{Hom}_B(B,f)} B$ defines a *companion* and the module $B \xrightarrow{\text{Hom}_B(f,B)} A$ defines a *conjoint* to the vertical arrow $f \colon A \to B$ (see Proposition 8.4.1, Theorem 8.4.4, and Corollary 8.4.6). These embeddings extend to natural transformations by Proposition 8.4.11: for any parallel pair of functors there are natural bijections between natural transformations in the homotopy 2-category

$$A \underset{g}{\overset{f}{\rightrightarrows}} {\Downarrow\alpha}\ B$$

and cells in the virtual equipment of modules:

$$
\begin{array}{ccccc}
A \xrightarrow{\operatorname{Hom}_B(B,f)} B & & A \xrightarrow{\operatorname{Hom}_A} A & & B \xrightarrow{\operatorname{Hom}_B(g,B)} A \\
\Big\| \quad \Downarrow\alpha_* \quad \Big\| & \rightsquigarrow & g\Big\downarrow \quad \Downarrow\tilde{\alpha} \quad \Big\downarrow f & \rightsquigarrow & \Big\| \quad \Downarrow\alpha^* \quad \Big\| \\
A \xrightarrow[\operatorname{Hom}_B(B,g)]{} B & & B \xrightarrow[\operatorname{Hom}_B]{} B & & B \xrightarrow[\operatorname{Hom}_B(f,B)]{} A
\end{array}
$$

As remarked upon in Definition 8.4.12, as a consequence of these results, there are three locally fully faithful homomorphisms from the homotopy 2-category $\mathfrak{h}\mathcal{K}$ into the virtual equipment $\mathsf{Mod}(\mathcal{K})$. The *vertical embedding* described by Propositions 8.3.18 sends the functors of $\mathfrak{h}\mathcal{K}$ to vertical arrows and the 2-cells to unary cells whose sources and targets are given by unit modules. The other two are the *covariant* and *contravariant embeddings*, respectively – embedding the homotopy 2-category into the substructure[1] of $\mathsf{Mod}(\mathcal{K})$ comprised only of unary cells whose vertical boundaries are identities. The modules in the image of the covariant embedding are the right representables and the modules in the image of the contravariant embedding are the left representables. We leave it to the reader to verify:

LEMMA 11.1.3. *The functor of virtual equipments* $F \colon \mathsf{Mod}(\mathcal{K}) \to \mathsf{Mod}(\mathcal{L})$ *induced by a cosmological functor* $F \colon \mathcal{K} \to \mathcal{L}$ *commutes with the covariant, contravariant, and vertical embeddings.*

Proof Exercise 11.1.ii. □

The theme of Chapter 9 could be summarized by saying that the virtual equipment of modules in an ∞-cosmos is a robust setting to develop the category theory of ∞-categories. On the one hand, it contains the homotopy 2-category of the ∞-cosmos, which is the setting for most of the results of Part I. It is also a very natural home to study ∞-categorical properties that are somewhat awkward to express in the homotopy 2-category. For instance, the weak 2-universal property of comma ∞-categories is now encoded by a bijection in Lemma 8.1.16: cells in the virtual equipment whose codomain is a comma module correspond bijectively to natural transformations of a particular form in the homotopy 2-category. Fibered equivalences of modules, as used to express

---

[1] This substructure is very nearly a bicategory, with horizontal composites of unary cells constructed as in Definition 8.3.16, except that compatible sequences of modules do not always admit a horizontal composite.

the universal properties of adjunctions, limits, and colimits in Chapter 4, are now vertical isomorphisms in the virtual equipment between parallel modules. The virtual equipment also cleanly encodes the universal property of pointwise left and right Kan extensions, which are used to define (weighted) limits and colimits of functors between ∞-categories.

We will show that when $F \colon \mathcal{K} \rightsquigarrow \mathcal{L}$ is a cosmological biequivalence, the corresponding functor of virtual equipments is a biequivalence in a suitable sense. In pursuit of this result, we first elaborate on Proposition 10.3.1(xiii).

Proposition 11.1.4. *A cosmological biequivalence* $F \colon \mathcal{K} \rightsquigarrow \mathcal{L}$ *preserves, reflects, and creates modules:*

> (i) *An isofibration* $(q, p) \colon E \twoheadrightarrow A \times B$ *is a module in* $\mathcal{K}$ *if and only if* $(Fq, Fp) \colon FE \twoheadrightarrow FA \times FB$ *is a module in* $\mathcal{L}$.
>
> (ii) *A pair of modules* $A \overset{E}{\nrightarrow} B$ *and* $A \overset{E'}{\nrightarrow} B$ *are equivalent in* $\mathcal{K}$ *if and only if the modules* $FA \overset{FE}{\nrightarrow} FB$ *and* $FA \overset{FE'}{\nrightarrow} FB$ *define equivalent modules in* $\mathcal{L}$.
>
> (iii) *For every module* $A' \overset{G}{\nrightarrow} B'$ *in* $\mathcal{L}$ *and every pair of* ∞-*categories* A, B *in* $\mathcal{K}$ *with specified equivalences* $FA \simeq A'$ *and* $FB \simeq B'$ *there is a module* $A \overset{E}{\nrightarrow} B$ *in* $\mathcal{K}$ *so that* FE *is equivalent to* G *over the pair of equivalences.*

Proof The result of (i) was proven already in Proposition 10.3.1(xiii). The result of (ii) follows from the fact that the induced cosmological biequivalence

$$
\begin{array}{ccc}
{}_{A\backslash}\mathcal{M}od(\mathcal{K})_{/B} & \overset{F}{\dashrightarrow} & {}_{FA\backslash}\mathcal{M}od(\mathcal{L})_{/FB} \\
\Downarrow & & \Downarrow \\
\mathcal{K}_{/A\times B} & \overset{F}{\underset{\sim}{\longrightarrow}} & \mathcal{L}_{/FA\times FB}
\end{array}
$$

preserves and reflects equivalences between objects.

For (iii), fix a pair of equivalences $FA \simeq A'$ and $FB \simeq B'$, defining an equivalence $e \colon A' \times B' \rightsquigarrow FA \times FB$, and consider the composite biequivalence

$$
\mathcal{K}_{/A\times B} \overset{F}{\underset{\sim}{\longrightarrow}} \mathcal{L}_{/FA\times FB} \overset{e^*}{\underset{\sim}{\longrightarrow}} \mathcal{L}_{/A'\times B'}
$$

given by Propositions 10.2.2 and 10.2.4. Consider a module $G \twoheadrightarrow A' \times B'$. By essential surjectivity, there is an isofibration $E \twoheadrightarrow A \times B$ whose image under this cosmological functor – the pullback of $FE \twoheadrightarrow FA \times FB$ along $e \colon A' \times B' \rightsquigarrow FA \times FB$ – defines an isofibration $(q, p) \colon E' \twoheadrightarrow A' \times B'$ that is equivalent to G in $\mathcal{L}_{/A'\times B'}$. It remains only to argue that E defines a module from A to B, which will follow, essentially as in the proof of (i), from the fact that $E' \simeq G$ defines a module from A' to B'.

As the image E' of E is equivalent to a discrete object, Proposition 10.3.6(v) tells us E is discrete in $\mathcal{K}_{/A \times B}$. The final step is to argue that the desired right adjoint to $E \to \mathrm{Hom}_B(B, p)$ is present in the image of the biequivalence $\mathcal{K}_{/A \times B} \overset{\approx}{\to} \mathcal{L}_{/A' \times B'}$, and apply Proposition 10.3.6(iii) to deduce its presence in $\mathcal{K}_{/A \times B}$; a similar argument of course applies to the functor $E \to \mathrm{Hom}_A(q, A)$. To see this note that $F \colon \mathcal{K}_{/A \times B} \overset{\approx}{\to} \mathcal{L}_{/FA \times FB}$ carries $\mathrm{Hom}_B(B, p) \twoheadrightarrow A \times B$ to $\mathrm{Hom}_{FB}(FB, Fp) \twoheadrightarrow FA \times FB$. By (i), it suffices to argue that this functor has a right adjoint over $FA \times FB$. Applying Proposition 10.3.6(iii) to the biequivalence $e^* \colon \mathcal{L}_{/FA \times FB} \overset{\approx}{\to} \mathcal{L}_{/A' \times B'}$, this follows from the fact that $E' \to \mathrm{Hom}_{B'}(B', p')$ has a right adjoint over $A' \times B'$. $\qquad\square$

PROPOSITION 11.1.5. *Let* $F \colon \mathcal{K} \overset{\approx}{\to} \mathcal{L}$ *be a cosmological biequivalence. Then a module* $A \overset{E}{\nrightarrow} B$ *in* $\mathcal{K}$ *is right representable if and only if the module* $FA \overset{FE}{\nrightarrow} FB$ *is right representable in* $\mathcal{L}$, *in which case,* F *carries the representing functor* $f \colon A \to B$ *in* $\mathcal{K}$ *to a representing functor* $Ff \colon FA \to FB$ *in* $\mathcal{L}$.

Proof To say that $A \overset{E}{\nrightarrow} B$ is right representable in $\mathcal{K}$ is to say that there exists a functor $f \colon A \to B$ together with an equivalence $E \simeq_{A \times B} \mathrm{Hom}_B(B, f)$ of modules over B. If this is the case then any cosmological functor $F \colon \mathcal{K} \to \mathcal{L}$ carries this to a fibered equivalence $FE \simeq_{FA \times FB} \mathrm{Hom}_{FB}(FB, Ff)$, and hence the module $FA \overset{FE}{\nrightarrow} FB$ is right represented by $Ff \colon FA \to FB$ in $\mathcal{L}$.

Conversely, if $FA \overset{FE}{\nrightarrow} FB$ is right represented by some functor $g \colon FA \to FB$, then by Corollary 10.3.2(ii), there exists a functor $f \colon A \to B$ in $\mathcal{K}$ so that $Ff \cong g$ in $\mathcal{L}$. By Proposition 8.4.11, naturally isomorphic functors represent equivalent modules; that is, $\mathrm{Hom}_{FB}(FB, g) \simeq_{FA \times FB} \mathrm{Hom}_{FB}(FB, Ff)$. Thus $FE \simeq_{FA \times FB} \mathrm{Hom}_{FB}(FB, Ff)$. By Proposition 11.1.4(ii), this fibered equivalence lifts along the cosmological functor $F \colon \mathcal{K}_{/A \times B} \to \mathcal{L}_{/FA \times FB}$ to a fibered equivalence $E \simeq_{A \times B} \mathrm{Hom}_B(B, f)$, which proves that E is right represented by $f \colon A \to B$ in $\mathcal{K}$. $\qquad\square$

We have seen that cosmological functors induce functors of virtual equipments, preserving all the structure. When a cosmological functor is a biequivalence, the induced functor of virtual equipments also creates structure and reflects universal properties, on account of bijections we now enumerate.

THEOREM 11.1.6 (model independence of ∞-category theory). *If* $F \colon \mathcal{K} \overset{\approx}{\to} \mathcal{L}$ *is a cosmological biequivalence, then the induced functor of virtual equipments*

$$F \colon \mathrm{Mod}(\mathcal{K}) \overset{\approx}{\to} \mathrm{Mod}(\mathcal{L})$$

*defines a **biequivalence of virtual equipments**: i.e., it is*

(i) *bijective on equivalence classes of objects;*

(ii) *locally bijective on isomorphism classes of parallel vertical functors extending the bijection of (i);*

(iii) *locally bijective on equivalence classes of parallel modules extending the bijection of (ii);*

(iv) *locally bijective on cells extending the bijections of (i), (ii), and (iii).*

Note further that if two ∞-cosmoi are connected by a finite zigzag of biequivalences, then the bijections described in Theorem 11.1.6 compose.

Proof Properties (i) and (ii) are restatements of Corollary 10.3.2(i) and (ii).

The local bijection (iii) follows immediately from Proposition 11.1.4 and the fact that for any pair of equivalences $e : A' \times B' \xrightarrow{\sim} FA \times FB$, the composite biequivalence

$$\mathcal{K}_{/A \times B} \xrightarrow{\;F\;} \mathcal{L}_{/FA \times FB} \xrightarrow{\;e^*\;} \mathcal{L}_{/A' \times B'}$$

preserves, reflects, and creates equivalences between objects, again by Corollary 10.3.2(i). Finally (iv) is an application of Corollary 10.3.2(ii) to this cosmological biequivalence. $\qquad\square$

Theorem 11.1.6 subsumes many of the model independence statements established thus far. For instance, the presence of an adjunctions between ∞-categories and the existence of limits and colimits inside an ∞-category can both be encoded as an equivalence invariant proposition in the virtual equipment of modules. The model independence of pointwise right and left extensions, first proven in Proposition 10.3.8, can also be established as an elementary corollary of Theorem 11.1.6, by an argument left to Exercise 11.1.iii. Along the same lines, a biequivalence of virtual equipments preserves, reflects, and creates composites of modules:

LEMMA 11.1.7. *Let $F : \mathcal{K} \xrightarrow{\sim} \mathcal{L}$ be a cosmological biequivalence.*

(i) *Then a compatible sequence of modules in $\mathcal{K}$ admits a composite in $\mathsf{Mod}(\mathcal{K})$ if and only if the image of this sequence admits a composite in $\mathsf{Mod}(\mathcal{L})$.*

(ii) *Hence, cosmological biequivalences preserve and reflect exact squares.*

Proof Via Definition 9.2.2, (ii) follows immediately from (i), so it remains only to show that a biequivalence of virtual equipments $F : \mathsf{Mod}(\mathcal{K}) \xrightarrow{\sim} \mathsf{Mod}(\mathcal{L})$ preserves, reflects, and creates composites of modules. To see that an n-ary composite cell $\mu : E_1 \times \cdots \times E_n \Rightarrow E$ in $\mathfrak{h}\mathcal{K}$ is preserved, note that by Theorem 11.1.6(iv), any cell in $\mathsf{Mod}(\mathcal{L})$ is isomorphic to a cell in the image of F: first replace the objects by equivalent ones in the image, then replace the vertical functors by naturally isomorphic ones in the image, then replace the modules by equivalent ones in the image over the specified equivalences between their ∞-categorical sources and targets, and then finally apply the local bijection (iv) to replace the cell in $\mathsf{Mod}(\mathcal{L})$ by a unique cell in the image of $\mathsf{Mod}(\mathcal{K})$ by composing with this data. Now, by local full and faithfulness and essential surjectivity, the universal property of the cocartesian cell $\mu \colon E_1 \times \cdots \times E_n \Rightarrow E$ implies that its image $F\mu \colon FE_1 \times \cdots \times FE_n \Rightarrow FE$ is again a cocartesian cell. Thus composites $E_1 \otimes \cdots \otimes E_n \simeq E$ are preserved by cosmological biequivalences.

Now if $F\mu \colon FE_1 \times \cdots \times FE_n \Rightarrow FE$ is a composite, since $F \colon \mathsf{Mod}(\mathcal{K}) \to \mathsf{Mod}(\mathcal{L})$ is locally fully faithful, then $\mu \colon E_1 \times \cdots \times E_n \Rightarrow E$ is also a composite; thus composites $FE_1 \otimes \cdots \otimes FE_n \simeq FE$ are reflected by cosmological biequivalences.

Finally, suppose the sequence $FE_1 \times \cdots \times FE_n$ of modules in $\mathsf{Mod}(\mathcal{L})$ admits a composite $FA_0 \xrightarrow{G} FA_n$; since the compatible sequence of modules is in the image of F the source and target ∞-categories of the composite module G are as well. By Theorem 11.1.6(iii) there exists a module $A_0 \xrightarrow{E} A_n$ in $\mathfrak{h}\mathcal{K}$ so that $FE \simeq G$ as modules from FA_0 to FA_n. The cocartesian cell $FE_1 \times \cdots \times FE_n \Rightarrow G$ that witnesses the composition relation composes with the unary cell of this equivalence to define a cocartesian cell $FE_1 \times \cdots \times FE_n \Rightarrow FE$. By Theorem 11.1.6(iv), this lifts to an n-ary cell $E_1 \times \cdots \times E_n \Rightarrow E$ in $\mathsf{Mod}(\mathcal{K})$. As we have just seen that cocartesianness of cells is reflected by biequivalences $F \colon \mathsf{Mod}(\mathcal{K}) \twoheadrightarrow \mathsf{Mod}(\mathcal{L})$, this completes the proof that composites are created by cosmological biequivalences. $\square$

The upshot of Theorem 11.1.6, and the reason that we consider this as a proof of the model independence of ∞-category theory, is that any statement about ∞-categories that can be encoded in the "language of virtual equipments" is invariant under change of model. Our experience gives us some informal understanding of this language. It includes statements that characterize an ∞-category up to equivalence (such as a comma or arrow ∞-category) or a functor up to natural isomorphism (such as a left or right adjoint). Other model independent statements are those which are expressible as an equivalence between modules (such as results concerning the left or right representability of modules) or in terms of the existence of a cell between modules with certain properties (such as in the case of pointwise extensions). But as we shall discover, this model independent language can be described much more formally by taking advantage of a logical framework that we now introduce.

Exercises

EXERCISE 11.1.i. Pick your favorite ∞-categorical notion and give an ad hoc proof of its model independence. Compare this with a proof using Theorem 11.1.6.

EXERCISE 11.1.ii. Prove Lemma 11.1.3.

EXERCISE 11.1.iii. Use Theorem 11.1.6 to prove a module-theoretic proof that cosmological biequivalences preserve, reflect, and create pointwise right and left extensions, and compare this argument with the proof of Proposition 10.3.8.

EXERCISE 11.1.iv. Prove that cosmological biequivalences between cartesian closed ∞ cosmoi preserve and reflect initial and final functors.

11.2 First-Order Logic with Dependent Sorts

There is an important caveat to the invariance of ∞-category theory under change of model. After passing from a complete Segal space to its underlying quasi-category and then back

$$\mathcal{CSS} \xrightarrow{\ (-)_0\ }_{\sim} \mathcal{QCat} \xrightarrow{\ \text{nerve}\ }_{\sim} \mathcal{CSS}$$

the resulting complete Segal space is equivalent, but likely not equal to the original one. But even in the classical strict case, not every statement about 1-categories is invariant under equivalence: "this category has a single object" is a famous counterexample. Similarly, not every statement about 2-categories is invariant under biequivalence. Consider, for instance, the statement that Proposition 1.4.5 proves for homotopy 2-categories: "this 2-category has a 2-terminal object." If $F: \mathcal{C} \xrightarrow{\sim} \mathcal{D}$ is a biequivalence between 2-categories and $\mathcal{D}$ has a 2-terminal object t, then by essential surjectivity there is an $s \in \mathcal{C}$ so that $Fs \simeq t$ in $\mathcal{D}$. But the local equivalence property only supplies an equivalence

$$\mathcal{C}(c, s) \simeq \mathcal{D}(Fc, Fs) \simeq \mathcal{D}(Fc, t) \cong 1,$$

making s into a *biterminal* object but not necessarily a 2-terminal object.

The Model Independence Theorem 11.1.6 supplies a biequivalence between the virtual equipments of modules defined for any pair of biequivalent ∞-cosmoi. Since the important statements about ∞-categories can be expressed as properties of the virtual equipment, this biequivalence gives a mechanism by which results proven with one model of ∞-categories can be transferred to another model. But how can we be sure when a statement about ∞-categories that has been expressed in the language of virtual equipments is in fact invariant under a biequivalence of virtual equipments?

This sort of problem has a long history. Blanc [19] and Freyd [43] characterize the properties of ordinary categories that are invariant under equivalence of categories (see also Preller [92]). Makkai's FOLDS was developed to extend these results to higher categorical structures, such as 2-categories and bicategories [82]. In future work, we hope to provide a complete characterization of the statements about ∞-categories that are invariant under biequivalence of ∞-cosmoi,[2] but our present aim is a result that is arguably more useful and much more easily obtained. In this section, we sketch the key ideas behind Makkai's FOLDS framework; all the main concepts and results that follow are due to him, though we have made a few minor modifications in the definitions and depart somewhat in notation and terminology, taking inspiration from conversations with Henry [52]. In §11.3, we apply this work to prove that any statement that can be expressed in the *language of virtual equipments* – a specialization of Makkai's formal language to a signature we introduce in 11.3.2 – gives rise to a model independent statement about ∞-categories.

Makkai's FOLDS provides a formal language for writing mathematical sentences whose variables are structured according to a given *signature*. As the name suggests, sentences in FOLDS closely resemble the sentences of first-order logic, with an important new ingredient. Each variable is typed to belong to a specific *sort* and these sorts may depend on a finite family of compatibly defined variables belonging to other sorts of lower degree. Universal and existential quantification is defined only over the variables of a single sort at a time, and the formula being quantified cannot contain any variables that depend on the variable being quantified over. Finally, relations, such as equality, are only permitted on the maximal sorts. For example, in the language of categories it is permissible to write "for all objects x and y and for all arrows f and g from x to y, f equals g," asserting that the category is a pre-order. But it is not permissible to write "there exists an object x so that for all arrows h, h equals 1_x" – asserting that the category is the terminal category – since the equality relation ranges over all arrows, rather than over the a single specified hom-set.

Before we introduce the FOLDS language, we must describe its signatures.

DEFINITION 11.2.1. A **simple inverse** category is an inverse category $\mathcal{J}$ (see Definition C.1.16) with "finite fan-out," meaning each object is the domain

---

[2] Note our sought-for result is more similar to Makkai's result about bicategories than Blanc and Freyd's result about 1-categories, since our concern has to do with invariance change of model of higher categories rather than with invariance under equivalence between ∞-categories in a given model.

of only finitely many arrows. The objects of such categories can be assigned canonical degrees by examining the graph of nonidentity arrows:

- objects which are not domains of any (nonidentity) arrow have degree 0, while
- an object for which every (nonidentity) "outgoing arrow" has codomain of degree 0 is assigned degree 1, and continuing inductively,
- any object is assigned the minimal degree that exceeds the degree of the codomain of each of its (nonidentity) outgoing arrows.

This defines a canonical identity-reflecting functor deg : $\mathcal{J} \to \omega^{op}$.

An object in a simple inverse category is **maximal** if it is not the codomain of any nonidentity arrow.

DEFINITION 11.2.2. A **FOLDS signature** is a simple inverse category with a distinguished (possibly empty) set of maximal objects called "relation symbols." The remaining objects are referred to as "kinds."

Each nonidentity arrow in a FOLDS signature indicates some dependency of a kind or relation symbol on other kinds, while commutativity conditions in the category encode compatibility conditions among the dependencies. In what follows, we indicate the relation symbols with a dot " ˙ ".

EXAMPLE 11.2.3. The FOLDS signature $\mathcal{J}_{\mathcal{C}at}$ for categories has kinds for the objects and arrows

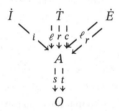

with the composition relations

$$s \cdot i = t \cdot i \qquad s \cdot c = s \cdot \ell \qquad s \cdot \ell = s \cdot r$$
$$t \cdot c = t \cdot r \qquad t \cdot \ell = t \cdot r$$
$$t \cdot \ell = s \cdot r$$

In this instance, every maximal object is a relation symbol. The unary relation symbol $\dot{I}$ encodes identity arrows, while $\dot{T}$ witnesses ternary composition relations, and $\dot{E}$ expresses the equality of parallel arrows.

EXAMPLE 11.2.4. The FOLDS signature $\mathcal{J}_{2\text{-}\mathcal{C}at}$ for 2-categories has kinds for the 0-, 1-, and 2-cells, as well as for identity 1-cells and composition of 1-cells. In addition there are relation symbols for equality of parallel 2-cells, identity 2-cells, and vertical and horizontal composition of 2-cells:

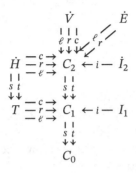

The inverse category $\mathcal{J}_{2\text{-}\mathcal{C}at}$ has the relations from the underlying 1-category

$$s \cdot i = t \cdot i \qquad s \cdot c = s \cdot \ell \qquad t \cdot c = t \cdot r \qquad t \cdot \ell = s \cdot r,$$

globularity relations

$$s \cdot s = s \cdot t \qquad\qquad t \cdot s = t \cdot t,$$

relations governing vertical composition, identities, and equality between parallel 2-cells

$$s \cdot i = t \cdot i \qquad\quad s \cdot c = s \cdot \ell \qquad\quad s \cdot \ell = s \cdot r$$
$$t \cdot c = t \cdot r \qquad\quad t \cdot \ell = t \cdot r$$
$$t \cdot \ell = s \cdot r$$

plus relations relating horizontal composition to composition of 1-cells

$$s \cdot \ell = \ell \cdot s \qquad\quad s \cdot r = r \cdot s \qquad\quad s \cdot c = c \cdot s$$
$$t \cdot \ell = \ell \cdot t \qquad\quad t \cdot r = r \cdot t \qquad\quad t \cdot c = c \cdot t.$$

DEFINITION 11.2.5. For a given FOLDS signature $\mathcal{J}$, an $\mathcal{J}$-**structure** is a functor $M : \mathcal{J} \to \mathcal{S}et$ so that for each relation symbol $\dot{R} \in \mathcal{J}$, the map induced by the family of nonidentity arrows with domain $\dot{R}$ is a monomorphism.[3]

$$M\dot{R} \rightarrowtail \prod_{p \,:\, \dot{R} \overset{\neq}{\to} K_p} MK_p$$

[3] An alternate, and arguably more useful, way to state this condition is to require that the components of the matching maps of Observation 11.2.8 for the relation symbols are monomorphisms (see Exercise 11.2.i.)

EXAMPLE 11.2.6. Consider the functor $D_{\mathcal{C}at} : \mathcal{J}^{\mathrm{op}}_{\mathcal{C}at} \to \mathcal{C}at$ whose image is given by

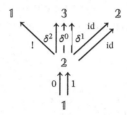

Mapping out of $D_{\mathcal{C}at}$ defines a functor $M_- : \mathcal{C}at \to \mathcal{S}et^{\mathcal{J}_{\mathcal{C}at}}$ whose image lies in the full subcategory of $\mathcal{J}_{\mathcal{C}at}$-structures.[4] For any small 1-category C, $M_C O$ is the set of objects and $M_C A$ is the set of arrows, with $M_C \dot{I}$, $M_C \dot{T}$, and $M_C \dot{E}$ encoding the relations that detect identity arrows, commutative triangles, and equality between parallel arrows, respectively.

EXAMPLE 11.2.7. Consider the functors $E_{2\text{-}\mathcal{C}at}, D_{2\text{-}\mathcal{C}at} : \mathcal{J}^{\mathrm{op}}_{2\text{-}\mathcal{C}at} \to 2\text{-}\mathcal{C}at$ defined by

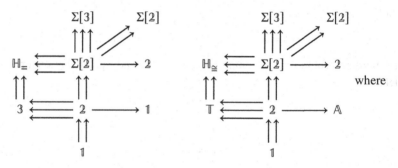

where

---

[4] This follows because the latching maps (see Definition C.4.14) for the diagram $D_{\mathcal{C}at}$ at the objects $\dot{I}$, $\dot{T}$, and $\dot{E}$ each define epimorphisms, surjective functors of categories.

are the free 2-categories generated by the depicted data. The diagram $D_{2\text{-}\mathcal{C}at}$ can be understood as a Reedy cofibrant replacement of $E_{2\text{-}\mathcal{C}at}$ relative to Lack's model structure for 2-categories (see Theorem C.5.14), in which the cofibrations are the 2-functors that are injective on objects and 1-cells and the trivial fibrations are the surjective equivalences [71].

Mapping out of $E_{2\text{-}\mathcal{C}at}$ or out of $D_{2\text{-}\mathcal{C}at}$ defines functors $N_-, M_- : 2\text{-}\mathcal{C}at \to \mathcal{S}et^{\mathcal{J}_{2\text{-}\mathcal{C}at}}$ whose images lie in the full subcategory of $\mathcal{J}_{2\text{-}\mathcal{C}at}$-structures.[5] We refer to $M_{\mathcal{C}}$ as the "saturated" $\mathcal{J}_{2\text{-}\mathcal{C}at}$-structure and $N_{\mathcal{C}}$ as the "naïve" $\mathcal{J}_{2\text{-}\mathcal{C}at}$-structure associated to a 2-category $\mathcal{C}$.

OBSERVATION 11.2.8 (the fibers of the matching map). For each $\mathcal{J}$-structure $M : \mathcal{J} \to \mathcal{S}et$ and object $K \in \mathcal{J}$, either a kind or a relation symbol, there is a canonical **matching map**

$$
\begin{array}{ccc}
MK & \xrightarrow{\hspace{4cm}} & \prod\limits_{p\,:\,K\overset{\neq}{\to}K_p} MK_p \\[2mm]
\quad {}^{m^K}\searrow & \partial^K M := \lim\limits_{p\,:\,K\overset{\neq}{\to}K_p} MK_p & \nearrow
\end{array}
$$

whose codomain is the matching object defined in Observation C.1.18. Note both the limit and the product are over nonidentity arrows with domain K, an implicit condition in similar constructions that follow. We think of an $\mathcal{J}$-structure M as a structured set in which each set MK is further partitioned into the fibers of the matching map $m^k : MK \to \partial^K M$.

We now describe Makkai's "dependent sorts," which are defined together with their variables by mutual recursion. The variables can be introduced purely syntactically, but we find it more intuitive to think of them as belonging to a **context**, this being an $\mathcal{J}$-structure Γ in which the sets ΓK associated to each kind $K \in \mathcal{J}$ are disjoint.

DEFINITION 11.2.9 (sorts and their variables). Fix a FOLDS signature $\mathcal{J}$ and a context $\Gamma : \mathcal{J} \to \mathcal{S}et$.

- Each kind K of degree zero defines a **sort**, also denoted by K, whose variables are the elements of the set ΓK. Write "$x : K$" to mean that $x \in \Gamma K$, i.e., that x is a variable belonging to the sort K in context Γ.
- For each kind K of degree one, the matching map m^K takes the form displayed below-right, where the product is over arrows in $\mathcal{J}$ with domain K and codomain of degree zero. For any family of variables $\{x_p : K_p\}_{p\,:\,K\overset{\neq}{\to}K_p}$,

---

[5] Again, the latching maps for each diagram at all four of the relation symbols are epimorphisms.

there is a sort $K\langle x_p\rangle$ whose variables are the elements of the fiber

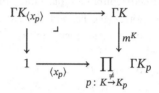

and are said to **depend on** the variables $x_p : K_p$.

- For a kind K of higher degree, a family of variables $\{x_p : K_p\langle x_{qp}\rangle\}_{p\,:\,K\xrightarrow{\neq}K_p}$ is **compatible** if $\langle x_p\rangle \in \prod_{p\,:\,K\xrightarrow{\neq}K_p} \Gamma K_p$ belongs to the matching object $\partial^K\Gamma$. In practice, this means that the higher degree variables in the list depend on the lower degree ones in the way prescribed by the dependency relations in the FOLDS signature. For any compatible family of variables, there is a sort $K\langle x_p\rangle$ whose variables are the elements of the fiber of the matching map m^K over $\langle x_p\rangle \in \partial^K\Gamma$.

$$
\begin{array}{ccc}
\Gamma K_{\langle x_p\rangle} & \longrightarrow & \Gamma K \\
\downarrow & \llcorner & \downarrow{\scriptstyle m^K} \\
1 & \xrightarrow{\langle x_p\rangle} & \partial^K\Gamma
\end{array}
$$

By convention, the only variables that are explicitly listed in the specification of a sort are those of highest degree; the lower-degree variables that these variables depend on can be deduced from the sorts to which the highest degree variables belong.

EXAMPLE 11.2.10. For the FOLDS signature $\mathcal{I}_{\mathcal{C}at}$:

- There is a sort O.
- There is a sort $A\langle x, y\rangle$ for any pair of variables $x, y : O$.

EXAMPLE 11.2.11. For the FOLDS signature $\mathcal{I}_{2\text{-}\mathcal{C}at}$:

- There is a sort C_0.
- There is a sort $C_1\langle x, y\rangle$ for any pair of variables $x, y : C_0$.
- There is a sort $C_2\langle x, y, f, g\rangle$ for any $x, y : C_0$ and $f, g : C_1\langle x, y\rangle$. This sort is typically abbreviated by $C_2\langle f, g\rangle$ with the implicit variables $x, y : C_0$ inferable from the sort in which the variables f and g live.

Using the same abbreviations:

- There is a sort $I_1\langle f\rangle$ for any $x : C_0$ and $f : C_1\langle x, x\rangle$.

- There is a sort $T\langle f, g, h \rangle$ for any $x, y, z : C_0$ and $f : C_1\langle x, y \rangle$, $g : C_1\langle y, z \rangle$, and $h : C_1\langle x, z \rangle$.

We are now prepared to introduce the formulas and sentences of the formal language associated to a FOLDS signature $\mathcal{J}$. These are defined by recursion starting from the atomic formulae, which are governed by the relation symbols. Again, we fix a FOLDS signature $\mathcal{J}$ and a context $\Gamma : \mathcal{J} \to \mathcal{S}et$.

DEFINITION 11.2.12. An **atomic formula** in the logic with dependent sorts is an entity of the form $\dot{R}\langle x_p \rangle$, where $\dot{R}$ is a relation symbol and the variables $\{x_p : K_p\langle x_{qp} \rangle\}_{p : \dot{R} \xrightarrow{\neq} K_p}$ must define a compatible family, meaning that they define an element $\langle x_p \rangle \in \partial^{\dot{R}}\Gamma$.

EXAMPLE 11.2.13. In the FOLDS signature $\mathcal{J}_{\mathcal{C}at}$, there is an atomic formula $\dot{T}\langle x, y, z, f, g, h \rangle$, abbreviated $\dot{T}\langle f, g, h \rangle$, for any $x, y, z : O$ and $f : A\langle x, y \rangle$, $g : A\langle y, z \rangle$, and $h : A\langle x, z \rangle$, which can be thought of as asserting the commutativity of the triangle formed by the arrows f, g, and h.

DIGRESSION 11.2.14. A key idea in the FOLDS philosophy is that it should express a logic with restricted equality. This is visible in the FOLDS signatures $\mathcal{J}_{\mathcal{C}at}$ and $\mathcal{J}_{2\text{-}\mathcal{C}at}$, which include binary equality relation symbols on the kinds A and C_2, but no similar equality predicates on the kinds O, C_0, and C_1 on which these depend. More precisely, the atomic formulae $\dot{E}\langle f, g \rangle$ for $\mathcal{J}_{\mathcal{C}at}$ and $\dot{E}\langle \alpha, \beta \rangle$ for $\mathcal{J}_{2\text{-}\mathcal{C}at}$ may be used to express the equality of a parallel pair of arrows, since compatibility requires $f, g : A\langle x, y \rangle$ for some $x, y : O$, or the equality of parallel 2-cells, since compatibility requires $\alpha, \beta : C_1\langle f, g \rangle$ for some $f, g : C_1\langle x, y \rangle$ and some $x, y : O$. This sort of restriction is essential for the main theorem: that the validity of a sentence expressed in the FOLDS language for a given signature is invariant under suitably defined equivalences between structures.

In FOLDS, one can quantify either universally or existentially only over the variables in a specified sort, provided the variables in the predicate under consideration do not depend on the variable being quantified over.

DEFINITION 11.2.15. **Formulae** ϕ and their sets of **free variables** $\mathrm{var}(\phi)$ are defined by simultaneous recursion:

- An atomic formula is a formula. The variables of $\dot{R}\langle x_p \rangle$ are the x_p.

Compound formulae are defined inductively from other formulae via the following procedures:

- $\top, \bot$ are formulae, with $\mathrm{var}(\top) := \mathrm{var}(\bot) := \varnothing$.

- Formulae may be combined using the sentential connectives $\wedge$, $\vee$, and $\rightarrow$, in which case the variables combine by unions: e.g., $\mathrm{var}(\phi \wedge \psi) = \mathrm{var}(\phi) \cup \mathrm{var}\,\psi$.
- When ϕ is a formula and x is a variable so that no variable in $\mathrm{var}(\phi)$ depends on x – though $x \in \mathrm{var}(\phi)$ is permitted – then $\forall x \phi$ and $\exists x \phi$ are well-formed formulas[6] whose variables are given by the set

$$\mathrm{var}(\forall x \phi) := \mathrm{var}(\exists x \phi) := (\mathrm{var}(\phi) - \{x\}) \cup \mathrm{dep}(x)$$

formed by removing x from the variables of ϕ if it appears and then adding all the variables on which x depends if they do not already appear.

When the set of variables in a formula is empty, we call the formula a **sentence**.

An **evaluation** of an $\mathcal{J}$-context Γ in an $\mathcal{J}$-structure M is given by a natural transformation $\alpha : \Gamma \rightarrow M$, which defines an interpretation of its variables, sending a variable $x : K\langle x_p \rangle$ to an element in the fiber of $MK \rightarrow \partial^K M$ over $\langle \alpha x_p \rangle$. To interpret a particular formula ϕ, it is not necessary to have specified an interpretation of the full context Γ. It suffices to merely specify the interpretation of its variables $\mathrm{var}(\phi) \subset \Gamma$, which may be regarded as a context in their own right (see Exercise 11.2.ii).

Inductively in the complexity of a formula, we define what it means for an $\mathcal{J}$-structure M to **satisfy** a formula ϕ under a given interpretation of its variables $\alpha : \mathrm{var}(\phi) \rightarrow M$, a property we denote by $M \vDash \phi[\alpha]$.

DEFINITION 11.2.16. An $\mathcal{J}$-structure M **satisfies** an atomic formula $\dot{R}\langle x_p \rangle$ under an interpretation α if and only if the tuple $\langle \alpha x_p \rangle \in \prod_{p \,:\, \dot{R} \rightarrow K_p} MK_p$ lies in the subset $M\dot{R}$, in which case one writes $M \vDash \dot{R}[\alpha]$.

The sentences $\top$ and $\bot$ have no variables so their semantics are independent of interpretation. Any $\mathcal{J}$-structure satisfies $\top$ and no $\mathcal{J}$-structure satisfies $\bot$.

DEFINITION 11.2.17. A $\mathcal{J}$-structure M **satisfies** compound formulas built from ϕ and ψ under an interpretation $\alpha : \mathrm{var}(\phi) \cup \mathrm{var}(\psi) \rightarrow M$ according to the rules:

- $M \vDash (\phi \wedge \psi)[\alpha]$ if and only if $M \vDash \phi[\alpha]$ and $M \vDash \psi[\alpha]$
- $M \vDash (\phi \vee \psi)[\alpha]$ if and only if $M \vDash \phi[\alpha]$ or $M \vDash \psi[\alpha]$
- $M \vDash (\phi \rightarrow \psi)[\alpha]$ if and only if whenever $M \vDash \phi[\alpha]$ then also $M \vDash \psi[\alpha]$.

DEFINITION 11.2.18. Consider a formula $\forall x \phi$ where $x : K\langle x_p \rangle$ together with an interpretation

$$\alpha : \mathrm{var}(\forall x \phi) \rightarrow M$$

---

[6] As we demonstrate in examples, the full syntax requires that each quantified variable is declared with its sort, which expresses the range of the quantification.

in an $\mathcal{J}$-structure M; note that $x \notin \text{var}(\forall x\phi)$ so this does not give an interpretation of the variable x itself. We say that M **satisfies** $\forall x\phi$ under the interpretation α if for all a in the fiber of $MK \to \partial^K M$ over $\langle \alpha x_p \rangle$, M ⊨ $\phi(a/x)[\alpha]$. That is, M ⊨ $\forall x\phi[\alpha]$ if any a with appropriate dependencies can be substituted for x in the interpretation of ϕ to yield a formula that M satisfies.

Similarly M **satisfies** $\exists x\phi$ under the interpretation α if there is some a in the fiber of $MK \to \partial^K M$ over $\langle \alpha x_p \rangle$, so that M ⊨ $\phi(a/x)[\alpha]$. That is, M ⊨ $\exists x\phi[\alpha]$ if some a with appropriate dependencies can be substituted for x in the interpretation of ϕ to yield a formula that M satisfies.

EXAMPLE 11.2.19. In the language for $\mathcal{J}_{\mathcal{C}at}$, in the context given by $x, y : O$, the formula

$$\exists f : A\langle x, y\rangle, \exists g : A\langle y, x\rangle, \forall 1_x : A\langle x, x\rangle, \forall 1_y : A\langle y, y\rangle,$$
$$(\dot{I}\langle 1_x\rangle \wedge \dot{I}\langle 1_y\rangle) \to (\dot{T}\langle f, g, 1_x\rangle \wedge \dot{T}\langle g, f, 1_y\rangle)$$

asserts the existence of an isomorphism between two objects specified by an interpretation.

EXAMPLE 11.2.20. Modulo a change in notation, there is a similar sentence in the language for $\mathcal{J}_{2\text{-}\mathcal{C}at}$ in the context given by $x, y : C_0$:

$$\exists f : C_1\langle x, y\rangle, \exists g : C_1\langle y, x\rangle, \forall 1_x : C_1\langle x, x\rangle, \forall 1_y : C_1\langle y, y\rangle,$$
$$\forall \mu : I_1\langle 1_x\rangle, \forall \nu : I_1\langle 1_y\rangle, \exists \alpha : T\langle f, g, 1_x\rangle \wedge \exists \beta : T\langle g, f, 1_y\rangle$$

where "$\exists \alpha : T\langle f, g, 1_x\rangle$" is shorthand for the formula "$\exists \alpha : T\langle f, g, 1_x\rangle.\mathsf{T}$."

In the naïve $\mathcal{J}_{2\text{-}\mathcal{C}at}$-structure $N_{\mathcal{C}}$ associated to a 2-category $\mathcal{C}$, the interpretation again gives an isomorphism between the specified objects, but in the saturated $\mathcal{J}_{2\text{-}\mathcal{C}at}$-structure $M_{\mathcal{C}}$ defined in Example 11.2.7, the interpretation now gives an equivalence. By the construction of Example 11.2.27, formulae in the language of 2-categories that are interpreted in the $\mathcal{J}_{2\text{-}\mathcal{C}at}$-structures $M_{\mathcal{C}}$ are invariant under biequivalence of 2-categories, though this result does not necessarily hold when they are interpreted in the naïve $\mathcal{J}_{2\text{-}\mathcal{C}at}$-structures $N_{\mathcal{C}}$ (see Exercise 11.2.vi and Digression B.1.7).

We now give a criterion under which two $\mathcal{J}$-structures are guaranteed to satisfy the same formulas.

DEFINITION 11.2.21. A natural transformation $\rho : M \to N$ of $\mathcal{J}$-structures is **fiberwise surjective** if for each $K \in \mathcal{J}$, either a relation symbol or a kind, the map to the pullback

$$
\begin{array}{ccc}
MK & \xrightarrow{\ \rho_K\ } & NK \\
m_K \downarrow & \quad & \downarrow m_K \\
\partial^K M & \xrightarrow[\ \partial^K\rho\ ]{} & \partial^K N
\end{array}
\tag{11.2.22}
$$

in the "matching square" for K is an epimorphism.

In the terminology of §C.5, the fiberwise surjective maps are exactly those natural transformations that define **Reedy epimorphisms**, i.e., for which the relative matching maps (11.2.22) are epimorphisms.

LEMMA 11.2.23. *Let* $\rho : M \to N$ *be fiberwise surjection between $\mathcal{J}$-structures. Then for each relation symbol $\dot{R}$, the matching square is a pullback:*

$$
\begin{array}{ccc}
M\dot{R} & \xrightarrow{\ \rho_{\dot{R}}\ } & N\dot{R} \\
m_{\dot{R}} \downarrow & \quad & \downarrow m_{\dot{R}} \\
\partial^{\dot{R}} M & \xrightarrow[\ \partial^{\dot{R}}\rho\ ]{} & \partial^{\dot{R}} N
\end{array}
$$

Proof Since M and N are $\mathcal{J}$-structures, the matching maps for the relation symbols of $\mathcal{J}$ are monomorphisms (see Exercise 11.2.i). Thus, the induced map to the pullback is a monomorphism and by the fiberwise surjectivity hypothesis also an epimorphism. Thus, this map is an isomorphism in the category of sets, and so the square is a pullback.   □

Recall the structures constructed in Examples 11.2.6 and 11.2.7.

LEMMA 11.2.24.

   (i) *A surjective equivalence of categories* $\mathcal{C} \twoheadrightarrow \mathcal{D}$ *induces a fiberwise surjection* $M_{\mathcal{C}} \twoheadrightarrow M_{\mathcal{D}}$ *of $\mathcal{J}_{\mathcal{C}at}$-structures.*

   (ii) *A surjective biequivalence of 2-categories* $\mathcal{C} \twoheadrightarrow \mathcal{D}$ *induces a fiberwise surjection* $M_{\mathcal{C}} \twoheadrightarrow M_{\mathcal{D}}$ *of $\mathcal{J}_{2\text{-}\mathcal{C}at}$-structures.*

Proof The two statements are special cases of a more general result and thus have a common proof [52]. On both $\mathcal{C}at$ and 2-$\mathcal{C}at$ there is a (cofibration, trivial fibration) weak factorization systems whose trivial fibrations are the surjective bi/equivalences referred to in the statement. The structures are defined by applying the hom bifunctors

$$
(\mathcal{C}at^{\mathcal{J}_{\mathcal{C}at}^{\mathrm{op}}})^{\mathrm{op}} \times \mathcal{C}at \xrightarrow{\ \mathrm{hom}\ } \mathcal{S}et^{\mathcal{J}_{\mathcal{C}at}} \qquad (2\text{-}\mathcal{C}at^{\mathcal{J}_{2\text{-}\mathcal{C}at}^{\mathrm{op}}})^{\mathrm{op}} \times 2\text{-}\mathcal{C}at \xrightarrow{\ \mathrm{hom}\ } \mathcal{S}et^{\mathcal{J}_{2\text{-}\mathcal{C}at}}
$$

to diagrams $D_{\mathcal{C}at}$ and $D_{2\text{-}\mathcal{C}at}$ that are Reedy cofibrant. By Lemma C.2.11 and Corollary C.5.16, these are right Leibniz bifunctors with respect to the Reedy weak factorization systems built from any weak factorization system on $\mathcal{C}at$ and 2-$\mathcal{C}at$ and the (monomorphism, epimorphism) weak factorization system on $\mathcal{S}et$, since they are the right adjoints in a two-variable adjunction involving the (unenriched) weighted colimit bifunctor. Consequently, the functors $M_- : \mathcal{C}at \to \mathcal{S}et^{\mathcal{J}_{\mathcal{C}at}}$ and $M_- : 2\text{-}\mathcal{C}at \to \mathcal{S}et^{\mathcal{J}_{2\text{-}\mathcal{C}at}}$ carry trivial fibrations, i.e., surjective bi/equivalences, to Reedy epimorphisms, i.e., fiberwise surjections. $\square$

DEFINITION 11.2.25. Let $\mathcal{J}$ be a FOLDS signature and let $\Gamma : \mathcal{J} \to \mathcal{S}et$ be a context with given interpretations $\alpha : \Gamma \to M$ and $\beta : \Gamma \to N$ in $\mathcal{J}$-structures $M, N : \mathcal{J} \to \mathcal{S}et$. The $\mathcal{J}$-structures M and N are $\mathcal{J}$-**equivalent in context** Γ, denoted $M \simeq_{\mathcal{J}}^{\Gamma} N$, just when there is a diagram of fiberwise surjections σ and ρ between $\mathcal{J}$-structures under Γ

$$
\begin{array}{ccccc}
 & & \Gamma & & \\
 & {}^{\alpha}\swarrow & \downarrow{\gamma} & \searrow^{\beta} & \\
M & & & & N \\
 & {}_{\sigma}\searrow & \downarrow & \swarrow_{\rho} & \\
 & & P & &
\end{array}
$$

The relation $\mathcal{J}$-equivalent in context Γ is manifestly reflexive and symmetric. In fact, it is also transitive, as the reader may verify in Exercise 11.2.v.

EXAMPLE 11.2.26. Consider an equivalence of categories $f : C \overset{\sim}{\to} D$ and form the iso-comma category of Definition 6.2.10 by the pullback.

$$
\begin{array}{ccc}
D \overset{\sim}{\underset{D}{\times}} C & \longrightarrow & D^{\mathbb{I}} \\
{\scriptstyle (q_1, q_0)}\downarrow & \lrcorner & \downarrow{\scriptstyle (q_1, q_0)} \\
D \times C & \underset{\mathrm{id} \times f}{\longrightarrow} & D \times D
\end{array}
$$

Since f is an equivalence of categories, the functors $D \overset{q_1}{\twoheadleftarrow} D \overset{\sim}{\underset{D}{\times}} C \overset{q_0}{\twoheadrightarrow} C$ are surjective equivalences. It follows from Lemma 11.2.24 that the corresponding natural transformations between the naïve $\mathcal{J}_{\mathcal{C}at}$-structures of Example 11.2.6 are fiberwise surjections

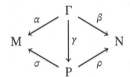

defining an $\mathcal{J}_{\mathcal{C}at}$-equivalence between M_C and M_D.

EXAMPLE 11.2.27. Consider a biequivalence of 2-categories $F \colon \mathcal{C} \to \mathcal{D}$ and form the **pseudo-comma 2-category** $\mathcal{P}$ whose

- objects are triples $(x \in \mathcal{C}, x' \in \mathcal{D}, a \colon Fx \xrightarrow{\sim} x' \in \mathcal{D})$ with a an equivalence in $\mathcal{D}$,
- 1-cells $(x, x', a) \to (y, y', b)$ are triples

$$
\left(
\begin{array}{ccc}
x & x' & Fx \xrightarrow{\ a\ }_{\sim} x' \\
{\scriptstyle f}\downarrow\ \in \mathcal{C}, & {\scriptstyle f'}\downarrow\ \in \mathcal{D}, & {\scriptstyle Ff}\downarrow \quad \cong\alpha \quad \downarrow{\scriptstyle f'}\ \in \mathcal{D} \\
y & y' & Fy \xrightarrow[\ b\ ]{\sim} y'
\end{array}
\right)
$$

- 2-cells $(f, f', \alpha) \Rightarrow (g, g', \beta)$ are given by a pair $(\gamma \colon f \Rightarrow g \in \mathcal{C}, \delta \colon f' \Rightarrow g' \in \mathcal{D})$ so that

$$
Ff\!\left(\!\!\begin{array}{c}{\scriptstyle F\gamma}\\ \Rightarrow\end{array}\!\!\right)\!Fg
\begin{array}{c}Fx \xrightarrow{\ a\ }_{\sim} x' \\ \cong\beta \\ Fy \xrightarrow[\ b\ ]{\sim} y'\end{array}\! g'
\ =\
Ff\left(\begin{array}{c}Fx \xrightarrow{\ a\ }_{\sim} x' \\ \cong\alpha \\ Fy \xrightarrow[\ b\ ]{\sim} y'\end{array}\right) f'\!\left(\!\!\begin{array}{c}{\scriptstyle \delta}\\ \Rightarrow\end{array}\!\!\right)\! g'
$$

The evident projections $\mathcal{C} \leftarrow \mathcal{P} \to \mathcal{D}$ define surjective biequivalences – 2-functors that are surjective on objects, full on 1-cells, and fully faithful on 2-cells – which by Lemma 11.2.24 induce fiberwise surjective natural transformations between the saturated $\mathcal{I}_{2\text{-}\mathcal{C}at}$-structures of Example 11.2.7 defining an $\mathcal{I}_{2\text{-}\mathcal{C}at}$-equivalence

$$
\begin{array}{ccc}
 & M_{\mathcal{P}} & \\
{\scriptstyle \sigma}\swarrow & & \searrow{\scriptstyle \rho} \\
M_{\mathcal{C}} & & M_{\mathcal{D}}
\end{array}
$$

Both of these constructions can be understood as instances of the Brown factorization C.1.6 applied in the folk model structures on $\mathcal{C}at$ and on 2-$\mathcal{C}at$. Our interest in these notions is on account of the following theorem of Makkai:

THEOREM 11.2.28. *If* M *and* N *are $\mathcal{I}$-equivalent in a context* Γ

$$
\begin{array}{ccccc}
 & & \Gamma & & \\
 & {\scriptstyle \alpha}\swarrow & \downarrow{\scriptstyle \gamma} & \searrow{\scriptstyle \beta} & \\
M & & \Downarrow & & N \\
 & {\scriptstyle \sigma}\searrow & \downarrow & \nearrow{\scriptstyle \rho} & \\
 & & P & &
\end{array}
$$

then $M \vDash \phi[\alpha]$ *if and only if* $N \vDash \phi[\beta]$ *for all formulae* ϕ *with* $\mathrm{var}(\phi) \subset \Gamma$.

Proof Since P is itself an $\mathcal{I}$-structure with an interpretation $\gamma : \Gamma \to$ P of the variables defined by the context, it suffices to show that for any fiberwise surjection $\sigma :$ P $\to$ M then P and M satisfy the same formulas: P $\vDash \phi[\gamma]$ if and only if M $\vDash \phi[\sigma\gamma]$. This is proven by an induction over the complexity of the formula ϕ.

For the base case, consider an atomic formula $\dot{R}\langle x_p \rangle$ with an interpretation $x_p \mapsto \gamma x_p$ in P. From the pullback of Lemma 11.2.23, $\langle \gamma x_p \rangle \in \partial^{\dot{R}}$P lies in P$\dot{R}$ if and only if $\langle \sigma\gamma x_p \rangle \in \partial^{\dot{R}}$M lies in M$\dot{R}$. Thus P $\vDash \dot{R}[\gamma]$ if and only if M $\vDash \dot{R}[\sigma\gamma]$.

Next consider the compound formulas $\phi \wedge \psi$, $\phi \vee \psi$, and $\phi \to \psi$ built from ϕ and ψ. Under the inductive hypothesis, we may assume that P $\vDash \phi[\gamma]$ if and only if M $\vDash \phi[\sigma\gamma]$ and similarly for ψ. Now it follows immediately that P $\vDash (\phi \wedge \psi)[\gamma]$ if and only if M $\vDash (\phi \wedge \psi)[\sigma\gamma]$ and similarly for compound formulas $\phi \vee \psi$ and $\phi \to \psi$.

Finally, consider a formula of the form $\forall x\phi$ where $x : K\langle x_p \rangle$. The interpretation $\gamma : \text{var}(\forall x\phi) \to$ P defines an element in the matching object $\langle \gamma x_p \rangle \in \partial^K$P. The fiberwise surjectivity condition tells us that the map from PK to the pullback in the matching square for K displayed below-right is a surjection:

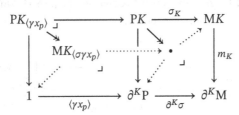

By composing pullbacks, it follows that we get a surjection from the fiber of PK $\to \partial^K$P over $\langle \gamma x_p \rangle$ to the fiber of MK $\to \partial^K$M over $\langle \sigma\gamma x_p \rangle$. In particular, if for all a in the fiber of PK $\to \partial^K$P, P $\vDash \phi(a/x)[\gamma]$, then by the inductive hypothesis M $\vDash \phi(\sigma a/x)[\sigma\gamma]$. By fiberwise surjectivity, every element b in the fiber of MK $\to \partial^K$M equals σa for some a in the fiber of PK $\to \partial^K$P, so this tells us P $\vDash (\forall x\phi)[\gamma]$ if and only if M $\vDash (\forall x\phi)[\sigma\gamma]$.

Similarly, for $\exists x\phi$, P $\vDash (\exists x\phi)[\gamma]$ if and only if there exists some a in the fiber so that P $\vDash \phi(a/x)[\gamma]$. But this holds if and only if M $\vDash \phi(\sigma a/x)[\sigma\gamma]$ and, by naturality of σ, σa lives in the appropriate fiber. So if P $\vDash (\exists x\phi)[\gamma]$ then M $\vDash (\exists x\phi)[\sigma\gamma]$. Conversely if M $\vDash (\exists x\phi)[\sigma\gamma]$, then there is some b in the fiber of MK $\twoheadrightarrow \partial^K$M over $\langle \sigma\gamma x_p \rangle$ so that M $\vDash \phi(b/x)[\sigma\gamma]$. By fiberwise surjectivity, there is some a in the fiber of PK $\twoheadrightarrow \partial^K$P over $\langle \gamma x_p \rangle$ so that $\sigma a = b$. By the inductive hypothesis P $\vDash \phi(a/x)[\gamma]$ if and only if M $\vDash \phi(b/x)[\sigma\gamma]$, so we see that if M $\vDash (\exists x\phi)[\sigma\gamma]$ then P $\vDash (\exists x\phi)[\gamma]$. $\square$

REMARK 11.2.29. An observation of Henry [52] provides a nice perspective on Makkai's Theorem 11.2.28. Henry proves that the collection of formulae for a fixed FOLDS signature $\mathcal{J}$ and define the initial functor from the category of contexts for that signature to boolean algebras, with the property that the restriction homomorphisms associated to display maps admit both left and right adjoints (existential and universal quantification) satisfying the Beck–Chevalley condition. By taking powersets, any $\mathcal{J}$-structure gives rise to a canonical functor from the category of contexts to boolean algebras, and the unique map from the initial object sends a formula to the subset of interpretations of its free variables that satisfy the formula. A map ρ of $\mathcal{J}$-structures defines a natural transformation between the corresponding boolean algebra valued functors – contravariantly, by reindexing – and this natural transformation respects the left and right adjoints if and only if ρ is fiberwise surjective. Thus, by initiality, we see that fiberwise surjections respect the interpretation of formulas.

Exercises

EXERCISE 11.2.i. Let $\mathcal{J}$ be a FOLDS signature with relation symbol $\dot{R} \in \mathcal{J}$ and let M : $\mathcal{J} \to \mathcal{S}et$ be any functor. Prove that the map of Definition 11.2.5 is a monomorphism if and only if the matching map of Observation C.1.18 is a monomorphism:

$$
\begin{array}{ccc}
\text{M}\dot{R} & \longrightarrow & \prod\limits_{p \,:\, \dot{R} \overset{\neq}{\to} K_p} \text{M}K_p \\
 & m^{\dot{R}} \searrow \quad \nearrow & p : \dot{R} \overset{\neq}{\to} K_p \\
 & \partial^{\dot{R}}\text{M} := \lim\limits_{p \,:\, \dot{R} \overset{\neq}{\to} K_p} \text{M}K_p &
\end{array}
$$

EXERCISE 11.2.ii.

(i) Show that if ϕ is any formula then $\mathrm{var}(\phi)$ is closed under dependences: if $x \in \mathrm{var}(\phi)$ then $\mathrm{dep}(x) \subset \mathrm{var}(\phi)$.

(ii) Show that the variables $\mathrm{var}(\phi)$ defined for a particular formula ϕ in context Γ define a subcontext $\mathrm{var}(\phi) \subset \Gamma$.

EXERCISE 11.2.iii. An object s in a 2-category $\mathcal{C}$ is **biterminal** if for every $c \in \mathcal{C}$, the hom-category $\mathcal{C}(c, s) \simeq \mathbb{1}$ is a contractible groupoid. Write a sentence in the language of 2-categories that asserts that the 2-category has a biterminal object.

EXERCISE 11.2.iv. Connect the result of Lemma 11.2.23 to the statement that $\text{M} \vDash \dot{R}[\alpha]$ if and only if $\text{N} \vDash \dot{R}[\rho\alpha]$ for some interpretation $\alpha : \Gamma \to \text{M}$.

EXERCISE 11.2.v. Verify that the relation defined by Definition 11.2.25 is transitive by composing the spans formed by the fiberwise surjections.

EXERCISE 11.2.vi. Use a sentence or formula along the lines considered in Example 11.2.20 and the result of Theorem 11.2.28 to show that there exist biequivalent 2-categories $\mathcal{C}$ and $\mathcal{D}$ so that the naïve $\mathcal{I}_{2\text{-}\mathcal{C}at}$-structures $N_{\mathcal{C}}$ and $N_{\mathcal{D}}$ are not $\mathcal{I}_{2\text{-}\mathcal{C}at}$-equivalent in an appropriate context.

11.3 A Language for Model Independent ∞-Category Theory

Our aim in this section is to apply Makkai's Theorem 11.2.28 to prove that statements about ∞-categories that are written in the language of virtual equipments, defined for a FOLDS signature $\mathcal{I}_{\mathcal{V}\mathcal{E}}$ introduced in Definition 11.3.2, are invariant under change-of-model. Most of the complexity in the simple inverse category $\mathcal{I}_{\mathcal{V}\mathcal{E}}$ is present already in the FOLDS signature for virtual double categories, which we introduce first.

DEFINITION 11.3.1. The FOLDS signature $\mathcal{I}_{\mathcal{V}\mathcal{D}bl\mathcal{C}at}$ for virtual double categories has kinds for the objects, vertical arrows, horizontal modules, and n-ary cells for each $n \geq 0$.

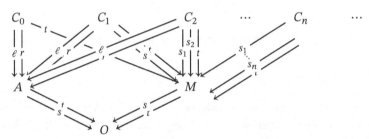

Each arrow and each module has a source object and target object. Each n-ary cell has a left boundary arrow, a right boundary arrow, a target module, and n source modules. The source and target objects of this boundary data must be compatible in the way specified by the composition relations

$$s \cdot \ell = s \cdot s_1 \qquad s \cdot r = t \cdot s_n \qquad t \cdot s_k = s \cdot s_{k+1}, \quad 1 \leq k < n$$
$$t \cdot \ell = s \cdot t \qquad t \cdot r = t \cdot t$$

in the case $n \geq 1$.[7]

Two further kinds I_A and T parametrize identity vertical arrows and composition of vertical arrows with the composition relations from the signature for

---

[7] For nullary cells, the relation $s \cdot \ell = s \cdot r$ replaces the relations involving the absent source modules.

1-categories:

$$s \cdot i = t \cdot i \qquad s \cdot c = s \cdot \ell \qquad t \cdot c = t \cdot r \qquad t \cdot \ell = s \cdot r$$

Each sort of n-ary cells supports a binary equality relation:

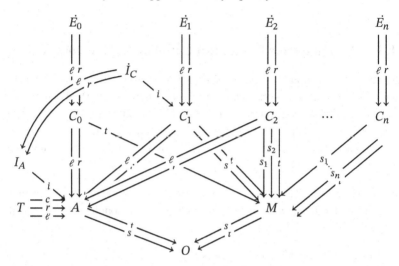

satisfying equations that demand that equal cells must have the same boundary type:

$$\ell \cdot \ell = \ell \cdot r \qquad r \cdot \ell = r \cdot r \qquad t \cdot \ell = t \cdot r \qquad s_k \cdot \ell = s_k \cdot r$$

In addition, in a virtual double category, each module has a specified unary identity cell whose vertical sources and targets are identity arrows. The relation symbol $\dot{I}_C$ satisfies the composition relation

$$s \cdot i = t \cdot i \qquad \ell \cdot i = i \cdot \ell \qquad r \cdot i = i \cdot r$$

which require that the source and target modules coincide and the left and right vertical arrows are identities.

Finally, for each partition $n_1 + \cdots + n_k = n$ there is a $(k + 4)$-ary relation symbols $\dot{T}_{n_1,\dots,n_k}$ with the indicated dependencies:

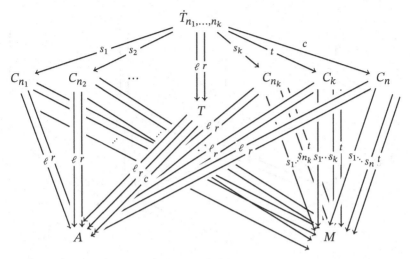

satisfying the composition relations

$$s_i \cdot s_j = s_{n_1+\cdots+n_{j-1}+i} \cdot c \quad \ell \cdot \ell = \ell \cdot s_1 \quad \ell \cdot r = r \cdot s_k \quad r \cdot s_j = \ell \cdot s_{j+1}$$

$$t \cdot s_j = s_j \cdot t \qquad\qquad r \cdot \ell = \ell \cdot t \qquad r \cdot r = r \cdot t$$

$$t \cdot t = t \cdot c \qquad\qquad c \cdot \ell = \ell \cdot c \qquad c \cdot r = r \cdot c$$

Here $\dot{T}_{n_1,\dots,n_k}$ is the relation that witnesses that a given n-cell is a composite of k specified source cells with arities $n_1, \dots, n_k$ into a given target k-cell. Such a composition depends also on the specification of left and right commutative triangles of vertical arrows.

DEFINITION 11.3.2. The FOLDS signature for a virtual equipment $\mathcal{J}_{\mathcal{VE}}$ extends the FOLDS signature for virtual double categories with two additional unary relation symbols $\dot{U}$ and $\dot{R}$ identifying the nullary unit modules and unary restriction modules:

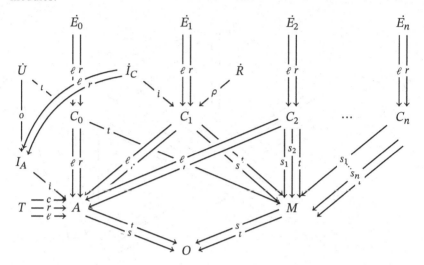

The left and right vertical arrows of a unit module are identities on the same object, as expressed by the composition relations

$$\ell \cdot \iota = i \cdot o = r \cdot \iota.$$

There are no composition relations imposed on the boundary of the unary restriction cell. This data comprises the FOLDS signature for virtual equipments a la Cruttwell and Shulman.

Any virtual equipment defines an $\mathcal{J}_{\mathcal{VE}}$-structure in a naïve manner, with the functor $\mathcal{J}_{\mathcal{VE}} \to \mathcal{S}et$ assigning each kind or relation symbol its intended interpretation.[8] But as was the case for 2-categories, the objects I_A and T are not relation symbols in the FOLDS signature for virtual equipments, so the matching maps associated to these objects need not be monomorphisms. This gives us the flexibility to convert a virtual equipment into an $\mathcal{J}_{\mathcal{VE}}$-structure in a more "saturated" manner. In fact, in analogy with Example 11.2.27 and Exercise 11.2.vi, it is necessary to "saturate" the $\mathcal{J}_{\mathcal{VE}}$-structures associated to virtual equipments in this manner to prove one of our desired results: that a biequivalence of virtual equipments gives rise to an $\mathcal{J}_{\mathcal{VE}}$-equivalence, in the sense of Definition 11.2.25. As in Example 11.2.7, it is possible to define the "saturated" $\mathcal{J}_{\mathcal{VE}}$-structure associated to a virtual equipment by defining a corresponding contravariant functor from $\mathcal{J}_{\mathcal{VE}}$ to the category of virtual equipments and structure-preserving functors, but in this case the virtual equipments that appear in the indexing dia-

---

[8] As the virtual equipments of greatest interest are large, the $\mathcal{J}_{\mathcal{VE}}$-structures are large as well.

gram are more complicated to describe, so instead we just describe the structure directly.

RECALL 11.3.3. Recall from Proposition 8.3.18 that any virtual equipment contains a *vertical 2-category* whose objects are the objects of the virtual equipment, whose arrows are the vertical arrows, and whose 2-cells are those unary cells

$$
\begin{array}{ccc}
A & \xrightarrow{\ \mathrm{Hom}_A\ } & A \\
{\scriptstyle g}\downarrow & \Downarrow\tilde{\alpha} & \downarrow{\scriptstyle f} \\
B & \xrightarrow[\ \mathrm{Hom}_B\ ]{} & B
\end{array}
\tag{11.3.4}
$$

whose horizontal boundary arrows are given by the unit modules. Confusingly, the "horizontal composition" of 2-cells is captured by the vertical composition of unary cells in the virtual equipment, while the "vertical composition" of 2-cells is captured by the horizontal composition operation defined by Definition 8.3.16. In particular, we describe a unary cell in the vertical 2-category as "invertible" when it is invertible in the usual sense for a 2-cell in a 2-category, that is for the horizontal composition operation of unary cells between unit modules in the virtual equipment; note that the vertical boundary 1-cells of an invertible cell in the vertical 2-category need not be invertible in any sense.

In fact, the construction described in Definition 8.3.16 generalizes as follows. Given any cell in the virtual equipment and pair of cells in the vertical 2-category as displayed

$$
\begin{array}{ccccccc}
A & \xrightarrow{\ \mathrm{Hom}_A\ } & A & \overset{\vec{E}}{\dashrightarrow} & C & \xrightarrow{\ \mathrm{Hom}_C\ } & C \\
{\scriptstyle g}\downarrow & \Downarrow\tilde{\alpha} & {\scriptstyle f}\downarrow & \Downarrow\beta & \downarrow{\scriptstyle h} & \Downarrow\tilde{\gamma} & \downarrow{\scriptstyle k} \\
B & \xrightarrow[\ \mathrm{Hom}_B\ ]{} & B & \xrightarrow[\ F\ ]{} & D & \xrightarrow[\ \mathrm{Hom}_D\ ]{} & D
\end{array}
\tag{11.3.5}
$$

there is a unique "horizontal composite" cell with boundary

$$
\begin{array}{ccc}
A & \overset{\vec{E}}{\dashrightarrow} & C \\
{\scriptstyle g}\downarrow & \Downarrow\tilde{\alpha}*\beta*\tilde{\gamma} & \downarrow{\scriptstyle k} \\
B & \xrightarrow[\ F\ ]{} & D
\end{array}
\tag{11.3.6}
$$

satisfying the analogue of the pasting equality (8.3.17) defined relative to the canonical composition cells of Lemma 8.3.10. In what follows, we abuse notation and write an expression like (11.3.5) to denote the composite cell (11.3.6).

In the virtual equipment $\mathbb{M}\mathrm{od}(\mathcal{K})$ of modules in an ∞-cosmos $\mathcal{K}$, this vertical 2-category is isomorphic to the homotopy 2-category $\mathfrak{h}\mathcal{K}$. Since our intention is to apply the following construction in that context, we borrow notation from the homotopy 2-category when it is simpler to read, for instance writing $\alpha : f \Rightarrow g$ for a unary cell (11.3.4) in the vertical 2-category whose right boundary is f and whose left boundary is g.

DEFINITION 11.3.7. For any virtual equipment $\mathbb{C}$, there is an $\mathcal{J}_{\mathcal{VE}}$-structure $M_{\mathbb{C}}$ in which:

- The sets $M_{\mathbb{C}}O$, $M_{\mathbb{C}}A$, $M_{\mathbb{C}}M$, and $M_{\mathbb{C}}C_n$ for each $n \geq 0$ are given their naïve interpretations as the sets of objects, vertical arrows, horizontal modules, and n-ary cells of $\mathbb{C}$.
- The binary relations $\dot{E}_n$ encode equality between parallel cells.
- The relation $\dot{R}$ encodes all unary cells that are cartesian, satisfying the universal property that characterizes the restriction cells.
- For any $i : M_{\mathbb{C}}A\langle X, X \rangle$, $M_{\mathbb{C}}I_A\langle i \rangle$ is the set of invertible cells $\alpha : \mathrm{id}_X \cong i$ in the vertical 2-category.
- For any $i : M_{\mathbb{C}}A\langle X, X \rangle$, $\alpha : M_{\mathbb{C}}I_A\langle i \rangle$, $j : M_{\mathbb{C}}A\langle Y, Y \rangle$, $\beta : M_{\mathbb{C}}I_A\langle j \rangle$, $E : M_{\mathbb{C}}M\langle X, Y \rangle$, and $\nu : M_{\mathbb{C}}C_1\langle i, j, E, E \rangle$, the set $M_{\mathbb{C}}\dot{I_C}\langle \nu \rangle$ is a singleton if the identity cell id_E associated to the module E equals the horizontal composite cell

$$
\begin{array}{ccccccc}
X & \xrightarrow{\mathrm{Hom}_X} & X & \xrightarrow{E} & Y & \xrightarrow{\mathrm{Hom}_Y} & Y \\
\Big\| & \Downarrow\overleftarrow{\alpha^{-1}}\ i\Big\downarrow & & \Downarrow\nu & \Big\downarrow j & \Downarrow\overleftarrow{\beta} & \Big\| \\
X & \xrightarrow[\mathrm{Hom}_X]{} & X & \xrightarrow[E]{} & Y & \xrightarrow[\mathrm{Hom}_Y]{} & Y
\end{array}
\quad = \quad
\begin{array}{ccc}
X & \xrightarrow{E} & Y \\
\Big\| & \Downarrow\mathrm{id}_E & \Big\| \\
X & \xrightarrow[E]{} & Y
\end{array}
$$

and is empty otherwise.

- For $i : M_{\mathbb{C}}A\langle X, X \rangle$, $\alpha : M_{\mathbb{C}}I_A\langle i \rangle$, $H : M_{\mathbb{C}}M\langle X, X \rangle$, and $\theta : M_{\mathbb{C}}C_0\langle i, i, H \rangle$, the set $M_{\mathbb{C}}\dot{U}\langle i, \theta \rangle$ is a singleton if the nullary composite cell

$$
\begin{array}{ccccccc}
X & \xrightarrow{\mathrm{Hom}_X} & X & = \!\!= & X & \xrightarrow{\mathrm{Hom}_X} & X \\
\Big\| & \Downarrow\overleftarrow{\alpha^{-1}}\ i\Big\downarrow & & \Downarrow\theta & \Big\downarrow i & \Downarrow\overleftarrow{\alpha} & \Big\| \\
X & \xrightarrow[\mathrm{Hom}_X]{} & X & \xrightarrow[H]{} & X & \xrightarrow[\mathrm{Hom}_X]{} & X
\end{array}
$$

is cocartesian and is empty otherwise.

- For any $f : M_{\mathbb{C}}A\langle X, Y \rangle$, $g : M_{\mathbb{C}}A\langle Y, Z \rangle$, and $h : M_{\mathbb{C}}A\langle X, Z \rangle$, the set $M_{\mathbb{C}}T\langle f, g, h \rangle$ is the set of invertible cells $\gamma : gf \cong h$ in the vertical 2-category.

- For any $n_1 + \cdots + n_k = n$, cells

$$
\begin{array}{ccccccc}
A_0 & \overset{\vec{E}_1}{\dashrightarrow} & A_1 & \overset{\vec{E}_2}{\dashrightarrow} & \cdots & \overset{\vec{E}_k}{\dashrightarrow} & A_k \\
{\scriptstyle f_0}\downarrow & {\scriptstyle \Downarrow\alpha_1} & {\scriptstyle f_1}\downarrow\; {\scriptstyle \Downarrow\alpha_2} & {\scriptstyle |\cdots|} & {\scriptstyle \Downarrow\alpha_k} & & \downarrow{\scriptstyle f_k} \\
B_0 & \overset{F_1}{\longrightarrow} & B_1 & \overset{F_2}{\longrightarrow} & \cdots & \overset{F_k}{\longrightarrow} & B_n \\
{\scriptstyle g}\downarrow & & {\scriptstyle \Downarrow\beta} & & & & \downarrow{\scriptstyle h} \\
C & & & \overset{G}{\longrightarrow} & & & D
\end{array}
$$

$$
\begin{array}{ccccccc}
A_0 & \overset{\vec{E}_1}{\dashrightarrow} & A_1 & \overset{\vec{E}_2}{\dashrightarrow} & \cdots & \overset{\vec{E}_k}{\dashrightarrow} & A_k \\
{\scriptstyle k}\downarrow & & & {\scriptstyle \Downarrow\epsilon} & & & \downarrow{\scriptstyle \ell} \\
C & & & \overset{G}{\longrightarrow} & & & D
\end{array}
$$

and any pair of isomorphisms $\gamma : T\langle f_0, g, k\rangle$ and $\delta : T\langle f_k, h, \ell\rangle$, the set $\mathrm{M_C}\dot{T}_{n_1,\dots,n_k}\langle \alpha_1, \dots, \alpha_k, \beta, \epsilon, \gamma, \delta\rangle$ is a singleton if ϵ equals the composite

$$
\begin{array}{ccccccccc}
A_0 & \overset{\mathrm{Hom}_{A_0}}{\longrightarrow} & A_0 & \overset{\vec{E}_1}{\dashrightarrow} & A_1 & \overset{\vec{E}_2}{\dashrightarrow} & \cdots \overset{\vec{E}_k}{\dashrightarrow} A_k & \overset{\mathrm{Hom}_{A_k}}{\longrightarrow} & A_k \\
{\scriptstyle k}\downarrow & {\scriptstyle \Downarrow\gamma} & {\scriptstyle f_0}\downarrow\; {\scriptstyle \Downarrow\alpha_1} & & {\scriptstyle f_1}\downarrow\; {\scriptstyle \Downarrow\alpha_2} & {\scriptstyle \Downarrow\alpha_k} & {\scriptstyle f_k}\downarrow & {\scriptstyle \Downarrow\delta^{-1}} & \downarrow{\scriptstyle \ell} \\
& & B_0 & \overset{F_1}{\longrightarrow} & B_1 & \overset{F_2}{\longrightarrow} & \cdots \overset{F_k}{\longrightarrow} B_n & & \\
& & {\scriptstyle g}\downarrow & & {\scriptstyle \Downarrow\beta} & & \downarrow{\scriptstyle h} & & \\
C & \overset{\mathrm{Hom}_C}{\longrightarrow} & C & & \overset{G}{\longrightarrow} & & D & \overset{\mathrm{Hom}_D}{\longrightarrow} & D
\end{array}
$$

and is empty otherwise.

We write $\mathrm{M}_{\mathcal{K}} : \mathcal{J}_{\mathcal{VE}} \to \mathit{Set}$ for the $\mathcal{J}_{\mathcal{VE}}$-structure obtained by applying the construction of Definition 11.3.7 to the virtual equipment of modules $\mathbb{Mod}(\mathcal{K})$.

REMARK 11.3.8. Recall that a functor of virtual equipments $F \colon \mathbb{C} \to \mathbb{D}$ is a map of virtual double categories that preserves all the structure, including unit and restriction cells. It follows that application of F directly defines the components of a natural transformation $F \colon \mathrm{M}_{\mathbb{C}} \to \mathrm{M}_{\mathbb{D}}$ between the corresponding $\mathcal{J}_{\mathcal{VE}}$-structures. By Proposition 11.1.2, a cosmological functor $F \colon \mathcal{K} \to \mathcal{L}$ induces a functor $F \colon \mathbb{Mod}(\mathcal{K}) \to \mathbb{Mod}(\mathcal{L})$ of virtual equipments, which thus induces a natural transformation $F \colon \mathrm{M}_{\mathcal{K}} \to \mathrm{M}_{\mathcal{L}}$ between $\mathcal{J}_{\mathcal{VE}}$-structures.

PROPOSITION 11.3.9. *Consider a pair of ∞-cosmoi connected by a cosmological functor* $F \colon \mathcal{K} \to \mathcal{L}$.

(i) If $F\colon \mathcal{K} \to \mathcal{L}$ is a cosmological biequivalence, then there exists a pair of fiberwise surjections between the corresponding $\mathcal{J}_{\mathcal{VE}}$-structures:

$$
\begin{array}{ccc}
 & P & \\
\overset{\sigma}{\swarrow} & & \overset{\rho}{\searrow} \\
M_{\mathcal{K}} & & M_{\mathcal{L}}
\end{array}
$$

(ii) Moreover, if the induced map between $\mathcal{J}_{\mathcal{VE}}$-structures preserves the given interpretations of a context Γ

$$
\begin{array}{ccc}
 & \Gamma & \\
\overset{\alpha}{\swarrow} & & \overset{\beta}{\searrow} \\
M_{\mathcal{K}} & \xrightarrow{\;\;F\;\;} & M_{\mathcal{L}}
\end{array}
$$

then the fiberwise surjections constructed in (i) defines an $\mathcal{J}_{\mathcal{VE}}$-equivalence in context Γ:

$$
\begin{array}{ccc}
 & \Gamma & \\
\overset{\alpha}{\swarrow} & \downarrow{\scriptstyle\gamma} & \overset{\beta}{\searrow} \\
M_{\mathcal{K}} & & M_{\mathcal{L}} \\
\overset{\sigma}{\searrow} & P & \overset{\rho}{\swarrow}
\end{array}
$$

Proof   We apply the same strategy used in Examples 11.2.26 and 11.2.27, forming a structure that we call the *pseudo-comma virtual equipment* $\mathbb{P}$ associated to the functor of virtual equipments $F\colon \mathsf{Mod}(\mathcal{K}) \to \mathsf{Mod}(\mathcal{L})$.[9] The $\mathcal{J}_{\mathcal{VE}}$-structure P is then defined to be $M_{\mathbb{P}}$ and σ and ρ are the evident projections. We then argue that these maps are fiberwise surjective under the hypothesis that F is a biequivalence of virtual equipments.

The **pseudo-comma virtual equipment** $\mathbb{P}$ associated to $F\colon \mathsf{Mod}(\mathcal{K}) \to \mathsf{Mod}(\mathcal{L})$ has

- objects given by triples $(A \in \mathcal{K}, A' \in \mathcal{L}, a\colon FA \overset{\sim}{\to} A' \in \mathcal{L})$ where a is an equivalence,
- vertical arrows $(A, A', a) \to (B, B', b)$ given by triples

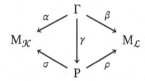

$$
\left(
\begin{array}{ccc}
A & A' & FA \xrightarrow{\;\overset{a}{\sim}\;} A' \\
{\scriptstyle f}\big\downarrow \in \mathcal{K}, & {\scriptstyle f'}\big\downarrow \in \mathcal{L}, & {\scriptstyle Ff}\big\downarrow \;{\cong\alpha}\; \big\downarrow{\scriptstyle f'} \in \mathfrak{h}\mathcal{L} \\
B & B' & FB \xrightarrow[\;\tilde{b}\;]{} B'
\end{array}
\right)
$$

---

[9]   A pseudo-comma virtual equipment can be defined for any functor (or more generally cospan of functors) between virtual equipments in exactly the same manner, but where it simplifies notation we refer also to structures in the ∞-cosmoi and their homotopy 2-categories.

- modules $(A, A', a) \longrightarrow (B, B', b)$ given by triples

$$
\left(
A \xrightarrow{E} B \ \in \mathcal{K}, \ A' \xrightarrow{E'} B' \ \in \mathcal{L}, \
\begin{array}{ccc}
FA & \xrightarrow{FE} & FB \\
a \downarrow & \Downarrow \epsilon & \downarrow b \\
A' & \xrightarrow[E']{} & B'
\end{array}
\ \in \mathsf{Mod}(\mathcal{L})
\right)
$$

where ϵ is a cartesian cell,
- n-ary cells given by a pair of n-ary cells μ in $\mathcal{K}$ and ν in $\mathcal{L}$ so that the composite cells formed from their boundary data are equal:

$$
\begin{array}{ccccccccc}
FA_0 & \xrightarrow{F\mathrm{Hom}_{A_0}} & FA_0 & \xrightarrow{FE_1} & FA_1 & \xrightarrow{FE_2} & \cdots & \xrightarrow{FE_n} & FA_n \\
a_0 \downarrow & & Ff \downarrow & & & \Downarrow F\mu & & & \downarrow Fg \\
A'_0 & \Downarrow \tilde{\alpha} & FB & & & \xrightarrow{FG} & & & FC \\
f' \downarrow & & b \downarrow & & & \Downarrow \epsilon & & & \downarrow c \\
B' & \xrightarrow[\mathrm{Hom}_{B'}]{} & B' & & & \xrightarrow{G'} & & & C'
\end{array}
\quad =
$$

$$
\begin{array}{ccccccccc}
FA_0 & \xrightarrow{FE_1} & FA_1 & \xrightarrow{FE_2} & \cdots & \xrightarrow{FE_n} & FA_n & \xrightarrow{F\mathrm{Hom}_{A_n}} & FA_n \\
a_0 \downarrow & \Downarrow \epsilon & a_1 \downarrow & \Downarrow \epsilon & |\cdots| & \Downarrow \epsilon & \downarrow a_n & & \downarrow Fg \\
A'_0 & \xrightarrow{E'_1} & A'_1 & \xrightarrow{E'_2} & \cdots & \xrightarrow{E'_n} & A'_n & \Downarrow \tilde{\beta} & FC \\
f' \downarrow & & & \Downarrow \nu & & & \downarrow g' & & \downarrow c \\
B' & & & \xrightarrow{G'} & & & C' & \xrightarrow[\mathrm{Hom}_{C'}]{} & C'
\end{array}
$$

We argue that $\mathbb{P}$ is a virtual double category. Note the objects and vertical arrows form a category: indeed it is in the underlying 1-category of the pseudo comma 2-category of Example 11.2.27 for the 2-functor $F \colon \mathfrak{h}\mathcal{K} \to \mathfrak{h}\mathcal{L}$. Each module $(E, E', \epsilon) \colon \ (A, A', a) \longrightarrow (B, B', b)$ has an identity cell whose components are the identity cells $\mathrm{id}_E \in \mathcal{K}$ and $\mathrm{id}_{E'} \in \mathcal{L}$. Composition of cells is inherited from $\mathsf{Mod}(\mathcal{K})$ and $\mathsf{Mod}(\mathcal{L})$, though we leave it to the reader to verify that the required pasting equation between the composite cells defined in this manner by drawing a very large diagram. This makes $\mathbb{P}$ into a virtual double category.

In fact $\mathbb{P}$ is a virtual equipment. Each object (A, A', a) admits a unit module

$(\mathrm{Hom}_A, \mathrm{Hom}_{A'}, \mathrm{Hom}_a)$ with a nullary cocartesian cell

$$
\begin{array}{ccc}
(A,A',a) \;=\!=\!=\!=\; (A,A',a) &
\begin{array}{ccc}
FA & =\!=\!= & FA \\
\| & \Downarrow F\iota & \| \\
FA & \xrightarrow{F\mathrm{Hom}_A} & FA \\
a\downarrow & \Downarrow \mathrm{Hom}_a & \downarrow a \\
A' & \xrightarrow[\mathrm{Hom}_{A'}]{} & A'
\end{array}
&
\begin{array}{ccc}
FA & =\!=\!= & FA \\
a\downarrow & & \downarrow a \\
A' & =\!=\!= & A' \\
\| & \Downarrow \iota & \| \\
A' & \xrightarrow[\mathrm{Hom}_{A'}]{} & A'
\end{array}
\end{array}
$$

(with vertical arrows $(\mathrm{id}_A,\mathrm{id}_{A'},\mathrm{id}_a)$ on the left diagram, cells $\Downarrow(\iota,\iota)$, bottom arrow $\xrightarrow[(\mathrm{Hom}_A,\mathrm{Hom}_{A'},\mathrm{Hom}_a)]{}$, $:=$ between the middle and right diagrams, $\rightleftharpoons$ linking)

whose vertical arrows are both taken to be the identity $(\mathrm{id}_A, \mathrm{id}_{A'}, \mathrm{id}_a)$ and whose component cells in $\mathcal{K}$ and $\mathcal{L}$ are the unit cells for A and A'. Since F preserves cocartesian cells, the universal property of $F\iota$ may be used to define a cell Hom_a satisfying the required pasting equality (compare with Definition 8.3.14). This cell is cocartesian in $\mathbb{P}$ since both components define cocartesian cells.

Finally, any diagram comprised of two vertical arrows and a horizontal module

$$
\begin{array}{ccc}
(X,X',x) & & (Y,Y',y) \\
(f,f',\alpha)\downarrow & & \downarrow (g,g',\beta) \\
(A,A',a) & \xrightarrow[(E,E',\epsilon)]{} & (B,B',b)
\end{array}
$$

can be completed to a unary cartesian cell

$$
\begin{array}{ccc}
FX \xrightarrow{F\mathrm{Hom}_X} FX \xrightarrow{FE(g,f)} FY & & FX \xrightarrow{FE(g,f)} FY \xrightarrow{F\mathrm{Hom}_Y} FY \\
x\downarrow \quad Ff\downarrow \;\Downarrow\rho \;\downarrow Fg & & x\downarrow \;\Downarrow\zeta\; \downarrow y \quad \downarrow Fg \\
X' \;\Downarrow\tilde\alpha\; FA \xrightarrow{FE} FB & = & X' \xrightarrow{E'(g',f')} Y' \;\Downarrow\tilde\beta\; FB \\
f'\downarrow \quad a\downarrow \;\Downarrow\epsilon\; \downarrow b & & f'\downarrow \;\Downarrow\rho'\; \downarrow g' \quad \downarrow b \\
A' \xrightarrow[\mathrm{Hom}_{A'}]{} A' \xrightarrow[E']{} B' & & A' \xrightarrow[E']{} B' \xrightarrow[\mathrm{Hom}_{B'}]{} B'
\end{array}
$$

where ρ and ρ' are the restriction cells in $\mathcal{K}$ and $\mathcal{L}$, respectively, and the cell ζ is defined by factoring the left-hand pasting diagram below through the cartesian cell ρ'.

$$
\begin{array}{cc}
\begin{array}{c}
FX \xrightarrow{F\mathrm{Hom}_X} FX \xrightarrow{FE(g,f)} FY \xrightarrow{F\mathrm{Hom}_Y} FY \\
x\downarrow \quad Ff\downarrow \;\Downarrow\rho\; \downarrow Fg \qquad \downarrow y \\
X' \;\Downarrow\tilde\alpha\; FA \xrightarrow{FE} FB \;\Downarrow\overline{\beta^{-1}}\; Y' \\
f'\downarrow \quad a\downarrow \;\Downarrow\epsilon\; \downarrow b \qquad \downarrow g' \\
A' \xrightarrow[\mathrm{Hom}_{A'}]{} A' \xrightarrow[E']{} B' \xrightarrow[\mathrm{Hom}_{B'}]{} B'
\end{array}
& =
\begin{array}{c}
FX \xrightarrow{FE(g,f)} FY \\
x\downarrow \;\exists!\Downarrow\zeta\; \downarrow y \\
X' \xrightarrow{E'(g',f')} Y' \\
f'\downarrow \;\Downarrow\rho'\; \downarrow g' \\
A' \xrightarrow[E']{} B'
\end{array}
\end{array}
$$

Since ϵ, $F\rho$, and ρ' are cartesian cells in $\mathcal{L}$, ζ is a cartesian cell as well, making

$(E(g, f), E'(g', f'), \zeta)$ into a module in $\mathbb{P}$. This makes the virtual double category $\mathbb{P}$ into a virtual equipment so that the evident forgetful functors to $\mathsf{Mod}(\mathcal{K})$ and $\mathsf{Mod}(\mathcal{L})$ define functors of virtual equipments.

By Remark 11.3.8 we obtain a diagram of $\mathcal{J}_{\mathcal{VE}}$-structures and natural transformations

Finally, when $F \colon \mathsf{Mod}(\mathcal{K}) \to \mathsf{Mod}(\mathcal{L})$ is a biequivalence of virtual equipments, satisfying the properties enumerated in Theorem 11.1.6, the forgetful functors

$$\mathsf{Mod}(\mathcal{K}) \longleftarrow \mathbb{P} \longrightarrow \mathsf{Mod}(\mathcal{L})$$

are surjective on objects, full on vertical arrows, full on horizontal modules, and fully faithful on cells, and reflect unit cells and restriction cells. It follows that the maps σ and ρ are fiberwise surjective.

Finally, if $F \colon \mathrm{M}_{\mathcal{K}} \to \mathrm{M}_{\mathcal{L}}$ strictly preserves the interpretations of a context Γ in $\mathsf{Mod}(\mathcal{K})$ and $\mathsf{Mod}(\mathcal{L})$, we claim that it is possible to simultaneously lift these interpretations along σ and ρ to an interpretation $\gamma \colon \Gamma \to \mathrm{P}$. This can easily be verified inductively in the $\mathcal{J}_{\mathcal{VE}}$-structure Γ. For a variable $x \colon O$, if $\alpha(x) = A$ and $\beta(x) = FA$, then define $\gamma(x) \coloneqq (A, FA, \mathrm{id}_{FA})$. The definition of γ at higher-degree variables is similar, with the missing data chosen to be the appropriate identities in $\mathcal{L}$. $\qquad\square$

Combining these results we obtain the desired corollary:

COROLLARY 11.3.10 (a language for model independent ∞-category theory). *Formulae in the language of virtual equipments are invariant under biequivalence of ∞-cosmoi.*

Proof Consider a formula ϕ in the language of virtual equipments with compatibly defined interpretations of its free variables, meaning that there is a commutative diagram

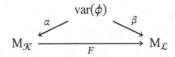

arising from the cosmological biequivalence $F \colon \mathcal{K} \rightsquigarrow \mathcal{L}$ and the $\mathcal{J}_{\mathcal{VE}}$-structures proscribed by Definition 11.3.7. By Proposition 11.3.9, these $\mathcal{J}_{\mathcal{VE}}$-structures are $\mathcal{J}_{\mathcal{VE}}$-equivalent in the context defined by the variables of ϕ. By Theorem 11.2.28, $\mathrm{M}_{\mathcal{K}}$ and $\mathrm{M}_{\mathcal{L}}$ then satisfy the same formulas. $\qquad\square$

We conclude with some examples.

EXAMPLE 11.3.11. In the context of a pair of objects $a, b : O$ and an arrow $u : A\langle a, b\rangle$, the formula

$$\exists f : A\langle b, a\rangle, \forall i_a : A\langle a, a\rangle, \forall \omega_a : I_A\langle i_a\rangle, \forall u_a : M\langle a, a\rangle,$$
$$\forall \iota_a : C_0\langle i_a, i_a, u_a\rangle, \dot{U}\langle \omega_a, \iota_a\rangle,$$
$$\forall i_b : A\langle b, b\rangle, \forall \omega_b : I_A\langle i_b\rangle, \forall u_b : M\langle b, b\rangle, \forall \iota_b : C_0\langle i_b, i_b, u_b\rangle, \dot{U}\langle \omega_b, \iota_b\rangle,$$
$$\exists m : M\langle a, b\rangle, \exists \rho_f : C_1\langle i_a, f, m, u_a\rangle, \exists \rho_u : C_1\langle u, i_b, m, u_b\rangle, \dot{R}\langle \rho_f\rangle \wedge \dot{R}\langle \rho_u\rangle,$$

which may be paraphrased as "there is an arrow f so that the covariant representation of u is equivalent to the contravariant representation of f," asserts the existence of a left adjoint to u.

EXAMPLE 11.3.12. In the context of three objects $a, b, c : O$ and a span of arrows $k : A\langle a, b\rangle$ and $f : A\langle a, c\rangle$, the formula

$$\exists r : A\langle b, c\rangle, \forall i_a : A\langle a, a\rangle, \forall \omega_a : I_A\langle i_a\rangle, \forall i_b : A\langle b, b\rangle, \forall \omega_b : I_A\langle i_b\rangle,$$
$$\forall i_c : A\langle c, c\rangle, \forall \omega_c : I_A\langle i_c\rangle,$$
$$\forall u_b : M\langle b, b\rangle, \forall \iota_b : C_0\langle i_b, i_b, u_b\rangle, \dot{U}\langle \omega_b, \iota_b\rangle,$$
$$\forall u_c : M\langle c, c\rangle, \forall \iota_c : C_0\langle i_c, i_c, u_c\rangle, \dot{U}\langle \omega_c, \iota_c\rangle,$$
$$\forall k_* : M\langle a, b\rangle, \forall f_* : M\langle a, c\rangle, \forall r_* : M\langle b, c\rangle,$$
$$\forall \rho_k : C_1\langle k, i_b, k_*, u_b\rangle, \forall \rho_f : C_1\langle f, i_c, f_*, u_c\rangle, \forall \rho_r : C_1\langle r, i_c, r_*, u_c\rangle,$$
$$\dot{R}\langle \rho_k\rangle \wedge \dot{R}\langle \rho_f\rangle \wedge \dot{R}\langle \rho_r\rangle,$$
$$\forall \tau_a : T\langle i_a, i_a, i_a\rangle, \forall \tau_c : T\langle i_c, i_c, i_c\rangle, \forall \operatorname{id}_{k_*} : C_1\langle i_a, i_b, k_*, k_*\rangle, \dot{I}_c\langle \omega_a, \omega_b, \operatorname{id}_{k_*}\rangle,$$
$$\exists \nu : C_2\langle i_a, i_c, k_*, r_*, f_*\rangle,$$
$$\forall e : O, \forall m : M\langle b, e\rangle, \forall n : M\langle e, c\rangle, \forall \alpha : C_3\langle i_a, i_c, k_*, m, n, f_*\rangle,$$
$$(\exists \mu : C_2\langle i_b, i_c, m, n, r_*\rangle, \dot{T}_{1,2}\langle \tau_a, \tau_c, \operatorname{id}_{k_*}, \mu, \nu, \alpha\rangle) \wedge (\forall \zeta, \xi : C_2\langle i_b, i_c, m, n, r_*\rangle,$$
$$\dot{T}_{1,2}\langle \tau_a, \tau_c, \operatorname{id}_{k_*}, \zeta, \nu, \alpha\rangle \wedge \dot{T}_{1,2}\langle \tau_a, \tau_c, \operatorname{id}_{k_*}, \xi, \nu, \alpha\rangle \to \dot{E}_2\langle \zeta, \xi\rangle))$$

asserts the existence of a pointwise right extension $\nu : rk \Rightarrow f$.

Here we have taken advantage of a simplification provided in the virtual equipment of modules. Recall from Theorem 9.3.3 that a natural transformation $\nu : rk \Rightarrow f$ defines a pointwise right extension if the corresponding binary cell, also denoted ν in the formula above, defines a right extension in the virtual equipment of modules. By Definition 9.1.2, this means that the binary cell ν must enjoy a unique factorization property for all cells whose target module is the covariant representation of f and whose source sequence of modules starts from the covariant representation of k.

In the above formula we do not quantify over a compatible sequence of modules from "b" to "c" of arbitrary length, instead quantifying over a single intermediate object "e" and a pair of modules "$m : M\langle b, e \rangle$" and "$n : M\langle e, c \rangle$." We can make this simplification on account of Remark 8.3.12, which tells us that any compatible sequence of modules $E_1 \times \cdots \times E_\ell$ in the source of a cell can be replaced by a compatible sequence of two modules. Here the object "e" is the summit of the composite two-sided fibration formed by composing the spans that encode the modules $E_1, \ldots, E_\ell$, while the modules "m" and "n" are, respectively, the contravariant and covariant representable modules associated to the legs of that composite span.

Special cases of this statement and its dual define limits and colimits of diagrams of ∞-categories, as well as many other concepts.

Exercises

EXERCISE 11.3.i. Express the axioms for a virtual equipment in the language of virtual equipments defined relative to the FOLDS signature $\mathcal{J}_{\mathcal{VE}}$. For instance, the axiom that asserts that for every module and compatible pair of vertical arrows there exists a unary restriction cell is expressed by the sentence:

$$\forall x, y, a, b, : O, \forall f : A\langle x, a \rangle, \forall g : A\langle y, b \rangle, \forall m : M\langle a, b \rangle,$$
$$\exists m(g, f) : M\langle x, y \rangle, \exists \rho : C_1\langle f, g, m(g, f), m \rangle, \dot{R}\langle \rho \rangle$$

12

Applications of Model Independence

In this chapter, we establish some special properties of a certain class of ∞-cosmoi we call ∞-**cosmoi of** $(\infty, 1)$-**categories**, by which we mean ∞-cosmoi that are biequivalent to the ∞-cosmos of quasi-categories. By Proposition 10.2.1 an ∞-cosmos $\mathcal{K}$ is an ∞-cosmos of $(\infty, 1)$-categories if and only if its underlying quasi-category functor $(-)_0 := \mathrm{Fun}(1, -) : \mathcal{K} \to \mathcal{QC}at$ is a biequivalence – meaning that every quasi-category is equivalent to the underlying quasi-category of an ∞-category in $\mathcal{K}$ and that for any $A, B \in \mathcal{K}$ the map on functor spaces defined by transposing the composition map in $\mathcal{K}$

$$\mathrm{Fun}(A, B) \xrightarrow{(-)_0} \mathrm{Fun}(A_0, B_0) \quad \rightsquigarrow \quad \mathrm{Fun}(A, B) \times \mathrm{Fun}(1, A) \xrightarrow{\circ} \mathrm{Fun}(1, B)$$

is an equivalence of quasi-categories. A few examples of ∞-cosmoi of this form are established in §E.2.

A secondary aim is to illustrate how the model independence theorem can be used to combine synthetic and analytic techniques to prove results concerning any family of biequivalent ∞-cosmoi. In what follows we appeal to the explicit model of $(\infty, 1)$-categories as quasi-categories to supply analytic proofs of certain key results – for instance, that a functor defines an equivalence of quasi-categories just when it is fully faithful and essential surjective in a suitable sense. We then explain how the model independence theorem can be used to transfer these results to biequivalent ∞-cosmoi, even when we cannot translate the specific proof used in the quasi-categorical case. We then apply some of our analytically proven theorems to further develop the synthetic theory of ∞-cosmoi of $(\infty, 1)$-categories.

Many of the results in this chapter have been alluded to previously in this text and indeed their proofs could have appeared earlier. The reason for the delay is that in the presence of the results of Chapters 10 and 11 their conclusions apply more broadly, to all ∞-cosmoi of $(\infty, 1)$-categories, not just in the quasi-categorical case. In particular, we discuss some special features of the ∞-cosmos of quasi-categories, proving in particular that universal properties in this ∞-cosmos are determined pointwise, again appealing to model independence to generalize this result to other ∞-cosmoi of $(\infty, 1)$-categories.

To warm up in §12.1, we define opposite $(\infty, 1)$-categories and the ∞-groupoid core of an $(\infty, 1)$-category. In practice, these notions are easily accessible in any model of choice, but our aim is to illustrate the general procedure for transferring ∞-categorical structures between biequivalent ∞-cosmoi in a relatively elementary setting. In §12.2, we establish a large suite of results which combine to express the pointwise generation of various universal properties in an ∞-cosmos of $(\infty, 1)$-categories. Finally, in §12.3, we cite a more sophisticated result from the $(\infty, 1)$-categorical literature concerning the exponentiability of cartesian and cocartesian fibrations and use this to tie up a lose end from Chapter 9: namely we reduce the existence of pointwise right and left extensions to the presence of certain limits or colimits. Since the indexing shapes for these limits and colimits vary with the elements in the domain of the pointwise extension, this result relies on the pointwise determination of universal properties established in §12.2. Along the way we also extend the calculus of modules in ∞-cosmoi of $(\infty, 1)$-categories, showing that the right extensions and right liftings of Definition 9.1.2 always exist, defining "homs" between modules to complement their "tensor products."

12.1 Opposite $(\infty, 1)$-Categories and ∞-Groupoid Cores

The construction of the co-dual of an ∞-cosmos in Definition 1.2.25 makes use of the construction of the opposite of a simplicial set. Recall that the opposite of a simplicial set $X \colon \Delta^{\mathrm{op}} \to Set$ is obtained by precomposing with the identity-on-objects involution $(-)^\circ \colon \Delta \to \Delta$ that reverses the ordering of the elements in each ordinal $[n] \in \Delta$. Precomposition with $(-)^\circ$ defines a functor $(-)^{\mathrm{op}} \colon sSet \to sSet$ which carries a simplicial set X to its **opposite simplicial set** X^{op}. We start by exploring the role played by this operation in the ∞-cosmos of quasi-categories and then investigate a related operation on other ∞-cosmoi of ∞-categories.

LEMMA 12.1.1. *If X is a quasi-category, then X^{op} is a quasi-category.*

Proof The lifting problem below-left is solved by the lifting problem below-right

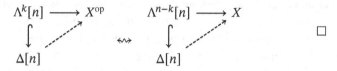

DEFINITION 12.1.2. For a quasi-category A, its **opposite quasi-category** A^{op} is the simplicial set defined by "reversing the ordering of the vertices in each simplex."

LEMMA 12.1.3. *The opposite quasi-category construction defines an involutive cosmological biequivalence* $(-)^{\mathrm{op}} : \mathcal{QC}at \Rightarrow \mathcal{QC}at^{co}$ *that acts on functor spaces via a natural isomorphism*

$$\mathrm{Fun}(A, B) \xrightarrow{\;\cong\;} \mathrm{Fun}(A^{\mathrm{op}}, B^{\mathrm{op}})^{\mathrm{op}}.$$

Proof The isomorphism $\mathrm{Fun}(A, B) \cong \mathrm{Fun}(A^{\mathrm{op}}, B^{\mathrm{op}})^{\mathrm{op}}$ is best understood at the level of simplices: the simplicial maps $A \times \Delta[n] \to B$ that define n-simplices in the functor space $\mathrm{Fun}(A, B)$ map via the isomorphism $(-)^{\mathrm{op}} : s\mathcal{S}et \Rightarrow s\mathcal{S}et$ to simplicial maps $A^{\mathrm{op}} \times \Delta[n]^{\mathrm{op}} \to B^{\mathrm{op}}$, and these define the n-simplices in the functor space $\mathrm{Fun}(A^{\mathrm{op}}, B^{\mathrm{op}})^{\mathrm{op}}$.

By an extension of the proof of Lemma 12.1.1, the opposite of an isofibration is an isofibration. The conical limits in $\mathcal{QC}at$, being defined pointwise in $\mathcal{S}et$, are preserved by restriction along the functor $(-)^{\circ} : \Delta \to \Delta$. Simplicial cotensors are also preserved: for a quasi-category B and a simplicial set X, $(B^X)^{\mathrm{op}} \cong (B^{\mathrm{op}})^{X^{\mathrm{op}}}$, which accords with the general construction of the cotensor of $B^{\mathrm{op}} \in \mathcal{QC}at^{co}$ with a simplicial set X as noted in Definition 1.2.25. This proves that $(-)^{\mathrm{op}} : \mathcal{QC}at \to \mathcal{QC}at^{co}$ defines a cosmological functor. $\square$

Lemma 12.1.3 extends the usual construction of the opposite of a 1-category and the corresponding 2-functor $(-)^{\mathrm{op}} : \mathcal{C}at \to \mathcal{C}at^{co}$ (see Exercise B.2.iii). On account of the explicitness of the construction given in Definition 12.1.2, the opposite of a quasi-category is defined up to isomorphism. By contrast, without any additional hypotheses, we are only be able to define the opposite of an ∞-category in a biequivalent ∞-cosmos up to equivalence.[1] While at first this may seem undesirable, it is arguably morally correct to give the definition in this manner, since from the model independent point of view, the ∞-category itself ought only be considered up to equivalence.

---

[1] In every model of $(\infty, 1)$-categories that we are aware of, there is in fact a strictly defined op-involution, and one can verify that these analytically defined opposite ∞-categories are compatible with the standard change-of-model functors. However, the benefits of this additional strictness might not be worth the effort in verifying the strict compatibility of the opposite ∞-category construction in each instance.

DEFINITION 12.1.4. Let A be an ∞-category in an ∞-cosmos $\mathcal{K}$ of $(\infty, 1)$-categories. Define the **opposite ∞-category** A^{op} to be any ∞-category in $\mathcal{K}$ whose underlying quasi-category is A_0^{op}.

We now argue that Definition 12.1.4 is well-defined up to equivalence:

PROPOSITION 12.1.5. *In an ∞-cosmos of $(\infty, 1)$-categories $\mathcal{K}$, any choices of opposite ∞-categories assemble into a quasi-pseudofunctor and biequivalence* $(-)^{op} : \mathcal{K} \rightsquigarrow \mathcal{K}^{co}$. *In particular, for any ∞-categories $A, B \in \mathcal{K}$, there is an equivalence* $\mathrm{Fun}(A, B) \simeq \mathrm{Fun}(A^{op}, B^{op})^{op}$ *that is quasi-pseudonatural in A and in B.*

Proof The quasi-pseudofunctorial biequivalence is defined as the composite of the zigzag of cosmological biequivalences

$$
\begin{array}{ccc}
\mathcal{K} & \xrightarrow[\sim]{\;(-)^{op}\;} & \mathcal{K}^{co} \\
{\scriptstyle(-)_0}\Big\downarrow\;\simeq & \simeq & \simeq\Big\downarrow{\scriptstyle(-)_0} \\
\mathcal{Q}\mathcal{C}at & \xrightarrow[(-)^{op}]{\;\sim\;} & \mathcal{Q}\mathcal{C}at^{co}
\end{array}
$$

We first choose opposite ∞-categories for every $A \in \mathcal{K}$ together with specified adjoint equivalences $(A^{op})_0 \simeq (A_0)^{op}$. Composing with these equivalences, the biequivalence $(-)_0 : \mathcal{K} \to \mathcal{Q}\mathcal{C}at$ provides local equivalences of quasi-categories:

$$
\begin{array}{ccc}
\mathrm{Fun}(A, B) & \dashrightarrow{\simeq}\dashrightarrow & \mathrm{Fun}(A^{op}, B^{op})^{op} \\
\Big\downarrow{\scriptstyle\wr} & \simeq & \Big\downarrow{\scriptstyle\wr} \\
\mathrm{Fun}(A_0, B_0) & \xrightarrow{\;\sim\;} & \mathrm{Fun}((A_0)^{op}, (B_0)^{op})^{op} \simeq \mathrm{Fun}((A^{op})_0, (B^{op})_0)^{op}
\end{array}
$$

which compose to define the desired equivalence in such a way that the square commutes up to a homotopy coherent isomorphism. In order for these equivalences to define the action on functor spaces of a quasi-pseudofunctor, we follow the construction of Corollary 10.4.17 described in the proof of Proposition 10.4.16 and choose adjoint equivalence inverses to the lower-left-hand horizontal simplicial natural equivalence, applying Lemma 10.4.15. Lemma 10.4.12 then proves the quasi-pseudonaturality statement.   □

On account of the equivalence $\mathrm{Fun}(A, B) \simeq \mathrm{Fun}(A^{op}, B^{op})^{op}$, a functor between $(\infty, 1)$-categories $f : A \to B$ has an opposite functor $f^{op} : A^{op} \to B^{op}$, well-defined up to isomorphism once the domain and codomain ∞-categories have been fixed. Similarly, a natural transformation between parallel functors has an opposite

$$A \underset{g}{\overset{f}{\rightrightarrows}} {\scriptstyle\Downarrow\alpha}\, B \qquad \rightsquigarrow \qquad A^{\mathrm{op}} \underset{g^{\mathrm{op}}}{\overset{f^{\mathrm{op}}}{\rightrightarrows}} {\scriptstyle\Uparrow\alpha^{\mathrm{op}}}\, B^{\mathrm{op}}$$

obtained by applying the pseudofunctor $(-)^{\mathrm{op}} : \mathfrak{h}\mathcal{K} \rightsquigarrow \mathfrak{h}\mathcal{K}^{\mathrm{op}}$ that underlies the quasi-pseudofunctor of Proposition 12.1.5. Furthermore:

LEMMA 12.1.6. *Let $\mathcal{K}$ be an ∞-cosmos of $(\infty, 1)$-categories.*

(i) *For any ∞-category A and simplicial set U, $(A^U)^{\mathrm{op}} \simeq (A^{\mathrm{op}})^{U^{\mathrm{op}}}$.*

(ii) *For any functors $f : B \to A$ and $g : C \to A$, there is an equivalence $\mathrm{Hom}_A(f, g)^{\mathrm{op}} \simeq \mathrm{Hom}_{A^{\mathrm{op}}}(g^{\mathrm{op}}, f^{\mathrm{op}})$ over $B^{\mathrm{op}} \times C^{\mathrm{op}}$.*

Proof We have a quasi-pseudonatural equivalence:

$$
\begin{array}{ll}
\mathrm{Fun}(X, (A^U)^{\mathrm{op}}) \simeq \mathrm{Fun}(X^{\mathrm{op}}, A^U)^{\mathrm{op}} & \text{by 12.1.5} \\
\cong (\mathrm{Fun}(X^{\mathrm{op}}, A)^U)^{\mathrm{op}} & \text{by (1.2.7)} \\
\cong (\mathrm{Fun}(X^{\mathrm{op}}, A)^{\mathrm{op}})^{U^{\mathrm{op}}} & \text{by 12.1.3} \\
\simeq \mathrm{Fun}(X, A^{\mathrm{op}})^{U^{\mathrm{op}}} & \text{by 12.1.5} \\
\cong \mathrm{Fun}(X, (A^{\mathrm{op}})^{U^{\mathrm{op}}}) & \text{by (1.2.7).}
\end{array}
$$

Hence, by Lemma 10.4.14, $(A^U)^{\mathrm{op}} \simeq (A^{\mathrm{op}})^{U^{\mathrm{op}}}$.

The second statement is a consequence of a more general result: that any quasi-pseudofunctorial biequivalence, such as $(-)^{\mathrm{op}} : \mathcal{K} \rightsquigarrow \mathcal{K}^{\mathrm{co}}$ established in Proposition 12.1.5, preserves and reflects comma ∞-categories. Alternatively, since the quasi-pseudofunctorial biequivalence under consideration here is defined as a zigzag of cosmological biequivalences, the result follows from the fact that cosmological biequivalences themselves preserve and reflect comma ∞-categories, as observed in Proposition 10.3.6(vi). $\qquad\square$

The next result provides another perspective on "appeals to duality" where facts about colimits of diagrams in an ∞-cosmos $\mathcal{K}$ were deduced from corresponding proofs about limits in $\mathcal{K}^{\mathrm{co}}$, and similarly results about cartesian fibrations were interpreted in $\mathcal{K}^{\mathrm{co}}$ to conclude the corresponding results about cocartesian fibrations in $\mathcal{K}$.

PROPOSITION 12.1.7. *Let $\mathcal{K}$ be an ∞-cosmos of $(\infty, 1)$-categories.*

(i) *A J-shaped family of diagrams in A has a colimit if and only if the corresponding J^{op}-shaped family of diagrams in A^{op} has a limit.*

(ii) A functor $p : E \twoheadrightarrow B$ defines a cartesian fibration if and only if the functor $p^{op} : E^{op} \twoheadrightarrow B^{op}$ defines a cocartesian fibration.[2]

Note that if $p : E \twoheadrightarrow B$ is an isofibration, it is always possible to choose a functor $p^{op} : E^{op} \twoheadrightarrow B^{op}$ that is again an isofibration, perhaps by changing the choice of total space E^{op}.

Proof By Lemma 12.1.6, a J-shaped family of diagrams $d : D \to A^J$ defines a J^{op}-shaped family of diagrams $d^{op} : D^{op} \to (A^{op})^{J^{op}}$. By Proposition 4.3.1, d admits a colimit in A if and only if there is an equivalence of comma ∞-categories

$$\mathrm{Hom}_{A^J}(d, \Delta) \simeq_{A \times D} \mathrm{Hom}_A(c, A),$$

in which case the representing functor $c : D \to A$ defines the colimit functor.

By Lemma 12.1.6, such an equivalence exists if and only if there is an equivalence

$$\mathrm{Hom}_{(A^{op})^{J^{op}}}(\Delta, d^{op}) \simeq_{D^{op} \times A^{op}} \mathrm{Hom}_{A^{op}}(A^{op}, c^{op}),$$

which, by Proposition 4.3.1, characterizes the limit functor $c^{op} : D^{op} \to A^{op}$.

The second statement is proven similarly. By Theorem 5.2.8, $p : E \twoheadrightarrow B$ defines a cartesian fibration if and only if the induced functor $i_1 \widehat{\pitchfork} p : E^2 \twoheadrightarrow \mathrm{Hom}_B(B, p)$ admits a right adjoint right inverse. By applying the quasi-pseudo-functorial biequivalence $(-)^{op} : \mathcal{K} \leadsto \mathcal{K}^{co}$, this adjunction exists if and only if the opposite functor admits a left adjoint right inverse. This functor need not be isomorphic to $i_0 \widehat{\pitchfork} p^{op} : (E^{op})^2 \twoheadrightarrow \mathrm{Hom}_{B^{op}}(p^{op}, B^{op})$, but by Lemma 12.1.6 it is equivalent to it, which by the equivalence invariance of adjunctions is good enough (see Exercise B.4.i). By the dual of Theorem 5.2.8, such an adjunction exists if and only if $p^{op} : E^{op} \twoheadrightarrow B^{op}$ is a cocartesian fibration. $\qquad\square$

We now turn our attention to the construction of the Kan complex core of a quasi-category and discuss its analogue in other ∞-cosmoi of $(\infty, 1)$-categories.

DEFINITION 12.1.8. The ∞-**groupoid core** of a quasi-category A is the largest Kan complex $\mathrm{core}A \subset A$, which may be constructed as the simplicial subset containing

- all of the vertices of A,
- only those edges that define isomorphisms in A (see Definition 1.1.13),
- every higher simplex whose edges are all isomorphisms.

---

[2] As noted in the introduction to Part III, the homotopy coherent diagram encoded by p^{op} is *not* the same as the homotopy coherent diagram encoded by p – a classical observation of Bénabou extended to the $(\infty, 1)$-categorical context by Barwick, Glasman, and Nardin [7].

LEMMA 12.1.9. *The ∞-groupoid core of a quasi-category is a Kan complex, and indeed is the largest Kan complex contained in the quasi-category A.*

Proof　The inclusion core$A$ ⊂ A constructed in Definition 12.1.8 is full on simplices of all dimensions except dimension one. Thus, to see that core$A$ is a quasi-category, we need only argue that it admits extensions along the horn $\Lambda^1[2]$ ↪ $\Delta[2]$. By construction, a horn $\Lambda^1[2]$ → core$A$ picks outs two isomorphisms in A. The filler $\Delta[2]$ → A witnesses a composition relation in the homotopy category h(A); thus the composite edge is also an isomorphism, and by fullness this filler lifts to $\Delta[2]$ → core$A$.

By construction h(core$A$) is a groupoid; indeed, it is the maximal subgroupoid contained in hA. So by Corollary 1.1.15, core$A$ is a Kan complex.

Finally, an intermediate simplicial subset core$A$ ⊊ K ⊂ A would necessarily contain an additional edge $f : x → y$. If K were a Kan complex, then it would have to admit fillers for $\Lambda^0[2]$- and $\Lambda^2[2]$-horns whose 2nd or 0th faces, respectively, were the 1-simplex f, and whose remaining face is degenerate. The fillers would construct left and right inverses to f in h(A). Hence, f is an isomorphism in A and already lives in core$A$. □

The inclusion of the ∞-cosmos of Kan complexes defines a cosmological embedding $\mathcal{K}an$↪$\mathcal{Q}\mathcal{C}at$ by Proposition 6.1.6. Functors of quasi-categories preserve isomorphisms, so a functor $f : A → B$ restricts to $f :$ core$A$ → core$B$. In this way the ∞-groupoid core construction acts functorially on the underlying category of $\mathcal{Q}\mathcal{C}at$ and, as an unenriched functor, is right adjoint to the inclusion. Note, however, as discussed in Example 1.3.7, that the core construction is not simplicial, at least not with respect to the usual quasi-categorical enrichment of $\mathcal{Q}\mathcal{C}at$. Indeed, a natural transformation between functors of quasi-categories only restricts to ∞-groupoid cores if each of its components is invertible.

The ∞-groupoid core does, however, define a simplicial functor with respect to a new enrichment that we now introduce. An ∞-cosmos is a type of $(∞, 2)$-category since it is a category enriched over a model of $(∞, 1)$-categories. We now introduce the $(∞, 1)$-**categorical core** of an ∞-cosmos. In the following definition, note that since core$(-) : \mathcal{Q}\mathcal{C}at → \mathcal{K}an$ is an (unenriched) right adjoint, it preserves products, so we may apply it to the functor spaces of a quasi-categorically enriched category to construct a Kan complex enriched subcategory (see Proposition A.7.3) that we now introduce.

DEFINITION 12.1.10 ($(∞, 1)$-core of an ∞-cosmos). For an ∞-cosmos $\mathcal{K}$, write core$_*\mathcal{K}$ ⊂ $\mathcal{K}$ for the subcategory with the same objects and with homs defined to be the ∞-groupoid cores of the functor spaces of $\mathcal{K}$. We refer to core$_*\mathcal{K}$ as the $(\infty, 1)$-**core** of $\mathcal{K}$ and think of it as being the core $(\infty, 1)$-category inside this $(\infty, 2)$-category.

REMARK 12.1.11. The $(\infty, 1)$-categorical core is not an ∞-cosmos in the strict sense axiomatized in Definition 1.2.1. It inherits its class of isofibrations and the conical limits from the original ∞-cosmos, but simplicial cotensors exist only weakly: the cotensor of an ∞-category A in $\mathrm{core}_*\mathcal{K}$ by a simplicial set U is constructed by the cotensor in $\mathcal{K}$ of A by a Kan complex replacement $\tilde{U}$ of U, defined by "freely inverting" its edges and adding fillers for horns. This results in an equivalence $\mathrm{core}(\mathrm{Fun}(X, A))^U \simeq \mathrm{core}(\mathrm{Fun}(X, A^{\tilde{U}}))$ in place of the usual isomorphism. Alternatively, Exercise 12.1.iii suggests an alternate approach to defining the enrichment of an ∞-cosmos in such a way that the $(\infty, 1)$-core remains an ∞-cosmos.

LEMMA 12.1.12. *The natural inclusion* $\mathcal{K}an \hookrightarrow \mathcal{QC}at$ *factors through the inclusion* $\mathrm{core}_*\mathcal{QC}at \subset \mathcal{QC}at$ *and this latter functor admits a simplicially enriched right adjoint left inverse, namely the functor that sends each quasi-category to its* ∞-*groupoid core.*

$$\mathcal{K}an \underset{\mathrm{core}}{\overset{\perp}{\rightleftarrows}} \mathrm{core}_*\mathcal{QC}at$$

Proof If K and L are Kan complexes, then so is $\mathrm{Fun}(K, L)$. Hence the natural inclusion $\mathcal{K}an \hookrightarrow \mathcal{QC}at$ factors through the $(\infty, 1)$-categorical core.

The right adjoint $\mathrm{core} : \mathrm{core}_*\mathcal{QC}at \to \mathcal{QC}at$ acts on objects by the construction of Definition 12.1.8. To define its action on functor spaces, we must supply a canonical map

$$\mathrm{core}(\mathrm{Fun}(A, B)) \to \mathrm{Fun}(\mathrm{core}A, \mathrm{core}B),$$

for any pair of quasi-categories A and B. By Corollary 1.1.22, the isomorphisms in $\mathrm{Fun}(A, B) \cong B^A$ are simplicial maps $\alpha : A \times \Delta[1] \to B$ whose components $\alpha_a : \Delta[1] \to B$, indexed by vertices a of A, define isomorphisms in B. Combining this observation with Definition 12.1.8, we see that an n-simplex in $\mathrm{core}(\mathrm{Fun}(A, B))$ is a simplicial map $\phi : A \times \Delta[n] \to B$ with the property that upon restriction to any vertex of A and any edge of $\Delta[n]$, the resulting edge in B is an isomorphism. When A is restricted to its Kan complex core, the edges of $\mathrm{core}A$ are also isomorphisms. It follows that $\phi : \mathrm{core}A \times \Delta[n] \to B$ carries every edge of the domain to an isomorphism in B, and hence factors through $\mathrm{core}B \hookrightarrow B$, since this inclusion is full on the invertible edges.[3] Thus

---

[3] In the language of marked simplicial sets, a map in $\mathrm{core}(\mathrm{Fun}(A, B))$ is a marked map $A^{\natural} \times \Delta[n]^{\sharp} \to B^{\natural}$. Upon restriction along $\mathrm{core}A^{\sharp} \hookrightarrow A^{\natural}$, the domain $\mathrm{core}A^{\sharp} \times \Delta[n]^{\sharp}$ is maximally marked, and hence factors through the maximally marked core $\mathrm{core}B^{\sharp} \hookrightarrow B^{\natural}$ (see §D.4).

the n-simplex ϕ restricts to define an n-simplex $\phi : \text{core}A \times \Delta[n] \to \text{core}B$. This defines the canonical map.

Now for a Kan complex K and quasi-category A, the simplicial natural isomorphism

$$\text{core}(\text{Fun}(K, A)) \cong \text{Fun}(K, \text{core}A)$$

is easily verified. The correspondence on vertices expresses the unenriched adjunction, while the correspondence on higher simplices follows for the reason just discussed and the isomorphism $\text{core}(K) \cong K$. $\qquad\square$

COROLLARY 12.1.13. *If A and B are equivalent quasi-categories, then $\text{core}A$ and $\text{core}B$ are equivalent Kan complexes.*

Proof An equivalence of quasi-categories is specified by a pair of 0-arrows together a pair of invertible 1-arrows. As such it is contained in the $(\infty, 1)$-categorical core $\text{core}_*\mathcal{QCat} \hookrightarrow \mathcal{QCat}$ and preserved by the simplicial functor $\text{core} : \text{core}_*\mathcal{QCat} \to \mathcal{Kan}$. $\qquad\square$

The core of an ∞-category in a general ∞-cosmos of $(\infty, 1)$-categories can be defined in a similar manner to Definition 12.1.4, but this notion also has an up-to-equivalence universal property that we prefer to use as the definition.

DEFINITION 12.1.14. Let $\mathcal{K}$ be an ∞-cosmos of $(\infty, 1)$-categories and let A be an ∞-category in $\mathcal{K}$. Its ∞-**groupoid core** is an ∞-category $\text{core}A$ equipped with a map $\iota : \text{core}A \to A$ so that

- $\text{core}A$ is a discrete ∞-category, meaning that $\text{Fun}(X, A)$ is a Kan complex for all X
- if G is a discrete ∞-category, then ι defines an equivalence

$$\text{Fun}(G, \text{core}A) \xrightarrow{\;\iota\circ-\;}_{\sim} \text{core}(\text{Fun}(G, A))$$

In practice, an ∞-cosmos of $(\infty, 1)$-categories frequently comes with an explicit core functor, but as in the case of opposites, one can also be defined by transferring the core construction along a suitable change-of-model functor:

PROPOSITION 12.1.15. *In an ∞-cosmos of $(\infty, 1)$-categories $\mathcal{K}$, any choices of ∞-groupoid cores assemble into a quasi-pseudofunctor* $\text{core} : \text{core}_*\mathcal{K} \to \mathcal{Disc}(\mathcal{K})$. *In particular, for any ∞-categories $A, B \in \mathcal{K}$, there is an map* $\text{core}(\text{Fun}(A, B)) \to \text{Fun}(\text{core}A, \text{core}B)$ *that is quasi-pseudonatural in A and in B as objects of* $\text{core}_*\mathcal{K}$.

Proof   The quasi-pseudofunctor is defined as the composite of the zigzag in which the backwards map is a cosmological biequivalence by Corollary 10.3.7:

$$
\begin{array}{ccc}
\mathrm{core}_*\mathcal{K} & \xrightarrow{\ \mathrm{core}(-)\ } & \mathcal{D}isc(\mathcal{K}) \\
{\scriptstyle (-)_0}\downarrow{\scriptstyle \wr} & \simeq & \downarrow{\scriptstyle (-)_0} \\
\mathrm{core}_*\mathcal{QC}at & \xrightarrow[\mathrm{core}(-)]{} & \mathcal{K}an \cong \mathcal{D}isc(\mathcal{QC}at)
\end{array}
$$

We leave it to Exercise 12.1.iv to verify that $\mathcal{K} \rightsquigarrow \mathcal{L}$ descends to a simplicially enriched biequivalence $\mathrm{core}_*\mathcal{K} \rightsquigarrow \mathrm{core}_*\mathcal{L}$. By Proposition 10.4.16 the inverse to the cosmological biequivalence $(-)_0 : \mathcal{D}isc(\mathcal{K}) \rightsquigarrow \mathcal{K}an$, defines a quasi-pseudofunctor and biequivalence $\mathcal{K}an \rightsquigarrow \mathcal{D}isc(\mathcal{K})$, which composes with the simplicial functor $\mathrm{core}((-)_0) : \mathrm{core}_*\mathcal{K} \to \mathcal{K}an$ to define the quasi-pseudo-functor $\mathrm{core} : \mathrm{core}_*\mathcal{K} \to \mathcal{D}isc(\mathcal{K})$. By Lemma 10.4.12, the action on homs of this quasi-pseudofunctor defines a quasi-pseudonatural transformation.

It remains only to verify that the action on objects of the quasi-pseudofunctor satisfies the conditions of Definition 12.1.14. By construction, $\mathrm{core}A$ is a discrete ∞-category for any $A \in \mathcal{K}$. The map $\iota : \mathrm{core}A \to A$ is defined by whiskering the corresponding inclusion of the Kan complex core of quasi-category with the underlying quasi-category functor and its quasi-pseudofunctorial inverse:

$$
\mathrm{core}_*\mathcal{K} \xrightarrow[\sim]{(-)_0} \mathrm{core}_*\mathcal{QC}at \underset{\Downarrow\iota}{\overset{\mathrm{core}}{\rightrightarrows}} \mathrm{core}_*\mathcal{QC}at \xrightarrow{(-)_0^{-1}} \mathrm{core}_*\mathcal{K}
$$

Now if G is a discrete ∞-category, then $G_0 = \mathrm{Fun}(1, G)$ is a Kan complex, so by Lemma 12.1.12 $\iota_{A_0} \circ - : \mathrm{Fun}(G_0, \mathrm{core}(A_0)) \rightsquigarrow \mathrm{core}(\mathrm{Fun}(G_0, A_0))$ is an equivalence. By construction $\mathrm{core}A$ is defined so that $(\mathrm{core}A)_0 \simeq \mathrm{core}(A_0)$. Note that since $(-)_0 : \mathcal{K} \rightsquigarrow \mathcal{QC}at$ is a biequivalence, this shows that the core of an ∞-category in an ∞-cosmos of $(\infty, 1)$-categories is well-defined up to equivalence. Since the simplicial functor $(-)_0 : \mathrm{core}_*\mathcal{K} \to \mathrm{core}_*\mathcal{QC}at$ is an equivalence on homs, the functor defined by post-composition with ι_A is equivalent to this functor

$$
\begin{array}{ccc}
\mathrm{Fun}(G, \mathrm{core}A) & \xrightarrow{\ \iota_A \circ -\ } & \mathrm{core}(\mathrm{Fun}(G, A)) \\
{\scriptstyle (-)_0}\downarrow{\scriptstyle \wr} & \cong & \downarrow{\scriptstyle \wr}{\scriptstyle (-)_0} \\
\mathrm{Fun}(G_0, \mathrm{core}A_0) & \xrightarrow[\iota_{A_0} \circ -]{\sim} & \mathrm{core}(\mathrm{Fun}(G_0, A_0))
\end{array}
$$

Thus post-composition with ι_A induces the required equivalence $\mathrm{Fun}(G, \mathrm{core}A) \simeq \mathrm{core}(\mathrm{Fun}(G, A))$, completing the proof. $\square$

Exercises

EXERCISE 12.1.i. Prove that the homotopy category of the opposite of an ∞-category A is equivalent to the opposite of the homotopy category of A.

EXERCISE 12.1.ii. Prove that a functor between ∞-categories is an equivalence if and only if its opposite functor is an equivalence.

EXERCISE 12.1.iii. In consultation with §D.4 and §D.5:

(i) Redefine the notion of an ∞-cosmos from Definition 1.2.1 to be a category enriched over marked simplicial sets, whose functor spaces are naturally marked quasi-categories.

(ii) Describe the construction of the ∞-groupoid core of a naturally marked quasi-category and of the $(\infty, 1)$-categorical core of an ∞-cosmos with this enrichment.

(iii) Show that $(\infty, 1)$-categorical cores are cotensored over simplicial sets, although these cotensors are not preserved by the inclusion $\mathrm{core}_* \mathcal{K} \hookrightarrow \mathcal{K}$.

(iv) Show that the $(\infty, 1)$-categorical core of an ∞-cosmos is an ∞-cosmos in the new sense, although the functor $\mathrm{core}_* \mathcal{K} \hookrightarrow \mathcal{K}$ is still not cosmological.

EXERCISE 12.1.iv. Let $F \colon \mathcal{K} \to \mathcal{L}$ be a cosmological functor. Prove that F induces a simplicial functor $F \colon \mathrm{core}_* \mathcal{K} \to \mathrm{core}_* \mathcal{L}$ that is a biequivalence if the original functor is.

EXERCISE 12.1.v.

(i) Prove that any adjunction between quasi-categories

$$B \underset{u}{\overset{f}{\rightleftarrows}} A$$

restricts to define an adjoint equivalence between the full sub-quasi-categories spanned by those elements $b \colon 1 \to B$ and $a \colon 1 \to A$ for which the unit and counit components, respectively, are invertible.

(ii) State and prove an analogous result about adjoint equivalences derived from adjunctions in an arbitrary ∞-cosmos of $(\infty, 1)$-categories.

12.2 Pointwise Universal Properties

In an ∞-cosmos of $(\infty, 1)$-categories, the terminal ∞-category 1 plays a special role which can be summarized by the slogan that "universal properties are detected pointwise." In this section, we collect together a number of results that encapsulate this slogan, which are proven through a combination of synthetic and analytic techniques.

For instance, Corollary 1.1.22 states and Corollary D.4.19 proves that a natural transformation between functors between quasi-categories $X \underset{g}{\overset{f}{\underset{\Downarrow\alpha}{\rightrightarrows}}} A$ is a natural isomorphism if and only if it is a **pointwise isomorphism**, meaning that each of its components $1 \xrightarrow{\ x\ } X \underset{g}{\overset{f}{\underset{\Downarrow\alpha}{\rightrightarrows}}} A$ is invertible. Consequently:

LEMMA 12.2.1. *In an ∞-cosmos of $(\infty, 1)$-categories, a natural transformation is a natural isomorphism if and only if it is a pointwise isomorphism.*

Proof Any natural isomorphism is clearly a pointwise isomorphism. For the converse, we deploy the 2-categorical biequivalence $(-)_0 : \mathfrak{h}\mathcal{K} \overset{\approx}{\to} \mathfrak{h}\mathcal{QC}at$ of Propositions 10.2.1 and 10.3.1. Suppose $X \underset{g}{\overset{f}{\underset{\Downarrow\alpha}{\rightrightarrows}}} A$ is a pointwise natural isomorphism in $\mathfrak{h}\mathcal{K}$ and consider the underlying natural transformation between underlying quasi-categories $X_0 \underset{g_0}{\overset{f_0}{\underset{\Downarrow\alpha_0}{\rightrightarrows}}} A_0$. By construction, vertices x of $X_0 \cong \mathrm{Fun}(1, X)$ correspond bijectively to elements of X, and the components

$$1 \xrightarrow{\ x\ } X \underset{g}{\overset{f}{\underset{\Downarrow\alpha}{\rightrightarrows}}} A \quad \leftrightsquigarrow \quad 1 \xrightarrow{\ x\ } X_0 \underset{g_0}{\overset{f_0}{\underset{\Downarrow\alpha_0}{\rightrightarrows}}} A_0$$

define corresponding arrows in $hA \cong h(A_0)$. Thus, the underlying natural transformation α_0 is a pointwise natural isomorphism in $\mathfrak{h}\mathcal{QC}at$ as well, and Corollary D.4.19 applies to prove that α_0 admits an inverse $X_0 \underset{f_0}{\overset{g_0}{\underset{\Downarrow\alpha_0^{-1}}{\rightrightarrows}}} A_0$. By the full and faithfulness of the local equivalence $h\mathrm{Fun}(X, A) \overset{\approx}{\to} h\mathrm{Fun}(X_0, A_0)$ established in Proposition 10.3.1, this 2-cell lifts to define an inverse natural transformation $X \underset{f}{\overset{g}{\underset{\Downarrow\alpha^{-1}}{\rightrightarrows}}} A$ witnessing the invertibility of α. $\qquad\square$

It is worth calling attention to a special feature of the cosmological biequivalence $(-)_0 : \mathcal{K} \overset{\approx}{\to} \mathcal{QC}at$ used in the proof of Lemma 12.2.1.

OBSERVATION 12.2.2 (on the elements of the underlying quasi-category). By Corollary 10.3.2(ii), a cosmological biequivalence $F \colon \mathcal{K} \xrightarrow{\sim} \mathcal{L}$ induces a bijection between isomorphism classes of elements of an ∞-category $A \in \mathcal{K}$ and isomorphism classes of elements of $FA \in \mathcal{L}$, and in fact induces an equivalence of homotopy categories $hA \xrightarrow{\sim} hFA$, the objects of which are exactly these elements (see Exercise 10.3.i). In particular, any element $x \colon 1 \to FA$ is naturally isomorphic to an element $Fa \colon 1 \to FA$ that is the image of an element $a \colon 1 \to A$. Since "pointwise" ∞-categorical notions – invertibility of the components of a natural isomorphism, possession of a terminal element in the fibers of a cocartesian fibration – are invariant under isomorphism, if A satisfies some pointwise criterion in $\mathcal{K}$, then FA satisfies the corresponding pointwise criterion in $\mathcal{L}$.

But in the case of ∞-cosmoi of $(\infty, 1)$-categories, Proposition 10.2.1 supplies a cosmological biequivalence $(-)_0 \colon \mathcal{K} \to Q\mathcal{C}at$ that acts bijectively on elements of ∞-categories. By construction, vertices of $A_0 \cong \mathrm{Fun}(1, A)$ correspond bijectively to elements of A. Consequently, as illustrated by the proof of Lemma 12.2.1, "pointwise" properties may be transferred even more readily.

Using Lemma 12.2.1, we can show that an isofibration in an ∞-cosmos of $(\infty, 1)$-categories is a discrete object of the slice ∞-cosmos if and only if its fibers are discrete ∞-categories. It follows that cocartesian, cartesian, or two-sided fibrations of $(\infty, 1)$-categories are discrete if and only if they have discrete fibers.

PROPOSITION 12.2.3. *Let* $p \colon E \twoheadrightarrow B$ *be an isofibration in an* ∞-*cosmos* $\mathcal{K}$ *of* $(\infty, 1)$-*categories. Then* p *is discrete as an object of* $\mathcal{K}_{/B}$ *if and only if the fibers of* p *are discrete* ∞-*categories in* $\mathcal{K}$.

Proof Any element $b \colon 1 \to B$ induces a cosmological functor $b^* \colon \mathcal{K}_{/B} \to \mathcal{K}$ which preserves discrete ∞-categories by Remark 1.3.3. So for any ∞-cosmos $\mathcal{K}$, if $p \colon E \twoheadrightarrow B$ is a discrete object in $\mathcal{K}_{/B}$, then its fibers are discrete ∞-categories.

For the converse we assume that $\mathcal{K}$ is an ∞-cosmos of $(\infty, 1)$-categories and appeal to Lemma 12.2.1. To show that $p \colon E \twoheadrightarrow B$ is discrete in $\mathcal{K}_{/B}$ we must argue that the quasi-category defined by the pullback

$$
\begin{array}{ccc}
\mathrm{Fun}_B(f \colon X \to B, p \colon E \twoheadrightarrow B) & \longrightarrow & \mathrm{Fun}(X, E) \\
\downarrow & \lrcorner & \downarrow{\scriptstyle p_*} \\
1 & \xrightarrow{\quad f \quad} & \mathrm{Fun}(X, B)
\end{array}
$$

is a Kan complex. By Corollary 1.1.15, it suffices to show that its homotopy category is a groupoid, and for this we use the smothering functor

$$h(\mathrm{Fun}_B(f\colon X \to B, p\colon E \twoheadrightarrow B)) \to \mathbb{1} \underset{h\mathrm{Fun}(X,B)}{\times} h\mathrm{Fun}(X,E)$$

of Lemma 3.1.5, which, in particular, reflects isomorphisms. The arrows of $h(\mathrm{Fun}_B(f,p))$ are represented by 1-simplices $\alpha\colon e \to e' \in \mathrm{Fun}(X,E)$ with the property that the whiskered composite $p\alpha \in \mathrm{Fun}(X,B)$ is the degenerate 1-simplex at f.[4] By Lemma 12.2.1, the natural transformation $\alpha\colon e \Rightarrow e' \in h\mathrm{Fun}(X,E)$ is invertible if and only if its components αx are invertible for every $x\colon 1 \to X$. Since the 1-simplex $p\alpha$ is degenerate in $\mathrm{Fun}(X,B)$, the 1-simplex $p\alpha x$ is degenerate at the vertex $fx \in \mathrm{Fun}(1,B)$, which says that αx lies in the fiber over $fx\colon 1 \to B$. Since the fibers of $p\colon E \twoheadrightarrow B$ are discrete, this tells us that αx is invertible, so we conclude by Lemma 12.2.1 that α is invertible as claimed. □

Our next series of results shows that universal properties can be detected pointwise in ∞-cosmoi of $(\infty, 1)$-categories. The key technical ingredient is an analytical result about quasi-categories in the style of Joyal and Lurie.

PROPOSITION 12.2.4. *A cocartesian fibration* $q\colon E \twoheadrightarrow A$ *of quasi-categories admits a right adjoint right inverse* $t\colon A \to E$ *if and only if for each* $a\colon 1 \to A$ *the fiber* E_a *has a terminal element.*

Proof By Lemma 3.6.9, a right adjoint right inverse to q can be interpreted as defining a terminal element in $q\colon E \twoheadrightarrow A$, considered as an object in the sliced ∞-cosmos $\mathcal{Q}\mathcal{C}at_{/A}$. By Lemma 3.6.6(i), this fibered adjunction may be pulled back along any element $a\colon 1 \to A$ to define a terminal element in the fiber E_a.

For the converse, let $ta\colon 1 \to E_a$ denote a chosen terminal element in the fiber E_a over $a\colon 1 \to A$. Lemma F.3.1 characterizes those isofibrations between quasi-categories that admit a right adjoint right inverse in terms of a lifting property. In this case, it suffices to show that any lifting problem

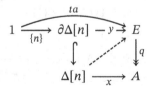

for $n \geq 1$ has a solution. To that end, consider the simplicial map $k\colon \Delta[n] \times \Delta[1] \to \Delta[n]$ defined on vertices by $k(i,0) := i$ and $k(i,1) := n$. The composite $xk\colon \Delta[n] \times \Delta[1] \to A$ restricts to define a map $xk\colon \partial\Delta[n] \times \Delta[1] \to A$ which

---

[4] This implies, but is stronger than, the property that the whiskered composite in the homotopy 2-category equals the 2-cell id_f (see §3.6 for a discussion).

represents a 2-cell whose codomain, defined by evaluating at the vertex $\{1\}$ of $\Delta[1]$, is constant at a and whose domain factors through q along y. This yields a new lifting problem

$$
\begin{array}{ccc}
\partial\Delta[n] & \xrightarrow{\ y\ } & E \\
{\scriptstyle i_0}\big\downarrow & \nearrow{\scriptstyle z} & \big\downarrow{\scriptstyle q} \\
\partial\Delta[n] \times \Delta[1] & \xrightarrow[\ xk\ ]{} & A
\end{array}
$$

which Lemma F.4.8 enables us to solve. By Lemma F.4.8, the lift z represents a q-cocartesian lift ζ of the 2-cell κ represented by the restriction of xk:

$$
\begin{array}{ccccc}
\partial\Delta[n] \xrightarrow{\ y\ } E & & & \partial\Delta[n] \xrightarrow{\ y\ } E & \\
{\scriptstyle !}\big\downarrow \quad {\scriptstyle \Downarrow\kappa} \quad \big\downarrow{\scriptstyle q} & = & & {\scriptstyle !}\big\downarrow \quad {\scriptstyle \Downarrow\zeta} \quad \big\downarrow{\scriptstyle q} & \\
1 \xrightarrow{\ a\ } A & & & 1 \xrightarrow{\ a\ } A &
\end{array}
$$

By construction, the codomain functor of the q-cocartesian lift displayed above right lands in the fiber over a. Now the component of κ at the final vertex $\{n\}\colon 1 \to \partial\Delta[n]$ is id_a, so by Lemma 5.1.6(ii), the component $z\{n\}$ representing the 2-cell $\zeta\{n\}$ is an isomorphism. In particular, the element $u\{n\}\colon 1 \to E$ is isomorphic to the terminal element $y\{n\} = ta$ of E_a, so we may apply the universal property of Proposition F.1.1(vi) to extend u to a simplex:

$$
\begin{array}{ccc}
\partial\Delta[n] & \xrightarrow{\ u\ } & E_a \\
\big\downarrow & \nearrow{\scriptstyle v} & \\
\Delta[n] & &
\end{array}
$$

This data defines a new lifting problem

$$
\begin{array}{ccccc}
\partial\Delta[n] \xhookrightarrow{\ i_0\ } & \partial\Delta[n]\times\Delta[1] \cup \Delta[n]\times\{1\} & \xrightarrow{\ z\cup v\ } & E & \\
\big\downarrow & \big\downarrow & & \big\downarrow{\scriptstyle q} & \quad (12.2.5) \\
\Delta[n] \xhookrightarrow{\ i_0\ } & \Delta[n]\times\Delta[1] & \xrightarrow[\ xk\ ]{} & A &
\end{array}
$$

which we solve inductively by choosing lifts of the $n+1$ $(n+1)$-simplices in $\Delta[n]\times\Delta[1]$ not present in $\partial\Delta[n]\times\Delta[1] \cup \Delta[n]\times\{1\}$, starting from the $n+1$-simplex that contains the face $\Delta[n]\times\{1\}$. All but the last of these can be lifted by means of lifting inner horns against the isofibration q. For the final simplex, we must solve an outer horn lifting problem

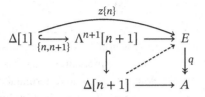

but in this case the final edge of the outer horn is the isomorphism $z\{n\}$, so Proposition 1.1.18 permits its solution as well. Now the lift (12.2.5) restricts to define the sought-for solution to the original lifting problem, proving that $q : E \twoheadrightarrow A$ admits a right adjoint right inverse. $\qquad\square$

Via the cosmological biequivalence $(-)^{\mathrm{op}} : \mathcal{QCat} \simeq \mathcal{QCat}^{\mathrm{co}}$ of Lemma 12.1.3, the proof of Proposition 12.2.4 dualizes to prove that a cartesian fibration of quasi-categories admits a left adjoint right inverse if and only if each fiber has an initial element (see Exercise 12.2.i).

The proof of Proposition 12.2.4 relied heavily on "analytic" techniques. Nevertheless its conclusion transfers to any ∞-cosmos that is biequivalent to the ∞-cosmos of quasi-categories.

PROPOSITION 12.2.6. *In an ∞-cosmos of* $(\infty, 1)$*-categories a cocartesian fibration* $q : E \twoheadrightarrow A$ *of admits a right adjoint right inverse* $t : A \to E$ *if and only if for each* $a : 1 \to A$ *the fiber* E_a *has a terminal element.*

Proof The argument that the fibers of an isofibration with right adjoint right inverse admit terminal elements is the same as given in the proof of Proposition 12.2.4. For the converse, suppose $q : E \twoheadrightarrow A$ is a cocartesian fibration in an ∞-cosmos $\mathcal{K}$ of $(\infty, 1)$-categories with the property that for each element $a : 1 \to A$ of the base, the fiber E_a has a terminal element. By Proposition 10.2.1, we may use the underlying quasi-category biequivalence $(-)_0 : \mathcal{K} \twoheadrightarrow \mathcal{QCat}$ to conclude that q admits a right adjoint right inverse once we show that $q_0 : E_0 \twoheadrightarrow A_0$ satisfies the hypotheses of Proposition 12.2.4.

Cosmological functors preserve cocartesian fibrations, so the underlying map $q_0 : E_0 \twoheadrightarrow A_0$ defines a cocartesian fibration of quasi-categories. By Observation 12.2.2, elements of the underlying quasi-category A_0 correspond bijectively to elements of the ∞-category A. By hypothesis, for every $a : 1 \to A$ the ∞-category E_a admits a terminal element, so by Proposition 10.1.4 the underlying quasi-category $(E_a)_0$ does as well. In this way, we see that every fiber of the cocartesian fibration of quasi-categories $q_0 : E_0 \twoheadrightarrow A_0$ admits a terminal element. By Proposition 12.2.4, q_0 admits a right adjoint right inverse.

Now by Proposition 10.3.6, we may conclude that $q : E \twoheadrightarrow A$ admits a right adjoint right inverse in $\mathcal{K}$, as desired. $\square$

An important special case of Proposition 12.2.6 proves a result promised in the discussion surrounding Proposition 4.1.6: in an ∞-cosmos of $(\infty, 1)$-categories a functor $f : B \to A$ admits a right adjoint just when for each element $a : 1 \to A$, the ∞-category $\mathsf{Hom}_A(f, a)$ admits a terminal element: namely the counit component $\ulcorner \epsilon a \urcorner : 1 \to \mathsf{Hom}_A(f, a)$. In the terminology used by Mac Lane [81, §III.1], this universal property exhibits $\ulcorner \epsilon a \urcorner$ as a "universal arrow" from the functor f to the element a.

COROLLARY 12.2.7. *In an ∞-cosmos of $(\infty, 1)$-categories, a functor $f : B \to A$ admits a right adjoint if and only if for each element $a : 1 \to A$, the comma ∞-category $\mathsf{Hom}_A(f, a)$ admits a terminal element.*

Proof Proposition 4.1.6 demonstrates that in any ∞-cosmos, $f : B \to A$ admits a right adjoint if and only if $\mathsf{Hom}_A(f, A)$ admits a terminal element over A, meaning that $p_1 : \mathsf{Hom}_A(f, A) \twoheadrightarrow A$ admits a right adjoint right inverse. By Corollary 5.5.13, this functor is a cocartesian fibration, so Proposition 12.2.6 tells us that p_1 admits a right adjoint right inverse if and only if each fiber $\mathsf{Hom}_A(f, a)$ admits a terminal element. $\square$

Proposition 12.2.6 implies that modules between $(\infty, 1)$-categories admit an analogous "pointwise" representability condition, characterizing those modules that are covariantly or contravariantly represented by a functor in the sense of Definition 7.4.7.

COROLLARY 12.2.8. *In an ∞-cosmos of $(\infty, 1)$-categories, a module $A \xrightarrow{E} B$ is covariantly represented if and only if for each $a : 1 \to A$, the module $1 \xrightarrow{E(1,a)} B$ is covariantly represented, which is the case if and only if each ∞-category $E(1, a)$ admits a terminal element.*

Proof By Proposition 8.4.10, a module $A \xrightarrow{E} B$ encoded by $(q, p) : E \twoheadrightarrow A \times B$ is covariantly represented if and only if its left leg $q : E \twoheadrightarrow A$ admits a right adjoint right inverse $r : A \to E$, in which case $E \simeq \mathsf{Hom}_B(B, pr)$. By Lemma 7.4.3, the left left $q : E \twoheadrightarrow A$ defines a cocartesian fibration, so by Proposition 12.2.6, q admits a right adjoint right inverse if and only if the fiber E_a over each element $a : 1 \to A$ admits a terminal element. For each $a : 1 \to A$, the module $1 \xrightarrow{E(1,a)} B$ is given by the pullback along $a \times \mathrm{id} : 1 \times B \to A \times B$ and hence is isomorphic to the module $1 \xrightarrow{E_a} B$. Applying Proposition 8.4.10 again, this module is covariantly represented by some element if and only if $! : E_a \twoheadrightarrow 1$ admits a right adjoint right inverse, which is the case if and only if the ∞-category $E_a \cong E(1, a)$ admits a terminal element. $\square$

Our next result concerns absolute lifting diagrams, used in Definition 2.3.8 to define the limit or colimit of a family of diagrams. There is a certain elegance in considering families of diagrams, rather than individual diagrams, when it comes to stating results such as Proposition 2.3.15, but it is also useful to establish the existence of limits and colimits one diagram at a time. In ∞-cosmoi of $(\infty, 1)$-categories, this sort of reduction is possible, on account of the following:

PROPOSITION 12.2.9. *In an ∞-cosmos of $(\infty, 1)$-categories:*

   (i) *A functor $g : C \to A$ admits an absolute right lifting through a functor*
       $f : B \to A$ if and only if for all $c : 1 \to C$, the comma ∞-category
       $\mathrm{Hom}_A(f, gc)$ admits a terminal element.
  (ii) *A triangle as below-left*

       displays r as an absolute right lifting of g through f if and only if for
       all $c : 1 \to C$, the restricted triangle as above-right displays rc as an
       absolute right lifting of gc through f.
 (iii) *A functor $g : C \to A$ admits an absolute right lifting through a functor*
       $f : B \to A$ if and only if for all $c : 1 \to C$ there exists an absolute right
       lifting of gc through f as below-left

$$\begin{array}{ccc} & B & \\ {\scriptstyle r_c}\nearrow & \Downarrow\rho_c & \downarrow{\scriptstyle f} \\ 1 \xrightarrow[gc]{} & & A \end{array} \qquad \begin{array}{ccc} & B & \\ {\scriptstyle r}\nearrow & \Downarrow\rho & \downarrow{\scriptstyle f} \\ C \xrightarrow[g]{} & & A \end{array}$$

       in which case the components of the above-right absolute right lifting
       of g through f are isomorphic to the corresponding pointwise absolute
       liftings: $r_c \cong rc$ and $\rho_c \cong \rho c$.

Proof Theorems 3.5.8 and 3.5.12 demonstrate that in any ∞-cosmos, a functor $g : C \to A$ admits a right lifting through $f : B \to A$ if and only if the codomain projection functor $p_1 : \mathrm{Hom}_A(f, g) \twoheadrightarrow C$ admits a right adjoint right inverse. By Corollary 5.5.13 this functor is a cocartesian fibration. Proposition 12.2.6 proves that in an ∞-cosmos of $(\infty, 1)$-categories, $p_1 : \mathrm{Hom}_A(f, g) \twoheadrightarrow C$ admits a right adjoint right inverse if and only if each fiber $\mathrm{Hom}_A(f, gc)$ over an element $c : 1 \to C$ admits a terminal element. This proves the first statement.

Since absolute lifting diagrams are stable under restriction, it is immediately clear that any absolute right lifting diagram as above-left, restricts to define a pointwise absolute right lifting diagram as above-right. For the converse, suppose $(rc, \rho c)$ defines an absolute right lifting of gc through f for any $c : 1 \to C$. By Theorem 3.5.12, it follows that each comma ∞-category $\mathrm{Hom}_A(f, gc)$ has a terminal element and so by (i), $g : C \to A$ must admit an absolute right lifting (s, σ) through $f : B \to A$. By its universal property, the pair (r, ρ) factors through (s, σ) via a 2-cell $\tau : r \Rightarrow s$ so that $\rho = \sigma \cdot f\tau$. Since $(rc, \rho c)$ and $(sc, \sigma c)$ are both absolute right lifting diagrams, we know that each component τc is an isomorphism. Hence, by Lemma 12.2.1, τ is an isomorphism, and thus (r, ρ) is also an absolute right lifting diagram, as desired.

The third statement is a convenient summary of the first two. If gc admits an absolute right lifting through f then $\mathrm{Hom}_A(f, gc)$ has a terminal element and (i) guarantees the existence of an absolute right lifting of g through f. The component of (r, ρ) at c defines a second absolute right lifting of gc through f inducing the claimed isomorphisms. $\square$

As a corollary, we may justify the claim made in Remark 2.3.9.

COROLLARY 12.2.10. *In an ∞-cosmos of $(\infty, 1)$-categories if A is an ∞-category that admits limits of every diagram $d : 1 \to A^J$ of shape J then A admits all limits of shape J: that is, the constant diagram functor admits a right adjoint*

$$A^J \underset{\lim}{\overset{\Delta}{\rightleftarrows}} A$$

Proof By Proposition 12.2.9(iii), the functor id : $A^J \to A^J$ admits an absolute right lifting through $\Delta : A \to A^J$ if and only if each diagram $d : 1 \to A^J$ admits an absolute right lifting through $\Delta : A \to A^J$. By Definition 2.3.8 the latter condition encodes what it means for A to admit a limit of the diagram d, while by Lemma 2.3.7 the former condition encodes what it means for $\Delta : A \to A^J$ to admit a right adjoint. $\square$

Another proof of Proposition 12.2.9(ii) is possible. By Theorem 3.5.8, a 2-cell $\rho : fr \Rightarrow g$ defines an absolute right lifting if and only if the induced functor $\ulcorner\rho\urcorner : \mathrm{Hom}_B(B, r) \to \mathrm{Hom}_A(f, g)$ is an equivalence over $C \times B$. As we shall now discover, equivalences between co/cartesian fibrations or modules can be detected fiberwise in ∞-cosmoi of $(\infty, 1)$-categories. As was our strategy for constructing right adjoint right inverses to cocartesian fibrations, we first prove this result for quasi-categories, using analytic techniques. Alternate analytic proofs can be found in [78, 3.3.1.5] and [5, 2.9]. We then use model independence to conclude that the same result holds true in arbitrary ∞-cosmoi of $(\infty, 1)$-categories.

PROPOSITION 12.2.11. *A cartesian functor*

$$E \xrightarrow{\quad g \quad} F$$
$$p \searrow \quad \swarrow q$$
$$B$$

between cocartesian fibrations of quasi-categories is a fibered equivalence if and only if it is a fiberwise equivalence, meaning that for each $b : 1 \to B$, the induced functor between fibers $g_b : E_b \to F_b$ is an equivalence.

Note the subtle difference in terminology between "fibered equivalences" – equivalences over B – and "fiberwise equivalences" – maps inducing equivalences on fibers over elements of B. This result and Proposition 12.2.12 to follow show that in fact these two notions coincide for cartesian functors in ∞-cosmoi of $(\infty, 1)$-categories.

Proof Fibered equivalences are stable under pullback to fibers, so the content is in the converse implication: that any cartesian functor between cocartesian fibrations that induces a fiberwise equivalence is necessarily an equivalence.

The cartesian functor g can be factored in the slice $\mathcal{QCat}_{/B}$ as an equivalence followed by an isofibration. By Corollary 5.3.1, the intermediate object of that factorization is again a cocartesian fibration and the isofibration from it to $q : F \twoheadrightarrow B$ is again a cartesian functor. Replacing $p : E \twoheadrightarrow B$ by the equivalent cocartesian fibration, it therefore suffices to assume that $g : E \twoheadrightarrow B$ is an isofibration and a cartesian functor and postulate that each induced map $g_b : E_b \twoheadrightarrow F_b$ is a trivial fibration. Under these assumptions, we must show that g is itself is a trivial fibration.

To that end, suppose that we are given a lifting problem

$$
\begin{array}{ccc}
\partial\Delta[n] & \xrightarrow{\quad e \quad} & E \\
\downarrow & & \downarrow g \\
\Delta[n] & \xrightarrow{\quad f \quad} & F
\end{array}
$$

over $b : \Delta[n] \to B$. Consider the retract diagram

$$\Delta[n] \xrightarrow{\mathrm{id} \times \{0\}} \Delta[n] \times \Delta[1] \xrightarrow{\quad r \quad} \Delta[n]$$

$$i \longmapsto (i, 0)$$

$$(i, j) \longmapsto \begin{cases} i & \text{if } j = 0 \\ n & \text{if } j = 1 \end{cases}$$

and choose a pointwise p-cocartesian lift

$$
\begin{array}{ccc}
\partial\Delta[n] \times \{0\} & \xrightarrow{\ e\ } & E \\
\Big\downarrow & \nearrow{\scriptstyle \chi} & \Big\downarrow{\scriptstyle p} \\
\partial\Delta[n] \times \Delta[1] & \xrightarrow{\ br\ } & B
\end{array}
$$

as permitted by Lemma F.4.8. Applying g we obtain a hollow cylinder

$$
\partial\Delta[n] \times \Delta[1] \xrightarrow{\ \chi\ } E \xrightarrow{\ g\ } F
$$

and since g is a cartesian functor and χ is pointwise p-cocartesian it follows that $g\chi$ is pointwise q-cocartesian. Now by construction the simplex $f : \Delta[n]\times\{0\} \to F$ agrees with $g\chi : \partial\Delta[n] \times \Delta[1] \to F$ on the subset $\partial\Delta[n] \times \{0\}$ where they are both defined. It follows that they combine to give a well-defined simplicial map on the union of their domains and so provide us with a second lifting problem:

$$
\begin{array}{ccc}
\Delta[n] \times \{0\} \cup \partial\Delta[n] \times \Delta[1] & \xrightarrow{\ f \cup g\chi\ } & F \\
\Big\downarrow & \nearrow{\scriptstyle \rho} & \Big\downarrow{\scriptstyle q} \\
\Delta[n] \times \Delta[1] & \xrightarrow{\ br\ } & B
\end{array}
$$

which can again be solved to give a pointwise q-cocartesian lift ρ by Lemma F.4.8. Note now that the retraction $r : \Delta[n] \times \Delta[1] \to \Delta[n]$ was constructed to map the subset $\Delta[n] \times \{1\}$ onto the vertex $\{n\}$, from which it follows that the n-simplex

$$
\Delta[n] \times \{1\} \hookrightarrow \Delta[n] \times \Delta[1] \xrightarrow{\ br\ } B
$$

is a degenerate image of the final vertex $b_n := b \cdot \{n\}$. Now observe that the cylinders χ and ρ were defined to lie over $br : \Delta[n] \times \Delta[1] \to B$, so it follows that the restricted maps

$$
\partial\Delta[n] \times \{1\} \hookrightarrow \partial\Delta[n] \times \Delta[1] \xrightarrow{\ \chi\ } E
$$

$$
\Delta[n] \times \{1\} \hookrightarrow \Delta[n] \times \Delta[1] \xrightarrow{\ \rho\ } F
$$

land in the fibers E_{b_n} and F_{b_n} of p and q, respectively. Thus, χ and ρ define a lifting problem

$$
\begin{array}{ccc}
\partial\Delta[n] & \xrightarrow{\ \chi|_{\{1\}}\ } & E_{b_n} \\
\Big\downarrow & \nearrow{\scriptstyle \gamma} & \Big\downarrow{\scriptstyle g_{b_n}} \\
\Delta[n] & \xrightarrow[\ \rho|_{\{1\}}\ ]{} & F_{b_n}
\end{array}
$$

which we may solve since the map of fibers on the right is, by assumption, a trivial fibration. Now the upper left triangle tells us that χ and γ agree on the subset $\partial\Delta[n] \times \{1\}$ where they are both defined. Thus, these maps combine to give a well-defined simplicial map on the union of their domains depicted as the upper-horizontal in the lifting problem on the right of the following diagram:

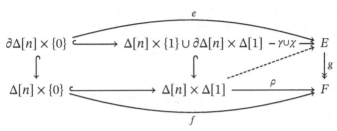

A standard argument shows that the lifting problem in the right-hand square can be solved by filling a sequence of inner horns and a single outer horn of shape $\Lambda^{n+1}[n+1]$ whose final edge is a cocartesian lift of the degeneracy at b_n. Lemma 5.1.6 observes that cocartesian lifts of degenerate simplices are isomorphisms, so this last horn is actually a "special outer horn" with first edge invertible. Consequently, by Theorem D.5.1 it can therefore be lifted against the isofibration p. This construction fills the sphere at the domain end of the hollow cylinder, solving the original lifting problem. □

PROPOSITION 12.2.12 (fiberwise equivalences of cartesian fibrations). *In an ∞-cosmos of* $(\infty, 1)$*-categories, a cartesian functor*

$$E \xrightarrow{\ g\ } F$$
$$p \searrow \quad \swarrow q$$
$$B$$

between cocartesian fibrations is a fibered equivalence if and only if it is a fiberwise equivalence, meaning that for each $b : 1 \to B$*, the induced functor between fibers* $g_b : E_b \to F_b$ *is an equivalence.*

Proof An equivalence of cocartesian fibrations is necessarily a fiberwise equivalence, so we need only prove the converse. By Propositions 10.2.2 and 10.2.1, if $\mathcal{K}$ is an ∞-cosmos of $(\infty, 1)$-categories and B is an ∞-category in $\mathcal{K}$, then the underlying quasi-category functor induces a cosmological biequivalence $(-)_0 : \mathcal{K}_{/B} \overset{\sim}{\to} \mathcal{QCat}_{/B_0}$. By Corollary 10.3.2, this cosmological biequivalence reflects equivalences, and by Proposition 10.1.4 $(-)_0 : \mathcal{K} \to \mathcal{QCat}$ preserves cocartesian fibrations and cartesian functors between them. Hence, to show that a cartesian functor and fiberwise equivalence $g : E \to F$ is an equivalence over

B, it suffices to show that $g_0 : E_0 \to F_0$ is an equivalence over B_0, which we do by verifying that this functor satisfies the hypotheses of Proposition 12.2.11.

By Observation 12.2.2, elements of the underlying quasi-category B_0 correspond bijectively to elements of the ∞-category B, and the cosmological functor $(-)_0 : \mathcal{K} \to \mathcal{QC}at$ preserves both fibers and equivalences. In this way we see that $g_0 : E_0 \to F_0$ is a fiberwise equivalence for all $b : 1 \to B_0$. By Proposition 12.2.11 this functor defines a fibered equivalence, and hence g does as well.   □

COROLLARY 12.2.13. *In an ∞-cosmos of $(\infty, 1)$-categories, a discrete cartesian fibration is a trivial fibration if and only if its fibers are contractible.*

Proof If $p : E \twoheadrightarrow B$ is a discrete cartesian fibration, then p defines a cartesian functor

$$
\begin{array}{ccc}
E & \xrightarrow{\ \ p\ \ } & B \\
& {}_{p}\searrow \quad \nearrow\!\!\!\!/\!/ & \\
& B &
\end{array}
$$

whose codomain is the identity discrete fibration. The conclusion now arises as a special case of the result of Proposition 12.2.12.   □

The result of Proposition 12.2.12 can be extended to modules:

COROLLARY 12.2.14 (equivalences of modules are determined fiberwise). *In an ∞-cosmos of $(\infty, 1)$-categories, a map*

$$
\begin{array}{ccc}
E & \xrightarrow{\ \ g\ \ } & F \\
{}_{(q,p)}\searrow & & \swarrow {}_{(s,r)} \\
& A \times B &
\end{array}
$$

between modules E and F from A to B is an equivalence if and only if it is a fiberwise equivalence, meaning that for each $a : 1 \to A$ and $b : 1 \to B$, the induced functor between the fibers $g_{a,b} : E(b, a) \to F(b, a)$ is an equivalence.

Proof Recall from Lemma 7.4.3 that the left-hand legs $q : E \twoheadrightarrow A$ and $s : F \twoheadrightarrow A$ are cocartesian fibrations and the functor g defines a cartesian functor between them. It follows, by Proposition 12.2.12, that g is an equivalence if and only if for each $a : 1 \to A$, the pullback $g_a : E(1, a) \to F(1, a)$ is an equivalence. Each of these ∞-categories defines a module from 1 to B, so by Lemma 7.4.3 again, the pulled back projections $p : E(1, a) \twoheadrightarrow B$ and $r : F(1, a) \twoheadrightarrow B$ are cartesian fibrations and g_a defines a cartesian functor between them. By the dual of Proposition 12.2.12, this functor is an equivalence if and only if for each $b : 1 \to B$, the induced action $g_{a,b} : E(b, a) \to F(a, b)$ on fibers is an equivalence. This proves the stated result.   □

A prototypical special case of this result characterizes units or counits of adjunctions between $(\infty, 1)$-categories:

COROLLARY 12.2.15. *In an ∞-cosmos of $(\infty, 1)$-categories, a 2-cell $\epsilon: fu \Rightarrow$ id_A defines the counit of an adjunction $f \dashv u$ if and only if the induced functor* $\ulcorner \epsilon \cdot f(-) \urcorner : \mathrm{Hom}_B(B, u) \to \mathrm{Hom}_A(f, A)$ *over $A \times B$ defines equivalences of mapping spaces* $\mathrm{Hom}_B(b, ua) \simeq \mathrm{Hom}_A(fb, a)$ *for each each $a: 1 \to A$ and $b: 1 \to B$.* □

Corollary 12.2.14 may also be applied to characterize fully faithful functors and equivalences between $(\infty, 1)$-categories.

PROPOSITION 12.2.16. *In an ∞-cosmos of $(\infty, 1)$-categories, a functor $f: A \to B$ is fully faithful if and only if for all elements $a, a': 1 \to A$, its action on mapping spaces $f_{a,a'}: \mathrm{Hom}_A(a, a') \to \mathrm{Hom}_B(fa, fa')$ defines an equivalence of discrete ∞-categories.*

Proof Corollary 3.5.6 defines a functor $f: A \to B$ to be fully faithful if and only if the induced functor $A^2 \to \mathrm{Hom}_B(f, f)$ between modules from A to A is an equivalence. By Corollary 12.2.14, this is the case if and only if this map defines a fiberwise equivalence, which means exactly that for all elements $a, a': 1 \to A$, its action on mapping spaces $f_{a,a'}: \mathrm{Hom}_A(a, a') \to \mathrm{Hom}_B(fa, fa')$ defines an equivalence of discrete ∞-categories. □

We now show that equivalences of $(\infty, 1)$-categories are precisely those functors that are pointwise fully faithful and essentially surjective in a suitable sense. In [102], Rezk refers to this result as "the fundamental theorem of quasi-category theory." Our proof mixes synthetic and analytic techniques:

THEOREM 12.2.17 (fundamental theorem of $(\infty, 1)$-category theory). *A functor $f: A \to B$ in an ∞-cosmos of $(\infty, 1)$-categories is an equivalence if and only if it is*

  (i) *fully faithful: in the sense that for all elements $a, a': 1 \to A$, the induced map*

$$\mathrm{Hom}_A(a, a') \to \mathrm{Hom}_B(fa, fa')$$

  is an equivalence and
  (ii) *essentially surjective in the sense that for all $b: 1 \to B$ there exists $a: 1 \to A$ and an isomorphism $fa \cong b$ in the homotopy category of B.*

Proof An equivalence of ∞-categories $f: A \xrightarrow{\sim} B$ induces an equivalence of homotopy categories $hf : hA \twoheadrightarrow hB$ as well as an equivalence $A^2 \simeq_{A \times A}$ $\mathrm{Hom}_B(f, f)$ between modules from A to A, by Propositions 3.3.3 and 3.4.5:

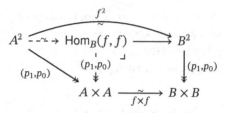

Essential surjectivity is a consequence of the first equivalence, while fully faithfulness follows by pulling the second equivalence back along a pair of elements of A.

To prove the converse, we start by factoring f as an equivalence followed by an isofibration. Both factors are easily seen to be pointwise fully faithful and essentially surjective, so it suffices to assume that $f : A \twoheadrightarrow B$ is an isofibration. Our task is now to show that f is a trivial fibration. In an ∞-cosmos $\mathcal{K}$ of $(\infty, 1)$-categories, the cosmological biequivalence $(-)_0 : \mathcal{K} \twoheadrightarrow \mathcal{QC}at$ of Proposition 10.2.1 preserves isofibrations and reflects equivalences. So to show that an isofibration $f : A \twoheadrightarrow B$ is a trivial fibration, it suffices to show that the underlying isofibration $f_0 : A_0 \twoheadrightarrow B_0$ is a trivial fibration, i.e., that we can solve lifting problems of the form

$$
\begin{array}{ccc}
\partial\Delta[n] & \longrightarrow & \mathrm{Fun}(1, A) \\
\downarrow & \nearrow & \downarrow f_0 \\
\Delta[n] & \longrightarrow & \mathrm{Fun}(1, B)
\end{array}
\tag{12.2.18}
$$

for $n \geq 0$.

In the case $n = 0$, this lifting property asserts that when $f_0 : A_0 \twoheadrightarrow B_0$ is an isofibration, the hypothesis that $f : A \twoheadrightarrow B$ is essentially surjective in fact implies that $f_0 : A_0 \twoheadrightarrow B_0$ is surjective on vertices. By essential surjectivity, for any $b : 1 \to B$, there is some $a : 1 \to A$ so that $fa \cong b$ in $hB := h\mathrm{Fun}(1, B) := h(B_0)$. Now Corollary 1.1.16 implies that any isomorphism in the homotopy category of B can be represented by a homotopy coherent isomorphism $\mathbb{I} \to \mathrm{Fun}(1, B)$. A choice of $a : 1 \to A$ and a homotopy coherent isomorphism $\beta : \mathbb{I} \to \mathrm{Fun}(1, B)$ representing $fa \cong b$ defines a lifting problem

$$
\begin{array}{ccccc}
\varnothing & \longrightarrow & 1 & \overset{a}{\longrightarrow} & \mathrm{Fun}(1, A) \\
\downarrow & & \downarrow & \nearrow & \downarrow f_0 \\
1 & \lhook\joinrel\longrightarrow & \mathbb{I} & \overset{\beta}{\dashrightarrow} & \mathrm{Fun}(1, B) \\
& & & \underset{b}{\searrow} &
\end{array}
$$

which can be solved by lifting the isomorphism along the isofibration. This solves the lifting problems (12.2.18) in the case $n = 0$.

By applying Proposition 3.4.5 to the commutative diagram below-left, we see that the induced map between modules is an isofibration.

$$
\begin{array}{ccccc}
A & = & A & = & A \\
\| & & \downarrow{\scriptstyle f} & & \| \\
A & \xrightarrow{\quad f \quad} & B & \xleftarrow{\quad f \quad} & A
\end{array}
\qquad \rightsquigarrow \qquad
\begin{array}{c}
A^2 \\
{}^{p_1}\diagup \quad \vdots \quad \diagdown{}^{p_0} \\
A \xleftarrow{\ } \mathrm{Hom}_f(A,A) \xrightarrow{\ } A \\
{}^{p_1}\diagup \quad \downarrow \quad \diagdown{}^{p_0} \\
\mathrm{Hom}_A(f,f)
\end{array}
$$

By Proposition 12.2.16 the hypothesis that f is pointwise fully faithful, inducing equivalences between fibers $f_{a,a'} : \mathrm{Hom}_A(a, a') \twoheadrightarrow \mathrm{Hom}_B(fa, fa')$, implies that the induced map $\mathrm{Hom}_A(f, f) : A^2 \twoheadrightarrow \mathrm{Hom}_B(f, f)$ is an equivalence and hence under present hypotheses a trivial fibration. The cosmological functor $(-)_0 : \mathcal{K} \twoheadrightarrow \mathcal{QC}at$ carries this map to a trivial fibration between quasi-categories, which then enjoys the lifting property below-left for $n \geq 0$:

$$
\begin{array}{ccc}
\partial\Delta[n] & \longrightarrow & (A_0)^2 \\
{\scriptstyle\lceil} & \diagup\nearrow & \downarrow{\scriptstyle \mathrm{Hom}_{f_0}(A_0,A_0)} \\
\Delta[n] & \longrightarrow & \mathrm{Hom}_{B_0}(f_0, f_0)
\end{array}
\quad \rightsquigarrow \quad
\begin{array}{ccc}
\partial\Delta[n] \times \Delta[1] \underset{\partial\Delta[n]\times\partial\Delta[1]}{\cup} \Delta[n] \times \partial\Delta[1] & \longrightarrow & A_0 \\
{\scriptstyle\lceil} & & \downarrow{\scriptstyle f_0} \\
\Delta[n] \times \Delta[1] & \longrightarrow & B_0
\end{array}
$$

Via the description of the comma construction as a weighted limit giving in Example A.6.14, the lifting property above-left transposes across the Leibniz version of the weighted limit two-variable adjunction of Definition A.6.5 to the lifting property displayed above-right, again for $n \geq 0$.

We have already shown that $f_0 : A_0 \twoheadrightarrow B_0$ also possesses the right lifting property with respect to the inclusion $\varnothing \hookrightarrow \Delta[0]$. Since this map and the Leibniz product inclusions $(\partial\Delta[n] \hookrightarrow \Delta[n]) \,\hat{\times}\, (\partial\Delta[1] \hookrightarrow \Delta[1])$ generate the class of monomorphisms of simplicial sets under transfinite composition, pushout, and retract, it follows now from the fact that the map $\mathrm{Hom}_{f_0}(A_0, A_0)$ is a trivial fibration that $f_0 : A_0 \twoheadrightarrow B_0$ is a trivial fibration. Hence $f : A \twoheadrightarrow B$ is a trivial fibration, which is what we wanted to show. $\qquad\square$

Exercises

EXERCISE 12.2.i. Use Lemma 12.1.3 to prove the duals of Propositions 12.2.4 and 12.2.11.

EXERCISE 12.2.ii. Use the duals established in Exercise 12.2.i to state and prove the duals of other results in this section.

EXERCISE 12.2.iii. Use Corollary 12.2.14 to give a second proof of Proposition 12.2.9(ii).

12.3 Existence of Pointwise Kan Extensions

The vast and rapidly expanding literature on $(\infty, 1)$-category theory greatly exceeds the scope of this book. We close with a demonstration that illustrates how results whose proofs are found elsewhere can be integrated into the "model independent" framework of ∞-cosmoi. In this section, we borrow one result from our colleagues – concerning the "exponentiability" of cartesian and cocartesian fibrations between quasi-categories – and extract considerable mileage from it.

The category of simplicial sets, as a category of set-valued presheaves, is **locally cartesian closed**, meaning that pullback along any map $f : A \to B$ of simplicial sets admits a right adjoint, called "pushforward," as well as a left adjoint, defined by composition:

$$sSet_{/B} \xrightarrow[\underset{\Pi_f}{\overset{\Sigma_f}{\longleftarrow}}]{\underset{\longleftarrow}{\overset{\perp}{\underset{f^*}{\overset{\perp}{\longrightarrow}}}}} sSet_{/A} \tag{12.3.1}$$

For some, but not all, isofibrations $f : A \twoheadrightarrow B$ between quasi-categories, the adjunctions restrict to isofibration-preserving functors

$$\mathcal{Q}\mathcal{C}at_{/B} \xrightarrow[\underset{\Pi_f}{\overset{\Sigma_f}{\longleftarrow}}]{\underset{\longleftarrow}{\overset{\perp}{\underset{f^*}{\overset{\perp}{\longrightarrow}}}}} \mathcal{Q}\mathcal{C}at_{/A}$$

in which case f^* and Π_f are both cosmological. Such isofibration $f : A \twoheadrightarrow B$ are called **exponentiable**.[5]

A characterization of the exponentiable functors between quasi-categories due to Lurie – who calls them "flat fibrations" – appears in [80, §B.3]. A model independent characterization is given by Ayala and Francis in [5, 2.2.8]. For our purposes here, we require only that the cocartesian fibrations and cartesian fibrations between quasi-categories are among the exponentiable functors, and moreover preserve cartesian and cocartesian fibrations respectively. In fact, as we note in [114], the resulting pushforward constructions are well-adapted to the cosmological setting:

---

[5] A famously nonexponentiable functor appears in Exercise 12.3.i.

THEOREM 12.3.2 ([114, 6.2.9–10]). *For a cocartesian fibration* $q : E \twoheadrightarrow B$ *or cartesian fibration* $p : E \twoheadrightarrow B$ *between quasi-categories, pullback along* q *or* p *admits a right adjoint, which restrict to the cosmologically embedded* ∞-*cosmoi*

$$
\begin{array}{ccc}
\mathcal{Q}\mathcal{C}at_{/E} \xleftarrow[\Pi_q]{\overset{q^*}{\perp}} \mathcal{Q}\mathcal{C}at_{/B} & \qquad & \mathcal{Q}\mathcal{C}at_{/E} \xleftarrow[\Pi_p]{\overset{p^*}{\perp}} \mathcal{Q}\mathcal{C}at_{/B} \\
\uparrow\ \ \ \ \ \ \ \uparrow & & \uparrow\ \ \ \ \ \ \ \uparrow \\
\mathcal{C}art(\mathcal{Q}\mathcal{C}at)_{/E} \xleftarrow[\Pi_q]{\overset{q^*}{\perp}} \mathcal{C}art(\mathcal{Q}\mathcal{C}at)_{/B} & & co\mathcal{C}art(\mathcal{Q}\mathcal{C}at)_{/E} \xleftarrow[\Pi_p]{\overset{p^*}{\perp}} co\mathcal{C}art(\mathcal{Q}\mathcal{C}at)_{/B}
\end{array}
$$

and moreover all of these functors are cosmological.

Lurie's proof that cartesian and cocartesian fibrations are exponentiable can be found in [80, B.3.11, B.4.5]; see also [78, 3.2.2.12] and [80, B.4.2]. A specialization of these results to a setting much closer to the case under consideration here appears as [115, 2.24].

We now extend these results to arbitrary ∞-cosmoi of $(\infty, 1)$-categories.

THEOREM 12.3.3. *For a cocartesian fibration* $q : E \twoheadrightarrow B$ *or cartesian fibration* $p : E \twoheadrightarrow B$ *in an* ∞-*cosmos of* $(\infty, 1)$-*categories* $\mathcal{K}$, *pullback along* q *or* p *admits a quasi-pseudofunctorial right biadjoint, which restrict to the cosmologically embedded* ∞-*cosmoi*

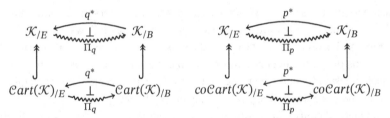

The "biadjointness" of the statement – again appropriating a term from weak 2-category theory – refers to a quasi-pseudonatural equivalence:

$$
\mathcal{K}_{/E} \xleftarrow[\Pi_p]{\overset{p^*}{\perp}} \mathcal{K}_{/B} \qquad \mathrm{Fun}_E(p^*(r), q) \simeq \mathrm{Fun}_B(r, \Pi_p q)
$$

Proof For a co/cartesian fibration $p : E \twoheadrightarrow B$, the pushforward quasi-pseudofunctor $\Pi_p : \mathcal{K}_{/E} \rightsquigarrow \mathcal{K}_{/B}$ is defined as the composite of the pushforward associated to the underlying co/cartesian fibration $p_0 : E_0 \twoheadrightarrow B_0$ between quasi-categories with the zigzag of cosmological biequivalences

$$
\begin{array}{ccc}
\mathcal{K}_{/E} & \xrightarrow{\;\;\Pi_p\;\;} & \mathcal{K}_{/B} \\
{\scriptstyle(-)_0}\Big\downarrow{\scriptstyle\wr} & \simeq & \Big\downarrow{\scriptstyle\wr}\,{\scriptstyle(-)_0} \\
\mathcal{QCat}_{/E_0} & \xrightarrow[\;\;\Pi_{p_0}\;\;]{} & \mathcal{QCat}_{/B_0}
\end{array}
\tag{12.3.4}
$$

Note that both pullback and the pushforward commute up to equivalence with the sliced underlying quasi-category functors $(-)_0 : \mathcal{K}_{/B} \rightsquigarrow \mathcal{QCat}_{/B_0}$.

The composite quasi-pseudofunctor can be understood as a pointwise right biadjoint on account of the composite equivalences

$$
\mathrm{Fun}_E(p^*(r), q) \xrightarrow[\sim]{(-)_0} \mathrm{Fun}_{E_0}((p_0)^*(r_0), q_0)
$$
$$
\wr\|
$$
$$
\mathrm{Fun}_{B_0}(r_0, \Pi_{p_0}(q_0)) \xleftarrow[\;\;\sim\;\;]{(-)_0} \mathrm{Fun}_B(r, \Pi_p q)
$$

that are quasi-pseudonaturally defined in $q : F \twoheadrightarrow E$ and $r : G \twoheadrightarrow B$. This proves the first statement.

Now let $p : E \twoheadrightarrow B$ be a cartesian fibration and let $q : F \twoheadrightarrow E$ be a cocartesian fibration in $\mathcal{K}$. Then by Proposition 10.3.6, $\Pi_p q$ is a cocartesian fibration in $\mathcal{K}$ if and only if its underlying functor is a cocartesian fibration on quasi-categories. By the essential commutativity of (12.3.4), this functor is equivalent to $\Pi_{p_0}(q_0)$, which is a cocartesian fibration of quasi-categories by Theorem 12.3.2. Since cocartesian fibrations are replete up to equivalence, the preservation result follows. The proof that pushforward preserves cartesian functors follows similarly.

Finally, Theorem 12.3.2 proves that cartesian functors between cocartesian fibrations of quasi-categories transpose to cartesian functors under $p^* \dashv \Pi_p$. Since cartesian functors are preserved and reflected by cosmological biequivalences, the same holds in any ∞-cosmos of $(\infty, 1)$-categories. Thus, the quasi-pseudonatural equivalence $\mathrm{Fun}_E(p^*(r), q) \simeq \mathrm{Fun}_B(r, \Pi_p q)$ restricts to define a quasi-pseudonatural equivalence $\mathrm{Fun}_E^{\mathrm{cart}}(p^*(r), q) \simeq \mathrm{Fun}_B^{\mathrm{cart}}(r, \Pi_p q)$. $\qquad\square$

LEMMA 12.3.5 (Beck–Chevalley). *Consider a pullback diagram in an ∞-cosmos of $(\infty, 1)$-categories,*

$$
\begin{array}{ccc}
F & \xrightarrow{\;\;g\;\;} & E \\
{\scriptstyle q}\Big\downarrow{\;\;\lrcorner} & & \Big\downarrow{\scriptstyle p} \\
A & \xrightarrow[\;\;f\;\;]{} & B
\end{array}
$$

and suppose that p is a cocartesian fibration or cartesian fibration. Then for all $r : G \twoheadrightarrow E$, the maps $f^*\Pi_p r \simeq \Pi_q g^* r$ are equivalent over A.

Proof Applying the cosmological biequivalence $(-)_0 : \mathcal{K} \twoheadrightarrow \mathcal{QC}at$ it suffices to prove this for quasi-categories, in which case the functors in question have left adjoints:

$$
\begin{array}{ccc}
\mathcal{QC}at_{/F_0} \xleftarrow{\ g_0^* \ } \mathcal{QC}at_{/E_0} & \qquad & sSet_{/F_0} \underset{\Sigma_{g_0}}{\overset{g_0^*}{\rightleftarrows}} sSet_{/E_0} \\[2ex]
\Pi_{q_0} \downarrow \qquad \qquad \downarrow \Pi_{p_0} & & \Pi_{q_0}\left(\dashv\right) q_0^* \qquad \qquad p_0^*\left(\dashv\right)\Pi_{p_0} \\[2ex]
\mathcal{QC}at_{/A_0} \xleftarrow[\ f_0^* \ ]{} \mathcal{QC}at_{/B_0} & & sSet_{/A_0} \underset{f_0^*}{\overset{\Sigma_{f_0}}{\rightleftarrows}} sSet_{/B_0}
\end{array}
$$

For any map of simplicial sets $a : X \to A_0$, by examining the pullback rectangle

$$
\begin{array}{ccc}
\bullet \longrightarrow & F_0 \xrightarrow{\ g_0\ } & E_0 \\
\downarrow \qquad & q_0 \downarrow \qquad & \downarrow p_0 \\
X \xrightarrow{\ a\ } & A_0 \xrightarrow{\ f_0\ } & B_0
\end{array}
$$

it is clear that $\Sigma_{g_0} q_0^* a \cong p_0^* \Sigma_{f_0} a$. Thus, the right adjoints $f_0^* \Pi_{p_0} \cong \Pi_{q_0} g_0^*$ also commute up to isomorphism, and conservativity provides the claimed equivalence in $\mathcal{K}$. $\qquad\qquad\square$

Now we explain our interest in these results. Recall from Definition 9.1.2 that a **right extension** of a module $A \xrightarrow{G} C$ along a module $A \xrightarrow{E} B$ consists of a pair given by a module $B \xrightarrow{\hom_A(E,G)} C$ together with a binary cell

$$
\begin{array}{ccccc}
A & \xrightarrow{E} & B & \xrightarrow{\hom_A(E,F)} & C \\
\| & & \Downarrow\epsilon & & \| \\
A & & \xrightarrow{\qquad G \qquad} & & C
\end{array}
$$

that is universal among the cells in the virtual equipment of modules in a sense detailed there. Dually, a **right lifting** of $A \xrightarrow{G} C$ through $B \xrightarrow{F} C$ consists of a pair given by a module $A \xrightarrow{\hom(F,G)_C} B$ together with a universal binary cell

$$
\begin{array}{ccccc}
A & \xrightarrow{\hom(F,G)_C} & B & \xrightarrow{F} & C \\
\| & & \Downarrow\epsilon & & \| \\
A & & \xrightarrow{\qquad G \qquad} & & C
\end{array}
$$

Here we introduce different notation for these notions with the aim of rebranding them. We refer to the module $\hom_A(E, G)$ as the **left hom** from E to F and refer to $\hom(F, G)_C$ as the **right hom** from F and G. Extending the "module" metaphor, we think of the left hom as the object of homomorphisms from E to G that respect the left A-actions, while the right hom is the object of homomorphisms from F to G that respect the right C-actions. Our first aim in this section is to demonstrate that the virtual equipment of modules between quasi-categories has left and right homs for all suitable pairs of modules. By model independence, this result extends to any ∞-cosmos of $(\infty, 1)$-categories.

THEOREM 12.3.6. *The virtual equipment of modules between quasi-categories is biclosed, admitting left and right homs that satisfy an enriched universal property: for modules $A \xrightarrow{E} B$, $B \xrightarrow{F} C$, and $A \xrightarrow{G} C$,*

$$\mathsf{Fun}_{A \times B}(E, \hom(F, G)_C) \cong \mathsf{Fun}_{A \times C}(E \times_B F, G) \cong \mathsf{Fun}_{B \times C}(F, \hom_A(E, G)).$$

Proof The cases of left and right homs are dual so we focus on the former. Fixing a module $A \xleftarrow{q} E \xrightarrow{p} B$, define $\hom_A(E, G)$ to be the image of the module $A \xrightarrow{G} C$ under the composite right adjoint functor

$$\mathcal{Q}\mathcal{C}at_{/A \times C} \underset{(q \times C)^*}{\overset{\Sigma_{q \times C}}{\underset{\perp}{\rightleftarrows}}} \mathcal{Q}\mathcal{C}at_{/E \times C} \underset{\Pi_{p \times C}}{\overset{(p \times C)^*}{\underset{\perp}{\rightleftarrows}}} \mathcal{Q}\mathcal{C}at_{/B \times C}$$

Note that the left adjoint carries a two-sided isofibration $B \xleftarrow{s} F \xrightarrow{r} C$ to the composite two-sided isofibration:

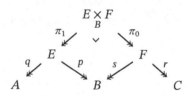

and simplicially enriched adjointness adjointness provides natural equivalences (or in this case isomorphisms) of quasi-categories

$$\mathsf{Fun}_{A \times C}(E \times_B F, G) \cong \mathsf{Fun}_{B \times C}(F, \hom_A(E, G)).$$

These equivalences induce bijections on the sets of isomorphism classes of objects in these quasi-categories. By Proposition 8.1.6, this gives rise to the universal property required by the right hom in the virtual equipment of modules (see Exercise 12.3.ii) – at least once we prove that $\hom_A(E, G)$ defines a module when E and G are modules.

Thus it remains only to show that when $A \xrightarrow{E} B$ is a module, then the functor

$$\hom_A(E, -): \mathcal{Q}\mathcal{C}at_{/A \times C} \xrightarrow{(q \times C)^*} \mathcal{Q}\mathcal{C}at_{/E \times C} \xrightarrow{\Pi_{p \times C}} \mathcal{Q}\mathcal{C}at_{/B \times C}$$

preserves modules. Proposition 7.4.5 tells us that modules are stable under pullback, so we need only demonstrate that $\Pi_{p \times C}$ carries a module $E \xleftarrow{v} M \xrightarrow{u} C$ to a module from B to C. Proposition 7.4.2 characterizes modules from E to C as those two-sided isofibrations $(v, u) \colon M \twoheadrightarrow E \times C$ so that

(i) $(v, u) \colon M \twoheadrightarrow E \times C$ is a cocartesian fibration in $Q\mathcal{C}at_{/C}$,
(ii) $(v, u) \colon M \twoheadrightarrow E \times C$ is a cartesian fibration in $Q\mathcal{C}at_{/E}$,
(iii) $(v, u) \colon M \twoheadrightarrow E \times C$ is discrete as an object in $Q\mathcal{C}at_{/E \times C}$.

We will show that $\Pi_{p \times C}$ preserves each of these properties, thus concluding that $\Pi_{p \times C}M$ defines a module from B to C. The preservation of (iii) is automatic: Theorem 12.3.2 demonstrates that the pushforward is a cosmological functor and Remark 1.3.3 observes that cosmological functors preserve discrete objects.

For (ii), we must show that $\Pi_{p \times C}M \twoheadrightarrow B \times C$ is a cartesian fibration in $Q\mathcal{C}at_{/B}$. To begin, note by Proposition 10.1.4 that the cosmological functor $\Pi_p \colon Q\mathcal{C}at_{/E} \to Q\mathcal{C}at_{/B}$ preserves cartesian fibrations,[6] so we conclude that $\Pi_p M \twoheadrightarrow \Pi_p(B \times C)$ is a cartesian fibration in $Q\mathcal{C}at_{/B}$. However, this functor does not coincide with $\Pi_{p \times C}M \twoheadrightarrow B \times C$. Rather, as demonstrated by Lemma 12.3.7 below, $\Pi_{p \times C}M \twoheadrightarrow B \times C$ is a pullback of $\Pi_p M \twoheadrightarrow \Pi_p(B \times C)$ in $Q\mathcal{C}at_{/B}$, so by pullback stability of cartesian fibrations, this proves (ii).

For the remaining property (i), recall from Lemma 7.1.1 that to say that $(v, u) \colon M \twoheadrightarrow E \times C$ defines a cocartesian fibration in $Q\mathcal{C}at_{/C}$ is equivalently to say that $(v, u) \colon M \twoheadrightarrow E \times C$ defines a morphism in the sub ∞-cosmos $co\mathcal{C}art(Q\mathcal{C}at)_{/E} \subset Q\mathcal{C}at_{/E}$. By Theorem 12.3.2, pushforward restricts to define a cosmological functor $\Pi_p \colon co\mathcal{C}art(Q\mathcal{C}at)_{/E} \to co\mathcal{C}art(Q\mathcal{C}at)_{/B}$, so $\Pi_p M \twoheadrightarrow \Pi_p(E \times C)$ lies in the sub cosmos $co\mathcal{C}art(Q\mathcal{C}at)_{/B}$. Since pullbacks are created by the cosmological embedding $co\mathcal{C}art(Q\mathcal{C}at)_{/B} \hookrightarrow Q\mathcal{C}at_{/B}$, by Lemma 12.3.7 $\Pi_{p \times C}M \twoheadrightarrow B \times C$ also lies in the sub cosmos $co\mathcal{C}art(Q\mathcal{C}at)_{/B}$. Lemma 7.1.1 now tells us that $\Pi_{p \times C}M \twoheadrightarrow B \times C$ is a cocartesian fibration in $Q\mathcal{C}at_{/C}$ as required. $\qquad\square$

LEMMA 12.3.7. *Let $p \colon E \twoheadrightarrow B$ be a cartesian fibration or cocartesian fibration between quasi-categories. Then the image of an isofibration $M \twoheadrightarrow E \times C$ under $\Pi_{p \times C} \colon Q\mathcal{C}at_{/E \times C} \to Q\mathcal{C}at_{/B \times C}$ is the left vertical isofibration defined by the*

---

[6] This is not an application of Theorem 12.3.3, which in this context would tell us that Π_p carries a *cocartesian* fibration over E to a cocartesian fibration over B. Rather, we note that Π_p, simply by virtue of being cosmological, carries cartesian fibrations between isofibration with codomain E to cartesian fibrations between isofibrations with codomain B.

pullback

$$
\begin{array}{ccc}
\Pi_{p\times C}M & \longrightarrow & \Pi_p M \\
\downarrow & \lrcorner & \downarrow \\
B \times C & \underset{\eta}{\longrightarrow} & \Pi_p(E \times C)
\end{array}
$$

whose right-hand vertical morphism is the image of the map $M \twoheadrightarrow E \times C$ under
$\Pi_p : \mathcal{QC}at_{/E} \to \mathcal{QC}at_{/B}.$

The conclusion of Lemma 12.3.7 holds, up to equivalence, in any ∞-cosmos of $(\infty, 1)$-categories $\mathcal{K}$ (see Exercise 12.3.iii). However, we find it easier to work with the strict adjunction of Theorem 12.3.2 rather than the weaker adjunctions of Theorem 12.3.3.

Proof To make sense of the statement note that p^* carries $\pi : B \times C \twoheadrightarrow B$ to $\pi : E \times C \twoheadrightarrow E$. Hence the map $\eta : B \times C \to \Pi_p(E \times C)$ is a component of the unit of the adjunction $p^* \dashv \Pi_p$. Now the claim follows directly by verifying that the displayed pullback has the universal property that defines $\Pi_{p\times C}M \in \mathcal{QC}at_{/B\times C}$. A cone over the pullback diagram may be interpreted as a diagram

$$
\begin{array}{ccc}
X & \longrightarrow & \Pi_p M \\
\downarrow & & \downarrow \\
B \times C & \underset{\eta}{\longrightarrow} & \Pi_p(E \times C)
\end{array}
$$

in $\mathcal{QC}at_{/B}$, which then transposes to define a commutative square

$$
\begin{array}{ccc}
X \times_B E & \longrightarrow & M \\
\downarrow & & \downarrow \\
E \times C & =\!=\!= & E \times C
\end{array}
$$

in $\mathcal{QC}at_{/E}$. In fact, this diagram lies in $\mathcal{QC}at_{/E\times C}$ and the left-hand vertical map $X \times_B E \to E \times C$ is isomorphic to $(p \times C)^*(X \to B \times C)$. Hence, this square transposes along $(p \times C)^* \dashv \Pi_{p\times C}$ to a commutative square

$$
\begin{array}{ccc}
X & \longrightarrow & \Pi_{p\times C}M \\
\downarrow & & \downarrow \\
B \times C & =\!=\!= & B \times C
\end{array}
$$

which proves the claim. $\qquad\qquad\qquad\qquad\qquad\qquad\qquad\qquad\qquad\square$

The result of Theorem 12.3.6 extends to biequivalent ∞-cosmoi:

COROLLARY 12.3.8. *In an ∞-cosmos of* (∞, 1)*-categories, the virtual equipment of modules is biclosed, admitting left and right homs that satisfy a weak enriched universal property: for modules* $A \xrightarrow{E} B$, $B \xrightarrow{F} C$, *and* $A \xrightarrow{G} C$,

$$\mathsf{Fun}_{A \times B}(E, \mathrm{hom}(F, G)_C) \simeq \mathsf{Fun}_{A \times C}(E \times_B F, G) \simeq \mathsf{Fun}_{B \times C}(F, \mathrm{hom}_A(E, G)).$$

There are two strategies for proving this. The first is to prove the model independence of the ingredients used in the proof of Theorem 12.3.6 (see Exercise 12.3.iii). A more expedient, though ultimately less informative strategy, is to simply make use of the biequivalence of virtual equipments, which is the approach we take here.

Proof Consider modules $A \xrightarrow{E} B$ and $A \xrightarrow{G} C$ in an ∞-cosmos of (∞, 1)-categories. By Theorem 11.1.6, there exists a module $B \xrightarrow{R} C$ whose underlying quasi-category is equivalent to the left hom $\mathrm{hom}_{A_0}(E_0, G_0)$ of Theorem 12.3.6, and binary cell $\nu : E \times R \Rightarrow G$ whose image under the underlying quasi-category functor composes with this equivalence to the binary map of modules between quasi-categories that defines the left hom from E_0 to G_0. It remains only to argue that this data has the universal property of a right extension in $\mathsf{Mod}(\mathcal{K})$, but this follows from the biequivalence of virtual equipments $(-)_0 : \mathsf{Mod}(\mathcal{K}) \rightsquigarrow \mathsf{Mod}(\mathcal{QC}at)$. □

Left and right homs can be used to define modules that encode hypothetical universal properties in ∞-category theory. When the modules so constructed are covariantly or contravariantly represented by a functor, as appropriate, then this data "satisfies the universal property," as we now illustrate. Proposition 9.4.9 observes that the value of a pointwise right extension of a functor $f : A \to C$ along a functor $k : A \to B$ at an element $b : 1 \to B$ is the limit of the composite diagram

$$\mathrm{Hom}_B(b, k) \xrightarrow{p_1} A \xrightarrow{f} C$$

The dual result expresses the value of a pointwise left extension of f along k at b as the colimit of the restriction of the functor f along $p_0 : \mathrm{Hom}_B(k, b) \to A$.

We now prove that in an ∞-cosmos of (∞, 1)-categories, if an ∞-category C admits such limits, then the pointwise right extension of f along k exists. The dual result reduces the question of the existence of pointwise left extensions to the existence of certain colimits.

THEOREM 12.3.9 (existence of pointwise right extensions). *Let* $k : A \to B$ *be a functor in an* ∞*-cosmos of* (∞, 1)*-categories and let* C *be an* ∞*-category that admits* $\mathrm{Hom}_B(b, k)$*-shaped limits for all elements* $b : 1 \to B$. *Then* C *admits* *pointwise right extensions along k*

$$A \xrightarrow{\ f\ } C$$
$$k \Big\downarrow \quad \Uparrow \nu \quad \nearrow^{\,\nearrow}$$
$$\text{ran}_k f$$
$$B$$

Proof Consider the module $\mathrm{Hom}_B(B, k)$ from A to B covariantly represented by the functor $k : A \to B$. Corollary 12.3.8 proves that any ∞-cosmos of $(\infty, 1)$-categories admits left and right homs, defining right extensions and right liftings in the virtual equipment of modules. In particular given any diagram $f : A \to C$, there exists a right extension in the virtual equipment of modules

$$A \xrightarrow{\ \mathrm{Hom}_B(B,k)\ } B \xdashrightarrow{\ \hom_A(\mathrm{Hom}_B(B,k),\mathrm{Hom}_C(C,f))\ } C$$
$$\Big\| \qquad\qquad \Downarrow \qquad\qquad \Big\|$$
$$A \xrightarrow[\mathrm{Hom}_C(C,f)]{} C$$

Now the ∞-category C admits a pointwise right Kan extension of $f : A \to C$ along $k : A \to B$ just when the module $\hom_A(\mathrm{Hom}_B(B, k), \mathrm{Hom}_C(C, f))$ is covariantly represented by a functor $r : B \to C$. By Corollary 12.2.8, this module is covariantly represented if and only if its pullbacks along each vertex $b : 1 \to B$ are covariantly represented, which is the case just when the fiber of the module $\hom_A(\mathrm{Hom}_B(B, k), \mathrm{Hom}_C(C, f)) \twoheadrightarrow B \times C$ over b has a terminal element.

By Lemma 9.1.4 the fiber $\hom_A(\mathrm{Hom}_B(B, k), \mathrm{Hom}_C(C, f))(1, b)$ arises as the the right extension

$$B \xrightarrow{\ \mathrm{Hom}_B(b,B)\ } 1 \xdashrightarrow{\ \hom_A(\mathrm{Hom}_B(B,k),\mathrm{Hom}_C(C,f))(1,b)\ } C$$
$$\Big\| \qquad\qquad \Downarrow\rho \qquad\qquad \Big\|$$
$$B \xrightarrow[\hom_A(\mathrm{Hom}_B(B,k),\mathrm{Hom}_C(C,f))]{} C$$

Since $\mathrm{Hom}_B(B, k) \otimes \mathrm{Hom}_B(b, B) \simeq \mathrm{Hom}_B(b, k)$ by Proposition 8.4.7, Proposition 9.1.6 allows us to combine these two right extensions into a single one

$$A \xrightarrow{\ \mathrm{Hom}_B(b,k)\ } 1 \xdashrightarrow{\ \hom_A(\mathrm{Hom}_B(B,k),\mathrm{Hom}_C(C,f))(1,b)\ } C$$
$$\Big\| \qquad\qquad \Downarrow \qquad\qquad \Big\|$$
$$A \xrightarrow[\mathrm{Hom}_C(C,f)]{} C$$

By exactness of the comma square

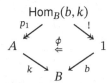

we also have a composition relation $\mathrm{Hom}_A(p_1, A) \otimes \mathrm{Hom}_1(1, !) \simeq \mathrm{Hom}_B(b, k)$.
By Lemma 9.1.4, there is a right extension diagram

$$
\begin{array}{ccccc}
A & \xrightarrow{\mathrm{Hom}_A(p_1,A)} & \mathrm{Hom}_B(b,k) & \dashrightarrow{\mathrm{Hom}_C(C,fp_1)} & C \\
\| & & \Downarrow & & \| \\
A & \xrightarrow[\mathrm{Hom}_C(C,f)]{\hspace{5cm}} & & & C
\end{array}
$$

Since $\mathrm{Hom}_C(C, f)$ admits a right extension along $\mathrm{Hom}_A(p_1, A) \otimes \mathrm{Hom}_1(1, !)$, by
Proposition 9.1.6 the right extension of $\mathrm{Hom}_C(C, fp_1)$ along $\mathrm{Hom}_1(1, !)$ exists
and is given by

$$
\begin{array}{ccccc}
\mathrm{Hom}_B(b,k) & \xrightarrow{\mathrm{Hom}_1(1,!)} & 1 & \dashrightarrow{\mathrm{hom}_A(\mathrm{Hom}_B(B,k),\mathrm{Hom}_C(C,f))(1,b)} & C \\
\| & & \Downarrow & & \| \\
\mathrm{Hom}_B(b,k) & \xrightarrow[\mathrm{Hom}_C(C,fp_1)]{\hspace{5cm}} & & & C
\end{array}
$$

In summary, the fiber of the module $\mathrm{hom}_A(\mathrm{Hom}_B(B, k), \mathrm{Hom}_C(C, f))$ over
$b : 1 \to B$ is the module defined by the right extension of $\mathrm{Hom}_C(C, fp_1)$
along $\mathrm{Hom}_1(1, !)$. Thus, we see that if this module is represented by an element $c : 1 \to C$, that element defines a pointwise right extension of the diagram

$$
\begin{array}{ccc}
\mathrm{Hom}_B(b,k) & \xrightarrow{p_1} A \xrightarrow{f} & C \\
{\scriptstyle !}\downarrow \quad {\scriptstyle \Uparrow} & \nearrow_{c} & \\
1 & &
\end{array}
$$

By Definition 9.4.7, this exists if and only if the diagram $fp_1 : \mathrm{Hom}_B(b, k) \twoheadrightarrow C$
has a limit in C. As we have assumed that this limit exists in C, so does the
pointwise right extension of f along k, as claimed. $\qquad\square$

Recall from Observation 10.3.4 that an ∞-cosmos of $(\infty, 1)$-categories admits
weakly defined exponentials.

COROLLARY 12.3.10. *Let* $k : A \to B$ *be a functor in an ∞-cosmos of $(\infty, 1)$-cat-
egories and let C be an ∞-category that admits* $\mathrm{Hom}_B(b, k)$*-shaped limits for*

all elements $b : 1 \to B$. Then pointwise right extension defines a right adjoint to the restriction functor:

$$C^A \underset{\mathrm{ran}_k}{\overset{-\circ k}{\rightleftarrows}} C^B$$

Proof By Lemma 2.3.7, to define a right adjoint to restriction it suffices to define an absolute right lifting of the identity through the functor $- \circ k : C^B \to C^A$. By Proposition 12.2.9, to achieve this it suffices to define absolute right lifting diagrams

$$
\begin{array}{ccc}
 & & C^B \\
 & \overset{\mathrm{ran}_k f}{\nearrow} \Downarrow \nu & \downarrow {\scriptstyle -\circ k} \\
1 & \underset{f}{\longrightarrow} & C^A
\end{array}
$$

for all elements $f : 1 \to C^A$ of C^A. As our notation suggests, we will demonstrate that the pointwise right extension $\nu : \mathrm{ran}_k f \circ k \Rightarrow f$ of Theorem 12.3.9 transposes across the equivalence $\mathrm{hFun}(X, C^A) \simeq \mathrm{hFun}(X \times A, C)$ of Observation 10.3.4 to define an absolute right lifting diagram.

The universal property that characterizes the absolute right lifting diagram

$$
\begin{array}{ccc}
X & \overset{e}{\longrightarrow} & C^B \\
{\scriptstyle !}\downarrow \quad \Downarrow \chi & & \downarrow {\scriptstyle C^k} \\
D & \underset{f}{\longrightarrow} & C^A
\end{array}
\quad = \quad
\begin{array}{ccc}
X & \overset{e}{\longrightarrow} & C^B \\
{\scriptstyle !}\downarrow \overset{\Downarrow \exists ! \zeta}{\underset{r}{\nearrow}} \Downarrow \nu & & \downarrow {\scriptstyle C^k} \\
1 & \underset{f}{\longrightarrow} & C^A
\end{array}
$$

where we have written $r := \mathrm{ran}_k f$ transposes to

$$
\begin{array}{ccc}
A \times X & \overset{\pi}{\twoheadrightarrow} A \overset{f}{\longrightarrow} C \\
{\scriptstyle k \times X}\downarrow \quad \Uparrow \chi & \nearrow \\
B \times X & \underset{e}{\longrightarrow}
\end{array}
\quad = \quad
\begin{array}{ccc}
A \times X & \overset{\pi}{\twoheadrightarrow} A \overset{f}{\longrightarrow} C \\
{\scriptstyle k \times X}\downarrow \quad \lrcorner \quad k\downarrow \overset{\Uparrow \nu}{\nearrow} {\scriptstyle r} \\
B \times X & \underset{\pi}{\twoheadrightarrow} B \; {\scriptstyle \Uparrow \exists ! \zeta} \\
& \underset{e}{\longrightarrow}
\end{array}
$$

Lemma 9.2.8 proves that the pullback square is exact. Now if ν is a pointwise right extension, then by Corollary 9.3.4 so is $\nu \cdot \pi$. The transposed universal property of this right extension diagram proves that ν defines an absolute right lifting. $\qquad\square$

The argument just given shows something further. By Corollary 9.3.4, the pasted composite of a pointwise right extension ν with an exact square ϕ gives another pointwise right extension as below-left, which we have just shown transposes to define an absolute right-lifting diagram below-right:

$$
\begin{array}{ccc}
E \xrightarrow{p} A \xrightarrow{f} C \\
q \downarrow \quad \Downarrow\phi \quad k \downarrow \quad \Uparrow\nu \quad \nearrow \\
D \xrightarrow{h} B \qquad \qquad r
\end{array}
\qquad \leftrightsquigarrow \qquad
\begin{array}{ccc}
& C^B \xrightarrow{C^h} C^D \\
r \nearrow \quad \Downarrow\nu \quad \downarrow C^k \quad \Downarrow C^\phi \quad \downarrow C^q \\
1 \xrightarrow{f} C^A \xrightarrow{C^p} C^E
\end{array}
$$

Thus, the absolute right lifting diagrams defined by transposing pointwise right extension diagrams are stable under pasting with exponentiated exact squares. In particular, if C admits pointwise right extensions along $k : A \to B$ then for any exact square ϕ we have an absolute right lifting diagram

$$
\begin{array}{ccc}
& C^B \xrightarrow{C^h} C^D \\
\text{ran}_k \nearrow \quad \Downarrow\nu \quad \downarrow C^k \quad \Downarrow C^\phi \quad \downarrow C^q \\
C^A = \!\!= \!\!= C^A \xrightarrow{C^p} C^E
\end{array}
$$

since by Proposition 12.2.9 the universal property of this absolute right lifting diagram can be checked pointwise at elements $f : 1 \to C^A$. Using this, we can establish a derivator-style "Beck–Chevalley" result as follows:

LEMMA 12.3.11 (Beck–Chevalley). *For any exact square*

$$
\begin{array}{ccc}
& D & \\
q \swarrow & & \searrow p \\
C & \phi \Leftarrow & B \\
g \searrow & & \swarrow f \\
& A &
\end{array}
$$

in an ∞-cosmos of $(\infty, 1)$-categories and any ∞-category E, the mates $\phi^!$ and $\phi_!$ of the induced natural transformation ϕ^* are both isomorphisms whenever these pointwise right and left extensions exist

$$
\begin{array}{ccc}
& E^A & \\
g^* \swarrow & & \searrow f^* \\
E^C & \phi^* \Leftarrow & E^B \\
q^* \searrow & & \swarrow p^* \\
& E^D &
\end{array}
\qquad
\begin{array}{ccc}
& E^A & \\
\text{ran}_g \nearrow & & \searrow f^* \\
E^C & \Downarrow\phi^! & E^B \\
q^* \searrow & & \nearrow \text{ran}_p \\
& E^D &
\end{array}
\qquad
\begin{array}{ccc}
& E^A & \\
g^* \swarrow & & \nwarrow \text{lan}_f \\
E^C & \Uparrow\phi_! & E^B \\
\text{lan}_q \nwarrow & & \swarrow p^* \\
& E^D &
\end{array}
$$

Proof By Corollary 12.3.10, the pointwise right extensions assemble into adjoint functors ran_g and ran_p which define absolute right lifting diagrams

Moreover, the mate $\phi^!$ of ϕ^* defines a factorization of the left-hand diagram through the right-hand diagram:

Immediately from the universal property of the absolute right liftings of q^* through p^* we have that $\phi^!$ is an isomorphism. The proof for $\phi_!$ is similar using the absolute left lifting diagrams arising from the pointwise left extensions defining lan_f and lan_q □

Exercises

EXERCISE 12.3.i. Conclude from the pullback diagram of simplicial sets

$$
\begin{array}{ccc}
1 + 1 & \longrightarrow & \Lambda^1[2] \\
\downarrow & \lrcorner & \downarrow \\
2 & \xrightarrow{\ d^1\ } & 3
\end{array}
$$

that the functor $d^1 : 2 \to 3$ is not exponentiable.

EXERCISE 12.3.ii. Suppose $A \xrightarrow{E} B$, $A \xrightarrow{G} C$, and $B \xrightarrow{R} C$ are modules equipped with a natural equivalence $\text{Fun}_{A \times C}(E \times_B F, G) \simeq \text{Fun}_{B \times C}(F, R)$ for all two-sided fibrations F from B to C. Extract a binary cell $\nu : E \times R \Rightarrow G$ from this natural equivalence and prove that ν exhibits R as a right extension of G along E.

EXERCISE 12.3.iii. Adapt Lemma 12.3.7 to a model independent statement that applies in any ∞-cosmos of $(\infty, 1)$-categories and then prove this result. Then use this to adapt the proof of Theorem 12.3.6 to directly demonstrate that an ∞-cosmos of $(\infty, 1)$-categories has all left and right homs.

APPENDIX OF ABSTRACT NONSENSE

Appendix A

Basic Concepts of Enriched Category Theory

Enriched category theory exists because enriched categories exist in nature. To explain, consider the data of a 1-category $\mathcal{C}$, given by:

- a collection of objects
- for each pair of objects $x, y \in \mathcal{C}$, a set $\mathcal{C}(x, y)$ of arrows in $\mathcal{C}$ from x to y
- for each $x \in \mathcal{C}$, a specified identity element $\mathrm{id}_x : 1 \to \mathcal{C}(x, x)$, and for each $x, y, z \in \mathcal{C}$, a specified composition map $\circ : \mathcal{C}(y, z) \times \mathcal{C}(x, y) \to \mathcal{C}(x, z)$ satisfying the associativity and unit[1] conditions:

$$\begin{array}{ccc} \mathcal{C}(y,z) \times \mathcal{C}(x,y) \times \mathcal{C}(w,x) & \xrightarrow{\circ \times \mathrm{id}} & \mathcal{C}(x,z) \times \mathcal{C}(w,x) \\ {\scriptstyle \mathrm{id} \times \circ} \downarrow & & \downarrow {\scriptstyle \circ} \\ \mathcal{C}(y,z) \times \mathcal{C}(w,y) & \xrightarrow{\quad \circ \quad} & \mathcal{C}(w,z) \end{array}$$

$$(A.0.1)$$

$$\begin{array}{ccc} \mathcal{C}(x,y) & \xrightarrow{\mathrm{id} \times \mathrm{id}_x} & \mathcal{C}(x,y) \times \mathcal{C}(x,x) \\ {\scriptstyle \mathrm{id}_y \times \mathrm{id}} \downarrow & \searrow {\scriptstyle \mathrm{id}} & \downarrow {\scriptstyle \circ} \\ \mathcal{C}(y,y) \times \mathcal{C}(x,y) & \xrightarrow{\quad \circ \quad} & \mathcal{C}(x,y) \end{array}$$

In many mathematical examples of interest, the set $\mathcal{C}(x, y)$ can be given additional structure, in which case it would be strange not to take it into account when performing further categorical constructions.

Perhaps there exists a specified zero arrow $0_{x,y} \in \mathcal{C}(x, y)$ in each hom-set, defining a two-sided ideal for composition: $g \circ 0 \circ f = 0$. Or extending this, perhaps each $\mathcal{C}(x, y)$ is an abelian group and composition is $\mathbb{Z}$-bilinear. Or in another direction, perhaps the set of arrows from x to y in $\mathcal{C}$ form the objects of

---

[1] Confusingly, this diagram contains two different sorts of identity arrows: e.g., the top horizontal arrow is the product of the specified identity element $\mathrm{id}_x : 1 \to \mathcal{C}(x, x)$ with the identity arrow associated with the object $\mathcal{C}(x, y)$.

a 1-category. In this setting, we regard the objects of $\mathcal{C}(x,y)$ as "1-dimensional" morphisms from x to y and the arrows of $\mathcal{C}(x,y)$ as "2-dimensional" morphisms from x to y in $\mathcal{C}$; here, it is natural to ask that the composition map defines a functor. Or perhaps the set of arrows from x to y in $\mathcal{C}$ form the vertices of a simplicial set, whose higher simplices now provide arrows in each positive dimension; here, it is natural to ask that composition defines a simplicial map. In all of these contexts, one says that the 1-category $\mathcal{C}$ can be *enriched over the category* $\mathcal{V}$ in which the objects $\mathcal{C}(x,y)$ and diagrams (A.0.1) live[2] – with $\mathcal{V}$ equal to the category of pointed sets, abelian groups, categories, or simplicial sets in the examples just described.

An alternate point of view on enriched category theory is often emphasized – adopted, for instance, in the classic textbook [68] from which we stole the title of this chapter. To borrow a distinction used by Peter May, the term "enriched" can be used as a compound noun – *enriched categories* – or as an adjective – *enriched* categories. In the noun form, an enriched category $\mathcal{C}$ has no preexisting underlying ordinary category, although we shall see in Definition A.2.2 that the underlying unenriched 1-category can always be identified a posteriori. When used as an adjective, an enriched category $\mathcal{C}$ is perhaps most naturally an ordinary category, whose hom-sets can be given additional structure.[3] While the noun perspective is arguably more elegant when discussing the general theory of enriched categories, the adjective perspective dominates when discussing examples, so we choose to emphasize the adjective form and focus on enriching unenriched categories here.

Before giving a precise definition of enriched category and the enriched functors between them in §A.2, in §A.1 we study the category $\mathcal{V}$ that defines the base for enrichment in which the hom-objects ultimately live. The primary examples appearing in this text are $\mathcal{V} = \mathcal{C}at$ and $\mathcal{V} = s\mathcal{S}et$ – as well as the unenriched case $\mathcal{V} = \mathcal{S}et$ – each of which has the special property of being a cartesian closed category. Since there are some simplifications in enriching over a cartesian closed category, we grant ourselves the luxury of working explicitly with these basis.[4]

We continue in §A.3 with an introduction to enriched natural transformations and the enriched Yoneda lemma. These notions allow us to correctly state the

---

[2] To interpret the diagrams (A.0.1) in $\mathcal{V}$ one needs to specify an interpretation for the monoidal product "$\times$" and its unit object 1 (which is not displayed in the diagram). In the examples we consider, this product is the cartesian product and this unit is the terminal object.

[3] To quote [84] "Thinking from the two points of view simultaneously, it is essential that the constructed ordinary category be isomorphic to the ordinary category that one started out with. Either way, there is a conflict of notation between that preferred by category theorists and that in common use by 'working mathematicians' (to whom [81] is addressed)."

[4] More generally, a category can be enriched over a monoidal category [68], a bicategory [17], or a virtual double category [75], each generalizing the preceding bases for enrichment.

universal properties, involving an enriched natural isomorphism, that character-ize *cotensors* and *tensors* in §A.4 and *conical limits* and *colimits* in §A.5. These are each special cases of *weighted limits* and *colimits*, which are introduced in §A.6. We conclude in §A.7 with a general theory of "change of base" – the one part of the theory of enriched categories that is not covered in encyclopedic detail in [68], the original reference instead being [41] – which allows us to be more precise about the procedure by which a 2-category may be regarded as a simplicial category or by which a simplicial category may be quotiented to define a 2-category, as alluded to in Digression 1.4.2.

A.1 Cartesian Closed Categories

Throughout this text, the base category for enrichment is always taken to be a complete and cocomplete cartesian closed category:

DEFINITION A.1.1. A category $\mathcal{V}$ is **cartesian closed** when it

- admits finite products, or equivalently, a terminal object $1 \in \mathcal{V}$ and binary products and
- for each $v \in \mathcal{V}$, the functor $v \times - : \mathcal{V} \to \mathcal{V}$ admits a right adjoint $(-)^v : \mathcal{V} \to \mathcal{V}$.

LEMMA A.1.2. *In a cartesian closed category $\mathcal{V}$, the product bifunctor is the left adjoint of a two-variable adjunction, this being captured by a commutative triangle of natural isomorphisms*

$$
\begin{array}{ccc}
 & \mathcal{V}(a \times b, c) & \\
{\scriptstyle \cong} \swarrow & & \searrow {\scriptstyle \cong} \\
\mathcal{V}(a, c^b) & \underset{\cong}{} & \mathcal{V}(b, c^a)
\end{array}
\tag{A.1.3}
$$

Proof The family of functors $(-)^a : \mathcal{V} \to \mathcal{V}$ extend to bifunctors

$$(-)^- : \mathcal{V}^{op} \times \mathcal{V} \to \mathcal{V}$$

in a unique way so that the isomorphisms defining each adjunction $a \times - \dashv (-)^a$

$$\mathcal{V}(a \times b, c) \cong \mathcal{V}(b, c^a)$$

become natural in a (as well as b and c). The details are left as Exercise A.1.i or to [104, 4.3.6]. This defines the natural isomorphism on the right-hand side of (A.1.3). The natural isomorphism on the left-hand is defined by composing with the symmetry isomorphism $a \times b \cong b \times a$. The third natural isomorphism is taken to be the composite of these two. $\square$

Example A.1.4 (cartesian closed categories).

(i) The category of sets is cartesian closed, with B^A defined to be the set of functions from A to B. Transposition across the natural isomorphism (A.1.3) is referred to as "currying."

(ii) The category $\mathcal{C}at$ of small[5] categories is cartesian closed, with B^A defined to be the category of functors and natural transformations from A to B.

$$\mathcal{C}at(A, C^B) \cong \mathcal{C}at(A \times B, C) \cong \mathcal{C}at(B, C^A)$$

identifies natural transformations, which are arrows $2 \to C^A$ in the category of functors, with "directed homotopies" $A \times 2 \to C$.

(iii) For any small category $\mathcal{C}$, the category $\mathcal{S}et^{\mathcal{C}^{op}}$ is cartesian closed. By the Yoneda lemma for $F, G \in \mathcal{S}et^{\mathcal{C}^{op}}$, the value of G^F at $c \in \mathcal{C}$ must be defined by

$$G^F(c) \cong \mathcal{S}et^{\mathcal{C}^{op}}(\mathcal{C}(-, c), G^F) \cong \mathcal{S}et^{\mathcal{C}^{op}}(F \times \mathcal{C}(-, c), G)$$

to be the set of natural transformations $F \times \mathcal{C}(-, c) \Rightarrow G$. As proscribed by Lemma A.1.2, the action of G^F on a morphism $f : c \to c' \in \mathcal{C}$ is defined by precomposition with the corresponding natural transformation $f \circ - : \mathcal{C}(-, c) \Rightarrow \mathcal{C}(-, c')$. This defines the functor G^F. Since any functor $H \in \mathcal{S}et^{\mathcal{C}^{op}}$ is canonically a colimit of representables, this definition extends to the required natural isomorphism $\mathcal{S}et^{\mathcal{C}^{op}}(H, G^F) \cong \mathcal{S}et^{\mathcal{C}^{op}}(F \times H, G)$.

(iv) In particular taking $\mathcal{C} = \Delta$, the category of simplicial sets $s\mathcal{S}et := \mathcal{S}et^{\Delta^{op}}$ is cartesian closed.

The exponential b^a is frequently referred to as an *internal hom*. As this name suggests, the internal hom b^a can be viewed as a lifting of the hom-set $\mathcal{V}(a, b)$ along a functor that we now introduce.

Definition A.1.5. For any cartesian closed category $\mathcal{V}$, the **underlying set functor** is the functor

$$\mathcal{V} \xrightarrow{(-)_0 := \mathcal{V}(1, -)} \mathcal{S}et$$

represented by the terminal object $1 \in \mathcal{V}$.

Lemma A.1.6. *For any pair of objects $a, b \in \mathcal{V}$ in a cartesian closed category, the underlying set of the internal hom b^a is $\mathcal{V}(a, b)$, i.e.:*

$$(b^a)_0 \cong \mathcal{V}(a, b).$$

---

[5] In general, the category of categories whose sets of morphisms are bounded by a fixed inaccessible cardinal is cartesian closed in that Grothendieck universe.

Proof Combining Definition A.1.5 with (A.1.3):

$$(b^a)_0 := \mathcal{V}(1, b^a) \cong \mathcal{V}(1 \times a, b) \cong \mathcal{V}(a, b)$$

since there is a natural isomorphism $1 \times a \cong a$. ∎

It makes sense to ask whether an isomorphism of underlying sets can be "enriched" to lie in $\mathcal{V}$, that is, lifted along the underlying set functor $(-)_0 : \mathcal{V} \to Set$.

LEMMA A.1.7. *The natural isomorphisms* (A.1.3) *characterizing the defining two-variable adjunction of a cartesian closed category lift to* $\mathcal{V}$: *for any* $a, b, c \in \mathcal{V}$

$$
\begin{array}{ccc}
 & c^{a \times b} & \\
(c^b)^a & \cong & (c^a)^b
\end{array}
\tag{A.1.8}
$$

Proof This follows from Lemma A.1.2, the associativity of finite products, and the Yoneda lemma. To prove (A.1.8), it suffices to show that $c^{a \times b}$, $(c^b)^a$, and $(c^a)^b$ represent the same functor. By composing the sequence of natural isomorphisms

$$\mathcal{V}(x, (c^b)^a) \cong \mathcal{V}(x \times a, c^b) \cong \mathcal{V}((x \times a) \times b, c) \cong \mathcal{V}(x \times (a \times b), c)$$
$$\cong \mathcal{V}(x, c^{a \times b})$$
$$\cong \mathcal{V}((a \times b) \times x, c) \cong \mathcal{V}(a \times (b \times x), c) \cong \mathcal{V}(b \times x, c^a)$$
$$\cong \mathcal{V}(x, (c^a)^b),$$

we see that

$$\mathcal{V}(x, (c^b)^a) \cong \mathcal{V}(x, c^{a \times b}) \cong \mathcal{V}(x, (c^a)^b). \qquad \square$$

REMARK A.1.9. Note $(-)^1 : \mathcal{V} \to \mathcal{V}$ is naturally isomorphic to the identity functor – i.e., $b^1 \cong b$, – since it is right adjoint to a functor $- \times 1 : \mathcal{V} \to \mathcal{V}$ that is naturally isomorphic to the identity.

A complete and cocomplete cartesian closed category is a special case of a complete and cocomplete closed symmetric monoidal category, this being deemed a **cosmos** by Jean Bénabou, to signify that such bases are an ideal setting for enriched category theory. For obvious reasons, we will not use this term here and instead refer to "complete and cocomplete cartesian closed categories" to highlight some common features of the categories appearing in Example A.1.4.[6]

---

[6] There is a competing 2-categorical notion of (fibrational) "cosmos" due to Street [117] that is more similar to the notion we consider here, which was the direct inspiration for the terminology we introduce in Definition 1.2.1.

Exercises

EXERCISE A.1.i. Prove that in a cartesian closed category $\mathcal{V}$, the family of functors $(-)^a : \mathcal{V} \to \mathcal{V}$ extend to bifunctors

$$(-)^- : \mathcal{V}^{\mathrm{op}} \times \mathcal{V} \to \mathcal{V}$$

in a unique way so that the isomorphism defining each adjunction $a \times - \dashv (-)^a$

$$\mathcal{V}(a \times b, c) \cong \mathcal{V}(b, c^a)$$

becomes natural in a (as well as b and c).

EXERCISE A.1.ii. The data of a **closed symmetric monoidal category** generalizes Definition A.1.1 by replacing finite products by an arbitrary bifunctor $- \otimes - : \mathcal{V} \times \mathcal{V} \to \mathcal{V}$, replacing the terminal object by an object $I \in \mathcal{V}$, and requiring the additional specification of natural isomorphisms

$$a \otimes (b \otimes c) \underset{\alpha}{\cong} (a \otimes b) \otimes c \qquad I \otimes a \underset{\lambda}{\cong} a \underset{\rho}{\cong} a \otimes I \qquad a \otimes b \underset{\gamma}{\cong} b \otimes a$$

satisfying various coherence axioms [41, 66] (see also [69]). Provide a canonical construction of these isomorphisms in the special case of a cartesian closed category and explain why the coherence conditions are automatic.

A.2 Enriched Categories and Enriched Functors

We now briefly switch perspectives and explain the meaning of the noun phrase "enriched category" before discussing what is required to "enrich" an ordinary 1-category. From here through §A.6, we fix a complete and cocomplete cartesian closed category $(\mathcal{V}, \times, 1)$ to serve as the base for enrichment.

DEFINITION A.2.1. A $\mathcal{V}$-**enriched category** or $\mathcal{V}$-**category** $\mathcal{C}$ is given by:

- a collection of **objects**
- for each pair of objects $x, y \in \mathcal{C}$, an **hom-object** $\mathcal{C}(x, y) \in \mathcal{V}$
- for each $x \in \mathcal{C}$, a specified **identity element** encoded by a map $\mathrm{id}_x : 1 \to \mathcal{C}(x, x) \in \mathcal{V}$, and for each $x, y, z \in \mathcal{C}$, a specified **composition map** $\circ : \mathcal{C}(y, z) \times \mathcal{C}(x, y) \to \mathcal{C}(x, z) \in \mathcal{V}$ satisfying the associativity and unit

conditions, both commutative diagrams lying in $\mathcal{V}$:[7]

$$\mathcal{C}(y,z) \times \mathcal{C}(x,y) \times \mathcal{C}(w,x) \xrightarrow{\circ \times \mathrm{id}} \mathcal{C}(x,z) \times \mathcal{C}(w,x)$$

$$\mathrm{id} \times \circ \downarrow \qquad\qquad\qquad\qquad\qquad \downarrow \circ$$

$$\mathcal{C}(y,z) \times \mathcal{C}(w,y) \xrightarrow{\qquad\circ\qquad} \mathcal{C}(w,z)$$

$$\mathcal{C}(x,y) \xrightarrow{\mathrm{id} \times \mathrm{id}_x} \mathcal{C}(x,y) \times \mathcal{C}(x,x)$$

$$\mathrm{id}_y \times \mathrm{id} \downarrow \qquad\searrow \mathrm{id} \qquad\qquad \downarrow \circ$$

$$\mathcal{C}(y,y) \times \mathcal{C}(x,y) \xrightarrow{\qquad\circ\qquad} \mathcal{C}(x,y)$$

Evidently from the diagrams of (A.0.1), a locally small 1-category defines a category enriched in *Set*. The underlying set functor of Definition A.1.5 can be used to define the "underlying category" of an enriched category.

DEFINITION A.2.2. If $\mathcal{C}$ is a $\mathcal{V}$-category, its **underlying category** $\mathcal{C}_0$ is the 1-category with the same collection of objects and with hom-sets defined by applying the underlying set functor $(-)_0 : \mathcal{V} \to$ *Set* to the hom-objects $\mathcal{C}(x,y) \in \mathcal{V}$. For the most part, we write "$\mathcal{C}(x,y)$" for both the hom-object and the hom-set and use our words to disambiguate, but when necessary "$\mathcal{C}(x,y)_0$" is also commonly used notation for the hom-set of the underlying category.

Note the identity arrow $\mathrm{id}_x : 1 \to \mathcal{C}(x,x)$ of the $\mathcal{V}$-category is by definition an element of the hom-set $\mathcal{C}(x,x)_0 := \mathcal{V}(1, \mathcal{C}(x,x))$. The composite of two arrows $f : 1 \to \mathcal{C}(x,y)$ and $g : 1 \to \mathcal{C}(y,z)$ in the underlying category is defined to be the arrow constructed as the composite

$$1 \xrightarrow{g \times f} \mathcal{C}(y,z) \times \mathcal{C}(x,y) \xrightarrow{\circ} \mathcal{C}(x,z)$$

In analogy with the discussion around Definition A.1.5, when one speaks of "enriching" an a priori unenriched category $\mathcal{C}$ over $\mathcal{V}$, the task is to define a $\mathcal{V}$-enriched category as in Definition A.2.1 whose underlying category recovers $\mathcal{C}$. When $\mathcal{V} = \mathcal{C}at$, the task is to define a 2-category whose underlying 1-category is the one given. When $\mathcal{V} = s\mathcal{S}et$, the task is to define simplicial hom-sets of n-arrows so that the 0-arrows are the ones given. When a simplicially enriched category $\mathcal{C}$ is encoded as a simplicial object $\mathcal{C}_{\bullet}$ in $\mathcal{C}at$ as explained in Digression 1.2.4, its underlying category is the category $\mathcal{C}_0$, further justifying the notion introduced in Definition A.2.2.

---

[7] These diagrams suppress the associativity and unit natural isomorphisms involving the product bifunctors $\times$ and its unit object 1. In a cartesian closed category these are *canonical* – rather than given by extra data, as is the case in the more general *closed symmetric monoidal category* (see Exercise A.1.ii).

For example, a cartesian closed category $\mathcal{V}$ as in Definition A.1.1 can be enriched to define a $\mathcal{V}$-category.

LEMMA A.2.3. *A cartesian closed category $\mathcal{V}$ defines a $\mathcal{V}$-category whose:*

- *objects are the objects of $\mathcal{V}$,*
- *hom-object in $\mathcal{V}$ from a to b is the internal hom b^a, and*
- *the identity map $\mathrm{id}_a : 1 \to a^a$ and composition map $\circ : c^b \times b^a \to c^a$ are defined to be the transposes of*

$$1 \times a \overset{\simeq}{\to} a \qquad \text{and} \qquad c^b \times b^a \times a \xrightarrow{\mathrm{id} \times \mathrm{ev}} c^b \times b \xrightarrow{\mathrm{ev}} c$$

the latter defined using the counit ev of the cartesian closure adjunction.

Proof   The task is to verify the commutative diagrams of (A.2.1) in $\mathcal{V}$ and then observe that Lemma A.1.6 reveals that the underlying category of the $\mathcal{V}$-category defined by the statement is the 1-category $\mathcal{V}$. We leave the identity conditions to the reader and verify associativity.

The definition of the composition map as an adjoint transpose implies that *its* adjoint transpose, the top-right composite below, is given by the left-bottom composite:

$$
\begin{array}{ccc}
c^b \times b^a \times a & \xrightarrow{\circ \times \mathrm{id}} & c^a \times a \\
{\scriptstyle \mathrm{id} \times \mathrm{ev}} \downarrow & & \downarrow {\scriptstyle \mathrm{ev}} \\
c^b \times b & \xrightarrow[\mathrm{ev}]{} & c
\end{array}
\qquad (A.2.4)
$$

The associativity diagram below-left commutes if and only if the transposed diagram appearing as the outer boundary composite below-right commutes:

$$
\begin{array}{ccc}
d^c \times c^b \times b^a & \overset{\circ \times \mathrm{id}}{\to} & d^b \times b^a \\
{\scriptstyle \mathrm{id} \times \circ} \downarrow & & \downarrow {\scriptstyle \circ} \\
d^c \times c^a & \xrightarrow[\circ]{} & d^a
\end{array}
$$

$$
\begin{array}{ccccc}
d^c \times c^b \times b^a \times a & \xrightarrow{\circ \times \mathrm{id} \times \mathrm{id}} & & d^b \times b^a \times a \\
{\scriptstyle \mathrm{id} \times \mathrm{id} \times \mathrm{ev}} \searrow & & & \downarrow {\scriptstyle \mathrm{id} \times \mathrm{ev}} \\
{\scriptstyle \mathrm{id} \times \circ \times \mathrm{id}} \downarrow & d^c \times c^b \times b & \overset{\circ \times \mathrm{id}}{\to} & d^b \times b \\
& {\scriptstyle \mathrm{id} \times \mathrm{ev}} \downarrow & & \downarrow {\scriptstyle \mathrm{ev}} \\
d^c \times c^a \times a & \xrightarrow[\mathrm{id} \times \mathrm{ev}]{} & d^c \times c & \xrightarrow[\mathrm{ev}]{} & d
\end{array}
$$

which follows from bifunctoriality of $\times$ and two instances of the commutative square above.   □

DEFINITION A.2.5. The **free $\mathcal{V}$-category** on a 1-category $\mathcal{C}$ has the same collection of objects with the hom-objects defined to be coproducts $\amalg_{\mathcal{C}(x,y)} 1$ of the terminal object $1 \in \mathcal{V}$ indexed by the hom-set $\mathcal{C}(x,y)$. The identity map $\mathrm{id}_x : 1 \to \amalg_{\mathcal{C}(x,x)} 1$ is given by the inclusion of the component indexed by the identity arrow, the composition map defined by acting by the composition function on the indexing sets:

$$\amalg_{\mathcal{C}(y,z)}1 \times \amalg_{\mathcal{C}(x,y)}1 \cong \amalg_{\mathcal{C}(y,z)\times\mathcal{C}(x,y)}1 \xrightarrow{\amalg_\circ 1} \amalg_{\mathcal{C}(x,z)}1$$

For example, free $\mathcal{C}at$-enriched categories are those with no nonidentity 2-cells, while free $s\mathcal{S}et$-enriched categories are those with no nondegenerate arrows in positive dimensions. We use the same notation for the 1-category $\mathcal{C}$ and the free $\mathcal{V}$-category it generates, using language to disambiguate.

DEFINITION A.2.6. A $\mathcal{V}$-**enriched functor** or $\mathcal{V}$-**functor** $F: \mathcal{C} \to \mathcal{D}$ is given by

- a mapping on objects that carries each $x \in \mathcal{C}$ to some $Fx \in \mathcal{D}$
- for each pair of objects $x, y \in \mathcal{C}$, an internal action on the hom-objects given by a morphism $F_{x,y}: \mathcal{C}(x,y) \to \mathcal{D}(Fx, Fy) \in \mathcal{V}$ so that the $\mathcal{V}$-functoriality diagrams commute:

$$
\begin{array}{ccc}
\mathcal{C}(y,z) \times \mathcal{C}(x,y) & \xrightarrow{\;\circ\;} & \mathcal{C}(x,z) \\
{\scriptstyle F_{y,z}\times F_{x,y}}\downarrow & & \downarrow{\scriptstyle F_{x,z}} \\
\mathcal{D}(Fy,Fz) \times \mathcal{D}(Fx,Fy) & \xrightarrow{\;\circ\;} & \mathcal{D}(Fx,Fz)
\end{array}
\qquad
\begin{array}{ccc}
1 & \xrightarrow{\;\mathrm{id}_x\;} & \mathcal{C}(x,x) \\
& {\scriptstyle \mathrm{id}_{Fx}}\searrow & \downarrow{\scriptstyle F_{x,x}} \\
& & \mathcal{D}(Fx,Fx)
\end{array}
$$

A prototypical example is given by the representable functors:

EXAMPLE A.2.7. For any $\mathcal{V}$-category $\mathcal{C}$ and object $c \in \mathcal{C}$, the **enriched representable** $\mathcal{V}$-**functor** $\mathcal{C}(c, -): \mathcal{C} \to \mathcal{V}$ is defined on objects by the assignment $x \in \mathcal{C} \mapsto \mathcal{C}(c, x) \in \mathcal{V}$ and whose internal action hom-objects is defined to be the adjoint transpose of the internal composition map for $\mathcal{C}$.

$$\mathcal{C}(x,y) \xrightarrow{\mathcal{C}(c,-)_{x,y}} \mathcal{C}(c,y)^{\mathcal{C}(c,x)} \quad\rightsquigarrow\quad \mathcal{C}(x,y) \times \mathcal{C}(c,x) \xrightarrow{\;\circ\;} \mathcal{C}(c,y)$$

The $\mathcal{V}$-functoriality diagrams are transposes of associativity and identity diagrams in $\mathcal{C}$. The contravariant enriched representable functors are defined similarly (see Exercise A.2.iii).

REMARK A.2.8. An enriched representable functor can be thought of as a "two-step" enrichment of the corresponding unenriched representable functor: the first step enriches the hom-sets to hom-objects in $\mathcal{V}$ and the second step enriches the composition function to an internal composition map in $\mathcal{V}$. To enrich a 1-category $\mathcal{C}$ to a $\mathcal{V}$-category with $\mathcal{C}$ as its underlying 1-category requires more than simply a lift of the hom bifunctor:

$$
\begin{array}{ccc}
\mathcal{C}^{\mathrm{op}} \times \mathcal{C} & \dashrightarrow{\mathcal{C}(-,-)} & \mathcal{V} \\
{\scriptstyle \mathcal{C}(-,-)}\searrow & {\scriptstyle \cong} & \swarrow{\scriptstyle (-)_0} \\
& \mathcal{S}et &
\end{array}
$$

In addition, the $\mathcal{V}$-valued representable $\mathcal{C}(-,-)\colon \mathcal{C}^{\mathrm{op}} \times \mathcal{C} \to \mathcal{V}$ must be a $\mathcal{V}$-bifunctor in the sense of Exercise A.2.iv, which tells us that the composition map may be defined internally to $\mathcal{V}$ and the identity and associativity laws hold there.

Both of the constructions of underlying unenriched categories and free categories are functorial (see Exercise A.2.v). The relationship between these constructions is summarized by the following proposition, whose proof is left as an exercise because it is strengthened by Corollary A.7.6.

PROPOSITION A.2.9. *The free $\mathcal{V}$-category functor defines a fully faithful left adjoint to the underlying category functor. Consequently, a $\mathcal{V}$-category is free just when it is isomorphic to the free category on its underlying category via the counit of this adjunction.*

Proof   Exercise A.2.vi.   □

Exercises

EXERCISE A.2.i.   Verify that the underlying category of an enriched category described in Definition A.2.2 is indeed a category.

EXERCISE A.2.ii.   Verify the unit condition left to the reader in the proof of Lemma A.2.3.

EXERCISE A.2.iii.   Define the **opposite** of a $\mathcal{V}$-category and dualize Example A.2.7 to define contravariant enriched representable functors.

EXERCISE A.2.iv.

(i) Define the **cartesian product** of two $\mathcal{V}$-categories.
(ii) Define a **multivariable $\mathcal{V}$-functor**.
(iii) Use these notions to show that any $\mathcal{V}$-category $\mathcal{C}$ comes equipped with a canonical $\mathcal{V}$-bifunctor $\mathcal{C}(-,-)\colon \mathcal{C}^{\mathrm{op}} \times \mathcal{C} \to \mathcal{V}$ that restricts to the co- and contravariant representable functors.

EXERCISE A.2.v.

(i) Define the underlying functor of an enriched functor.
(ii) Prove that the passage from enriched functors to underlying unenriched functors is functorial.
(iii) Define the free enriched functor on an unenriched functor.
(iv) Prove the passage from unenriched functors to free enriched functors is functorial.

EXERCISE A.2.vi (A.7.6).   Prove Proposition A.2.9.

A.3 Enriched Natural Transformations and the Enriched Yoneda Lemma

Recall an (unenriched) natural transformation $\alpha : F \Rightarrow G$ between parallel functors $F, G : \mathcal{C} \to \mathcal{D}$ is given by:

- the data of an arrow $\alpha_x \in \mathcal{D}(Fx, Gx)$ for each $x \in \mathcal{C}$
- subject to the condition that for each morphism $f \in \mathcal{C}(x, y)$, the diagram

$$
\begin{array}{ccc}
Fx & \xrightarrow{\alpha_x} & Gx \\
{\scriptstyle Ff}\downarrow & & \downarrow{\scriptstyle Gf} \\
Fy & \xrightarrow[\alpha_y]{} & Gy
\end{array}
\qquad (A.3.1)
$$

commutes in $\mathcal{D}$.

This enriches to the notion of a $\mathcal{V}$-natural transformation whose data is exactly the same – a family of arrows in the underlying category of $\mathcal{D}$ indexed by the objects of $\mathcal{C}$ – but with a stronger $\mathcal{V}$-naturality condition expressed by internalizing the naturality condition (A.3.1).

DEFINITION A.3.2. A $\mathcal{V}$-**enriched natural transformation** or $\mathcal{V}$-**natural transformation** $\alpha : F \Rightarrow G$ between $\mathcal{V}$-enriched functors $F, G : \mathcal{C} \to \mathcal{D}$ is given by:

- an arrow $\alpha_x : 1 \to \mathcal{D}(Fx, Gx)$ for each $x \in \mathcal{C}$
- so that for each pair of objects $x, y \in \mathcal{C}$, the following $\mathcal{V}$-naturality square commutes in $\mathcal{V}$:

$$
\begin{array}{ccc}
\mathcal{C}(x, y) & \xrightarrow{\alpha_y \times F_{x,y}} & \mathcal{D}(Fy, Gy) \times \mathcal{D}(Fx, Fy) \\
{\scriptstyle G_{x,y} \times \alpha_x}\downarrow & & \downarrow{\scriptstyle \circ} \\
\mathcal{D}(Gx, Gy) \times \mathcal{D}(Fx, Gx) & \xrightarrow[\circ]{} & \mathcal{D}(Fx, Gy)
\end{array}
\qquad (A.3.3)
$$

EXAMPLE A.3.4. An arrow $f : 1 \to \mathcal{C}(x, y)$ in the underlying category of a $\mathcal{V}$-category $\mathcal{C}$ defines a $\mathcal{V}$-natural transformation $- \circ f : \mathcal{C}(y, -) \Rightarrow \mathcal{C}(x, -)$ between the enriched representable functors whose component at $z \in \mathcal{C}$ is defined by evaluating the adjoint transpose of the composition map at f:

$$
1 \xrightarrow{\;f\;} \mathcal{C}(x, y) \xrightarrow{\mathcal{C}(-, z)_{x,y}} \mathcal{C}(x, z)^{\mathcal{C}(y, z)}
$$

The required $\mathcal{V}$-naturality square is obtained by evaluating one component of the associativity diagram for $\mathcal{C}$ at f.

$\mathcal{V}$-natural transformations compose as unenriched natural transformations do:

DEFINITION A.3.5. The **vertical composite** $\beta \cdot \alpha$ of $\mathcal{V}$-natural transformations $\alpha \colon F \Rightarrow G$ and $\beta \colon G \Rightarrow H$, both from $\mathcal{C}$ to $\mathcal{D}$, has component $(\beta \cdot \alpha)_x$ at $x \in \mathcal{C}$ defined by the composite

$$1 \xrightarrow{\beta_x \times \alpha_x} \mathcal{D}(Gx, Hx) \times \mathcal{D}(Fx, Gx) \xrightarrow{\quad \circ \quad} \mathcal{D}(Fx, Hx)$$

The **horizontal composite** $\gamma * \alpha$ of $\alpha \colon F \Rightarrow G$ from $\mathcal{C}$ to $\mathcal{D}$ and $\gamma \colon H \Rightarrow K$ from $\mathcal{D}$ to $\mathcal{E}$ has component $(\gamma * \alpha)_x$ at $x \in \mathcal{C}$ defined by the common composite

$$
\begin{array}{ccc}
1 \xrightarrow{\quad \alpha_x \quad} \mathcal{D}(Fx, Gx) & \xrightarrow{\gamma_{Gx} \times H_{Fx,Gx}} & \mathcal{E}(KGx, HGx) \times \mathcal{E}(HFx, HGx) \\
{\scriptstyle K_{Fx,Gx} \times \gamma_{Fx}} \downarrow & & \downarrow {\scriptstyle \circ} \\
\mathcal{E}(KFx, KGx) \times \mathcal{E}(HFx, KFx) & \xrightarrow{\quad \circ \quad} & \mathcal{D}(HFx, KGx)
\end{array}
$$

which is well-defined by $\mathcal{V}$-naturality of γ. The $\mathcal{V}$-naturality of these constructions is left to Exercise A.3.i.

The data of the **underlying natural transformation** of a $\mathcal{V}$-natural transformation is given by the same family of arrows. The unenriched naturality condition (A.3.1) is obtained by evaluating the enriched naturality condition (A.3.3) at an underlying arrow $f \colon 1 \to \mathcal{C}(x, y)$. In particular, the middle four interchange rule (see Definition B.1.1) for horizontal and vertical composition of $\mathcal{V}$-natural transformations follows from the middle four interchange rule for horizontal and vertical composition of unenriched natural transformations for the data of the latter determines the data of the former. Consequently, Exercise A.3.i implies that:

COROLLARY A.3.6. *For any cartesian closed category $\mathcal{V}$, there is a 2-category $\mathcal{V}\text{-}\mathcal{C}at$ of $\mathcal{V}$-categories, $\mathcal{V}$-functors, and $\mathcal{V}$-natural transformations.*

We now turn our attention to the $\mathcal{V}$-enriched Yoneda lemma, which we present in several forms. One role of the Yoneda lemma is to give a representable characterization of isomorphic objects in $\mathcal{C}$. When $\mathcal{C}$ is a $\mathcal{V}$-category, this has several forms. The notion of $\mathcal{V}$-**natural isomorphism** referred to in the following result is defined to be a $\mathcal{V}$-natural transformation $\alpha \colon F \Rightarrow G$ with an inverse $\alpha^{-1} \colon G \Rightarrow F$ for vertical composition.

LEMMA A.3.7. *For objects x, y in a $\mathcal{V}$-category $\mathcal{C}$ the following are equivalent:*

  (i) *x and y are isomorphic as objects of the underlying category of $\mathcal{C}$.*
 (ii) *The $\mathcal{S}et$-valued unenriched representables $\mathcal{C}(x, -), \mathcal{C}(y, -) \colon \mathcal{C} \to \mathcal{S}et$ are naturally isomorphic.*
(iii) *The $\mathcal{V}$-valued unenriched representables $\mathcal{C}(x, -), \mathcal{C}(y, -) \colon \mathcal{C} \to \mathcal{V}$ are naturally isomorphic.*

(iv) *The V-valued V-functors $\mathcal{C}(x, -), \mathcal{C}(y, -) \colon \mathcal{C} \to V$ are V-naturally isomorphic.*

Proof Applying the underlying category 2-functor $(-)_0 \colon V\text{-}\mathcal{C}at \to \mathcal{C}at$, the fourth statement implies the third. The third statement implies the second by whiskering with the underlying set functor $(-)_0 \colon V \to \mathcal{S}et$. The second statement implies the first by the unenriched Yoneda lemma; this is still the main point. Finally, the first statement implies the last by a direct construction: if $f \colon 1 \to \mathcal{C}(x, y)$ and $g \colon 1 \to \mathcal{C}(y, x)$ define an isomorphism in the underlying category of $\mathcal{C}$, the corresponding representable V-natural transformations of Example A.3.4 define a V-natural isomorphism. $\square$

Lemma A.3.7 defines a common notion of **isomorphism** between two objects of an enriched category, which turns out to be no different than the usual unenriched notion of isomorphism. This can be thought of as defining a "cheap" form of the enriched Yoneda lemma. The full form of the V-Yoneda lemma enriches the usual statement – a natural isomorphism between the set of natural transformations whose domain is a representable functor to the set defined by evaluating the codomain at the representing object – to an isomorphism in V. The first step to make this precise is to enrich the set of V-natural transformations between a parallel pair of V-functors to an object of V.

DEFINITION A.3.8. Let V be a complete cartesian closed category and consider a parallel pair of V-functors $F, G \colon \mathcal{C} \to \mathcal{D}$, with $\mathcal{C}$ a small V-category. Then the V-**object of** V-**natural transformations** is defined by the equalizer diagram

$$\mathcal{D}^{\mathcal{C}}(F, G) \rightarrowtail \prod_{z \in \mathcal{C}} \mathcal{D}(Fz, Gz) \rightrightarrows \prod_{x, y \in \mathcal{C}} \mathcal{D}(Fx, Gy)^{\mathcal{C}(x, y)}$$

where one map to $\mathcal{D}(Fx, Gy)^{\mathcal{C}(x, y)}$ in the equalizer diagram is defined by projecting to $\mathcal{D}(Fx, Gx)$, applying the internal action of G on arrows, and then composing, while the other is defined by projecting to $\mathcal{D}(Fy, Gy)$, applying the internal action of F on arrows, and then composing:

$$\prod_{z \in \mathcal{C}} \mathcal{D}(Fz, Gz) \;\; \overset{\pi}{\nearrow} \;\; \begin{array}{c} \mathcal{D}(Fx, Gx) \overset{\circ}{\to} \mathcal{D}(Fx, Gy)^{\mathcal{D}(Gx, Gy)} \;\xrightarrow{- \circ G_{x,y}}\; \\ \qquad\qquad\qquad\qquad\qquad \mathcal{D}(Fx, Gy)^{\mathcal{C}(x, y)} \\ \overset{\pi}{\searrow} \; \mathcal{D}(Fy, Gy) \underset{\circ}{\to} \mathcal{D}(Fx, Gy)^{\mathcal{D}(Fx, Fy)} \;\xrightarrow{- \circ F_{x,y}}\; \end{array}$$

LEMMA A.3.9. *The underlying set of the V-object of V-natural transformations $V^{\mathcal{C}}(F, G)$ is the set of V-natural transformations from F to G.*

Proof By its defining universal property, elements of the underlying set of

$\mathcal{V}^{\mathcal{C}}(F, G)$ correspond to maps $\alpha: 1 \to \prod_{z \in \mathcal{C}} \mathcal{V}(Fz, Gz)$ that equalize the parallel pair of maps described in Definition A.3.8. The map α defines the components of a $\mathcal{V}$-natural transformation $\alpha: F \Rightarrow G$ and the commutativity condition transposes to (A.3.3). $\qquad\qquad\square$

The Yoneda lemma can be expressed by the slogan "evaluation at the identity is an isomorphism," but since in the enriched context the enriched object of natural transformations is defined via a limit, it is easier to define the map that induces a natural transformation instead. Given an object $a \in \mathcal{A}$ in a small $\mathcal{V}$-category $\mathcal{A}$ and a $\mathcal{V}$-functor $F: \mathcal{A} \to \mathcal{V}$, the internal action of F on arrows transposes to define a map that equalizes the parallel pair

$$Fa \overset{F_{a,-}}{\dashrightarrow} \prod_{z \in \mathcal{A}} Fz^{\mathcal{A}(a,z)} \rightrightarrows \prod_{x,y \in \mathcal{A}} Fy^{\mathcal{A}(a,x) \times \mathcal{A}(x,y)} \qquad (A.3.10)$$

and thus induces a canonical map $Fa \to \mathcal{V}^{\mathcal{A}}(\mathcal{A}(a,-), F)$ in $\mathcal{V}$.

THEOREM A.3.11 (enriched Yoneda lemma). *For any small $\mathcal{V}$-category $\mathcal{A}$, object $a \in \mathcal{A}$, and $\mathcal{V}$-functor $F: \mathcal{A} \to \mathcal{V}$, the canonical map defines an isomorphism in $\mathcal{V}$*

$$Fa \overset{\cong}{\to} \mathcal{V}^{\mathcal{A}}(\mathcal{A}(a,-), F)$$

that is $\mathcal{V}$-natural in both a and F.

Proof To prove the isomorphism, it suffices to verify that (A.3.10) is a limit cone. To that end consider another cone over the parallel pair

$$\upsilon \overset{\lambda}{\to} \prod_{z \in \mathcal{A}} Fz^{\mathcal{A}(a,z)} \rightrightarrows \prod_{x,y \in \mathcal{A}} Fy^{\mathcal{A}(a,x) \times \mathcal{A}(x,y)}$$

and define a candidate factorization by evaluating the transpose of the component λ_a at id_a:

$$\lambda_a(\mathrm{id}_a) := \upsilon \overset{\mathrm{id}_a \times \upsilon}{\longrightarrow} \mathcal{A}(a, a) \times \upsilon \overset{\lambda_a}{\longrightarrow} Fa$$

To see that $\lambda_a(\mathrm{id}_a): \upsilon \to Fa$ indeed defines a factorization of λ through the limit cone, it suffices to show commutativity at each component $Fz^{\mathcal{A}(a,z)}$ of the product, which we verify in transposed form:

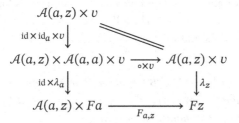

The upper triangle commutes by the identity law for $\mathcal{A}$ while the bottom square commutes because λ defines a cone over the parallel pair. Uniqueness of the factorization $\lambda_a(\mathrm{id}_a)$ follows from the same diagram by taking $z = a$ and evaluating at id_a.

The verification of the $\mathcal{V}$-naturality of $Fa \cong \mathcal{V}^{\mathcal{A}}(\mathcal{A}(a, -), F)$ is left to the reader or to [68, §2.4]. $\qquad\square$

Passing to underlying sets:

COROLLARY A.3.12. *For any small $\mathcal{V}$-category $\mathcal{A}$, object $a \in \mathcal{A}$, and $\mathcal{V}$-functor $F \colon \mathcal{A} \to \mathcal{V}$, there is a natural bijection between $\mathcal{V}$-natural transformations $\alpha \colon \mathcal{A}(a, -) \Rightarrow F$ and elements $u \colon 1 \to Fa$ in the underlying set of Fa implemented by evaluating the component at $a \in \mathcal{A}$ at the identity id_a.* $\qquad\square$

This gives a criterion for establishing the representability of a $\mathcal{V}$-functor by presenting the minimal data required to establish the defining $\mathcal{V}$-natural isomorphism.

COROLLARY A.3.13. *For a $\mathcal{V}$-functor $F \colon \mathcal{A}^{\mathrm{op}} \to \mathcal{V}$ and an object $a \in \mathcal{A}$ the following are equivalent and define what it means for a to **represent** F:*

(i) *There exists an isomorphism $\mathcal{A}(x, a) \cong Fx$ in $\mathcal{V}$ that is $\mathcal{V}$-natural in $x \in \mathcal{A}$.*

(ii) *There exists an element $u \colon 1 \to Fa$ in the underlying set of Fa so that the composite map*

$$\mathcal{A}(x, a) \xrightarrow{\ F_{x,a}\ } \mathcal{V}(Fa, Fx) \xrightarrow{\ -\circ u\ } \mathcal{V}(1, Fx) \cong Fx$$

defines an isomorphism in $\mathcal{V}$ for all $x \in \mathcal{A}$.

Proof By Corollary A.3.12 the element $u \colon 1 \to Fa$ in the underlying set of Fa determines a unique $\mathcal{V}$-natural transformation $\mathcal{A}(-, a) \Rightarrow F$ whose component at $x \in \mathcal{A}$ is the map of the statement. Thus, the universal element u defines a $\mathcal{V}$-natural isomorphism and not just a $\mathcal{V}$-natural transformation just when the map of the statement is an isomorphism. $\qquad\square$

Since we have assumed our bases for enrichment to be cartesian closed, the 2-category $\mathcal{V}\text{-}\mathcal{C}at$ admits finite products, allowing us to define multivariable $\mathcal{V}$-functors (see Exercise A.2.iv). The following result implies that the structures characterized by $\mathcal{V}$-natural isomorphisms in §A.4 and §A.5 assemble into $\mathcal{V}$-functors.

PROPOSITION A.3.14. *Let* $M \colon \mathcal{B}^{\mathrm{op}} \times \mathcal{A} \to \mathcal{V}$ *be a $\mathcal{V}$-functor so that for each* $a \in \mathcal{A}$, *the $\mathcal{V}$-functor* $M(-, a) \colon \mathcal{B}^{\mathrm{op}} \to \mathcal{V}$ *is represented by some* $Fa \in \mathcal{B}$, *meaning there exists a $\mathcal{V}$-natural isomorphism*

$$\mathcal{B}(b, Fa) \cong M(b, a).$$

Then there is a unique way of extending the mapping $a \in \mathcal{A} \mapsto Fa \in \mathcal{B}$ *to a $\mathcal{V}$-functor* $F \colon \mathcal{A} \to \mathcal{B}$ *so that the isomorphisms are $\mathcal{V}$-natural in* $a \in \mathcal{A}$ *as well as* $b \in \mathcal{B}$.

Proof   By the Yoneda lemma in the form of Corollary A.3.12, to define a family of isomorphisms $\alpha_{b,a} \colon \mathcal{B}(b, Fa) \cong M(b, a)$ for each $a \in \mathcal{A}$ that are $\mathcal{V}$-natural in $b \in \mathcal{B}$ is to define a family of elements $\eta_a \colon 1 \to M(Fa, a)$ for each $a \in \mathcal{A}$ that satisfy the condition of Corollary A.3.13(ii). By the $\mathcal{V}$-naturality statement in the Yoneda lemma, for the isomorphism $\alpha_{b,a}$ to be $\mathcal{V}$-natural in a is equivalent to the family of elements $\eta_a \colon 1 \to M(Fa, a)$ being "extraordinarily" $\mathcal{V}$-natural in a. What this means is that for any pair of objects $a, a' \in \mathcal{A}$, the outer square commutes:

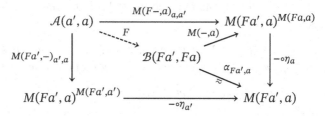

As exhibited by the top triangle, the top-horizontal map $M(F-, a)_{a',a}$ factors through the internal action of F on arrows, which we seek to define, but note that the composite of the other factor with the right vertical map is the natural isomorphism $\alpha_{Fa',a}$, as exhibited by the right triangle. Thus, there is a unique way to define $F_{a',a}$ making the extraordinary $\mathcal{V}$-naturality square commute, namely, by ensuring that the lower left quadrilateral commutes, which is exactly the claim. $\mathcal{V}$-functoriality of these internal action maps for F follows from $\mathcal{V}$-functoriality of M in the $\mathcal{A}$ variable.   □

We close this section with some applications of the enriched Yoneda lemma. The correct notions of $\mathcal{V}$-enriched equivalence or $\mathcal{V}$-enriched adjunction are given by interpreting the standard 2-categorical notions of equivalence and adjunction in the 2-category $\mathcal{V}$-$\mathcal{C}at$. For sake of contrast, we present both notions in an alternate form here and leave it to the reader to apply Theorem A.3.11 to relate these to the 2-categorical notions.

DEFINITION A.3.15. A pair of $\mathcal{V}$-categories $\mathcal{C}$ and $\mathcal{D}$ are $\mathcal{V}$-**equivalent** if there exists a $\mathcal{V}$-functor $F \colon \mathcal{C} \to \mathcal{D}$ that is

- $\mathcal{V}$-**fully faithful**: each $F_{x,y} \colon \mathcal{C}(x, y) \to \mathcal{D}(Fx, Fy)$ is an isomorphism in $\mathcal{V}$ and
- essentially surjective on objects: each $d \in \mathcal{D}$ is isomorphic to Fc for some $c \in \mathcal{C}$.

DEFINITION A.3.16. A $\mathcal{V}$-**adjunction** is given by a pair of $\mathcal{V}$-functors $F \colon \mathcal{B} \to \mathcal{A}$ and $U \colon \mathcal{A} \to \mathcal{B}$ together with isomorphisms

$$\mathcal{A}(Fb, a) \cong \mathcal{B}(b, Ua)$$

that are $\mathcal{V}$-natural in both $a \in \mathcal{A}$ and $b \in \mathcal{B}$.

REMARK A.3.17. By Proposition A.3.14, a $\mathcal{V}$-functor $U \colon \mathcal{A} \to \mathcal{B}$ admits a $\mathcal{V}$-left adjoint if and only if each $\mathcal{B}(b, U-) \colon \mathcal{A} \to \mathcal{V}$ is represented by some object $Fb \in \mathcal{A}$, in which case the data of the $\mathcal{V}$-natural isomorphism $\mathcal{A}(Fb, -) \cong \mathcal{B}(b, U-)$ equips $b \in \mathcal{B} \mapsto Fb \in \mathcal{A}$ with the structure of a $\mathcal{V}$-functor. Dual remarks construct enriched right adjoints to a given $\mathcal{V}$-functor.

Exercises

EXERCISE A.3.i. Verify that the vertical and horizontal composites of Definition A.3.5 are $\mathcal{V}$-natural.

EXERCISE A.3.ii. Suppose $F \colon \mathcal{C} \to \mathcal{D}$ is a $\mathcal{V}$-functor. Prove that the map

$$\mathcal{C}(x, y) \xrightarrow{\ F_{x,y}\ } \mathcal{D}(Fx, Fy)$$

is $\mathcal{V}$-natural in both x and y.

EXERCISE A.3.iii. When $\mathcal{V}$ is complete, show for any pair of $\mathcal{V}$-categories $\mathcal{C}$ and $\mathcal{D}$, with $\mathcal{C}$ small, that Definition A.3.8 makes $\mathcal{D}^{\mathcal{C}}$ into a $\mathcal{V}$-category.

EXERCISE A.3.iv. Verify the $\mathcal{V}$-naturality statement in Theorem A.3.11.

EXERCISE A.3.v. Use Corollary A.3.12 to show that the notions of $\mathcal{V}$-equivalence and $\mathcal{V}$-adjunction given in Definitions A.3.15 and A.3.16 are equivalent to the 2-categorical notions in $\mathcal{V}$-$\mathcal{C}at$.

A.4 Tensors and Cotensors

A $\mathcal{V}$-category $\mathcal{C}$ admits *tensors* just when for all $c \in \mathcal{C}$, the covariant representable functor $\mathcal{C}(c, -) \colon \mathcal{C} \to \mathcal{V}$ admits a left $\mathcal{V}$-adjoint $- \otimes c \colon \mathcal{V} \to \mathcal{C}$. Dually, a $\mathcal{V}$-category $\mathcal{C}$ admits *cotensors* just when the contravariant representable functor $\mathcal{C}(-, c) \colon \mathcal{C}^{\mathrm{op}} \to \mathcal{V}$ admits a mutual right $\mathcal{V}$-adjoint $c^- \colon \mathcal{V}^{\mathrm{op}} \to \mathcal{C}$. The aim in this section is to introduce both constructions formally. As we shall discover, the presence of tensors or cotensors is useful when enriching functors between the underlying categories of enriched categories.

DEFINITION A.4.1. A $\mathcal{V}$-category $\mathcal{C}$ is **cotensored** if, for all $v \in \mathcal{V}$ and $c \in \mathcal{C}$, the $\mathcal{V}$-functor $\mathcal{C}(-, c)^v \colon \mathcal{C}^{\mathrm{op}} \to \mathcal{V}$ is represented by an object $c^v \in \mathcal{C}$, i.e., there exists an isomorphism

$$\mathcal{C}(x, c^v) \cong \mathcal{C}(x, c)^v$$

in $\mathcal{V}$ that is $\mathcal{V}$-natural in x. By Proposition A.3.14, the cotensor product defines a unique $\mathcal{V}$-functor

$$\mathcal{C} \times \mathcal{V}^{\mathrm{op}} \xrightarrow{\ (-)^-\ } \mathcal{C}$$

making the defining isomorphism $\mathcal{V}$-natural in all three variables.

DEFINITION A.4.2. Dually, a $\mathcal{V}$-category $\mathcal{C}$ is **tensored** if, for all $v \in \mathcal{V}$ and $c \in \mathcal{C}$, the $\mathcal{V}$-functor $\mathcal{C}(c, -)^v \colon \mathcal{C} \to \mathcal{V}$ is represented by an object $v \otimes c \in \mathcal{C}$, i.e., there exists an isomorphism

$$\mathcal{C}(v \otimes c, x) \cong \mathcal{C}(c, x)^v$$

in $\mathcal{V}$ that is $\mathcal{V}$-natural in x. By Proposition A.3.14, the tensor product defines a unique $\mathcal{V}$-functor

$$\mathcal{V} \times \mathcal{C} \xrightarrow{\ - \otimes -\ } \mathcal{C}$$

making the defining isomorphism $\mathcal{V}$-natural in all three variables.

Immediately from these definitions:

LEMMA A.4.3. *A $\mathcal{V}$-category $\mathcal{C}$ is tensored and cotensored if and only if the $\mathcal{V}$-functor $\mathcal{C}(-, -) \colon \mathcal{C}^{\mathrm{op}} \times \mathcal{C} \to \mathcal{V}$ is part of a two-variable $\mathcal{V}$-adjunction*

$$\begin{array}{ccc} & \mathcal{C}(v \times a, b) & \\ {}^{\nearrow\!\!\!\nearrow} & & {}^{\nwarrow\!\!\!\nwarrow} \\ \mathcal{C}(a, b^v) & \cong & \mathcal{C}(a, b)^v \end{array}$$

as expressed by the commutative triangle of $\mathcal{V}$-natural isomorphisms.   □

The $\mathcal{V}$-naturality of the defining natural isomorphisms has the following consequence:

Lemma A.4.4. *In any category* $\mathcal{C}$ *that is enriched and cotensored over* $\mathcal{V}$, *there are* $\mathcal{V}$-*natural isomorphisms*

$$c^1 \cong c \qquad and \qquad \begin{array}{ccc} & c^{u \times v} & \\ {}_{\not\cong} \nearrow & & \nwarrow_{\cong} \\ (c^v)^u & \cong & (c^u)^v \end{array}$$

for $u, v \in \mathcal{V}$ *and* $c \in \mathcal{C}$.

Dually if $\mathcal{C}$ *is enriched and tensored over* $\mathcal{V}$, *there are* $\mathcal{V}$-*natural isomorphisms*

$$1 \otimes c \cong c \qquad and \qquad \begin{array}{ccc} & (u \times v) \otimes c & \\ {}_{\not\cong} \nearrow & & \nwarrow_{\cong} \\ u \otimes (v \otimes c) & \cong & v \otimes (u \otimes c) \end{array}$$

for $u, v \in \mathcal{V}$ *and* $c \in \mathcal{C}$.

There are various coherence relations between these $\mathcal{V}$-natural isomorphisms that derive from the coherences of the cartesian closed category $\mathcal{V}$.

Proof By Lemma A.3.7, to define the displayed isomorphisms, it suffices to prove that these objects represent the same $\mathcal{V}$-functors $\mathcal{C}^{op} \to \mathcal{V}$. By the defining universal property of the cotensor, for any $x \in \mathcal{C}$, there are $\mathcal{V}$-natural isomorphisms

$$\mathcal{C}(x, c^1) \cong \mathcal{C}(x, c)^1 \cong \mathcal{C}(x, c),$$

the last isomorphism by Remark A.1.9. Since these isomorphisms are $\mathcal{V}$-natural in c as well as x, the isomorphism $c^1 \cong c$ is $\mathcal{V}$-natural as well.

Similarly, there are $\mathcal{V}$-natural isomorphisms in each of the three vertices of the triangle below

$$\begin{array}{ccc} & \mathcal{C}(x, c^{u \times v}) \cong \mathcal{C}(x, c)^{u \times v} & \\ {}_{\not\cong} \nearrow & & \nwarrow_{\cong} \\ \mathcal{C}(x, (c^v)^u) \cong (\mathcal{C}(x, c)^v)^u & \cong & (\mathcal{C}(x, c)^u)^v \cong \mathcal{C}(x, (c^u)^v) \end{array}$$

with the connecting $\mathcal{V}$-natural isomorphisms given by Lemma A.1.7. Again the $\mathcal{V}$-naturality of these isomorphisms in u, v, and c gives the $\mathcal{V}$-naturality of the statement. $\qquad \square$

Extending Lemma A.2.3:

Lemma A.4.5. *A cartesian closed category* $(\mathcal{V}, \times, 1)$ *is enriched, tensored, and cotensored over itself, with tensors defined by the cartesian product and cotensors defined by the internal hom*[8]:

$$v \otimes w := v \times w \qquad and \qquad w^v := w^v.$$

---

[8] This excuses the abuse of exponential notation for both internal homs and cotensors.

Proof   Lemma A.1.7 establishes the required isomorphisms (A.1.8) in $\mathcal{V}$. The proof of their $\mathcal{V}$-naturality is left to the reader or to [68, §1.8].   □

Now consider a $\mathcal{V}$-functor $F \colon \mathcal{C} \to \mathcal{D}$. If $\mathcal{C}$ and $\mathcal{D}$ are tensored over $\mathcal{V}$, then the composite map

$$v \xrightarrow{\eta_v} \mathcal{C}(c, v \otimes c) \xrightarrow{F_{c, v \otimes c}} \mathcal{D}(Fc, F(v \otimes c))$$

transposes to define a canonical $\mathcal{V}$-natural transformation $\tau_{v,c} \colon v \otimes Fc \to F(v \otimes c)$. The $\mathcal{V}$-functor F **preserves tensors** if this map is an isomorphism in $\mathcal{D}$.

The presence of tensors and cotensors provides a convenient mechanism for testing whether an a priori unenriched adjunction may be enriched to a $\mathcal{V}$-adjunction.

PROPOSITION A.4.6.   *Suppose $\mathcal{A}$ and $\mathcal{B}$ are $\mathcal{V}$-categories.*

  (i) *If $\mathcal{A}$ and $\mathcal{B}$ are tensored over $\mathcal{V}$ then a $\mathcal{V}$-functor $F \colon \mathcal{B} \to \mathcal{A}$ is a left $\mathcal{V}$-adjoint if and only if F admits an unenriched right adjoint and preserves tensors.*

  (ii) *If $\mathcal{A}$ and $\mathcal{B}$ are cotensored over $\mathcal{V}$ then a $\mathcal{V}$-functor $U \colon \mathcal{A} \to \mathcal{B}$ is a right $\mathcal{V}$-adjoint if and only if U admits an unenriched left adjoint and preserves cotensors.*

Proof   We prove the sufficiency of the conditions in the first of the dual pair of statements and leave their necessity to Exercise A.4.i. Suppose F admits an unenriched right adjoint U with counit ϵ. By Example A.3.4 and Exercise A.3.ii, the composite map of hom-objects in $\mathcal{V}$

$$\mathcal{B}(b, Ua) \xrightarrow{F_{b, Ua}} \mathcal{A}(Fb, FUa) \xrightarrow{\epsilon_a \circ -} \mathcal{A}(Fb, a) \qquad (\mathrm{A.4.7})$$

which lifts the hom-set bijection of the unenriched adjunction, is $\mathcal{V}$-natural in b. Our task is to show that (A.4.7) is an isomorphism in $\mathcal{V}$. Then by Proposition A.3.14, there is a unique way to enrich the right adjoint U so that this map is $\mathcal{V}$-natural in a as well as b.

After applying the unenriched representable functor $\mathcal{V}(v, -) \colon \mathcal{V} \to Set$, we claim that this map fits into a commutative diagram

$$
\begin{array}{ccc}
\mathcal{V}(v, \mathcal{B}(b, Ua)) \xrightarrow{F_{b, Ua} \circ -} \mathcal{V}(v, \mathcal{A}(Fb, FUa)) \xrightarrow{(\epsilon_a \circ -) \circ -} \mathcal{V}(v, \mathcal{A}(Fb, a)) \\[2pt]
\cong\!\downarrow \qquad\qquad\qquad\qquad\qquad\qquad\qquad\qquad\qquad \downarrow\!\cong \\[2pt]
\mathcal{B}(v \otimes b, Ua) \xrightarrow{\;\cong\;} \mathcal{A}(F(v \otimes b), a) \xrightarrow{\; - \circ \tau_{v,b}\;} \mathcal{A}(v \otimes Fb, a)
\end{array}
$$

where the three isomorphisms are defined by transposing across the unenriched adjunctions. Since this is a diagram of functions between sets, to verify its commutativity, it suffices to consider an element $f : \upsilon \to \mathcal{B}(b, Ua)$; the image under both composites is the composite map

$$\upsilon \otimes Fb \xrightarrow{f \otimes Fb} B(b, ua) \otimes Fb \xrightarrow{F_{b,ua} \otimes Fb} A(Fb, FUa) \otimes Fb \xrightarrow{\text{ev}} FUa \xrightarrow{\epsilon_a} a$$

in the underlying category of $\mathcal{A}$. Since F preserves tensors, the map $\tau_{\upsilon,b}$ is an isomorphism, and thus, by the unenriched Yoneda lemma, (A.4.7) is an isomorphism as well. $\square$

Exercises

EXERCISE A.4.i. Prove that enriched left adjoints preserve tensors and define unenriched left adjoints.

EXERCISE A.4.ii. Let $\mathcal{J}$ be a small unenriched category and let $\mathcal{C}$ be a $\mathcal{V}$-category. Prove that if $\mathcal{C}$ is tensored or cotensored then so is $\mathcal{C}^{\mathcal{J}}$.

A.5 Conical Limits and Colimits

Consider a diagram $F \colon \mathcal{J} \to \mathcal{C}$ indexed by a 1-category $\mathcal{J}$ and valued in a $\mathcal{V}$-category $\mathcal{C}$. A cone λ over the diagram F with summit $\ell \in \mathcal{C}$ is a limit cone if composition with λ induces a bijection of hom-sets

$$\mathcal{C}(x, \ell)_0 \xrightarrow[\approx]{\lambda \circ -} \lim_{j \in \mathcal{J}} \mathcal{C}(x, Fj)_0$$

for all $x \in \mathcal{C}$. If $\mathcal{V}$ admits $\mathcal{J}$-shaped limits, then the composition map lifts to $\mathcal{V}$, and it is natural to assert a stronger version of the usual universal property of a limit cone, demanding that composition with λ induces an isomorphism of $\mathcal{V}$-objects and not just of hom-sets. This gives the notion of a *conical limit* of a diagram valued in an enriched category. *Conical colimits* are defined dually (see Definition A.5.2).

The underlying set functor $(-)_0 := \mathcal{V}(1, -) \colon \mathcal{V} \to \mathcal{S}et$ preserves limits and carries isomorphisms in $\mathcal{V}$ to bijections between sets. Thus, conical limits necessarily define 1-categorical limits, so we pay particular attention to what is required to enrich a 1-categorical limit to a conical limit. Our first observation along these lines is that limits and colimits of diagrams valued in a cartesian closed category $\mathcal{V}$ always enrich to define conical limits and conical colimits.

LEMMA A.5.1. *If $\mathcal{V}$ is a cartesian closed category, then any 1-categorical limit* *cone* $\lambda \colon \Delta \lim F \Rightarrow F$ *or 1-categorical colimit cone* $\gamma \colon F \Rightarrow \Delta \operatorname{colim} F$ *over or under a diagram* $F \colon \mathcal{J} \to \mathcal{V}$ *give rise to isomorphisms*

$$(\lim F)^{\upsilon} \cong \lim_{j \in \mathcal{J}} (F j^{\upsilon}) \quad and \quad \upsilon^{\operatorname{colim} F} \cong \lim_{j \in \mathcal{J}^{\mathrm{op}}} \upsilon^{F j}$$

in $\mathcal{V}$ *that are* $\mathcal{V}$-*natural in* $\upsilon \in \mathcal{V}$.

Proof   For any $\upsilon \in \mathcal{V}$, the exponential $(-)^{\upsilon} \colon \mathcal{V} \to \mathcal{V}$ is right adjoint to the product functor $- \times \upsilon \colon \mathcal{V} \to \mathcal{V}$; as such it preserves limits, giving rise to the first isomorphism of the statement, while $\upsilon^{-} \colon \mathcal{V}^{\mathrm{op}} \to \mathcal{V}$, as a mutual right adjoint, carries colimits to limits, establishing the second. $\square$

For the remainder of this section, suppose that $\mathcal{V}$ is a complete cartesian closed category, so that the limits in $\mathcal{V}$ that encode the universal properties of conical limits and colimits are assumed to exist.

DEFINITION A.5.2.   Let $\mathcal{C}$ be a $\mathcal{V}$-category and let $\mathcal{J}$ be a small 1-category. The **conical limit** of an unenriched diagram $F \colon \mathcal{J} \to \mathcal{C}$ is given by an object $\ell \in \mathcal{C}$ and a cone $\lambda \colon \Delta\ell \Rightarrow F$ inducing a $\mathcal{V}$-natural[9] isomorphism of hom-objects in $\mathcal{V}$

$$\mathcal{C}(x, \ell) \xrightarrow[\approx]{\lambda \circ -} \lim_{j \in \mathcal{J}} \mathcal{C}(x, F j) \quad \in \quad \mathcal{V}$$

for all $x \in \mathcal{C}$. Dually, the **conical colimit** is given by an object $c \in \mathcal{C}$ and a cone $\gamma \colon F \Rightarrow \Delta c$ inducing a $\mathcal{V}$-natural isomorphism

$$\mathcal{C}(c, x) \xrightarrow[\approx]{\gamma^{*}} \lim_{j \in \mathcal{J}^{\mathrm{op}}} \mathcal{C}(F j, x) \quad \in \quad \mathcal{V}$$

for all $x \in \mathcal{C}$.

The isomorphisms that characterize conical limits and colimits closely resemble the usual isomorphisms that characterize 1-categorical limits and colimits except for one very important difference: they postulate isomorphisms in $\mathcal{V}$ rather than isomorphisms in $\mathcal{S}et$. In the case where $\mathcal{V} = s\mathcal{S}et$, the isomorphism of vertices that underlies this isomorphism of simplicial sets describes the usual 1-categorical universal property. To say that the limit is *conical* and not merely *1-categorical* is to assert that this universal property extends to all positive dimensions.

Inspecting Definition A.5.2, we see immediately that:

---

[9] For any cone $\lambda \colon \Delta\ell \Rightarrow F$, not necessarily a limit cone, composition defines a map $\lambda \circ - \colon \mathcal{C}(x, \ell) \to \lim_{j \in \mathcal{J}} \mathcal{C}(x, F j)$ that is $\mathcal{V}$-natural in x (see Exercise A.5.i). This follows by essentially the same argument used in Example A.3.4 to demonstrate that arrows in the underlying category of an enriched category induce enriched natural transformations between representable functors. So the content of the universal property that characterizes conical limits is that this $\mathcal{V}$-natural map is invertible.

PROPOSITION A.5.3. *A 1-categorical limit cone is conical just when it is preserved by all $\mathcal{V}$-valued representable functors $\mathcal{C}(x,-)\colon \mathcal{C} \to \mathcal{V}$, while a 1-categorical colimit cone is conical just when it is preserved by all $\mathcal{V}$-valued representable functors $\mathcal{C}(-,x)\colon \mathcal{C}^{\mathrm{op}} \to \mathcal{V}$.* □

The proof of Lemma A.5.1 generalizes to show:

PROPOSITION A.5.4. *In an enriched category with tensors all 1-categorical limits in $\mathcal{C}$ are conical, while in an enriched category with cotensors all 1-categorical colimits are conical.*

Proof By Proposition A.5.3, to show that a 1-categorical limit in a $\mathcal{V}$-category $\mathcal{C}$ is conical, it suffices to show that it is preserved by the $\mathcal{V}$-valued representable functors $\mathcal{C}(x,-)\colon \mathcal{C} \to \mathcal{V}$ for all $x \in \mathcal{C}$. If $\mathcal{C}$ admits tensors, then each of these functors admits a left adjoint; as right adjoints, they necessarily preserve the 1-categorical limits of the statement. □

In analogy with Proposition A.5.3, we have:

PROPOSITION A.5.5. *Let $\mathcal{C}$ be enriched and cotensored over $\mathcal{V}$. A limit of an unenriched diagram $F\colon \mathcal{J} \to \mathcal{C}$ is a conical limit if and only if it is preserved by cotensors with all objects $v \in \mathcal{V}$.*

Proof Cotensors are $\mathcal{V}$-enriched right adjoints, which preserve conical limits. The content is that preservation by cotensors suffices to enrich a 1-categorical limit to a $\mathcal{V}$-categorical one.

By Proposition A.5.3, a 1-categorical limit cone $\lambda\colon \Delta\ell \Rightarrow F$ is conical just when it is preserved by all $\mathcal{V}$-valued representable functors $\mathcal{C}(x,-)\colon \mathcal{C} \to \mathcal{V}$. To see that the natural map

$$\mathcal{C}(x,\ell) \xrightarrow{\ \lambda\circ-\ } \lim_{j\in\mathcal{J}} \mathcal{C}(x,Fj)$$

is an isomorphism in $\mathcal{V}$ we appeal to the unenriched Yoneda lemma and argue that this map induces an isomorphism upon applying an unenriched representable functor $\mathcal{V}(v,-)\colon \mathcal{V} \to \mathcal{S}et$ for any $v \in \mathcal{V}$. To see this invertibility, note that the induced map of hom-sets fits into a commutative diagram

$$
\begin{array}{ccc}
\mathcal{V}(v,\mathcal{C}(x,\ell)) & \xrightarrow{\ \cong\ } & \mathcal{C}(x,\ell^v) \\
{\scriptstyle (\lambda\circ-)\circ-}\downarrow & & \downarrow{\scriptstyle \lambda^v\circ-} \\
\mathcal{V}(v,\lim_{j\in\mathcal{J}}\mathcal{C}(x,Fj)) \cong \lim_{j\in\mathcal{J}}\mathcal{V}(v,\mathcal{C}(x,Fj)) \cong \lim_{j\in\mathcal{J}}\mathcal{C}(x,Fj^v) \cong \mathcal{C}(x,\lim_{j\in\mathcal{J}}(Fj^v))
\end{array}
$$

where the horizontal isomorphisms express the unenriched universal property of cotensors and the fact that unenriched representable functors preserve limits. To say that cotensors preserve the limit $\ell \cong \lim_{j \in \mathcal{J}} Fj$ means that $\ell^{\upsilon} \cong (\lim_{j \in \mathcal{J}} Fj)^{\upsilon} \cong \lim_{j \in \mathcal{J}} (Fj^{\upsilon})$, so by the Yoneda lemma, the right-vertical map is an isomorphism, and thus the left-vertical map is as well. $\square$

Exercises

EXERCISE A.5.i. Consider a diagram $F: \mathcal{J} \to \mathcal{C}$ indexed by a 1-category and valued in a $\mathcal{V}$-category and a cone $\lambda: \Delta\ell \Rightarrow F$. Assuming $\mathcal{V}$ has $\mathcal{J}$-shaped limits, define a canonical map $\lambda \circ -: \mathcal{C}(x, \ell) \to \lim_{j \in \mathcal{J}} \mathcal{C}(x, Fj)$ and show, by arguing along the lines of Example A.3.4, that it is $\mathcal{V}$-natural.

EXERCISE A.5.ii ([67, p. 306]). Specialize the result of Proposition A.5.4 to prove the following: in any 2-category $\mathcal{C}$ that admits tensors with the walking-arrow category 2, any 1-categorical limits that $\mathcal{C}$ admits are automatically conical.[10]

A.6 Weighted Limits and Colimits

The cotensors of §A.4 and conical limits of §A.5 are both instances of a more general notion of *weighted limit* that we now introduce. We continue in the context of a complete and cocomplete cartesian closed category $(\mathcal{V}, \times, 1)$. The examples we have in mind are $(s\mathcal{S}et, \times, 1)$, its cartesian closed subcategory $(\mathcal{C}at, \times, \mathbb{1})$, or its further cartesian closed subcategory $(\mathcal{S}et, \times, 1)$.

Ordinary limits and colimits are objects representing the functor of cones with a given apex over or under a fixed diagram. Weighted limits and colimits are defined analogously, except that the cones over or under a diagram might have exotic "shapes," which are allowed to vary with the objects indexing the diagram. More formally, in the $\mathcal{V}$-enriched context, the *weight*, defining the "shape" of a cone over a diagram indexed by $\mathcal{A}$ or under a diagram indexed by $\mathcal{A}^{op}$, takes the form of a functor in $\mathcal{V}^{\mathcal{A}}$; note the indexing category $\mathcal{A}$ may be $\mathcal{V}$-enriched, unlike the diagrams considered in §A.5.

We develop the general notions of weighted limit and weighted colimit from three different viewpoints that we introduce in the reverse of the logical order, because we find this route to be the most intuitive. We first describe the axioms that characterize the weighted limit and colimit bifunctors, whenever they exist. We then explain how weighted limits and colimits can be constructed, assuming

---

[10] The statement asserts that the presence of tensors with 2 implies that the universal property of 1-dimensional limits automatically has an additional 2-dimensional aspect, such as illustrated by the discussion around Proposition 1.4.5.

certain other limits and colimits exist. We then finally introduce the general universal property that defines a particular weighted limit or colimit, which stipulates exactly what is required for the notions just introduced to in fact exist.

DEFINITION A.6.1 (weighted limits and colimits, axiomatically). For a small $\mathcal{V}$-enriched category $\mathcal{A}$ and a large $\mathcal{V}$-enriched category $\mathcal{C}$, the **weighted limit** and **weighted colimit bifunctors**

$$\lim_{-} - : (\mathcal{V}^{\mathcal{A}})^{\mathrm{op}} \times \mathcal{C}^{\mathcal{A}} \to \mathcal{C} \qquad \text{and} \qquad \mathrm{colim}_{-} - : \mathcal{V}^{\mathcal{A}} \times \mathcal{C}^{\mathcal{A}^{\mathrm{op}}} \to \mathcal{C}$$

are characterized by the following pair of axioms whenever they exist:

(i) Weighted co/limits with representable weights evaluate at the representing object:

$$\lim_{\mathcal{A}(a,-)} F \cong Fa \qquad \text{and} \qquad \mathrm{colim}_{\mathcal{A}(-,a)} G \cong Ga.$$

(ii) The weighted co/limit bifunctors are cocontinuous in the weight: for any diagrams $F \in \mathcal{C}^{\mathcal{A}}$ and $G \in \mathcal{C}^{\mathcal{A}^{\mathrm{op}}}$, the functor $\mathrm{colim}_{-} G$ preserves colimits, while the functor $\lim_{-} F$ carries colimits to limits.[11]

We interpret axiom (ii) to mean that weights can be "made-to-order": a weight constructed as a colimit of representables – as all $\mathcal{V}$-valued functors are – will stipulate the expected universal property.

DEFINITION A.6.2 (weighted limits and colimits, constructively). The **limit** of $F \in \mathcal{C}^{\mathcal{A}}$ **weighted by** $W \in \mathcal{V}^{\mathcal{A}}$ is computed by the functor cotensor product:

$$\lim_{W} F := \int_{a \in \mathcal{A}} Fa^{Wa} := \mathrm{eq}\left(\prod_{a \in \mathcal{A}} Fa^{Wa} \rightrightarrows \prod_{a,b \in \mathcal{A}} Fb^{\mathcal{A}(a,b) \times Wa} \right),$$

$$(\text{A.6.3})$$

where the product and equalizer should be interpreted as conical limits in the sense of Definition A.5.2. The maps in the equalizer diagram are induced by the actions $\mathcal{A}(a,b) \times Wa \to Wb$ and $Fa \to Fb^{\mathcal{A}(a,b)}$ of the hom-object $\mathcal{A}(a,b)$ on the $\mathcal{V}$-functors W and F.

Dually, the **colimit** of $G \in \mathcal{C}^{\mathcal{A}^{\mathrm{op}}}$ **weighted by** $W \in \mathcal{V}^{\mathcal{A}}$ is computed by the

---

[11] More precisely, as proven in Proposition A.6.10, the weighted colimit functor $\mathrm{colim}_{-} G$ preserves weighted colimits, while the weighted limit functor $\lim_{-} F$ carries weighted colimits to weighted limits.

functor tensor product:

$$\mathrm{colim}_W G := \int^{a \in \mathcal{A}} Wa \otimes Ga$$

$$:= \mathrm{coeq}\left(\coprod_{a,b \in \mathcal{A}} (Wa \times \mathcal{A}(a,b)) \otimes Gb \rightrightarrows \coprod_{a \in \mathcal{A}} Wa \otimes Ga \right),$$

$$(A.6.4)$$

where the coproduct and coequalizer should be interpreted as conical colimits. One of the maps in the coequalizer diagram is induced by the action $\mathcal{A}(a,b) \otimes Gb \to Ga$ of $\mathcal{A}(a,b)$ on the contravariant $\mathcal{V}$-functor G, while the other uses the covariant action of $\mathcal{A}(a,b)$ on W as before.

DEFINITION A.6.5 (weighed limits and colimits, the universal property). The **limit** $\lim_W F$ of the diagram $F \in \mathcal{C}^{\mathcal{A}}$ weighted by $W \in \mathcal{V}^{\mathcal{A}}$ and the **colimit** $\mathrm{colim}_W G$ of $G \in \mathcal{C}^{\mathcal{A}^{\mathrm{op}}}$ weighted by $W \in \mathcal{V}^{\mathcal{A}}$ are characterized by the universal properties:

$$\mathcal{C}(x, \lim_W F) \cong \mathcal{V}^{\mathcal{A}}(W, \mathcal{C}(x, F)) \text{ and } \mathcal{C}(\mathrm{colim}_W G, x) \cong \mathcal{V}^{\mathcal{A}}(W, \mathcal{C}(G, x)),$$

$$(A.6.6)$$

each of these defining an isomorphism between objects of $\mathcal{V}$ that is $\mathcal{V}$-natural in x.

When the indexing category is not clear from context, we may add it as a superscript to the notation for the weighted limit and weighted colimit. Proposition A.6.10 shows that these three definitions characterize the same objects. Along the way to proving it, we obtain results of interest in their own right, that we record separately.

LEMMA A.6.7. *A complete cartesian closed category $\mathcal{V}$ admits all weighted limits, as defined by the formula of (A.6.3) satisfying the natural isomorphism of (A.6.6) and the axioms of Definition A.6.1. Explicitly, for a weight $W : \mathcal{A} \to \mathcal{V}$ and a diagram $F : \mathcal{A} \to \mathcal{V}$, the weighted limit*

$$\lim_W F := \mathcal{V}^{\mathcal{A}}(W, F),$$

is the $\mathcal{V}$-object of $\mathcal{V}$-natural transformations from W to V.

Proof The $\mathcal{V}$-functor $\mathcal{V}(1, -) : \mathcal{V} \to \mathcal{V}$ represented by the terminal object is naturally isomorphic to the identity functor. So taking $x = 1$ in the universal property of (A.6.6) in the case where the diagram $F \in \mathcal{V}^{\mathcal{A}}$ is valued in the $\mathcal{V}$-category $\mathcal{V}$, we have

$$\lim_W F \cong \mathcal{V}^{\mathcal{A}}(W, F).$$

Simultaneously, the formula (A.6.3) computes the V-object $V^{\mathcal{A}}(W, F)$ of V-natural transformations from W to F introduced in Definition A.3.8. The enriched Yoneda lemma of Theorem A.3.11 proves the first axiom, while the second axiom follows from the universal property of (A.6.6). □

The V-object of V-natural transformations satisfies the natural isomorphism

$$V(v, V^{\mathcal{A}}(W, F)) \cong V^{\mathcal{A}}(W, V(v, F))$$

for any $v \in V$. By applying the observation that W-weighted limits of V-valued functors are V-objects of natural transformations to the functors $\mathcal{C}(x, F-)$ and $\mathcal{C}(G-, x)$, we may re-express the natural isomorphism (A.6.6) as:

COROLLARY A.6.8. *The weighted limits and weighted colimits of* (A.6.6) *are representably defined as weighted limits in* V: *for* $W \in V^{\mathcal{A}}$ *and* $F \in \mathcal{C}^{\mathcal{A}}$ *and* $G \in \mathcal{C}^{\mathcal{A}^{\mathrm{op}}}$ *the weighted limit and colimit are characterized by* V-*natural isomorphisms in* x:

$$\mathcal{C}(x, \lim_W F) \cong \lim_W \mathcal{C}(x, F) \qquad and \qquad \mathcal{C}(\mathrm{colim}_W G, x) \cong \lim_W \mathcal{C}(G, x)$$
$$(A.6.9)$$
□

We now unify the Definitions A.6.1, A.6.2, and A.6.5.

PROPOSITION A.6.10. *When the limits and colimits of* (A.6.3) *and* (A.6.4) *exist they define objects satisfying the universal properties* (A.6.6) *or equivalently* (A.6.9). *The* V-*bifunctors defined by these universal properties satisfy the axioms of Definition A.6.1.*

Proof The equivalence of Definitions A.6.2 and A.6.5 – for either weighted limits or weighted colimits – is a direct consequence of the special case of this implication for weighted limits valued in $\mathcal{C} = V$ proven as Lemma A.6.7 and Corollary A.6.8. The limits of (A.6.3) in $\mathcal{C}$ are also defined representably in terms of the analogous limits in V. So the objected defined by (A.6.3) represents the V-functor $\lim_W \mathcal{C}(-, F)$ that defines the weighted limit $\lim_W F$.

It remains to prove that the weighted limits of Definitions A.6.2 and A.6.5 satisfy the axioms of Definition A.6.1. In the case of a V-valued diagram $F \in V^{\mathcal{A}}$, axiom (i) is the V-Yoneda lemma: $V^{\mathcal{A}}(\mathcal{A}(a, -), F) \cong Fa$ proven in Theorem A.3.11. Once again, the general case for $F \in \mathcal{C}^{\mathcal{A}}$ follows from the special case for V-valued diagrams. To demonstrate an isomorphism $\lim_{\mathcal{A}(a,-)} F \cong Fa$ in $\mathcal{C}$ it suffices to produce an isomorphism $\mathcal{C}(x, \lim_{\mathcal{A}(a,-)} F) \cong \mathcal{C}(x, Fa)$ in V for all $x \in \mathcal{C}$, and we have such a natural isomorphism by applying (A.6.9) and the observation just made to the functor $\mathcal{C}(x, F-) \in V^{\mathcal{A}}$.

For the axiom (ii), consider a diagram $W: \mathcal{J}^{\mathrm{op}} \to V^{\mathcal{A}}$ of weights and a weight $V \in \mathcal{V}^{\mathcal{J}}$ so that $\mathrm{colim}_V^{\mathcal{J}} W$ is an object in $\mathcal{V}^{\mathcal{A}}$. For any $F \in \mathcal{C}^{\mathcal{A}}$, we will show that the $\mathcal{V}$-functor $\lim_{-}^{\mathcal{A}} F \colon (\mathcal{V}^{\mathcal{A}})^{\mathrm{op}} \to \mathcal{C}$ carries the V-weighted colimit of the diagram of weights W to the V-weighted limit of the composite diagram $\lim_{W-}^{\mathcal{A}} F \colon \mathcal{J} \to \mathcal{C}$.

The universal property (A.6.6), applied first to the $\mathrm{colim}_V^{\mathcal{J}} W$-weighted limit of the diagram F and the object x, and then to the V-weighted colimit of the diagram W and the object $\mathcal{C}(x, F)$, supplies isomorphisms:

$$\mathcal{C}(x, \lim_{\mathrm{colim}_V^{\mathcal{J}} W}^{\mathcal{A}} F) \cong \mathcal{V}^{\mathcal{A}}(\mathrm{colim}_V^{\mathcal{J}} W, \mathcal{C}(x, F)) \cong \mathcal{V}^{\mathcal{J}}(V, \mathcal{V}^{\mathcal{A}}(W, \mathcal{C}(x, F))).$$

Applying (A.6.6) twice more, first for the weights Wj for each $j \in \mathcal{J}$ and then for the weight V and the diagram $\lim_W^{\mathcal{A}} F \colon \mathcal{J} \to \mathcal{C}$, we have

$$\cong \mathcal{V}^{\mathcal{J}}(V, \mathcal{C}(x, \lim_W^{\mathcal{A}} F)) \cong \mathcal{C}(x, \lim_V^{\mathcal{J}} \lim_W^{\mathcal{A}} F).$$

By the Yoneda lemma, this proves that

$$\lim_{\mathrm{colim}_V^{\mathcal{J}} W}^{\mathcal{A}} F \cong \lim_V^{\mathcal{J}} \lim_W^{\mathcal{A}} F,$$

i.e., that the weighted limit functor $\lim_{-}^{\mathcal{A}} F$ is carries a weighted colimit of weights to the analogous weighted limit of weights.   □

REMARK A.6.11 (for unenriched indexing categories). When the indexing category is unenriched, the limit and colimit formulas from Definition A.6.2 simplify as follows

$$\lim_W F \cong \mathrm{eq}\left(\prod_{a \in \mathcal{A}} Fa^{Wa} \Longrightarrow \prod_{\mathcal{A}(a,b)} Fb^{Wa} \right)$$

$$\mathrm{colim}_W G \cong \mathrm{coeq}\left(\coprod_{\mathcal{A}(a,b)} Wa \otimes Gb \Longrightarrow \coprod_{a \in \mathcal{A}} Wa \otimes Ga \right)$$

and in fact, it suffices to consider only nonidentity arrows or even just atomic arrows.

EXAMPLE A.6.12 (conical limits and colimits). For any small $\mathcal{V}$-category $\mathcal{A}$, the constant diagram at the terminal object of $\mathcal{V}$ defines the terminal weight $1 \in \mathcal{V}^{\mathcal{A}}$. For diagrams $F \in \mathcal{C}^{\mathcal{A}}$ and $G \in \mathcal{C}^{\mathcal{A}^{\mathrm{op}}}$, respectively, limits and colimits weighted by this weight satisfy the defining universal properties

$$\mathcal{C}(x, \lim_1 F) \cong \mathcal{V}^{\mathcal{A}}(1, \mathcal{C}(x, F)) \cong \lim_1 \mathcal{C}(x, F) \qquad \text{and}$$
$$\mathcal{C}(\mathrm{colim}_1 G, x) \cong \mathcal{V}^{\mathcal{A}}(1, \mathcal{C}(G, x)) \cong \lim_1 \mathcal{C}(G, x),$$

which say that $\lim F$ and $\mathrm{colim}\, G$ represent the functors of $\mathcal{V}$-enriched conical cones over F or under G, respectively.

When $\mathcal{A}$ is an unenriched category, this recovers the universal property that characterized the **conical limits** and **conical colimits** of Definition A.5.2, so we extend those terms to refer to general limits and colimits weighted by the terminal weight; it is common to use the simplified notation $\lim F := \lim_1 F$ and $\operatorname{colim} G := \operatorname{colim}_1 G$ for conical limits and colimits. This explains the name "conical": among the weighted limits, the conical limits are the ones with ordinary cone shapes, involving a single arrow in the underlying category pointing from the summit to each object in the diagram. Thus, conical limits and colimits arise as special cases of weighted limits and colimits whose weights are terminal.

EXAMPLE A.6.13 (tensors and cotensors). A diagram indexed by the terminal category $\mathbb{1}$ and valued in a $\mathcal{V}$-enriched category $\mathcal{C}$ is just an object $c \in \mathcal{C}$. A weight for such diagrams is just given by another object $v \in \mathcal{V}$. The v-weighted limit of the diagram c is defined by the universal property

$$\mathcal{C}(x, \lim_v^1 c) \cong \mathcal{C}(x, c)^v$$

that characterizes the **cotensor** c^v, while the v-weighted colimit of the diagram c is defined by the universal property

$$\mathcal{C}(\operatorname{colim}_v^1 c, x) \cong \mathcal{C}(c, x)^v$$

that characterizes the **tensor** $v \otimes c$. Thus, cotensors and tensors arise as special cases of weighted limits and colimits whose indexing categories are terminal.

EXAMPLE A.6.14 (commas). An ∞-cosmos $\mathcal{K}$ is, in particular, a category enriched over simplicial sets. The comma ∞-category is the limit in the ∞-cosmos $\mathcal{K}$ of the diagram $\lrcorner \to \mathcal{K}$ given by the cospan

$$C \xrightarrow{\ g\ } A \xleftarrow{\ f\ } B$$

weighted by the diagram $\lrcorner \to sSet$ given by the cospan

$$\mathbb{1} \xrightarrow{\ 1\ } 2 \xleftarrow{\ 0\ } \mathbb{1}$$

Under the simplification of Remark A.6.11, the formula for the weighted limit reduces to the equalizer of the pair of maps

$$C \times A^2 \times B \begin{array}{c} \xrightarrow{\ \pi\ } A^2 \xrightarrow{(p_1, p_0)} \\ \xrightarrow[\ \pi\ ]{} C \times B \xrightarrow[g \times f]{} \end{array} A \times A$$

which computes the pullback of (3.4.2). The universal property (A.6.6) provides a correspondence between functors $\ulcorner\alpha\urcorner : X \to \operatorname{Hom}_A(f, g)$ in $\mathcal{K}$ and simplicial natural transformations, the data of which is given by the three dashed vertical maps that fit into two commutative squares:

$$
\begin{array}{ccccc}
\mathbb{1} & \overset{1}{\hookrightarrow} & \mathbb{2} & \overset{0}{\hookleftarrow} & \mathbb{1} \\
{\scriptstyle c}\downarrow & & {\scriptstyle \ulcorner\alpha\urcorner}\downarrow & & \downarrow{\scriptstyle b} \\
\mathsf{Fun}(X,C) & \underset{g\circ-}{\longrightarrow} & \mathsf{Fun}(X,A) & \underset{f\circ-}{\longleftarrow} & \mathsf{Fun}(X,B)
\end{array}
$$

EXAMPLE A.6.15 (Kan extensions as weighted co/limits). The usual colimit or limit formula that computes the value of a pointwise left or right Kan extension of an unenriched functor $F\colon C \to E$ along $K\colon C \to D$ at an object $d \in D$ can be succinctly expressed by the weighted colimit or weighted limit

$$
\mathsf{lan}_K F(d) := \mathsf{colim}_{D(K-,d)} F \qquad \text{and} \qquad \mathsf{ran}_K F(d) := \lim_{D(d,K-)} F.
$$

The formulae of Definition A.6.2 give criteria under which weighted limits or colimits are guaranteed to exist.

COROLLARY A.6.16. *A $\mathcal{V}$-category that admits cotensors and conical limits of all small unenriched diagram shapes then admits all small weighted limits. Dually, a $\mathcal{V}$-category that admits tensors and conical colimits of all small unenriched diagram shapes then admits all small weighted colimits.*　□

DEFINITION A.6.17. A $\mathcal{V}$-category is $\mathcal{V}$-**complete** if it admits small $\mathcal{V}$-weighted limits, or equivalently, by Corollary A.6.16, if it admits cotensors by objects in $\mathcal{V}$ and small conical limits. Dually, a $\mathcal{V}$-category is $\mathcal{V}$-**cocomplete** if it admits small $\mathcal{V}$-weighted colimits, or equivalently, by Corollary A.6.16, if it admits tensors by objects in $\mathcal{V}$ and small conical colimits.

REMARK A.6.18 (on proving enriched completeness). In practice one often shows that a $\mathcal{V}$-category is complete by demonstrating that its underlying category is complete and that the $\mathcal{V}$-category has both cotensors and tensors, the latter being an instance of a weighted *colimit*. Then Proposition A.5.4 applies to establish that the unenriched limits are in fact conical limits.

We conclude with a few results from the general theory of weighted limits and colimits. Immediately from their defining universal properties, it can be verified that:

LEMMA A.6.19 (weighted limits of restricted diagrams). *Consider a $\mathcal{V}$-functor $K\colon \mathcal{A} \to \mathcal{B}$, a weight $W\colon \mathcal{A} \to \mathcal{V}$, and diagrams $F\colon \mathcal{B} \to \mathcal{C}$ and $G\colon \mathcal{B}^{\mathrm{op}} \to \mathcal{C}$. Then the W-weighted limit or colimit of the restricted diagram is isomorphic to the $\mathsf{lan}_K W$-weighted limit or colimit of the original diagram:*

$$
\lim_W^{\mathcal{A}}(F \circ K) \cong \lim_{\mathsf{lan}_K W}^{\mathcal{B}} F \qquad \text{and} \qquad \mathsf{colim}_W^{\mathcal{A}}(G \circ K) \cong \mathsf{colim}_{\mathsf{lan}_K W}^{\mathcal{B}} G.
$$

Proof   Exercise A.6.ii.   □

Recall from Definition A.3.16 that an enriched adjunction is comprised of
a pair of $\mathcal{V}$-functors $F\colon \mathcal{B} \to \mathcal{A}$ and $U\colon \mathcal{A} \to \mathcal{B}$ together with a family of
isomorphisms $\mathcal{A}(Fb, a) \cong \mathcal{B}(b, Ua)$ that are $\mathcal{V}$-natural in both variables. The
usual Yoneda-style argument enriches to show:

PROPOSITION A.6.20. *Enriched right adjoints preserve weighted limits and
enriched left adjoints preserve weighted colimits.*

Proof   Exercise A.6.iii.   □

Exercises

EXERCISE A.6.i. Taking the base for enrichment $\mathcal{V}$ to be *Set*, compute the
following weighted limits of a simplicial set X, regarded as a diagram in $Set^{\Delta^{\mathrm{op}}}$,
weighted by:

(i) the standard n-simplex $\Delta[n] \in Set^{\Delta^{\mathrm{op}}}$,
(ii) the **spine** of the n-simplex, the simplicial subset of $\Delta[n]$ obtained by
     gluing together the n edges from i to $i+1$ into a composable path,
(iii) the n-simplex boundary $\partial\Delta[n] \in Set^{\Delta^{\mathrm{op}}}$.[12]

EXERCISE A.6.ii. Prove Lemma A.6.19.

EXERCISE A.6.iii. Prove Proposition A.6.20.

EXERCISE A.6.iv.

(i) Extend the definitions of weighted limit and colimit to allow the weight
    W to be an **enriched profunctor** – i.e., a $\mathcal{V}$-functor $W\colon \mathcal{B}^{\mathrm{op}} \times \mathcal{A} \to \mathcal{V}$
    for small $\mathcal{V}$-categories $\mathcal{A}$ and $\mathcal{B}$ – in such a way that the weighted limit
    and colimit functors have the form

$$(\mathcal{V}^{\mathcal{B}^{\mathrm{op}} \times \mathcal{A}})^{\mathrm{op}} \times \mathcal{C}^{\mathcal{A}} \xrightarrow{\ \lim_{-} -\ } \mathcal{C}^{\mathcal{B}} \quad \text{and} \quad \mathcal{V}^{\mathcal{B}^{\mathrm{op}} \times \mathcal{A}} \times \mathcal{C}^{\mathcal{B}} \xrightarrow{\ \mathrm{colim}_{-} -\ } \mathcal{C}^{\mathcal{A}}$$

(ii) Show that the weighted limit and weighted colimit bifunctors from (i)
     form two thirds of a two-variable adjunction, with a $\mathcal{V}$-natural isomorph-
     ism

$$\mathcal{C}^{\mathcal{A}}(\mathrm{colim}_W G, F) \cong \mathcal{C}^{\mathcal{B}}(G, \lim_W F)$$

for $W \in \mathcal{V}^{\mathcal{B}^{\mathrm{op}} \times \mathcal{A}}$, $F \in \mathcal{C}^{\mathcal{A}}$, and $G \in \mathcal{C}^{\mathcal{B}}$.

(iii) What is the third bifunctor of the two-variable adjunction of (ii)?

---

[12] The limit of a simplicial object weighted by $\partial\Delta[n]$ is called the **nth-matching object** (see
Definition C.4.14).

A.7 Change of Base

"Change of base," first considered by Eilenberg and Kelly in [41], refers to a systematic procedure by which enrichment over one category $\mathcal{V}$ is converted into enrichment over another category $\mathcal{W}$. Corollary A.3.6 notes that for a cartesian closed category $\mathcal{V}$, there is a 2-category $\mathcal{V}\text{-}\mathcal{C}at$ of $\mathcal{V}$-categories, $\mathcal{V}$-functors, and $\mathcal{V}$-natural transformations. The first main result, appearing as Proposition A.7.3, gives conditions under which a functor $T: \mathcal{V} \to \mathcal{W}$ between cartesian closed categories induces a change-of-base 2-functor $T_*: \mathcal{V}\text{-}\mathcal{C}at \to \mathcal{W}\text{-}\mathcal{C}at$.

As the context we are working in here is less general than the one considered by Eilenberg and Kelly – our base categories are cartesian closed while theirs are closed symmetric monoidal – we take a shortcut which covers all of our examples and is easier to explain. In general, all that is needed to produce a change of base 2-functor is a *lax monoidal* functor between symmetric monoidal categories, but the lax monoidal functors we encounter between cartesian closed categories are in fact finite-product-preserving, so we content ourselves with explicating the results in that case instead.

Recall that a functor $T: \mathcal{V} \to \mathcal{W}$ between cartesian closed categories **preserves finite products** just when the natural maps defined for any $u, v \in \mathcal{V}$

$$T(u \times v) \overset{\sim}{\to} Tu \times Tv \qquad \text{and} \qquad T1 \overset{\sim}{\to} 1$$

are isomorphisms. For example:

EXAMPLE A.7.1. Since representable functors preserve products, for any cartesian closed category $\mathcal{V}$, the underlying set functor $(-)_0 : \mathcal{V} \to \mathcal{S}et$ is product-preserving

EXAMPLE A.7.2. In a cartesian closed category $\mathcal{V}$, finite products distribute over arbitrary coproducts. In particular, for any sets X and Y there is an isomorphism

$$\amalg_{X \times Y} 1 \cong (\amalg_X 1) \times (\amalg_Y 1)$$

between coproducts of the terminal object 1, which proves that the functor

$$\mathcal{S}et \xrightarrow{\ \amalg_-1\ } \mathcal{V}$$

is finite-product-preserving.

A finite-product-preserving functor may be used to change the base as follows

PROPOSITION A.7.3. *A finite-product-preserving functor $T: \mathcal{V} \to \mathcal{W}$ between cartesian closed categories induces a change-of-base 2-functor*

$$\mathcal{V}\text{-}\mathcal{C}at \xrightarrow{\ T_*\ } \mathcal{W}\text{-}\mathcal{C}at.$$

An early observation along these lines was first stated as [41, II.6.3], with the proof left to the reader. We adopt the same tactic and leave the diagram chases to the reader or to [31, 4.2.4] and instead just give the construction of the change-of-base 2-functor, which is the important thing.

Proof Let $\mathcal{C}$ be a $\mathcal{V}$-category and define a $\mathcal{W}$-category $T_*\mathcal{C}$ to have the same objects and to have mapping objects $T_*\mathcal{C}(x,y) := T\mathcal{C}(x,y)$. The composition and identity maps are given by the composites

$$T\mathcal{C}(y,z) \times T\mathcal{C}(x,y) \xrightarrow{\;\cong\;} T(\mathcal{C}(y,z) \times \mathcal{C}(x,y)) \xrightarrow{\;T\circ\;} T\mathcal{C}(x,z)$$

$$1 \xrightarrow{\;\cong\;} T1 \xrightarrow{\;T\,\mathrm{id}_x\;} T\mathcal{C}(x,x)$$

which make use of the inverses of the natural maps that arise when a finite-product-preserving functor is applied to a finite product. A straightforward diagram chase verifies that $T_*\mathcal{C}$ is a $\mathcal{W}$-category.

If $F\colon \mathcal{C} \to \mathcal{D}$ is a $\mathcal{V}$-functor, then we define a $\mathcal{W}$-functor $T_*F\colon T_*\mathcal{C} \to T_*\mathcal{D}$ to act on objects by $c \in \mathcal{C} \mapsto Fc \in \mathcal{D}$ and with internal action on arrows defined by

$$T\mathcal{C}(x,y) \xrightarrow{\;TF_{x,y}\;} T\mathcal{D}(Fx,Fy)$$

Again, a straightforward diagram chase verifies that T_*F is $\mathcal{W}$-functorial. It is evident from this definition that $T_*(GF) = T_*G \cdot T_*F$.

Finally, let $\alpha\colon F \Rightarrow G$ be a $\mathcal{V}$-natural transformation between $\mathcal{V}$-functors $F, G\colon \mathcal{C} \to \mathcal{D}$ and define a $\mathcal{W}$-natural transformation $T_*\alpha\colon T_*F \Rightarrow T_*G$ to have components

$$1 \xrightarrow{\;\cong\;} T1 \xrightarrow{\;T\alpha_c\;} T\mathcal{D}(Fc,Gc)$$

Another straightforward diagram chase verifies that $T_*\alpha$ is $\mathcal{W}$-natural.

It remains to verify this assignment is functorial for both horizontal and vertical composition of enriched natural transformations. Consulting Definition A.3.5, we see that the component of $T_*(\beta \cdot \alpha)$ is defined by the top-horizontal composite below while the component of the vertical composite of $T_*\alpha$ with $T_*\beta\colon T_*G \Rightarrow T_*H$ is defined by the bottom composite:

$$
\begin{array}{ccccc}
1 & \xrightarrow{\;\cong\;} & T1 & \xrightarrow{\;T(\beta_c \times \alpha_c)\;} T(\mathcal{D}(Gc,Hc) \times \mathcal{D}(Fc,Gc)) & \xrightarrow{\;T\circ\;} & T\mathcal{D}(Fc,Hc) \\
& \searrow_{\cong} & \uparrow_{\cong} & & \uparrow_{\cong} & \\
& & T1 \times T1 & \xrightarrow[\;T\beta_c \times T\alpha_c\;]{} T\mathcal{D}(Gc,Hc) \times T\mathcal{D}(Fc,Gc) & &
\end{array}
$$

The square commutates by the naturality of the isomorphism $T(u \times v) \cong$

$Tu \times Tv$, while the triangle commutes because 1 is terminal, so the inverses of the displayed isomorphisms form a commutative triangle. The argument for functoriality of horizontal composites is similar.   □

REMARK A.7.4. In fact, the "change of base" procedure $\mathcal{V} \mapsto \mathcal{V}\text{-}\mathcal{C}at$ is *itself* a 2-functor from the 2-category of cartesian closed categories, finite-product-preserving functors, and natural transformations to the 2-category of 2-categories, 2-functors, and 2-natural transformations. See [31, §4.3] for a discussion and proof.

As an immediate consequence of the 2-functoriality of Remark A.7.4:

PROPOSITION A.7.5. *Any adjunction between cartesian closed categories whose left adjoint preserves finite products induces a change-of-base 2-adjunction*

$$\mathcal{V} \underset{U}{\overset{F}{\rightleftarrows}} \mathcal{W} \quad\leadsto\quad \mathcal{V}\text{-}\mathcal{C}at \underset{U_*}{\overset{F_*}{\rightleftarrows}} \mathcal{W}\text{-}\mathcal{C}at$$

Proof Of course right adjoints always preserve products, so the adjoint pair of functors $F \dashv U$ defines an adjunction in the 2-category of cartesian closed categories and finite-product-preserving functors described in Remark A.7.4. The 2-functor $\mathcal{V} \mapsto \mathcal{V}\text{-}\mathcal{C}at$ then carries the adjunction displayed on the left to the adjunction displayed on the right.   □

As a special case:

COROLLARY A.7.6. *For any cartesian closed category $\mathcal{V}$ with coproducts, the underlying category construction and free category construction define adjoint 2-functors*

$$\mathcal{C}at \underset{(-)_0}{\overset{}{\rightleftarrows}} \mathcal{V}\text{-}\mathcal{C}at \qquad\qquad □$$

In light of Proposition A.7.5 and results to follow, an adjunction between cartesian closed categories whose left adjoint preserves finite products provides a **change-of-base adjunction**. While Proposition A.7.5 permits the change of base along either adjoint of a finite-product-preserving adjunction, the next series of results reveal that change of base along the right adjoint is somewhat better behaved.

LEMMA A.7.7. *Any adjunction comprised of finite-product-preserving functors between cartesian closed categories*

$$\mathcal{V} \underset{U}{\overset{F}{\rightleftarrows}} \mathcal{W} \quad\leadsto\quad \mathcal{V} \underset{U}{\overset{F}{\rightleftarrows}} U_*\mathcal{W}$$

defines a $\mathcal{V}$-enriched adjunction between the $\mathcal{V}$-categories V and U_*W; i.e., there exists a $\mathcal{V}$-natural isomorphism $UW(Fv, w) \cong \mathcal{V}(v, Uw)$.

Proof The internal action $U_{a,b} : UW(a, b) \to \mathcal{V}(Ua, Ub)$ of the $\mathcal{V}$-functor $U : U_*W \to \mathcal{V}$ is defined by the transpose of the map $U\mathrm{ev} : UW(a, b) \times Ua \to Ub$ defined by applying U to the counit of the cartesian closure adjunction of W. The commutative square (A.2.4) provides the $\mathcal{V}$-functoriality of this map.

By the $\mathcal{V}$-functoriality of $U : U_*W \to \mathcal{V}$, the map

$$UW(Fv, w) \xrightarrow{U_{Fv,w}} \mathcal{V}(UFv, Uw) \xrightarrow{-\circ \eta_v} \mathcal{V}(v, Uw)$$

is $\mathcal{V}$-natural in $w \in U_*W$ for all $v \in \mathcal{V}$. By Remark A.3.17, to construct a compatible $\mathcal{V}$-enrichment of F, we need only demonstrate that this map in an isomorphism in $\mathcal{V}$.

We do this by constructing an explicit inverse, namely

$$\mathcal{V}(v, Uw) \xrightarrow{\eta} UFV(v, Uw) \xrightarrow{U(F_{v,Uw})} UW(Fv, FUw) \xrightarrow{\epsilon_w \circ -} UW(Fv, w)$$

where the middle map is defined by applying the unenriched functor U to the action map from the W-functor $F : F_*V \to W$, which is defined similarly to the $\mathcal{V}$-functor $U : U_*W \to \mathcal{V}$.

The proof that these maps are inverses involves a pair of diagram chases, the first of which demonstrates that the top-right composite reduces to the left-bottom composite, which is the identity:

The only subtle point is the commutativity of the trapezoidal region, which expresses the fact that $\eta : \mathrm{id}_{\mathcal{V}} \Rightarrow UF$ is a *closed natural transformation* between product-preserving functors between cartesian closed categories. This region commutes because the transposed diagram does:

the right-hand square by naturality, and the left-hand square because any naturally transformation between product-preserving functors is automatically a monoidal natural transformation (see Exercise A.7.i). The other diagram chase is similar. □

PROPOSITION A.7.8. *Given an adjunction between cartesian closed categories*

$$V \underset{U}{\overset{F}{\rightleftarrows}} W$$

whose left adjoint preserves finite products then if $\mathcal{C}$ is co/tensored as a W-category, $U_*\mathcal{C}$ is co/tensored as V-category with the co/tensor of $c \in \mathcal{C}$ by $v \in V$ defined by

$$v \otimes c := Fv \otimes c \qquad and \qquad c^v := c^{Fv}.$$

Proof Suppose $\mathcal{C}$ admits cotensors as a W-category. To verify that $U_*\mathcal{C}$ admits cotensors as a V-category we must supply an isomorphism

$$U\mathcal{C}(x, c^{Fv}) \cong (U\mathcal{C}(x, c))^v$$

in V that is V-natural in x. By the enriched Yoneda lemma, we can extract this isomorphism from an isomorphism

$$V(u, U\mathcal{C}(x, c^{Fv})) \cong V(u, (U\mathcal{C}(x, c))^v)$$

that is V-natural in $u \in V$. To that end, by composing the V-natural isomorphisms of Lemma A.7.7, the enriched natural isomorphisms arising from the cartesian closed structure on V and on U_*W, and the isomorphisms that characterize the cotensor on $\mathcal{C}$ and express the fact that F preserves binary products, we have:

$$V(u, U\mathcal{C}(x, c^{Fv})) \cong UW(Fu, \mathcal{C}(x, c^{Fv})) \cong UW(Fu, \mathcal{C}(x, c)^{Fv})$$
$$\cong UW(Fu \times Fv, \mathcal{C}(x, c)) \cong UW(F(u \times v), \mathcal{C}(x, c))$$
$$\cong V(u \times v, U\mathcal{C}(x, c)) \cong V(u, (U\mathcal{C}(x, c))^v). \qquad \square$$

This theory of change of base is all well and good from the compound noun perspective on enriched categories, but an additional concern arises from the adjectival point of view. If the finite-product-preserving functor $T: V \to W$ commutes with the underlying set functors for V and W up to natural isomorphism, then by the 2-functoriality of Remark A.7.4, the change-of-base 2-functor $T_*: V\text{-}\mathcal{C}at \to W\text{-}\mathcal{C}at$ also preserves the underlying categories up to natural isomorphism. This happens in particular in the following setting.

LEMMA A.7.9. *Consider a finite-product-preserving adjunction between cartesian closed categories:*

$$\mathcal{V} \underset{U}{\overset{F}{\rightleftarrows}} \mathcal{W}$$

Then change of base along the right adjoint respects the underlying categories:

$$\mathcal{W}\text{-}\mathcal{C}at \xrightarrow{\;\; U_* \;\;} \mathcal{V}\text{-}\mathcal{C}at$$
$$\underset{(-)_0}{\searrow} \qquad \underset{(-)_0}{\swarrow}$$
$$\mathcal{C}at$$

Proof Let $\mathcal{C}$ be a $\mathcal{W}$ category. Then the hom-set in the underlying category of $U_*\mathcal{C}$ from x to y is isomorphic to the corresponding hom-set

$$U_*\mathcal{C}(x,y)_0 \cong \mathcal{V}(1, U\mathcal{C}(x,y))_0 \cong \mathcal{W}(F1, \mathcal{C}(x,y))_0 \cong \mathcal{W}(1, \mathcal{C}(x,y))_0 \cong \mathcal{C}(x,y)_0$$

in the underlying category of $\mathcal{C}$ and moreover this isomorphism respects the composition and identities in the underlying categories. Thus $\mathcal{C}_0 \cong U\mathcal{C}_0$. A similar argument shows that change of base along U respects underlying functors and natural transformations. □

We close this chapter by returning to an example previewed in Digression 1.4.2.

EXAMPLE A.7.10. Both adjoints of the adjunction

$$s\mathcal{S}et \underset{}{\overset{h}{\rightleftarrows}} \mathcal{C}at$$

of Proposition 1.1.11 preserve finite products. Hence, Proposition A.7.5 induces a change-of-base adjunction defined by the 2-functors

$$s\mathcal{S}et\text{-}\mathcal{C}at \underset{}{\overset{h_*}{\rightleftarrows}} 2\text{-}\mathcal{C}at$$

that act identically on objects and act by applying the homotopy category functor or nerve functor, respectively, on homs. By Lemma A.7.9, the right adjoint, which builds a simplicially enriched category from a 2-category, respects the underlying category: the underlying category of objects and 1-cells is identified with the underlying category of objects and 0-arrows. In this case, the functor $h\colon s\mathcal{S}et \to \mathcal{C}at$ commutes with the underlying set functors, so in fact both adjoints preserve underlying categories, as is evident from direct computation. In particular, the homotopy 2-category of an ∞-cosmos has the same underlying 1-category.

Exercises

Exercise A.7.i. Look up the definition of monoidal functor and monoidal transformation – sometimes referred to as a *lax monoidal functor* and a *lax monoidal transformation* – and show that

  (i) any product-preserving functor between categories with finite products is monoidal, and
  (ii) any natural transformation between such functors is a monoidal natural transformation.

Exercise A.7.ii. Let $\mathcal{V}$ be a cartesian closed category and suppose $\mathcal{C}$ is a tensored and cotensored $\mathcal{V}$-category. By Proposition A.7.8, the underlying category $\mathcal{C}_0$ is tensored and cotensored as an unenriched category. Describe these tensors and cotensors.

Appendix B

An Introduction to 2-Category Theory

An important special case of enriched category theory arises when the base for enrichment is the cartesian closed category of categories themselves. Categories *enriched* in $\mathcal{C}at$ – *2-categories* – and categories defined *internally* to $\mathcal{C}at$ – *double categories* – were first introduced by Charles Ehresmann. A notable early expository account appeared in [70], while comprehensive modern treatments include [73] and [59]. The basic definitions are given in §B.1, which pays particular attention to the composition of 2-cells in a 2-category by means of *pasting diagrams*.

In §B.2, we briefly answer the question: what do 2-categories form? We define three dimensions of morphisms between 2-categories – the 2-functors, 2-natural transformations, and modifications – and observe that these collectively assemble into a *3-category*, this being a category enriched over the cartesian closed category of 2-categories.

There are many aspects of the theory of 2-categories that fall outside the purview of enriched category theory. We meet some of these in §B.3, where we develop the calculus of adjunctions and mates in any 2-category, complementing the results of §2.1, and in §B.4, where we study the special case of right adjoint right inverse adjunctions. In §B.5, we prove a lemma that produces absolute lifting diagrams that are preserved by any 2-functor. Finally, in §B.6 we consider the representability of various 2-categorical structures and comment briefly on the bicategorical Yoneda lemma.

B.1 2-Categories and the Calculus of Pasting Diagrams

The category $\mathcal{C}at$ of small categories is cartesian closed with the exponential B^A defined to be the category of functors and natural transformations from A to B and a terminal object given by the terminal category $\mathbb{1}$. Exploiting the work in Appendix A, we can concisely define a 2-category to be a category enriched over this cartesian closed category. Lemma A.2.3 then provides our first example: since any cartesian closed category is enriched over itself, $\mathcal{C}at$ defines an example of 2-category – the prototypical example.

Unpacking Definition A.2.1, we see that a 2-category contains a considerable amount of structure:

DEFINITION B.1.1 (2-category). A **2-category** $\mathcal{C}$ is a category enriched in $\mathcal{C}at$. Explicitly it has:

- a collection of objects;
- for each pair of objects $a, b \in \mathcal{C}$, a collection of arrows $f : a \to b$, also known as **1-cells**, these being the objects of the hom-category $\mathcal{C}(a, b)$; and
- for each p air of 1-cells $f, g : a \to b$, a collection of arrows between arrows

$$a \overset{f}{\underset{g}{\Longrightarrow}} b \quad \Downarrow\alpha$$

, called **2-cells**,[1] these being the morphisms $\alpha : f \Rightarrow g$ of the hom-category $\mathcal{C}(a, b)$ from f to g

so that:

(i) For each fixed pair of objects $a, b \in \mathcal{C}$, the 1-cells and 2-cells form a category. In particular:

- A pair of 2-cells as below-left admits a **vertical composite** as below-right:

$$a \overset{f}{\underset{h}{\overset{\Downarrow\alpha}{\underset{\Downarrow\beta}{\longrightarrow g \longrightarrow}}}} b \quad =: \quad a \overset{f}{\underset{h}{\Longrightarrow}} b \quad \Downarrow\beta\cdot\alpha$$

- Each 1-cell $f : a \to b$ has an identity 2-cell $a \overset{f}{\underset{f}{\Longrightarrow}} b \quad \Downarrow\mathrm{id}_f$.

(ii) The objects and 1-cells define a category in the ordinary sense; in particular, each object has an identity arrow $\mathrm{id}_a : a \to a$.

(iii) The objects and 2-cells form a category. In particular:

- A pair of 2-cells as below-left admits a **horizontal composite** as

---

[1] Implicit in this graphical representation is the requirement that a 2-cell α has a domain 1-cell f and a codomain 1-cell g, and these 1-cells have a common domain object a and codomain object b, the 0-cell source and 0-cell target of α.

below-right:

$$a \underset{g}{\overset{f}{\rightrightarrows}}\Downarrow\alpha\; b \underset{k}{\overset{j}{\rightrightarrows}}\Downarrow\gamma\; c \quad =: \quad a \underset{kg}{\overset{jf}{\rightrightarrows}}\Downarrow\gamma*\alpha\; c$$

- The identity 2-cells on identity 1-cells

$$a \underset{\mathrm{id}_a}{\overset{\mathrm{id}_a}{\rightrightarrows}}\Downarrow\mathrm{id}_{\mathrm{id}_a}\; a$$

define identities for horizontal composition.

(iv) Finally, the horizontal composition is functorial with respect to the vertical composition:

- The horizontal composite of identity 2-cells is an identity 2-cell:

$$a \underset{g}{\overset{g}{\rightrightarrows}}\Downarrow\mathrm{id}_g\; b \underset{k}{\overset{k}{\rightrightarrows}}\Downarrow\mathrm{id}_k\; c \quad = \quad a \underset{kg}{\overset{kg}{\rightrightarrows}}\Downarrow\mathrm{id}_{kg}\; c$$

- In the situation below, the horizontal composite of the vertical composites coincides with the vertical composite of the horizontal composites, a property referred to as **middle-four interchange**:

$$a \xrightarrow{g} b \xrightarrow{k} c \qquad\qquad (\delta*\beta)\cdot(\gamma*\alpha) = (\delta\cdot\gamma)*(\beta\cdot\alpha)$$

A degenerate special case of horizontal composition, in which all but one of the 2-cells is an identity id_f on its boundary 1-cell f, is called "whiskering."

DEFINITION B.1.2 (whiskering). The **whiskered composite** $h\alpha k$ of a 2-cell $a \underset{g}{\overset{f}{\rightrightarrows}}\Downarrow\alpha\; b$ with a pair of 1-cells $k: x \to a$ and $h: b \to y$ is defined by the horizontal composite:

$$x \underset{hgk}{\overset{hfk}{\rightrightarrows}}\Downarrow h\alpha k\; y \quad := \quad x \xrightarrow{k} a \underset{g}{\overset{f}{\rightrightarrows}}\Downarrow\alpha\; b \xrightarrow{h} y$$

$$:= \quad x \underset{k}{\overset{k}{\rightrightarrows}}\Downarrow\mathrm{id}_k\; a \underset{g}{\overset{f}{\rightrightarrows}}\Downarrow\alpha\; b \underset{h}{\overset{h}{\rightrightarrows}}\Downarrow\mathrm{id}_h\; y$$

As the following lemma reveals, horizontal composition can be recovered from vertical composition and whiskering. Our primary interest in this result, however, has to with a rather prosaic consequence appearing as the final part of the statement, which is surprisingly frequently apposite

LEMMA B.1.3 (naturality of whiskering). *For any horizontally composable pair of 2-cells in a 2-category there is a commutative square in the hom-category from the domain object to the codomain object formed by the whiskered cells whose diagonal defines the horizontal composite:*

$$\mathcal{C} \ni \quad a \underset{g}{\overset{f}{\underset{\Downarrow\alpha}{\rightrightarrows}}} b \underset{k}{\overset{h}{\underset{\Downarrow\beta}{\rightrightarrows}}} c \quad \rightsquigarrow \quad \begin{array}{ccc} hf & \overset{\beta f}{\Longrightarrow} & kf \\ h\alpha \Big\downarrow & \overset{\searrow}{\beta*\alpha} & \Big\downarrow k\alpha \\ hg & \underset{\beta g}{\Longrightarrow} & kg \end{array} \quad \in \mathcal{C}(a,c)$$

In particular, if any three of the four whiskered 2-cells $h\alpha$, $k\alpha$, βf, and βg are invertible, so is the fourth.

Proof By middle-four interchange:

$$\beta g \cdot h\alpha = (\beta * \mathrm{id}_g) \cdot (\mathrm{id}_h *\alpha) = (\beta \cdot \mathrm{id}_h) * (\mathrm{id}_g \cdot\alpha) = \beta * \alpha$$
$$= (\mathrm{id}_k \cdot\beta) * (\alpha \cdot \mathrm{id}_f) = (\mathrm{id}_k *\alpha) \cdot (\beta * \mathrm{id}_f) = k\alpha \cdot \beta f. \qquad \square$$

The operations of horizontal and vertical composition are special cases of composition by *pasting*, an operation first introduced by Bénabou [9]. The main result, proven in the 2-categorical context by Power [91] is that a well-formed *pasting diagram* such as

(B.1.4)

has a unique 2-cell composite.[2] We leave the formal statement and proof of this result to the literature and instead describe informally how such pasting composites should be interpreted.

DIGRESSION B.1.5 (how to read a pasting diagram). A pasting diagram in a 2-category $\mathcal{C}$ represents a unique composite 2-cell, defining a morphism in one

---

[2] This result was generalized to the bicategorical context by Verity [130], in which case the composite 2-cell is well-defined once its source and target 1-cells are specified (see also [58]).

of the hom-categories between a pair of objects. To identify these objects, look at the underlying directed graph of objects and 1-cells in the pasting diagram. If the pasting diagram is well-formed, that graph will have a unique source object a and a unique target object z. This indicates that the pasting diagram defines a 2-cell in·the hom-category $\mathcal{C}(a, z)$. The object a is its **source 0-cell** and the object z is its **target 0-cell**.

The next step is to identify the source 1-cell and the target 1-cell of the pasting diagram. These will both be objects of $\mathcal{C}(a, z)$, i.e., 1-cells in the 2-category from a to z. Again if the pasting diagram is well-formed, the **source 1-cell** will be the unique composable path of 1-cells none of which occur as part of the codomain of any 2-cell in the pasting diagram. In the diagram (B.1.4), these are the 1-cells whose labels appear above the arrow, and their composite is hgf. Dually, the **target 1-cell** will be the unique composable path of 1-cells, none of which occur as part of the domain of any 2-cell in the pasting diagram. In (B.1.4), these are the 1-cells whose labels appear below the diagram, and their composite is trp.

The final step is to represent the pasting diagram as a vertical composite of 2-cells from a to z, each of which is a whiskered composite of one of the displayed "atomic" 2-cells. The source and target of the whiskered atomic 2-cells trace composable paths from a to z through the directed graph underlying the pasting diagram that differ only by substituting the source of the atomic 2-cell for its target. Each 2-cell in the pasting diagram will label precisely one of the 2-cells of this composite. The expressions of these vertical 2-cell composites are not necessarily unique and may not necessarily pass through every possible composable path of 1-cells, though there will be some vertical composite of 2-cells that does pass through each path of 1-cells.

To start, pick any 2-cell in the pasting diagram whose 1-cell source can be found as a subsequence of the source 1-cell; in the (B.1.4), either α or β can be chosen first. Whisker it so that it defines a 2-cell from the source 1-cell hgf to another path of composable 1-cells from a to z through the pasting diagram. Then this whiskered composite forms the first step in the sequence of composable 2-cells. Remove this part of the pasting diagram and repeat until you arrive at the target 1-cell. In the example above, there are four possible ways to express the composite pasted cell (B.1.4) as vertical composites of whiskered 2-cells, represented by the four paths through the following commutative diagram in the category $\mathcal{C}(a, z)$:

$$hgf \xRightarrow{h\alpha f} hkjif \xRightarrow{hkj\gamma} hkjn \xRightarrow{hk\delta} hkqp$$

$$\beta gf \Downarrow \qquad \beta kjif \Downarrow \qquad \Downarrow \beta kjn \qquad \Downarrow \beta kqp$$

$$m\ell gf \underset{m\ell\alpha a}{\Longrightarrow} m\ell kjia \underset{m\ell kj\gamma}{\Longrightarrow} m\ell kjn \underset{m\ell k\delta}{\Longrightarrow} m\ell kqp \xRightarrow{m\epsilon p} msrp \xRightarrow{\zeta rp} trp$$

A 2-category has four duals, including itself, each of which have the same objects, 1-cells, and 2-cells, but swapping the domains and codomains in some dimension.

DEFINITION B.1.6 (op and co duals). Let $\mathcal{C}$ be a 2-category.

- Its **op-dual** $\mathcal{C}^{\mathrm{op}}$ is the 2-category with $\mathcal{C}^{\mathrm{op}}(a, b) := \mathcal{C}(b, a)$. This reverses the direction of the 1-cells but not the 2-cells.
- Its **co-dual** $\mathcal{C}^{\mathrm{co}}$ is the 2-category with $\mathcal{C}^{\mathrm{co}}(a, b) := \mathcal{C}(a, b)^{\mathrm{op}}$. This reverses the direction of the 2-cells but not the 1-cells.
- Its **coop-dual** $\mathcal{C}^{\mathrm{coop}}$ is the 2-category with $\mathcal{C}^{\mathrm{coop}}(a, b) := \mathcal{C}(b, a)^{\mathrm{op}}$. This reverses the direction of both the 2-cells and the 1-cells.

Recall from Definition 1.4.6 that an **equivalence** in a 2-category is given by

- a pair of objects a and b
- a pair of 1-cells $f : a \to b$ and $g : b \to a$
- a pair of invertible 2-cells[3]

$$a \underset{gf}{\overset{}{\rightleftarrows}} \cong \Downarrow \alpha \; a \qquad \text{and} \qquad b \overset{fg}{\underset{}{\rightleftarrows}} \cong \Downarrow \beta \; b$$

DIGRESSION B.1.7 (notions of sameness inside a 2-category). From the point of view of 2-category theory, the most natural notion of "sameness" for two objects of a 2-category is *equivalence*: a and b are to be regarded as the same if there exists an equivalence between them.

The most natural notion of "sameness" for a parallel pair of morphisms in a 2-category is *isomorphism*: $h, k : a \to b$ are to be regarded as the same if there exists an invertible 2-cell $\gamma : h \cong k$.

The most natural notion of "sameness" for a pair of 2-cells with common boundary is *equality*. Because a 2-category lacks any higher dimensional morphisms to mediate between 2-cells, there is no weaker notion available.

---

[3] The default meaning of "invertibility" for 2-cells is invertibility for vertical composition. Note the boundary 1-cells of an invertible 2-cell need not be invertible.

In Digression 1.2.4, we saw that the data of a simplicial category could be expressed as a diagram of a particular type valued in $\mathcal{C}at$. A small 2-category can be similarly encoded – in fact in two different ways – as a category defined *internally* to the category of categories.

DEFINITION B.1.8 (internal category). Let $\mathcal{E}$ be any category with pullbacks. An **internal category** in $\mathcal{E}$ is given by the data

$$
\begin{array}{c}
C_1 \underset{C_0}{\times} C_1 \\
\pi_\ell \swarrow \quad \vee \quad \searrow \pi_r \\
C_1 \qquad\qquad C_1 \\
d \searrow \qquad \swarrow c \\
C_0
\end{array}
\qquad
C_1 \underset{C_0}{\times} C_1 \;-\!\circ\!\to\; C_1 \underset{\xrightarrow{\;\;d\;\;}}{\overset{\xleftarrow{\;i\;}}{\xrightarrow[\;\;c\;\;]{}}} C_0
$$

subject to commutative diagrams that define the domains and codomains of composites and identities

$$
\begin{array}{ccc}
C_1 & \xleftarrow{\;\pi_\ell\;} C_1 \underset{C_0}{\times} C_1 \xrightarrow{\;\pi_r\;} & C_1 \\
c \downarrow & \circ \downarrow & \downarrow d \\
C_0 & \xleftarrow{\;c\;} C_1 \xrightarrow{\;d\;} & C_0
\end{array}
\qquad
\begin{array}{c}
C_0 \\
\diagup \downarrow i \diagdown \\
C_0 \xleftarrow{\;c\;} C_1 \xrightarrow{\;d\;} C_0
\end{array}
$$

and encode the fact that composition is associative and unital.

$$
\begin{array}{ccc}
C_1 \underset{C_0}{\times} C_1 \underset{C_0}{\times} C_1 & \xrightarrow{\;\circ\times\mathrm{id}\;} & C_1 \underset{C_0}{\times} C_1 \\
\mathrm{id}\times\circ \downarrow & & \downarrow \circ \\
C_1 \underset{C_0}{\times} C_1 & \xrightarrow{\;\circ\;} & C_1
\end{array}
\qquad
\begin{array}{ccc}
C_1 & \xrightarrow{(\mathrm{id},i)} C_1 \underset{C_0}{\times} C_1 \xleftarrow{(i,\mathrm{id})} & C_1 \\
& \diagdown \; \circ \downarrow \; \diagup & \\
& C_1 &
\end{array}
$$

An internal category in Set is an ordinary small category. An internal category in $\mathcal{C}at$ defines a double category:

DEFINITION B.1.9 (double category). A **double category** $\mathcal{C}$ is an internal category

$$
C_{h,s} \underset{C_{0,v}}{\times} C_{h,s} \;-\!\circ\!\to\; C_{h,s} \underset{\xrightarrow{\;\;d\;\;}}{\overset{\xleftarrow{\;i\;}}{\xrightarrow[\;\;c\;\;]{}}} C_{0,v}
$$

in $\mathcal{C}at$. Explicitly it has:

- a category $C_{0,v}$ of **objects** and **vertical arrows**;
- a category $C_{h,s}$ of **horizontal arrows** and **squares**;

- functors $c, d : C_{h,s} \to C_{o,v}$ that assign a co/domain object to each horizontal arrow and a co/domain vertical arrow to each square;
- a functor $i : C_{o,v} \to C_{h,s}$ that assigns a horizontal identity arrow to each object and an identity square to each vertical arrow; and
- a composition functor $\circ : C_{h,s} \times_{C_{o,v}} C_{h,s} \to C_{h,s}$ that defines horizontal composition of horizontal arrows and squares that is functorial with respect to vertical composition in each variable

satisfying the axioms imposed by the commutative diagrams of an internal category.

A 2-category can be realized as a special case of this construction in the following two ways.

DIGRESSION B.1.10 (2-categories as category objects). A 2-category may be defined to be an internal category in $\mathcal{C}at$

$$C_{1,2} \underset{C_0}{\times} C_{1,2} \ -\circ\to\ C_{1,2} \ \overset{d}{\underset{c}{\leftarrow i -}}\ C_0$$

in which the category C_0 is a set, namely the set of objects of the 2-category. The 1- and 2-cells occur as the objects and arrows of the category $C_{1,2}$. The functors $d, c : C_{1,2} \to C_0$ send 1- and 2-cells to their domain and codomain 0-cells. The functor $i : C_0 \to C_{1,2}$ sends each object to its identity 1-cell. The action of the functor $\circ : C_{1,2} \times_{C_0} C_{1,2} \to C_{1,2}$ on objects defines composition of 1-cells and the action on morphisms defines the horizontal composition on 2-cells. Vertical composition on 2-cells is the composition inside the category $C_{1,2}$. Functoriality of this map encodes middle-four interchange.[4]

Dually, a 2-category may be defined to be an internal category in $\mathcal{C}at$

$$C_{0,2} \underset{C_{0,1}}{\times} C_{0,2} \ -\circ\to\ C_{0,2} \ \overset{d}{\underset{c}{\leftarrow i -}}\ C_{0,1}$$

in which the categories $C_{0,1}$ and $C_{0,2}$ have the same set of objects and all four functors are identity-on-objects. Here the common set of objects defines the objects of the 2-category and the arrows of $C_{0,1}$ and $C_{0,2}$ define the 1- and 2-cells, respectively. The functors $d, c : C_{0,2} \to C_{0,1}$ define the domain and codomain 1-cells for a 2-cell, which the functor $\circ : C_{0,2} \times_{C_{0,1}} C_{0,2} \to C_{0,2}$ encodes vertical composition of 2-cells. The composition inside the category $C_{0,2}$ defines horizontal composition of 2-cells. Functoriality of this map encodes middle-four interchange.

---

[4] This definition motivates the Segal category model of $(\infty, 1)$-categories described in Definition E.2.4.

Exercises

EXERCISE B.1.i. Relate the structures itemized in Definition B.1.1 to the structures itemized in Definition A.2.1 in the case where the base for enrichment is $\mathcal{C}at$.

EXERCISE B.1.ii. Define a duality involution on double categories that exchanges the two expressions of a 2-category as an internal category appearing in Digression B.1.10.

B.2 The 3-Category of 2-Categories

Ordinary 1-categories form the objects of a 2-category of categories, functors, and natural transformations. Similarly, 2-categories form the objects of a 3-category of 2-categories, 2-functors, 2-natural transformations, and modifications. In this section, we briefly introduce all of these notions.

Recall from Definition B.1.1, that a 2-category is a category enriched in $\mathcal{C}at$. Similarly, 2-functors and 2-natural transformations are precisely the $\mathcal{C}at$-enriched functors and $\mathcal{C}at$-enriched natural transformations of Appendix A. By Corollary A.3.6, 2-categories, 2-functors, and 2-natural transformations assemble into a 2-category. The 3-dimensional cells between 2-categories – the modifications – are defined using the 2-cells of a 2-category, like the 2-dimensional cells between 1-categories – the natural transformations – are defined using the 1-cells in a 1-category.

DEFINITION B.2.1. A **2-functor** $F \colon \mathcal{C} \to \mathcal{D}$ between 2-categories is given by:

- a mapping on objects $\mathcal{C} \ni x \mapsto Fx \in \mathcal{D}$;
- a functorial mapping on 1-cells $\mathcal{C} \ni f \colon x \to y \mapsto Ff \colon Fx \to Fy \in \mathcal{D}$ respecting domains and codomains; and
- a mapping on 2-cells

$$\mathcal{C} \ni \quad x \underset{g}{\overset{f}{\underset{\Downarrow\alpha}{\rightrightarrows}}} y \quad \mapsto \quad Fx \underset{Fg}{\overset{Ff}{\underset{\Downarrow F\alpha}{\rightrightarrows}}} Fy \quad \in \mathcal{D}$$

that respects 0- and 1-cell sources and targets that is functorial for both horizontal and vertical composition and horizontal and vertical identities.

DEFINITION B.2.2. A **2-natural transformation** $\mathcal{C} \underset{G}{\overset{F}{\underset{\Downarrow\phi}{\rightrightarrows}}} \mathcal{D}$ between a parallel pair of 2-functors F and G is given by a family of 1-cells $(\phi_c \colon Fc \to$

$Gc)_{c \in C}$ in $\mathcal{D}$ indexed by the objects of C that are natural with respect to the 1-cells $f : x \to y$ in C, in the sense that the square

$$
\begin{array}{ccc}
Fx & \xrightarrow{Ff} & Fy \\
{\scriptstyle \phi_x}\downarrow & & \downarrow{\scriptstyle \phi_y} \\
Gx & \xrightarrow[Gf]{} & Gy
\end{array}
$$

commutes in $\mathcal{D}$, and also natural with respect to the 2-cells $x \overset{f}{\underset{g}{\Rightarrow}}{\scriptstyle \Downarrow\alpha}\, y$ in C, in the sense that the whiskered composites $\phi_y \cdot F\alpha$ and $G\alpha \cdot \phi_x$ are equal:

$$
\begin{array}{c}
Fx \\
{\scriptstyle \phi_x}\downarrow \qquad \overset{Gf}{\underset{Gg}{\Longrightarrow}}{\scriptstyle \Downarrow G\alpha} \\
Gx \qquad\qquad Gy
\end{array}
\quad = \quad
\begin{array}{c}
Fx \overset{Ff}{\underset{Fg}{\Longrightarrow}}{\scriptstyle \Downarrow F\alpha}\, Fy \\
\qquad\qquad\quad \downarrow{\scriptstyle \phi_y} \\
\qquad\qquad\quad Gy
\end{array}
$$

The 3-dimensional morphisms between 2-categories are outside the purview of C*at*-enriched category theory:

DEFINITION B.2.3. A **modification** $\Xi : \phi \Rightarrow \psi$

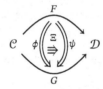

between parallel 2-natural transformations is given by a family of 2-cells in $\mathcal{D}$

$$
Fc \overset{\phi_c}{\underset{\psi_c}{\Longrightarrow}}{\scriptstyle \Downarrow\Xi_c}\, Gc
$$

indexed by the objects $c \in C$ with the property that for any 1-cell $f : x \to y$ in C, the whiskered composites $\Xi_y \cdot Ff = Gf \cdot \Xi_x$ are equal in $\mathcal{D}$ and for any 2-cell $\alpha : f \Rightarrow g$ in $\mathcal{D}$, the horizontal composites are equal in $\mathcal{D}$:

$$
Fx \overset{Ff}{\underset{Fg}{\Longrightarrow}}{\scriptstyle \Downarrow F\alpha}\, Fy \overset{\phi_y}{\underset{\psi_y}{\Longrightarrow}}{\scriptstyle \Downarrow\Xi_y}\, Gy \;=\; Fx \overset{\phi_x}{\underset{\psi_x}{\Longrightarrow}}{\scriptstyle \Downarrow\Xi_x}\, Gx \overset{Gf}{\underset{Gg}{\Longrightarrow}}{\scriptstyle \Downarrow G\alpha}\, Gy
$$

The category of 2-categories is cartesian closed, with internal hom $\mathcal{B}^{\mathcal{A}}$ defined to be the 2-category of 2-functors, 2-natural transformations, and modifications. So now we can define a **3-category** to be a category enriched in 2-categories.

COROLLARY B.2.4. *There is a 3-category* 2-$\mathcal{C}at$ *of 2-categories, 2-functors, 2-natural transformations, and modifications.*

Proof Lemma A.2.3 proves that any cartesian closed category is enriched over itself. Thus 2-$\mathcal{C}at$ defines a 3-category. □

Exercises

EXERCISE B.2.i. Relate the structures itemized in Definitions B.2.1 and B.2.2 to the structures itemized in Definition A.2.6 and A.3.2 in the case where the base for enrichment is $\mathcal{C}at$.

EXERCISE B.2.ii. For the reader who has a large writing surface, unpack the definition of a 3-category just given.

EXERCISE B.2.iii.

(i) Show that the functor that sends a 1-category to its opposite defines an involutive 2-functor $(-)^{\mathrm{op}} : \mathcal{C}at^{\mathrm{co}} \to \mathcal{C}at$ on the 2-category of categories, functors, and natural transformations.

(ii) Similarly the functors that send a 2-category to one of its three duals can be understood as involutions of the 3-category of 2-categories, 2-functors, 2-natural transformations, and modifications. What is the variance of each of these mappings?

B.3 Adjunctions and Mates

As discussed in Chapter 2, any 2-category has an internally defined notion of **adjunction**, comprised of:

- a pair of objects a and b,
- a pair of 1-cells $u : a \to b$ and $f : b \to a$,
- and a pair of 2-cells $\eta : 1_b \Rightarrow uf$ and $\epsilon : fu \Rightarrow 1_a$, called the **unit** and **counit** respectively,

so that the triangle equalities hold:

The 1-cell f is called the **left adjoint** and u is called the **right adjoint**, a relationship that is denoted symbolically in text by writing $f \dashv u$ or in a displayed diagram such as

$$a \underset{u}{\overset{f}{\rightleftarrows}} \perp \; b \qquad (B.3.1)$$

The basic 2-category theory of adjunctions is developed in §2.1, whose results specialize to prove theorems about adjunctions between ∞-categories. Here we extend that theory in a complementary direction by developing the calculus of *mates*, which generalize the more familiar adjoint transposes.

In the presence of an adjunction as in (B.3.1), certain 2-cells with codomain a "transpose" into 2-cells with codomain b; op-dually, certain 2-cells with domain a "transpose" into 2-cells with domain b: for any 1-cells $w \xrightarrow{x} b, w \xrightarrow{y} a, b \xrightarrow{h} z$, and $a \xrightarrow{k} z$

$$fx \overset{\alpha}{\Longrightarrow} y \;\rightsquigarrow\; x \overset{\beta}{\Longrightarrow} uy \qquad hu \overset{\gamma}{\Longrightarrow} k \;\rightsquigarrow\; h \overset{\delta}{\Longrightarrow} kf$$
$$(B.3.2)$$

Both of these transposition operations admit a common generalization due to Kelly and Street [70] referred to as the "mates correspondence" which describes a duality between 2-cells induced by a pair of adjunctions.

DEFINITION B.3.3 (mates). Given any pair of adjunctions and functors

$$
\begin{array}{ccc}
b & \xrightarrow{\;k\;} & b' \\
f\left(\dashv\right)u & & f'\left(\dashv\right)u' \\
a & \xrightarrow{\;h\;} & a'
\end{array}
$$

there is a bijection between 2-cells as below-left and 2-cells as below-right

$$
\begin{array}{ccc}
b \xrightarrow{\;k\;} b' & & b \xrightarrow{\;k\;} b' \\
f\downarrow \quad {\swarrow\alpha} \quad \downarrow f' & \;\rightsquigarrow\; & u\uparrow \quad {\searrow\beta} \quad \uparrow u' \\
a \xrightarrow[h]{} a' & & a \xrightarrow[h]{} a'
\end{array}
\qquad (B.3.4)
$$

implemented by pasting with the units and counits of the adjunctions:

$$
\begin{array}{ccc}
b \xrightarrow{\;k\;} b' & & b === b \xrightarrow{\;k\;} b' \\
f\downarrow \;\; {\swarrow\alpha} \;\; \downarrow f' & := & {\searrow\eta}\;\; u\uparrow \;\; {\searrow\beta} \;\; u'\uparrow \;\; {\searrow} f' \\
a \xrightarrow[h]{} a' & & f{\searrow}\;\; a \xrightarrow[h]{} a' === a' \quad {\searrow\epsilon'}
\end{array}
$$

$$b \xrightarrow{k} b' \qquad b \xrightarrow{k} b' = b'$$

$$u \uparrow \quad \searrow\beta \quad \uparrow u' \quad := \quad {}^{u}\nearrow \quad \begin{smallmatrix} | \\ f \\ \downarrow \end{smallmatrix} \quad \swarrow\alpha \quad \begin{smallmatrix} | \\ f' \\ \downarrow \end{smallmatrix} \quad \swarrow\eta' \quad \nearrow$$

$$a \xrightarrow{h} a' \qquad a = a \xrightarrow{h} a' \qquad u'$$

Corresponding 2-cells (B.3.4) under this bijection are referred to as **mates**.

EXAMPLE B.3.5. The mates correspondence specializes to define a bijection between 2-cells between a parallel pair of left adjoints and 2-cells pointing in the opposite direction between their right adjoints that Mac Lane refers to as **conjugates** [81, §IV.7]:

$$b = b \qquad b = b$$

$$f \downarrow \quad \swarrow\alpha \quad \downarrow f' \qquad \leftrightsquigarrow \qquad u \uparrow \quad \searrow\beta \quad \uparrow u'$$

$$a = a \qquad a = a$$

The mates correspondence is respected by horizontal and vertical composition of squares (B.3.4) in the sense made precise by the following result:

THEOREM B.3.6 (double-functoriality of the mates correspondence). *For any 2-category, there is a double isomorphism* $\mathbb{L}\mathsf{adj} \cong \mathbb{R}\mathsf{adj}$ *between the double categories whose*

- *objects and horizontal morphisms are the objects and 1-cells*
- *vertical morphisms are fully specified adjunctions* (f, u, η, ϵ) *pointing in the direction of the left adjoint*[5]
- *cells in* $\mathbb{L}\mathsf{adj}$ *are 2-cells of the form displayed below-left, while cells in* $\mathbb{R}\mathsf{adj}$ *are 2-cells of the form displayed below-right:*

$$b \xrightarrow{k} b' \qquad b \xrightarrow{k} b'$$

$$f \downarrow \quad \swarrow\alpha \quad \downarrow f' \qquad u \uparrow \quad \searrow\beta \quad \uparrow u'$$

$$a \xrightarrow{h} a' \qquad a \xrightarrow{h} a'$$

that acts as the identity on objects and on horizontal and vertical morphisms and acts on cells by taking mates.[6]

Proof The horizontal and vertical functoriality of the mates correspondence can be verified by a pasting diagram chase (or see [70, 2.2]). □

---

[5] The composition of vertical morphisms makes use of the construction given in the proof of Proposition 2.1.9.

[6] The isomorphism of double categories can be regarded as defining a component of a 2-natural isomorphism between 2-functors $\mathbb{L}\mathsf{adj}, \mathbb{R}\mathsf{adj} : 2\text{-}\mathcal{C}at \to Dbl\text{-}\mathcal{C}at$ from $2\text{-}\mathcal{C}at$ to the 2-category of double categories, double functors, and horizontal natural transformations.

WARNING B.3.7 (mates of isomorphisms need not be isomorphisms). In general it is possible for one of the 2-cells in a mate pair (B.3.4) to be invertible without the other being so. For instance, the unit and counit of an adjunction $f \dashv u$ are each mates with both the identity cells id_f and id_u, depending on which way these 2-cells are arranged to fit in squares (see Exercise B.3.i). However if both horizontal 1-cells h and k are equivalences, or if both adjunctions $f \dashv u$ and $f' \dashv u'$ are adjoint equivalences, then $\alpha : f'k \Rightarrow hf$ is invertible if and only if its mate $\beta : ku \Rightarrow u'h$ is invertible.

Elaborating upon Warning B.3.7 we have:

PROPOSITION B.3.8 (equivalence invariance of adjointness). *Suppose given an essentially commutative square whose horizontal arrows are equivalences:*

$$
\begin{array}{ccc}
b & \xrightarrow{\ k\ }_{\sim} & b' \\
u \uparrow{\scriptstyle\vdash}\ \Big\downarrow & f \cong_{\mathscr{l}} \alpha\ f' & \dashv u' \\
a & \xrightarrow[\ h\ ]{\sim} & a'
\end{array}
\quad\rightsquigarrow\quad
\begin{array}{ccc}
b & \xrightarrow{\ k\ }_{\sim} & b' \\
u\ \cong\,{\scriptstyle\searrow} u'h\epsilon\cdot u'\alpha u\cdot\eta' & & u' \\
a & \xrightarrow[\ h\ ]{\sim} & a'
\end{array}
$$

Then f admits a right adjoint u if and only f' admits a right adjoint u', in which case the mate of the isomorphism α is an isomorphism.

Proof Proposition 2.1.12 may be used to choose inverse adjoint equivalences $k' \dashv k$ and $h \dashv h'$. If f is a left adjoint, then by Proposition 2.1.9, $f' \cong hfk'$ is isomorphic to a left adjoint, and so by Proposition 2.1.10, f' is left adjoint to kuh'. If $f' \dashv u'$ is defined to be the composite adjunction as in the previous paragraph, the mate of α works out to be the whiskered composite of $h'h \cong \mathrm{id}_a$ with ku. By Proposition 2.1.10, any other choice of right adjoint to f' is isomorphic to this one, so the mate of α is still an isomorphism. $\qquad\square$

Exercise B.3.ii suggests a new proof that any pair of left adjoints $f' \dashv u$ and $f \dashv u$ to a common 1-cell are isomorphic (see Proposition 2.1.10) by applying the double isomorphism $\mathbb{L}\mathsf{adj} \cong \mathbb{R}\mathsf{adj}$. A more complicated argument along the same lines can be used to prove:

LEMMA B.3.9. *Suppose given a triple of adjoint functors $\ell \dashv i \dashv r$. Then the counit of $\ell \dashv i$ is invertible if and only if the unit of $i \dashv r$ is invertible.*

Proof Let $i : a \to b$ and write $\epsilon : \ell i \Rightarrow \mathrm{id}_a$ for the counit of $\ell \dashv i$ and $\eta : \mathrm{id}_a \Rightarrow ri$ for the unit of $i \dashv r$. If ϵ admits an inverse isomorphism $\epsilon^{-1} : \ell i \Rightarrow \mathrm{id}_a$, then the vertical composite in $\mathbb{L}\mathsf{adj}$ displayed below-left admits an inverse cell for horizontal composition in $\mathbb{L}$adj displayed below-right:

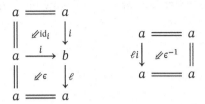

Applying the horizontal functoriality of the double isomorphism $\mathbb{L}$adj $\cong \mathbb{R}$adj, the mates of these cells must also compose horizontally in $\mathbb{R}$adj to identities.[7] Applying the vertical functoriality of the double isomorphism $\mathbb{L}$adj $\cong \mathbb{R}$adj, the mate of the vertical composite equals the composite

$$
\begin{array}{ccc}
a & = & a \\
\| & \searrow\eta & \uparrow r \\
a & \xrightarrow{\ i\ } & b \\
\| & \searrow\mathrm{id}_i & \uparrow i \\
a & = & a
\end{array}
$$

in $\mathbb{R}$adj, and thus the mate of ϵ^{-1} must define the inverse of η. In summary, we see that the counit of $\ell \dashv i$ is an isomorphism if and only if the unit of $i \dashv r$ is an isomorphism, in which case its inverse isomorphism $\eta^{-1} : ri \Rightarrow \mathrm{id}_a$ is the conjugate of $\epsilon^{-1} : \mathrm{id}_a \Rightarrow \ell i$ via the composite adjunction $\ell i \dashv ri$:

$$
\begin{array}{ccc}
a & = & a \\
\ell i\downarrow & \swarrow\epsilon^{-1} & \| \\
a & = & a
\end{array}
\qquad \leftrightsquigarrow \qquad
\begin{array}{ccc}
a & = & a \\
ri\uparrow & \searrow\eta^{-1} & \| \\
a & = & a
\end{array}
\qquad \square
$$

Exercises

EXERCISE B.3.i.

  (i) Explain how the bijections (B.3.2) may be realized as special cases of the mates correspondence.
 (ii) By choosing a suitable pair of adjunctions and functors, explain how the unit of an adjunction $f \dashv u$ is mates with id_u.
(iii) By choosing a suitable pair of adjunctions and functors, explain how the unit of an adjunction $f \dashv u$ is mates with id_f.

---

[7] Since the horizontal morphisms in the cells in question are all identities, the concern raised in Warning B.3.7 does not apply.

EXERCISE B.3.ii. Consider two adjunctions $f \dashv u$ and $f' \dashv u$ as vertical morphisms in $\mathbb{L}\mathsf{adj} \cong \mathbb{R}\mathsf{adj}$ and apply the double functoriality of the mates correspondence to prove that $f \cong f'$.

EXERCISE B.3.iii. Consider a 2-cell

$$
\begin{array}{ccc}
b & \xrightarrow{\ r\ } & b' \\
f \downarrow & \Swarrow \alpha & \downarrow f' \\
a & \xrightarrow[\ r'\ ]{} & a'
\end{array}
$$

in which the vertical 1-cells admit right adjoints $f \dashv u$ and $f' \dashv u'$ and the horizontal 1-cells admit left adjoints $\ell \dashv r$ and $\ell' \dashv r'$. Show that the mate of α with respect to the vertical adjunctions defines an isomorphism $ru \cong u'r'$ if and only if the mate of α with respect to the horizontal adjunctions defines an isomorphism $\ell' f' \cong f\ell$ – and these isomorphisms are themselves mates in the sense of Example B.3.5 with respect to the composite adjunctions.

EXERCISE B.3.iv. Show that any 2-functor $F \colon \mathcal{C} \to \mathcal{D}$ preserves equivalences, adjunctions, and mates.

B.4 Right Adjoint Right Inverse Adjunctions

An important class of adjunctions are those whose counits are invertible.

DEFINITION B.4.1. A 1-cell $f \colon b \to a$ in a 2-category admits a **right adjoint right inverse** if it admits a right adjoint $u \colon a \to b$ so that the counit of the adjunction $f \dashv u$ is an isomorphism. In this situation, f is **left adjoint left inverse** to u.

The co-dual defines the class **left adjoint right inverse** or **right adjoint left inverse** adjunctions with invertible unit.

When the counit of $f \dashv u$ is an isomorphism, the whiskered composites $f\eta$ and ηu of the unit must also be isomorphisms. Indeed, to construct an adjunction of this form it suffices to give 2-cells with these properties, as demonstrated by the following 2-categorical lemma.

LEMMA B.4.2. *Suppose we are given a pair of 1-cells $u \colon a \to b$ and $f \colon b \to a$ and a 2-isomorphism $fu \cong \mathrm{id}_a$ in a 2-category. If there exists a 2-cell $\eta' \colon \mathrm{id}_b \Rightarrow uf$ with the property that $f\eta'$ and $\eta'u$ are 2-isomorphisms, then f is left adjoint to u. Furthermore, in the special case where u is a section of f, then f is left adjoint to u with the counit of the adjunction an identity.*

Since $f\eta'$ and $\eta'u$ are isomorphisms and $fu \cong \mathrm{id}_a$, the "triangle equality composites" of Lemma 2.1.11 are invertible, so from that result we can conclude that $f \dashv u$. Indeed, from the construction given in that proof, we see that we can take the specified isomorphism $fu \cong \mathrm{id}_a$ to be the counit of the adjunction, so in particular when u is a section of f this counit may be taken to be the identity. The direct proof given below compiles out the argument just sketched.

Proof Let $\epsilon : fu \Rightarrow \mathrm{id}_a$ be the isomorphism, taken to be the identity in the case where u is a section of f. We will define an adjunction $f \dashv u$ with counit ϵ by modifying $\eta' : \mathrm{id}_b \Rightarrow uf$. The "triangle identity composite" $\theta := u\epsilon \cdot \eta'u : u \Rightarrow u$ defines an automorphism of u. Define

$$\eta := \mathrm{id}_b \xRightarrow{\eta'} uf \xRightarrow{\theta^{-1}f} uf := \mathrm{id}_b \xRightarrow{\eta'} uf \xRightarrow{(u\epsilon f)^{-1}} ufuf \xRightarrow{(\eta'uf)^{-1}} uf.$$

Immediately, $u\epsilon \cdot \eta u = \mathrm{id}_u$, as is verified by the calculation:

The other triangle identity composite $\phi := \epsilon f \cdot f\eta$ is an isomorphism, as a composite of isomorphisms, and also an idempotent:

By cancelation, any idempotent isomorphism is the identity, proving that $\epsilon f \cdot f\eta = \mathrm{id}_f$. $\square$

A generalized element $y : z \to b$ is said to be in the **essential image** of a 1-cell $u : a \to b$ if there exists a generalized element $x : z \to a$ and an invertible 2-cell $\beta : y \cong ux$. When the functor u is right adjoint right inverse to f, there is a convenient characterization of its essential image:

LEMMA B.4.3. *A generalized element $y : z \to b$ is in the essential image of the right adjoint right inverse $u : a \to b$ if and only if the unit component $\eta y : y \Rightarrow ufy$ is an isomorphism.*

Proof It remains only to argue that if given an invertible 2-cell $\beta : y \cong ux$, the unit component ηy is also an isomorphism. This follows from the banal final statement of Lemma B.1.3. From the horizontally composable pair below-left, naturality of whiskering defines the commutative diagram below-right:

$$
z \xrightarrow{\;y\;} b \;=\!=\!=\; b \qquad\qquad y \xrightarrow{\;\eta y\;} ufy
$$

$$
\underset{x}{\searrow}\; {\cong\Downarrow\beta}\; \underset{u}{\nearrow} \quad \underset{f}{\searrow}\; {\Downarrow\eta}\; \underset{u}{\nearrow} \qquad {\leftrightsquigarrow} \qquad \beta\Big\|\cong \qquad \cong\Big\|\beta
$$

$$
a \qquad\qquad a \qquad\qquad\qquad ux \xrightarrow[\eta ux]{\cong} ufux
$$

Since u is a right adjoint right inverse, ηu is invertible. Thus, if β is invertible so is ηy. $\qquad\qquad\square$

Sometimes it is more convenient to make use of a stricter notion of right adjoint right inverse adjunction in which the counit ϵ is required to be the identity $fu = \mathrm{id}_a$. In this case it follows from the triangle equalities that $f\eta = \mathrm{id}_f$, so that the unit is fibered over a. When the left adjoint is an isofibration in the following 2-categorical sense, a right adjoint right inverse up to isomorphism can always be replaced by a right adjoint right inverse up to identity (see Lemma 3.6.9).

Definition B.4.4 (isofibration). A 1-cell $f : b \to a$ in a 2-category defines an **isofibration** – in which case the arrow is typically denoted by "$\twoheadrightarrow$" – if given any invertible 2-cell $\alpha : fy \cong x$ abutting to a with a specified lift of one of its boundary 1-cells through f, there exists an invertible 2-cell $\beta : y \cong \bar{y}$ abutting to b with this boundary 1-cell that whiskers with f to the original 2-cell:

$$
\begin{array}{ccc}
z \xrightarrow{\;y\;} b & & z \xrightarrow[\bar{y}]{\overset{y}{\cong\Downarrow\beta}} b \\
\searrow {\cong\Downarrow\alpha}\;\Big\downarrow f & = & \Big\downarrow f \\
x \searrow\; a & & a
\end{array}
$$

Lemma B.4.5. *Let $f : b \twoheadrightarrow a$ be any isofibration in a 2-category $\mathcal{C}$ that admits a right adjoint $u' : a \to b$ with counit $\epsilon : fu' \cong \mathrm{id}_a$ an isomorphism. Then u' is isomorphic to a functor u that lies strictly over a and defines a strict right adjoint right inverse to f, in which case $f \dashv u$ defines an adjunction in the 2-category $\mathcal{C}_{/a}$ of isofibrations with codomain a, commutative triangles over a, and 2-cells that whisker to identities abutting to a.*

Proof Define the 1-cell $u : a \to b$ by lifting the counit isomorphism through the isofibration $f : b \twoheadrightarrow a$

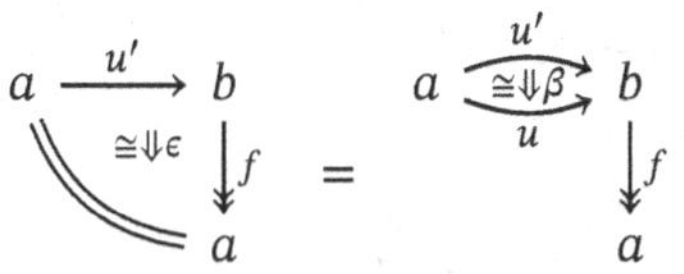

$$
\begin{array}{c}
a \xrightarrow{\; u' \;} b \\
\quad \cong \Downarrow \epsilon \quad \downarrow f \\
\qquad\qquad a
\end{array}
\quad = \quad
\begin{array}{c}
a \underset{u}{\overset{u'}{\cong \Downarrow \beta}} b \\
\qquad \downarrow f \\
\qquad a
\end{array}
$$

Note by construction that $fu = \mathrm{id}_a$, so u defines a 1-cell in $\mathcal{C}_{/a}$. By the triangle equalities for the adjunction $f \dashv u'$, the unit defines a 2-cell $\eta : \mathrm{id}_b \Rightarrow u'f$ with $\eta u'$ and $f\eta$ both invertible. The composite 2-cell

$$
\eta' := \mathrm{id}_b \xLongrightarrow{\;\eta\;} u'f \xLongrightarrow[\cong]{\;\beta f\;} uf
$$

lies in $\mathcal{C}_{/a}$ has the properties that $\eta' u$ and $f\eta$ are both invertible. Applying Lemma B.4.2 in the 2-category $\mathcal{C}_{/a}$, this 2-cell may then be modified to define the unit of an adjunction $f \dashv u$ with counit $fu = \mathrm{id}_a$. $\qquad\square$

Exercises

EXERCISE B.4.i. Consider a pair of equivalent adjunctions, satisfying the condition of Proposition B.3.8. Show that if either of these is a right adjoint right inverse adjunction then both are.

B.5 Absolute Absolute Lifting Diagrams

Recall from Definition 2.3.5 that for any cospan $c \xrightarrow{g} a \xleftarrow{f} b$ in a 2-category, an **absolute right lifting** of g through f is given by a 1-cell r and 2-cell ρ as displayed below-left

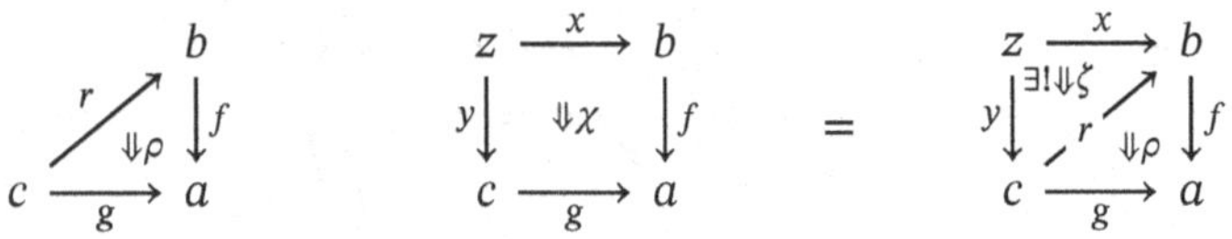

$$
\begin{array}{c}
\quad\quad b \\
{}^r\nearrow \Downarrow\rho \downarrow f \\
c \xrightarrow{\;g\;} a
\end{array}
\quad
\begin{array}{c}
z \xrightarrow{\;x\;} b \\
y\downarrow \quad \Downarrow\chi \quad \downarrow f \\
c \xrightarrow{\;g\;} a
\end{array}
\quad = \quad
\begin{array}{c}
z \xrightarrow{\;x\;} b \\
y\downarrow {}^{\exists!\Downarrow\zeta}\nearrow^r \Downarrow\rho \downarrow f \\
c \xrightarrow{\;g\;} a
\end{array}
$$

so that any 2-cell as displayed above-center factors uniquely through (r, ρ) as displayed above-right. The adjective "absolute" refers to the property that absolute right lifting diagrams are stable under restriction along any 1-cell $k : d \to c$.

The following lemma yields absolute right lifting diagrams which are absolute in a second sense: namely, they are preserved by any 2-functor.

LEMMA B.5.1. *Suppose* $(f \dashv u, \eta : \mathrm{id}_a \Rightarrow uf, \epsilon : fu \Rightarrow \mathrm{id}_c)$ *is an adjunction under* b *in the sense that*

- *the solid-arrow triangles involving both adjoints commute*

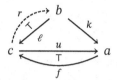

- *and* $\eta k = \mathrm{id}_k$ *and* $\epsilon \ell = \mathrm{id}_\ell$.

Then if ℓ *admits a right adjoint* r *with unit* $\iota : \mathrm{id}_b \Rightarrow r\ell$ *and counit* $\nu : \ell r \Rightarrow \mathrm{id}_c$, *then* $u\nu$ *exhibits* r *as an absolute right lifting of* u *through* k.

Moreover, such absolute right lifting diagrams are preserved by any 2-functor.

Proof　The argument is purely diagrammatic. Any 2-cell as below-left factors through $u\nu$ as below-right:

Conversely, if ζ defines a factorization of χ through $u\nu$, then

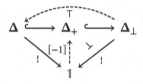

proving that the factorization constructed above is unique.

Finally, because the universal property of the absolute right lifting diagram is "equationally witnessed" by the presence of the adjunctions, it is preserved by any 2-functor. □

EXAMPLE B.5.2. For example, there is a diagram of adjoint functors

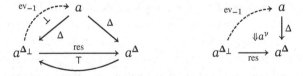

involving the categories introduced in Definition 2.3.13. The inclusion $\Delta \hookrightarrow \Delta_+ \hookrightarrow \Delta_\perp$ freely adjoins a bottom element to each ordinal, while its right adjoint can be identified with the inclusion $\Delta_\perp \hookrightarrow \Delta$ of the wide subcategory whose morphisms are bottom element preserving maps. The adjunction $[-1] \dashv\ !$ witnesses the fact that $[-1] \in \Delta_\perp$ defines an initial object with the counit ν defining the canonical natural transformation from the initial object to the identity functor.

This diagram satisfies the premises of Lemma B.5.1 in $\mathcal{C}at^{\mathrm{op}}$. Let a be an object of any 2-category $\mathcal{C}$ that is cotensored over $\mathcal{C}at$. Then the 2-functor $a^{(-)} : \mathcal{C}at^{\mathrm{op}} \to \mathcal{C}$ carries the given data to a diagram of adjoint functors in $\mathcal{C}$ as below-left and hence the triangle below-right is absolute right lifting:

This proves Proposition 2.3.15.

EXAMPLE B.5.3. There is a similar diagram of adjoint functors

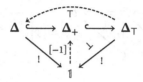

again involving the categories introduced in Definition 2.3.13. The inclusion $\Delta \hookrightarrow \Delta_+ \hookrightarrow \Delta_\top$ freely adjoins a top element and its right adjoint can be identified with the inclusion $\Delta_\top \hookrightarrow \Delta$ of the wide subcategory whose morphisms are top element preserving maps. The adjunction $[-1] \dashv \,!$ witnesses the fact that $[-1] \in \Delta_\top$ defines an initial object with the counit ν defining the canonical natural transformation from the initial object to the identity functor. This proves another version of Proposition 2.3.15 where the "splittings" occur on the other side of the co/simplicial objects.

Exercises

EXERCISE B.5.i. Search for other diagrams of categories satisfying the premises of Lemma B.5.1, such as the example implicit in Exercise 2.3.iv.

B.6 Representable Characterizations of 2-Categorical Notions

In the cartesian closed category of 2-categories, the hom 2-functor $\mathcal{C}(-,-)$: $\mathcal{C}^{\mathrm{op}} \times \mathcal{C} \to \mathcal{C}at$ associated to any 2-category $\mathcal{C}$ transposes to define a Yoneda embedding 2-functor $\bot : \mathcal{C} \hookrightarrow \mathcal{C}at^{\mathcal{C}^{\mathrm{op}}}$ whose codomain is the 2-category of 2-functors, 2-natural transformations, and modifications from $\mathcal{C}$ to $\mathcal{C}at$. By the enriched Yoneda lemma, this 2-functor is fully faithful in an enriched sense: the 2-category $\mathcal{C}$ is isomorphic to the full sub 2-category spanned by the representable 2-functors $\mathcal{C}(-,x)$: $\mathcal{C}^{\mathrm{op}} \to \mathcal{C}at$. On account of this enriched fully faithfulness, structures such as equivalences or adjunctions that are defined internally to the 2-category $\mathcal{C}$ are both preserved and reflected by the Yoneda embedding.

This result is less useful than one might expect. Indeed, none of the "representable characterizations" of equivalences, adjunctions, and absolute lifting diagrams proven in this section are consequences of it. The main point is that the 1-cells in $\mathcal{C}at^{\mathcal{C}}$ are 2-natural transformations, but as our proofs reveal, weaker naturality conditions suffice to detect 2-categorical structures that exist in $\mathcal{C}$.

We say more about the Yoneda perspective in Remark B.6.7 at the close of this section.

By analogy with Theorem 1.4.7, we have the following result which tells us that equivalences in a 2-category represent equivalences of categories.

PROPOSITION B.6.1 (equivalences are representably defined). *A 1-cell* $f : a \to b$ *in a 2-category* $\mathcal{C}$ *defines an equivalence if and only if for all* $x \in \mathcal{C}$ *the induced functor*

$$\mathcal{C}(x, a) \xrightarrow{\ f_* \ } \mathcal{C}(x, b)$$

defines an equivalence of categories.

Proof Each $x \in \mathcal{C}$ defines a 2-functor $\mathcal{C}(x, -): \mathcal{C} \to \mathcal{C}at$, so if $f : a \rightsquigarrow b$ is an equivalence in $\mathcal{C}$, then $f_* : \mathcal{C}(x, a) \rightsquigarrow \mathcal{C}(x, b)$ is an equivalence in $\mathcal{C}at$.

Conversely, by essential surjectivity of $f_* : \mathcal{C}(b, a) \rightsquigarrow \mathcal{C}(b, b)$, there exists some $g : b \to a$ and isomorphism $\beta : fg \cong \mathrm{id}_b$. By fully faithfulness of $f_* : \mathcal{C}(a, a) \rightsquigarrow \mathcal{C}(a, b)$ the isomorphism $\beta^{-1}f : f \cong fgf$ lifts to an isomorphism $\alpha : \mathrm{id}_a \cong gf$. $\qquad\qquad\square$

Similarly, an adjoint functor in a 2-category induces pointwise defined adjunctions between the hom-categories, but in this case, a further "exactness" condition is required to convert a representably defined adjunction into an adjunction in the 2-category.

PROPOSITION B.6.2 (adjunctions are representably defined). *A 1-cell* $u : a \to b$ *in a 2-category* $\mathcal{C}$ *admits a left adjoint if and only if:*

(i) *For all* $x \in \mathcal{C}$*, the induced functor admits a left adjoint*

$$\mathcal{C}(x, a) \underset{u_*}{\overset{f^x}{\rlap{\raisebox{-0.5ex}{$\xleftarrow{\hphantom{aaa}}$}}\underrightarrow{\hphantom{aaa}}}} \perp \mathcal{C}(x, b)$$

(ii) *For all* $k : y \to x \in \mathcal{C}$*, the mate of the identity 2-cell is an isomorphism:*

$$
\begin{array}{ccc}
\mathcal{C}(x, a) & \xrightarrow{\ k^* \ } & \mathcal{C}(y, a) \\
{\scriptstyle u_*}\downarrow & \nearrow\mathrm{id} & \downarrow{\scriptstyle u_*} \\
\mathcal{C}(x, b) & \xrightarrow{\ k^* \ } & \mathcal{C}(y, b)
\end{array}
\qquad \leftrightsquigarrow \qquad
\begin{array}{ccc}
\mathcal{C}(x, a) & \xrightarrow{\ k^* \ } & \mathcal{C}(y, a) \\
{\scriptstyle f^x}\uparrow & \cong\kern-0.3em\raisebox{0.2ex}{$\scriptstyle \epsilon^y k^* f^x \cdot f^y k^* \eta^x$} & \uparrow{\scriptstyle f^y} \\
\mathcal{C}(x, b) & \xrightarrow{\ k^* \ } & \mathcal{C}(y, b)
\end{array}
$$

Proof The Yoneda embedding 2-functor $よ : \mathcal{C} \hookrightarrow \mathcal{C}at^{\mathcal{C}op}$ preserves adjunctions, carrying an adjoint pair $f \dashv u$ in $\mathcal{C}$ to an adjunction between the representable 2-functors $\mathcal{C}(-, a)$ and $\mathcal{C}(-, b)$

$$\mathcal{C}(-, a) \underset{u_*}{\overset{f_*}{\rightleftarrows}} \bot \; \mathcal{C}(-, b)$$

whose left and right adjoints are the 2-natural transformations $f_* \dashv u_*$ and whose unit and counit are modifications. Evaluating at $x \in \mathcal{C}$, this defines a family of adjunction as in (i) and **strict adjunction morphisms**, i.e., so that any $k : y \to x$ induces a strictly commutative square with respect to the left and right adjoints inhabited by a mate pair of identity 2-cells.

The real content is in the converse. Assuming (i), define a candidate left inverse by $f := f^b(\mathrm{id}_b)$. By construction $uf := u_* f^b(\mathrm{id}_b)$ so we may define a candidate unit to be the component of the unit η^b of $f^b \dashv u_*$ at id_b:

$$\eta := \mathrm{id}_b \overset{\eta^b_{\mathrm{id}_b}}{\Longrightarrow} uf \quad \in \mathcal{C}(b, b).$$

Note that these definitions do not a priori give any information about the other composite $fu \in \mathcal{C}(a, a)$, but condition (ii) defines a natural isomorphism $\alpha : f^a u^* \cong u^* f^b$

$$\begin{array}{ccc}
\mathcal{C}(b, a) & \overset{u^*}{\longrightarrow} & \mathcal{C}(a, a) \\
{\scriptstyle f^b}\big\uparrow & {\scriptstyle \cong\searrow \epsilon^a u^* f^b \cdot f^a u^* \eta^b} & \big\uparrow{\scriptstyle f^a} \\
\mathcal{C}(b, b) & \underset{u^*}{\longrightarrow} & \mathcal{C}(a, b)
\end{array}$$

whose component at id_b defines an isomorphism

$$\alpha_{\mathrm{id}_b} := f^a(u) \overset{f^a(\eta u)}{\Longrightarrow} f^a(ufu) = f^a u_*(fu) \overset{\epsilon^a(fu)}{\Longrightarrow} fu \quad \in \mathcal{C}(a, a).$$

Using this, we define the counit to be the composite of the inverse of this isomorphism with the component of the counit ϵ^a of $f^a \dashv u_*$ at id_a:

$$\epsilon := fu \overset{\alpha^{-1}_{\mathrm{id}_b}}{\underset{\cong}{\Longrightarrow}} f^a(u) \overset{\epsilon^a_{\mathrm{id}_a}}{\Longrightarrow} \mathrm{id}_a \quad \in \mathcal{C}(a, a).$$

The commutative diagram

$$\begin{array}{ccc}
u & \overset{\eta u}{\Longrightarrow} & ufu \\
{\scriptstyle \eta^a(u)}\big\Vert\big\downarrow & & \big\Vert\big\downarrow{\scriptstyle \eta^a(ufu)} \quad\searrow \\
uf^a(u) & \overset{uf^a(\eta u)}{\Longrightarrow} uf^a(ufu) \overset{u\epsilon^a(fu)}{\Longrightarrow} & ufu \\
& \underset{u\alpha_{\mathrm{id}_b}}{\longrightarrow} &
\end{array}$$

reveals that $u\alpha_{\mathrm{id}_b} \cdot \eta^a u = \eta u$, so

$$u\epsilon \cdot \eta u = (u\epsilon^a_{\mathrm{id}_a} \cdot u\alpha^{-1}_{\mathrm{id}_b}) \cdot (u\alpha_{\mathrm{id}_b} \cdot \eta^a u) = u\epsilon^a_{\mathrm{id}_a} \cdot u\alpha_{\mathrm{id}_b} = \mathrm{id}_u,$$

which verifies one of the two triangle equalities.

It is somewhat delicate to prove that the other triangle equality composite

$$f \xrightarrow{\ f\eta\ } fuf \xrightarrow{\ \epsilon f\ } f \in \mathcal{C}(b,a)$$

is the identity because we do not have any way to understand the arrow $f\eta$. Note, however, that this arrow defines an endomorphism of the object $f^b(\mathrm{id}_b) \in \mathcal{C}(b,a)$, so if we verify that its transpose under the adjunction $f^b \dashv u_*$ is the unit component $\eta^b_{\mathrm{id}_b}$, then by uniqueness of adjoint transposition, we must have $\epsilon f \cdot f\eta = \mathrm{id}_f$ as desired. This can be verified by direct calculation: the adjoint transpose is computed by applying the functor u_* and then precomposing with $\eta^b_{\mathrm{id}_b} = \eta$, which yields the left-bottom composite below.

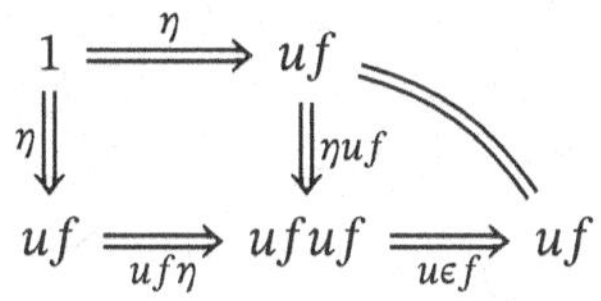

$$
\begin{array}{ccc}
1 & \xrightarrow{\ \eta\ } & uf \\
{\scriptstyle \eta}\downarrow & & \downarrow{\scriptstyle \eta uf} \quad \searrow \\
uf & \xrightarrow[\ uf\eta\ ]{} & ufuf \xrightarrow[\ u\epsilon f\ ]{} uf
\end{array}
$$

An easy diagram chase making use of the previously verified triangle equality completes the proof. $\qquad\square$

Condition (ii) of Proposition B.6.2 is referred to as a "Beck–Chevalley" or *exactness* condition. Another exactness condition appears in a representable characterization of absolute lifting diagrams.

DEFINITION B.6.3. A trio of functors (u, v, w) between a pair of absolute right lifting diagrams (r, ρ) and (r', ρ') as below defines a **right exact transformation** if and only if the 2-cell τ induced by the universal property of the absolute right lifting is invertible:

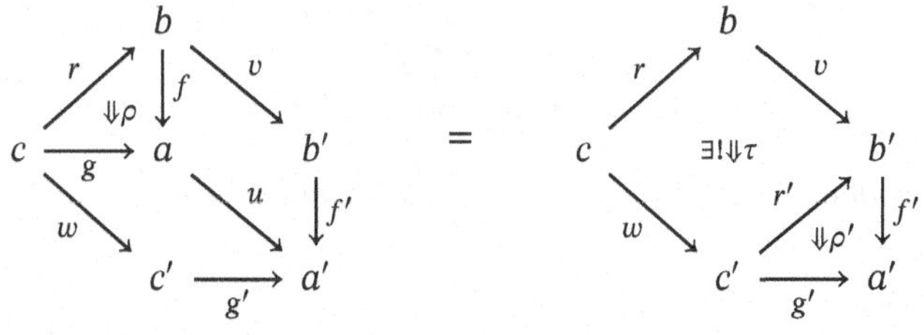

(B.6.4)

This right exactness condition holds if and only if, in the diagram on the left of (B.6.4), the whiskered 2-cell $u\rho$ displays vr as the absolute right lifting of

$g'w$ through f', which is to say that the right exact transformations are those that preserve absolute right lifting diagrams.

LEMMA B.6.5. *The mate of a commutative square between left adjoints as below*

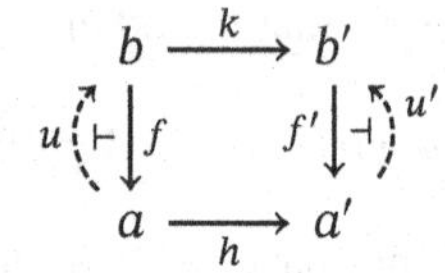

$$
\begin{array}{ccc}
b & \xrightarrow{\ k\ } & b' \\
u \uparrow \vdash \downarrow f & & f' \downarrow \dashv \uparrow u' \\
a & \xrightarrow{\ h\ } & a'
\end{array}
$$

is invertible if and only if (h, k, h) defines a right exact transformation between the absolute right lifting diagrams (u, ϵ) and (u', ϵ') of id_a through f and $\mathrm{id}_{a'}$ through f'.

Proof The unique 2-cell τ satisfying the pasting diagram below is the mate of $\mathrm{id} : f'k \Rightarrow hf$.

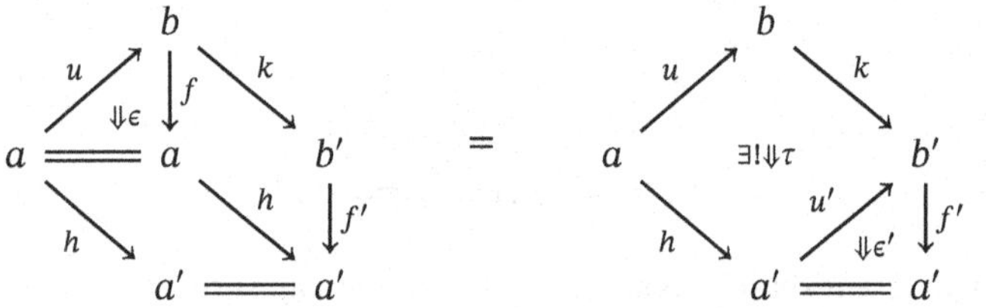

PROPOSITION B.6.6. *Consider a 2-cell in a 2-category $\mathcal{C}$*

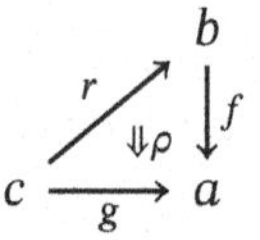

$$
\begin{array}{ccc}
 & & b \\
 & r \nearrow \quad \Downarrow\rho & \downarrow f \\
c & \xrightarrow{\ g\ } & a
\end{array}
$$

   (i) *If (r, ρ) defines an absolute right lifting diagram in $\mathcal{C}$, then*

     (a) *For all $x \in \mathcal{C}$,*

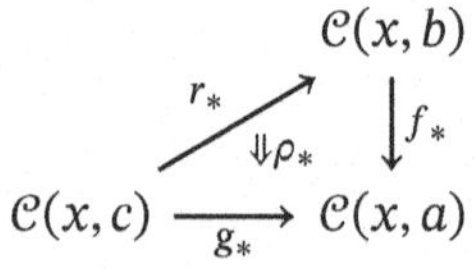

$$
\begin{array}{ccc}
 & & \mathcal{C}(x, b) \\
 & r_* \nearrow \quad \Downarrow\rho_* & \downarrow f_* \\
\mathcal{C}(x, c) & \xrightarrow{\ g_*\ } & \mathcal{C}(x, a)
\end{array}
$$

defines an absolute right lifting diagram in $\mathcal{C}\mathrm{at}$.

(b) For all k : $w \to x \in \mathcal{C}$, the induced transformation is right exact.

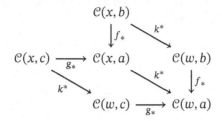

(ii) Conversely if ((a)) holds for each $x \in \mathcal{C}$, then (r, ρ) defines an absolute right lifting diagram in $\mathcal{C}$.

(iii) Moreover, if $g : c \to a$ and $f : b \to a$ are so that for all $x \in \mathcal{C}$, the functor $g_* : \mathcal{C}(x, c) \to \mathcal{C}(x, a)$ admits an absolute right lifting through $f_* : \mathcal{C}(x, b) \to \mathcal{C}(x, a)$ for which condition ((b)) holds, then g admits an absolute right lifting through f in $\mathcal{C}$.

Proof We leave the proof of the first statement, which is the most straightforward, to the reader with the hint that to verify the universal property of an absolute lifting diagram in $\mathcal{C}at$, it suffices to consider cones over the cospan (g_*, f_*) whose summit is the terminal category $\mathbb{1}$.

For the second assertion, consider a cone

$$
\begin{array}{ccc}
x & \xrightarrow{\;y\;} & b \\
{\scriptstyle z}\downarrow & \Downarrow\chi & \downarrow{\scriptstyle f} \\
c & \xrightarrow[g]{} & a
\end{array}
$$

over the cospan (g, f) in $\mathcal{C}$. This data defines a diagram of categories as below-left, which factors uniquely as below-right:

$$
\begin{array}{ccc}
\mathbb{1} & \xrightarrow{\;y\;} & \mathcal{C}(x, b) \\
{\scriptstyle z}\downarrow & \Downarrow\chi & \downarrow{\scriptstyle f_*} \\
\mathcal{C}(x, c) & \xrightarrow[g_*]{} & \mathcal{C}(x, a)
\end{array}
\quad = \quad
\begin{array}{ccc}
\mathbb{1} & \xrightarrow{\;y\;} & \mathcal{C}(x, b) \\
{\scriptstyle z}\downarrow & {\scriptstyle \exists!\Downarrow\zeta} \quad {\scriptstyle r_*} \;\nearrow\; \Downarrow\rho_* & \downarrow{\scriptstyle f_*} \\
\mathcal{C}(x, c) & \xrightarrow[g_*]{} & \mathcal{C}(x, a)
\end{array}
$$

defining the desired unique factorization

$$
\begin{array}{ccc}
x & \xrightarrow{\;y\;} & b \\
{\scriptstyle z}\downarrow & \Downarrow\chi & \downarrow{\scriptstyle f} \\
c & \xrightarrow[g]{} & a
\end{array}
\quad = \quad
\begin{array}{ccc}
x & \xrightarrow{\;y\;} & b \\
{\scriptstyle z}\downarrow & {\scriptstyle \Downarrow\zeta} \quad {\scriptstyle r}\;\nearrow\; \Downarrow\rho & \downarrow{\scriptstyle f} \\
c & \xrightarrow[g]{} & a
\end{array}
$$

For the final statement, we define the pair (r, ρ) by evaluating the functor and natural transformation of the postulated absolute right lifting (r^c, ρ^c) in the case

$x = c$ at $\mathrm{id}_c \in \mathcal{C}(c,c)$. To verify that $\rho : fr \Rightarrow g$ defines an absolute right lifting of g through f, consider a 1-cell $z : x \to c$. The hypothesis of right exactness tells us that the composite transformation

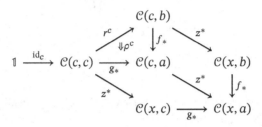

is absolute right lifting. By the proof of the second statement above, this tells us that $(rc, \rho c)$ is an absolute right lifting of gc through f, which proves that (r, ρ) is an absolute right lifting as required.   □

REMARK B.6.7. The results of Propositions B.6.2 and B.6.6 can be viewed as applications of the bicategorical Yoneda lemma, which defines a 2-fully faithful embedding of a bicategory $\mathcal{C}$ into the 2-category $[\mathcal{C}^{\mathrm{op}}, \mathcal{C}at]$ of pseudofunctors, pseudonatural transformations, and modifications (see Definitions 10.4.1, 10.4.2, and B.2.3). If a 1-cell $u : a \to b$ in $\mathcal{C}$ satisfies condition (i) of Proposition B.6.2, then by Theorem B.3.6, the left adjoints $f^x : \mathcal{C}(x,b) \to \mathcal{C}(x,a)$ define the components of an oplax natural transformation. Condition (ii) demands that this oplax natural transformation is a pseudonatural transformation. Now 2-fully faithfulness allows us to lift this to an arrow $f : b \to a$ in $\mathcal{C}$, which is left adjoint to u.

In the case of Proposition B.6.1, where $u : a \to b$ induces equivalences $\mathcal{C}(x,a) \xrightarrow{\simeq} \mathcal{C}(x,b)$, the inverses can be chosen to define adjoint equivalences, which automatically assemble into a pseudonatural transformation (see Lemma 10.4.15). This is why no additional hypothesis was required.

Exercises

EXERCISE B.6.i. Confirm the assertion made in the proof of Lemma B.6.5.

EXERCISE B.6.ii. Prove Proposition B.6.6(i).

Appendix C

Abstract Homotopy Theory

The underlying 1-category of an ∞-cosmos, together with its classes of isofibrations, equivalences, and trivial fibrations, defines a *category of fibrant objects*, a categorical setting for abstract homotopy theory first studied by Brown [23]. In §C.1, we develop some of the general theory of categories of fibrant objects in order to present some classical proofs that are omitted in the main text.

The remainder of this chapter develops material that is applied in later appendices. In Appendix E, we discover that examples of ∞-cosmoi can be found "in the wild" as *model categories* that are enriched as such over Joyal's model structure on the category of simplicial sets. To explain the notions that feature in the statements and proofs of these results, model categories, enriched model categories, and the various functors between them are introduced in §C.3.

A model category is an axiomatic framework for abstract homotopy theory developed by Quillen [93].[1] In the introduction to " Chapter I. Axiomatic Homotopy Theory" where the definition first appears, Quillen highlights the factorization and lifting axioms as being the most important. These are most clearly encapsulated in the categorical notion of a *weak factorization system* discussed in §C.2, the axioms for which were enumerated later.

Finally, some of the technical combinatorial proofs of Appendix D require inductive arguments involving the *Reedy category* Δ. Thus, we conclude in §C.4 and §C.5 with a brief introduction to Reedy category theory and the Reedy model structure following the presentation of [107].

---

[1] Similarly, an ∞-cosmos is an axiomatic framework for abstract ∞-category theory, which may productively be thought of as a categorification of a Quillen model category.

C.1 Abstract Homotopy Theory in a Category of Fibrant Objects

In this section, we work in an (unenriched) category of fibrant objects, a notion first introduced by Brown [23]. Examples include the underlying category of an ∞-cosmos or the full subcategory of fibrant objects in a Quillen model category (hence the name).

DEFINITION C.1.1 (category of fibrant objects). A **category of fibrant objects** consists of a category $\mathcal{M}$ together with two subcategories of morphisms W and F satisfying the following axioms:

(i) $\mathcal{M}$ has products and in particular a terminal object 1. Moreover, the classes F and F ∩ W are each closed under products.

(ii) W has the **2-of-3 property**: for any composable pair of morphisms, if any two of f, g, and gf is in W then so is the third.

(iii) Pullbacks of maps in F exist and lie in F, and the class F ∩ W is also stable under pullback.

(iv) Limits of countable towers[2] of maps in F exist and also lie in F, and the class F ∩ W is also closed under forming limits of towers.

(v) For every object B, there exists a **path object** PB together with a factorization of the diagonal into a map in W followed by a map in F:

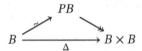

(vi) All objects are **fibrant**: for every $B \in \mathcal{M}$, the map $B \to 1$ lies in F.

REMARK C.1.2. The original definition only requires the existence finite products in axiom (i) and omits axiom (iv). The closure of the classes F and F ∩ W under finite products follows, by induction, from the closure under pullback assumed in axiom (iii) (see Corollary C.1.14). Here, we ask for these infinite limits to parallel the limit axiom 1.2.1(i) in our definition of an ∞-cosmos. In practice, the classes F and F ∩ W are frequently characterized by a right lifting property, in which case the closure axioms (i), (iii), and (iv) are automatic (see Lemma C.2.3).

---

[2] A (countable) **tower** is a diagram of shape ω^{op}. Closure under limits of towers asserts that if the images of each of the atomic arrows in the tower lie in F, then the map from the limit object to the terminal object in the diagram is also in F. The dual notion, a map from the initial object in an ω-shaped diagram to its colimit, is commonly referred to as a "countable composite" or a "transfinite composite" in the case of larger limit ordinals. More general towers, indexed by other limit ordinals, are considered in Lemma C.2.3 and beyond.

In general, it is customary to refer to the maps in W as "weak equivalences," the maps in F as "fibrations," and the maps in F ∩ W as "trivial fibrations" – unless the specific context dictates alternate names – and depict these classes by the decorated arrows, ⥲, ↠, and ⤳, respectively. Our primary interest in categories of fibrant objects is on account of the following two examples.

EXAMPLE C.1.3. The underlying category of an ∞-cosmos defines a category of fibrant objects with W the class of equivalences and F the class of isofibrations. Most of the axioms of Definition C.1.1 are subsumed by the limit and isofibration axioms of Definition 1.2.1. The remaining pieces are established in Lemmas 1.2.14, 1.2.17, and 1.2.19.

EXAMPLE C.1.4. The full subcategory of fibrant objects in a model category defines a category of fibrant objects with W the class of weak equivalences and F the class of fibrations between fibrant objects (see Definition C.3.2 and Exercise C.3.i).

REMARK C.1.5. Both of the examples just discussed have the additional property of being **right proper**, satisfying an additional axiom:

(vii) Pullbacks of maps in W along maps in F define maps in W:

$$
\begin{array}{ccc}
F & \overset{g}{\underset{\sim}{\longrightarrow}} & E \\
{\scriptstyle q}\Big\downarrow & \lrcorner & \Big\downarrow{\scriptstyle p} \\
A & \underset{f}{\overset{\sim}{\longrightarrow}} & B
\end{array}
$$

For ∞-cosmoi, this is proven in Proposition 3.3.3 and for fibrant objects in model categories, this was first observed by Reedy in [99, Theorem B] (see also [84, 15.4.2]).

The factorization axiom in a category of fibrant objects can be generalized to construct factorizations of any map (see Lemma 1.2.19).

LEMMA C.1.6 (Brown factorization lemma). *Any map $f : A \to B$ in a category of fibrant objects may be factored as a weak equivalence followed by a fibration, where the weak equivalence is constructed as a section of a trivial fibration.*

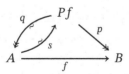

Moreover, f is a weak equivalence if and only if the fibration p is a trivial fibration.

Proof   The displayed factorization is constructed by the pullback of the path object factorization $B \rightsquigarrow PB \twoheadrightarrow B \times B$ of (v):

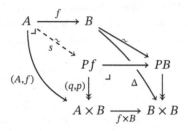

By the 2-of-3 property for the weak equivalences, both projections $PB \twoheadrightarrow B$ are trivial fibrations. Since the map q is a pullback of one of these projections along $f : A \to B$, it follows from axiom (iii) that q is a trivial fibration. Its section s, constructed by applying the universal property of the pullback to the displayed cone with summit A, is thus a weak equivalence. Finally, if either f or p are weak equivalences, the other must be as well by the 2-of-3 property.   □

In analogy with Proposition 1.2.22:

COROLLARY C.1.7.   *If $\mathcal{M}$ is a category of fibrant objects and $B \in \mathcal{M}$, then the category $\mathcal{M}_{/B}$ of fibrations in $\mathcal{M}$ with codomain B and maps over B becomes a category of fibrant objects with weak equivalences and fibrations created by the forgetful functor $\mathcal{M}_{/B} \to \mathcal{M}$.*

Proof   The construction of limits in the slice category $\mathcal{M}_{/B}$ is described in the proof of Proposition 1.2.22; note in particular, that id_B is the terminal object, so all objects in $\mathcal{M}_{/B}$, being fibrations in $\mathcal{M}$, are fibrant. Path objects for a fibration $f : A \twoheadrightarrow B$ are constructed by applying Lemma C.1.6 to the "diagonal" map $(f, f) : A \to A \times_B A$ from A to the pullback of f along itself.   □

The dual of a result of Blumberg and Mandell [20, 6.4] demonstrates that the equivalences in any ∞-cosmos satisfy the 2-of-6 property. The proof reveals that this holds in any category of fibrant objects in which the class $\mathbb{W}$ is closed under retracts.[3]

PROPOSITION C.1.8.   *Let $\mathcal{M}$ be a category with classes of maps $\mathbb{W}$ and $\mathbb{F}$ so that:*

- $\mathbb{W}$ *satisfies the 2-of-3 property, and is closed under retracts.*

---

[3] An arrow $f : A \to B$ is a **retract** of an arrow $g : C \to D$ if there exists a diagram:

$$
\begin{array}{ccc}
A & \xrightarrow{s} C \xrightarrow{r} & A \\
f\downarrow & \downarrow g & \downarrow f \\
B & \xrightarrow{u} D \xrightarrow{v} & B
\end{array}
$$

- *The pullback of a map in* F ∩ W *is in* F ∩ W *and these pullbacks always exist.*
- *Every map in* W *factors as a section of a map in* F ∩ W *followed by a map in* F ∩ W.

Then the class W *satisfies the* **2-of-6 property***: for any composable triple of morphisms*

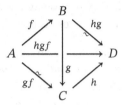

if gf and hg are in a class W *then f, g, h, and hgf are too.*

Proof Form the factorization of the weak equivalence hg displayed below-left, and form the pullback of p along h and the induced map t:

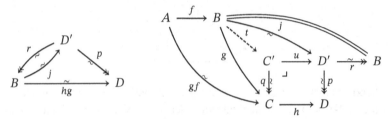

By pullback stability of the trivial fibrations, the map q is in W, so by the 2-of-3 property and the assumption that gf is in W, the composite $tf : A \to C'$ must be in W. The map f is a retract of this composite

$$
\begin{array}{ccc}
A & == & A & == & A \\
f\downarrow & & \downarrow tf & & \downarrow f \\
B & \xrightarrow{t} & C' & \xrightarrow{ru} & B
\end{array}
$$

so by retract closure of the class W, f is in W. Now it follows from the 2-of-3 property that g, h, and hgf lie in W as well. □

COROLLARY C.1.9. *The equivalences in an* ∞-*cosmos satisfy the 2-of-6 property.*

Proof It remains only to argue that the premises of Proposition C.1.8 hold for the classes of equivalences, isofibrations, and trivial fibrations in any ∞-cosmos.

Lemma 1.2.17 proves that the equivalences in an ∞-cosmos are also closed under retracts and have the 2-of-3 property. Lemma 1.2.14 proves that the class of trivial fibrations is stable under pullbacks, which exist in any ∞-cosmos. Lemma

1.2.19 constructed the desired factorization, which by the 2-of-3 property factors an equivalence as a section of a trivial fibration followed by a trivial fibration. Now Proposition C.1.8 applies to prove that the equivalences in any ∞-cosmos satisfy the stronger 2-of-6 property. $\square$

The following consequence of Lemma C.1.6, traditionally referred to as "Ken Brown's lemma," is the key to proving the invariance of various limit constructions in a category of fibrant objects under pointwise weak equivalence.

LEMMA C.1.10 (Ken Brown's lemma). *Consider a functor $F \colon \mathcal{M} \to \mathcal{N}$ whose domain is a category of fibrant objects and whose codomain is a category with a subcategory of "weak equivalences" satisfying the 2-of-3 property. If F carries trivial fibrations to weak equivalences, then F carries weak equivalences in $\mathcal{M}$ to weak equivalences in $\mathcal{N}$.*

Proof By Lemma C.1.6, any weak equivalence in a category of fibrant objects may be factored as a section of a trivial fibration followed by a trivial fibration.

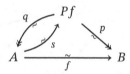

By hypothesis, the images of the maps q and p under F are weak equivalences. By the 2-of-3 property of the weak equivalences in $\mathcal{N}$, it follows that the image of s and thus also the image of f are weak equivalences. $\square$

The rest of this section is devoted to applications of Lemma C.1.10 to establish the weak equivalence invariance of limits in a category of fibrant objects.

LEMMA C.1.11. *In a category of fibrant objects, a weak equivalence between fibrations pulls back to a weak equivalence between fibrations:*

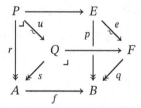

Proof By Corollary C.1.7, slices of a category $\mathcal{M}$ of fibrant objects define categories of fibrant objects and pullback along f defines a functor $f^* \colon \mathcal{M}_{/B} \to \mathcal{M}_{/A}$. Note that the map u in the displayed diagram is the pullback of the map e, so it follows directly from axiom (iii) of Definition C.1.1 that pullback preserves trivial fibrations. Now Lemma C.1.10 implies that it also preserves equivalences.   □

Other results in a similar vein require somewhat more delicate arguments. The proofs appearing below are originally due to Reedy in an unpublished manuscript [99] that implicitly gave birth to the notion of a "Reedy category" that we introduce in §C.4.

PROPOSITION C.1.12. *Consider a diagram in a category of fibrant objects:*

$$
\begin{array}{ccccc}
C & \xrightarrow{\ g\ } & A & \xleftarrow{\ f\ } & B \\
{\scriptstyle r}\downarrow & & \downarrow{\scriptstyle p} & & \downarrow{\scriptstyle q} \\
\bar{C} & \xrightarrow[\ \bar{g}\ ]{} & \bar{A} & \xleftarrow[\ \bar{f}\ ]{} & \bar{B}
\end{array}
$$

If the map r and the map $z : B \to A \times_{\bar{A}} \bar{B}$ *are both fibrations or both trivial fibrations then the induced map from the pullback of f along g to the pullback of* $\bar{f}$ *along* $\bar{g}$ *again a fibration or trivial fibration, respectively.*

Proof By pullback composition and cancelation, the induced map *t* factors as a pullback of *z* followed by a pullback of *r* as displayed below

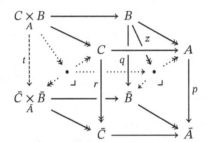

and is thus an fibration or trivial fibration if both of these maps are.   □

Similarly, we have the following result whose dual form is sometimes called the "gluing lemma."

PROPOSITION C.1.13. *In a category of fibrant objects, the induced map between the pullbacks of the horizontal rows of a diagram of the following form is again a weak equivalence:*

$$
\begin{array}{ccccc}
C & \xrightarrow{\ g\ } & A & \xleftarrow{\ f\ } & B \\
{\scriptstyle r}\downarrow\wr & & \downarrow\wr{\scriptstyle p} & & \downarrow\wr{\scriptstyle q} \\
\bar{C} & \xrightarrow[\ \bar{g}\ ]{} & \bar{A} & \xleftarrow[\ \bar{f}\ ]{} & \bar{B}
\end{array}
$$

The proof of Proposition 3.3.4 applies equally in any right proper category of fibrant objects, but as we shall discover, the hypothesis of right properness is not actually necessary.

Proof By Exercise C.1.i, for any category of fibrant objects $\mathcal{M}$ there is a category of fibrant objects $\mathcal{M}^{\lrcorner}$ whose

- objects are cospans $C \xrightarrow{g} A \xleftarrow{f} B$ whose right leg is a fibration in $\mathcal{M}$,
- weak equivalences are pointwise weak equivalences, and
- fibrations are diagrams

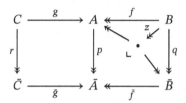

in which the maps r, p, and q are fibrations, as is the induced map $z : B \to A \times_{\bar{A}} \bar{B}$.

with the requisite limits inherited pointwise from $\mathcal{M}$. By Proposition C.1.12, the pullback functor $\lim : \mathcal{M}^{\lrcorner} \to \mathcal{M}$ carries trivial fibrations to trivial fibrations, so by Lemma C.1.10, it also preserves weak equivalences. □

COROLLARY C.1.14. *In a category of fibrant objects, finite products of fibrations, trivial fibrations, or weak equivalences are again fibrations, trivial fibrations, or weak equivalences.*

Proof The product of a finite family of maps $\{f_i : A_i \to \bar{A}_i\}_{i=1,\ldots,n}$ can be formed inductively as a pullback over the terminal object 1

$$
\begin{array}{ccccc}
\prod_{i=1}^{n-1} A_i & \longrightarrow & 1 & \longleftarrow & A_n \\
\scriptstyle\prod f_i \downarrow & & \| & & \downarrow \scriptstyle f_n \\
\prod_{i=1}^{n-1} \bar{A}_i & \longrightarrow & 1 & \longleftarrow & \bar{A}_n
\end{array}
$$

When each map f_i is a fibration or trivial fibration, by Proposition C.1.12 and an induction starting from the case of binary products, the same is true of the product of these maps. When each map f_i is a weak equivalence, the same conclusion follows from Proposition C.1.13. □

Similarly, limits of towers of fibrations are invariant under pointwise weak equivalence:

PROPOSITION C.1.15. *Consider a natural transformation between countable towers of fibrations in a category of fibrant objects*

$$
\begin{array}{ccccccc}
X_\omega := \lim_n X_n & \cdots\!\!\twoheadrightarrow \cdots \longrightarrow & X_2 & \xrightarrow{f_2} & X_1 & \xrightarrow{f_1} & X_0 \\
\alpha_\omega \downarrow & & \downarrow \alpha_2 & & \downarrow \alpha_1 & & \downarrow \alpha_0 \\
Y_\omega := \lim_n Y_n & \cdots\!\!\twoheadrightarrow \cdots \longrightarrow & Y_2 & \xrightarrow[g_2]{} & Y_1 & \xrightarrow[g_1]{} & Y_0
\end{array}
$$

(i) *If for each $n \geq 0$, the map $\langle \alpha_n, f_n \rangle \colon X_n \to Y_n \times_{Y_{n-1}} X_{n-1}$ is a fibration or trivial fibration,[4] then the induced map α_ω between the limits is as well.*

(ii) *If for each $n \geq 0$, the map α_n is a weak equivalence, then the induced map α_ω between the limits is as well.*

Proof When the hypotheses of (i) hold in a category of fibrant objects $\mathcal{M}$, the induced map $\alpha_\omega \colon X_\omega \to Y_\omega$ between the limits of the towers of fibrations is itself the limit composite of a tower of fibrations or trivial fibrations, respectively

$$
X_\omega \overset{}{\underset{}{\Longrightarrow}}\!\!\twoheadrightarrow \cdots \longrightarrow P_n \twoheadrightarrow P_{n-1} \longrightarrow\!\!\twoheadrightarrow \cdots \longrightarrow P_1 \twoheadrightarrow P_0 \twoheadrightarrow Y_\omega
$$
$$
\alpha_\omega
$$

where each layer is a pullback of the map

$$
\begin{array}{ccc}
P_n & \longrightarrow\!\!\!\!\twoheadrightarrow & P_{n-1} \\
\downarrow & \lrcorner & \downarrow \\
X_n & \xrightarrow[\langle \alpha_n, f_n \rangle]{} & Y_n \underset{Y_{n-1}}{\times} X_{n-1}
\end{array}
$$

assumed to be either a fibration or a trivial fibration. Starting from the bottom P_0 is defined to be the pullback of α_0 along the leg of the limit cone for Y_ω:

$$
\begin{array}{ccc}
P_0 & \longrightarrow\!\!\!\!\twoheadrightarrow & Y_\omega \\
\downarrow & \lrcorner & \downarrow \\
X_0 & \xrightarrow[\alpha_0]{} & Y_0
\end{array}
$$

By construction, P_0 admits a canonical map to the pullback $Y_1 \times_{Y_0} X_0$, and P_1 is

---

[4] To make sense of the case $n = 0$, declare X_{-1} and Y_{-1} to be terminal so that the map under consideration is α_0.

defined to be the pullback:

$$
\begin{array}{ccc}
P_1 & \longrightarrow\!\!\!\!\!\rightarrow & P_0 \\
\downarrow & \lrcorner & \downarrow \\
X_1 & \xrightarrow[\langle \alpha_1, f_1 \rangle]{} & Y_1 \underset{Y_0}{\times} X_0
\end{array}
$$

Continuing inductively, the limit of the tower of fibrations $P_n \twoheadrightarrow P_{n-1}$ can be seen to coincide with the limit of the $\omega^{\mathrm{op}} \times 2$ shaped diagram formed by the maps f_n, g_n, and α_n. Since the inclusion $\omega^{\mathrm{op}} \hookrightarrow \omega^{\mathrm{op}} \times 2$ of the top row of this diagram is initial, this limit recovers X_ω and the composite of the tower of fibrations recovers the map α_ω. Thus, by axioms (iii) and (iv) of Definition C.1.1, the induced map α_ω is again a fibration or trivial fibration, respectively.

The second statement now follows by applying Lemma C.1.10 to the limit functor $\lim : \mathcal{M}^{\omega^{\mathrm{op}}} \to \mathcal{M}$ whose domain is the category of towers of fibrations (see Exercise C.1.ii). $\qquad\square$

Again in this proof we have made use of the fact that the category of diagrams valued in a category of fibrant objects may itself be equipped with the structure of a category of fibrant objects, at least for certain types of diagrams and certain diagram shapes. We now establish this result more systematically for a particularly useful family of diagrams, namely those indexed by inverse categories.

DEFINITION C.1.16. A category $\mathcal{J}$ is a **inverse category** if there exists a functor $\deg : \mathcal{J} \to \omega^{\mathrm{op}}$ that reflects identities.[5]

The degree functor assigns a natural number degree to each object of $\mathcal{J}$ in such a way that all nonidentity morphisms "lower degree," in the sense that the degree of their domain object is strictly greater than the degree of their codomain object. The utility of the degree functor for an inverse category is that it allows us to define the data of an $\mathcal{J}$-indexed diagram or natural transformation by inductively specifying diagrams indexed by the full subcategories

$$
\mathcal{J}_{\leq 0} \hookrightarrow \cdots \hookrightarrow \mathcal{J}_{\leq n-1} \hookrightarrow \mathcal{J}_{\leq n} \hookrightarrow \cdots \hookrightarrow \mathrm{colim}_{n \in \omega} \mathcal{J}_{\leq n} \cong \mathcal{J}
$$

of objects with bounded degree. To extend $X \in \mathcal{M}^{\mathcal{J}_{\leq n-1}}$ to $\mathcal{M}^{\mathcal{J}_{\leq n}}$ requires the specification, for each object i with degree n of an object $X^i \in \mathcal{M}$ together with

---

[5] The mathematics does not change in any substantial way if ω is replaced by the category of ordinals. The reason we restrict to finite degrees is because Definition C.1.1 only asks for limits of ω^{op}-indexed towers (see Proposition C.1.21(i)).

a map $X^i \to \partial^i X$ from X^i to the limit

$$\partial^i X := \lim_{\substack{i \overset{\neq}{\to} j}} X_j := \lim\left(\,{}^{i\!\diagup}\mathcal{I}_{\leq n-1} \xrightarrow{\ \text{cod}\ } \mathcal{I}_{\leq n-1} \xrightarrow{\ X\ } \mathcal{M} \,\right) \qquad \text{(C.1.17)}$$

indexed by the nonidentity maps $i \to j$ in $\mathcal{I}$.

OBSERVATION C.1.18 (boundary data as a weighted limit). As the notation suggests, the object $\partial^i X$ should be thought of as the object of "boundary data" associated to X^i. This intuition can be made precise through the formalism of weighted limits, illustrating their utility even in unenriched contexts.

Recall from Definition A.6.1(i) that the limit of a diagram $X \in \mathcal{M}^{\mathcal{I}}$ weighted by the representable functor $\mathcal{I}(i, -) \in \mathcal{S}et^{\mathcal{I}}$ at an object $i \in \mathcal{I}$ is the object X^i. Define the **boundary** $\partial\mathcal{I}(i, -) \in \mathcal{S}et^{\mathcal{I}}$ of the representable functor to be the functor defined by

$$\partial\mathcal{I}(i, j) := \begin{cases} \mathcal{I}(i, j) & \deg(j) < \deg(i) \\ \varnothing & \deg(j) \geq \deg(i) \end{cases}$$

By comparing (C.1.17) with the formula of Remark A.6.11, we observe that $\lim_{\partial\mathcal{I}(i, -)} X \cong \partial^i X$.

Recall from Definition A.6.1, that weighted limits are contravariantly functorial in the weight. Thus, the natural inclusion $\partial\mathcal{I}(i, -) \hookrightarrow \mathcal{I}(i, -)$ induces a canonical map

$$X^i \cong \lim_{\mathcal{I}(i,-)} X \xrightarrow{\ m^i\ } \lim_{\partial\mathcal{I}(i,-)} X \cong \partial^i X$$

between the weighted limits. Anticipating the terminology of Definition C.4.14 we refer to $\partial^i X$ as the ith **matching object** of the diagram X and call $m^i : X^i \to \partial^i X$ the ith **matching map**.

For reasons that will momentarily become clear we define:

DEFINITION C.1.19. Let $\mathcal{M}$ be a category of fibrant objects and let $\mathcal{I}$ be an inverse category.

- A **fibrant diagram** is a diagram $X \in \mathcal{M}^{\mathcal{I}}$ with the property that for each $i \in \mathcal{I}$, the matching map $m^i : X^i \twoheadrightarrow \partial^i X$ is a fibration.
- A **fibrant natural transformation** is a natural transformation $\alpha : X \to Y \in \mathcal{M}^{\mathcal{I}}$ between fibrant diagrams so that for each $i \in \mathcal{I}$ the **relative matching map** $\hat{m}^i$ defined by the pullback in the square formed by the matching maps

is a fibration:

$$
\begin{array}{ccc}
X^i & \xrightarrow{\alpha^i} & Y^i \\
\end{array}
$$

(C.1.20)

For example, a fibrant diagram of shape ω is a tower of fibrations, as in Proposition C.1.15. The maps referenced in the first statement of that result are exactly the relative matching maps associated to a natural transformation α between fibrant diagrams, so the first hypothesis asserts exactly that α is a fibrant natural transformation. The following generalization of Proposition C.1.15 is an unenriched version of Proposition 6.2.8.

PROPOSITION C.1.21.

(i) *A category of fibrant objects $\mathcal{M}$ admits limits of any fibrant diagram $X \in \mathcal{M}^{\mathcal{J}}$ indexed by an inverse category $\mathcal{J}$, with $\lim_{\mathcal{J}} F \in \mathcal{M}$ constructed as the limit of a tower[6]*

$$
\lim_{\mathcal{J}} X := \lim_{\omega^{\mathrm{op}}} \left(\cdots \twoheadrightarrow \lim_{\mathcal{J}_{\leq n}} X \twoheadrightarrow \lim_{\mathcal{J}_{\leq n-1}} X \twoheadrightarrow \cdots \twoheadrightarrow \lim_{\mathcal{J}_{\leq 0}} X \right)
$$

each layer of which is a pullback

$$
\begin{array}{ccc}
\lim_{\mathcal{J}_{\leq n}} X & \longrightarrow & \lim_{\mathcal{J}_{\leq n-1}} X \\
\downarrow & & \downarrow \\
\prod_{\deg(i)=n} X^i & \longrightarrow & \prod_{\deg(i)=n} \partial^i X
\end{array}
$$

In particular, each leg of the limit cone $\lim_{\mathcal{J}} X \twoheadrightarrow X^i$ is a fibration as is each map in the image of the fibrant diagram X.

(ii) *For any fibrant natural transformation $\alpha \colon X \to Y \in \mathcal{M}^{\mathcal{J}}$ between fibrant diagrams, the induced map $\lim_{\mathcal{J}} X \to \lim_{\mathcal{J}} Y$ is the limit composite of a tower whose nth layer is a pullback of the map p_n constructed as a*

---

[6] The objects are the limits of the restricted diagrams, with the subscript indicating the indexing category.

pullback in the diagram below:

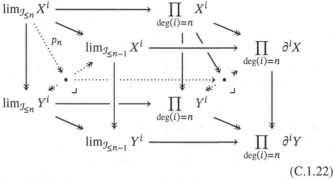

$$(C.1.22)$$

Moreover, each component map $\alpha^i : X^i \twoheadrightarrow Y^i$ is a fibration.

Proof Note that the slice category $^i\mathcal{J}_{n-1}$ is again an inverse category with degree functor

$$^i\mathcal{J}_{n-1} \xrightarrow{\text{cod}} \mathcal{J}_{n-1} \xrightarrow{\text{deg}} \omega^{\text{op}}$$

in which every object has degree at most $n-1$. In the case where i has degree 1, this category has only identity arrows, so by induction we may assume that the limit $\partial^i X$ defined by (C.1.17) exists. Now the result of (i) follows by direct inspection of the universal property of this construction, as in the proof of Proposition C.1.15. The final assertion follows from this construction and is left to Exercise C.1.iii.

By (i), it follows that the induced map between the inverse limits is defined as the limit of an ω^{op}-indexed diagram in the arrow category $\mathcal{M}^2$:

$$\lim_{\mathcal{J}} X := \lim_{n \in \omega^{\text{op}}} \lim_{\mathcal{J}_{\leq n}} X \cdots\twoheadrightarrow \cdots \to \lim_{\mathcal{J}_{\leq 2}} X \twoheadrightarrow \lim_{\mathcal{J}_{\leq 1}} X \to \lim_{\mathcal{J}_{\leq 0}} X$$

$$\lim_{\mathcal{J}} \alpha \downarrow \qquad\qquad\qquad\qquad \downarrow \qquad\qquad \downarrow \qquad \downarrow p_0$$

$$\lim_{\mathcal{J}} Y := \lim_{n \in \omega^{\text{op}}} \lim_{\mathcal{J}_{\leq n}} Y \cdots\twoheadrightarrow \cdots \to \lim_{\mathcal{J}_{\leq 2}} Y \twoheadrightarrow \lim_{\mathcal{J}_{\leq 1}} Y \twoheadrightarrow \lim_{\mathcal{J}_{\leq 0}} Y$$

Under the hypothesis of (ii), it follows by the proof of Proposition C.1.15(i), the map $\lim_{\mathcal{J}} \alpha$ then factors as the limit composite of a tower whose bottom layer is the pullback of the map p_0 along the lower-horizontal composite above, whose next layer is the pullback of the map p_1 appearing in the right-most square, whose next layer is the pullback of the map p_2 appearing in the second right-most square, and so on, where in each square p_n is the map from the upper left-hand corner to the pullback of the lower-right cospan. The map p_0 is a product of the relative matching maps indexed by objects of degree zero, and is thus a fibration. By applying pullback composition and cancelation in the cube

(C.1.22), it follows from (i) that the top and bottom faces are pullbacks, and consequently the map p_n from the initial vertex to the pullback in the left face is a pullback of the corresponding map from $\prod_{\deg(i)=n} X^i$ to the pullback in the right face. This map is the product of the relative matching maps indexed by objects of degree n, and in thus a fibration, so p_n is a fibration as well. Thus, by Proposition C.1.15(i), the induced map $\lim_{\mathcal{J}} \alpha \colon \lim_{\mathcal{J}} X \to \lim_{\mathcal{J}} Y$ is a fibration as well. This proves all but the final clause of (ii), which is also left to Exercise C.1.iii. □

We would like to conclude also that for any pointwise weak equivalence $\alpha \colon X \to Y \in \mathcal{M}^{\mathcal{J}}$ between fibrant diagrams indexed by inverse categories, the induced map $\lim_{\mathcal{J}} X \to \lim_{\mathcal{J}} Y$ is a weak equivalence. As in the proof of Proposition C.1.15(ii) this requires an intermediate result of independent interest, closely related to Exercise 6.1.iii.

PROPOSITION C.1.23. *Let $\mathcal{M}$ be a category of fibrant objects with fibrations* F *and weak equivalences* W *and let $\mathcal{J}$ be an inverse category. The category $\mathcal{M}^{\mathcal{J}}$ of fibrant diagrams and all natural transformations between them inherits the structure of a category of fibrant objects in which:*

- *the weak equivalences are those natural transformations whose components lie in* W
- *the fibrations are the fibrant natural transformations, those $\alpha \colon X \to Y \in \mathcal{M}^{\mathcal{J}}$ so that for each $i \in \mathcal{J}$ the relative matching map $X^i \xrightarrow{\hat{m}^i} Y^i \underset{\partial^i Y}{\times} \partial^i X$ is in* F.
- *the trivial fibrations are those natural transformations $\alpha \colon X \to Y \in \mathcal{M}^{\mathcal{J}}$ so that for each $i \in \mathcal{J}$ the relative matching map $X^i \xrightarrow[\sim]{\hat{m}^i} Y^i \underset{\partial^i Y}{\times} \partial^i X$ is in* F ∩ W.

Proof The proof is a very lengthy exercise for the reader, which only entails specializing the corresponding arguments from §C.4 to this "one-sided" case. A proof of a similar result using a mildly different axiomatization can be found in [94, 9.2.4]. □

The payoff for all this work is the following result.

PROPOSITION C.1.24. *Let $\mathcal{M}$ be a category of fibrant objects and let $\mathcal{J}$ be an inverse category. Then for any pointwise weak equivalence $\alpha \colon X \to Y \in \mathcal{M}^{\mathcal{J}}$ between fibrant diagrams, the induced map $\lim_{\mathcal{J}} X \to \lim_{\mathcal{J}} Y$ between the limits is a weak equivalence.*

Proof   We make use of Proposition C.1.23, which gives the category $\mathcal{M}^{\mathcal{J}}$ of fibrant diagrams the structure of a category of fibrant objects. Consider a map $\alpha : X \to Y \in \mathcal{M}^{\mathcal{J}}$ in F or F ∩ W. By Proposition C.1.21(ii), this map is the limit composite of a tower of maps, each layer of which is the pullback of a product of the maps that we have assumed lies in F or F ∩ W. Since the classes F and F ∩ W are closed under product, pullback, and limits of towers, it is now clear that the limit functor preserves these classes. The fact that it also proves the class W then follows from Lemma C.1.10. □

Exercises

EXERCISE C.1.i. Show that for any category of fibrant objects $\mathcal{M}$ there is a category of fibrant objects $\mathcal{M}^{\lrcorner}$ whose

- objects are cospans $C \xrightarrow{g} A \xleftarrow{f} B$ whose right leg is a fibration in $\mathcal{M}$,
- weak equivalences are pointwise weak equivalences, and
- fibrations are diagrams

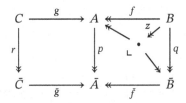

in which the maps r, p, and q are fibrations, as is the induced map $z : B \to A \times_{\bar{A}} \bar{B}$.

with the requisite limits inherited pointwise from $\mathcal{M}$.

EXERCISE C.1.ii. Specialize Proposition C.1.23 to describe the category of fibrant objects structure on the category $\mathcal{M}^{\omega^{op}}$ of towers of fibrations in a category of fibrant objects $\mathcal{M}$, and use this to complete the proof of Proposition C.1.15(ii).

EXERCISE C.1.iii (C.5.10).

(i) Verify that each leg of the limit cone constructed in Proposition C.1.21(i) is a fibration.
(ii) Conclude that each morphism in the image of a fibrant diagram is a fibration.
(iii) Arguing along the same lines, verify that each component of a fibrant natural transformation is a fibration.

EXERCISE C.1.iv. Prove Proposition C.1.23 and determine whether Exercises C.1.i and C.1.ii are special cases of this result.

C.2 Lifting Properties, Weak Factorization Systems, and Leibniz Closure

Fixing two arrows j and p in a category $\mathcal{M}$, we regard any commutative square of the form

as presenting a **lifting problem** between j and p, which is solved by constructing a **lift**: a diagonal morphism ℓ making both triangles commute. If every lifting problem between j and p has a solution, we say that j has the **left lifting property** with respect to p and, equivalently, that p has the **right lifting property** with respect to j. When this is the case, we use the suggestive symbol $j \boxslash p$ to assert this lifting property.

Frequently in abstract homotopy theory a class of maps of interest is characterized by a left or right lifting property with respect to another class or set of maps.

DEFINITION C.2.1. Let J be a class of maps in a category $\mathcal{M}$.

- Write $J^{\boxslash}$ for the class of maps in $\mathcal{M}$ that have the right lifting property with respect to every morphism in J.
- Write $^{\boxslash}J$ for the class of maps in $\mathcal{M}$ that have the left lifting property with respect to every morphism in J.

EXAMPLE C.2.2. Definitions 1.1.17 and 1.1.25 characterize the isofibrations and trivial fibrations between quasi-categories by right lifting properties against the sets of maps

$$\{\Lambda^k[n] \hookrightarrow \Delta[n]\}_{n \geq 2, 0 < k < n} \cup \{\mathbb{1} \hookrightarrow \mathbb{I}\} \qquad \text{and} \qquad \{\partial\Delta[n] \hookrightarrow \Delta[n]\}_{n \geq 0},$$

respectively.

Maps characterized by a right lifting property automatically satisfy various closure properties that may now be familiar.

LEMMA C.2.3. *Any class of maps* $J^{\boxslash}$ *characterized by a right lifting property contains the isomorphisms and is closed under composition, product, pullback, retract, and limits of towers.*

In the statement, "products" and "retracts" refer to limits formed in the category of arrows, while the "pullbacks" are of a map in $J^{\boxslash}$ along an arbitrary map. A "tower" refers to a diagram of shape α^{op}, where α is a limit ordinal (most likely ω).

Proof All of the arguments are similar. For instance, suppose q is a pullback of $p \in J^{\boxtimes}$. By juxtaposing a lifting problem as below-left with the pullback square as below-right, we may solve the composite lifting problem of j against p to obtain the dashed diagonal morphism ℓ, and then induce a solution s to the lifting problem of j against q via the cone formed by (v, ℓ) over the pullback diagram

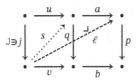

So q lifts against J and is therefore in $J^{\boxtimes}$. □

On account of the dual of Lemma C.2.3, any set of maps in a cocomplete category "cellularly generates" a larger class of maps with the same left lifting property.

DEFINITION C.2.4. Let J be a set of maps that we think of as "basic cells." A J-**cell complex** is a map built as a transfinite composite of pushouts of coproducts of maps in J:

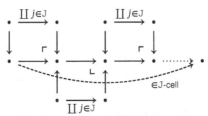

The class J-cell of J-cell complexes is said to be **cellularly generated** by a set of maps J. The class J-cof of maps **cofibrantly generated** by a set of maps J is comprised of those maps obtained as retracts of sequential composites of pushouts of coproducts of those maps.

DEFINITION C.2.5. A **weak factorization system** (L, R) on a category $\mathcal{M}$ is comprised of two classes of morphisms L and R so that

(i) Every morphism in $\mathcal{M}$ may be factored as a morphism in L followed by a morphism in R.

$$\begin{array}{ccc} \bullet & \xrightarrow{\ f\ } & \bullet \\ & {}_{L \ni \ell}\searrow \quad \nearrow_{r \in R} & \\ & \bullet & \end{array}$$

(ii) The classes L and R, respectively, have the left and right lifting properties

L ⊠ R with respect to each other: that is, any commutative square from a mall in L to a map in R admits a diagonal filler:

$$
\begin{array}{ccc}
\bullet & \longrightarrow & \bullet \\
{\scriptstyle L \ni \ell} \downarrow & \nearrow & \downarrow {\scriptstyle r \in R} \\
\bullet & \longrightarrow & \bullet
\end{array}
$$

(iii) Moreover $L = {}^{⊠}R$ and $R = L^{⊠}$.

As a consequence of axiom (iii), the right class of a weak factorization system enjoys the closure properties of Lemma C.2.3, while the left class is closed under the dual constructions.

In the presence of a pair of adjoint functors, lifting properties transpose.

LEMMA C.2.6. *In the presence of an adjunction:*

(i) *A solution to the lifting problem in $\mathcal{N}$ displayed below-left transposes to define a solution for the transposed lifting problem in $\mathcal{M}$ displayed below-right:*

$$
\begin{array}{ccc}
FA & \xrightarrow{f^{\sharp}} & X \\
{\scriptstyle F\ell} \downarrow & {\scriptstyle k^{\sharp}}\nearrow & \downarrow {\scriptstyle r} \\
FB & \xrightarrow{g^{\sharp}} & Y
\end{array}
\qquad
\begin{array}{ccc}
A & \xrightarrow{f^{\flat}} & UX \\
{\scriptstyle \ell} \downarrow & {\scriptstyle k^{\flat}}\nearrow & \downarrow {\scriptstyle Ur} \\
B & \xrightarrow{g^{\flat}} & UY
\end{array}
$$

(ii) *If $\mathcal{M}$ has a weak factorization system* (L, R) *and $\mathcal{N}$ has a weak factorization system* (L', R') *then F preserves the left classes if and only if U preserves the right classes:*

$$
FL \subset L' \qquad \leftrightsquigarrow \qquad R \supset UR'.
$$

The factorizations of Definition C.2.5 are completely irrelevant to (ii) but we have stated this result for weak factorization systems because that is the context in which it is typically applied.

Proof   Exercise C.2.iii.   □

Lemma C.2.6(ii) defines the notion of **adjunction of weak factorization systems**, this being an adjoint pair of functors between categories equipped with weak factorization systems so that the left adjoint preserves the left classes and the right adjoint preserves the right classes. Our aim is now to extend this notion to two-variable adjunctions,[7] which are given by a triple of bifunctors,

$$\mathcal{V} \times \mathcal{M} \xrightarrow{\otimes} \mathcal{N}, \quad \mathcal{V}^{\mathrm{op}} \times \mathcal{N} \xrightarrow{\{,\}} \mathcal{M}, \quad \mathcal{M}^{\mathrm{op}} \times \mathcal{N} \xrightarrow{\mathrm{hom}} \mathcal{V} \qquad (\mathrm{C.2.7})$$

written using notation that suggests the most common examples, equipped with a natural isomorphism

$$\mathcal{N}(V \otimes M, N) \cong \mathcal{M}(M, \{V, N\}) \cong \mathcal{V}(V, \mathrm{hom}(M, N)).$$

The "pushout product" of a bifunctor $\otimes : \mathcal{V} \times \mathcal{M} \to \mathcal{N}$ defines a bifunctor $\widehat{\otimes} : \mathcal{V}^2 \times \mathcal{M}^2 \to \mathcal{N}^2$ that we refer to as the "Leibniz tensor" (when the bifunctor $\otimes$ is called a "tensor"). The "Leibniz cotensor" and "Leibniz hom"

$$\widehat{\{,\}} : (\mathcal{V}^2)^{\mathrm{op}} \times \mathcal{N}^2 \to \mathcal{M}^2 \qquad \text{and} \qquad \widehat{\mathrm{hom}} : (\mathcal{M}^2)^{\mathrm{op}} \times \mathcal{N}^2 \to \mathcal{V}^2$$

are defined dually, using pullbacks in $\mathcal{M}$ and $\mathcal{V}$, respectively.

DEFINITION C.2.8 (Leibniz tensors and cotensors). Given a bifunctor $\otimes : \mathcal{V} \times \mathcal{M} \to \mathcal{N}$ valued in a category with pushouts, the **Leibniz tensor** of a map $k : I \to J$ in $\mathcal{V}$ and a map $\ell : A \to B$ in $\mathcal{M}$ is the map $k \mathbin{\widehat{\otimes}} \ell$ in $\mathcal{N}$ induced by the pushout diagram below-left:

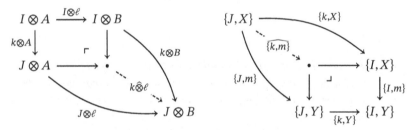

In the case of a bifunctor $\{,\} : \mathcal{V}^{\mathrm{op}} \times \mathcal{N} \to \mathcal{M}$ contravariant in one of its variables valued in a category with pullbacks, the **Leibniz cotensor** of a map $k : I \to J$ in $\mathcal{V}$ and a map $m : X \to Y$ in $\mathcal{N}$ is the map $\widehat{\{k, m\}}$ induced by the pullback diagram above right.

PROPOSITION C.2.9. *The Leibniz construction preserves:*

(i) structural isomorphisms: a natural isomorphism

$$X * (Y \otimes Z) \cong (X \times Y) \mathbin{\square} Z$$

between suitably composable bifunctors extends to a natural isomorphism

$$f \mathbin{\widehat{*}} (g \mathbin{\widehat{\otimes}} h) \cong (f \mathbin{\widehat{\times}} g) \mathbin{\widehat{\square}} h$$

---

[7] There is an analogous generalization to n-variable adjunctions that can be found in [26, §4].

between the corresponding Leibniz products;

(ii) *adjointness: if* $(\otimes, \{,\}, \mathrm{hom})$ *define a two-variable adjunction, then the Leibniz functors* $(\widehat{\otimes}, \widehat{\{,\}}, \widehat{\mathrm{hom}})$ *define a two-variable adjunction between the corresponding arrow categories;*

(iii) *retracts: if f is a retract of h and g is a retract of k then $f \widehat{\otimes} g$ is a retract of $h \widehat{\otimes} k$;*

(iv) *colimits in the arrow category: if $\otimes : \mathcal{V} \times \mathcal{M} \to \mathcal{N}$ is cocontinuous in either variable, then so is $\widehat{\otimes} : \mathcal{V}^2 \times \mathcal{M}^2 \to \mathcal{N}^2$;*

(v) *pushouts: if $\otimes : \mathcal{V} \times \mathcal{M} \to \mathcal{N}$ is cocontinuous in its second variable, and if g' is a pushout of g, then $f \widehat{\otimes} g'$ is a pushout of $f \widehat{\otimes} g$;*

(vi) *composition, in a sense: the Leibniz tensor $f \widehat{\otimes} (h \cdot g)$ factors as a composite of a pushout of $f \widehat{\otimes} g$ followed by $f \widehat{\otimes} h$*

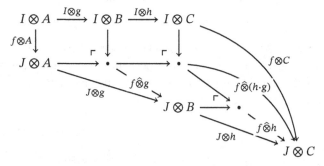

(vii) *cell complex structures: if f and g may be presented as cell complexes with cells f_α and g_β, respectively, and if $\otimes$ is cocontinuous in both variables, then $f \widehat{\otimes} g$ may be presented as a cell complex with cells $f_\alpha \widehat{\otimes} g_\beta$.*

Proof The components of the induced structural isomorphism between Leibniz products are instances of the given structural isomorphism and hence invertible, proving (i). For (ii), by naturality of the isomorphisms defining a two-variable adjunction $(\otimes, \{,\}, \mathrm{hom})$, each of the squares below commutes if and only if the other two do, under the hypothesis that the horizontal arrows given the same names in each diagram are transposes:

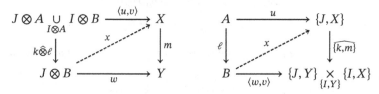

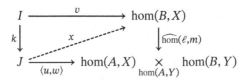

This transposition correspondence extends to solutions to the lifting problems presented by these squares (see Exercise C.2.v).

Property (iii) is a consequence of the bifunctoriality of $\hat{\otimes}$ in the arrow categories: any bifunctor preserves retracts. Property (iv) is nearly as immediate, since limits or colimits in arrow categories are computed pointwise. For (v), consider the commutative cube:

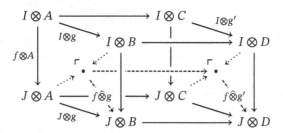

Since $I \otimes -$ and $J \otimes -$ preserve the pushout defining g' as a pushout of g, the top and bottom faces of the cube are pushouts. The squares defining the domains of the Leibniz tensors define pushouts inside the left and right-hand faces. It follows by pushout composition and cancelation that $f \hat{\otimes} g'$ is a pushout of $f \hat{\otimes} g$ as claimed.

The displayed diagram in (vi) proves the assertion made there, so it remains only to prove (vii). First note that pushouts of transfinite composites of pushouts are again transfinite composites of pushouts and transfinite composites of transfinite composites are transfinite composites, so it suffices to work one variable at a time and prove that $f \hat{\otimes} -$ preserves cell complex presentations for g. To that end, suppose g is a α-composite of maps g_i each of which are pushouts of a coproduct of maps $g'_i = \coprod_j g'_{i,j}$. We may promote this colimit to the arrow category to regard $g = g_{0,\alpha}$ as the colimit of the diagram $g_{-,\alpha} : \alpha \to \mathcal{M}^2$ with one-step maps

$$
\begin{array}{ccc}
\bullet & \xrightarrow{g_i} & \bullet \\
g_{i,\alpha} \downarrow & & \downarrow g_{i+1,\alpha} \\
\bullet & = & \bullet
\end{array}
$$

Similarly, the pushout square defining g_i from g'_i can be promoted to a pushout square in the arrow category:

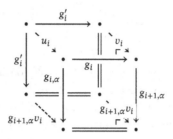

We interpret this cube as presenting the square in the front face as a pushout of the square in the back face, which decomposes as a coproduct of similar squares, one for each component of $g'_i = \coprod_j g'_{i,j}$. In this way we see that $g = g_{0,\alpha}$ is the domain component of the colimit of a diagram $\alpha \to \mathcal{M}^2$, each step of which is a pushout of a coproduct of maps in the arrow category. Now $f \mathbin{\widehat{\otimes}} - : \mathcal{M}^2 \to \mathcal{N}^2$ preserves colimits in the arrow category, and the domain functor dom : $\mathcal{N}^2 \to \mathcal{N}$ preserves colimits as well. Thus, $f \mathbin{\widehat{\otimes}} g$ is a colimit of an α-sequence of pushouts of coproducts of the maps $f \mathbin{\widehat{\otimes}} g_{i,j}$. $\square$

More details establishing these assertions are given in [107, §4–§5].

DEFINITION C.2.10. Let $\mathcal{V}$, $\mathcal{M}$, and $\mathcal{N}$ be cocomplete categories each equipped with weak factorization systems (L, R), (L', R'), and (L'', R''), respectively. A **left Leibniz bifunctor** is a bifunctor $\otimes : \mathcal{V} \times \mathcal{M} \to \mathcal{N}$ that is

  (i) cocontinuous in each variable separately, and
  (ii) has the **Leibniz property**: $\otimes$-pushout products of a map in L with a map in L' are in L''.

Dually, a bifunctor between complete categories equipped with weak factorization systems is a **right Leibniz bifunctor** if it is continuous in each variable separately and if pullback cotensors of maps in the right classes land in the right class. We most frequently apply this definition in the case of a bifunctor

$$\{,\} : \mathcal{V}^{\mathrm{op}} \times \mathcal{N} \to \mathcal{M}$$

that is contravariant in one of its variables, in which we case the relevant hypotheses are that $\mathcal{V}$ is cocomplete and colimits in the first variable are carried to limits in $\mathcal{M}$, and furthermore the Leibniz cotensor of a map in $\mathcal{L}$ with a map in $\mathcal{R}''$ defines a map in $\mathcal{R}'$. The nature of the duality between left and right Leibniz bifunctors is somewhat subtle to articulate, and left this as a puzzle for the reader (or see [26]).

LEMMA C.2.11. *If the bifunctors*

$$\mathcal{V} \times \mathcal{M} \xrightarrow{\otimes} \mathcal{N}, \quad \mathcal{V}^{\mathrm{op}} \times \mathcal{N} \xrightarrow{\{,\}} \mathcal{M}, \quad \text{and} \quad \mathcal{M}^{\mathrm{op}} \times \mathcal{N} \xrightarrow{\mathrm{hom}} \mathcal{V}$$

define a two-variable adjunction, and (L, R), (L', R'), *and* (L'', R'') *are three weak factorization systems on* $\mathcal{V}$, $\mathcal{M}$, *and* $\mathcal{N}$, *respectively, then the following are equivalent*

(i) $\otimes \colon \mathcal{V} \times \mathcal{M} \to \mathcal{N}$ *defines a left Leibniz bifunctor.*
(ii) $\{,\} \colon \mathcal{V}^{\mathrm{op}} \times \mathcal{N} \to \mathcal{M}$ *defines a right Leibniz bifunctor.*
(iii) $\mathrm{hom} \colon \mathcal{M}^{\mathrm{op}} \times \mathcal{N} \to \mathcal{V}$ *defines a right Leibniz bifunctor.*

When these conditions are satisfied, we say that $(\otimes, \{,\}, \mathrm{hom})$ *defines a* **Leibniz two-variable adjunction**.

Proof The presence of the adjoints ensures that each of the bifunctors satisfy the required (co)continuity hypotheses. Note that, for instance, $L \hat{\otimes} L' \subset L''$ if and only if $L \hat{\otimes} L' \boxvoid R''$. Now the equivalence of the three statements follows from the equivalence of the following three lifting properties:

$$L \hat{\otimes} L' \boxvoid R'' \quad \leftrightsquigarrow \quad L' \boxvoid \widehat{\{L, R''\}} \quad \leftrightsquigarrow \quad L \boxvoid \widehat{\mathrm{hom}}(L', R''),$$

the proof of which is left to Exercise C.2.v. $\square$

REMARK C.2.12. By Proposition C.2.9(vii), to show that a cocontinuous bifunctor $\otimes$ satisfies the Leibniz property, it suffices to show that $\otimes$-Leibniz products of generating morphisms are in the left class of the codomain weak factorization system.

LEMMA C.2.13. *For any category* $\mathcal{M}$ *with products and coproducts that is equipped with a weak factorization system* (L, R) *the set-tensor, set-cotensor, and hom*

$$* \colon \mathcal{S}et \times \mathcal{M} \to \mathcal{M}, \quad \{,\} \colon \mathcal{S}et^{\mathrm{op}} \times \mathcal{M} \to \mathcal{M}, \quad \text{and} \quad \mathrm{hom} \colon \mathcal{M}^{\mathrm{op}} \times \mathcal{M} \to \mathcal{S}et$$

respectively define a Leibniz two-variable adjunction relative to the (monomorphism, epimorphism) weak factorization system on $\mathcal{S}et$.

Proof By Lemma C.2.11, it suffices to prove any one of these bifunctors is Leibniz. When $A \hookrightarrow B$ is a monomorphism in $\mathcal{S}et$, the Leibniz tensor with

$f : X \to Y$ decomposes as a coproduct of maps that are manifestly in L.

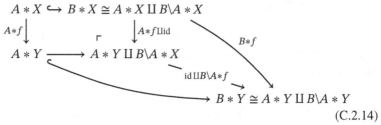

$$\text{(C.2.14)}$$

A slicker proof is also possible. Because every monomorphism may be presented as a cell complex built from a single cell $\varnothing \hookrightarrow 1$, it suffices, by Proposition C.2.9(vii), to consider Leibniz tensor with the generating monomorphism $\varnothing \hookrightarrow 1$. But note that the functor

$$\mathcal{M}^2 \xrightarrow{(\varnothing \hookrightarrow 1)\hat{*}-} \mathcal{M}^2$$

is naturally isomorphic to the identity, which certainly preserves the left class L.   □

REMARK C.2.15. By Lemma C.2.13, hom : $\mathcal{M}^{\mathrm{op}} \times \mathcal{M} \to \mathcal{S}et$ is right Leibniz, meaning that for any $\ell \in$ L and $r \in$ R, the morphism

$$\mathcal{M}(\operatorname{cod}\ell, \operatorname{dom}r) \xrightarrow{r\circ - \circ\ell} \mathcal{M}(\operatorname{dom}\ell, \operatorname{dom}r) \underset{\mathcal{M}(\operatorname{dom}\ell,\operatorname{cod}r)}{\times} \mathcal{M}(\operatorname{cod}\ell, \operatorname{cod}r)$$

is an epimorphism. The target of this map is the set of commutative squares in $\mathcal{M}$ from ℓ to r, while the fiber over any element is the set of solutions to the lifting problem so-presented. The fact that this is an epimorphism re-expresses the lifting property L ⊠ R.

Exercises

EXERCISE C.2.i. Finish the proof of Lemma C.2.3.

EXERCISE C.2.ii.

(i) Prove the "retract argument": Suppose $f = r \circ \ell$ and f has the left lifting property with respect to its right factor r. Then f is a retract of its left factor ℓ.

(ii) Conclude that in the presence of axioms (i) and (ii) of Definition C.2.5, that axiom (iii) may be replaced by the hypothesis that the classes L and R are closed under retracts.

EXERCISE C.2.iii. Prove Lemma C.2.6

EXERCISE C.2.iv.

(i) Suppose $\mathcal{M}$ is a category with products, pullbacks, and limits of towers equipped with a weak factorization system (L, R). Prove that for any inverse category $\mathcal{J}$, the category of diagrams $\mathcal{M}^{\mathcal{J}}$ has a weak factorization system whose left class is comprised of those maps whose components are in L and whose right class is comprised of those maps $\alpha : X \to Y$ so that for each $i \in I$, the relative matching map $\hat{m}_i : X_i \to \partial_i X \times_{\partial_i Y} Y_i$ lies in R.

(ii) Give a new proof that $\lim : \mathcal{M}^{\mathcal{J}} \to \mathcal{M}$ preserves fibrations and trivial fibrations under the additional hypothesis that the classes F and $F \cap W$ of $\mathcal{M}$ are the right classes of weak factorization systems.

EXERCISE C.2.v. Given a two variable adjunction (C.2.7) and classes of maps A, B, C in $\mathcal{V}, \mathcal{M}, \mathcal{N}$, respectively, prove that the following lifting properties are equivalent

$$A \otimes B \boxtimes C \quad \Leftrightarrow \quad B \boxtimes \widehat{\{A, C\}} \quad \Leftrightarrow \quad A \boxtimes \widehat{\hom}(B, C).$$

C.3 Model Categories and Quillen Functors

The following reformulation of Quillen's definition of a "closed model category" [93, I.5.1] was given by Joyal and Tierney [64, 7.7], who prove that a category $(\mathcal{M}, W)$ with weak equivalences satisfying the 2-of-3 property admits a model structure just when there exist classes C and F that define a pair of weak factorization systems as follows:

DEFINITION C.3.1 (model category). A **model structure** on a complete and cocomplete category $\mathcal{M}$ consists of three classes of maps – the **weak equivalences** W denoted "$\overset{\sim}{\to}$" which must satisfy the 2-of-3 property, the **cofibrations** C denoted "$\rightarrowtail$," and the **fibrations** F denoted "$\twoheadrightarrow$" – so that $(C, F \cap W)$ and $(C \cap W, F)$ each define weak factorization systems on $\mathcal{M}$.[8]

A **model category** is a complete and cocomplete category equipped with a model structure. The better-behaved objects in a model category are either "fibrant" or "cofibrant" or both:

DEFINITION C.3.2. In a model category $\mathcal{M}$ an object X is **fibrant** just when the unique map $X \to 1$ to the terminal object is a fibration and **cofibrant** just when

---

[8] There is one axiom in standard definition of a model category – the closure of weak equivalences under retracts – that is not obviously packaged into these hypotheses, but this is a consequence of the axioms given here [64, 7.8].

Abstract Homotopy Theory

the unique map $\varnothing \to X$ from the initial object is a cofibration. By factoring the unique maps, any object X has a **cofibrant replacement** QX and a **fibrant replacement** RX constructed as follows:

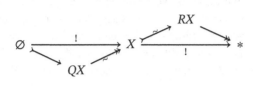

Note that Definitions C.2.5 and C.3.1 are self-dual: if (L, R) defines a weak factorization system on $\mathcal{M}$ then (R, L) defines a weak factorization system on $\mathcal{M}^{\mathrm{op}}$. Thus all general theorems about the right classes F of fibrations and $F \cap W$ of **trivial fibrations** "$\twoheadrightarrow$," imply the dual results involving the left classes C of cofibrations and $C \cap W$ of **trivial cofibrations** "$\hookrightarrow$." In particular, by Example C.1.4, all of the results proven in §C.1 about a category of fibrant objects hold for the fibrations, trivial fibrations, and weak equivalence between the eponymous fibrant objects in a model category, and the duals of these results hold for the cofibrations, trivial cofibrations, and weak equivalences between cofibrant objects.

Note also that since either class of a weak factorization system determines the other, the trivial cofibrations can be defined without reference to either the cofibrations or weak equivalences as those maps that have the left lifting property with respect to the fibrations, and dually the trivial fibrations are precisely those maps that have the right lifting property with respect to the cofibrations.

DEFINITION C.3.3. A functor between model categories is

- **left Quillen** if it preserves cofibrations, trivial cofibrations, and cofibrant objects, and
- **right Quillen** if it preserves fibrations, trivial fibrations, and fibrant objects.

Left Quillen functors admit left derived functors while right Quillen functors admit right derived functors. We leave a full account of this to other authors [103, §2.1–2] so as to avoid defining these terms, but an important component of the "derivability" of Quillen functors is captured by the following result:

LEMMA C.3.4. *Any left Quillen functor between model categories preserves weak equivalences between cofibrant objects, while any right Quillen functor preserves weak equivalences between fibrant objects.*

Proof For right Quillen functors this follows directly from Lemma C.1.10 and Example C.1.4. The result for left Quillen functors is dual. $\qquad \square$

Most left Quillen functors are "cocontinuous," preserving all colimits, while most right Quillen functors are "continuous," preserving all limits; when this is the case there is no need to separately assume that cofibrant or fibrant objects are preserved. In fact, Quillen functors commonly occur as an adjoint pair:

DEFINITION C.3.5. Consider an adjunction between a pair of model categories.

$$\mathcal{M} \underset{U}{\overset{F}{\underset{\perp}{\rightleftarrows}}} \mathcal{N}$$

By Lemma C.2.6 the following are equivalent, defining a **Quillen adjunction**.

(i) The left adjoint F is left Quillen.
(ii) The right U is right Quillen.
(iii) The left adjoint preserves cofibrations and the right adjoint preserves fibrations.
(iv) The left adjoint preserves trivial cofibrations and the right adjoint preserves trivial fibrations.

LEMMA C.3.6. *For a Quillen adjunction*

$$\mathcal{M} \underset{U}{\overset{F}{\underset{\perp}{\rightleftarrows}}} \mathcal{N}$$

the following are equivalent and characterize those Quillen adjunctions that define **Quillen equivalences***:*

(i) *For every cofibrant object $M \in \mathcal{M}$ and every fibrant object $N \in \mathcal{N}$, a map $f^{\sharp} : FM \to N$ is a weak equivalence in $\mathcal{N}$ if and only if its transpose $f^{\flat} : M \to UN$ is a weak equivalence in $\mathcal{M}$.*

(ii) *For every cofibrant object $M \in \mathcal{M}$, the derived unit*

$$M \xrightarrow{\eta} UFM \longrightarrow URFM$$

defined by composing the unit with fibrant replacement is a weak equivalence, and for every fibrant object $N \in \mathcal{N}$, the derived counit

$$FQUN \longrightarrow FUN \xrightarrow{\epsilon} N$$

defined by composing the counit with cofibrant replacement is a weak equivalence.

Proof Assume (i) and consider a cofibrant object $M \in \mathcal{M}$ and a fibrant object

$N \in \mathcal{N}$. The derived unit and counit are, respectively, transposes of the fibrant replacement and cofibrant replacement maps

$$FM \rightarrowtail\xrightarrow{\sim} RFM \qquad QUN \xrightarrow{\sim}\twoheadrightarrow UN$$

By Definition C.3.2 these maps are weak equivalences, and since M and QUN are cofibrant while RFM and N are fibrant, (i) tells us that the derived unit and counit must be weak equivalences as well.

Now assume that the derived unit is a weak equivalence and consider a weak equivalence $f^{\sharp} : FM \xrightarrow{\sim}\twoheadrightarrow N$ where M is cofibrant and N is fibrant. Applying the right adjoint and fibrant replacement, we obtain a commutative diagram

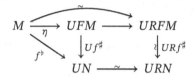

By Lemma C.3.4, the right and bottom maps are weak equivalences and by hypothesis the derived unit appearing as the top composite is as well. By the 2-of-3 property, the transpose $f^{\flat} : M \xrightarrow{\sim}\twoheadrightarrow UN$ is a weak equivalence as well. Dually, if the derived counit is a weak equivalence then weak equivalences $f^{\flat} : M \xrightarrow{\sim}\twoheadrightarrow UN$ transpose to weak equivalences $f^{\sharp} : FM \xrightarrow{\sim}\twoheadrightarrow N$. $\square$

We now introduce a pair of model structures on diagram categories that are designed to ensure that the diagonal functor $\Delta : \mathcal{M} \to \mathcal{M}^{\mathcal{J}}$ is, respectively, right or left Quillen, so that the colimit and limit functors, respectively, are left or right Quillen. The corresponding left and right derived functors then define the **homotopy colimit** and **homotopy limit** functors.

DEFINITION C.3.7. Let $\mathcal{M}$ be a model category and let $\mathcal{J}$ be a small category.

(i) The **projective model structure** on $\mathcal{M}^{\mathcal{J}}$ has weak equivalences and fibrations defined pointwise in $\mathcal{M}$.

(ii) The **injective model structure** on $\mathcal{M}^{\mathcal{J}}$ has weak equivalences and cofibrations defined pointwise in $\mathcal{M}$.

When the model category $\mathcal{M}$ is *combinatorial* or more generally *accessible*, the projective and injective model structures always exist [54]. Of course, the projective and injective model structures might happen to exist on $\mathcal{M}^{\mathcal{J}}$, perhaps for particular diagram shapes $\mathcal{J}$, in the absence of these hypotheses.

A Quillen two-variable adjunction is a two-variable adjunction in which the left adjoint is a left Quillen bifunctor while the right adjoints are both right

Quillen bifunctors. By Exercise C.2.v, any one of these conditions implies the other two:

DEFINITION C.3.8. A two-variable adjunction

$$V \times \mathcal{M} \xrightarrow{\otimes} \mathcal{N}, \quad V^{\mathrm{op}} \times \mathcal{N} \xrightarrow{\{,\}} \mathcal{M}, \quad \mathcal{M}^{\mathrm{op}} \times \mathcal{N} \xrightarrow{\mathrm{hom}} V$$

between model categories V, $\mathcal{M}$, and $\mathcal{N}$ defines a **Quillen two-variable adjunction** if any, and hence all, of the following equivalent conditions are satisfied:

(i) The functor $\hat{\otimes} \colon V^2 \times \mathcal{M}^2 \to \mathcal{N}^2$ carries a cofibration in V and a cofibration in $\mathcal{M}$ to a cofibration in $\mathcal{N}$ and furthermore this cofibration is a weak equivalence if either of the domain cofibrations are.

(ii) The functor $\widehat{\{,\}} \colon (V^2)^{\mathrm{op}} \times \mathcal{N}^2 \to \mathcal{M}^2$ carries a cofibration in V and a fibration in $\mathcal{N}$ to a fibration in $\mathcal{N}$ and furthermore this fibration is a weak equivalence if either of the domain maps are.

(iii) The functor $\widehat{\mathrm{hom}} \colon (\mathcal{M}^2)^{\mathrm{op}} \times \mathcal{N}^2 \to V^2$ carries a cofibration in $\mathcal{M}$ and a fibration in $\mathcal{N}$ to a fibration in V and furthermore this fibration is a weak equivalence if either of the domain maps are.

REMARK C.3.9. By Definition C.3.8, a two-variable adjunction is Quillen if and only if its left adjoint $\otimes \colon V \times \mathcal{M} \to \mathcal{N}$ is a **left Quillen bifunctor**: a bifunctor that is left Leibniz with respect to seven of the eight possible choices of constituent weak factorization systems, the exception the choice of $(C, F \cap W)$ for both V and $\mathcal{M}$ and $(C \cap W, F)$ for $\mathcal{N}$.

Quillen's axiomatization of the additional properties enjoyed by his model structure on the category of simplicial sets has been generalized by Hovey [57, §4.2] to define the notions of monoidal model category and enriched model category. We specialize the former to the cartesian closed categories of §A.1 as those are the only cases needed here. If V has a model structure and also is a cartesian closed category it is natural to ask that these structures be compatible in the following sense:

DEFINITION C.3.10. A **cartesian closed model category** is a cartesian closed category $(V, \times, 1)$ with a model structure so that

(i) the cartesian product and internal hom define a Quillen two-variable adjunction and

(ii) the map $Q1 \times v \to 1 \times v \cong v$ defined by the cofibrant replacement of the terminal object is a weak equivalence whenever v is cofibrant.

DEFINITION C.3.11. If V is a cartesian closed model category a V-**model category** is a model category $\mathcal{M}$ that is tensored, cotensored, and V-enriched and so that

(i) $(\otimes, \{,\}, \mathrm{hom})$ is a Quillen two-variable adjunction and
(ii) the map $Q1 \otimes m \to 1 \otimes m \cong m$ defined by the cofibrant replacement of the terminal object is a weak equivalence whenever m is cofibrant.

If $1 \in \mathcal{V}$ is cofibrant, the second condition in both Definition C.3.10 and C.3.11 follows from the first one (see Exercise C.3.iii).

LEMMA C.3.12. *If $\mathcal{M}$ is a $\mathcal{V}$-model category, then for any cofibrant object M and fibrant object N in $\mathcal{M}$, $\mathrm{hom}(M, N)$ is a fibrant object in $\mathcal{V}$. More generally, for any cofibrant object M and fibration $p: N \twoheadrightarrow P$, the induced map $p_*: \mathrm{hom}(M, N) \to \mathrm{hom}(M, P)$ is a fibration in $\mathcal{V}$.*

Proof By Proposition A.5.4 – which implies, for the terminal object $1 \in \mathcal{M}$ and any $M \in \mathcal{V}$, that $\mathrm{hom}(M, 1) \cong 1$ is terminal in $\mathcal{V}$ – the second statement subsumes the first. By Exercise C.2.v, the lifting problem below-left for any trivial cofibration i in $\mathcal{V}$ transposes to the lifting problem below-right

By Exercise C.3.iii, since M is cofibrant, $- \otimes M: \mathcal{V} \to \mathcal{M}$ is left Quillen, so $i \otimes M$ is a trivial cofibration in $\mathcal{M}$ and since $p: N \twoheadrightarrow P$ is a fibration, a solution to the lifting problem exists. $\square$

The next result was formulated by Gambino [45] in the context of a model category enriched over Quillen's cartesian closed model structure on simplicial sets, but its proof applies in greater generality.

THEOREM C.3.13. *If $\mathcal{M}$ is a $\mathcal{V}$-model category and $\mathcal{J}$ is a small category, then the weighted colimit functor*

$$\mathrm{colim}_- -: \mathcal{V}^{\mathcal{J}} \times \mathcal{M}^{\mathcal{J}^{\mathrm{op}}} \to \mathcal{M}$$

is left Quillen if the domain categories have the (injective, projective) or (projective, injective) model structure. Similarly, the weighted limit functor

$$\lim_- -: (\mathcal{V}^{\mathcal{J}})^{\mathrm{op}} \times \mathcal{M}^{\mathcal{J}} \to \mathcal{M}$$

is right Quillen if the domain categories have the (projective, projective) or (injective, injective) model structure.

Proof By Definition C.3.8 we can prove both statements in adjoint form. The weighted colimit bifunctor of Definition A.6.5 has a right adjoint (used to express the defining universal property of the weighted colimit)

$$\hom : (\mathcal{M}^{\mathcal{I}^{\mathrm{op}}})^{\mathrm{op}} \times \mathcal{M} \to \mathcal{V}^{\mathcal{I}}$$

which sends $F \in \mathcal{M}^{\mathcal{I}^{\mathrm{op}}}$ and $m \in \mathcal{M}$ to $\hom(F-, m) \in \mathcal{V}^{\mathcal{I}}$.

To prove the statement when $\mathcal{V}^{\mathcal{I}}$ has the projective and $\mathcal{M}^{\mathcal{I}^{\mathrm{op}}}$ has the injective model structure, we must show that this is a right Quillen bifunctor with respect to the pointwise (trivial) cofibrations in $\mathcal{M}^{\mathcal{I}^{\mathrm{op}}}$, (trivial) fibrations in $\mathcal{M}$, and pointwise (trivial) fibrations in $\mathcal{V}^{\mathcal{I}}$. Because the limits involved in the definition of right Quillen bifunctors are also formed pointwise, this follows immediately from the corresponding property of the enriched hom bifunctor, which was part of the definition of an enriched model category. The other cases are similar. $\square$

DIGRESSION C.3.14 (on the construction of homotopy colimits). In a model category, the terms *homotopy colimit* and *homotopy limit* refer to the derived functors of the colimit and limit functors. The upshot of Theorem C.3.13 is that there are two approaches to constructing a homotopy colimit: "fattening up the diagram" – for instance, by requiring that its objects are cofibrant and its morphisms are cofibrations – or "fattening up the weight" – typically by taking a cofibrant replacement of the terminal weight [103, §11.5]. Lemmas 6.2.14 and 6.2.18 can be understood as examples of the general equivalence between these two approaches.

We single out one of many consequences of Theorem C.3.13, of interest because the flexible weights of Definition 6.2.1 are precisely the cofibrant objects in the projective model structure on the category of weights defines relative to the Joyal model structure on simplicial sets.

COROLLARY C.3.15. *If $\mathcal{M}$ is a $\mathcal{V}$-model category, then for any diagram $F \in \mathcal{M}^{\mathcal{I}}$ whose objects are all fibrant and any projective cofibrant weight $W \in \mathcal{V}^{\mathcal{I}}$, the weighted limit is a fibrant object.*

Proof By Theorem C.3.13, the weighted limit bifunctor $\lim_{-} - : (\mathcal{V}^{\mathcal{I}})^{\mathrm{op}} \times \mathcal{M}^{\mathcal{I}} \to \mathcal{M}$ is right Quillen with respect to the projective model structure on the category of weights and the projective model structure on the category of diagrams. Since right Quillen bifunctors preserve fibrant objects, it follows that the limit of a pointwise fibrant diagram weighted by a projective cofibrant weight is fibrant. $\square$

Finally, we make use of the following theorem which enables the change of base of enrichment for model categories extending the results of §A.7. The premises of Theorem C.3.16 are the obvious extension of the premises of Proposition A.7.5 to the enriched model category context, but the conclusion only allows us to transfer enrichments in the direction of the right adjoint because an enriched model category must also be tensored and cotensored and these properties only transfer in that direction.

The result below is a specialization of a more general theorem proven in [50, 3.8] to cartesian closed bases for enrichment.

THEOREM C.3.16. *Consider a Quillen adjunction, in which the left adjoint preserves finite products, between cartesian closed model categories:*

$$\mathcal{V} \underset{U}{\overset{F}{\underset{\perp}{\rightleftarrows}}} \mathcal{W}$$

Then any $\mathcal{W}$-model category admits the structure of a $\mathcal{V}$-model category with the same underlying unenriched model category with enriched homs, cotensors, and tensors defined by:

$$\hom_{\mathcal{V}}(M, N) := U \hom_{\mathcal{W}}(M, N), \quad V \otimes M := FV \otimes M, \quad \text{and} \quad M^V := M^{FV}.$$

Proof By Proposition A.7.8 these definitions make $\mathcal{W}$ into a tensored and cotensored $\mathcal{V}$-enriched category. Lemma A.7.9 observes that change of base along the right adjoint of a finite-product-preserving adjunction preserves underlying 1-categories. It remains only to verify that the functors underlying the $\mathcal{V}$-enriched hom, tensor, and cotensor define a Quillen two-variable adjunction, but this follows easily from the cartesian closure of the model categories $\mathcal{V}$ and $\mathcal{W}$ and the fact that $F \dashv U$ is Quillen.   □

Exercises

EXERCISE C.3.i. Verify that the full subcategory of fibrant objects in a model category defines a category of fibrant objects with W the class of weak equivalences and F the class of fibrations between fibrant objects.

EXERCISE C.3.ii. Verify that a model structure on $\mathcal{M}$, if it exists, is uniquely determined by any of the following data:

   (i) the cofibrations and weak equivalences,
   (ii) the fibrations and weak equivalences,
   (iii) the cofibrations and fibrations, or
   (iv) the trivial cofibrations and trivial fibrations.[9]

EXERCISE C.3.iii.

---

[9] By a more delicate observation of Joyal [63, E.1.10], a model structure is also uniquely determined by (v) the cofibrations and fibrant objects, or (vi) the fibrations and cofibrant objects.

(i) Prove that if $\otimes: \mathcal{V} \times \mathcal{M} \to \mathcal{N}$ is a left Quillen bifunctor and $V \in \mathcal{V}$ is cofibrant then $V \otimes -: \mathcal{M} \to \mathcal{N}$ is a left Quillen functor.

(ii) Conclude that the second conditions of Definitions C.3.10 and C.3.11 are unnecessary if $1 \in \mathcal{V}$ is cofibrant.

EXERCISE C.3.iv. In a locally small category $\mathcal{M}$ with products and coproducts the hom bifunctor is part of a two-variable adjunction:

$$*: Set \times \mathcal{M} \to \mathcal{M}, \quad \{,\}: Set^{op} \times \mathcal{M} \to \mathcal{M}, \quad \hom: \mathcal{M}^{op} \times \mathcal{M} \to Set.$$

Equipping Set with the model structure whose weak equivalences are all maps, whose cofibrations are the monomorphisms, and whose fibrations are the epimorphisms, prove that

(i) Set is a cartesian closed model category.

(ii) Any model category $\mathcal{M}$ is a Set-model category.

C.4 Reedy Categories as Cell Complexes

In this section, we describe a structure borne by certain small categories $\mathcal{A}$ first exploited by Reedy to prove homotopical results about the category of $\mathcal{A}$-indexed diagrams [99]. Our primary examples – the ordinal category Δ_+, inverse categories, their opposites, and products of these – are all (strict) *Reedy categories* as defined by Daniel Kan,[10] so we confine our attention to this special case.[11] Our presentation follows [107].

DEFINITION C.4.1. A **Reedy structure** on a small category $\mathcal{A}$ consists of a **degree function** $\deg: \mathrm{obj}\,\mathcal{A} \to \omega$ together with a pair of wide subcategories $\vec{\mathcal{A}}$ and $\overleftarrow{\mathcal{A}}$ of **degree-increasing** and **degree-decreasing** arrows, respectively, so that

(i) For each nonidentity morphism in $\vec{\mathcal{A}}$, the degree of its domain is strictly less than the degree of its codomain, and for each nonidentity morphism in $\overleftarrow{\mathcal{A}}$, the degree of its domain is strictly greater than the degree of its domain.

---

[10] The original written reference for this definition, as cited by the canonical model category texts [55] and [57], was an early draft of the book that became [39], though by the time this manuscript appeared in print, it in turn referenced those sources in order to "review the notion of a Reedy category."

[11] This theory has been usefully extended by Berger and Moerdijk in such a way as to encompass "generalized Reedy categories" in which objects are permitted to have nonidentity automorphisms [12].

(ii) Every morphism f in $\mathcal{A}$ may be uniquely factored as

$$
\begin{array}{ccc}
\bullet & \xrightarrow{\quad f \quad} & \bullet \\
& \searrow_{\overleftarrow{\mathcal{A} \ni f}} \quad \bullet \quad \nearrow_{\overrightarrow{f} \in \mathcal{A}} &
\end{array}
\tag{C.4.2}
$$

Axiom (i) implies that $\overrightarrow{\mathcal{A}} \cap \overleftarrow{\mathcal{A}} = \mathrm{obj}(\mathcal{A})$, while both conditions together imply that $\mathcal{A}$ contains no nonidentity automorphisms (see Exercise C.4.i).

EXAMPLE C.4.3. Any inverse category $\mathcal{I}$ (see Definition C.1.16) is a Reedy category, with $\overleftarrow{\mathcal{I}} = \mathcal{I}$ and $\overrightarrow{\mathcal{I}} = \mathrm{obj}\,\mathcal{I}$. Conversely, any Reedy category $\mathcal{A}$ with $\mathcal{A} = \overleftarrow{\mathcal{A}}$ is an inverse category, in which case its degree *function* extends to a degree *functor* that reflects identity arrows.

EXAMPLE C.4.4. The category $\mathbf{\Delta}_+$ is a Reedy category with $\overrightarrow{\mathbf{\Delta}}_+$ the subcategory of monomorphisms and $\overleftarrow{\mathbf{\Delta}}_+$ the subcategory of epimorphisms. Here it is convenient to take advantage of the order isomorphism $\mathbb{1} + \omega \cong \omega$ to define $\deg[n] := n$. The subcategories $\mathbf{\Delta}$, $\mathbf{\Delta}_\top$, and $\mathbf{\Delta}_\perp$ all inherit analogous Reedy category structures.

REMARK C.4.5. If $\mathcal{A}$ is a Reedy category, then so is $\mathcal{A}^{\mathrm{op}}$: its Reedy structure has the same degree function but has the degree-increasing and degree-decreasing arrows interchanged. In particular, the Reedy categories of Example C.4.3 dualize to define **direct categories**, with an identity-reflecting functor $\deg : \mathcal{A} \to \omega$.

EXAMPLE C.4.6. If $\mathcal{A}$ and $\mathcal{B}$ are Reedy categories, so is $\mathcal{A} \times \mathcal{B}$, with $\deg(a, b) := \deg(a) + \deg(b)$ (see Exercise C.4.ii).

For the remainder of this section, we fix a Reedy category $\mathcal{A}$. We refer to the unique factorization (C.4.2) as the **Reedy factorization** of the map f and the degree of the object $\mathrm{cod}\,\overleftarrow{f} = \mathrm{dom}\,\overrightarrow{f}$ as the **degree** of f. Our next aim is to show that:

(i) The degree of f is the minimal degree of an object through which f factors.

(ii) The only factorization of f through an object with this degree is the Reedy factorization.

To prove these assertions, consider the category $\mathcal{F}\mathrm{act}f := (^{a/}\mathcal{A})_{/f} \cong {}^{f/}(\mathcal{A}_{/b})$ whose objects are factorizations $a \xrightarrow{g} c \xrightarrow{h} b$ of f and whose morphisms

$h \cdot g \to h' \cdot g'$ are maps $k : c \to c'$ so that the front triangles

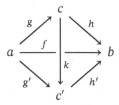

commute. Write $\mathcal{F}act_{\leq n}f \subset \mathcal{F}actf$ for the full subcategory of factorizations through an object of degree at most n.

LEMMA C.4.7. *The category $\mathcal{F}actf$ is connected, and each full subcategory $\mathcal{F}act_{\leq n}f$ is either empty or connected. The minimal n with $\mathcal{F}act_{\leq n}f$ nonempty is the degree of f, and $\mathcal{F}act_{\leq \deg(f)}f \cong \mathbb{1}$.*

Proof Consider $h \cdot g \in \mathcal{F}actf$ and their Reedy factorizations:

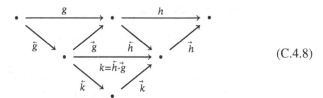

$$(C.4.8)$$

In this way, we define a zigzag of morphisms in $\mathcal{F}actf$ connecting $h \cdot g$ to $\vec{hk} \cdot \overleftarrow{kg}$, which by axiom (ii) must be the Reedy factorization of f. This shows that $\mathcal{F}actf$ is connected.

Moreover, axiom (i) implies that the degree of $\mathrm{cod}(g) = \mathrm{dom}(h)$ is at least the degree of f. In particular, if $h \cdot g \in \mathcal{F}act_{\leq n}f$, then each of the factorizations in (C.4.8) is as well, proving that $\mathcal{F}act_{\leq n}f$ is connected if it is nonempty. This diagram also shows that each nonempty category $\mathcal{F}act_{\leq n}f$ contains the Reedy factorization. Hence, the minimal such n is the degree of f.

Finally, if the degree of $\mathrm{cod}(g) = \mathrm{dom}(h)$ equals the degree of f, then $\vec{g}$ and $\overleftarrow{h}$ (and $\vec{k}$ and $\overleftarrow{k}$) must be identities, from which we deduce that $g \in \overleftarrow{\mathcal{A}}$ and $h \in \vec{\mathcal{A}}$: i.e., that $h \cdot g$ is the Reedy factorization. Hence $\mathcal{F}act_{\leq \deg(f)}f \cong \mathbb{1}$ is the terminal category as claimed. $\qquad\square$

Lemma C.4.7 is used to establish a "cellular decomposition" for the hom bifunctor, elsewhere denoted by $\mathcal{A}(-, -)$ but here abbreviated by $\mathcal{A} \in \mathcal{S}et^{\mathcal{A}^{\mathrm{op}} \times \mathcal{A}}$. The Reedy structure allows us to present the bifunctor $\mathcal{A}$ as a cell complex in the sense of Definition C.2.4: a sequential composite of pushouts of coproducts of basic "cells" that have a particular form. Lemma C.4.7 implies that the subset of arrows of degree at most n assembles into a sub*functor* of the hom-bifunctor.

DEFINITION C.4.9 (n-skeleton of the hom bifunctor). For any Reedy category $\mathcal{A}$, the n-**skeleton** is the subfunctor of arrows of degree at most n.

$$\mathrm{sk}_n \mathcal{A} \hookrightarrow \mathcal{A} \in \mathcal{S}et^{\mathcal{A}^{\mathrm{op}} \times \mathcal{A}}$$

There are obvious inclusions $\mathrm{sk}_{n-1} \mathcal{A} \hookrightarrow \mathrm{sk}_n \mathcal{A}$. The colimit of the sequence

$$\varnothing \hookrightarrow \mathrm{sk}_0 \mathcal{A} \hookrightarrow \cdots \hookrightarrow \mathrm{sk}_{n-1} \mathcal{A} \hookrightarrow \mathrm{sk}_n \mathcal{A} \hookrightarrow \cdots \hookrightarrow \mathrm{colim}\, \mathrm{sk}_n \mathcal{A} \cong \mathcal{A}$$

is the hom bifunctor $\mathcal{A}$. The morphisms of degree n first appear in $\mathrm{sk}_n \mathcal{A}$. It remains to express each inclusion $\mathrm{sk}_{n-1} \mathcal{A} \hookrightarrow \mathrm{sk}_n \mathcal{A}$ as a pushout of a coproduct of basic "cells" we now describe.

The external (pointwise) product defines a bifunctor $\square \colon \mathcal{S}et^{\mathcal{A}} \times \mathcal{S}et^{\mathcal{A}^{\mathrm{op}}} \to \mathcal{S}et^{\mathcal{A}^{\mathrm{op}} \times \mathcal{A}}$. For any $a \in \mathcal{A}$, there is a natural "composition" map $\circ \colon \mathcal{A}_a \square \mathcal{A}^a \to \mathcal{A}$ whose domain is the external product of the contravariant $\mathcal{A}^a$ and covariant $\mathcal{A}_a$ representables; here we write $\mathcal{A}^a$ to abbreviate $\mathcal{A}(-, a)$ and $\mathcal{A}_a$ to abbreviate $\mathcal{A}(a, -)$. By Lemma C.4.7, the composite of any pair of maps that factor through an object a of degree n lies in $\mathrm{sk}_n \mathcal{A}$, defining a map:

$$\coprod_{\deg(a)=n} \mathcal{A}_a \square \mathcal{A}^a \xrightarrow{\ \circ\ } \mathrm{sk}_n \mathcal{A}$$

Our next task is to describe the subfunctor of the domain that factors through $\mathrm{sk}_{n-1} \mathcal{A} \hookrightarrow \mathrm{sk}_n \mathcal{A}$, for which we require some new notation.

DEFINITION C.4.10 (boundaries of representable functors). If $a \in \mathcal{A}$ has degree n, write

$$\partial \mathcal{A}_a := \mathrm{sk}_{n-1} \mathcal{A}_a \qquad \in \mathcal{S}et^{\mathcal{A}} \qquad \text{and}$$
$$\partial \mathcal{A}^a := \mathrm{sk}_{n-1} \mathcal{A}^a \qquad \in \mathcal{S}et^{\mathcal{A}^{\mathrm{op}}}.$$

By Lemma C.4.7, $\partial \mathcal{A}_a \hookrightarrow \mathcal{A}_a$ is the subfunctor of arrows in $\mathcal{A}$ with domain a that do *not* lie in $\vec{\mathcal{A}}$, while $\partial \mathcal{A}^a \hookrightarrow \mathcal{A}^a$ is the subfunctor of arrows with codomain a that do *not* lie in $\overleftarrow{\mathcal{A}}$.

In particular, the exterior Leibniz product

$$\mathcal{A}_a \square \partial \mathcal{A}^a \underset{\partial \mathcal{A}_a \square \partial \mathcal{A}^a}{\cup} \partial \mathcal{A}_a \square \mathcal{A}^a \xrightarrow{(\partial \mathcal{A}_a \hookrightarrow \mathcal{A}_a) \hat{\square} (\partial \mathcal{A}^a \hookrightarrow \mathcal{A}^a)} \mathcal{A}_a \square \mathcal{A}^a$$

defines the subfunctor of pairs of morphisms $h \cdot g$ with $\mathrm{dom}\, h = \mathrm{cod}\, g = a$ in which at least one of the morphisms g and h has degree less than the degree of a.

PROPOSITION C.4.11. *The displayed commutative square is both a pullback and a pushout in* $\mathcal{S}et^{\mathcal{A}^{op} \times \mathcal{A}}$.

$$
\begin{array}{ccc}
\coprod_{\deg(a)=n} \partial \mathcal{A}_a \square \mathcal{A}^a \cup \mathcal{A}_a \square \partial \mathcal{A}^a & \hookrightarrow & \coprod_{\deg(a)=n} \mathcal{A}_a \square \mathcal{A}^a \\
{\scriptstyle \circ} \downarrow & & \downarrow {\scriptstyle \circ} \\
\mathrm{sk}_{n-1}\mathcal{A} & \hookrightarrow & \mathrm{sk}_n\mathcal{A}
\end{array}
\tag{C.4.12}
$$

The fact that (C.4.12) is a pullback is used to facilitate the proof that it is also a pushout.

Proof An element of the pullback consists of $f \in \mathrm{sk}_{n-1}\mathcal{A}$ together with a factorization $f = h \cdot g$ through an object a of degree n. If both h and g have degree n, then Lemma C.4.7 tells us that $h \cdot g$ is a Reedy factorization, contradicting the fact that f has degree at most $n - 1$. So we must have either $h \in \partial \mathcal{A}_a$ or $g \in \partial \mathcal{A}^a$, which tells us that the map from the upper left corner of (C.4.12) surjects onto the pullback. Because the top-horizontal map is monic, the comparison is therefore an isomorphism; i.e., (C.4.12) is a pullback square.

To see that it is a pushout, it suffices now to show that the right-hand vertical is one-to-one on the complement of $\mathrm{sk}_{n-1}\mathcal{A} \hookrightarrow \mathrm{sk}_n\mathcal{A}$. This follows from Lemma C.4.7, which argued that any morphism of degree n has a unique factorization through an object of that degree: namely its Reedy factorization. $\square$

As a corollary of Proposition C.4.11, the hom bifunctor $\mathcal{A}$ has a canonical presentation as a cell complex.

THEOREM C.4.13. *The inclusion* $\varnothing \hookrightarrow \mathcal{A} \in \mathcal{S}et^{\mathcal{A}^{op} \times \mathcal{A}}$ *has a canonical presentation as a cell complex:*

$$
\begin{array}{ccc}
\coprod_{\deg(a)=n} \partial \mathcal{A}_a \square \mathcal{A}^a \cup \mathcal{A}_a \square \partial \mathcal{A}^a & \hookrightarrow & \coprod_{\deg(a)=n} \mathcal{A}_a \square \mathcal{A}^a \\
{\scriptstyle \circ} \downarrow & & \downarrow {\scriptstyle \circ} \\
\varnothing \hookrightarrow \mathrm{sk}_0\mathcal{A} \dashrightarrow \mathrm{sk}_{n-1}\mathcal{A} & \xrightarrow{\ \ulcorner\ } & \mathrm{sk}_n\mathcal{A} \dashrightarrow \mathrm{colim}_n \mathrm{sk}_n\mathcal{A} \cong \mathcal{A}
\end{array}
$$

i.e., is a sequential composite of pushouts of coproducts of cells defined as exterior Leibniz products

$$
(\partial \mathcal{A}_a \hookrightarrow \mathcal{A}_a) \,\hat{\square}\, (\partial \mathcal{A}^a \hookrightarrow \mathcal{A}^a),
$$

where the cell for each $a \in \mathcal{A}$ of degree n is attached at stage n. $\square$

As a corollary of Theorem C.4.13, any morphism $f \in \mathcal{M}^{\mathcal{A}}$ is itself a cell complex: the cellular decomposition of $\mathcal{A}$ is translated into a cellular decomposition for f by taking weighted colimits. Taking weighted limits instead transforms the cellular decomposition of $\mathcal{A}$ into a "Postnikov presentation" for f as the limit of a countable tower of pullbacks of products of particular maps. This sort of result is exemplary of the slogan of [107] that "it is all in the weights." Before proving this corollary, we require notation for the maps appearing as the basic cells and basic layers.

DEFINITION C.4.14 (latching and matching objects). Let $a \in \mathcal{A}$. The **latching** and **matching objects** of diagram $X \in \mathcal{M}^{\mathcal{A}}$ at a are defined to be the colimits and limits, respectively, weighted by the boundary representables of appropriate variance:

$$L^a X := \operatorname{colim}_{\partial \mathcal{A}^a} X \qquad M^a X := \lim_{\partial \mathcal{A}_a} X.$$

The boundary inclusions $\partial \mathcal{A}^a \hookrightarrow \mathcal{A}^a$ and $\partial \mathcal{A}_a \hookrightarrow \mathcal{A}_a$ induce the **latching** and **matching maps** $\ell^a : L^a X \to X^a$ and $m^a : X^a \to M^a X$, on account of the isomorphisms $\operatorname{colim}_{\mathcal{A}^a} X \cong X^a \cong \lim_{\mathcal{A}_a} X$ of Definition A.6.1(i).

DEFINITION C.4.15 (relative latching and matching maps). The **relative latching** and **relative matching maps** of a natural transformation $f : X \to Y \in \mathcal{M}^{\mathcal{A}}$ are defined to be the Leibniz weighted colimits and limits

$$\hat{\ell}^a f := \widehat{\operatorname{colim}}_{\partial \mathcal{A}^a \hookrightarrow \mathcal{A}^a} f \qquad \hat{m}^a f := \widehat{\lim}_{\partial \mathcal{A}_a \hookrightarrow \mathcal{A}_a} f,$$

i.e., by the pullbacks and pushouts:

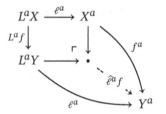 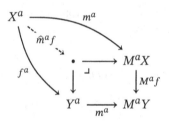

of the maps $L^a f := \operatorname{colim}_{\partial \mathcal{A}^a} f$ and $M^a f := \lim_{\partial \mathcal{A}_a} f$.

NOTATION C.4.16. For any diagram $X \in \mathcal{M}^{\mathcal{A}}$ let

$$\operatorname{sk}_n X := \operatorname{colim}_{\operatorname{sk}_n \mathcal{A}} X \qquad \text{and} \qquad \operatorname{cosk}_n X := \lim_{\operatorname{sk}_n \mathcal{A}} X$$

denote the results of applying the weighted colimit and weighted limit bifunctors $\operatorname{colim}_- - : \mathcal{S}et^{\mathcal{A}^{op} \times \mathcal{A}} \times \mathcal{M}^{\mathcal{A}} \to \mathcal{M}^{\mathcal{A}}$ and $\lim_- - : (\mathcal{S}et^{\mathcal{A}^{op} \times \mathcal{A}})^{op} \times \mathcal{M}^{\mathcal{A}} \to \mathcal{M}^{\mathcal{A}}$ of Exercise A.6.iv to the diagram X with weight $\operatorname{sk}_n \mathcal{A}$.

Recall the set-tensor $* : \mathcal{S}et \times \mathcal{M} \to \mathcal{M}$ and set-cotensor $\{,\} : \mathcal{S}et^{op} \times \mathcal{M} \to \mathcal{M}$ bifunctors defined for category with coproducts and products.

COROLLARY C.4.17. *Let $\mathcal{A}$ be a Reedy category and let $\mathcal{M}$ be complete and cocomplete. Any morphism $f : X \to Y \in \mathcal{M}^{\mathcal{A}}$ is a cell complex*

$$X \to X \underset{\mathrm{sk}_0 X}{\cup} \mathrm{sk}_0 Y \to \cdots \to X \underset{\mathrm{sk}_{n-1} X}{\cup} \mathrm{sk}_{n-1} Y \to X \underset{\mathrm{sk}_n X}{\cup} \mathrm{sk}_n Y \to \cdots \to \mathrm{colim} \cong Y$$

whose nth stage attaches the coproduct of the cells

$$(\partial \mathcal{A}_a \hookrightarrow \mathcal{A}_a) \hat{*} \hat{\ell}^a f$$

indexed by objects a of degree n, and also a "Postnikov tower"

$$X \cong \lim \to \cdots \to \mathrm{cosk}_n X \underset{\mathrm{cosk}_n Y}{\times} Y \to \mathrm{cosk}_{n-1} X \underset{\mathrm{cosk}_{n-1} Y}{\times} Y \to \cdots \to \mathrm{cosk}_0 X \underset{\mathrm{cosk}_0 Y}{\times} Y \to Y$$

whose nth layer is the product of the maps

$$\{\partial \mathcal{A}^a \hookrightarrow \widehat{\mathcal{A}^a}, \widehat{m}^a f\}$$

indexed by the objects a of degree n.

Proof These dual results follow immediately by applying the weighted colimit and weighted limit bifunctors of Exercise A.6.iv

$$\mathrm{colim}_{-} - : \mathcal{S}et^{\mathcal{A}^{\mathrm{op}} \times \mathcal{A}} \times \mathcal{M}^{\mathcal{A}} \to \mathcal{M}^{\mathcal{A}} \quad \text{and} \quad \lim_{-} - : (\mathcal{S}et^{\mathcal{A}^{\mathrm{op}} \times \mathcal{A}})^{\mathrm{op}} \times \mathcal{M}^{\mathcal{A}} \to \mathcal{M}^{\mathcal{A}}$$

to the cell complex presentations of Theorem C.4.13; recall from Definition A.6.1(ii) that both bifunctors are cocontinuous in the weight.

To see that the cell complex presentation for f has the asserted form, note that for any diagram $X \in \mathcal{M}^{\mathcal{A}}$ and weight defined by an exterior product of $U \in \mathcal{S}et^{\mathcal{A}}$ and $V \in \mathcal{S}et^{\mathcal{A}^{\mathrm{op}}}$, there is a natural isomorphism $\mathrm{colim}_{U \boxdot V} X \cong U * \mathrm{colim}_V X$, which extends to a natural isomorphism between Leibniz products (Proposition C.2.9(i)).

By the coYoneda lemma, $f \cong \mathrm{colim}_{\mathcal{A}} f \cong \widehat{\mathrm{colim}}_{\emptyset \hookrightarrow \mathcal{A}} f$. By Proposition C.2.9(vii), the Leibniz weighted colimit functor $\widehat{\mathrm{colim}}_{-} f$ preserves cell structures. It follows that f admits a canonical presentation as a cell complex with cells

$$\widehat{\mathrm{colim}}_{(\partial \mathcal{A}_a \hookrightarrow \mathcal{A}_a) \boxdot (\partial \mathcal{A}^a \hookrightarrow \mathcal{A}^a)} f \cong (\partial \mathcal{A}_a \hookrightarrow \mathcal{A}_a) \hat{*} \widehat{\mathrm{colim}}_{\partial \mathcal{A}^a \hookrightarrow \mathcal{A}^a} f \cong (\partial \mathcal{A}_a \hookrightarrow \mathcal{A}_a) \hat{*} \hat{\ell}^a f. \qquad \square$$

This presentation is most familiar for the Reedy category $\boldsymbol{\Delta}^{\mathrm{op}}$. Here we write $\boldsymbol{\Delta}^n$ for the standard n-simplex $\Delta[n]$ and $\partial \boldsymbol{\Delta}^n$ for its boundary, common notational conventions in the literature that are consistent with the notation of Definition C.4.10.

EXAMPLE C.4.18. A simplicial object Y taking values in any cocomplete category admits a skeletal filtration

$$\varnothing \to \mathrm{sk}_0\, Y \to \cdots \to \mathrm{sk}_{n-1}\, Y \to \mathrm{sk}_n\, Y \to \cdots \to \operatorname*{colim}_n \mathrm{sk}_n\, Y \cong Y$$

in which the step from stage $n - 1$ to stage n is given by a pushout

$$
\begin{array}{ccc}
\Delta^n * L_n Y \cup \partial\Delta^n * Y_n & \longrightarrow & \Delta^n * Y_n \\
\downarrow & & \downarrow \\
\mathrm{sk}_{n-1}\, Y & \xrightarrow{\quad\ulcorner\quad} & \mathrm{sk}_n\, Y
\end{array}
$$

where $L_n Y \to Y_n$ is the object of "degenerate n-simplices."

Considering the Yoneda embedding as a simplicial object $\Delta \in (Set^\Delta)^{\Delta^{op}}$, this specializes to the "canonical cell complex presentation" of the hom bifunctor of Theorem C.4.13

$$
\begin{array}{ccc}
\Delta^n \times \partial\Delta_n \cup \partial\Delta^n \times \Delta_n & \lhook\joinrel\longrightarrow & \Delta^n \times \Delta_n \\
\downarrow & & \downarrow \\
\varnothing \lhook\joinrel\longrightarrow \cdots \lhook\joinrel\longrightarrow \mathrm{sk}_{n-1}\Delta & \xlhook{\quad\ulcorner\quad}\joinrel\longrightarrow \mathrm{sk}_n \Delta \lhook\joinrel\longrightarrow \cdots \lhook\joinrel\longrightarrow \Delta
\end{array}
$$

In summary, Corollary C.4.17 tells us that we may express a generic natural transformation between diagrams of shape $\mathcal{A}$ valued in a complete and cocomplete category as

(i) a cell complex whose cells are Leibniz tensors built from boundary inclusions of covariant representables and relative latching maps, and dually

(ii) as a Postnikov tower whose layers are Leibniz cotensors built from boundary inclusions of contravariant representables and relative matching maps.

This explains the importance of these maps to Reedy category theory, as we shall discover in the next section.

Exercises

EXERCISE C.4.i. Show that any isomorphism in a (strict) Reedy category is an identity.

EXERCISE C.4.ii. Show that the product of two Reedy categories is a Reedy category, with the degree of an object defined to be the sum of the degrees.

C.5 The Reedy Model Structure

Our aim in this section is to explain how any weak factorization system on $\mathcal{M}$ gives rise to a *Reedy weak factorization system* on $\mathcal{M}^{\mathcal{A}}$, and moreover that the Reedy weak factorization systems associated to a model structure on $\mathcal{M}$ define the *Reedy model structure* on $\mathcal{M}^{\mathcal{A}}$. Both results derive from an analysis of what is required to inductively define functors and natural transformations indexed by a Reedy category. Finally, we prove that the weighted limit and weighted colimit bifunctors define Quillen bifunctors – an important special case of a more general algebraic result – and discuss the implications of this result for the theory of homotopy limits and homotopy colimits indexed by Reedy categories.

This work requires one preliminary: a discussion of how the skeleta and coskeleta introduced in the previous section feature in the inductive definition of Reedy-shaped diagrams. For a Reedy category $\mathcal{A}$, write

$$\mathcal{A}_{\leq 0} \subset \mathcal{A}_{\leq 1} \subset \cdots \subset \mathcal{A}_{\leq n-1} \subset \mathcal{A}_{\leq n} \subset \cdots \subset \mathcal{A}$$

for the full subcategories of objects with degree at most the ordinal appearing in the subscript. These categories give us a new way to understand the skeleton and coskeleton functors of C.4.16.

LEMMA C.5.1. *For any complete and cocomplete category $\mathcal{M}$, restriction and left and right Kan extension define an adjoint triple of functors*

$$\mathcal{M}^{\mathcal{A}} \underset{\underset{\mathrm{ran}_n}{\overset{\bot}{\longleftarrow}}}{\overset{\overset{\mathrm{lan}_n}{\longleftarrow}}{\underset{\mathrm{res}_n}{\overset{\bot}{\longrightarrow}}}} \mathcal{M}^{\mathcal{A}_{\leq n}}$$

with induced comonad $\mathrm{sk}_n := \mathrm{lan}_n \circ \mathrm{res}_n$ *and monad* $\mathrm{cosk}_n := \mathrm{ran}_n \circ \mathrm{res}_n$ *that are adjoint* $\mathrm{sk}_n \dashv \mathrm{cosk}_n$ *and naturally isomorphic to the functors defined by weighted colimit and weighted limit*

$$\mathrm{lan}_n\,\mathrm{res}_n(-) \cong \mathrm{colim}_{\mathrm{sk}_n\mathcal{A}} - \qquad and \qquad \mathrm{ran}_n\,\mathrm{res}_n(-) \cong \lim_{\mathrm{sk}_n\mathcal{A}} -.$$

Proof Exercise C.5.i. □

For example:

DEFINITION C.5.2. Specializing the notation above, write $\Delta_{\leq n} \subset \Delta$ for the full subcategory of the simplex category of 1.1.1 spanned by the ordinals $[0], \ldots, [n]$. Restriction and left and right Kan extension define adjunctions

$$\mathcal{S}et^{\Delta^{\mathrm{op}}} \underset{\underset{\mathrm{ran}_n}{\overset{\bot}{\longleftarrow}}}{\overset{\overset{\mathrm{lan}_n}{\longleftarrow}}{\underset{\mathrm{res}_n}{\overset{\bot}{\longrightarrow}}}} \mathcal{S}et^{\Delta_{\leq n}^{\mathrm{op}}}$$

inducing an idempotent comonad $\mathrm{sk}_n := \mathrm{lan}_n \circ \mathrm{res}_n$ and an idempotent monad $\mathrm{cosk}_n := \mathrm{ran}_n \circ \mathrm{res}_n$ on $s\mathcal{S}et$ that are adjoint $\mathrm{sk}_n \dashv \mathrm{cosk}_n$. The counit and unit of this comonad and monad define canonical maps

$$\mathrm{sk}_n X \xrightarrowtail{\;\epsilon\;} X \xrightarrow{\;\eta\;} \mathrm{cosk}_n X$$

relating a simplicial set X with its n-**skeleton** and n-**coskeleton**. We say X is n-**skeletal** or n-**coskeletal** if the former or latter of these maps, respectively, is an isomorphism.

The following lemma records special properties of an adjoint triple of functors $\mathrm{lan}_n \dashv \mathrm{res}_n \dashv \mathrm{ran}_n$ arising from a fully faithful inclusion $\mathrm{sk}_n \mathcal{A} \hookrightarrow \mathcal{A}$. In particular, the canonical map from the n-skeleton of a simplicial set to its n-coskeleton can be defined more generally, not just for diagrams defined as restrictions:

LEMMA C.5.3. *For any fully faithful inclusion $\mathcal{B} \hookrightarrow \mathcal{A}$ and complete and co-complete category $\mathcal{M}$, consider the associated adjoint triple:*

$$\mathcal{M}^{\mathcal{A}} \underset{\underset{\mathrm{ran}}{\overset{\bot}{\xleftarrow{\hspace{1cm}}}}}{\overset{\overset{\mathrm{lan}}{\xleftarrow{\hspace{1cm}}}}{\underset{\bot}{\xrightarrow[\mathrm{res}]{\hspace{1cm}}}}} \mathcal{M}^{\mathcal{B}}$$

(i) *The functors $\mathrm{lan}, \mathrm{ran} : \mathcal{M}^{\mathcal{B}} \to \mathcal{M}^{\mathcal{A}}$ are fully faithful; that is, the unit of $\mathrm{lan} \dashv \mathrm{res}$ and the counit of $\mathrm{res} \dashv \mathrm{ran}$ are isomorphisms.*

(ii) *The common composite in the commutative square below defines a canon-ical natural transformation*

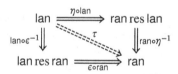

While we find the commutative square in (ii) amusing, since the inclusion $\mathcal{B} \hookrightarrow \mathcal{A}$ is fully faithful, that τ can be defined more simply using the universal properties of left and right Kan extensions as the initial and terminal functors that extend a given diagram.

Proof It is well-known that a right adjoint functor is fully faithful if and only if the counit is an isomorphism and that the counit of a pointwise right Kan extension along a fully faithful functor is an isomorphism; for proofs, specialize the results of Lemma 9.4.4 and Proposition 9.4.5 to the ∞-cosmos $\mathcal{C}at$. These statements and their duals prove (i).

In (ii), τ is defined to be the adjoint transpose of $\eta^{-1} : \mathrm{res}\,\mathrm{lan} \Rightarrow \mathrm{id}$ under res ⊣ ran and also to be the adjoint transpose of ϵ^{-1} : id ⇒ res ran under lan ⊣ res. To see that these definitions agree, observe that the former asserts that the composite of the right two morphisms below is the unique right inverse of the left morphism, while the latter asserts that the composite of the left two morphisms below is the unique left inverse of the right morphism:

$$\text{id} \xrightarrow[\cong]{\eta} \text{res lan} \xrightarrow{\text{res } \tau} \text{res ran} \xrightarrow[\cong]{\epsilon} \text{id}$$

In other words, both definitions assert exactly that the displayed triple composite is the identity. □

In this way, we obtain a natural transformation τ_n : $\text{sk}_n \Rightarrow \text{cosk}_n$ between the comonad and monad introduced in Lemma C.5.1. These structures allow us to inductively define Reedy diagrams:

PROPOSITION C.5.4 (inductive definition of diagrams).

(i) *A diagram $X \in \mathcal{M}^{\mathcal{A}_{\leq n-1}}$ together with a family of factorizations for each object $a \in \mathcal{A}$ of degree n*

$$\text{sk}_{n-1} X^a \xrightarrow{\tau_{n-1} X^a} \text{cosk}_{n-1} X^a$$
$$\overset{i^a}{\dashrightarrow} X^a \overset{p^a}{\dashrightarrow}$$

uniquely determines an extension of X to a diagram $X \in \mathcal{M}^{\mathcal{A}_{\leq n}}$.

(ii) *A natural transformation $f : X \to Y \in \mathcal{M}^{\mathcal{A}_{\leq n-1}}$ together with a family of factorizations for each object $a \in \mathcal{A}$ of degree n*

$$
\begin{array}{ccccc}
 & \tau_{n-1} X^a & & & \\
\text{sk}_{n-1} X^a & \overset{i^a}{\dashrightarrow} X^a & \overset{p^a}{\dashrightarrow} & \text{cosk}_{n-1} X^a & \\
\text{sk}_{n-1} f^a \downarrow & & f^a \downarrow & & \downarrow \text{cosk}_{n-1} f^a \\
\text{sk}_{n-1} Y^a & \overset{i^a}{\dashrightarrow} Y^a & \overset{p^a}{\dashrightarrow} & \text{cosk}_{n-1} Y^a & \\
 & \tau_{n-1} Y^a & & & \\
\end{array}
$$

uniquely determines an extension of f to a map $f : X \to Y \in \mathcal{M}^{\mathcal{A}_{\leq n}}$.

Proof For (i), it remains to define the action of X on nonidentity morphisms whose domain or codomain has degree n. The Reedy factorization of any such morphism $k : a \to a'$ is through an object b of degree less than n. By composing the maps in the upper-right or lower-left square, there exist unique dotted-arrow maps making the following diagram commute

$$
\begin{array}{ccccc}
\mathrm{sk}_{n-1}X^a & \xrightarrow{\ i^a\ } & X^a & \xrightarrow{\ p^a\ } & \mathrm{cosk}_{n-1}X^a \\
{\scriptstyle \mathrm{sk}_{n-1}X^{\vec{k}}}\downarrow & & \downarrow{\scriptstyle X^{\vec{k}}} & & \downarrow{\scriptstyle \mathrm{cosk}_{n-1}X^{\vec{k}}} \\
\mathrm{sk}_{n-1}X^b & \xleft=\xright[\ i^b\ ] & X^b & \xleft=\xright[\ p^b\ ] & \mathrm{cosk}_{n-1}X^b \\
{\scriptstyle \mathrm{sk}_{n-1}X^{\vec{f}}}\downarrow & & \downarrow{\scriptstyle X^{\vec{k}}} & & \downarrow{\scriptstyle \mathrm{cosk}_{n-1}X^{\vec{k}}} \\
\mathrm{sk}_{n-1}X^{a'} & \xrightarrow[\ i^{a'}\ ]{} & X^{a'} & \xrightarrow[\ p^{a'}\ ]{} & \mathrm{cosk}_{n-1}X^{a'}
\end{array}
$$

and these compose to define the action of X on k. The functoriality of this construction in a composable pair of morphisms $k \cdot h$ follows from connectedness of the category $\mathcal{F}act_{\leq n-1}(k \cdot h)$.

For (ii), apply (i) to the diagram $a \mapsto f^a : \mathcal{A}_{\leq n-1} \to \mathcal{M}^2$. $\qquad\qquad\square$

Now we turn our attention to the main subject of this section. Let $\mathcal{M}$ be a category with a weak factorization system (L, R) and let $\mathcal{A}$ be a Reedy category.

DEFINITION C.5.5. The **Reedy weak factorization system** $(L[\mathcal{A}], R[\mathcal{A}])$ on $\mathcal{M}^{\mathcal{A}}$ defined relative to the weak factorization system (L, R) on $\mathcal{M}$ has:

- as left class $L[\mathcal{A}]$ those maps $f : X \to Y \in \mathcal{M}^{\mathcal{A}}$ whose relative latching maps $\hat{\ell}^a f : \ell^a f \to Y^a \in \mathcal{M}$ are in L, and
- as right class $R[\mathcal{A}]$ those maps $f : X \to Y \in \mathcal{M}^{\mathcal{A}}$ whose relative matching maps $\hat{m}^a f : X^a \to m^a f \in \mathcal{M}$ are in R.

We say a map $f : X \to Y \in \mathcal{M}^{\mathcal{A}}$ is **Reedy in** L or **Reedy in** R if its relative latching or relative matching maps are in L or R, respectively.

The classes $L[\mathcal{A}]$ and $R[\mathcal{A}]$ are lifts, respectively, of the classes L and R along the $\mathcal{A}$-indexed family of functors $\hat{\ell}^a : (\mathcal{M}^{\mathcal{A}})^2 \to \mathcal{M}^2$ or $\hat{m}^a : (\mathcal{M}^{\mathcal{A}})^2 \to \mathcal{M}^2$. By functoriality, we see that these classes are closed under retracts. Now Exercise C.2.ii combines with the following pair of lemmas to imply that these two classes indeed define a weak factorization system on the category $\mathcal{M}^{\mathcal{A}}$.

LEMMA C.5.6. *The maps $i \in L[\mathcal{A}]$ have the left lifting property with respect to the maps $p \in R[\mathcal{A}]$.*

$$
\begin{array}{ccc}
A & \longrightarrow & K \\
{\scriptstyle i}\downarrow & \nearrow & \downarrow{\scriptstyle p} \\
B & \longrightarrow & L
\end{array}
$$

Proof By Lemma C.2.3 and Corollary C.4.17, to show that $i \boxtimes p$ for any pair of morphisms $i, p \in \mathcal{M}^{\mathcal{A}}$, it suffices to solve the lifting problems below-left

$$(\partial \mathcal{A}_a \hookrightarrow \mathcal{A}_a) \hat{*} \hat{\ell}^a i \downarrow \quad\nearrow\quad \downarrow p \qquad\leftrightsquigarrow\qquad \hat{\ell}^a i \downarrow \quad\nearrow\quad \downarrow \widehat{\lim}_{\partial \mathcal{A}_a \hookrightarrow \mathcal{A}_a} p \cong \hat{m}^a p$$

in $\mathcal{M}^{\mathcal{A}}$ for each $a \in \mathcal{A}$. By adjunction, it suffices to solve the transposed lifting problem in $\mathcal{M}$ above-right (see Lemma C.2.6). If $i \in L[\mathcal{A}]$ and $p \in R[\mathcal{A}]$, then by definition $\hat{\ell}^a i \in L$ and $\hat{m}^a p \in R$, so a solution exists. $\qquad\square$

LEMMA C.5.7. *Every map $f : X \to Y \in \mathcal{M}^{\mathcal{A}}$ can be factored as a map in $L[\mathcal{A}]$ followed by a map in $R[\mathcal{A}]$.*

Proof We define the components of the factorization of $f^a : X^a \to Y^a$ inductively in the degree of a. To start, we use the factorization of (L, R) to factor all components indexed by objects at degree zero. Since the full subcategory $\mathcal{A}_{\leq 0}$ spanned by these objects has only identity arrows, this defines a factorization of the subdiagram $f \in \mathcal{M}^{\mathcal{A}_{\leq 0}}$.

Continuing inductively, suppose we have factored the restriction $f \in \mathcal{M}^{\mathcal{A}_{<n}}$ as

$$X \xrightarrow{\;\;f\;\;} Y \qquad \in \mathcal{M}^{\mathcal{A}_{<n}}$$
$$X \xrightarrow{\;g\;} Z \xrightarrow{\;h\;} Y$$

with the relative latching maps $\hat{\ell}^a g \in L$ and $\hat{m}^a h \in R$ for all objects a of degree less than n. By Proposition C.5.4, to define the attendant factorization of f^a, it suffices to define an object Z^a of $\mathcal{M}$ together with the dotted arrow maps.

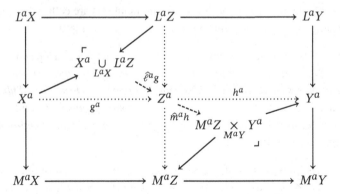

We factor the dashed diagonal map from the pushout to the pullback using (L, R) to define the object Z^a. The left and right factors become the ath relative latching map and matching map of the composite morphisms g^a and h^a so-defined. Note by construction that these maps lie in the classes L and R, respectively. It follows from the universal properties of the pushout and the pullback that $f^a = h^a \cdot g^a$. By Proposition C.5.4 these definitions extend the natural transformations g and h to degree n. □

It follows from Proposition C.2.9 and Corollary C.4.17 that if the left class of a weak factorization system (L, R) on $\mathcal{M}$ is cofibrantly or cellularly generated, as in Definition C.2.4, then the left class of the Reedy weak factorization system is too:

PROPOSITION C.5.8. *If (L, R) is a weak factorization system on $\mathcal{M}$ that is cellularly or cofibrantly generated by a set of maps J, then the Reedy weak factorization system $(L[\mathcal{A}], R[\mathcal{A}])$ on $\mathcal{M}^{\mathcal{A}}$ is cellularly or cofibrantly generated, respectively, by the set*

$$\{(\partial \mathcal{A}_a \hookrightarrow \mathcal{A}_a) \mathbin{\hat{*}} j\}_{a \in \mathcal{A}, j \in \mathcal{J}}.$$

Proof By Corollary C.4.17, any morphism $f : X \to Y \in \mathcal{M}^{\mathcal{A}}$ may be presented as a cell complex built from cells

$$\{(\partial \mathcal{A}_a \hookrightarrow \mathcal{A}_a) \mathbin{\hat{*}} \hat{\ell}^a f\}_{a \in \mathcal{A}}.$$

If $f \in L[\mathcal{A}]$, then $\hat{\ell}^a f \in L$ for each a, and by hypothesis these relative latching maps may be presented as cell complexes or retracts of cell complexes built from the maps in the generating set J. By Proposition C.2.9(vii), the Leibniz tensors $(\partial \mathcal{A}_a \hookrightarrow \mathcal{A}_a) \mathbin{\hat{*}} \hat{\ell}^a f$ may then be presented as (retracts of) cell complexes built from the Leibniz tensors of the boundary inclusions and the maps in J, exactly as claimed in the statement. □

For example, the monomorphisms of simplicial sets are cellularly generated by the simplex boundary inclusions $\partial \Delta[n] \hookrightarrow \Delta[n]$ for $n \geq 0$.

LEMMA C.5.9. *The Reedy weak factorization system $(mono[\Delta^{\mathrm{op}}], epi[\Delta^{\mathrm{op}}])$ on $s\mathcal{S}et = \mathcal{S}et^{\Delta^{\mathrm{op}}}$ defined relative to the (monomorphism, epimorphism) weak factorization system on $\mathcal{S}et$ coincides with the (monomorphism, trivial fibration) weak factorization system. Consequently, any monomorphism of simplicial sets decomposes canonically as a sequential composite of pushouts of coproducts of the maps $\partial \Delta[n] \hookrightarrow \Delta[n]$ for $n \geq 0$.*

Proof Monomorphisms of sets are cellularly generated by a single map, the inclusion $! : \varnothing \hookrightarrow *$. Consequently, by Proposition C.5.8, the Reedy weak factorization system is cellularly generated as well. In this case, the pushout product functor $- \mathbin{\hat{*}} ! : \mathcal{S}et^{\Delta^{\mathrm{op}}} \to \mathcal{S}et^{\Delta^{\mathrm{op}}}$ is the identity, so the set of generating maps are the familiar simplex boundary inclusions $\{\partial \Delta[n] \hookrightarrow \Delta[n]\}_{[n] \in \Delta}$, the right lifting property against which characterizes the trivial fibrations (see Definition 1.1.25). □

PROPOSITION C.5.10.

(i) *If $f : X \to Y \in \mathcal{M}^{\mathcal{A}}$ is Reedy in L, that is, if the relative latching maps $\hat{\ell}^a f$ are in L, then each of the components $f^a : X^a \to Y^a$ and each of the latching maps $L^a f : L^a X \to L^a Y$ are also in L.*

(ii) *If $f : X \to Y \in \mathcal{M}^{\mathcal{A}}$ is Reedy in R, that is, if the relative matching maps $\hat{m}^a f$ are in R, then each of the components $f^a : X^a \to Y^a$ and each of the matching maps $M^a f : M^a X \to M^a Y$ are also in R.*

Proof The maps f^a and $L^a f$ are the Leibniz weighted colimits of f with the maps $\varnothing \hookrightarrow \mathcal{A}^a$ and $\varnothing \hookrightarrow \partial \mathcal{A}^a$, respectively. Evaluating the covariant variable of the cell complex presentation of Theorem C.4.13 at $a \in \mathcal{A}$, we see that $\varnothing \hookrightarrow \mathcal{A}^a$ is a cell complex whose cells have the form

$$((\partial \mathcal{A}_x)^a \hookrightarrow \mathcal{A}_x^a) \, \hat{\Box} \, (\partial \mathcal{A}^x \hookrightarrow \mathcal{A}^x), \qquad (C.5.11)$$

indexed by the objects $x \in \mathcal{A}$. In fact, it suffices to consider those objects with $\deg(x) \le \deg(a)$ because when $\deg(x) > \deg(a)$ the inclusion $(\partial \mathcal{A}_x)^a \hookrightarrow \mathcal{A}_x^a$, and hence the cell (C.5.11), is an isomorphism. Similarly, since $\partial \mathcal{A}^a = \mathrm{sk}_{\deg(a)-1} \mathcal{A}^a$, Theorem C.4.13 implies that $\varnothing \hookrightarrow \partial \mathcal{A}^a$ is a cell complex whose cells have the form (C.5.11) with $\deg(x) < \deg(a)$.

By Proposition C.2.9(vii), the maps f^a and $L^a f$ are then cell complexes whose cells, indexed by the objects $x \in \mathcal{A}$ with the $\deg(x) \le \deg(a)$ and $\deg(x) < \deg(a)$, respectively, have the form

$$\overline{\mathrm{colim}}_{((\partial \mathcal{A}_x)^a \hookrightarrow \mathcal{A}_x^a) \hat{\Box} (\partial \mathcal{A}^x \hookrightarrow \mathcal{A}^x)} f \cong ((\partial \mathcal{A}_x)^a \hookrightarrow \mathcal{A}_x^a) \, \hat{\ast} \, \hat{\ell}^x f, \qquad (C.5.12)$$

the isomorphism arising from Proposition C.2.9(i). By Lemma C.2.13, the Leibniz tensor of a monomorphism with a map in the left class of a weak factorization system is again in the left class. Thus, since $(\partial \mathcal{A}_x)^a \hookrightarrow \mathcal{A}_x^a$ is a monomorphism and $\hat{\ell}^x f$ is in L, these cells, and thus the maps f^a and $L^a f$ are in L as well. □

Recall from Definition C.3.1 that a model structure on a category $\mathcal{M}$ with a class of weak equivalences W satisfying the 2-of-3 property is given by two classes of maps C and F so that $(C \cap W, F)$ and $(C, F \cap W)$ define weak factorization systems. To show that the Reedy weak factorization systems on $\mathcal{M}^{\mathcal{A}}$ relative to a model structure on $\mathcal{M}$ define a model structure on $\mathcal{M}^{\mathcal{A}}$ with the weak equivalences defined pointwise, one lemma is needed.

LEMMA C.5.13. *Let the classes (W, C, F) define a model structure on $\mathcal{M}$. Then a map $f : X \to Y \in \mathcal{M}^{\mathcal{A}}$*

(i) is Reedy in C ∩ W *if and only if f is Reedy in* C *and a pointwise weak equivalence, and*

(ii) is Reedy in F ∩ W *if and only if f is Reedy in* F *and a pointwise weak equivalence.*

Proof We prove the first of these dual statements. If f is Reedy in C ∩ W, then it is obviously Reedy in C, and Proposition C.5.10 implies that its components f^a are also in C ∩ W. Thus f is a pointwise weak equivalence.

For the converse, we make use of the diagram

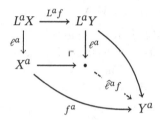

which relates the maps $L^a f$, $\hat{\ell}^a f$, and f^a for any $a \in \mathcal{A}$; this is an instance of Proposition C.2.9(vi) applied to $(\varnothing \hookrightarrow \partial A^a \hookrightarrow A^a) \,\hat{*}_A\, f$. Suppose that f is Reedy in C and a pointwise weak equivalence. By Proposition C.5.10, it follows that $L^a f$ is in C. We will show that $L^a f$ is in fact in C ∩ W and then apply pushout stability of the left class of a weak factorization system and the 2-of-3 property, to conclude that $\hat{\ell}^a f \in$ W and hence that f is Reedy in C ∩ W. We argue by induction. If a has degree zero, then $L^a f$ is the identity at the initial object, which is certainly a weak equivalence, and $\hat{\ell}^a f = f^a$ is in C ∩ W. If a has degree n, we may now assume that $\hat{\ell}^x f \in$ C ∩ W for any x with degree less than the degree of a. By the proof of Proposition C.5.10, $L^a f$ may be presented as a cell complex whose cells (C.5.12) are Leibniz tensors of monomorphisms with maps in C ∩ W, and thus lie in C ∩ W. Thus, we conclude that $L^a f \in$ C ∩ W, completing the proof.   □

Lemmas C.5.6, C.5.7, and C.5.13 assemble to prove:

THEOREM C.5.14 (the Reedy model structure). *If $\mathcal{A}$ is a Reedy category and* (W, C, F) *define a model structure on* $\mathcal{M}$, *then the Reedy weak factorization systems* (C ∩ W[$\mathcal{A}$], F[$\mathcal{A}$]) *and* (C[$\mathcal{A}$], F ∩ W[$\mathcal{A}$]) *define a model structure on* $\mathcal{M}^{\mathcal{A}}$ *with pointwise weak equivalences.*   □

One reason the Reedy model structure is important is because it is equipped with convenient Quillen bifunctors arising in the following manner.

THEOREM C.5.15. *Let $\mathcal{A}$ be a Reedy category and let $\otimes : \mathcal{V} \times \mathcal{M} \to \mathcal{N}$ be a* *left Leibniz bifunctor with respect to weak factorization systems* (L, R), (L', R'), *and* (L'', R''). *Then the functor tensor product*

$$\otimes_{\mathcal{A}} \colon \mathcal{V}^{\mathcal{A}^{\mathrm{op}}} \times \mathcal{M}^{\mathcal{A}} \to \mathcal{N}$$

is left Leibniz with respect to (L'', R'') *and the Reedy weak factorization systems* $(L[\mathcal{A}^{\mathrm{op}}], R[\mathcal{A}^{\mathrm{op}}])$ *and* $(L'[\mathcal{A}], R'[\mathcal{A}])$.

The **functor tensor product** of $F \in \mathcal{V}^{\mathcal{A}^{\mathrm{op}}}$ and $G \in \mathcal{M}^{\mathcal{A}}$ is defined by the coend

$$F \otimes_{\mathcal{A}} G := \int^{a \in \mathcal{A}} Fa \otimes Ga := \operatorname{coeq}\left(\coprod_{a,b \in \mathcal{A}} \coprod_{\mathcal{A}(a,b)} Fa \otimes Gb \rightrightarrows \coprod_{a \in \mathcal{A}} Fa \otimes Ga \right).$$

Proof As a construction built from a cocontinuous functor and colimits, the functor tensor product is cocontinuous in both variables. We argue that $\otimes_{\mathcal{A}}$ has the Leibniz property. Corollary C.4.17 asserts that the maps $f \in \mathcal{V}^{\mathcal{A}^{\mathrm{op}}}$ can be built as cell complexes whose cells are Leibniz products

$$(\partial \mathcal{A}^a \hookrightarrow \mathcal{A}^a) \mathbin{\hat{*}} \hat{\ell}_a f,$$

and the maps $g \in \mathcal{M}^{\mathcal{A}}$ can be built as cell complexes whose cells are Leibniz products

$$(\partial \mathcal{A}_b \hookrightarrow \mathcal{A}_b) \mathbin{\hat{*}} \hat{\ell}^b g.$$

By Proposition C.2.9(vii), $f \mathbin{\hat{\otimes}}_{\mathcal{A}} g$ is then a cell complex whose cells have the form

$$\left((\partial \mathcal{A}^a \hookrightarrow \mathcal{A}^a) \mathbin{\hat{*}} \hat{\ell}_a f \right) \mathbin{\hat{\otimes}}_{\mathcal{A}} \left((\partial \mathcal{A}_b \hookrightarrow \mathcal{A}_b) \mathbin{\hat{*}} \hat{\ell}^b g \right)$$
$$\cong \left((\partial \mathcal{A}^a \hookrightarrow \mathcal{A}^a) \mathbin{\hat{\times}}_{\mathcal{A}} (\partial \mathcal{A}_b \hookrightarrow \mathcal{A}_b) \right) \mathbin{\hat{*}} (\hat{\ell}_a f \mathbin{\hat{\otimes}} \hat{\ell}^b g).$$

To say that f is Reedy in L and g is Reedy in L' means that $\hat{\ell}_a f \in L$ and $\hat{\ell}^b g \in L'$. Since $\otimes$ is left Leibniz, it follows that $\hat{\ell}_a f \mathbin{\hat{\otimes}} \hat{\ell}^b g \in L''$. The Leibniz functor tensor product

$$(\partial \mathcal{A}^b \hookrightarrow \mathcal{A}^b) \mathbin{\hat{\times}}_{\mathcal{A}} (\partial \mathcal{A}_a \hookrightarrow \mathcal{A}_a)$$

of the maps in $\mathcal{S}et^{\mathcal{A}^{\mathrm{op}}}$ and in $\mathcal{S}et^{\mathcal{A}}$ amounts to the inclusion into the hom-set $\mathcal{A}^a_b = \mathcal{A}(b, a)$ of the subset of morphisms from b to a that factor through an object of degree strictly less than a or strictly less than b; in particular, this map is a monomorphism. Now Lemma C.2.13 applies to the weak factorization system (L'', R'') on $\mathcal{N}$ to prove that the Leibniz tensor of this monomorphism with $\hat{\ell}_a f \mathbin{\hat{\otimes}} \hat{\ell}^b g$ remains in L'', completing the proof. $\square$

For example, by (A.6.4), the weighted colimit bifunctor is the functor tensor product defined from the set-tensor bifunctor of Lemma C.2.13. Applying Theorem C.5.15 with (monomorphism, epimorphism) taken as the default weak factorization system on $\mathcal{S}et$, we conclude:

COROLLARY C.5.16. *For any complete and cocomplete category $\mathcal{M}$ with a weak factorization system (L, R) and any Reedy category, the weighted colimit and weighted limit*

$$\mathrm{colim}_{-} - : \mathcal{S}et^{\mathcal{A}} \times \mathcal{M}^{\mathcal{A}^{\mathrm{op}}} \to \mathcal{M} \quad \text{and} \quad \lim_{-} - : (\mathcal{S}et^{\mathcal{A}})^{\mathrm{op}} \times \mathcal{M}^{\mathcal{A}} \to \mathcal{M}$$

define left and right Leibniz bifunctors relative to the Reedy weak factorization systems.   □

In the setting of a model category, a cartesian closed model category, or a $\mathcal{V}$-model category (which subsumes the previous two cases by taking $\mathcal{V}$ to be $\mathcal{S}et$ with the model structure of Exercise C.3.iv or taking $\mathcal{V}$ to be the model category itself), Corollary C.5.16 specializes to the following result, which helps us understand homotopy limits and colimits of diagrams of Reedy shape.

COROLLARY C.5.17. *Let $\mathcal{M}$ be a $\mathcal{V}$-model category and let $\mathcal{A}$ be a Reedy category. Then for any weight W in $\mathcal{V}^{\mathcal{A}}$ that is Reedy cofibrant,[12] the weighted colimit and weighted limit functors*

$$\mathrm{colim}_{W} - : \mathcal{M}^{\mathcal{A}^{\mathrm{op}}} \to \mathcal{M} \quad \text{and} \quad \lim_{W} - : \mathcal{M}^{\mathcal{A}} \to \mathcal{M}$$

are respectively left and right Quillen with respect to the Reedy model structure on $\mathcal{M}^{\mathcal{A}}$.

Proof By Exercise C.3.iii, the Quillen bifunctors give rise to Quillen functors when plugging a cofibrant object into the appropriate variable.   □

EXAMPLE C.5.18 (homotopy limits and colimits). Taking the terminal weight 1 in $\mathcal{S}et^{\mathcal{A}}$, the weighted limit reduces to the ordinary limit functor. The functor $1 \in \mathcal{S}et^{\mathcal{A}}$ is Reedy monomorphic just when, for each $a \in \mathcal{A}$, the category of elements for the weight $\partial \mathcal{A}^{a}$ is either empty or connected. This is the case if and only if $\mathcal{A}$ has **cofibrant constants**, meaning that the constant $\mathcal{A}$-indexed diagram at any cofibrant object in any model category is Reedy cofibrant. Thus, we conclude that if $\mathcal{A}$ has cofibrant constants, then the limit functor $\lim : \mathcal{M}^{\mathcal{A}} \to \mathcal{M}$ is right Quillen.

Dually, the colimit functor $\mathrm{colim} : \mathcal{M}^{\mathcal{A}} \to \mathcal{M}$ is a special case of the weighted colimit functor with the terminal weight $1 \in \mathcal{S}et^{\mathcal{A}^{\mathrm{op}}}$. This is Reedy monomorphic

_____

[12] In the case of $\mathcal{V} = \mathcal{S}et$, "Reedy cofibrant" should be read as "Reedy monomorphic."

just when each category of elements for the weights $\partial \mathcal{A}_a$ is either empty or connected, which is the case if and only if $\mathcal{A}$ has **fibrant constants**, meaning that the constant $\mathcal{A}$-indexed diagram at any fibrant object in any model category is Reedy fibrant. Thus, we conclude that if $\mathcal{A}$ has fibrant constants, then the colimit functor colim : $\mathcal{M}^{\mathcal{A}} \to \mathcal{M}$ is left Quillen (see [107, §9] for more discussion).

In presheaf categories, the cofibrations are often the monomorphisms, and with a bit of elbow grease, we can identify Reedy monomorphic weights for use in applications of Corollary C.5.17. For instance, any bisimplicial set is Reedy cofibrant as an object of $s\mathcal{S}et^{\Delta^{\mathrm{op}}}$ (see [103, 14.3.7]). Reedy monomorphic cosimplicial objects can also be identified, on account of the following lemma.

LEMMA C.5.19.

(i) *Let* $X \colon \Delta \to \mathcal{S}et^{\mathcal{I}^{\mathrm{op}}}$ *be a cosimplicial object in a presheaf category. If* X *is unaugmentable, in the sense that the equalizer of the pair of coface maps* $\delta^0, \delta^1 \colon X^0 \to X^1$ *is empty, then the latching maps of* X *are all monomorphisms.*

(ii) *If* X *is an unaugmentable cosimplicial object in a slice category of a presheaf category, then the latching maps of* X *are all monomorphisms.*

Proof Since latching objects are defined in terms of certain colimits in $\mathcal{S}et^{\mathcal{I}^{\mathrm{op}}}$ computed pointwise in $\mathcal{S}et$, we may reduce this result to the corresponding one for cosimplicial sets $X \colon \Delta \to \mathcal{S}et$. A simplex in a cosimplicial set is "nondegenerate" if it is not in the image of a monomorphism from Δ. The nth latching map $L^n X \to X^n$ is a monomorphism just when each expression of an n-simplex x as the image of a nondegenerate simplex z under a monomorphism $\sigma \colon [k] \rightarrowtail [n]$ is unique.

So suppose we have two such representations $x = \sigma \cdot z = \sigma' \cdot z'$. Any monomorphism $\sigma \in \Delta$ has a left inverse τ, so we see that $z = \tau\sigma' \cdot z'$. The map $\tau\sigma'$ can be factored as an epimorphism followed by a monomorphism. Because z is nondegenerate, this monomorphism must be the identity, so $\tau\sigma'$ is an epimorphism. Repeating this argument with an left inverse τ' for σ' we see that $\tau'\sigma$ is an epimorphism, so z and z' have the same degree and both epimorphisms are identities. This proves that $z = z'$.

If the set of left inverses for a monomorphism in Δ uniquely characterized that monomorphism, then we could conclude that σ and σ' must be equal, and hence that such decompositions would be fully unique. This is true for nearly all monomorphisms in Δ, the only exceptions being the face maps $\delta^0, \delta^1 \colon [0] \rightarrowtail [1]$. However, X is assumed to be unaugmentable, so there is no $z \in Z^0$ with $\delta^0 \cdot z = \delta^1 \cdot z$, and thus $L^n X \to X^n$ is a monomorphism for all n. This proves the first statement.

For the second statement, we only need to consider slice categories $^{A/}Set^{\mathcal{I}^{op}}$ under a presheaf $A \in Set^{\mathcal{I}^{op}}$, since the slice category $Set^{\mathcal{I}^{op}}{}_{/A}$ is equivalent to the category of presheaves on the category of elements of A, and so this case is covered by (i). The forgetful functor $^{A/}Set^{\mathcal{I}^{op}} \to Set^{\mathcal{I}^{op}}$ creates monomorphisms and connected colimits, which tells us that the latching maps $L^n X \to X^n$ of a complicial object X in $^{A/}Set^{\mathcal{I}^{op}}$ are calculated in $Set^{\mathcal{I}^{op}}$ for all $n > 1$. The direct calculation given above proves that these are monomorphisms, so it remains only to consider the cases $n = 0$ and $n = 1$. If X is unaugmentable, then by hypothesis the equalizer of the map $\delta^0, \delta^1 : X^0 \to X^1$ in $Set^{\mathcal{I}^{op}}$ is A. Thus, the 0th latching map $A \to X^0$ is an equalizer, and so it must be a monomorphism. Finally, we claim that the 1st latching map $(\delta^0, \delta^1) : L^1 X \cong X^0 + X^0 \to X^1$ is a monomorphism: arguing in $Set^{\mathcal{I}^{op}}$ and then ultimately in Set, it is easy to see that the pullback of this map along itself is A. This completes the proof that X is a Reedy monomorphism. $\square$

EXAMPLE C.5.20 (geometric realization and totalization). By Lemma C.5.19, the Yoneda embedding defines a Reedy cofibrant weight $\Delta \in sSet^{\Delta}$. The weighted colimit and weighted limit functors

$$\mathrm{colim}_{\Delta} - : \mathcal{M}^{\Delta^{op}} \to \mathcal{M} \qquad \text{and} \qquad \lim_{\Delta\cdot} - : \mathcal{M}^{\Delta} \to \mathcal{M}$$

typically go by the names of **geometric realization** and **totalization**. Corollary C.5.17 proves that if $\mathcal{M}$ is a simplicial model category, then these functors are left and right Quillen.

Exercises

EXERCISE C.5.i. Prove Lemma C.5.1.

EXERCISE C.5.ii ([103, 14.3.8]). Prove the relative analog of Lemma C.5.19: if X and Y are both unaugmentable cosimplicial objects in a presheaf category, then any pointwise monomorphism $X \to Y$ is also a Reedy monomorphism, i.e., its relative latching maps are monic.

APPENDIX OF CONCRETE CONSTRUCTIONS

Appendix D

The Combinatorics of (Marked) Simplicial Sets

In this appendix we explore the combinatorics of simplicial sets, proving results stated in Chapters 1 and 4. Certain of these results, namely those involving isomorphisms in quasi-categories, are more easily proved in the closely related category of "marked" simplicial sets, where the quasi-categories are identified with those marked simplicial sets that are *1-complicial*. Because the corresponding *n-complicial sets* provide one of the families of examples of ∞-cosmoi appearing in Appendix E, we prove the necessary combinatorial results in that more general context.

The category of marked simplicial sets is introduced in §D.1. Certain objects in this category that have composites of simplices in all dimensions define *complicial sets*, which are characterized by a right lifting property that also defines the *complicial isofibrations*. In §D.2, we begin our combinatorial work by revisiting the join and slice constructions from §4.2, redeveloping these notions from the viewpoint of augmented simplicial sets. This work is completed in §D.6, where we prove that various models for the quasi-category of cones over or under a diagram are equivalent.

In §D.3, we prove the Leibniz stability of complicial isofibrations under exponentiation with monomorphisms of marked simplicial sets. This closely resembles some of the unproven results of §1.1, but before making the connection, we must relate complicial sets and complicial isofibrations to quasi-categories and isofibrations. This task occupies the remaining two sections. In §D.4 we establish the connection between the theory of complicial sets and the theory of quasi-categories, showing that any quasi-category can be equipped with a canonically defined "natural" marking in such a way that it defines a complicial set with all simplices above dimension 1 marked. The proof involves a careful study of the data that define an isomorphism in a quasi-category. Applying this analysis in §D.5, we prove Joyal's "special outer horn filling" Proposition 1.1.14, which amounts to the observation that isofibrations between naturally marked quasi-categories coincide with the complicial isofibrations. With this connection established, the previous results can be assembled to supply proofs of the claims made in §1.1.

D.1 Complicial Sets

When a quasi-category is regarded as an $(\infty, 1)$-category, its vertices play the role of the objects and its edges represent the morphisms, with the degenerate edge at a vertex representing its identity. The n-simplices then witness n-ary composition relations. When a complicial set is regarded as an (∞, ∞)-category, its n-simplices must play a dual role: both serving as witnesses for lower dimensional composition relations and representing a priori noninvertible n-dimensional cells in their own right. To disambiguate between these two interpretations, certain positive dimensional simplices in a complicial set are *marked* as "thin," indicating that they should be interpreted as equivalences witnessing a weak composition relation between their boundary faces. Thus the ambient category in which complicial sets are defined is not the category of ordinary simplicial sets but a closely related category of *marked simplicial sets*[1] that we now introduce.

DEFINITION D.1.1 (marked simplicial sets). A **marked simplicial set** is a simplicial set with a designated subset of **marked** or **thin** positive dimensional simplices that includes all degenerate simplices. A map of marked simplicial sets is a simplicial map that preserves marked simplices.

DEFINITION D.1.2 (minimal and maximal marking). The category $sSet^+$ of marked simplicial sets is equipped with an evident forgetful functor to $sSet$ admitting both left and right adjoints:

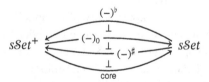

The left adjoint $(-)^{\flat}$ defines the **minimal marking** of a simplicial set, in which only the degeneracies are marked, while the right adjoint $(-)^{\sharp}$ defines the **maximal marking**, with all simplices marked. This functor has a further right adjoint,

---

[1] In the original sources [128, 129], marked simplicial sets are called *stratified simplicial sets*. To avoid confusing with the increasingly prominent unrelated notion of stratified spaces, we have elected to change the name. In [78], Lurie uses the term *marked simplicial sets* for a special case of the more general notion we presently introduce.

which takes a marked simplicial set to its **core**, the simplicial set with the same vertices comprised of those marked simplices whose faces are also marked.

On various occasions, it is convenient to identify *sSet* with either of the fully faithful embeddings into *sSet*⁺ just introduced. Unless otherwise specified, the default convention is to identify simplicial sets with their minimal markings. In particular, with this convention, we are free to regard the standard simplices and their subspaces as minimally marked simplicial sets.

To succinctly introduce other marked simplicial sets, the following terminology is convenient:

DEFINITION D.1.3. An inclusion $U \hookrightarrow V$ of marked simplicial sets is:

- **regular**, denoted $U \hookrightarrow_r V$, if thin simplices in U are created in V and
- **entire**, denoted $U \hookrightarrow_e V$, if the map is an isomorphism (or more commonly the identity) on underlying simplicial sets, in which case the only difference between U and V is that V has more marked simplices.

For example, for each $n \geq 1$, we define the **marked simplex** $\Delta[n]_t$ to be the entire extension of minimally marked simplex $\Delta[n]$ that also marks the top nondegenerate simplex.

NOTATION D.1.4 (the marked simplex category). Let $t\Delta \hookrightarrow sSet^+$ denote the full subcategory spanned by the minimally marked simplices $\Delta[n]$, for $n \geq 0$, together with the marked simplices $\Delta[n]_t$, for $n \geq 1$. It can be built from the simplex category Δ of Notation 1.1.1 by:

- adjoining objects $[n]_t$ for $n \geq 1$,
- adjoining maps $\phi \colon [n] \to [n]_t$ for $n \geq 1$ and $\zeta^i \colon [n+1]_t \to [n]$ for $n \geq 0$ and $0 \leq i \leq n$, and
- imposing relations $\zeta^i \phi = \sigma^i$ and $\sigma^i \zeta^{j+1} = \sigma^j \zeta^i$ for $i \leq j$.[2]

For a marked simplicial set X, the maps $\Delta[n] \to X$ and $\Delta[n]_t \to X$, respectively, parametrize n-simplices in X and marked n-simplices in X. This defines a canonical embedding $sSet^+ \hookrightarrow Set^{t\Delta^{op}}$, which is easily seen to be fully faithful. Moreover:

PROPOSITION D.1.5. *There is a reflective embedding* $sSet^+ \xleftarrow{\quad\perp\quad} Set^{t\Delta^{op}}$ *whose*

(i) *essential image consists of those presheaves F for which the component maps* $- \circ \phi \colon F_{[n]_t} \to F_{[n]}$ *are monomorphisms, and*

---

[2] Viktoriya Ozornova and Martina Rovelli pointed out to us that this last family of relations was omitted from the original source [122] but should have been included. A corrected definition appears in [89, 1.1].

(ii) left adjoint is constructed by replacing the set $F_{[n]_t}$ with the image of the map $- \circ \phi : F_{[n]_t} \to F_{[n]}$.

Consequently, $sSet^+$ is a locally finitely presentable category, and in particular is complete and cocomplete, with limits constructed pointwise as presheaves and with colimits constructed by applying the reflector to the pointwise colimit of presheaves.

Put in more elementary terms, limits and colimits of marked simplicial sets are created by the underlying simplicial set functor $(-)_0 : sSet^+ \to sSet$. A simplex in the limit is marked if and only if *each* of its components, defined by composing with the legs of the limit cone, are marked simplices. A simplex in a colimit is marked if *any* of its lifts along any leg of the colimit cone are marked simplices. The reflection in (ii) is a sort of "propositional truncation," remembering which simplices should be marked while forgetting the data that indicates why.

Proof The right action of the operators in $t\Delta$ on a presheaf $F \in Set^{t\Delta^{op}}$ gives the sets of elements of F the structure of a marked simplicial set with the exception of one condition: namely that the marked n-simplices form a subset of the n-simplices. This explains the condition appearing in (i) and the construction appearing in (ii). It follows that marked simplicial sets are the category of models for a finite limit sketch, and hence form a locally finitely presentable category. Any reflective full subcategory of a complete and cocomplete category inherits limits in the manner constructed in the statement (see, e.g., [104, 4.5.15]). □

Lemma C.5.9 extends to marked simplicial sets as follows:

LEMMA D.1.6. *The momomorphisms in $sSet^+$ are cellularly generated by*

$$\{\partial\Delta[n] \hookrightarrow_r \Delta[n]\}_{n \geq 0} \cup \{\Delta[n] \hookrightarrow_e \Delta[n]_t\}_{n \geq 1}.$$

Proof Exercise D.1.i. □

A marked simplicial set is a simplicial set with enough structure to talk about composition of simplices in all dimensions. A complicial set is a marked simplicial set in which composites exist and in which thin witnesses to composition compose to define thin simplices, an associativity condition that ultimately implies that thin simplices are equivalences in a sense that is made explicit in Lemma D.4.2 and Digression D.4.21. The following form of the definition of a (née *weak*) *complicial set*, due to Verity [129], modifies an earlier equivalent presentation due to Street [120]. Verity's modification focuses on a particular set of *k-admissible n-simplices*, which are thin n-simplices that exhibit their kth face as a composite of their $(k + 1)$th and $(k - 1)$th faces, in the case where

$0 < k < n$. In the cases $k = 0$ or $k = n$, a k-admissible n-simplex witnesses an equivalence between the first or last pair of faces, respectively.

DEFINITION D.1.7 (k-admissible n-simplex). For $n \geq 1$ and $0 \leq k \leq n$, the **k-admissible n-simplex** $\Delta^k[n]$ is the entire superset of the standard n-simplex with certain additional faces marked thin: a nondegenerate m-simplex in $\Delta^k[n]$ is thin if and only if it contains all of the vertices in $\{k - 1, k, k + 1\} \cap [n]$. Thin faces include in particular:

- the top dimensional n-simplex
- all codimension-one faces except for the $(k - 1)$th, kth, and $(k + 1)$th
- the 2-simplex spanned by $\{k - 1, k, k + 1\}$ when $0 \leq k \leq n$ or the edge spanned by $\{k - 1, k, k + 1\} \cap [n]$ when $k = 0$ or $k = n$.

When drawing pictures of marked simplicial sets, we use the symbol "$\simeq$" to decorate marked simplices and "$\sim$" to decorate marked edges. Our diagrams also adopt a convention for the direction of the cells inhabiting an unmarked n-simplex. Following the combinatorics introduced by Street in his "Algebra of oriented simplexes" [120], we regard an n-simplex as an n-cell from the pasted composite of its odd-numbered faces to the pasted composite of its even-numbered faces. Note this is compatible with the convention already in use for depicting a 1-simplex in a simplicial set as an arrow from its 1st face (the 0th vertex) to its 0th face (the 1st vertex).

EXAMPLE D.1.8 (admissible simplices in low dimensions).

(i) For $n = 1$, both admissible simplices $\Delta^0[1]$ and $\Delta^1[1]$ equal the thin 1-simplex $\Delta[1]_t = \Delta[1]^\sharp$. A map $\Delta[1]^\sharp \to A$ is interpreted as defining an equivalence between the two vertices in its image.

(ii) For $n = 2$, the admissible simplex $\Delta^1[2]$ equals the thin 2-simplex $\Delta[2]_t$. A map $\Delta^1[2] \to A$ is interpreted as specifying that the image of the $\{02\}$-edge is a composite of the images of the $\{01\}$- and $\{12\}$-edges.

By contrast, $\Delta^0[2]$ and $\Delta^2[2]$ each have a marked edge, as well as a marked 2-simplex as indicated by the diagrams:

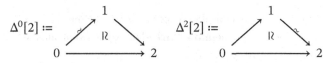

A map $\Delta^0[2] \to A$ witnesses an equivalence between the image of the $\{12\}$ edge and the image of the $\{02\}$ edge.

(iii) For $n = 3$, the admissible simplices $\Delta^1[3]$ and $\Delta^2[3]$ have their 3rd and 0th faces marked, respectively, as well as the top dimensional 3-simplex, with no other nondegenerate faces marked. We choose to draw admissible 3-simplices in such a way that allows us to see all of their codimension-one faces:

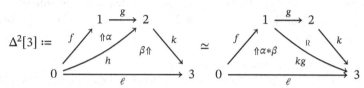

Here we label the faces in order to better describe the interpretation of a map $\Delta^2[3] \to A$. Its 0th face, which is itself an admissible simplex $\Delta^1[2]$, witnesses that the edge $\{13\}$ is a composite of the edges $\{12\}$ and $\{23\}$. Note that because the 0th face is thin, its 1st edge is interpreted as a composite kg of g and k, which is needed so that the boundary of the 2-cell appearing in the 2nd face agrees with the boundary of the pasted composite of β and α. On account of this boundary condition and the thin 3-simplex, we interpret the 2nd face as the pasted composite of the 1st and 3rd faces depicted on the right.

The admissible simplex $\Delta^0[3]$ has both its 2nd and 3rd faces marked, as well as the top dimensional 3-simplex, and the edge $\{01\}$. Dually, $\Delta^3[3]$ has its 0th and 1st faces marked, as well as the top dimensional 3-simplex, and the edge $\{23\}$.

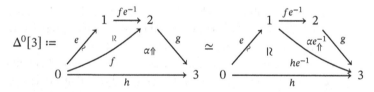

A map $\Delta^0[3] \to A$ is interpreted as witnessing an equivalence between the pair of nonthin 2-simplices occupying the 0th and 1st faces, respectively.

Definition D.1.9. A **complicial set** is a marked simplicial set A that admits extensions along the **elementary marked anodyne extensions**, which are given by the following two sets of maps:

(i) The **complicial horn extensions**

$$\Lambda^k[n] \longrightarrow A$$
$$r \downarrow \qquad \qquad \text{for} \quad n \geq 1,\ 0 \leq k \leq n \qquad \text{(D.1.10)}$$
$$\Delta^k[n]$$

are regular inclusions of k-**admissible** n-**horns**. An inner admissible n-horn parametrizes "admissible composition" of a pair of $(n-1)$-simplices. The extension defines a composite $(n-1)$-simplex together with a thin n-simplex witness.

(ii) The **complicial thinness extensions**

$$\Delta^k[n]' \longrightarrow A$$
$$\int \qquad \qquad \text{for} \quad n \geq 2,\ 0 \leq k \leq n \qquad \text{(D.1.11)}$$
$$\Delta^k[n]''$$

are entire inclusions of two entire supersets of $\Delta^k[n]$. The marked simplicial set $\Delta^k[n]'$ is obtained from $\Delta^k[n]$ by also marking the $(k-1)$th and $(k+1)$th faces, while $\Delta^k[n]''$ has all codimension-one faces marked. This extension problem demands that whenever the composable pair of simplices in an admissible horn are thin, then so is any composite.

EXAMPLE D.1.12 (complicial horn extensions). For $\Lambda^2[4] \hookrightarrow_r \Delta^2[4]$ the nonthin codimension-one faces in the horn define the two 3-simplices with a common face displayed on the left, while their composite is a 3-simplex as displayed on the right.

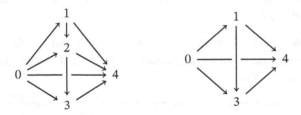

It makes sense to interpret the right hand simplex, the 2nd face of the 2-admissible 4-simplex, as a composite of the 3rd and 1st faces because the following simplices are thin:

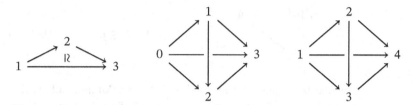

We refer to the maps that are cellularly generated by the elementary marked anodyne extensions as **marked anodyne extensions**. For instance, by a mild extension of the argument that solves Exercise 1.1.v:

LEMMA D.1.13. *Either inclusion $2^\sharp \hookrightarrow \mathbb{I}^\sharp$ of the marked 1-simplex into the maximally marked isomorphism is an marked anodyne extension, as is the injection $\mathbb{1} \hookrightarrow \mathbb{I}^\sharp$.*

Proof   Exercise D.1.iii.   □

DEFINITION D.1.14. A map of marked simplicial sets is a **complicial isofibration** if it has the right lifting property with respect to the elementary marked anodyne extensions and if its domain and codomain are complicial sets.

By Exercise D.1.iv, a marked map between complicial sets is a complicial isofibration if and only if it lifts against the complicial horn extensions – the complicial thinness extensions come for free. By Lemma C.2.3, complicial isofibrations then enjoy the right lifting property against all marked anodyne extensions. Among the complicial isofibrations are the **trivial fibrations**, defined to be those maps of marked simplicial sets that lift against the monomorphisms, as characterized by Lemma D.1.6.

The original meaning of "complicial sets" referred to a particular variety that we now call *strict*.

DEFINITION D.1.15. A **strict complicial set** is a marked simplicial set that admits unique extensions along the elementary marked anodyne extensions (D.1.10) and (D.1.11).

In the manuscript [128], Verity proves that the strict complicial sets are precisely those marked simplicial sets that are Street nerves of strict ω-categories, resolving a conjecture of Roberts and Street. In this manuscript, we primarily utilize marked simplicial sets to streamline the proofs of results concerning isomorphisms in quasi-categories and equivalences between quasi-categories. We discuss this topic more explicitly in §D.4 and §D.5 after developing some combinatorial constructions required in the interim.

DEFINITION D.1.16. A marked simplicial set X is n-**trivial** if all r-simplices are marked for $r > n$.

The full subcategory of n-trivial marked simplicial sets is reflective and coreflective

$$n\text{-}s\mathcal{S}et^+ \overset{\mathrm{trv}_n}{\underset{\mathrm{core}_n}{\overset{\perp}{\underset{\perp}{\rightleftarrows}}}} s\mathcal{S}et^+$$

in the category of marked simplicial sets. That is n-**trivialization** defines an idempotent monad on $s\mathcal{S}et^+$ with unit the entire inclusion $X \hookrightarrow_e \mathrm{trv}_n X$ of a marked simplicial set X into the marked simplicial set $\mathrm{trv}_n X$ with the same marked simplices in dimensions $1, \dots, n$, and with all higher simplices "made thin." A complicial set is n-**trivial** if this map is an isomorphism.

The n-**core** $\mathrm{core}_n X$, defined by restricting to those simplices whose faces above dimension n are all thin in X, defines an idempotent comonad with counit the regular inclusion $\mathrm{core}_n X \hookrightarrow_r X$. Again, a complicial set is n-**trivial** just when this map is an equivalence. As is always the case for a monad–comonad pair arising in this way, these functors are adjoints: $\mathrm{trv}_n \dashv \mathrm{core}_n$.

The inclusions of the subcategories of n-trivial marked simplicial sets have adjoints

$$s\mathcal{S}et \xrightarrow[\approx]{(-)^\sharp} 0\text{-}s\mathcal{S}et^+ \hookrightarrow \cdots \hookrightarrow (n\text{-}1)\text{-}s\mathcal{S}et^+ \overset{\mathrm{trv}_{n-1}}{\underset{\mathrm{core}_{n-1}}{\overset{\perp}{\underset{\perp}{\rightleftarrows}}}} n\text{-}s\mathcal{S}et^+ \hookrightarrow \cdots \hookrightarrow s\mathcal{S}et^+$$

The n-core of a complicial set is a complicial set, but the n-trivialization functor, which just marks simplices in the appropriate dimension without changing the underlying simplicial set, does not necessarily preserve complicial sets (see Exercise D.1.v).

REMARK D.1.17 (the odd dual). Recall that the **opposite** of a simplicial set X is the simplicial set obtained by reindexing along the involution $(-)^{\mathrm{op}} : \Delta \to \Delta$ that reverses the ordering in each ordinal. This operation may be extended to marked simplicial sets in a natural way: marking an n-simplex in X^{op} just when the corresponding n-simplex in X is marked. Note, however, that under Street's interpretation of an n-simplex as encoding an n-dimensional morphism from the composite of its odd $(n-1)$-dimensional faces to the composite of its even $(n-1)$-dimensional faces, this operation does not simply "reverse the direction of all the cells" in a marked simplicial set. Rather, it reverses the direction of all the simplices in the *odd* dimensional cells, while preserving the direction in all of the even dimensional cells. Thus, we refer to the vertex reordering construction as defining the **odd dual** of a marked simplicial set.

We close with a discussion of the equivalences between complicial sets.

DEFINITION D.1.18 (equivalences of complicial sets). A map $f : A \to B$ between complicial sets is an **equivalence** if it extends to the data of a homotopy equivalence with the marked 1-simplex $\Delta[1]^\sharp$ serving as the interval:[3] that is, if there exist maps $g : B \to A$,

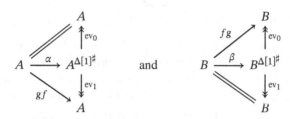

and

The data of an equivalence of complicial sets transposes to define an instance of a more general notion of "marked homotopy equivalence," for which we drop the requirement that the marked simplicial sets are complicial sets.

DEFINITION D.1.19. A **marked homotopy** between a pair of maps $f, g : X \to Y$ is given by a map $\alpha : X \times \Delta[1]^\sharp \to Y$ that restricts along the endpoint inclusions $X + X \hookrightarrow_r X \times \Delta[1]^\sharp$ to the maps f and g, respectively. In the case where X and Y are minimally marked simplicial sets, a map $X \times \Delta[1] \to Y$ extends to a map $X \times \Delta[1]^\sharp \to Y$ just when for each 0-simplex $x \in X$, the 1-simplex $(x\sigma^0, \mathrm{id}_{[1]}) \in X \times \Delta[1]^\sharp$ maps to a degenerate and hence marked 1-simplex of Y.[4]

A **marked homotopy equivalence** consists of:

- a pair of marked maps $f : X \to Y$ and $g : Y \to X$ and
- a pair of marked homotopies $\alpha : X \times \Delta[1]^\sharp \to X$ and $\beta : Y \times \Delta[1]^\sharp \to Y$ between id_X and gf and fg and id_Y, respectively.

When X and Y are complicial sets, marked homotopies can be reversed and composed. Even when this is not the case, we permit ourselves the reverse the direction of the marked homotopies that comprise a marked homotopy equivalence without comment.

DIGRESSION D.1.20 (the Verity model structure for complicial sets). The category of marked simplicial sets bears a cartesian closed, cofibrantly generated

---

[3] Essentially by Lemma D.1.13 equivalences of complicial sets could be defined using the marked homotopy coherent isomorphism $\mathbb{I}^\sharp$ instead (see Corollary D.3.13). But one of the advantages of the complicial sets model of higher categories is that the correct notion of equivalence can be defined with the simpler data of the marked 1-simplex.

[4] The cartesian product of marked simplicial sets is described in more detail in Proposition D.3.3.

model structure whose fibrant objects are exactly the complicial sets and whose cofibrations are the monomorphisms [129, §6.2-4]. The fibrations and weak equivalences between fibrant objects are precisely the classes of complicial isofibrations and equivalences defined above. In the following sections, we verify many of these properties for the category of fibrant objects directly, leaving only the verification of the actual model structure, which follows from Jeff Smith's theorem, to the literature.

Exercises

EXERCISE D.1.i. Prove Lemma D.1.6.

EXERCISE D.1.ii. Prove that a maximally marked simplicial set defines a complicial set if and only if the underlying simplicial set is a Kan complex.

EXERCISE D.1.iii. Prove Lemma D.1.13.

EXERCISE D.1.iv. Let $f : A \to B$ be any map of marked simplicial sets whose domain A is a complicial set. Prove that f has the (unique) right lifting property against the complicial thinness extensions.

EXERCISE D.1.v.

(i) Prove that the n-core of a complicial set is a complicial set.
(ii) Find an example of a complicial set whose n-trivialization for some n is no longer a complicial set.

D.2 The Join and Slice Constructions

In this section, we revisit Joyal's join and slice constructions in considerably more detail than given in Definition 4.2.4 and discuss their extension to marked simplicial sets. We prove that Leibniz joins of monomorphisms and various classes of anodyne maps again define monomorphisms of the same type. The combinatorics are slightly easier if we work with *augmented* simplicial sets in place of ordinary simplicial sets, an approach that follows the original definition of the simplicial join by Ehlers and Porter [40].

DEFINITION D.2.1 (ordinal sum). The algebraists' skeletal category Δ_+ of finite ordinals and order-preserving maps – with objects $[n] = \{0 \leq 1 \leq \cdots \leq n\}$ and $[-1] = \varnothing$ – supports a strict (nonsymmetric) monoidal structure $(\Delta_+, \oplus, [-1])$ in which $\oplus$ denotes the **ordinal sum** defined

- for objects $[n], [m] \in \Delta_+$ by $[n] \oplus [m] := [n + 1 + m]$,
- for arrows $\alpha \colon [n] \to [n'], \beta \colon [m] \to [m']$ by $\alpha \oplus \beta \colon [n + 1 + m] \to [n' + 1 + m']$ where

$$\alpha \oplus \beta(i) = \begin{cases} \alpha(i) & \text{if } i \leq n, \\ \beta(i - n - 1) + n' + 1 & \text{otherwise.} \end{cases}$$

By Day convolution [33], the join bifunctor $\oplus \colon \Delta_+ \times \Delta_+ \to \Delta_+$ extends to a (nonsymmetric) monoidal closed structure

$$(sSet_+, \star, \Delta[-1], dec_l, dec_r)$$

on the category of augmented simplicial sets $sSet_+ := Set^{\Delta_+^{op}}$.

DEFINITION D.2.2 (join of augmented simplicial sets). The **join** $X \star Y$ of augmented simplicial sets X and Y may be described explicitly as follows:

- Its simplices are pairs $(x, y) \in (X \star Y)_{r+1+s}$ with $x \in X_r$, $y \in Y_s$.
- If (x, y) is a simplex of $X \star Y$ with $x \in X_r$ and $y \in Y_s$ and $\alpha \colon [n] \to [r + 1 + s]$ is a simplicial operator in Δ_+, then α may be uniquely decomposed as $\alpha = \alpha_1 \oplus \alpha_2$ with $\alpha_1 \colon [n_1] \to [r]$ and $\alpha_2 \colon [n_2] \to [s]$, and $(x, y) \cdot \alpha := (x \cdot \alpha_1, y \cdot \alpha_2)$.

Note by construction that $\Delta[n] \star \Delta[m] \cong \Delta[n + 1 + m]$, since $[n] \oplus [m] = [n + 1 + m]$.[5]

Note that $\Delta[-1]$, the marked simplicial set with a single -1-simplex and no other simplices, is a two-sided unit for the join bifunctor.

DEFINITION D.2.3 (décalage of augmented simplicial sets). The closures dec_l and dec_r, known as the **left and right décalage constructions**, respectively, are defined as the parametrized right adjoints to the join:

$$sSet_+ \underset{dec_l(X,-)}{\overset{X\star-}{\rightleftarrows}} sSet_+ \quad \text{and} \quad sSet_+ \underset{dec_r(X,-)}{\overset{-\star X}{\rightleftarrows}} sSet_+$$

so that there is a two-variable adjunction formed by the bifunctors:

$$sSet_+ \times sSet_+ \xrightarrow{-\star-} sSet_+ ,$$

$$sSet_+^{op} \times sSet_+ \xrightarrow{dec_l(-,-)} sSet_+, \quad sSet_+^{op} \times sSet_+ \xrightarrow{dec_r(-,-)} sSet_+ .$$

---

[5] A general feature of the Day convolution product is that the Yoneda embedding
よ $\colon \Delta_+ \hookrightarrow Set^{\Delta_+^{op}}$ defines a strong monoidal functor.

OBSERVATION D.2.4 (simplicial sets vs augmented simplicial sets). The evident functor that forgets the augmentation $U\colon s\mathcal{S}et_+ \to s\mathcal{S}et$ admits both left and right adjoints

$$s\mathcal{S}et_+ \underset{\underset{*}{\overset{\pi_0}{\underset{\bot}{\overset{\top}{\longleftarrow}}}}}{\overset{\longleftarrow}{\underset{U}{\longrightarrow}}} s\mathcal{S}et$$

where the left adjoint augments a simplicial set X with its set of path components $\pi_0 X$, defined by the coequalizer

$$X_1 \underset{\delta^0}{\overset{\delta^1}{\rightrightarrows}} X_0 \longrightarrow\!\!\!\!\!\rightarrow \pi_0 X$$

and the right adjoint augments a simplicial set "terminally" by adding a single -1-simplex. The unit of $\pi_0 \dashv U$ and counit of $U \dashv *$ are both isomorphisms; hence either adjoint defines a fully faithful embedding $s\mathcal{S}et \hookrightarrow s\mathcal{S}et_+$.

Any augmented simplicial set is canonically a coproduct of its terminally augmented "components":

LEMMA D.2.5. *Let X be an augmented simplicial set.*

　(i) *For each $i \in X_{-1}$, the subset $X_{\langle i\rangle}$ comprised of those simplices in any dimension whose -1-simplex face is i forms a terminally augmented simplicial subset of X.*

　(ii) *The disjoint union $\coprod_{i\in X_{-1}} X_{\langle i\rangle}$ of these components is isomorphic to X.*

Proof　Exercise D.2.i.　　　　　　　　　　　　　　　　　□

DEFINITION D.2.6 (join of simplicial sets). By convention, the **join** of a pair of simplicial sets is defined to be the underlying simplicial set of the join of the trivially augmented simplicial sets. Thus, the join bifunctor is the composite

$$
\begin{array}{ccc}
s\mathcal{S}et \times s\mathcal{S}et & \overset{-\star-}{\dashrightarrow} & s\mathcal{S}et \\[4pt]
{\scriptstyle *\times*}\big\downarrow & & \big\uparrow{\scriptstyle U} \\[4pt]
s\mathcal{S}et_+ \times s\mathcal{S}et_+ & \overset{-\star-}{\longrightarrow} & s\mathcal{S}et_+
\end{array}
$$

Explicitly, n-simplices of $X \star Y$ are pairs comprised of a j-simplex of X and a k-simplex of Y where $j + k = n - 1$, where in the case $j = -1$ such a "pair" consists of a single n-simplex of Y and in the case $k = -1$ such a "pair" consists of a single n-simplex of X. This recovers Definition 4.2.4.

As observed in Definition 4.2.4, the join of simplicial sets X and Y admits canonical embeddings

$$X \hookrightarrow X \star Y \hookleftarrow Y$$

which can be understood as the maps obtained by applying $X \star -$ or $- \star Y$ respectively to the maps $\Delta[-1] \to Y$ and $\Delta[-1] \to X$ in $sSet_+$ that pick out the unique -1-simplices in the trivial augmentations.

LEMMA D.2.7.

(i) *The join bifunctor* $- \star - :$ $sSet \times sSet \to sSet$ *preserves connected colimits in each variable.*

(ii) *For any simplicial set* X*, the join functors*

$$sSet \xrightarrow{\;X\star-\;} {}^{X/}sSet \qquad and \qquad sSet \xrightarrow{\;-\star X\;} {}^{X/}sSet$$

preserve all colimits.

Proof   In Definition D.2.6, the join of simplicial sets is defined as the composite of three functors, two of which possess right adjoints and hence preserve all colimits. The third functor $* :$ $sSet \to sSet_+$ does not possess a right adjoint but nevertheless preserves *connected* colimits as is clear from the following definition: an indexing 1-category J is connected just when the colimit of the constant J-indexed diagram valued at the singleton set is a singleton. This proves (i).

Now the forgetful functor ${}^{X/}sSet \to sSet$ strictly creates connected colimits [104, 3.3.8], so the join functors of (ii) preserve connected colimits. Arbitrary colimits may be built from connected colimits and coproducts, so to prove (ii) it remains only to argue that these functors preserve coproducts. While $X \star (\amalg_i Y_i) \ncong \amalg_i (X \star Y_i)$ if the latter coproduct is interpreted in $sSet$, it can be directly verified that $X \star (\amalg_i Y_i)$ is the quotient of $\amalg_i (X \star Y_i)$ modulo the identification of the images of each inclusion $X \hookrightarrow X \star Y_i$ with a single copy of X, which is exactly the construction of the coproduct in the category ${}^{X/}sSet$. $\square$

DEFINITION D.2.8 (slice of simplicial sets).   The categories $sSet$ and ${}^{X/}sSet$ are locally presentable, so the cocontinuous functors of Lemma D.2.7(ii) have right adjoints (see [1, 1.57]) the values of which at $f : X \to A$ define Joyal's sliced simplicial sets ${}^{f/}A$ and $A_{/f}$ characterized by the universal properties described in Proposition 4.2.5:

$$
sSet \underset{-/\_}{\overset{X\star-}{\rightleftarrows}} \bot \ X/sSet \quad \left\{ Y \to {}^{f}/A \right\} := \left\{ \begin{array}{c} X \\ {}^{\nearrow} \searrow^{f} \\ X \star Y \longrightarrow A \end{array} \right\}
$$

$$
sSet \underset{-/\_}{\overset{-\star Y}{\rightleftarrows}} \bot \ Y/sSet \quad \left\{ X \to A_{/g} \right\} := \left\{ \begin{array}{c} Y \\ {}^{\nearrow} \searrow^{g} \\ X \star Y \longrightarrow A \end{array} \right\}.
$$

We think of the slice ${}^{f}/A$ as being the simplicial set of **cones under the diagram** f and we think of the dual slice $A_{/f}$ as being the simplicial set of **cones over the diagram** f. This terminology is reconciled with the terminology of Definition 4.2.1 in Proposition D.6.4. We can also recover these sliced simplicial sets from the décalage construction of Definition D.2.3 via Lemma D.2.5:

LEMMA D.2.9. *For any simplicial map* $f : X \to A$, *the simplicial sets* ${}^{f}/A$ *and* $A_{/f}$ *are the terminally augmented components of* $\mathrm{dec}_l(X,A)$ *and* $\mathrm{dec}_r(X,A)$, *respectively, indexed by the* -1*-simplex* $f : X \to A$.

Proof We identify simplicial sets X and A with their terminally augmented simplicial sets. Recall that $\Delta[-1]$ is the monoidal unit for the join bifunctor on $sSet_+$. Consequently, by adjunction, maps $\Delta[-1] \to \mathrm{dec}_l(X,A)$ or $\Delta[-1] \to \mathrm{dec}_r(X,A)$ correspond to maps $X \to A$. For another terminally augmented simplicial set Y, transposing across the adjunction of Definition D.2.3 provides a correspondence:

$$
\left\{ \begin{array}{c} X \star \Delta[-1] \cong X \\ {}^{X\star!}{\nearrow} \qquad \searrow^{f} \\ X \star Y \longrightarrow A \end{array} \right\} \cong \left\{ \begin{array}{c} \Delta[-1] \\ {}^{!}{\nearrow} \qquad \searrow^{f} \\ Y \longrightarrow \mathrm{dec}_l(X,A) \end{array} \right\}
$$

which shows that the simplicial subset of $\mathrm{dec}_l(X,A)$ comprised of those simplices whose -1-simplex face is f has the universal property that defines ${}^{f}/A$. The dual argument proves that the simplicial subset of $\mathrm{dec}_r(X,A)$ comprised of those simplices whose -1-simplex face is f has the universal property that defines $A_{/f}$. In other words, these décalages admit the following canonical decompositions as disjoint unions of (terminally augmented) slices:

$$
\mathrm{dec}_r(X,A) = \coprod_{f\,:\,X\to A} A_{/f} \qquad \mathrm{dec}_l(X,A) = \coprod_{f\,:\,X\to A} {}^{f}/A \qquad \square
$$

DEFINITION D.2.10 ((left-/right-/inner-)anodyne extensions).

- The set of **horn inclusions** $\Lambda^k[n] \hookrightarrow \Delta[n]$ for $n \geq 1$ and $0 \leq k \leq n$ cellularly generates the **anodyne extensions**.

- The set of **left horn inclusions** $\Lambda^k[n] \hookrightarrow \Delta[n]$ for $n \geq 1$ and $0 \leq k < n$ cellularly generates the **left anodyne extensions**.
- The set of **right horn inclusions** $\Lambda^k[n] \hookrightarrow \Delta[n]$ for $n \geq 1$ and $0 < k \leq n$ cellularly generates the **right anodyne extensions**.
- The set of **inner horn inclusions** $\Lambda^k[n] \hookrightarrow \Delta[n]$ for $n \geq 2$ and $0 < k < n$ cellularly generates the **inner anodyne extensions**.

We refer to the right classes that lift against these maps as **Kan**, **left**, **right**, and **inner fibrations**, respectively. By an easy direct calculation:

LEMMA D.2.11. *The Leibniz join of a horn inclusion and a boundary inclusion is isomorphic to a single horn inclusion:*

$$(\Lambda^k[n] \hookrightarrow \Delta[n]) \, \hat{\star} \, (\partial\Delta[m] \hookrightarrow \Delta[m]) \cong \Lambda^k[n+1+m] \hookrightarrow \Delta[n+1+m]$$
$$(\partial\Delta[n] \hookrightarrow \Delta[n]) \, \hat{\star} \, (\Lambda^k[m] \hookrightarrow \Delta[m]) \cong \Lambda^{n+k+1}[n+1+m] \hookrightarrow \Delta[n+1+m]$$

Proof Since the join bifunctor is the Day convolution of the ordinal sum $[n] \oplus [m] = [n+1+m]$, $\Delta[n] \star \Delta[m] \cong \Delta[n+1+m]$. The domain of the first Leibniz tensor is the simplicial set

$$\Lambda^k[n] \star \Delta[m] \underset{\Lambda^k[n] \star \partial\Delta[m]}{\cup} \Delta[n] \star \partial\Delta[m].$$

We use Definition D.2.2 to identify this with a simplicial subset of $\Delta[n+1+m]$. Since $\partial\Delta[m]$ contains all the codimension-one faces of $\Delta[m]$, the $\Delta[n] \star \partial\Delta[m]$ component contains the jth face of $\Delta[n+1+m]$ for each index $j > n$. Similarly, since $\Lambda^k[n]$ contains all codimension-one faces of $\Delta[n]$-except one, the $\Lambda^k[n] \star \Delta[m]$ component contains the ith face of $\Delta[n+1+m]$ for each index $i \leq n$ except $i = k$. Thus, we see that only the kth face and the $n+1+m$-simplex are missing, which allows us to identify the domain of this Leibniz join with the horn $\Lambda^k[n+1+m]$.   □

Consequently:

COROLLARY D.2.12. *If $f : X \to A$ is any simplicial map and A is a quasi-category, then $f^{/}A$ and $A_{/f}$ are quasi-categories.*

Proof By Lemma D.2.9, to prove that $f^{/}A$ is a quasi-category, it suffices to show that the augmented simplicial set $\mathrm{dec}_l(X, A)$ admits fillers for all inner horns, considered as trivially augmented simplicial sets. For this, it suffices to solve the transposed lifting problem, and argue that A admits extensions along the Leibniz join $(\varnothing \hookrightarrow X) \,\widehat{\star}\, (\Lambda^k[n] \hookrightarrow \Delta[n])$ for all $n \geq 2$ and $0 < k < n$:

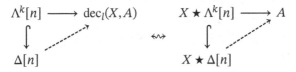

By Lemma C.5.9, the inclusion $\varnothing \hookrightarrow X$ is cellularly generated by the inclusions $\partial\Delta[m] \hookrightarrow \Delta[m]$ for $m \geq 0$, and by Lemma D.2.11 the Leibniz joins of these maps with inner horn inclusions are inner anodyne extensions. By Proposition C.2.9(vii), the Leibniz join $(\varnothing \hookrightarrow X) \,\widehat{\star}\, (\Lambda^k[n] \hookrightarrow \Delta[n])$ is then inner anodyne, so the lifting problem above-right admits a solution, which transpose to define a solution to the lifting problem above-left. □

Our next aim is to prove that the slice quasi-categories are equivalent to the quasi-categories of cones introduced in §4.2. As sketched there, this result hinges on a suitable equivalence between the join construction and the so-called "fat join" construction of Definition 4.2.2, which we now extend to augmented simplicial sets. Recall from Lemma D.2.5 that an augmented simplicial set X canonically decomposes into a coproduct $X \cong \amalg_{i \in X_{-1}} X_{\langle i \rangle}$ of terminally augmented simplicial sets, indexed by the set of -1-simplices.

DEFINITION D.2.13 (fat join and décalage of augmented simplicial sets). For augmented simplicial sets $X \cong \amalg_{i \in X_{-1}} X_{\langle i \rangle}$ and $Y \cong \amalg_{j \in Y_{-1}} Y_{\langle j \rangle}$ their **fat join** is constructed by the pushout:

$$\begin{array}{ccc} (X \times Y) \sqcup (X \times Y) & \xrightarrow{\pi_X \sqcup \pi_Y} & (X \times Y_{-1}) \sqcup (X_{-1} \times Y) \\ \downarrow & & \downarrow \\ X \times 2 \times Y & \longrightarrow & X \diamond Y \end{array} \qquad X \diamond Y := \coprod_{(i,j) \in X_{-1} \times Y_{-1}} X_{\langle i \rangle} \diamond Y_{\langle j \rangle}$$

where $X_{\langle i \rangle} \diamond Y_{\langle j \rangle}$ is the terminally augmented simplicial set defined by the fat join of Definition 4.2.2. This construction is arranged so that the bifunctor $- \diamond - : sSet_+ \times sSet_+ \to sSet_+$ preserves all colimits in each variable, not simply the connected ones preserved by the bifunctor $- \diamond - : sSet \times sSet \to sSet$.

Explicitly, the set of n-simplices $(X \diamond Y)_n$ is the quotient of the set $X_n \times \Delta([n], [1]) \times Y_n$ of n-simplices of $X \times 2 \times Y$ modulo the relation that identifies triples

- $(x, 0, y) \sim (x, 0, y')$ where $0 : [n] \to [1]$ is the constant operator and y and y' are in the same component of $Y \cong \amalg_{j \in Y_{-1}} Y_{\langle j \rangle}$ and
- $(x, 1, y) \sim (x', 1, y)$ where $1 : [n] \to [1]$ is the constant operator and x and x' are in the same component of $X \cong \amalg_{i \in X_{-1}} X_{\langle i \rangle}$.

By cocontinuity and the adjoint functor theorems, the fat join bifunctor on $sSet_+$ has both left and right closures $\mathrm{fatdec}_l(X,A)$ and $\mathrm{fatdec}_r(X,A)$, called **left and right fat décalage**, respectively, whose handedness we fix by declaring that if X is an augmented simplicial set then $X \diamond - \dashv \mathrm{fatdec}_l(X,-)$ and $- \diamond X \dashv \mathrm{fatdec}_r(X,-)$.

There is a canonical comparison map from the fat join to the join as previewed in the discussion surrounding Proposition 4.2.7.

LEMMA D.2.14. *There exists a canonical map of augmented simplicial sets*

$$s^{X,Y} : X \diamond Y \to X \star Y$$

natural in X and Y that in particular defines a natural transformation

$$s^{n,m} : \Delta[n] \diamond \Delta[m] \to \Delta[n] \star \Delta[m] \quad \in sSet_+^{\Delta_+ \times \Delta_+}$$

that is an isomorphism[6] *if n or m equals -1 and otherwise arises as a quotient of the map defined by its order-preserving action on vertices:*

$$\Delta[n] \times \Delta[1] \times \Delta[m] \xrightarrow{\bar{s}^{n,m}} \Delta[n+1+m]$$
$$(i,0,k) \longmapsto i$$
$$(i,1,k) \longmapsto k+n+1$$

Note that $\bar{s}^{n,m}$ takes simplices related under the congruence described in Definition D.2.13 to the same simplex and thus induces a unique map $s^{n,m} : \Delta[n] \diamond \Delta[m] \to \Delta[n] \star \Delta[m]$ on the quotient simplicial set. In the proof, we give a general construction of the map $s^{X,Y}$ that can be shown to coincide with this description in the case where X and Y are standard simplices.

Proof Identifying the set X_{-1} with the augmented simplicial set $\amalg_{X_{-1}} \Delta[-1]$, the Yoneda lemma supplies a canonical map $X_{-1} \to X$ of augmented simplicial sets, which gives rise to a canonical map

$$(X \times Y) \sqcup (X \times Y) \xrightarrow{\pi \sqcup \pi} (X \times Y_{-1}) \sqcup (X_{-1} \times Y)$$

$$X \times 2 \times Y \xrightarrow{\ulcorner} X \diamond Y \xrightarrow{s^{X,Y}} X \star Y$$

$$\pi \searrow \quad \downarrow \quad \swarrow$$
$$2$$

(D.2.15)

Note that the fibers of both $X \diamond Y$ and $X \star Y$ over the endpoints $0, 1$ of 2 are $X \times Y_{-1}$

---

[6] Note that $X \diamond \Delta[-1] \cong X \star \Delta[-1] \cong X \cong \Delta[-1] \diamond X \cong \Delta[-1] \star X$.

and $X_{-1} \times Y$, respectively, and the map $s^{X,Y}$ commutes with the inclusions of these fibers.

The map $s^{X,Y}$ is defined on those n-simplices of $X \diamond Y$ that map surjectively onto 2 by sending a triple $(\sigma \in X_n, \alpha : [n] \twoheadrightarrow [1], \tau \in Y_n)$ representing an n-simplex of $X \diamond Y$ to the pair $(\sigma|_{\{0,\ldots,k\}} \in X_k, \tau|_{\{k+1,\ldots,n\}} \in Y_{n-k-1})$ representing an n-simplex of $X \star Y$, where $k \in [n]$ is the maximal vertex in $\alpha^{-1}(0)$. $\square$

We now prove that the natural comparison between the fat join of simplices and the join of simplices defines a component of a marked homotopy equivalence (see Definition D.1.19).

PROPOSITION D.2.16. *For each $n, m \geq -1$, the map of augmented simplicial sets*

$$s^{n,m} : \Delta[n] \diamond \Delta[m] \longrightarrow \Delta[n] \star \Delta[m] \quad \in s\mathcal{S}et_+^{\Delta_+ \times \Delta_+}$$

is a marked homotopy retract equivalence which is an isomorphism in the case $n = -1$ or $m = -1$.

Proof To define a section and left homotopy inverse to $s^{n,m}$, we consider a map determined by its order-preserving action on vertices:

$$\Delta[n+1+m] \xrightarrow{\bar{\imath}^{n,m}} \Delta[n] \times \Delta[1] \times \Delta[m]$$

$$i \longmapsto \begin{cases} (i,0,0) & \text{if } i \leq n, \\ (n,1,i-n-1) & \text{if } i > n. \end{cases}$$

and note immediately that that $\bar{s}^{n,m} \circ \bar{\imath}^{n,m} = \text{id}$. The obverse composite is given by the explicit formula:

$$\Delta[n] \times \Delta[1] \times \Delta[m] \xrightarrow{\bar{\imath}^{n,m} \circ \bar{s}^{n,m}} \Delta[n] \times \Delta[1] \times \Delta[m]$$

$$(i,0,k) \longmapsto (i,0,0)$$
$$(i,1,k) \longmapsto (n,1,k)$$

Now we may define a related order-preserving endo-map $\bar{u}^{n,m}$ on $[n] \times [1] \times [m]$ by

$$\Delta[n] \times \Delta[1] \times \Delta[m] \xrightarrow{\bar{u}^{n,m}} \Delta[n] \times \Delta[1] \times \Delta[m]$$

$$(i,0,k) \longmapsto (i,0,0)$$
$$(i,1,k) \longmapsto (i,1,k)$$

which is of interest because in the pointwise ordering on such maps we have

$\bar{u}^{n,m} \leq \bar{t}^{n,m} \circ \bar{s}^{n,m}$ and $\bar{u}^{n,m} \leq \text{id}_{[n]\times[1]\times[m]}$, representing simplicial homotopies:

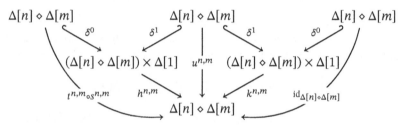

Passing to quotients under the congruence defined in Definition D.2.13, this induces simplicial maps $t^{n,m} : \Delta[n] \star \Delta[m] \to \Delta[n] \diamond \Delta[m]$ and $u^{n,m} : \Delta[n] \diamond \Delta[m] \to \Delta[n] \diamond \Delta[m]$ so that $s^{n,m} \circ t^{n,m} = \text{id}_{\Delta[n]\star\Delta[m]}$, and simplicial homotopies $h^{n,m}$ and $k^{n,m}$ that assemble into a diagram:

$$
\begin{array}{ccc}
\Delta[n] \diamond \Delta[m] & \Delta[n] \diamond \Delta[m] & \Delta[n] \diamond \Delta[m]
\end{array}
$$

$$
\begin{array}{c}
(\Delta[n] \diamond \Delta[m]) \times \Delta[1] \quad u^{n,m} \quad (\Delta[n] \diamond \Delta[m]) \times \Delta[1]
\end{array}
$$

$$
\Delta[n] \diamond \Delta[m]
$$

To see that the maps $h^{n,m}$ and $k^{n,m}$ define marked homotopies, Definition D.1.19 tells us that we must verify, for each 0-simplex $[i, j, k]_{\sim}$ of $\Delta[n] \diamond \Delta[m]$, that the 1-simplex $([i, j, k]_{\sim} \cdot \sigma^0, \text{id}_{[1]})$ of $(\Delta[n] \diamond \Delta[m]) \times \Delta[1]$ is mapped by $h^{n,m}$ and $k^{n,m}$ to degenerate, and thus marked, simplices in $\Delta[n] \diamond \Delta[m]$. We argue by cases in the index j. If $j = 0$, then $\bar{u}^{n,m}(i, 0, k) = (i, 0, 0) \sim (i, 0, k) = \bar{t}^{n,m} \circ \bar{s}^{n,m}(i, 0, k)$, so the components of both $h^{n,m}$ and $k^{n,m}$ are degenerate. If $j = 1$, then $\bar{u}^{n,m}(i, 1, k) = (i, 1, k) \sim (n, i, k) = \bar{t}^{n,m} \circ \bar{s}^{n,m}(i, 1, k)$, so again the components of both $h^{n,m}$ and $k^{n,m}$ are degenerate. Thus, $s^{n,m}$ extends to a marked homotopy retract equivalence with equivalence inverse $t^{n,m}$. $\square$

The marked simplicial homotopy equivalence of Proposition D.2.16 witnesses a pointwise weak equivalence in a suitable sense between two diagrams in $s\mathcal{S}et_+^{\Delta_+ \times \Delta_+}$ considered in Lemma D.2.14. This is the key ingredient in the proof that the canonical map of augmented simplicial sets $s^{X,Y} : X \diamond Y \to X \star Y$ is also a weak equivalence in a suitable sense, but this conclusion requires an exploration of the connection between the homotopy theory of marked simplicial sets and the homotopy theory of quasi-categories. We make this connection in §D.4 and then resume this line of reasoning in §D.6.

We close this section with one final result, a marked analogue of Lemma D.2.11. The join construction of Definition D.2.6 is extended to marked simplicial sets in [128].

DEFINITION D.2.17 (join of marked simplicial sets). The simplicial join lifts to a join bifunctor

$$sSet^+ \times sSet^+ \xrightarrow{\ \star\ } sSet^+$$

in which a simplex $(\alpha, \beta) \colon \Delta[n] \to A \star B$, with components $\alpha \colon \Delta[k] \to A$ and $\beta \colon \Delta[n - k - 1] \to B$, is marked in $A \star B$ if and only if at least one of the simplices α or β is marked in A or B.

LEMMA D.2.18.

(i) *The Leibniz join of a complicial horn extension and a boundary inclusion is isomorphic to a single complicial horn extension:*

$$(\Lambda^k[n] \hookrightarrow_r \Delta^k[n]) \widehat{\star} (\partial\Delta[m] \hookrightarrow_r \Delta[m]) \cong \Lambda^k[n+1+m] \hookrightarrow_r \Delta^k[n+1+m]$$

unless $k = n$, in which case $(\Lambda^n[n] \hookrightarrow_r \Delta^n[n]) \widehat{\star} (\partial\Delta[m] \hookrightarrow_r \Delta[m])$ is a pushout of $\Lambda^n[n + 1 + m] \hookrightarrow_r \Delta^n[n + 1 + m]$.

(ii) *The Leibniz joins below are complicial thinness extensions*

$$(\Lambda^k[n] \hookrightarrow_r \Delta^k[n]) \widehat{\star} (\Delta[m] \hookrightarrow_e \Delta[m]_t) \cong$$
$$\Delta^k[n + 1 + m]' \hookrightarrow_e \Delta^k[n + 1 + m]''$$
$$(\Delta^k[n]' \hookrightarrow_e \Delta^k[n]'') \widehat{\star} (\partial\Delta[m] \hookrightarrow_r \Delta[m]) \cong$$
$$\Delta^k[n + 1 + m]' \hookrightarrow_e \Delta^k[n + 1 + m]''$$

unless $k = n$, in which case the Leibniz joins are instead pushouts of $\Delta^k[n + 1 + m]' \hookrightarrow_e \Delta^k[n + 1 + m]''$, while $(\Delta^k[n]' \hookrightarrow_e \Delta^k[n]'') \widehat{\star}$ $(\Delta[m] \hookrightarrow_e \Delta[m]_t)$ is the identity map.

Since the join is antisymmetric – with $(A \star B)^{op} \cong B^{op} \star A^{op}$ – the cases where the left and right maps of each Leibniz join are exchanged are easily deduced from the cases considered here.

Proof The underlying map of simplicial sets in (i) is identified in Lemma D.2.11, so it remains only to consider the markings. Similarly, in each of the three Leibniz joins considered in (ii), the underlying map of simplicial sets is a Leibniz join of a monomorphism with an identity, and is thus an identity, so it remains only to consider the markings in the resulting entire inclusion. Since a simplex in a join of marked simplicial sets is marked if and only if either of its components are, this description lends itself readily to a case analysis. We leave the details to Exercise D.2.ii or to [129, 38]. □

The slice construction of Definition D.2.8 also extends to marked simplicial sets. We adopt new notation for this construction because it is not typically the case that the underlying simplicial set of the marked join is the slice of corresponding map of underlying simplicial sets.

LEMMA D.2.19. *For any map of marked simplicial sets $f : X \to A$, there exist marked simplicial sets $^{f}\!/\!\!/A$ and $A_{/\!\!/f}$ characterized by the universal properties*

$$\left\{ \begin{array}{c} X \\ \diagup \quad \searrow^{f} \\ X \star Y \longrightarrow A \end{array} \right\} \cong \left\{ Y \to {}^{f}\!/\!\!/A \right\} \quad \left\{ \begin{array}{c} X \\ \diagup \quad \searrow^{f} \\ Y \star X \longrightarrow A \end{array} \right\} \cong \left\{ Y \to A_{/\!\!/f} \right\}.$$

Proof Exercise D.2.iv.   □

Exercises

EXERCISE D.2.i. Prove Lemma D.2.5.

EXERCISE D.2.ii ([129, 38]). Finish the proof of Lemma D.2.18.

EXERCISE D.2.iii. Generalize the last case of Lemma D.2.18(ii) by showing that the Leibniz join of two entire inclusions is an identity.

EXERCISE D.2.iv. Prove Lemma D.2.19.

D.3 Leibniz Stability of Cartesian Products

We now turn our attention to analogous Leibniz constructions defined with respect to the cartesian product, which in the context of marked simplicial sets is called the *Gray tensor product* in [129]. We warm up with a basic result about the geometry of the Leibniz product.

LEMMA D.3.1. *The Leibniz product of any pair of monomorphisms of simplicial sets is a monomorphism.*

Proof Products, pushouts, and monomorphisms in *sSet* are determined pointwise in the category of sets, so this result follows from the fact that for monomorphisms $S \hookrightarrow T$ and $U \hookrightarrow V$ of sets, the Leibniz product

$$(S \hookrightarrow T) \,\hat{\times}\, (U \hookrightarrow V) \cong (S \times V \underset{S \times U}{\cup} T \times U \hookrightarrow T \times V)$$

is a monomorphism, which is clear by inspection.   □

REMARK D.3.2 (why "Leibniz"). The inclusion defined by the Leibniz product of a pair of simplex boundary inclusions

$$(\partial\Delta[n] \hookrightarrow \Delta[n]) \,\hat{\times}\, (\partial\Delta[m] \hookrightarrow \Delta[m]) \cong$$
$$\partial\Delta[n] \times \Delta[m] \underset{\partial\Delta[n]\times\partial\Delta[m]}{\cup} \Delta[n] \times \partial\Delta[m] \hookrightarrow \Delta[n] \times \Delta[m]$$

corresponds geometrically to the inclusion of the boundary $\partial(\Delta[n] \times \Delta[m]) \hookrightarrow \Delta[n] \times \Delta[m]$ of the prism. The identification

$$\partial(\Delta[n] \times \Delta[m]) \cong \partial\Delta[n] \times \Delta[m] \underset{\partial\Delta[n]\times\partial\Delta[m]}{\cup} \Delta[n] \times \partial\Delta[m]$$

is formally similar to various identities that are commonly called "the Leibniz rule."

To prove an analogous result for marked simplicial sets, we first require the following observation:

PROPOSITION D.3.3. *The category of marked simplicial sets is cartesian closed with*

- *cartesian product defined by marking a simplex in the cartesian product of the underlying simplicial sets just when both components are marked simplices*
- *internal hom Y^X defined to be the simplicial set whose n-simplices are maps of marked simplicial sets $\sigma: X \times \Delta[n] \to Y$, where σ is marked just when this map extends to a map of marked simplicial sets:*

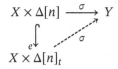

Proof It is clear from the universal property of the product and its closure that the cartesian product and internal hom must be defined in this way if these objects exist. To verify the adjunction, recall from Proposition D.1.5 that marked simplicial sets embed as a reflexive full subcategory of a category of presheaves $Set^{t\Delta^{op}}$. By Example A.1.4, the category $Set^{t\Delta^{op}}$ is cartesian closed and moreover this embedding preserves the products and internal homs as just defined. Now we conclude that these define the functors of a two-variable adjunction on $sSet^+$ by restricting the corresponding natural isomorphisms from $Set^{t\Delta^{op}}$ to its full subcategory. $\square$

The result of Lemma D.3.1 extends to the marked context. Our proof uses a simple observation that is also deployed elsewhere.

LEMMA D.3.4. *The Leibniz product of any map of marked simplicial sets with an entire inclusion is an entire inclusion.*

Proof   By definition the Leibniz product of $f : X \to Y$ and $i : U \hookrightarrow_e V$ is the induced map of marked simplicial sets

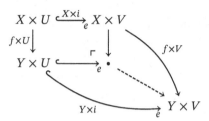

Note that the forgetful functor $(-)_0 : sSet^+ \to sSet$ preserves products and pushouts, and recall that a map of marked simplicial sets is entire just when the underlying map is an isomorphism. Since the product of a simplicial set with an isomorphism is an isomorphism, the maps $X \times U \hookrightarrow_e X \times V$ and $Y \times U \hookrightarrow_e Y \times V$ are entire. Since pushouts of isomorphisms are isomorphisms, it follows that the remaining horizontal map is also entire. Finally, since isomorphisms obey the 2-of-3 property, the Leibniz product map must also be entire.   □

LEMMA D.3.5. *The Leibniz product of two regular inclusions is again a regular inclusion.*

Proof   By Lemma D.3.1, the underlying simplicial set of the Leibniz product of two regular inclusions $A \hookrightarrow_r B$ and $C \hookrightarrow_r D$ is the monomorphism

$$A \times D \underset{A \times C}{\cup} B \times C \hookrightarrow B \times D.$$

Note, in particular, that the inclusions of the components $A \times D$ and $B \times C$ jointly surject onto the domain of this map. Our task is to show that any n-simplex in $A \times D \underset{A \times C}{\cup} B \times C$ that is marked in $B \times D$ is marked in $A \times D \underset{A \times C}{\cup} B \times C$. We argue by cases and assume without loss of generality that the n-simplex is in the image of the inclusion from $B \times C$. In this case, the regularity of the map $B \times C \hookrightarrow_r B \times D$ implies that it is marked in $A \times D \underset{A \times C}{\cup} B \times C$ as claimed.   □

A generic monomorphism of marked simplicial sets is neither regular nor entire, but the generating monomorphisms of marked simplicial sets have one of these properties. Thus, by Lemmas D.3.4 and D.3.5, we immediately conclude:

LEMMA D.3.6. *For any $n, m \geq 0$, the Leibniz products*

$$(\partial\Delta[n] \hookrightarrow_r \Delta[n]) \hat{\times} (\partial\Delta[m] \hookrightarrow_r \Delta[m]), (\partial\Delta[n] \hookrightarrow_r \Delta[n]) \hat{\times} (\Delta[m] \hookrightarrow_e \Delta[m]_t),$$

$$and \quad (\Delta[n] \hookrightarrow_e \Delta[n]_t) \,\hat{\times}\, (\Delta[m] \hookrightarrow_e \Delta[m]_t)$$

are monomorphisms of marked simplicial sets. □

As a corollary, we can characterize Leibniz products of arbitrary monomorphisms of marked simplicial sets by invoking a very general argument that is repeated throughout this chapter.

COROLLARY D.3.7.

(i) *The Leibniz product of any pair of monomorphisms of marked simplicial sets is again a monomorphism.*

(ii) *The Leibniz exponential of any trivial fibration of marked simplicial sets with any monomorphism is again a trivial fibration.*

Proof In the terminology of Lemma C.2.11, the two statements combine to assert that the two-variable adjunction defined by the cartesian product and internal hom is a Leibniz two-variable adjunction with respect to the (monomorphism, trivial fibration) weak factorization system. By Remark C.2.12 this follows from the properties established in Lemmas D.1.6 and D.3.6. □

Considerably harder is to show the Leibniz stability of the class of marked anodyne extensions with the class of marked monomorphisms. We prove a slightly more specific result that also describes the cases of inner, left, or right marked anodyne extensions, which restrict the inequalities $0 \leq k \leq n$ to $0 < k < n, k < n$, or $0 < k$, respectively.

PROPOSITION D.3.8. *For $n \geq 1$, $m \geq 0$, and $0 < k \leq n$ each of the Leibniz products is a right marked anodyne extension and is an inner marked anodyne extension if $k < n$.*

$$(\Lambda^k[n] \hookrightarrow_r \Delta^k[n]) \,\hat{\times}\, (\partial\Delta[m] \hookrightarrow_r \Delta[m])$$
$$(\Lambda^k[n] \hookrightarrow_r \Delta^k[n]) \,\hat{\times}\, (\Delta[m] \hookrightarrow_e \Delta[m]_t)$$
$$(\Delta^k[n]' \hookrightarrow_e \Delta^k[n]'') \,\hat{\times}\, (\partial\Delta[m] \hookrightarrow_r \Delta[m])$$
$$(\Delta^k[n]' \hookrightarrow_e \Delta^k[n]'') \,\hat{\times}\, (\Delta[m] \hookrightarrow_e \Delta[m]_t)$$

On account of the symmetry of the cartesian product, it is immaterial whether the horn inclusion or the simplex boundary inclusion appears on the left or right. The proof of this result requires some special notation to describe the cartesian product of simplices.

DIGRESSION D.3.9 (on shuffles). Since $\Delta[n]$ is the nerve of the poset $[n]$, an r-simplex $i : \Delta[r] \to \Delta[n]$ may equally be encoded by the ordered sequence of vertices $i_0 \leq \cdots \leq i_r \in [n]$ appearing in its image.

By the universal property of the product, an r-simplex in $\Delta[n] \times \Delta[m]$ is given by an r-simplex $i \colon \Delta[r] \to \Delta[n]$ and an r-simplex $j \colon \Delta[r] \to \Delta[m]$. Since $\Delta[n] \times \Delta[m]$ is the nerve of the poset $[n] \times [m]$, such simplices correspond bijectively to ordered sequences of pairs

$$(i_0, j_0) \le (i_1, j_1) \le \cdots \le (i_r, j_r) \tag{D.3.10}$$

with each $i_s \in [n]$ and each $j_t \in [m]$, and is degenerate just when one of these inequalities is an equality.

The nondegenerate $n + m$-simplices of the simplicial set $\Delta[n] \times \Delta[m]$ are called **shuffles**; these are the simplices of maximal dimension. An $n+m$-simplex (i, j) defines a shuffle just when $i_t + j_t = t$ for all $t \in [n + m]$. If the objects of $[n] \times [m]$ are arranged in a rectangular grid, the shuffles are those maximal-length nondegenerate paths that start from $(0, 0)$ and end with (n, m), by taking steps which add one to exactly one coordinate at a time.

The first case of the following proof is an adaptation of an argument of Dugger and Spivak [36, A.1] to the marked context.

Proof   By Lemma D.3.5, the Leibniz product $(\Lambda^k[n] \hookrightarrow_r \Delta^k[n]) \,\hat{\times}\, (\partial\Delta[m] \hookrightarrow_r \Delta[m])$ is the regular inclusion

$$\Lambda^k[n] \times \Delta[m] \underset{\Lambda^k[n] \times \partial\Delta[m]}{\cup} \Delta^k[n] \times \partial\Delta[m] \overset{\ \ \ }{\hookrightarrow}_r \Delta^k[n] \times \Delta[m].$$

A nondegenerate r-simplex (D.3.10) of $\Delta^k[n] \times \Delta[m]$ is missing from the domain of the Leibniz product inclusion just when

- its component $\{i_0, \dots, i_r\} \supset [n]\backslash\{k\}$ and
- its component $\{j_0, \dots, j_r\} \supset [m]$.

We filter this inclusion as a sequence of regular inclusions

$$\Lambda^k[n] \times \Delta[m] \underset{\Lambda^k[n] \times \partial\Delta[m]}{\cup} \Delta^k[n] \times \partial\Delta[m] =: Y^{-1} \hookrightarrow_r \cdots \hookrightarrow_r Y^m = \Delta^k[n] \times \Delta[m]$$

and argue that each $Y^t \hookrightarrow Y^{t+1}$ is right or inner marked anodyne, as appropriate.

Starting from $Y^{-1} := \Lambda^k[n] \times \Delta[m] \underset{\Lambda^k[n] \times \partial\Delta[m]}{\cup} \Delta^k[n] \times \partial\Delta[m]$, we define Y^t to be the smallest regular simplicial subset of $\Delta^k[n] \times \Delta[m]$ that contains Y^{t-1} together with every simplex (D.3.10) that contains the vertex (k, t). Since every missing simplex is a face of a simplex that contains one of the vertices $(k, 0), \dots, (k, m)$, it is clear from this description that $Y^m = \Delta^k[n] \times \Delta[m]$.

It remains only to analyze the regular inclusions $Y^{t-1} \hookrightarrow_r Y^t$, which we do by producing another filtration

$$Y^{t-1} = Y^{t,n-1} \hookrightarrow_r Y^{t,n} \hookrightarrow_r \cdots \hookrightarrow_r Y^{t,n+m} = Y^t.$$

Note that every simplex of $\Delta^k[n] \times \Delta[m]$ that contains the vertex (k, t) and has dimension $n - 1$ or less is contained in Y^{-1}, so the simplices containing the vertex (k, t) that are attached to Y^{t-1} to form Y^t have dimensions between n and $n + m$. With this in mind, we define $Y^{t,r}$ to be the smallest regular simplicial subset of $\Delta^k[n] \times \Delta[m]$ containing $Y^{t,r-1}$ and all simplices of dimension r that contain the vertex (k, t). In particular, $Y^{t,n-1} = Y^{t-1}$ and $Y^{t,n+m} = Y^t$.

We now argue that each regular inclusion in this filtration is a pushout of a coproduct of complicial horn extensions followed by complicial thinness extensions

indexed by the sets $S^{t,r}$ of r-simplices containing the vertex (k, t) and not already present in $Y^{t,r-1}$ and $T^{t,r} \subset S^{t,r}$ defined to be the subset of those r-simplices τ so that $\tau\delta^{\ell_\tau}$ is marked in $\Delta^k[n] \times \Delta[m]$. Moreover, for each $\tau \in S^{t,r}$, we will see that that $0 < \ell_\tau$ and if $k < n$ then $\ell_\tau < r$. This will show that the Leibniz product is a right marked anodyne extension, which is an inner marked anodyne extension if $k < n$.

To see this, let $\tau \in S^{t,r}$ be the r-simplex

$$(i_0, j_0) \leq \cdots \leq (i_{\ell_\tau}, j_{\ell_\tau}) = (k, t) \leq \cdots \leq (i_r, j_r)$$

containing (k, t) as its ℓ_τth vertex; for readability we write ℓ for ℓ_τ going forward. Since the set $\{i_0, \ldots, i_r\} \supset [n]$ and $0 < k$ we must also have $0 < \ell$, and if $k < n$, we must also have $\ell < r$. We will argue that:

- Each face of τ except the ℓth is contained in $Y^{t,r-1}$.
- The ℓth face of τ is not in $Y^{t,r-1}$.
- The r-simplex τ is a ℓ-admissible simplex of $\Delta^k[n] \times \Delta[m]$.
- If $\tau\delta^\ell$ is marked in $\Delta^k[n] \times \Delta[m]$ then so is $\tau\delta^{\ell-1}$ and $\tau\delta^{\ell+1}$ (in the case $\ell < r$).

Thus, the union of τ with $Y^{t,r-1}$ may be formed as a pushout of a complicial horn extension $\Lambda^\ell[r] \hookrightarrow_r \Delta^\ell[r]$ as claimed.

For the first item, note that each codimension-one face except for $\tau\delta^\ell$ has dimension $r - 1$ and contains the vertex (k, t) and thus lies in $Y^{t,r-1}$ as claimed. To see that $Y^{t,r-1}$ does not also contain the face $\tau\delta^\ell$, we consider the vertex $(i_{\ell-1}, j_{\ell-1})$. If $i_{\ell-1} = k$ then by nondegeneracy, $j_{\ell-1} < t$, in which case we would have $\tau \in Y^{t-1}$, a contradiction. Thus $i_{\ell-1} < k$. Now if $i_{\ell-1} \le k - 2$, then we would have $\tau \in Y^{-1}$, again a contradiction. So it must be that $i_{\ell-1} = k - 1$. Now if $j_{\ell-1} < t$, then τ would be a face of the $r + 1$-dimensional simplex

$$(i_0, j_0) \le \cdots \le (i_{\ell-1}, j_{\ell-1}) \le (k, t - 1) \le (i_\ell, j_\ell) = (k, t) \le \cdots \le (i_r, j_r)$$

which is contained in Y^{t-1}, a contradiction. So we conclude that $(i_{\ell-1}, j_{\ell-1}) = (k - 1, t)$.

From this computation we see that the vertices of $\tau\delta^\ell$ satisfy $\{i_0, \dots, i_{\ell-1} = k - 1, i_{\ell+1}, \dots, i_r\} \supset [n]\backslash\{k\}$ and $\{j_0, \dots, j_{\ell-1}, j_{\ell+1}, \dots, j_r\} \supset [m]$. Thus, $\tau\delta^\ell$ is not in Y^{-1}. Furthermore, $\tau\delta^\ell$ was not added in along the way to $Y^{t,r-1}$, since it is not a face of a simplex containing the vertex (k, s) for any $s < t$. This completes our second task.

We have shown that it is possible to attach τ to $Y^{t,r-1}$ along with its ℓth face by filling a suitable horn. It remains only to argue that the horn $\Lambda^\ell[r] \to Y^{t,r-1}$ along which we are attaching τ is admissible. Since the inclusion $Y^{t,r-1} \hookrightarrow_r \Delta^k[n] \times \Delta[m]$ is regular it suffices to show that each simplex containing the vertices $(i_{\ell-1}, j_{\ell-1})$, (i_ℓ, j_ℓ), and $(i_{\ell+1}, j_{\ell+1})$ – or just the first two of these in the case $\ell = r$ – is marked in $\Delta^k[n] \times \Delta[m]$. We have seen above that $(i_{\ell-1}, j_{\ell-1}) = (k-1, t)$ and $(i_\ell, j_\ell) = (k, t)$. In the case $\ell < r$, since τ is missing from Y^{-1}, $i_{\ell+1}$ must equal k or $k + 1$. But now the component in $\Delta[m]$ of this simplex is degenerate, containing the sequence $t \le t$, while the component in $\Delta^k[n]$ is either degenerate, contains the sequence $k - 1 \le k \le k+1$, or contains the sequence $k - 1 \le k$ in the case $\ell = r$ in which case $k = n$. Thus both components are marked simplices, which means that their product is marked in $\Delta^k[n] \times \Delta[m]$ as required.

Finally, we must argue that the simplices attached by the pushout contain all the markings present in the regular subset $Y^{t,r} \hookrightarrow_r \Delta^k[n] \times \Delta[m]$. The only simplices present in $Y^{t,r}$ but not $Y^{t,r-1}$ are in dimensions r and $r - 1$. The newly attached r-simplices are all marked, so we need only concern ourselves with the $(r - 1)$-simplex

$$(i_0, j_0) \le \cdots \le (i_{\ell-1}, j_{\ell-1}) = (k - 1, t) \le (i_{\ell+1}, j_{\ell+1}) \le \cdots \le (i_r, j_r)$$

arising as the ℓ-face for each $\tau \in S^{t,r}$ when this simplex is marked in $\Delta^k[n] \times \Delta[m]$.

There are two cases depending on whether $i_{\ell+1} = k + 1$ or $i_{\ell+1} = k$. In the former case, the fact that this simplex is marked tells us that there is a duplication present in the sequence $i_0 \le \cdots \le i_{\ell_\tau-1} \le i_{\ell_\tau+1} \le \cdots \le i_r$ and also in the sequence $j_0 \leq \cdots \leq j_{\ell_\tau - 1} \leq j_{\ell_\tau + 1} \leq \ldots \leq j_r$. When we substitute $i_\ell = k$ in the first sequence for $i_{\ell - 1}$ or $i_{\ell + 1}$ either the duplication remains or we now have the subsequence $k - 1 \leq k \leq k + 1$. Either way, this tells us that the component of the faces $\tau \delta^{\ell - 1}$ and $\tau \delta^{\ell + 1}$ is marked in $\Delta^k[n]$. Similarly, when we substitute $j_\ell = t$ for $j_{\ell - 1} = t$, the sequence is unchanged, and when we substitute for $j_{\ell + 1}$ our sequence now contains a duplication $t \leq t$. Either way, this tells us that the component of the faces $\tau \delta^{\ell - 1}$ and $\tau \delta^{\ell + 1}$ is marked in $\Delta[m]$. In conclusion, the simplex $\tau \colon \Delta^\ell[r] \to \Delta^k[n] \times \Delta[m]$ extends along $\Delta^k[n] \hookrightarrow_e \Delta^k[n]'$, so we obtain the desired marking of its ℓth face by extending along the entire inclusion $\Delta^k[n]' \hookrightarrow_e \Delta^k[n]''$ included in the second pushout.

In the case where $i_{\ell + 1} = k$, the sequence of vertices for $\tau \delta^{\ell - 1}$ contains $k \leq k$ in its first component and the same sequence of vertices as $\tau \delta^\ell$ in its second component. Thus, $\tau \delta^{\ell - 1}$ is marked. The sequence of vertices for $\tau \delta^{\ell + 1}$ contains $k - 1 \leq k \leq k + 1$ in its first component and $t \leq t$ in its second component. Thus, $\tau \delta^{\ell + 1}$ is marked, and once more we obtain the desired marking of its ℓth face by extending along the entire inclusion $\Delta^k[n]' \hookrightarrow_e \Delta^k[n]''$ included in the second pushout. This completes the proof that the first Leibniz product is a marked anodyne extension.

By Lemma D.3.4, the remaining three Leibniz products are entire inclusions, so all that is required is to verify that the additional markings present in the codomains are the results of complicial thinness extensions. We treat simultaneously the two cases involving Leibniz products

$$(\Delta^k[n]' \hookrightarrow_e \Delta^k[n]'') \hat{\times} (A \hookrightarrow B)$$

of a complicial horn extension with a marked monomorphism. The only marked simplex of $\Delta^k[n]''$ that is not marked in $\Delta^k[n]'$ is the face $\delta^k \colon \Delta[n-1]_t \to \Delta^k[n]''$, which implies that we have a pullback and pushout square

$$
\begin{array}{ccc}
\coprod\limits_{\tau \in S} \Delta[n-1] & \longrightarrow & \Delta^k[n]' \times B \underset{\Delta^k[n]' \times A}{\cup} \Delta^k[n]'' \times A \\
{\scriptstyle e}\Big\downarrow & \ulcorner & \Big\downarrow{\scriptstyle e} \\
\coprod\limits_{\tau \in S} \Delta[n-1]_t & \xrightarrow{\quad (\delta^k, \tau) \quad} & \Delta^k[n]'' \times B
\end{array}
$$

where S is the set of marked $(n-1)$-simplices in B that are not present or not marked in A. We argue that for any marked $(n-1)$-simplex $\tau \in B$, the degenerate n-simplex $\tau\sigma^{k-1}$ admits the indicated markings:

$$
\begin{array}{ccc}
\Delta[n] & \overset{e}{\longrightarrow} & \Delta^k[n]'' \\
{\scriptstyle \sigma^{k-1}}\downarrow & & \downarrow \\
\Delta[n-1]_t & \underset{\tau}{\longrightarrow} & B
\end{array}
$$

because the $k-1$th and kth faces equal τ, and so are marked, and any face that contains the vertices $k-1$ and k is degenerate and so is also marked; these conditions cover all of the required marked faces. Now it is clear that the pushout square above factors through the left-hand pushout diagram

$$
\begin{array}{ccccc}
\coprod\limits_{\tau\in S}\Delta[n-1] & \longrightarrow & \coprod\limits_{\tau\in S}\Delta^k[n]' & \longrightarrow & \Delta^k[n]'\times B \underset{\Delta^k[n]'\times A}{\cup} \Delta^k[n]''\times A \\
{\scriptstyle e}\downarrow & & {\scriptstyle e}\downarrow & & {\scriptstyle e}\downarrow \\
\coprod\limits_{\tau\in S}\Delta[n-1]_t & \underset{\delta^k}{\longrightarrow} & \coprod\limits_{\tau\in S}\Delta^k[n]'' & \underset{(\mathrm{id},\tau\sigma^{k-1})}{\longrightarrow} & \Delta^k[n]''\times B
\end{array}
$$

demonstrating that the Leibniz product inclusion is a pushout of coproducts of suitable complicial thinness extensions.

The final case of

$$
(\Lambda^k[n] \hookrightarrow_r \Delta^k[n]) \,\hat{\times}\, (\Delta[m] \hookrightarrow_e \Delta[m]_t) \cong \\
\Lambda^k[n]\times\Delta[m]_t \underset{\Lambda^k[n]\times\Delta[m]}{\cup} \Delta^k[n]\times\Delta[m] \hookrightarrow_e \Delta^k[n]\times\Delta[m]_t
$$

is again an entire inclusion. Since the only simplex that is marked in $\Delta[m]_t$ but not in $\Delta[m]$ is the top dimensional m-simplex, the only simplices that are marked in the codomain but not in the domain are m-simplices $(\tau,\mathrm{id}) : \Delta[m]_t \to \Delta^k[n]\times\Delta[m]_t$ in which the image of τ either

- contains $[n]$ or
- contains $[n]\backslash\{k\}$ but not $\{k\}$ and is degenerate.

In particular, this Leibniz product inclusion is an identity if $m < n$. In the case $m = n$, there are $m + 1$ simplices that are marked in $\Delta^k[m]\times\Delta[m]_t$ but not in the domain, corresponding to the m m-simplices that are degenerate on the kth face of $\Delta^k[m]$ and the top dimensional m-simplex.

We factor the inclusion as a finite composite of pushouts of coproducts of maps $\Delta^s[m+1]' \hookrightarrow_e \Delta^s[m+1]''$ for varying $0 < s \le m+1$, where each $s < m+1$ if $k < n$. This will prove that this Leibniz product is a complicial thinness extension of the appropriate kind.

We can classify the missing marked simplices in terms of their component $\tau : \Delta[m]_t \to \Delta^k[n]$, which we may represent as a sequence $i_0, \ldots, i_m$ of vertices of $[n]$ that either contains $[n]$ or contains $[n]\backslash\{k\}$ and has repetitions. We induct over a partial ordering of these simplices in decreasing order of the sum $\sum_{t=0}^{m} i_t$.[7]

For a simplex τ with maximal vertex sum $\sum_{t=0}^{m} i_t$ among those simplices that remain to be marked, let s be minimal so that $i_s \geq k$; when $k = n$ it is possible that all $i_t < k = n$, which gives a second case that we will consider in a moment. Then we consider the $m + 1$-simplex:

$$\Delta^s[m+1]'' \xrightarrow{(\tau\sigma^s, \sigma^{s-1})} \Delta^k[n] \times \Delta[m]_t.$$

By construction, the $s + 1$th face is marked in $\Delta^k[n] \times \Delta[m]$, while the $s - 1$th face has strictly greater vertex sum, and so is marked by the inductive hypothesis. The faces containing the $s - 1$, s, and $s + 1$ vertices are all degenerate and thus marked. This proves that the face τ can be marked by forming an extension $\Delta^s[m+1]' \hookrightarrow_e \Delta^s[m+1]''$.

In the case where τ is a simplex where all $i_t < k = n$, then we consider the $m + 1$-simplex

$$\Delta^{m+1}[m+1]'' \xrightarrow{(\chi, \sigma^m)} \Delta^n[n] \times \Delta[m]_t.$$

where $\chi : \Delta[m+1] \to \Delta^n[n]$ is the simplex spanned by the vertices $i_0, \dots, i_m = n - 1, i_{m+1} = n$. Here the mth face has strictly greater sum, and so is marked by the inductive hypothesis. The faces containing the mth and $m + 1$th vertices have a degenerate component in $\Delta[m]$ and have a component in $\Delta^n[n]$ that contains the last two vertices $n - 1$ and n. Thus, all such simplices are marked. This proves that the face τ can be marked by forming an extension $\Delta^{m+1}[m+1]' \hookrightarrow_e \Delta^{m+1}[m+1]''$. $\qquad\square$

As in the proof of Corollary D.3.7, by Remark C.2.12 the result of Proposition D.3.8 extends to Leibniz products of marked anodyne extensions with marked monomorphisms. We derive a few consequences of this in Corollary D.3.12, but first explain how this result implies its unmarked analogue.

COROLLARY D.3.11. *Let $i: A \hookrightarrow B$ and $j: K \hookrightarrow L$ be monomorphisms of simplicial sets. If either i or j is also anodyne (or, respectively, left-, right-, or inner-anodyne), then so is the Leibniz product*

$$A \times L \cup_{A \times K} B \times K \xhookrightarrow{\ i \hat{\times} j\ } B \times L$$

---

[7] Here the vertex sum of an m-simplex τ is greater than the vertex sum of an m-simplex τ' if and only if τ has greater "depth" in the sense defined in [129, 68]. The inductive argument of [129, §5.2] involves Leibniz products of inner or left horn inclusions and starts by considering simplices of lowest depth; ours involves an inner or right horn inclusion and starts by considering simplices of highest depth.

Proof   The maximal marking functor $(-)^{\sharp} : s\mathcal{S}et \to s\mathcal{S}et^{+}$ sends (left/right/inner) horn inclusions to (left/right/inner) marked anodyne extensions, and as a left adjoint, therefore carries all (left/right/inner) anodyne extensions of simplicial sets to (left/right/inner) marked anodyne extensions. As a left and right adjoint, the maximal marking functor also preserves Leibniz products. Similarly, the forgetful functor $(-)_{0} : s\mathcal{S}et^{+} \to s\mathcal{S}et$ preserves products and colimits and carries the (left/right/inner) marked anodyne extensions to the classes of anodyne extensions introduced in Definition D.2.10. Thus, this result follows immediately from the first case of Proposition D.3.8 and Remark C.2.12. $\square$

Recall from Definition D.1.14 that a marked map between complicial sets is a **complicial isofibration** if it has the right lifting property with respect to the complicial horn extensions and complicial thinness extensions of Definition D.1.9.

COROLLARY D.3.12.

  (i) *For any quasi-category A and simplicial set X, A^{X} is again a quasi-category.*

  (ii) *For any complicial set A and marked simplicial set X, A^{X} is again a complicial set.*

  (iii) *For any complicial isofibration $p : E \twoheadrightarrow B$ and any monomorphism of marked simplicial sets $i : X \hookrightarrow Y$, the Leibniz exponential $\widehat{\{i, p\}} : E^{Y} \twoheadrightarrow E^{X} \times_{B^{X}} B^{Y}$ is a complicial isofibration.*

  (iv) *For any complicial isofibration $p : E \twoheadrightarrow B$ and any marked anodyne extension $i : X \hookrightarrow Y$, the Leibniz exponential $\widehat{\{i, p\}} : E^{Y} \twoheadrightarrow E^{X} \times_{B^{X}} B^{Y}$ is a trivial fibration of complicial sets.*

Proof   The second statement is a special case of the third statement, which, together with the fourth statement, follows by transposing the result of Proposition D.3.8 across the Leibniz version of the two variable adjunction of Proposition D.3.3 (see Proposition C.2.9). The first statement follows similarly by applying Corollary D.3.11 in place of Proposition D.3.8. $\square$

In particular, by Lemma D.1.13, for any complicial set A, $A^{\mathbb{I}^{\sharp}} \to A^{\Delta[1]^{\sharp}}$ is a trivial fibration of complicial sets. Thus the notion of equivalence of complicial sets can be redefined as follows:

COROLLARY D.3.13. *If $f : A \to B$ and $g : B \to A$ are inverse equivalences of complicial sets, then there exists a homotopy equivalence with the marked* *homotopy coherent isomorphism serving as the interval:*

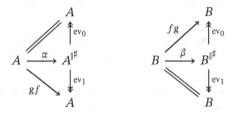

Conversely, such data restricts to exhibit an equivalence of complicial sets.

Proof The data of Definition D.1.18 can be lifted as follows:

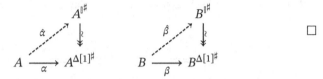

We would like to prove the analogous statement to Corollary D.3.12(iii) and (iv) for isofibrations between quasi-categories, which requires analogous statements to Proposition D.3.8 and Corollary D.3.11 analyzing Leibniz products of monomorphisms of simplicial sets with the inclusion $\mathbb{1} \hookrightarrow \mathbb{I}$. We shall deduce this by considering the relationship between isomorphisms in quasi-categories and marked edges in complicial sets, which is the subject of the next section.

Exercises

EXERCISE D.3.i. State and prove the unmarked analogue of Corollary D.3.7.

EXERCISE D.3.ii. The Leibniz product of a regular inclusion with a noninvertible entire inclusion is entire but not necessarily regular. Find examples that illustrate both possibilities and state and prove a characterization of those Leibniz products of this form that are necessarily regular. Would this have simplified the proof of Proposition D.3.8?

EXERCISE D.3.iii. If $f : X \to Y$ is a marked homotopy equivalence and A is a complicial set, prove that the restriction map $f^* : A^Y \to A^X$ is an equivalence of complicial sets.

D.4 Isomorphisms in Naturally Marked Quasi-Categories

By Exercise D.1.ii, Kan complexes can be regarded as a special case of complicial sets: namely Kan complexes coincide with the complicial sets that are maximally marked. In this section, we discover that quasi-categories can similarly be identified with certain 1-trivial complicial sets (see Definition D.1.16) whose marked 1-simplices are determined by the underlying simplicial set. In Theorem D.4.14, we prove that the category of quasi-categories may be identified with the 1-trivial complicial sets whose marked 1-simplices are *saturated* in a precise sense to be introduced.

The markings on the 1-simplices in a complicial set cannot be arbitrarily assigned: every marked edge must be an equivalence in a sense that we now introduce.

DEFINITION D.4.1. A 1-simplex f in a marked simplicial set is an **equivalence** if there exist a pair of thin 2-simplices as displayed

$$
\begin{array}{ccc}
& x & \\
{}^{g}\nearrow \;{}_{\cong} & & \searrow^{f} \\
y & =\!=\!=\!= & y
\end{array}
\qquad
\begin{array}{ccc}
& y & \\
{}^{f}\nearrow \;{}_{\cong} & & \searrow^{g'} \\
x & =\!=\!=\!= & x
\end{array}
$$

Note the notion of equivalence is defined relative to the choice of markings on the 2-simplices. A very similar notion is defined for the edges of a quasi-category in Definition 1.1.13 under the name "isomorphism."

LEMMA D.4.2. *Every marked edge in a complicial set is an equivalence.*

Proof If f is a marked edge in any complicial set A, then the $\Lambda^2[2]$-horn with 0th face f and 1st face degenerate is admissible, so f has a right equivalence inverse. A dual construction involving a $\Lambda^0[2]$-horn shows that f has a left equivalence inverse:[8]

$$
\begin{array}{ccc}
& x & \\
{}\nearrow \;{}_{\cong} & & \searrow^{f} \\
y & =\!=\!=\!= & y
\end{array}
\qquad
\begin{array}{ccc}
& y & \\
{}^{f}\nearrow \;{}_{\cong} & & \searrow \\
x & =\!=\!=\!= & x
\end{array}
\qquad\qquad \square
$$

This observation suggests two ways to mark the edges in the nerve of a 1-category.

LEMMA D.4.3. *The nerve of a 1-category defines a complicial set by marking all simplices in dimension greater than one and then either defining:*

 (i) the marked edges to be the identity arrows only or
 (ii) the marked edges to be all isomorphisms.

Proof Exercise D.4.ii. $\square$

---

[8] Note also that the complicial thinness extensions imply further that these one-sided inverses are also marked, so they admit further inverses of their own.

The first of these options defines the minimal 1-trivial marking that makes the nerve of a 1-category into a complicial set. By Lemma D.4.2, the latter option defines the maximal marking that makes the nerve of a 1-category into a complicial set. We now introduce terminology to describe "maximally marked" 1-trivial complicial sets.

DEFINITION D.4.4. A complicial set is 1-**saturated** if every equivalence it contains among its edges is marked.[9]

This motivates a definition of the canonical marking of a quasi-category, which is called the "natural marking" in [78]. Namely, we assign a quasi-category its unique 1-saturated 1-trivial marking:

DEFINITION D.4.5. For any quasi-category A, its **natural marking** is defined by:

- marking all simplices in dimension greater than one
- marking exactly those edges that are isomorphisms, in the sense of Definition 1.1.13, or equivalently marking all those edges that are equivalences, in the sense of Definition D.4.1.

The natural marking for quasi-categories is convenient for stating and proving an important combinatorial result due to Joyal:

PROPOSITION D.4.6. *Any naturally marked quasi-category A admits fillers for outer complicial horn extensions for $n \geq 1$:*

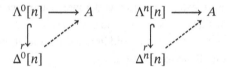

In the original [61], the result is stated without reference to markings as follows: a quasi-category admits fillers for **special outer horns**, left horns $\Lambda^0[n] \to A$ whose initial $\{01\}$-edge is mapped to an isomorphism in A and right horns $\Lambda^n[n] \to A$ whose final $\{n-1\,n\}$-edge is mapped to an isomorphism in A.

Many proofs of Proposition D.4.6 are possible; for instance, see [36, §B] or the original [61]. We choose to use combinatorial results of Verity [129, §4.2], which we present in stages, that use an alternate (a posteriori equivalent) notion

---

[9] Since the characterization of equivalences among the 1-simplices requires prior agreement about which 2-simplices are marked, we typically apply Definition D.4.4 to 1-trivial complicial sets, in which case we say a 1-saturated 1-trivial complicial set is simply "saturated." This is a special case of a general notion of saturated complicial set discussed in Digression D.4.21.

of homotopy coherent isomorphism, which a homotopy type theorist would recognize by the name "half adjoint equivalence" [125, §4.2].

NOTATION D.4.7 (subcomplexes of the coherent isomorphism). Recall the **coherent isomorphism** is the simplicial set $\mathbb{I}$ defined as the nerve of the free-living isomorphism. It has exactly two nondegenerate simplices in each dimension. If we label its vertices as "$-$" and "$+$," then its remaining nondegenerate simplices are uniquely determined by their vertices, which are given by alternating sequences of "$-$" and "$+$" starting from either vertex. As marked simplicial sets, we give $\mathbb{I}$ and its subcomplexes the maximal marking.

Following the notation introduced in [129, 42], we write $E_n^+, E_n^- \subset \mathbb{I}$ for the simplicial subsets generated by the n-simplices $- + - \cdots \pm$ and $+ - + \cdots \mp$, respectively. Both simplicial subsets include uniquely into both E_{n+1}^+ and E_{n+1}^- and these inclusions factor as follows

$$
\begin{array}{ccc}
\Lambda^0[n+1] \longrightarrow E_n^{\pm} & \qquad & \Lambda^{n+1}[n+1] \longrightarrow E_n^{\pm} \\
\end{array}
$$

$$
\begin{array}{ccc}
\Delta^0[n+1] \longrightarrow \bullet \longleftarrow \Delta^0[n+1]' & \qquad & \Delta^{n+1}[n+1] \longrightarrow \bullet \longleftarrow \Delta^{n+1}[n+1]' \\
\end{array}
$$

$$
\begin{array}{ccc}
E_{n+1}^{\pm} \longleftarrow \Delta^0[n+1]'' & \qquad & E_{n+1}^{\mp} \longleftarrow \Delta^{n+1}[n+1]'' \\
\end{array}
$$

proving that any inclusion $E_n^{\pm} \hookrightarrow E_m^{\pm}$ or $E_n^{\pm} \hookrightarrow E_m^{\mp}$ with $m > n$ is a marked anodyne extension.

The following result gives a criterion under which an **inner complicial fibration** – a marked map that is only assumed to have the right lifting property against the inner complicial horn extensions and inner complicial thinness extensions – in fact defines a complicial isofibration.

PROPOSITION D.4.8. *Let $p : A \twoheadrightarrow B$ be an inner complicial fibration whose codomain B is a complicial set. Then p is a complicial isofibration if and only if p admits lifts against the inclusions $E_0^- \hookrightarrow E_1^-$ and $E_1^- \hookrightarrow E_3^-$.*

Proof Any complicial isofibration admits lifts against the marked anodyne extensions $E_0^- \hookrightarrow E_1^-$ and $E_1^- \hookrightarrow E_3^-$. Thus, the heart of this result, and the only part that remains to be proven, is the assertion that any inner complicial fibration that admits lifts against this pair of inclusions and whose codomain is a complicial set necessarily also admits fillers for outer complicial horn extensions and outer complicial thinness extensions.

We begin by arguing that an inner complicial fibration $p : A \twoheadrightarrow B$ satisfying the conditions of the statement also admits lifts against the dual inclusion $E_0^- \hookrightarrow$

E_1^+. Given a lifting problem such as presented by the maps a and b in the square below:

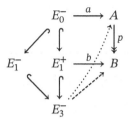

there exists the dashed extension of b along $E_1^+ \hookrightarrow E_3^-$, since this inclusion is marked anodyne and B is a complicial set. Now the inclusion $E_0^- \hookrightarrow E_3^-$ factors as indicated $E_0^- \hookrightarrow E_1^- \hookrightarrow E_3^-$ and since p is assumed to lift against both maps, the dotted lift exists as well, which restricts to define a solution to the original lifting problem.

With this result in hand, it follows that the odd dual of $p : A \twoheadrightarrow B$ satisfies the same lifting properties as $p : A \twoheadrightarrow B$ does, since the odd dual of $E_0^- \hookrightarrow E_1^-$ is $E_0^- \hookrightarrow E_1^+$, while the odd dual of $E_1^- \hookrightarrow E_3^-$ is isomorphic to itself. So it suffices to show that p admits lifts against left complicial horn extensions $\Lambda^0[n] \hookrightarrow_r \Delta^0[n]$ and left complicial thinness extensions $\Delta^0[n]' \hookrightarrow_e \Delta^0[n]''$ since its odd dual will then share these properties, which implies that p also admits lifts against right complicial horn extensions and right complicial thinness extensions. The case $\Lambda^0[1] \hookrightarrow \Delta^0[1]$ is the map $E_0^- \hookrightarrow E_1^-$ so it suffices also to assume $n > 1$, in which case we have an isomorphism

$$\Lambda^0[n] \hookrightarrow \Delta^0[n] \cong (\Lambda^0[1] \hookrightarrow \Delta^0[1]) \,\widehat{\star}\, (\partial\Delta[n-2] \hookrightarrow \Delta[n-2])$$

by Lemma D.2.18. Writing $m = n - 2$ for concision, consider a lifting problem as presented by the maps a and b:

$$
\begin{array}{ccc}
E_0^- \star \Delta[m] \cup E_1^- \star \partial\Delta[m] & \xrightarrow{\quad\quad a \quad\quad} & A \\
\downarrow & & \downarrow{\scriptstyle p} \\
E_1^- \star \Delta[m] \xhookrightarrow{\ } E_2^- \star \Delta[m]_t \cup E_3^- \star \partial\Delta[m] \xhookrightarrow{\ } E_3^- \star \Delta[m] & \dashrightarrow & B
\end{array}
$$

with maps i, j, b

By Lemma D.2.18, the map j is a marked anodyne extension, so since B is a complicial set, the dashed extension exists. To show that the dotted lift exists as well, we argue that the map i is cellularly generated by the inner complicial horn extensions and the map $E_1^- \hookrightarrow E_3^-$. To see this, factor the map i as the composite of the three vertical maps in the middle column of the diagram of

pushout squares

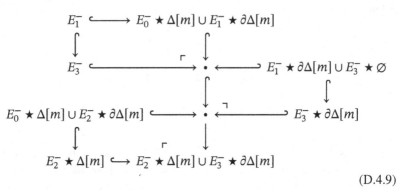

$$(D.4.9)$$

The first attached cell is the map $E_1^- \hookrightarrow E_3^-$ itself. As observed in D.4.7, $E_1^- \hookrightarrow E_3^-$ is a right anodyne extension, so by Lemma D.2.18 the second attached cell $(E_1^- \hookrightarrow E_3^-) \widehat{\star} (\emptyset \hookrightarrow \partial\Delta[m])$ is an inner anodyne extension. Similarly $E_0^- \hookrightarrow E_2^-$ is a right anodyne extension, so the final attached cell $(E_0^- \hookrightarrow E_2^-) \widehat{\star} (\partial\Delta[m] \hookrightarrow \Delta[m])$ is also an inner anodyne extension. Thus, p admits lifts against left complicial horn extensions as claimed.

To see that $p \colon A \twoheadrightarrow B$ also admits outer complicial thinness extensions we make use of the isomorphism

$$(\Lambda^0[1] \hookrightarrow \Delta^0[1]) \widehat{\star} (\Delta[m] \hookrightarrow \Delta[m]_t) \cong \Delta^0[m+2]' \hookrightarrow \Delta^0[m+2]''$$

of Lemma D.2.18, recalling that $\Lambda^0[1] \hookrightarrow \Delta^0[1]$ is the inclusion $E_0^- \hookrightarrow E_1^-$. So we may consider a lifting problem as presented by the maps a and b:

$$
\begin{array}{ccc}
E_0^- \star \Delta[m]_t \cup E_1^- \star \Delta[m] & \xrightarrow{\quad a \quad} & A \\
\downarrow {\scriptstyle i} & & \downarrow {\scriptstyle p} \\
E_1^- \star \Delta[m]_t \hookrightarrow E_2^- \star \Delta[m] \cup E_3^- \star \Delta[m] \hookrightarrow E_3^- \star \Delta[m]_t & \xrightarrow{\quad b \quad} & B
\end{array}
$$

By Lemma D.2.18, the map j is a marked anodyne extension, so since B is a complicial set, the dashed extension exists. To show that the dotted lift exists as well, we argue that the map i is cellularly generated by the inner complicial horn extensions, the inner complicial thinness extensions, and the map $E_1^- \hookrightarrow E_3^-$. To see this, factor the map i as the composite of the three vertical maps in the middle column of the diagram of pushout squares

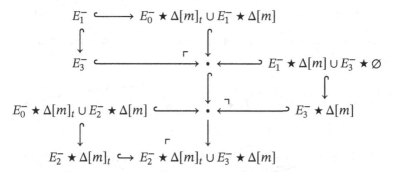

The first attached cell is the map $E_1^- \hookrightarrow E_3^-$ itself. As observed in D.4.7, $E_1^- \hookrightarrow E_3^-$ is a right marked anodyne extension, so by Lemma D.2.18 the second attached cell $(E_1^- \hookrightarrow E_3^-) \,\hat{\star}\, (\varnothing \hookrightarrow \Delta[m])$ is an inner marked anodyne extension. Similarly $E_0^- \hookrightarrow E_2^-$ is a right anodyne extension, so the final attached cell $(E_0^- \hookrightarrow E_2^-) \,\hat{\star}\, (\Delta[m] \hookrightarrow \Delta[m]_t)$ is also an inner complicial thinness extension. Thus, p admits lifts against left complicial thinness extensions as claimed. □

Since the inclusions $E_0^- \hookrightarrow E_1^-$ and $E_1^- \hookrightarrow E_3^-$ are left marked anodyne extensions, Proposition D.4.8 has the following immediate corollary.

COROLLARY D.4.10. *Let* $p \colon A \twoheadrightarrow B$ *be an inner complicial fibration whose codomain B is a complicial set. Then if p is a left complicial fibration or a right complicial fibration, then p is a complicial isofibration.* □

Note that in the proof of Proposition D.4.8, lifts against the inclusion $E_0^- \hookrightarrow E_1^-$ are only needed to construct lifts for the outer complicial horn extensions $\Lambda^0[1] \hookrightarrow \Delta^0[1]$ and $\Lambda^1[1] \hookrightarrow \Delta^1[1]$. Thus, even if this lifting condition is dropped, the outer complicial horn extensions in higher dimensions can still be constructed. The argument just given supplies a proof of a special case of "special outer horn filling" that is useful in proving the general version.

LEMMA D.4.11. *Let* $p \colon A \twoheadrightarrow B$ *be an inner complicial fibration whose codomain B is a complicial set. Then p admits fillers for left horns*

$$
\begin{array}{ccc}
\Lambda^0[n] & \xrightarrow{\ a\ } & A \\
\big\downarrow & \nearrow & \big\downarrow{\scriptstyle p} \\
\Delta^0[n] & \xrightarrow[\ b\ ]{} & B
\end{array}
$$

with $n > 1$ *provided a carries the* $\{01\}$ *edge of the horn* $\Lambda^0[n]$ *to a degenerate simplex in A.*

Proof   Writing $m = n - 2$, by Lemma D.2.18, we have an isomorphism

$$\Lambda^0[m+2] \hookrightarrow \Delta^0[m+2] \cong (\Lambda^0[1] \hookrightarrow \Delta^0[1]) \,\widehat{\star}\, (\partial\Delta[m] \hookrightarrow \Delta[m])$$

so once more we are asked to consider a lifting problem as presented by the maps a and b:

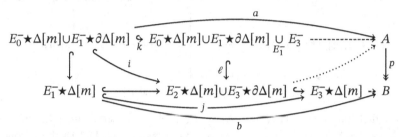

The map j is a marked anodyne extension, so since B is a complicial set, the lower dashed extension exists, and we are left to solve a lifting problem between the map i and the map p. To do so, we factor the map i as a composite of the three middle vertical morphisms displayed in (D.4.9) and let k denote the first of these morphisms while ℓ denotes the composite of the second two. We next solve the lifting problem between k and p by defining the image of the attached E_3^- to be a degenerate 3-simplex; note that k is constructed by attaching this E_3^- to the E_1^- that corresponds to the initial edge of the horn $\Lambda^0[m+2]$, which a maps to a degenerate edge.

Now to construct the dotted lift, it remains only to solve the lifting problem between ℓ and p, and the diagram (D.4.9) reveals that this can be done, as it expresses the map ℓ as the composite of pushouts of inner complicial horn extensions.   □

COROLLARY D.4.12.   *A marked simplicial set A that admits fillers for inner complicial horn extensions and inner complicial thinness extensions is a complicial set if and only if it admits extensions along $\Delta[1]^\sharp \hookrightarrow \mathrm{sk}_2\, \mathbb{I}^\sharp$.*

Proof   Applying Proposition D.4.8 to the inner complicial fibration $A \to 1$, to conclude that A is a complicial set, we need only construct extensions along the maps $E_0^- \hookrightarrow E_1^-$ and $E_1^- \hookrightarrow E_3^-$. The former is automatic, since $E_0^- \hookrightarrow E_1^-$ is isomorphic to $\Delta[0] \hookrightarrow \Delta[1]^\sharp$, which admits a retraction, so to complete our proof we will show that if A admits extensions along the map $E_1^- \hookrightarrow E_3^-$ under the stated hypotheses.

The domain map $\Delta[1]^\sharp \cong E_1^- \to A$ defines a marked 1-simplex $f : x \to y$ in A, which by hypothesis we may extend to a map $\mathrm{sk}_2\, \mathbb{I}^\sharp \to A$, which specifies marked 2-simplices

$$f \nearrow^{y} \searrow^{f^{-1}} \quad f^{-1} \nearrow^{x} \searrow^{f}$$
$$x \xrightarrow{\quad\alpha\quad} x \quad\quad y \xrightarrow{\quad\alpha'\quad} y$$

The 2-simplices α and α' can be understood as defining 2-dimensional equivalences $\alpha\colon \mathrm{id}_x \simeq f^{-1}f$ and $\alpha'\colon \mathrm{id}_y \simeq ff^{-1}$. It is not necessarily be the case that the pair of 2-simplices α and α' form the two nondegenerate 2-simplex faces of a map $E_3^- \to A$, which would amount to the additional requirement that the triangle identity composite $(\alpha')^{-1}f \cdot f\alpha$ is equivalent to id_f,[10] but we will construct a replacement β of α' so that α and β form the nondegenerate 2-simplices of $E_3^- \to A$.[11]

The 3-simplex E_3^- will be constructed as the 2nd face of the filler to a horn $\Lambda^2[4] \to A$ that we now build. As orientation for the construction given below, we first summarize the end result:

- the 0th, 2nd, and 3rd vertices will be x, while the 1st and 4th vertex will be y;
- all of the edges will be either id_x, id_y, f, or f^{-1}, with the positioning of these determined uniquely by the vertices;
- the faces $\{0, 1, 2\}$ and $\{0, 1, 3\}$ are α, while the face $\{1, 2, 4\}$ is α';
- the faces $\{0, 2, 3\}$, $\{0, 1, 4\}$, $\{0, 3, 4\}$, and $\{1, 2, 3\}$ are degenerate; and
- the missing faces $\{0, 2, 4\}$ called γ, $\{2, 3, 4\}$ called $\bar\gamma$, and $\{1, 3, 4\}$ called β will be filled in in this order, with the desired 3-simplex E_3^- appearing as the 2nd face of the horn.

The 4th face is the degenerated 3-simplex $\alpha\sigma^2$. The 3rd face is constructed by filling the horn $\Lambda^1[3] \to A$ depicted below:

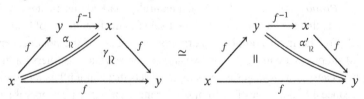

By a complicial thinness extension, the face γ defined by filling this horn is marked. Next, the 1st face is constructed by filling a horn $\Lambda^0[3] \to A$, as

---

[10] This is why the marked simplicial sets E_3^- and E_3^+ might be referred to as "half adjoint equivalences."

[11] This should be compared with the proof of Proposition 2.1.12 via Lemma 2.1.11.

permitted by Lemma D.4.11.

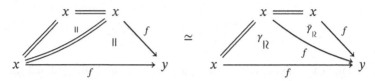

This produces another marked 3-simplex $\bar{\gamma}$ defined by filling this horn. The 0th face is constructed by filling the horn $\Lambda^1[3] \to A$ depicted below:

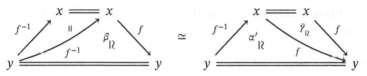

The face defined by filling this horn is the replacement 2-simplex β. These four 3-simplices define a map $\Lambda^2[4] \to A$ whose filler defines a 3-simplex face as depicted below:

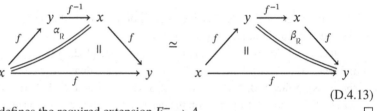

$$(D.4.13)$$

This defines the required extension $E_3^- \to A$.   □

With Corollary D.4.12 in hand, we can now prove Joyal's special outer horn filling result.

Proof of Proposition D.4.6   Let A be a naturally marked quasi-category. We demonstrate that the unique map $! : A \to 1$ satisfies the hypotheses of Corollary D.4.12. For the inner complicial horn extensions, all of the nondegenerate marked simplices are in dimension two and higher, so A admits fillers for these, simply because quasi-categories are simplicial sets that admit fillers for all inner horns. Since the composite of a pair of isomorphisms in a quasi-category is again an isomorphism, A admits extensions along $\Delta^1[2]' \hookrightarrow_e \Delta^1[2]''$. The remaining complicial thinness extensions of (D.1.11) are entire inclusions that differ only in markings of simplices in dimension at least two; since all such simplices are thin in the natural marking, A admits these extensions as well. Thus, $A \to 1$ is an inner complicial fibration

To conclude, we need only argue that A admits extensions of the form

$$\Delta[1]^\sharp \xrightarrow{\ f\ } A$$

$$\downarrow \qquad \nearrow$$

$$\mathrm{sk}_2 \, \mathbb{I}^\sharp$$

Since A is naturally marked, the attaching map $f : x \to y$ defines an isomorphism in A. By Definition 1.1.13, this means there exist 2-simplices

$$\begin{array}{c} \overset{y}{\underset{f \nearrow \ \alpha \ \searrow f^{-1}}{}} \\ x =\!=\!=\!= x \end{array} \qquad \begin{array}{c} \overset{x}{\underset{f^{-1} \nearrow \ \alpha' \ \searrow f}{}} \\ y =\!=\!=\!= y \end{array}$$

which provide the data of the required extension $\mathrm{sk}_2 \, \mathbb{I}^\sharp \to A$. $\qquad\square$

This proves the hard direction of the following characterization of quasi-categories as complicial sets:

THEOREM D.4.14. *The natural marking of a quasi-category is a complicial set and indeed is the maximal marking that turns a quasi-category into a complicial set. Conversely, the underlying simplicial set of any complicial set with all simplices above dimension one marked is a quasi-category.*

Proof Proposition D.4.6 demonstrates that a naturally marked quasi-category admits fillers for outer complicial horn extensions and its proof demonstrates that it admits fillers for inner complicial horn extensions as well. As argued there, since naturally marked quasi-categories are 1-trivial, the complicial thinness extensions in dimension greater than 1 are automatic, while the three complicial thinness extensions in dimension 1 ask that the isomorphisms in a quasi-category satisfy the 2-of-3 property, which is true. Thus, naturally marked quasi-categories are complicial sets. By Lemma D.4.2, it is not possible to mark any additional edges in A and retain the property of being a complicial set, so the natural marking is the maximal one.

The converse is elementary, and left to the reader in Exercise D.4.i. $\qquad\square$

We now give a few sample applications of Theorem D.4.14, revisiting some results that were proven in §1.1 using Proposition D.4.6. For instance, we return to Corollary 1.1.15:

COROLLARY D.4.15. *A quasi-category A is a Kan complex if and only if its homotopy category is a groupoid.*

Proof It is clear that the homotopy category of a Kan complex is a groupoid, so we focus our attention on the converse. By Theorem D.4.14, a quasi-category may be regarded as a complicial set, with every simplex above dimension 1 marked, and where the marked edges are exactly the isomorphisms defined to be those 1-simplices that represent isomorphisms in the homotopy category. If the homotopy category of A is a groupoid, then this tells us that A is maximally marked, and Exercise D.1.ii observes that a maximally marked complicial set defines a Kan complex. □

For our next result, we revisit Corollary 1.1.16 and eliminate the reference to the maximal Kan complex spanned by the isomorphisms in a quasi-category – the existence of which follows from special outer horn lifting – from the proof given there.

COROLLARY D.4.16. *An arrow f in a quasi-category A is an isomorphism if and only if it extends to a homotopy coherent isomorphism*

Proof When the quasi-category A is regarded as a naturally marked complicial set, the isomorphism f defines a marked map $f : 2^\sharp \to A$. By Lemma D.1.13, the injection $2^\sharp \hookrightarrow \mathbb{I}^\sharp$ is a marked anodyne extension, and thus A lifts against this map. Forgetting markings, this proves that every isomorphism in A extends to a homotopy coherent isomorphism. The converse is obvious from Definition 1.1.13. □

Next, we prove the two statements appearing in Corollary 1.1.22.

LEMMA D.4.17. *If A is a naturally marked quasi-category and X is a minimally marked simplicial set, then A^X is a naturally marked quasi-category.*

Proof By Proposition D.3.3, n-simplices in A^X correspond to maps $X \times \Delta[n] \to A$. Since the domain is minimally marked, it follows that the underlying simplicial set of A^X coincides with the exponential of the underlying simplicial sets and hence, by Corollary D.3.12(i) defines a quasi-category.

To see that every simplex of dimension greater than one is marked in A^X consider an extension problem

$$
\begin{array}{ccc}
X \times \Delta[n] & \longrightarrow & A \\
{\scriptstyle e}\downarrow & \nearrow & \\
X \times \Delta[n]_t & &
\end{array}
$$

for $n > 1$. By Proposition D.3.3, the only simplices that are marked in $X \times \Delta[n]_t$ but not in $X \times \Delta[n]$ are n-simplices, and since all n-simplices in A are marked, it is clear that the desired extension exists.

By Corollary D.3.12(ii), A^X is a complicial set, so by Lemma D.4.2 every marked edge in A^X is an isomorphism. It remains only to show that every isomorphism $f : \Delta[1] \to A^X$ in the quasi-category A^X is marked, admitting an extension as indicated below-right:

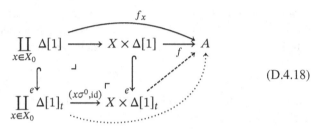

$$(D.4.18)$$

By Proposition D.3.3, the only simplices that are marked in $X \times \Delta[1]_t$ but not in $X \times \Delta[1]$ are 1-simplices whose component in X is degenerate and whose component in $\Delta[1]_t$ is the nondegenerate 1-simplex, as indicated by the square above-left that is both a pullback and a pushout. The images of such simplices in A define the "components" of f, as indicated by the top composite above.

If f is an isomorphism in A^X then each of its components f_x, the image of f under the evaluation function $\mathrm{ev}_x : A^X \to A$, are clearly also isomorphisms, which is the case if and only if each f_x is marked in A. These components f_x are marked in A if and only if the dotted lift exists, and by the universal property of the pushout, this is equivalent to the existence of the dashed lift, as required. $\square$

As a consequence of Lemma D.4.17, we can prove an oft-cited result:

COROLLARY D.4.19. *For any quasi-category A and simplicial set X, an edge in A^X is an isomorphism if and only if each of its components in A, indexed by the vertices of X, are isomorphisms.*

Proof Regarding A as a naturally marked quasi-category and X as a minimally marked simplicial set, by Lemma D.4.17 a 1-simplex in A^X is marked if and only if it defines an isomorphism in the quasi-category A^X. So the statement asserts that a 1-simplex is marked in A^X if and only if its components are marked in A. By the definition, given in Proposition D.3.3, of the markings in the exponential, a 1-simplex $f : \Delta[1] \to A^X$ is marked if and only if the dashed extension of (D.4.18) exists. By the pushout given there, this is equivalent to the existence of the dotted extensions, which say exactly that each component f_x is marked in A. $\square$

We close by observing that marked homotopy equivalences induce equivalences of naturally marked quasi-categories.

LEMMA D.4.20. *Let A be a quasi-category and let $I \to J$ be a map of simplicial sets that extends to a marked homotopy equivalence between minimally marked simplicial sets. Then the induced map $A^J \twoheadrightarrow A^I$ is an equivalence of quasi-categories.*

Proof Equip A with its natural marking so that by Theorem D.4.14 it defines a complicial set. By Exercise D.3.iii, $A^J \twoheadrightarrow A^I$ is then an equivalence of complicial sets. By Corollary D.3.13 the data of this equivalence may be given by the maps

$$A^J \to A^I, \quad A^I \to A^J, \quad A^I \to (A^I)^{\natural}, \quad A^J \to (A^J)^{\natural},$$

and upon forgetting the markings, this data defines an equivalence of quasi-categories in the sense of Definition 1.1.23.[12]   □

DIGRESSION D.4.21 (on equivalences and saturation for higher simplices). Lemma D.4.2 demonstrates that the marked edges in a complicial set should be interpreted as equivalences, in a suitable sense. A similar interpretation is appropriate for the higher dimensional marked simplices as well. Consequently, we may interpret the condition that a complicial set is n-trivial as demanding that all simplices in dimension $r > n$ are weakly invertible.

Having understood that every marked simplex in a complicial set is an equivalence, we are lead to consider complicial sets that satisfy the converse of this condition, in which every equivalence is marked. Such *saturated* complicial sets are especially important, and we suggest the terminology n-**complicial set** to describe an n-trivial saturated complicial set (see [105] and [89] for the precise definition and some discussion). The Kan complexes are precisely the 0-*complicial sets* by Exercise D.1.ii, while the quasi-categories are precisely the 1-*complicial sets* by Theorem D.4.14. As this pattern suggests, the n-*complicial sets* define a well-behaved model for (∞, n)-categories in the sense that the full subcategory of such defines a cartesian closed ∞-cosmos, as we prove Proposition E.3.9.

Exercises

EXERCISE D.4.i. Prove that the underlying simplicial set of any 1-trivial complicial set is a quasi-category.[13]

---

[12] Note that $A^{\natural} \cong A^{\natural\natural}$ when A is a naturally marked quasi-category.
[13] A converse of sorts to this result appears in Theorem D.4.14.

Exercise D.4.ii. Prove Lemma D.4.3.

Exercise D.4.iii. Let $\Delta[3]_{eq}$ denote the "2-of-6 3-simplex," in which the edges $\{02\}$ and $\{13\}$ and all simplices in dimension greater than 1 are marked. Show that:

(i) A 1-simplex in a complicial set A is an equivalence if and only if it defines the $\{12\}$-edge of a 3-simplex $\Delta[3]_{eq} \to A$.
(ii) A complicial set is 1-saturated if and only if it admits extensions along the entire inclusion

D.5 Isofibrations between Quasi-Categories

Our aim in this section is to explain the relevance of Proposition D.3.8 to the theory of quasi-categories. In particular, this finally enables us to complete the combinatorial work required to supply proofs of the Leibniz stability results stated in §1.1. The missing ingredient is a relative version of Theorem D.4.14, which admits a similar proof:

Theorem D.5.1. *A map between quasi-categories defines an isofibration if and only if it defines a complicial isofibration when those quasi-categories are given their natural markings. In particular, an isofibration between naturally marked quasi-categories admits fillers for outer complicial horn extensions for $n \geq 1$:*

$$
\begin{array}{ccc}
\Lambda^0[n] & \longrightarrow & A \\
{\scriptstyle r}\downarrow & \nearrow & \downarrow \\
\Delta^0[n] & \longrightarrow & B
\end{array}
\qquad
\begin{array}{ccc}
\Lambda^n[n] & \longrightarrow & A \\
{\scriptstyle r}\downarrow & \nearrow & \downarrow \\
\Delta^n[n] & \longrightarrow & B
\end{array}
$$

Proof We leave it to the reader to verify that, upon forgetting the markings, a complicial isofibration between naturally marked quasi-categories defines an isofibration between quasi-categories. The content is in the converse, so consider an isofibration $p : A \twoheadrightarrow B$ between quasi-categories which have been given their natural markings. By Theorem D.4.14, the quasi-categories A and B are complicial sets, so by Proposition D.4.8, to prove that $p : A \twoheadrightarrow B$ is a complicial isofibration it suffices to show that it has the right lifting property against the inner complicial horn extensions, inner complicial thinness extensions, and the two maps $E_0^- \hookrightarrow E_1^-$ and $E_1^- \hookrightarrow E_3^-$.

The inner complicial horn and thinness extensions are straightforward, as in the proof of Proposition D.4.6. To construct a lift

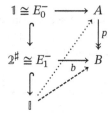

we use Lemma D.1.13, recalled as Corollary D.4.16, to extend the codomain to a homotopy coherent isomorphism and then solve the composite lifting problem.

The construction of the lift against $E_1^- \hookrightarrow E_3^-$ is considerably more laborious. To begin, we argue that since A and B are complicial sets and p is an inner complicial fibration, then p admits lifts against the complicial horn extension $\Lambda^0[2] \to \Delta^0[2]$. To see this, we identify the codomain $\Delta^0[2]$ with the second face of the 3-simplex $E_2^- \star \Delta[0]$ and consider the lifting problem presented by the exterior diagram

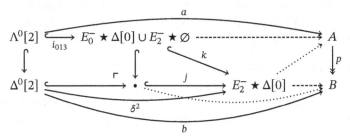

The inclusion i_{013} is marked anodyne, so the top dashed extension exists since A is a complicial set. This induces the dotted map by the universal property of the pushout. Since the map j is also marked anodyne, the bottom dashed extension exists since B is a complicial set. These dashed maps define a new lifting problem between the composite map k and p and since k is an inner marked anodyne extension, the dotted lift exists, solving the original lifting problem.

Now we can use the fact that p admits lifts along $\Lambda^0[2] \hookrightarrow \Delta^0[2]$ to construct lifts along the horizontal composite map

$$
\begin{array}{ccc}
E_1^+ & \lhook\joinrel\longrightarrow & E_2^+ \\
\big\uparrow & & \big\uparrow \\
E_1^- \lhook\joinrel\longrightarrow E_2^- & \lhook\joinrel\longrightarrow & E_2^- \underset{E_1^+}{\cup} E_2^+
\end{array}
$$

since the observation made in Notation D.4.7 reveals that both maps $E_1^\pm \hookrightarrow E_2^\pm$ are composites of pushouts of $\Lambda^0[2] \hookrightarrow \Delta^0[2]$ and $\Delta^0[2]' \hookrightarrow \Delta^0[2]''$; since A is a complicial set, the complicial thinness extension comes for free. Now lifts against the composite $E_1^- \hookrightarrow E_2^- \underset{E_1^+}{\cup} E_2^+$ have the effect of giving the data of a right inverse and also a right inverse to the right inverse to an isomorphism in A, lifting the corresponding data in B. To solve a lifting problem against $E_1^- \hookrightarrow E_3^-$

$$
\begin{array}{ccccc}
& & E_1^- & \xrightarrow{\ f\ } & A \\
& \nearrow & \big\downarrow{\scriptstyle (\alpha,\alpha')} & & \big\downarrow{\scriptstyle p} \\
E_2^- \underset{E_1^+}{\cup} E_2^+ & \longrightarrow & E_3^- & \xrightarrow{\ \tau\ } & B
\end{array}
$$

we first construct the outer lift, defining a pair of 2-simplices in A:

$$
\begin{array}{cc}
\begin{array}{c}
\overset{y}{x \nearrow\ \searrow x} \\
f \quad \alpha \quad f^{-1} \\
x =\!=\!=\!= x
\end{array}
&
\begin{array}{c}
\overset{x}{y \nearrow\ \searrow y} \\
f^{-1} \quad \alpha' \quad (f^{-1})^{-1} \\
y =\!=\!=\!= y
\end{array}
\end{array}
$$

We now extend the data of the lifted $E_2^- \underset{E_1^+}{\cup} E_2^+ \to A$ to construct a 3-simplex $E_3^- \to A$ lifting $\tau \colon E_3^- \to B$ using a mild modification of the construction in the proof of Corollary D.4.12. We define a complicial inner horn extension

$$
\begin{array}{ccc}
\Lambda^2[4] & \longrightarrow & A \\
\big\downarrow & \nearrow & \big\downarrow{\scriptstyle p} \\
\Delta^2[4] & \xrightarrow[\tau\sigma^2]{} & B
\end{array}
$$

so that the lift of the 2nd face defines the simplex $E_3^- \to A$ that we seek. It remains only to define an appropriate horn $\Lambda^2[4] \to A$ over $\tau\sigma^2 \colon \Delta^2[4] \to B$. The 4th face is $\alpha\sigma^2$. The 3rd face is constructed by lifting along a $\Lambda^1[3]$-horn whose 3rd face is α, whose 0th face is α', and whose 2nd face is degenerate. Writing γ for the face defined by filling this horn, the 1st face is constructed by filling a $\Lambda^0[3]$-horn whose 2nd face is γ and 1st and 3rd faces are degenerate; this lift is permitted by Lemma D.4.11. The 0th face is constructed by filling the $\Lambda^1[3]$-horn whose faces have all already been described. The face defined by filling this horn is the replacement 2-simplex β – a replacement for α' – which witnesses that f is a right inverse to its right inverse f^{-1}. These four 3-simplices define a map $\Lambda^2[4] \to A$ over $\tau\sigma^2$ whose filler defines a 3-simplex face of the form displayed in (D.4.13). This defines the required lift along $E_1^- \hookrightarrow E_3^-$. Now Proposition D.4.8 completes the proof that isofibrations between quasi-categories are complicial isofibrations. $\square$

With this result in hand, we may now integrate the class of isofibrations between quasi-categories into the results of §D.3, proving Propositions 1.1.20 and 1.1.29, restated here for convenience.

PROPOSITION D.5.2.

(i) *There is a solution to any lifting problem between the Leibniz product of a monomorphism* $i : X \hookrightarrow Y$ *and the map* $\mathbb{1} \hookrightarrow \mathbb{I}$ *and any isofibration* $f : A \twoheadrightarrow B$.

(ii) *If* $i : X \hookrightarrow Y$ *is a monomorphism and* $f : A \twoheadrightarrow B$ *is an isofibration, then the induced Leibniz exponential map* $i \widehat{\pitchfork} f : A^Y \twoheadrightarrow B^Y \times_{B^X} A^X$ *is again an isofibration.*

(iii) *If* $i : X \hookrightarrow Y$ *is a monomorphism and* $f : A \twoheadrightarrow B$ *is a trivial fibration, then the induced Leibniz exponential map* $i \widehat{\pitchfork} f : A^Y \twoheadrightarrow B^Y \times_{B^X} A^X$ *is again a trivial fibration.*

(iv) *If* $i : X \hookrightarrow Y$ *is in the class cellularly generated by the inner horn inclusions and the map* $\mathbb{1} \hookrightarrow \mathbb{I}$ *and* $f : A \twoheadrightarrow B$ *is an isofibration, then the induced Leibniz exponential map* $i \widehat{\pitchfork} f : A^Y \twoheadrightarrow B^Y \times_{B^X} A^X$ *is a trivial fibration.*

Proof It suffices to construct the lift of (i) in marked simplicial sets and then forget the markings. By Lemma D.1.13, $\mathbb{1} \hookrightarrow \mathbb{I}$ is a marked anodyne extension, when $\mathbb{I}$ is assigned its natural maximal marking. Thus by Proposition D.3.8, the Leibniz product of the minimally marked monomorphism i with this map is again a marked anodyne extension. By Theorem D.5.1, an isofibration defines a complicial isofibration between naturally marked quasi-categories, so the postulated lift exists.

Parts (ii) and (iv) follow from the conclusion of (i) and a similar result, Corollary D.3.11, by transposing lifting problems across the two-variable adjunction between the Leibniz product and the Leibniz exponential.

Part (iii) follows by a similar argument from an easier observation made in Lemma D.3.1: that the Leibniz product of two monomorphisms is again a monomorphism. $\square$

Theorems D.4.14 and D.5.1 permit the use of complicial techniques to solve lifting problems involving isofibrations between quasi-categories. The results of this section suggest that these techniques are particularly fruitful when isomorphisms are involved. We now develop a few specific applications of this principle, which are used to prove the final missing result from §1.1.

To that end we consider a pair of cosimplicial marked simplicial sets

$$\Delta[\bullet]^\sharp \hookrightarrow \mathbb{I}[\bullet]^\sharp \quad \in (s\mathcal{S}et^+)^\Delta$$

the former of which is given by the maximally marked simplices $\Delta[n]^\sharp$ and the latter of which is given by the maximally marked contractible groupoids $\mathbb{I}[n]^\sharp$ on objects $0, 1, \dots, n$.

LEMMA D.5.3. *The natural inclusion $\Delta[\bullet]^\sharp \hookrightarrow \mathbb{I}[\bullet]^\sharp$ is a Reedy monomorphism in $(s\mathcal{S}et^+)^\Delta$ between Reedy monomorphic cosimplicial objects that is moreover a pointwise weak equivalence in the Verity model structure.*

Proof When $s\mathcal{S}et^+$ is identified with its image in $\mathcal{S}et^{t\Delta^{op}}$, the first half of the statement follows the relative version of Lemma C.5.19 stated as Exercise C.5.ii: any pointwise monomorphism between "unaugmentable" cosimplicial objects is a Reedy monomorphism between Reedy monomorphic cosimplicial objects. As the equalizers of the face maps $\delta^0, \delta^1 : \Delta[0]^\sharp \to \Delta[1]^\sharp$ and $\delta^0, \delta^1 : \mathbb{I}[0]^\sharp \to \mathbb{I}[1]^\sharp$ are empty, both $\Delta[\bullet]^\sharp$ and $\mathbb{I}[\bullet]^\sharp$ are unaugmentable, so we conclude that these simplicial objects are Reedy monomorphic and the natural inclusion is a Reedy monomorphism.

Finally to prove that $\Delta[n]^\sharp \to \mathbb{I}[n]^\sharp$ is a pointwise weak equivalence, we appeal to the 2-of-3 property and argue that both $\Delta[n]^\sharp$ and $\mathbb{I}[n]^\sharp$ are marked homotopy equivalent to $\mathbb{1} \in s\mathcal{S}et^+$. The inverse equivalences are given by $0 : \mathbb{1} \to \Delta[n]^\sharp$ and $0 : \mathbb{1} \to \mathbb{I}[n]^\sharp$ and the marked homotopies $\Delta[n]^\sharp \times \Delta[1]^\sharp \to \Delta[n]^\sharp$ and $\mathbb{I}[n]^\sharp \times \Delta[1]^\sharp \to \mathbb{I}[n]^\sharp$ are both defined by the map on objects $(i, 0) \mapsto 0$ and $(i, 1) \mapsto i$. $\qquad\qquad\square$

Our intent is to use the simplicial objects $\Delta[\bullet]^\sharp$ and $\mathbb{I}[\bullet]^\sharp$ to "freely invert" the simplices of a simplicial set K. To see how this works, consider also the cosimplicial object $\Delta[\bullet] \in s\mathcal{S}et^\Delta$ defined by the Yoneda embedding. By the coYoneda lemma, the weighted colimit $\text{colim}_K \Delta[\bullet] \cong K$ recovers the original simplicial set K. Similarly, since the maximal marking functor $(-)^\sharp : s\mathcal{S}et \to s\mathcal{S}et^+$ is a left adjoint, the weighted colimit $\text{colim}_K \Delta[\bullet]^\sharp \cong K^\sharp$ equips the simplicial set K with the maximal marking. Finally, we define $\tilde{K}^\sharp := \text{colim}_K \mathbb{I}[\bullet]^\sharp$ using the weighted colimit bifunctor. The idea of this functor is that it replaces each n-simplex of K by $\mathbb{I}[n]^\sharp$, a "homotopy coherent composite of n isomorphisms." As the notation suggests, $\tilde{K}^\sharp$ is also maximally marked.

PROPOSITION D.5.4. *For any simplicial set K, the natural map $K^\sharp \to \tilde{K}^\sharp$ is a trivial cofibration in the Verity model structure .*

Proof Any simplicial set K is Reedy monomorphic when considered as an object of $Set^{\Delta^{op}}$. Hence, by Corollary C.5.17, the weighted colimit functor

$$\mathrm{colim}_K - : \; (s Set^+)^\Delta \to s Set^+$$

is left Quillen with respect to the Reedy model structure on $(s Set^+)^\Delta$. Lemmas C.5.13 and D.5.3 prove that $\Delta[\bullet]^\sharp \hookrightarrow \mathbb{I}[\bullet]^\sharp$ is a Reedy trivial cofibration, so it follows that $K^\sharp \to \tilde{K}^\sharp$ is a trivial cofibration as claimed.   □

We refer to the functor $(\tilde{-}): \; s Set \to s Set$ defined by applying $\mathrm{colim}_- \mathbb{I}[\bullet]^\sharp$ and forgetting the markings as the "free-inversion functor."

EXAMPLE D.5.5. Proposition D.5.4 can be applied to the anodyne extension $\Lambda^1[2] \hookrightarrow \Delta[2]$ to prove that composites of homotopy coherent isomorphisms can be lifted along isofibrations of quasi-categories. By definition $\widetilde{\Delta[2]} \cong \mathbb{I}[2]$, the contractible groupoid on three vertices 0, 1, and 2, so we adopt similar notation $\widetilde{\Lambda^1[2]} =: \Lambda^1[\mathbb{I}[2]]$ for the freely inverted horn. Since $\Lambda^1[2]$ is built by gluing two 1-simplices along a common vertex and the free inversion functor preserves colimits, we see that $\Lambda^1[\mathbb{I}[2]]$ is the union of two homotopy coherent isomorphisms between 0 and 1 and between 1 and 2. Giving these simplicial sets their maximal markings, it follows from the 2-of-3 property applied to the square

$$
\begin{array}{ccc}
\Lambda^1[2]^\sharp & \hookrightarrow & \Lambda^1[\mathbb{I}[2]]^\sharp \\
\cap\downarrow & & \cap\downarrow \\
\Delta[2]^\sharp & \hookrightarrow & \mathbb{I}[2]^\sharp
\end{array}
$$

that the inclusion $\Lambda^1[\mathbb{I}[2]]^\sharp \hookrightarrow \mathbb{I}[2]^\sharp$ is a trivial cofibration of marked simplicial sets. Applying Theorem D.5.1, we conclude that composites of homotopy coherent isomorphisms can be lifted along isofibrations between quasi-categories

$$
\begin{array}{ccc}
\Lambda^1[\mathbb{I}[2]] & \longrightarrow & A \\
\cap\downarrow & \nearrow & \downarrow \\
\mathbb{I}[2] & \longrightarrow & B
\end{array}
$$

A similar example is left as Exercise D.5.ii. To complete the verification of the results claimed in §1.1, it remains only to prove Proposition 1.1.28:

PROPOSITION D.5.6. *For an isofibration $f : A \twoheadrightarrow B$ of quasi-categories the following are equivalent:*

(i) f is at trivial fibration
(ii) f is both an isofibration and an equivalence
(iii) f is a **split fiber homotopy equivalence**: *an isofibration admitting a section s that is also an equivalence inverse via a homotopy from id_A to sf that composes with f to the constant homotopy from f to f.*

Proof For (i)$\Rightarrow$(ii), observe that the simplex boundary inclusions generate the monomorphisms of simplicial sets under coproduct, pushout, and sequential composition (see Lemma C.5.9), so the lifting property of (1.1.26) implies that the trivial fibrations lift against all monomorphisms of simplicial sets, and in particular against the monomorphisms that detect the class of isofibrations. Thus, trivial fibrations are isofibrations. By the same lifting property, every trivial fibration admits a section

To show that s defines an inverse equivalence to f, observe that the outer rectangle built from the constant homotopy $\pi : A \times \mathbb{I} \to A$

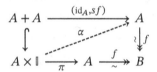

commutes since $fsf = f$. The lift defines a homotopy between id_A and sf completing the proof that trivial fibrations are equivalences. And note in fact that the equivalence just constructed is a split fiber homotopy equivalence, proving that (i)$\Rightarrow$(iii).

To prove (ii)$\Rightarrow$(iii), suppose that f is an isofibration with equivalence inverse g. By Lemma D.5.7 below, the homotopies α from id_A to gf and β from fg to id_B may be chosen so as to define a "half adjoint equivalence," meaning that there exists a map $\Phi : A \times \mathbb{I}[2] \to B$, where $\mathbb{I}[2]$ is the contractible groupoid on three objects, whose boundary is formed by $f\alpha$, βf, and the constant homotopy $\mathrm{id}_f := f\pi$.[14]

Applying Proposition D.5.2(i) to the monomorphism $\varnothing \hookrightarrow B$, we find that

---

[14] This is similar but not isomorphic to the notion of half adjoint equivalence encoded by the marked simplicial sets E_3^- and E_3^+.

we can lift the homotopy β between fg and id_B along f

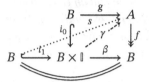

The composite map s defines a strict section of f, while the lift defines a homotopy γ from g to s. Applying Proposition D.3.8, Theorem D.5.1, and Example D.5.5, we can solve the lifting problem

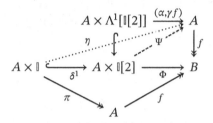

The lift defines a composite homotopy η from id_A to sf so that $f\eta = f\pi$ is the constant homotopy. This data exhibits f as a split fiber homotopy equivalence.

Finally, for (iii)$\Rightarrow$(i) note that the data of a split fiber homotopy equivalence defines a retract diagram

$$
\begin{array}{ccccc}
A & \xrightarrow{\ \alpha\ } & A^{\mathbb{I}} & \xrightarrow{\ \mathrm{ev}_0\ } & A \\
{\scriptstyle f}\downarrow & & {\scriptstyle\langle f^{\mathbb{I}},\mathrm{ev}_1\rangle}\downarrow & & \downarrow{\scriptstyle f} \\
B & \xrightarrow[\langle\Delta,s\rangle]{} & B^{\mathbb{I}}\underset{B}{\times}A & \xrightarrow[\mathrm{ev}_0\,\pi]{} & B
\end{array}
$$

The central map is the Leibniz cotensor of $\{1\}\hookrightarrow\mathbb{I}$ with the isofibration f and so is a trivial fibration by Proposition D.5.2. Since the trivial fibrations are characterized by a right lifting property, Lemma C.2.3 tells us that they are closed under retracts. Thus f is a trivial fibration as desired. □

LEMMA D.5.7. *Any equivalence of quasi-categories*

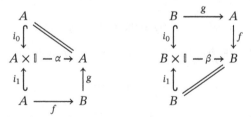

can be extended to a **half adjoint equivalence of quasi-categories**, with an additional coherence homotopy $\Phi : A \times \mathbb{I}[2] \to B$ whose boundary is comprised of the three homotopy coherent isomorphisms:

$$
\begin{array}{ccc}
 & fgf & \\
f\alpha \nearrow & \Phi & \searrow \beta f \\
f \xrightarrow{\quad \mathrm{id}_f \quad} & & f
\end{array}
$$

at the cost of replacing one of the homotopies α or β.

The proof is by a simplicial reinterpretation of the 2-categorical argument that proves Proposition 2.1.12.

Proof Consider an equivalence of quasi-categories as in the statement. By Example D.5.5, the homotopies $f\alpha$ and βf admit some composite defined by solving the lifting problem

$$
\begin{array}{ccc}
A \times \Lambda^1[\mathbb{I}[2]] & \xrightarrow{(f\alpha,\beta f)} & B \\
{\scriptstyle \psi} \downarrow {\scriptstyle \ulcorner} & \nearrow {\scriptstyle \Psi} & \\
A \times \mathbb{I} & \xhookrightarrow{\;\delta^1\;} & A \times \mathbb{I}[2]
\end{array}
$$

Restriction along the nonidentity involution $\mathbb{I} \to \mathbb{I}$ defines the inverse of any homotopy, denoted by $(-)^{-1}$. We will replace α by the composite $\alpha' := g\psi^{-1} \cdot \alpha$ defined by solving the lifting problem

$$
\begin{array}{ccc}
A \times \Lambda^1[\mathbb{I}[2]] & \xrightarrow{(\alpha,g\psi^{-1})} & A \\
{\scriptstyle \alpha'} \downarrow {\scriptstyle \ulcorner} & \nearrow {\scriptstyle \Xi} & \\
A \times \mathbb{I} & \xhookrightarrow{\;\delta^1\;} & A \times \mathbb{I}[2]
\end{array}
$$

and show that this homotopy defines a half adjoint equivalence with β.

The witness to the half adjoint equivalence is obtained by solving a final lifting problem

$$
\begin{array}{ccc}
A \times \Lambda^1[\mathbb{I}[3]] & \xrightarrow{\;\Gamma\;} & B \\
{\scriptstyle \Phi} \downarrow {\scriptstyle \ulcorner} & \nearrow & \\
A \times \mathbb{I}[2] & \xhookrightarrow{\;\delta^1\;} & A \times \mathbb{I}[3]
\end{array}
$$

involving an extension along $\Lambda^1[\mathbb{I}[3]] \hookrightarrow \mathbb{I}[3]$, whose codomain is the contractible groupoid with four objects $0, 1, 2, 3$ and whose domain is the union of the three faces $\mathbb{I}[2]$ that contain the vertex 1. This is permitted by Exercise D.5.ii.

It remains to define the faces of the "horn of homotopy coherent isomorphisms" Γ, which is built from the homotopy coherent isomorphisms

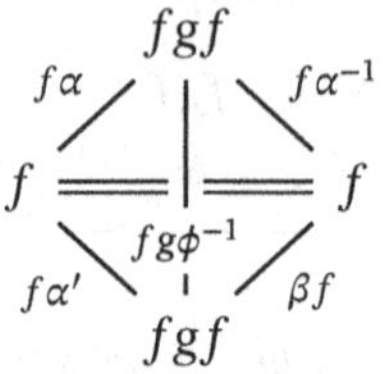

$$
\begin{array}{ccccc}
 & & fgf & & \\
 & {}^{f\alpha}\nearrow & \big| & \searrow^{f\alpha^{-1}} & \\
f & =\!=\!= & \big| & =\!=\!= & f \\
 & {}_{f\alpha'}\searrow & {\scriptstyle fg\phi^{-1}} & \nearrow_{\beta f} & \\
 & & fgf & &
\end{array}
$$

The 3rd face is $f\Xi : A \times \mathbb{1}[2] \to B$ while the second face is the composite

$$
A \times \mathbb{1}[2] \xrightarrow{A \times q} A \times \mathbb{1} \xrightarrow{f\alpha} B
$$

where $q : \mathbb{1}[2] \to \mathbb{1}$ is the unique map defined by $0, 2 \mapsto 0$ and $1 \mapsto 1$. It remains to define the 0th face. For this, we first extend along another horn $\Lambda^1[\mathbb{1}[3]] \hookrightarrow \mathbb{1}[3]$ of homotopy coherent isomorphisms in B^A as depicted below

$$
\begin{array}{ccccc}
 & & f & & \\
 & {}^{\beta f}\nearrow & \big| & \searrow^{\psi^{-1}} & \\
fgf & \xrightarrow{\ \delta\ } & \big| & \xrightarrow{\quad} & f \\
 & \searrow & {\scriptstyle \beta^{-1}f} & \nearrow_{f\alpha^{-1}} & \\
 & & fgf & &
\end{array}
$$

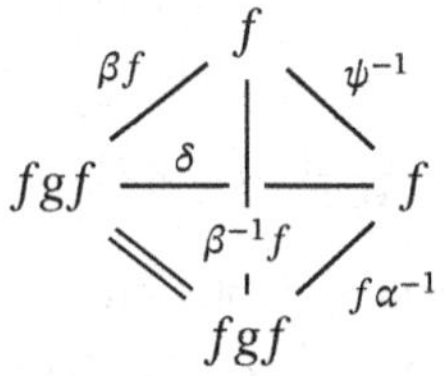

Here the 0th face is Ψ^{-1}, the 3rd face is the composite

$$
A \times \mathbb{1}[2] \xrightarrow{A \times q} A \times \mathbb{1} \xrightarrow{\beta f} B
$$

and the 2nd face is the composite

$$
A \times \mathbb{1}[2] \xrightarrow{A \times d} A \times \mathbb{1}[2] \times \mathbb{1} \xrightarrow{\Psi \times \mathbb{1}} B \times \mathbb{1} \xrightarrow{\beta} B
$$

where $d : \mathbb{1}[2] \to \mathbb{1}[2] \times \mathbb{1}$ is the unique map defined by $0 \mapsto (2, 0)$, $1 \mapsto (2, 1)$, and $2 \mapsto (0, 1)$. The δ^1-face of this horn can be used to define a final horn $\Lambda^1[\mathbb{1}[3]] \hookrightarrow \mathbb{1}[3]$ of homotopy coherent isomorphisms in B^A as depicted below

$$
\begin{array}{ccccc}
 & & fgf & & \\
 & {}^{\!}\nearrow\!\!\nearrow & \big| & \searrow^{\delta} & \\
fgf & \xrightarrow{\ f\alpha^{-1}\ } & \big| & \xrightarrow{\quad} & f \\
 & {}_{fg\psi^{-1}}\searrow & {\scriptstyle fg\psi^{-1}} & \nearrow_{\beta f} & \\
 & & fgf & &
\end{array}
$$

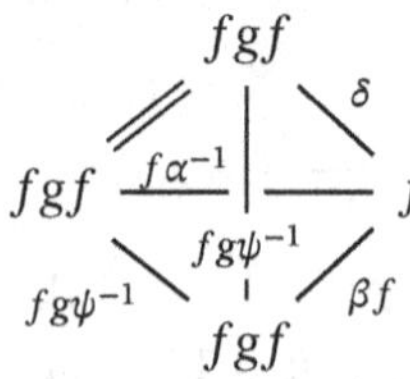

whose 3rd face is degenerate, whose 2nd face is an inversion that swaps the first two vertices of the δ^1-face $A \times \mathbb{I}[2] \to B$ just defined, and whose 0th face is the composite

$$A \times \mathbb{I}[2] \xrightarrow{\ A \times c\ } A \times \mathbb{I}[2] \times \mathbb{I} \xrightarrow{\ \Psi \times \mathbb{I}\ } B \times \mathbb{I} \xrightarrow{\ \beta\ } B$$

where $c : \mathbb{I}[2] \to \mathbb{I}[2] \times \mathbb{I}$ is the unique map defined by $0 \mapsto (2,0)$, $1 \mapsto (0,0)$, and $2 \mapsto (0,1)$. The filler defines the desired 0th face

that completes the horn $\Gamma : A \times \Lambda^1[\mathbb{I}[3]] \to B$ whose filler defines the witness for the half adjoint equivalence between α' and β. $\square$

Exercises

EXERCISE D.5.i. Verify that the underlying map defined by a complicial isofibration between naturally marked quasi-categories is an isofibration of quasi-categories.

EXERCISE D.5.ii. Extend the result of Example D.5.5 to show that the maximally marked "$\Lambda^1[3]$-horn of homotopy coherent isomorphisms" $\Lambda^1[\mathbb{I}[3]] \hookrightarrow \mathbb{I}[3]$, whose codomain is the contractible groupoid on four vertices $0, 1, 2, 3$ and whose domain is the union of the three copies of $\mathbb{I}[2]$ spanned by the subsets of three of these four vertices that include the vertex 1, is a marked anodyne extension.

D.6 Equivalence of Slices and Cones

Theorems D.4.14 and D.5.1 also enable us to finally prove the results sketched in §4.2, which now follow easily from the combinatorial work done in §D.2. Recall in particular Proposition D.2.16, which shows that the map of augmented simplicial sets

$$s^{n,m} : \Delta[n] \diamond \Delta[m] \longrightarrow \Delta[n] \star \Delta[m] \quad \in sSet_+^{\Delta_+ \times \Delta_+}$$

is a marked homotopy retract equivalence. By Lemma D.4.20, we now know that such maps induce equivalences upon mapping into quasi-categories.

In this section, we use these observations to verify that for any augmented simplicial sets X and Y the canonical map of augmented simplicial sets $s^{X,Y} : X \diamond$

$Y \to X \star Y$ also induces an equivalence on mapping into quasi-categories. For this, we use the Reedy categorical homotopy theory developed in §C.5. To begin:

LEMMA D.6.1. *The latching maps for the diagrams $F_\diamond, F_\star \in sSet_+^{\Delta_+ \times \Delta_+}$ defined by*

$$F_\diamond^{n,m} := \Delta[n] \diamond \Delta[m] \qquad and \qquad F_\star^{n,m} := \Delta[n] \star \Delta[m]$$

are the maps

$$(\partial\Delta[n] \hookrightarrow \Delta[n]) \hat{\diamond} (\partial\Delta[m] \hookrightarrow \Delta[m]) and (\partial\Delta[n] \hookrightarrow \Delta[n]) \hat{\star} (\partial\Delta[m] \hookrightarrow \Delta[m]),$$
(D.6.2)

which are both monomorphisms. Hence, both $F_\diamond$ and $F_\star$ are Reedy monomorphic.

Proof For any pair of augmented simplicial sets X and Y we define their **external product** $X \square Y \in Set^{\Delta_+^{op} \times \Delta_+^{op}}$ to be the functor that takes an object $([n], [m])$ to the set $X_n \times Y_m$. In particular, the functor represented by $([n], [m])$ is $\Delta[n] \square \Delta[m]$. We can view an external product $X \square Y$ as a weight for the diagrams $F_\diamond$ and $F_\star$ and use Definition A.6.2 and cocontinuity of the join and fat join bifunctors to compute the weighted colimits:

$$\text{colim}_{X \square Y} F_\diamond \cong \int^{([n],[m]) \in \Delta_+ \times \Delta_+} \coprod_{X_n \times Y_m} \Delta[n] \diamond \Delta[m]$$

$$\cong \left(\int^{[n] \in \Delta_+} \coprod_{X_n} \Delta[n] \right) \diamond \left(\int^{[m] \in \Delta_+} \coprod_{Y_m} \Delta[m] \right)$$

$$\cong X \diamond Y$$

Similarly, $\text{colim}_{X \square Y} F_\star \cong X \star Y$.

By Definition C.4.14, the latching map at the object $([n], [m]) \in \Delta_+ \times \Delta_+$ is the map on weighted colimits induced by the map

$$(\partial\Delta[n] \hookrightarrow \Delta[n]) \hat{\square} (\partial\Delta[m] \hookrightarrow \Delta[m])$$

of weights. By the weighted colimit calculation just given, we see that the latching maps for $F_\diamond$ and $F_\star$ are the maps (D.6.2).

It remains to verify that these latching maps are monomorphisms. By direct calculation,

$$(\partial\Delta[n] \hookrightarrow \Delta[n]) \hat{\star} (\partial\Delta[m] \hookrightarrow \Delta[m]) \cong \partial\Delta[n+1+m] \hookrightarrow \Delta[n+1+m],$$

which is clearly a monomorphism, so $F_\star$ is Reedy monomorphic.

For the analogous result for the fat join, observe first that

$$(\partial\Delta[-1] \hookrightarrow \Delta[-1]) \hat{\diamond} (X \hookrightarrow Y) \cong (X \hookrightarrow Y),$$

since $\partial\Delta[-1]$ is the initial object and $\Delta[-1]$ is the unit for the fat join. Thus, it suffices to consider Leibniz products of simplex boundary inclusions with $n, m \geq 0$, in which case the formula for the fat join simplifies since $\partial\Delta[n]$ and $\Delta[n]$ are both terminally augmented. Recall from Definition 4.2.2 that for terminally augmented simplicial sets X and Y and for $n \geq 0$, $(X \diamond Y)_n \cong X_n \sqcup (\bigsqcup_{[n] \twoheadrightarrow [1]} X_n \times Y_n) \sqcup Y_n$. Thus, we see that for any monomorphisms of terminally augmented simplicial sets $X \hookrightarrow Y$ and $U \hookrightarrow V$, the square

$$
\begin{array}{ccc}
(U \diamond X)_n & \longhookrightarrow & (V \diamond X)_n \\
\Big\uparrow & \lrcorner & \Big\uparrow \\
(U \diamond Y)_n & \longhookrightarrow & (V \diamond Y)_n
\end{array}
$$

is a pullback in the category of sets. For any pullback square comprised of monomorphisms in the category of sets, the pushout inside the square is constructed by the joint image of the lower and right-hand legs: in particular the map $(U \diamond Y)_n \cup_{(U \diamond X)_n} (V \diamond X)_n \hookrightarrow (V \diamond Y)_n$ is a monomorphism. Thus, the Leibniz fat join $(U \hookrightarrow V) \hat{\diamond} (X \hookrightarrow Y)$ of two monomorphisms of terminally augmented simplicial sets is a monomorphism, and in particular the latching maps of $F_\diamond$ are monomorphisms as claimed. □

Recall the natural comparison map of Lemma D.2.14 from the fat join of a pair of simplicial sets to the join of the pair of simplicial sets. We are now in the position to prove Proposition 4.2.7.

PROPOSITION D.6.3. *For all simplicial sets X and Y, the natural map $s^{X,Y} : X \diamond Y \to X \star Y$ induces an equivalence of quasi-categories $A^{X \star Y} \twoheadrightarrow A^{X \diamond Y}$ for all quasi-categories A.*

Lurie gives an alternate proof of this result in [78, 4.2.1.2].

Proof A direct verification shows that the latching maps of the augmented bisimplicial set $X \square Y \in Set^{\Delta_+^{op} \times \Delta_+^{op}}$ are monomorphisms.[15] Hence by Corollary C.5.17, the weighted colimit functor

$$
\mathrm{colim}_{X \square Y} - : sSet^{\Delta_+ \times \Delta_+} \longrightarrow sSet
$$

is a left Quillen bifunctor from the Reedy version of the Joyal model structure to the Joyal model structure on simplicial sets.

By Proposition D.2.16 and Lemma D.4.20, the components $s^{n,m} : \Delta[n] \diamond \Delta[m] \to \Delta[n] \star \Delta[m]$ are weak equivalences in the Joyal model structure, as

---

[15] This follows because $\Delta_+ \times \Delta_+$ is an **elegant Reedy category**: every element of a presheaf indexed by this category is a degeneracy of some nondegenerate element in a unique way [16], and hence all presheaves are Reedy monomorphic.

these are characterized as those maps that induce equivalences upon mapping into an arbitrary quasi-category. So Proposition D.2.16 and Lemma D.6.1 establish that the natural transformation $s: F_\diamond \to F_\star$ is a pointwise weak equivalence between Reedy cofibrant objects in $sSet^{\Delta_+ \times \Delta_+}$. By Lemma C.3.4, the induced map on weighted colimits $s^{X,Y}: X \diamond Y \to X \star Y$ is then a weak equivalence in the Joyal model structure, which means exactly that it induces an equivalence of quasi-categories $A^{X\star Y} \overset{\sim}{\Longrightarrow} A^{X \diamond Y}$ for all quasi-categories A as claimed.   □

In particular, for any quasi-category A, there are natural equivalences $A^{1\star J} \overset{\sim}{\Longrightarrow} A^{1\diamond J}$ and $A^{J\star 1} \overset{\sim}{\Longrightarrow} A^{J \diamond 1}$ over $A \times A^J$. By Lemma 4.2.3 the codomains pullback to define the quasi-categories of cones under or over a diagram $d: J \to A$, respectively. However, as discussed in Warning 4.2.10, the domains do not pull back to the slice quasi-categories $A_{/d}$ and $^{d/}A$ of Definition D.2.8. Nonetheless, we can use the equivalence between the join and fat join constructions to prove that $A_{/d} \simeq \mathrm{Hom}_{A^J}(\Delta, d)$ and $^{d/}A \simeq \mathrm{Hom}_{A^J}(d, \Delta)$ over A and do so now.

PROPOSITION D.6.4. *For any diagram* $d: J \to A$ *indexed by a simplicial set J and valued in a quasi-category A, there are natural equivalences*

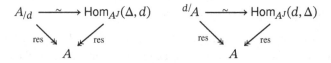

between the slice quasi-categories and the quasi-categories of cones.

Proof Our proof again uses Reedy category theory. To begin, recall the adjunction of Definition D.2.8

$$sSet \underset{-/-}{\overset{-\star J}{\rightleftarrows}} {}^{J/}sSet$$

which gives a correspondence, for a simplicial set I and a map $d: J \to A$, between maps $I \star J \to A$ under J and maps of simplicial sets $I \to A_{/d}$. A right adjoint to the fat join functor $- \diamond J: sSet \to {}^{J/}sSet$ can be calculated similarly. By the defining pushout of Definition 4.2.2, the data of a map $I \diamond J \to A$ under J displayed below-left transposes to the data displayed below-right

$$
\begin{array}{ccc}
(I \times J) \sqcup (I \times J) & \xrightarrow{\pi_I \sqcup \pi_J} & I \sqcup J \\
\big\uparrow & & \big\downarrow \\
I \times 2 \times J & \xrightarrow{\quad\quad} & I \diamond J \dashrightarrow A
\end{array}
\qquad
\begin{array}{ccc}
I \dashrightarrow \mathrm{Hom}_{A^J}(\Delta, d) \longrightarrow A^{2\times J} \\
\big\downarrow \quad\quad \big\downarrow \\
1 \times A \xrightarrow{d \times \Delta} A^J \times A^J
\end{array}
$$

whence we see that the value of the right adjoint to $- \diamond J$

$$sSet \underset{\mathrm{Hom}_{(-)^J}(\Delta, -)}{\overset{- \diamond J}{\underset{\perp}{\rightleftarrows}}} {}^{J/}sSet$$

at $d : J \to A$ defines the ∞-category of cones $\mathrm{Hom}_{A^J}(\Delta, d)$ over d.

The natural map $s^{X,Y} : X \diamond Y \to X \star Y$ of Lemma D.2.14 defines a natural transformation

$$\Delta \underset{\Delta[\bullet] \star J}{\overset{\Delta[\bullet] \diamond J}{\underset{\Downarrow s}{\rightrightarrows}}} {}^{J/}sSet \quad = \quad \Delta \overset{\downarrow}{\longrightarrow} sSet \underset{- \star J}{\overset{- \diamond J}{\underset{\Downarrow s}{\rightrightarrows}}} {}^{J/}sSet$$

that by Proposition D.6.3 is a pointwise weak equivalence in the (sliced) Joyal model structure. Note that by adjunction, $\mathrm{Hom}_{A^J}(\Delta, d)$ is isomorphic to the simplicial set defined by mapping from the cosimplicial object $\Delta[\bullet] \diamond J$ to $d : J \to A$ in ${}^{J/}sSet$, and $A_{/d}$ is isomorphic to the simplicial set defined by mapping from the cosimplicial object $\Delta[\bullet] \star J$ to $d : J \to A$ in ${}^{J/}sSet$. In particular, the mate of the natural transformation s defines a natural comparison map $\hat{s} : A_{/d} \to \mathrm{Hom}_{A^J}(\Delta, d)$. Moreover, since $s^{\Delta[n], \varnothing}$ is the identity, this natural comparison map lies over A. Observe also that the cosimplicial objects $\Delta[\bullet] \diamond J$ and $\Delta[\bullet] \star J$ are both unaugmentable and thus Reedy cofibrant by Lemma C.5.19.

The hom-bifunctor in the category ${}^{J/}sSet$ defines a bifunctor

$$(({}^{J/}sSet)^\Delta)^{\mathrm{op}} \times {}^{J/}sSet \overset{\mathrm{hom}}{\longrightarrow} Set^{\Delta^{\mathrm{op}}} =: sSet$$

$$\begin{array}{ccc} (\Delta[\bullet] \diamond J, d) & \longmapsto & \mathrm{Hom}_{A^J}(\Delta, d) \\ {\scriptstyle (s,\mathrm{id})} \downarrow & & \uparrow {\scriptstyle \hat{s}} \\ (\Delta[\bullet] \star J, d) & \longmapsto & A_{/d} \end{array} \qquad (D.6.5)$$

which carries the map s and the object $d : J \to A$ to the map $\hat{s} : A_{/d} \to \mathrm{Hom}_{A^J}(\Delta, d)$ over A.

By Lemma C.2.13, the hom bifunctor for any model category, such as ${}^{J/}sSet$, is a right Quillen bifunctor relative to the model structure on Set of Exercise C.3.iv, with both weak factorization systems taken to be (monomorphism, epimorphism). Now since Δ is a Reedy category, Theorem C.5.15 tells us that the bifunctor (D.6.5) is again right Quillen, with respect to the Reedy model structure on $({}^{J/}sSet)^\Delta$ and the model structure on $Set^{\Delta^{\mathrm{op}}}$ for which both weak factorization systems are taken to be (Reedy monomorphism, Reedy epimorphism). By Lemma C.5.9, this latter weak factorization system coincides with the familiar (monomorphism, trivial fibration) weak factorization system on *sSet*. In particular, this bifunctor carries a Reedy trivial cofibration in its first variable and a fibrant object $d : J \to A$ in its second variable to a trivial fibration of simplicial sets. By Ken Brown's Lemma C.1.10, it follows that (D.6.5) carries a pointwise weak equivalence between Reedy cofibrant objects to an equivalence of quasi-categories over A, proving that $\hat{s}$ defines the claimed fibered equivalence of quasi-categories. $\square$

In the case $J = \mathbb{1}$, a diagram $a : \mathbb{1} \to A$ defines an element of the quasi-category A, and we have the same result with different notion.

COROLLARY D.6.6. *For any element* $a : \mathbb{1} \to A$ *of a quasi-category* A, *there are canonical equivalences* $A_{/a} \twoheadrightarrow \mathrm{Hom}_A(A, a)$ *and* $^{a/}A \twoheadrightarrow \mathrm{Hom}_A(a, A)$ *over* A.

Proof   When $J = \mathbb{1}$, the constant diagram functor $\Delta : A \to A^J$ appearing in Proposition D.6.4 reduces to the identity functor on A. $\square$

Exercises

EXERCISE D.6.i.   In [78, §1.2.2], Lurie defines the right and left mapping spaces between a pair of elements x and y in a quasi-category A by the pullbacks:

$$
\begin{array}{ccc}
\mathrm{Hom}_A^R(x, y) & \longrightarrow & A_{/y} \\
\downarrow & \lrcorner & \downarrow{\scriptstyle p_0} \\
\mathbb{1} & \xrightarrow{\ x\ } & A
\end{array}
\qquad
\begin{array}{ccc}
\mathrm{Hom}_A^L(x, y) & \longrightarrow & {}^{x/}A \\
\downarrow & \lrcorner & \downarrow{\scriptstyle p_1} \\
\mathbb{1} & \xrightarrow{\ y\ } & A
\end{array}
$$

Show that these simplicial sets are Kan complexes, which are equivalent to the mapping spaces of Definition 3.4.9:

$$
\mathrm{Hom}_A^R(x, y) \simeq \mathrm{Hom}_A(x, y) \simeq \mathrm{Hom}_A^L(x, y).
$$

Appendix E

∞-Cosmoi Found in Nature

In this appendix we establish concrete examples of ∞-cosmoi found in nature. Typically, the objects of these ∞-cosmoi are infinite-dimensional categories as instantiated by some particular nonalgebraic model and the functors between them are the morphisms of such. In most cases, there is an accompanying model structure known to the higher categories literature, which lends us appropriate classes of isofibrations, equivalences, and trivial fibrations. After indicating where a proof of the existence of a suitable model structure may be found, the only work that remains for us is to transfer previously established enrichments to an enrichment over Joyal's model structure for quasi-categories on simplicial sets.

The general theory of what we might call "quasi-categorically enriched model categories" is discussed in §E.1. In particular, we prove that any "quasi-categorically enriched category of fibrant objects" – a combination of Definitions C.1.1 and C.3.11 – defines an ∞-cosmos and describe a change-of-base result that helps produce examples.

In §E.2, we apply these results to establish the ∞-cosmoi of (∞, 1)-categories defined using the complete Segal space, Segal categories, and 1-complicial set models. These complement the ∞-cosmos of quasi-categories of Proposition 1.2.10. We also prove that the change-of-model functors displayed in (10.0.1) define cosmological biequivalences.

Finally, in §E.3, we turn our attention to what might be called *higher ∞-categories*, establishing ∞-cosmoi whose objects are (∞, n)- or even (∞, ∞)-categories in various models.

E.1 Quasi-Categorically Enriched Model Categories

Many examples of ∞-cosmoi arise as categories of fibrant objects in a model category that is enriched over Joyal's model structure simplicial sets – at least if all fibrant objects are cofibrant as is surprisingly often the case.[1] For example, the ∞-cosmos of quasi-categories itself arises in this manner, as Joyal's model structure is cartesian closed (see Digression 1.1.31). The fibrant objects in the Joyal model structure are exactly the quasi-categories, the fibrations between fibrant objects are precisely the isofibrations of Definition 1.1.17, and the weak equivalences between fibrant objects are exactly the equivalences of Definition 1.1.23.

PROPOSITION E.1.1. *Let* $\mathcal{M}$ *be any model category that is enriched over the Joyal model structure and in which every fibrant object is cofibrant. Then the full subcategory of fibrant objects* $\mathcal{M}_f$ *inherits the structure of an* ∞-*cosmos in which the isofibrations are the fibrations between fibrant objects, the equivalences are the weak equivalences between fibrant objects, and the trivial fibrations are the trivial fibrations between fibrant objects.*

Proof   Since the fibrant objects in $\mathcal{M}$ are also cofibrant, Lemma C.3.12 implies the simplicially enriched homs between fibrant–cofibrant objects of $\mathcal{M}$ are quasi-categories, which we denote by $\mathsf{Fun}(A, B)$. The same result also implies that for any fibration $f : A \twoheadrightarrow B$ between fibrant objects, the induced map $f_* : \mathsf{Fun}(X, A) \twoheadrightarrow \mathsf{Fun}(X, B)$ is an isofibration of quasi-categories.

By Example C.1.4, the fibrant objects and fibrations and weak equivalences between them define a category of fibrant objects (see Definition C.1.1). In particular, the unenriched category $\mathcal{M}_f$ has a terminal object, small products, pullbacks of isofibrations, and limits of countable towers of isofibrations, with each of these limits created in $\mathcal{M}$. Since $\mathcal{M}$ admits simplicial tensors, Proposition A.5.4 implies that these 1-categorical limits are conical, and thus $\mathcal{M}_f$ possesses the conical limits of axiom 1.2.1(i). The stability of the isofibrations under the 1-categorical limits of axiom 1.2.1(ii) then is also part of the category of fibrant objects structure of Example C.1.4, though in many examples the stability of the class of isofibrations can also be established via Lemma C.2.3.

It remains to verify the axioms concerning the simplicial cotensors. By hypothesis $\mathcal{M}$ is also cotensored over simplicial sets, and since $\mathcal{M}$ is an enriched model category, the cotensor bifunctor is a right adjoint of a Quillen two-variable adjunction. Directly from the defining axiom of Definition C.3.8, the fibrant

---

[1] This hypothesis – that all fibrant objects are cofibrant – is not essential for the development of ∞-category theory, though it does streamline various proofs. Indeed, the first definition of an "∞-cosmos" to appear in the literature did not include this requirement [110].

objects are closed under cotensor with cofibrant objects, but since all objects in the Joyal model structure are cofibrant, the fibrant objects are closed under simplicial cotensors. Thus $\mathcal{M}_f$ possesses all the limits of 1.2.1(i). Leibniz stability of the isofibrations is a special case of Definition C.3.8, proving that $\mathcal{M}_f$ is an ∞-cosmos.

Since all fibrant objects are cofibrant, $\mathrm{Fun}(X, -) \colon \mathcal{M}_f \to \mathcal{QC}at$ is the action on fibrant objects of a right Quillen functor (see Exercise C.3.iii), and thus by Lemma C.3.4 weak equivalences in $\mathcal{M}_f$ are sent to the equivalences of Definition 1.2.2. Conversely, by Lemma 1.2.15 each equivalence in the ∞-cosmos admits the structure of a homotopy equivalence, defined using what Quillen would call "right homotopies." Quillen proves that any homotopy equivalence is necessarily a weak equivalence in the model category [93, §I.1].   □

Furthermore:

Corollary E.1.2. *Any simplicially enriched right Quillen adjoint between quasi-categorically enriched model categories with all fibrant objects cofibrant defines a cosmological functor that is a cosmological biequivalence whenever the Quillen adjoint defines a Quillen equivalence.*

Proof A right Quillen adjoint preserves fibrant objects, so a simplicially enriched right Quillen adjoint defines a simplicial functor between the subcategories of fibrant objects, satisfying the first requirement of a cosmological functor. As a right Quillen functor, it preserves fibrations between fibrant objects, and thus preserves the isofibrations in the ∞-cosmoi defined by Proposition E.1.1. Finally, a simplicially enriched right adjoint preserves both conical and weighted limits, by Proposition A.6.20. This verifies all of the axioms defining a cosmological functor.

Now consider a simplicially enriched right Quillen equivalence $U \colon \mathcal{N} \to \mathcal{M}$, with left adjoint F. The action on homs of the functor U is isomorphic to the map defined by precomposition with a counit component:

$$\mathcal{N}(X, Y) \xrightarrow{U_{X,Y}} \mathcal{M}(UX, UY)$$
$$\mathcal{N}(FUX, Y)$$

with $-\circ\epsilon_X$ on the lower-left diagonal and $\cong$ on the right vertical.

where the displayed vertical isomorphism is adjoint transposition. By Lemma C.3.6, the counit component ϵ_X is a weak equivalence when X is fibrant. When Y is fibrant, the functor $\mathcal{N}(-, Y) \colon \mathcal{N}^{\mathrm{op}} \to s\mathcal{S}et$ is right Quillen by Exercise C.3.iii. Hence, by Lemma C.3.4, this functor carries weak equivalences between cofibrant objects in $\mathcal{N}$ to equivalences between quasi-categories. Putting all this together, we see that when X and Y are fibrant, the diagonal map defines an equivalence of quasi-categories, and thus $U_{X,Y} \colon \mathcal{N}(X, Y) \to \mathcal{M}(UX, UY)$ must be an equivalence of quasi-categories as well. This verifies the local equivalence property of cosmological biequivalences.

Essential surjectivity is also a consequence of Lemma C.3.6, since the derived unit supplies a weak equivalence $M \overset{\sim}{\to} URFM$ for any fibrant object $M \in \mathcal{M}$. Since RFM is a fibrant object in $\mathcal{N}$, this defines an equivalence in the ∞-cosmos $\mathcal{M}_f$ involving an object in the image of $U \colon \mathcal{N}_f \to \mathcal{M}_f$. □

With Proposition E.1.1 in hand, the next question is where do model categories enriched over the Joyal model structure come from? This question has not attracted much attention in the literature, but the community has done us a considerable favor, in many cases, by providing model categories of infinite-dimensional categories that are enriched over some other cartesian closed model category. This allows us to apply Theorem C.3.16 to convert a known enrichment to an enrichment over Joyal's model structure for quasi-categories. Combining that result with Proposition E.1.1, we obtain a useful change-of-base result that produces model categories enriched over the Joyal model structure.

PROPOSITION E.1.3. *Let $\mathcal{V}$ be a cartesian closed model category equipped with a Quillen adjunction whose right adjoint is valued in the Joyal model structure and whose left adjoint preserves finite products.*

$$sSet \underset{U}{\overset{F}{\underset{\perp}{\rightleftarrows}}} \mathcal{V}$$

(i) *Then for any $\mathcal{V}$-model category $\mathcal{M}$ in which every fibrant object is cofibrant, the full subcategory of fibrant objects $\mathcal{M}_f$ defines an ∞-cosmos in which the isofibrations are the fibrations between fibrant objects, the equivalences are the weak equivalences between fibrant objects, the trivial fibrations are the trivial fibrations between fibrant objects, the functor spaces are defined by $\mathsf{Fun}(M, N) := U\mathcal{M}(M, N)$, where $\mathcal{M}(M, N)$ is the hom-object in $\mathcal{V}$, and the simplicial cotensor of $M \in \mathcal{M}_f$ with $S \in sSet$ are defined by the $\mathcal{V}$-cotensor M^{FS}.*

(ii) *Moreover, any right Quillen $\mathcal{V}$-adjoint between $\mathcal{V}$-model categories of this form defines a cosmological functor that is a cosmological biequivalence whenever the Quillen adjoint is a Quillen equivalence.*

Proof   The adjunction $F \dashv U$ is assumed to be Quillen adjunction between cartesian closed model categories in which the left adjoint preserves finite products. Thus, by Theorem C.3.16, any $\mathcal{V}$-model category $\mathcal{M}$ admits the structure of a model category enriched over the Joyal model structure with the same underlying unenriched model category with enriched homs and cotensors defined by:

$$\mathrm{Fun}(M,N) := U\mathcal{M}(M,N) \qquad \text{and} \qquad M^S := M^{FS},$$

as claimed. Since the underlying model category is unchanged, every fibrant object in $\mathcal{M}$ is still cofibrant. Thus, by Proposition E.1.1, the full subcategory of fibrant objects defines an ∞-cosmos.

Again because the change-of-base result does not affect the underlying model categories, any right Quillen adjoint between $\mathcal{V}$-model categories remains a right Quillen adjoint. By Proposition A.7.3, a $\mathcal{V}$-functor becomes a simplicially enriched functor under the new enrichments. Corollary E.1.2 now tells us that a right Quillen $\mathcal{V}$-adjoint gives rise to a cosmological functor that is a cosmological biequivalence when that Quillen adjoint is a Quillen equivalence. □

We often make use of this result in the following special case:

COROLLARY E.1.4. *Let $\mathcal{V}$ be a cartesian closed model category in which all fibrant objects are cofibrant that is equipped with a Quillen adjunction whose right adjoint is valued in the Joyal model structure and whose left adjoint preserves finite products.*

$$s\mathcal{S}et \underset{U}{\overset{F}{\underset{\perp}{\rightleftarrows}}} \mathcal{V}$$

Then the fibrant objects of $\mathcal{V}$ define a cartesian closed ∞-cosmos $\mathcal{V}_f$ and the right adjoint defines a cosmological functor $U\colon \mathcal{V}_f \to \mathcal{QC}at$ that is naturally isomorphic to the underlying quasi-category functor and is a cosmological biequivalence whenever $F \dashv U$ is a Quillen equivalence.

Proof When the fibrant objects of $\mathcal{V}$ are cofibrant, $\mathcal{V}$ itself satisfies the hypotheses of Proposition E.1.3(i) and thus $\mathcal{V}_f$ defines an ∞-cosmos. By Lemma A.7.7, $F \dashv U$ defines a simplicially enriched adjunction between simplicial sets and the simplicially enriched category $U_*\mathcal{V}$, so in this way we obtain a simplicially enriched right Quillen functor $U\colon U_*\mathcal{V} \to s\mathcal{S}et$ between model categories enriched over the Joyal model structure. By Corollary E.1.2, U then defines a cosmological functor $U\colon \mathcal{V}_f \to \mathcal{QC}at$ that is a biequivalence if $F \dashv U$ is a Quillen equivalence. Combining Remark A.1.9 with the definition of the functor spaces of $\mathcal{V}_f$ given in Proposition E.1.3, we obtain the following natural isomorphism for any $X \in \mathcal{V}_f$

$$U(X) \cong U(X^1) \cong \mathrm{Fun}(1,X)$$

proving that the cosmological functor $U \colon \mathcal{V}_f \to \mathcal{QC}at$ is naturally isomorphic to the underlying quasi-category functor associated to the ∞-cosmos $\mathcal{V}_f$ (see also Remark 1.3.11).   □

Proposition E.1.3 inspires the following trivial examples of ∞-cosmoi.

EXAMPLE E.1.5 (1-categories as ∞-cosmoi). Any complete locally small 1-category $\mathcal{C}$ can be made into an ∞-cosmos in which $\mathsf{Fun}(A, B)$ is just the set of morphisms from A to B. By the Yoneda lemma, the equivalences are then the isomorphisms in $\mathcal{C}$ and so by Lemma 1.2.19 all maps must necessarily be isofibrations. The cotensor of an object $A \in \mathcal{C}$ with a simplicial set S is defined by

$$A^S := A^{\pi_0 S} := \prod_{\pi_0 S} A.$$

Ignoring the fact that model categories are typically assumed to have colimits as well as limits, this construction can be seen as a special case of Proposition E.1.3 applied to the adjunction

$$\mathcal{S}et \underset{sk_0}{\overset{\pi_0}{\rightleftarrows}} \mathcal{S}et$$

whose right adjoint embeds $\mathcal{S}et \hookrightarrow s\mathcal{S}et$ as the subcategory of 0-skeletal simplicial sets (see Definition C.5.2). Here the cartesian closed model structure on $\mathcal{S}et$ is not the one considered in Exercise C.3.iv but rather the one in which the weak equivalences are the isomorphisms and all maps are taken to be both cofibrations and fibrations. To see that this adjunction is Quillen, note that π_0 vacuously preserves cofibrations, while sk_0 carries any map to an isofibration of quasi-categories: This latter claim follows by adjunction since the defining lifting properties below-left transpose to the lifting properties below-right:

$$
\begin{array}{ccc}
\Lambda^k[n] \longrightarrow sk_0\, A & & \pi_0 \Lambda^k[n] \longrightarrow A \\
\Big\downarrow \quad \nearrow \quad \Big\downarrow & \leftrightsquigarrow & \Big\| \quad \overset{\exists!}{\nearrow} \quad \Big\downarrow \\
\Delta[n] \longrightarrow sk_0\, B & & \pi_0 \Delta[n] \longrightarrow B
\end{array}
$$

$$
\begin{array}{ccc}
\mathbb{1} \longrightarrow sk_0\, A & & \pi_0 \mathbb{1} \longrightarrow A \\
\Big\downarrow \quad \nearrow \quad \Big\downarrow & \leftrightsquigarrow & \Big\| \quad \overset{\exists!}{\nearrow} \quad \Big\downarrow \\
\mathbb{I} \longrightarrow sk_0\, B & & \pi_0 \mathbb{I} \longrightarrow B
\end{array}
$$

Famously, π_0 preserves finite products, so the conditions of the change-of-base theorem apply. The homotopy 2-category of an ∞-cosmos arising in this way has only identity 2-cells.

EXAMPLE E.1.6 (2-categories as ∞-cosmoi). Categorifying the previous example, any 2-category $\mathcal{C}$ with sufficient limits defines an ∞-cosmos where $\mathrm{Fun}(A, B)$ is the nerve of the hom-category of morphisms from A to B in $\mathcal{C}$. By Theorem 1.4.7 and Proposition B.6.1, the equivalences are necessarily the equivalences in the 2-category. Inspired by Proposition 1.4.9, we take the isofibrations to be the isofibrations in the 2-category.

Interpreting "sufficient limits" to mean the limits of axiom 1.2.1(i), the remaining axiom 1.2.1(ii) can be verified by hand. Alternatively, again ignoring the fact that model categories are typically assumed to have colimits as well as limits, we may apply Proposition E.1.3 to the homotopy category ⊣ nerve adjunction of Proposition 1.1.11

$$s\mathcal{S}et \underset{}{\overset{\mathsf{h}}{\rightleftarrows}} \mathcal{C}at$$

which is a Quillen adjunction between Joyal's model structure and the usual "folk" model structure on categories. Any 2-category admitting suitable finite limits and colimits is canonically enriched over the folk model structure on categories, when it is given the "trivial" $\mathcal{C}at$-enriched model structure described by Lack [72], with weak equivalences and fibrations are exactly the equivalences and isofibrations just described.

It remains to unpack the meaning of the weaselly phrase "sufficient limits." By Proposition 6.2.8, the 2-category $\mathcal{C}$ is required to have all PIE limits, that is 2-categorical products, inserters, and equifiers discussed in Digression 6.2.7. This implies that $\mathcal{C}$ admits pseudopullbacks of all maps, by the construction of Definition 6.2.10, but this does not quite imply that $\mathcal{C}$ admits 2-pullbacks of isofibrations. Instead, the proof of Lemma 6.2.14 constructs a bipullback of an isofibration, with the usual hom-category isomorphism replaced by a hom-category equivalence. Similar remarks apply to limits of towers of isofibrations. But in practice, the 2-categories that admit PIE limits such as those considered in [18] do seem to admit 2-pullbacks of isofibrations and 2-limits of towers of isofibrations and thus define examples of ∞-cosmoi.

In particular, Example E.1.6 specializes to recover the ∞-cosmos structure on $\mathcal{C}at$ discussed in Proposition 1.2.11. Intriguingly, it also defines an ∞-cosmos structure on $\mathcal{C}at^{\mathrm{op}}$ in which the "isofibrations" are those functors that are injective on objects.[2] Combining these observations with the dual ∞-cosmos

---

[2] In the "folk" model structure on $\mathcal{C}at$, the fibrations are the isofibrations, the weak equivalences are the equivalences, and the cofibrations are the injective-on-objects functors.
Injective-on-objects functors satisfy an isomorphism extension property dual to the isomorphism lifting property that defines the 2-categorical notion of isofibration.

construction of Definition 1.2.25, we see that the four 2-categorical duals $\mathcal{C}at$, $\mathcal{C}at^{op}$, $\mathcal{C}at^{co}$, and $\mathcal{C}at^{coop}$ are all ∞-cosmoi.

The ∞-cosmoi of Example E.1.6 admit an abstract characterization as those ∞-cosmoi that are isomorphic (as quasi-categorically enriched categories) to their homotopy 2-categories. In this case, the weak 2-limits of Chapter 3 are actually strict and many of our results specialize to known theorems in the 2-categorical literature.

EXAMPLE E.1.7 (simplicial model categories as ∞-cosmoi). The identity functor id : $s\mathcal{S}et \to s\mathcal{S}et$ defines a right Quillen adjoint from Quillen's model structure for Kan complexes to Joyal's model structure for quasi-categories; evidently its left adjoint preserves products. Hence, any Kan complex enriched model category – or **simplicial model category** in the usual parlance – may be regarded as a quasi-categorically enriched model category in which each of the mapping spaces between fibrant–cofibrant objects happens to be a Kan complex. Thus, any simplicial model category whose fibrant objects are cofibrant may be regarded as presenting an ∞-cosmos.

The homotopy 2-categories arising in this manner are all $(2, 1)$-categories, with every natural transformation defining a natural isomorphism.

DIGRESSION E.1.8 (on accessible ∞-cosmoi). Many of the model categories one meets in practice, including all of the examples considered in this text, are *combinatorial*, meaning that the underlying category is locally presentable and the model structure is cofibrantly generated. When a model category $\mathcal{M}$ satisfying the hypotheses of Proposition E.1.1 is combinatorial, the resulting ∞-cosmos $\mathcal{M}$ is an **accessible ∞-cosmos**, a notion being studied by Bourke and Lack, based on their earlier work [22] with Vokřínek on homotopical adjoint functor theorems. In fact, it seems likely that all of the constructions of Chapter 6 preserve accessibility, which would mean that every ∞-cosmos considered in this text is accessible.

There are innumerable applications of this observation that will be explored in future work, stemming from the specialization of the main theorems of [22] to this setting. First any accessible ∞-cosmos has all **flexible weighted homotopy colimits**, which are defined by a simplicially enriched universal property of the form of Definition A.6.5 except expressed by an equivalence, rather than an isomorphism, of quasi-categories.[3] This result itself can be understood as a consequence of a second theorem which says that any cosmological functor $U : \mathcal{K} \to \mathcal{L}$ between accessible ∞-cosmoi that is accessible as an unenriched functor admits a homotopical left adjoint: for every $A \in \mathcal{L}$, there exists $FA \in \mathcal{K}$

---

[3] The fact that all of the simplicial sets of (A.6.6) are quasi-categories is a consequence of the flexibility of the weight (see Definition 6.2.1 and Proposition 6.2.8).

and a map $\eta_A : A \to UFA$ inducing an equivalence of quasi-categories

$$\mathrm{Fun}(FA, B) \xrightarrow{\sim} \mathrm{Fun}(A, UB)$$

for all $B \in \mathcal{K}$. For instance, the cosmological embedding $\mathcal{D}isc(\mathcal{K}) \hookrightarrow \mathcal{K}$ is accessible whenever $\mathcal{K}$ is an accessible ∞-cosmos, so this result defines a weak reflection, that "freely inverts" all of the arrows in an ∞-category.

Exercises

EXERCISE E.1.i. State and prove a version of Proposition E.1.3 that applies to enriched categories of fibrant objects, such as considered in Examples E.1.5 and E.1.6, that may have few colimits.

E.2 ∞-**Cosmoi of** (∞, 1)-**Categories**

The ∞-cosmos of quasi-categories is established in Proposition 1.2.10. In this section, we establish three other ∞-cosmoi whose objects define (∞, 1)-categories – modeled as complete Segal spaces, Segal categories, or 1-complicial sets – and construct the following biequivalences between them:

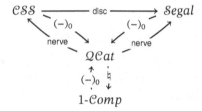

Before giving the formal definition of a complete Segal space, introduced by Charles Rezk in [100], we explain the idea. To start, a complete Segal space is a bisimplicial set $X \in \mathcal{S}et^{\Delta^{\mathrm{op}} \times \Delta^{\mathrm{op}}}$. It is conventional to regard the simplicial sets $X_m := X_{m,\bullet}$ as the "columns" of the bisimplicial set X, while the simplicial sets $X_{\bullet,n}$ define the "rows." In a complete Segal space, the diagram

$$X_\bullet := X_0 \rightrightarrows X_1 \rightrightarrows X_2 \qquad \cdots$$

defines a simplicial object in the category of Kan complexes. Moreover, for each $m \geq 0$, the matching map $X_m \to M_m X$ whose codomain is the space of "boundary data" associated with the m-simplex is a Kan fibration. The spaces X_0 and X_1 are the "spaces of objects and arrows" for the complete Segal space. The so-called "Segal condition" implies that the space X_n may be regarded as the "space of n-composable arrows." A Segal space, satisfying the conditions enumerated thus far, is then something like an "internal category up to homotopy" (compare with Definition B.1.8). The final "completeness" condition relates the spatial structure of X_0 with the categorical structure just defined, expressing the idea that paths in X_0 should correspond to isomorphisms in $X_\bullet$.

The formal definition of a complete Segal space has three conditions, which are most easily described in terms of the weighted limit bifunctor

$$(\mathcal{S}et^{\Delta^{op}})^{op} \times s\mathcal{S}et^{\Delta^{op}} \xrightarrow{\lim_- -} s\mathcal{S}et$$

where $s\mathcal{S}et$ is regarded as a $\mathcal{S}et$-enriched category. Note that the weights for Δ^{op}-indexed diagrams in $s\mathcal{S}et$ are Δ^{op}-indexed diagrams in $\mathcal{S}et$, i.e., simplicial sets. In more detail:

DEFINITION E.2.1 (complete Segal space).

(i) A simplicial object $X_\bullet \in s\mathcal{S}et^{\Delta^{op}}$ is **Reedy fibrant** just when the induced map on weighted limits

$$X_m \cong \lim_{\Delta[m]} X \to \lim_{\partial\Delta[m]} X =: M_m X$$

is a Kan fibration of simplicial sets for all $m \geq 0$.

(ii) A Reedy fibrant simplicial object $X_\bullet$ is a **Segal space** just when the induced map on weighted limits

$$X_n \cong \lim_{\Delta[n]} X \to \lim_{\Lambda^k[n]} X$$

is a trivial fibration of simplicial sets for all $n \geq 2$ and $0 < k < n$.[4]

(iii) A Segal space $X_\bullet$ is a **complete Segal space**, just when the induced map on weighted limits

$$\lim_{\mathbb{I}} X \to \lim_{\Delta[0]} X \cong X_0$$

is a trivial fibration of simplicial sets, asserting that the "space of isomorphisms in X"[5] is equivalent to the space X_0.[6]

---

[4] By Reedy fibrancy, the induced map is already a Kan fibration, so to demand that it is a trivial fibration is equivalent to demanding that it is a weak homotopy equivalence. A priori, this definition is stronger than the usual **Segal condition**, which requires that the map induced on weighted limits by the inclusion of the spine of the n-simplex for each $n \geq 2$ is a trivial fibration. The spine inclusions are in the class cellularly generated by the inner horn inclusions, so by Exercise C.2.v applied to the two-variable adjunction involving the weighted limit, our condition clearly implies the classical Segal condition. The proof of the converse is more subtle and can be found as [64, 3.4].

[5] Other weights may be used to define the "space of isomorphisms" such as the pushout of $\Delta[2] \xleftarrow{\delta^0} \Delta[1] \xrightarrow{\delta^2} \Delta[2]$. See [100, §11] for a discussion.

[6] By the 2-of-3 property, this is equivalent to the arguably more natural condition that the map $\Delta : X_0 \to \lim_{\mathbb{I}} X$, induced by $! : \mathbb{I} \to \Delta[0]$ is a weak homotopy equivalence.

The category of bisimplicial sets, as a presheaf category, is cartesian closed and hence enriched over itself. Among the great supply of product-preserving functors $Set^{\Delta^{op} \times \Delta^{op}} \to Set^{\Delta^{op}}$ that may be used to convert this to a simplicial enrichment, there are two of particular interest: $\mathrm{column}_0 : Set^{\Delta^{op} \times \Delta^{op}} \to Set^{\Delta^{op}}$, which sends a bisimplicial set X to its space X_0 of 0-simplices and $\mathrm{row}_0 : Set^{\Delta^{op} \times \Delta^{op}} \to Set^{\Delta^{op}}$, which passes to set the set of vertices in each space in the simplicial object. As observed by Joyal and Tierney [64], the former construction carries a complete Segal space to a Kan complex, while the latter construction carries a complete Segal space to a quasi-category and will be used to prove:

PROPOSITION E.2.2. *The full subcategory $\mathcal{CSS} \hookrightarrow Set^{\Delta^{op} \times \Delta^{op}}$ of complete Segal spaces defines a cartesian closed ∞-cosmos in which the functor space $\mathrm{Fun}(A, B)$ is defined to be the underlying quasi-category, formed by the vertices in each internal hom B^A. With respect to this ∞-cosmos structure:*

(i) *The underlying quasi-category functor $(-)_0 := \mathrm{row}_0 : \mathcal{CSS} \twoheadrightarrow \mathcal{QCat}$ is a cosmological biequivalence.*

(ii) *A second cosmological biequivalence* $\mathrm{nerve} : \mathcal{QCat} \twoheadrightarrow \mathcal{CSS}$ *carries a quasi-category A to the bisimplicial set whose (m, n)-simplices are simplicial maps $\Delta[m] \times \mathbb{I}[n] \to A$ indexed by the product of the ordinal category with the ordinal groupoid.*

Proof By a theorem of Rezk, the complete Segal spaces form the fibrant objects in a cartesian closed model structure borne by the category of bisimplicial sets in which all objects are cofibrant [100]. Precomposition with the adjoint pair of functors defined by $\pi_1([m] \times [n]) := [m]$ and $\iota_0([m]) := [m] \times [0]$ induces an adjunction as below-right:

$$
\Delta \overset{\pi_1}{\underset{\iota_0}{\rightleftarrows}} \Delta \times \Delta \quad \rightsquigarrow \quad Set^{\Delta^{op}} \overset{\pi_1^*}{\underset{\iota_0^*}{\rightleftarrows}} Set^{\Delta^{op} \times \Delta^{op}} \tag{E.2.3}
$$

Joyal and Tierney prove that this pair of functors defines a Quillen equivalence between the model structure for quasi-categories and the model structure for complete Segal spaces [64, 4.11]. By inspection, the left adjoint preserves finite products, so Corollary E.1.4 applies to create a cartesian closed ∞-cosmos structure on the full subcategory $\mathcal{CSS}$, as detailed in Proposition E.1.3, for which $(-)_0 := \mathrm{row}_0 := \iota_0^*$ is a cosmological biequivalence.

A second adjunction between simplicial sets and bisimplicial sets pointing in the opposite direction has a left adjoint defined as the left Kan extension of the functor

$$\Delta \times \Delta \longrightarrow \mathcal{S}et^{\Delta^{op}}$$

$$[m] \times [n] \longmapsto \Delta[m] \times \mathbb{I}[n]$$

along the Yoneda embedding $\Delta \times \Delta \hookrightarrow \mathcal{S}et^{\Delta^{op} \times \Delta^{op}}$; here $\mathbb{I}[n]$ is the nerve of the groupoid with $n+1$ objects and one exactly morphism in each hom-set, obtained by freely inverting the morphisms in the ordinal category $\mathbb{n+1}$. The right adjoint is the corresponding "nerve" functor described in the statement of (ii). Joyal and Tierney also prove that the adjunction

$$\mathcal{S}et^{\Delta^{op} \times \Delta^{op}} \underset{\text{nerve}}{\overset{\text{lan}}{\underset{\bot}{\rightleftarrows}}} \mathcal{S}et^{\Delta^{op}}$$

is a Quillen equivalence with respect to the model structures for complete Segal spaces and quasi-categories [64, 4.12]. To conclude from Corollary E.1.2 that nerve : $\mathcal{Q}\mathcal{C}at \to \mathcal{C}\mathcal{S}\mathcal{S}$ is a cosmological biequivalence it remains only to show that this functor is simplicially enriched and preserves simplicial cotensors, or equivalently, by Proposition A.4.6, that the adjunction lan ⊣ nerve is simplicially enriched.

To verify this, we make use of the external product bifunctor:

$$\mathcal{S}et^{\Delta^{op}} \times \mathcal{S}et^{\Delta^{op}} \xrightarrow{\quad \square \quad} \mathcal{S}et^{\Delta^{op} \times \Delta^{op}}$$

$$(A, B) \longmapsto (A \,\square\, B)_{m,n} := A_m \times B_n$$

Since any bisimplicial set X may be recovered as a canonical colimit of representables, it suffices to consider maps from a representable bisimplicial set $\Delta[m] \,\square\, \Delta[n]$ to a simplicial set A. In the simplicial enrichment of $\mathcal{S}et^{\Delta^{op} \times \Delta^{op}}$ just defined, the simplicial set of maps from $\Delta[m] \,\square\, \Delta[n]$ to nerve(A) has k-simplices defined to be $(k, 0)$-simplices in nerve(A)$^{\Delta[m]\square\Delta[n]}$. Now

$$(\text{nerve}(A)^{\Delta[m]\square\Delta[n]})_{k,0} := \mathcal{S}et^{\Delta^{op} \times \Delta^{op}}((\Delta[m] \,\square\, \Delta[n]) \times (\Delta[k] \,\square\, \Delta[0]), \text{nerve}(A))$$

by the definition of the cartesian closed structure on bisimplicial sets, which is

$$\cong \mathcal{S}et^{\Delta^{op} \times \Delta^{op}}((\Delta[m] \times \Delta[k]) \,\square\, \Delta[n], \text{nerve}(A))$$

by the definition of the external product, which is

$$\cong \mathcal{S}et^{\Delta^{op}}(\text{lan}((\Delta[m] \times \Delta[k]) \,\square\, \Delta[n]), A)$$

by adjunction. Joyal and Tierney prove in [64, 2.11] that the left Kan extension acts on the external tensor product by $\text{lan}(B \,\square\, \Delta[n]) \cong B \times \mathbb{I}[n]$. So we have

$$\cong \mathcal{S}et^{\Delta^{op}}((\Delta[m] \times \Delta[k]) \times \mathbb{I}[n], A)$$

$$\cong \mathcal{S}et^{\Delta^{op}}((\Delta[m] \times \mathbb{I}[n]) \times \Delta[k], A)$$

$$\cong \mathcal{S}et^{\Delta^{op}}(\text{lan}(\Delta[m] \,\square\, \Delta[n]) \times \Delta[k], A)$$

$$=: (A^{\text{lan}(\Delta[m] \square \Delta[n])})_k$$

by the definition of the cartesian closed structure on simplicial sets. This proves that the adjunction is compatible with the simplicial enrichments, so it follows from Corollary E.1.2 and [64, 4.12] that nerve : $\mathcal{QC}at \to \mathcal{CSS}$ is a cosmological biequivalence. □

A second model of $(\infty, 1)$-categories is closely related.

DEFINITION E.2.4 (Segal categories). A **Segal precategory** is a bisimplicial set $X_{\bullet} \in s\mathcal{S}et^{\Delta^{op}}$ whose space of 0-simplices X_0 is 0-skeletal on the set $X_{0,0}$ of its vertices. A **Segal category** is a Segal category that is Reedy fibrant and satisfies the Segal condition of Definition E.2.1.

Definition E.2.4 is mildly stronger than the usual definition first introduced by Dwyer, Kan, and Smith [38] and further developed by Hirschowitz and Simpson [56]. The usual convention defines a **Segal category** to be a Segal precategory $X_{\bullet}$ so that for each $n \geq 2$, the map induced on weighted limits by the inclusion $\Gamma[n] \hookrightarrow \Delta[n]$ of the spine of the n-simplex

$$X_n \cong \lim_{\Delta[n]} X \to X_1 \underset{X_0}{\times} \cdots \underset{X_0}{\times} X_1 \cong \lim_{\Gamma[n]} \Delta[n]$$

is a weak homotopy equivalence of simplicial sets – without requiring Reedy fibrancy. We prefer to include Reedy fibrancy in our notion of Segal category so that the Segal categories are precisely the fibrant objects in an appropriate model structure on the category $\mathcal{PC}at$ of Segal precategories, which then gives rise to an ∞-cosmos.

Before we introduce the ∞-cosmos *Segal*, we explain how to transform a complete Segal space into a Segal category.

LEMMA E.2.5. *There is a functor* disc : $s\mathcal{S}et^{\Delta^{op}} \to s\mathcal{S}et^{\Delta^{op}}$ *called **discretization** defined by the pullback*

$$
\begin{array}{ccc}
\text{disc}(X) & \lrcorner\!\!\!\hookrightarrow & X \\
\downarrow & & \downarrow \\
\text{cosk}_0(X_{0,0}) & \hookrightarrow & \text{cosk}_0(X_0)
\end{array}
$$

that lands in the subcategory of Segal precategories, and indeed is right adjoint to the inclusion $\mathcal{P}Cat \hookrightarrow sSet^{\Delta^{op}}$. *Moreover, the discretization of a Reedy fibrant Segal space is a Segal category.*

Proof Since the "vertex evaluation" map $X \to \mathrm{cosk}_0(X_0)$ is bijective on the 0th column, the pullback $\mathrm{disc}(X) \to \mathrm{cosk}_0(X_{0,0})$ must be as well. Hence $\mathrm{disc}(X)_0 \cong X_{0,0}$, which proves that $\mathrm{disc}(X)$ is a Segal precategory. To prove the adjointness, note that for any Segal precategory Y and bisimplicial map $f : Y \to X$, the component $f_0 : Y_0 \to X_0$ factors uniquely through $X_{0,0} \hookrightarrow X_0$ by discreteness of Y. This induces the required unique factorization of f through $\mathrm{disc}(X) \hookrightarrow X$.

Finally, any simplicial space that is 0-coskeletal is automatically Reedy fibrant and a Segal space since the maps of Definition E.2.1(i) and (ii) are both isomorphisms. When X is Reedy fibrant, the map $X_n \to \mathrm{cosk}_0(X_0)_n \cong X_0^n$ is a Kan fibration, so the pullback that defines the simplicial set $\mathrm{disc}(X)_n$ is a homotopy pullback. Applying Lemma C.1.11 to Quillen's model structure for Kan complexes on simplicial sets, the Segal maps (ii) for X pull back to define analogous weak homotopy equivalences for $\mathrm{disc}(X)$. □

PROPOSITION E.2.6. *The full subcategory* $Segal \hookrightarrow \mathcal{P}Cat$ *of Segal categories defines a cartesian closed* ∞-*cosmos in which the functor space* $\mathrm{Fun}(A, B)$ *is defined to be the underlying quasi-category, formed by the vertices in each internal hom* B^A. *With respect to this* ∞-*cosmos structure:*

(i) *The underlying quasi-category functor* $(-)_0 := \mathrm{row}_0 : Segal \twoheadrightarrow QCat$ *is a cosmological biequivalence.*

(ii) *There is a cosmological biequivalence* $\mathrm{disc} : CSS \twoheadrightarrow Segal$ *that "discretizes" a complete Segal space into a Segal category.*

(iii) *Another cosmological biequivalence* $\mathrm{nerve} : QCat \twoheadrightarrow Segal$ *carries a quasi-category* A *to the bisimplicial set whose* (m, n)-*simplices are simplicial maps* $\Delta[m] \times \Delta[n] \to A$ *whose components at each vertex of* $\Delta[m]$ *are constant.*

Proof By Pellissier and Bergner [90, 13, 14], the (Reedy fibrant) Segal categories form the fibrant objects in a cartesian closed model structure borne by the category of Segal precategories in which all objects are cofibrant. The cartesian closed structure on $\mathcal{P}Cat$ can be defined explicitly, or deduced from the observation that $\mathcal{P}Cat$ is a category of presheaves (see Exercise E.2.i).

The adjoint functors of (E.2.3) restrict to an adjunction between simplicial sets and Segal precategories, which Joyal and Tierney show define a Quillen equivalence between the model structure for quasi-categories and the model structure for Segal categories [64, 5.6]. Again by inspection, the left adjoint preserves finite products, so Corollary E.1.4 applies to create a cartesian closed

∞-cosmos structure on the full subcategory $Segal$ for which $(-)_0 := \mathrm{row}_0 := \iota_0^*$ is a cosmological biequivalence.

By a theorem of Bergner, the inclusion $\dashv$ discretization adjunction of Lemma E.2.5 defines a Quillen equivalence between the model structure for complete Segal spaces and the model structure for Segal categories [14, §6]. To conclude from Corollary E.1.2 that $\mathsf{disc} \colon \mathcal{CSS} \to \mathcal{S}egal$ is a cosmological biequivalence it remains only to show that this functor is simplicially enriched and preserves simplicial cotensors, or equivalently, by Proposition A.4.6, that the adjunction is simplicially enriched. This follows from the fact that this adjunction commutes with the underlying quasi-category adjunctions for $\mathcal{CSS}$ and $\mathcal{S}egal$ (see Remark E.2.7). In particular, since the inclusion $\mathcal{PC}at \hookrightarrow s\mathcal{S}et^{\Delta^{\mathrm{op}}}$ preserves binary products, for any bisimplicial set C and Segal precategory S, $\mathsf{disc}(C^S) \cong \mathsf{disc}(C)^S$. A similar argument shows that the simplicial cotensors are preserved. Passing to underlying quasi-categories, this induces the desired simplicially enriched adjunction, which makes $\mathsf{disc} \colon \mathcal{CSS} \to \mathcal{S}egal$ simplicial and hence cosmological.

A second adjunction between simplicial sets and Segal precategories pointing in the opposite direction has left adjoint defined by restriction along the diagonal functor $\Delta \colon \Delta^{\mathrm{op}} \to \Delta^{\mathrm{op}} \times \Delta^{\mathrm{op}}$ and right adjoint, which we call "nerve," given by right Kan extension along the same followed by discretization. Joyal and Tierney also prove that this adjunction defines a Quillen equivalence with respect to the model structures for complete Segal spaces and quasi-categories [64, 5.7]. As above, to conclude that $\mathsf{nerve} \colon \mathcal{QC}at \to \mathcal{S}egal$ is a cosmological biequivalence it remains only to argue that this adjunction is simplicially enriched. Since the functor nerve is the composite of the right adjoint to the diagonal functor $\mathsf{diag} \colon \mathcal{S}et^{\Delta^{\mathrm{op}} \times \Delta^{\mathrm{op}}} \to \mathcal{S}et^{\Delta^{\mathrm{op}}}$ followed by discretization and we have already argued that the latter adjunction is simplicially enriched, it suffices to show that $\mathsf{diag} \dashv \mathsf{ran}$ is simplicially enriched.

To that end, consider a bisimplicial set X and a simplicial set A. By definition

$$(\mathsf{ran}(A)^X)_k := (\mathsf{ran}(A)^X)_{k,0} := \mathcal{S}et^{\Delta^{\mathrm{op}} \times \Delta^{\mathrm{op}}}(X \times (\Delta[k] \,\square\, \Delta[0]), \mathsf{ran}(A))$$

$$\cong \mathcal{S}et^{\Delta^{\mathrm{op}} \times \Delta^{\mathrm{op}}}(\mathsf{diag}(X \times (\Delta[k] \,\square\, \Delta[0])), A)$$

$$\cong \mathcal{S}et^{\Delta^{\mathrm{op}} \times \Delta^{\mathrm{op}}}(\mathsf{diag}(X) \times \Delta[k], A) =: (A^{\mathsf{diag}(X)})_k,$$

which is what we wanted to show. $\square$

REMARK E.2.7. This discretization functor commutes with the underlying quasi-category functors:

$$\mathcal{CSS} \xrightarrow{\quad \text{disc} \quad} \mathcal{S}egal$$
$$(-)_0 \searrow \quad \swarrow (-)_0$$
$$\mathcal{QCat}$$

as can most easily be seen by considering the left adjoints to these functors at the level of model categories. However, discretization does not commute with the nerve constructions on the nose, only up to equivalence. For a quasi-category A, nerve(A) is the Segal category with (m, n)-simplices given by the set of simplicial maps $\Delta[m] \times \Delta[n] \to A$ whose components at each vertex of $\Delta[m]$ are constant. By contrast, disc(nerve(A)) is the Segal category with (m, n)-simplices given by the set of simplicial maps $\Delta[m] \times \mathbb{I}[n] \to A$ whose components at each vertex of $\Delta[m]$ are constant.

Recall from Digression D.4.21 that a 1-complicial set is a complicial set that is 1-trivial and saturated (see Definitions D.1.9, D.1.16, and D.4.4). Theorem D.4.14 identifies quasi-categories with 1-complicial sets – there called "naturally marked quasi-categories" – so unsurprisingly:

PROPOSITION E.2.8. *The full subcategory* 1-$\mathcal{Comp} \hookrightarrow$ 1-$s\mathcal{Set}^+$ *of 1-complicial sets defines a cartesian closed ∞-cosmos in which the functor space* Fun(A, B) *is defined to be the underlying quasi-category of the internal hom* B^A. *With respect to this ∞-cosmos structure, both the underlying quasi-category functor* $(-)_0 :$ 1-$\mathcal{Comp} \twoheadrightarrow \mathcal{QCat}$ *and the natural marking functor* $(-)^\natural : \mathcal{QCat} \twoheadrightarrow$ 1-$\mathcal{Comp}$ *are cosmological.*

Proof By independent theorems of Lurie [78, §3.1.3–4] and Verity [129, §6.5], the naturally marked quasi-categories, which we call **1-complicial sets**, form the fibrant objects in a cartesian closed model structure borne by the category of marked simplicial sets in which all objects are cofibrant. There is an adjunction

$$s\mathcal{Set} \underset{(-)_0}{\overset{(-)^\flat}{\rightleftarrows}} \text{1-}s\mathcal{Set}^+$$

in which the right adjoint forgets the marking and the left adjoint assigns each simplicial set the minimal 1-trivial marking, which Lurie proves is a Quillen equivalence between the model structure for quasi-categories and the model structure for 1-complicial sets [78, 3.1.5.1]. By inspection, the left adjoint preserves finite products,[7] so Corollary E.1.4 applies to create a cartesian closed

---

[7] Note $(\Delta[1] \times \Delta[1])^\flat \neq \Delta[1]^\flat \times \Delta[1]^\flat$ as marked simplicial sets so for the finite-product-preservation property to hold it is essential that the minimal marking functor lands in the category of 1-trivial marked simplicial sets.

∞-cosmos structure on the full subcategory 1-$\mathcal{C}omp$ so that the forgetful functor defines a cosmological biequivalence $(-)_0 : 1\text{-}\mathcal{C}omp \twoheadrightarrow \mathcal{Q}\mathcal{C}at$ that coincides with the underlying quasi-category functor. The fact that the model category of 1-complicial sets is enriched over the model structure for quasi-categories via this construction is observed already in [78, 3.1.4.5].

For any quasi-categories A and B, observe that there is a natural isomorphism $\mathrm{Fun}(A, B) \cong \mathrm{Fun}(A^\natural, B^\natural)$ between the functor quasi-category in $\mathcal{Q}\mathcal{C}at$ and the just-defined functor space in 1-$\mathcal{C}omp$ between their natural markings; the point is that simplicial maps $A \to B$ preserve isomorphisms and hence the natural markings. Verity shows that the natural marking functor $(-)^\natural : \mathcal{Q}\mathcal{C}at \to$ 1-$\mathcal{C}omp$ creates the fibrations between fibrant objects [129, 114–118]. Since limits in $\mathcal{Q}\mathcal{C}at$ and 1-$\mathcal{C}omp$ are both created in $s\mathcal{S}et$, it follows that the functor $(-)^\natural : \mathcal{Q}\mathcal{C}at \to 1\text{-}\mathcal{C}omp$ is a cosmological biequivalence, and indeed an inverse isomorphism to $(-)_0 : 1\text{-}\mathcal{C}omp \to \mathcal{Q}\mathcal{C}at$. $\qquad\qquad\Box$

Exercises

EXERCISE E.2.i. Joyal and Tierney identify the subcategory $\mathcal{P}\mathcal{C}at \hookrightarrow \mathcal{S}et^{\Delta^{op} \times \Delta^{op}}$ with the category of presheaves indexed by the 1-categorical quotient $\Delta|_2$ of $\Delta \times \Delta$ defined by inverting the maps in the image of the functor $[0] \times \Delta \hookrightarrow \Delta \times \Delta$ [64, 5.4]. Redefine the three adjunctions between $\mathcal{P}\mathcal{C}at$, $\mathcal{S}et^{\Delta^{op} \times \Delta^{op}}$, and $\mathcal{S}et^{\Delta^{op}}$ appearing in the proof of Proposition E.2.6 from this point of view.

EXERCISE E.2.ii. Verify that the cosmological biequivalences nerve : $\mathcal{Q}\mathcal{C}at \twoheadrightarrow \mathcal{C}\mathcal{S}\mathcal{S}$ and nerve : $\mathcal{Q}\mathcal{C}at \twoheadrightarrow \mathcal{S}egal$ each define sections of the respective underlying quasi-category functors.

E.3 ∞-Cosmoi of (∞, n)-Categories

In this section, we introduce a variety of ∞-cosmoi whose objects are models of (∞, n)-categories for $1 < n \le \infty$. These ∞-cosmoi describe the $(\infty, 2)$-categories of ∞-categories, ∞-functors, and ∞-natural transformations, omitting higher-dimensional transformations, though generalized elements and internal homs allow access to higher-dimensional noninvertible morphisms.

Because the combinatorics entailed in fully specifying a model of (∞, n)-categories can be rather involved, to save space, we do not define every one of the higher categorical notions discussed here, instead providing external references to where such definitions can be found.

A few of our models of (∞, n)-categories are defined as presheaves indexed by a 1-category Θ_n first introduced by Joyal in an unpublished note [60], which we present in an equivalent form due to Berger [11].

DEFINITION E.3.1. For $0 \leq n \leq \infty$, define a family of 1-categories Θ_n inductively as follows.

- $\Theta_0 := \mathbb{1}$ is the terminal category and $\Theta_1 := \Delta$ is the category of finite nonempty ordinals and order-preserving maps.
- $\Theta_n := \Delta \wr \Theta_{n-1}$, where $\Delta \wr -: \mathcal{C}at \to \mathcal{C}at$ is the **categorical wreath product** construction. Explicitly, for a 1-category C, $\Delta \wr C$ is the category whose:
  - objects are tuples $[n](c_1, \dots, c_n)$ where $[n] \in \Delta$ and $c_i \in C$.
  - morphisms $(\alpha; \vec{f}): [n](c_1, \dots, c_n) \to [m](c'_1, \dots, c'_m)$ are given by a simplicial map $\alpha: [n] \to [m] \in \Delta$ together with morphisms $f_{i,j}: c_i \to c'_j \in C$ for all $0 < i \leq n$ and $\alpha(i-1) < j \leq \alpha(i)$.

The objects of Θ_n define pasting diagrams of k-cells for $0 \leq k \leq n$ while the morphisms define projection, composition, and degeneracy maps. The functor $\Theta_n \hookrightarrow n\text{-}\mathcal{C}at$ that sends a pasting diagram to the free strict n-category that it generates is full and faithful [11, 3.7].

For instance, the morphism $(\delta^2; (\delta^1, !, \mathrm{id})): [2]([1], [1]) \to [3]([2], [0], [1])$ in Θ_2 corresponds to the 2-functor between the free 2-categories generated by the pasting diagrams

$$0 \underset{\Downarrow}{\overset{}{\rightrightarrows}} 1 \underset{\Downarrow}{\overset{}{\rightrightarrows}} 2 \qquad \mapsto \qquad 0 \underset{\Downarrow}{\overset{\Downarrow}{\rightrightarrows}} 1 \longrightarrow 2 \underset{\Downarrow}{\overset{}{\rightrightarrows}} 3$$

that sends 0 to 0, 1 to 1, and 2 to 3, and sends the left 2-cell of the domain to the vertical composite of the leftmost 2-cells of the codomain and the right 2-cell of the domain to the whiskered composite of the rightmost 2-cell of the codomain with the central 1-cell.

LEMMA E.3.2. *For any 1-category with a terminal element t, the adjunction below-left induces an adjunction below-right:*

$$\mathbb{1} \underset{t}{\overset{!}{\underset{\longrightarrow}{\rightleftarrows}}} C \qquad \rightsquigarrow \qquad \Delta \cong \Delta \wr \mathbb{1} \underset{\Delta \wr t}{\overset{\Delta \wr !}{\underset{\longrightarrow}{\rightleftarrows}}} \Delta \wr C$$

Proof The categorical wreath product defines a 2-functor $\Delta \wr -: \mathcal{C}at \to \mathcal{C}at$. $\qquad \square$

Ara introduced a model of (∞, n)-categories for each $1 \leq n < \infty$ called n-**quasi-categories** as presheaves on Θ_n characterized by a particular right lifting property described in [4, §5]. Ara's 1-quasi-categories coincide with the usual quasi-categories.

PROPOSITION E.3.3. *For each $n \geq 1$, the full subcategory n-$\mathcal{QCat} \hookrightarrow \mathcal{Set}^{\Theta_n^{op}}$ of n-quasi-categories defines a cartesian closed ∞-cosmos.*

Proof Ara constructs a cartesian closed model structure on the category $\mathcal{Set}^{\Theta_n^{op}}$, generalizing the Joyal model structure in the case $n = 1$, in which the fibrant objects are exactly the n-quasi-categories and in which the cofibrations are the monomorphisms [4]; in particular, all objects are cofibrant. Hence, to induce a cartesian closed ∞-cosmos structure on the full subcategory n-$\mathcal{QCat}$ it suffices to find a Quillen adjunction from this model structure to the model structure for quasi-categories whose left adjoint $\mathcal{Set}^{\Delta^{op}} \to \mathcal{Set}^{\Theta_n^{op}}$ preserves binary products.

To that end, note that $[0] \in \Theta_{n-1}$ is terminal for all $n > 1$, so Lemma E.3.2 provides an adjunction as below-left and hence an adjunction as below-right

$$\Delta \underset{\Delta\wr[0]}{\overset{\Delta\wr!}{\rightleftarrows}} \perp \Theta_n \qquad \rightsquigarrow \qquad \mathcal{Set}^{\Delta^{op}} \underset{(\Delta\wr[0])^*}{\overset{(\Delta\wr!)^*}{\rightleftarrows}} \perp \mathcal{Set}^{\Theta_n^{op}}$$

The right adjoint $\Delta \wr [0] : \Delta \hookrightarrow \Theta_n$ includes Δ as the subcategory of "pasting diagrams comprised of only 1-cells"; hence, restriction along this functor $(\Delta \wr [0])^* : \mathcal{Set}^{\Theta_n^{op}} \to \mathcal{Set}^{\Delta^{op}}$ forgets higher-dimensional cells. The other adjoint $\Delta\wr! : \Theta_n \to \Delta$ projects onto the first component of the categorical wreath product. Note that the corresponding restriction functor between presheaf categories has its own left adjoint, defined by left Kan extension, and so clearly preserves products.

Indeed, for the same reason, the left adjoint preserves all limits and hence also preserves monomorphisms (which can be characterized as those maps whose kernel pair is given by identities). By a result of Joyal and Tierney [64, 7.15], to prove that an adjunction is Quillen, it suffices to show that the left adjoint preserves cofibrations, as we have just done, and the right adjoint preserves fibrations between fibrant objects. By Lemma C.2.6, this means that we need only verify that the left adjoint carries the inner horn inclusions $\{\Lambda^k[n] \hookrightarrow \Delta[n]\}_{n\geq 2, 0<k<n}$ and the map $\mathbb{1} \hookrightarrow \mathbb{I}$ to trivial cofibrations in Ara's model structure. In fact, by [64, 3.5], it suffices to consider the spine inclusions $\{\Gamma[n] \hookrightarrow \Delta[n]\}_{n\geq 2}$ in place of the inner horn inclusions, which we shall.

To see this, it is helpful to note, as observed in [4, §6], that the left adjoint commutes with the nerve embeddings of strict 1-categories and strict n-categories:

$$\begin{array}{ccc} \mathcal{C}at & \longhookrightarrow & n\text{-}\mathcal{C}at \\ \uparrow & & \downarrow \\ \mathcal{S}et^{\Delta^{op}} & \xrightarrow{(\Delta\iota!)^*} & \mathcal{S}et^{\Theta_n^{op}} \end{array}$$

In particular, the left adjoint carries the 1-categorical nerve of $\mathbb{1} \hookrightarrow \mathbb{I}$ to the strict n-categorical nerve of this map and, since the left adjoint also preserves colimits, it carries the inner horn inclusion $\Gamma[n] \hookrightarrow \Delta[n]$ to the corresponding "spine inclusion" for the object $[n]([0], \dots, [0]) \in \Theta_n$. As both types of maps are among Ara's "localizer of n-quasi-categories" of [4, 5.17], they are certainly trivial cofibrations. Hence, the adjunction is Quillen, as claimed, and Corollary E.1.4 applies to create a cartesian closed ∞-cosmos structure on n-$\mathcal{QC}at$. □

Another model of (∞, n)-categories, for $0 \leq n < \infty$ is due to Rezk [101]. A Θ_n-**space** is a simplicial presheaf on Θ_n-satisfying Reedy fibrancy, Segal, and completeness conditions analogous to those of Definition E.2.1. A Θ_1-space is exactly a complete Segal space, while a Θ_0-space is just a Kan complex.

PROPOSITION E.3.4. *For each $n \geq 1$, the full subcategory Θ_n-$\mathcal{S}p \hookrightarrow s\mathcal{S}et^{\Theta_n^{op}}$ of Θ_n-spaces defines a cartesian closed ∞-cosmos for which the underlying complete Segal space functor $U : \Theta_n$-$\mathcal{S}p \to \Theta_1$-$\mathcal{S}p \cong \mathcal{CSS}$ is cosmological.*

Proof Rezk constructs a cartesian closed model structure on the category $s\mathcal{S}et^{\Theta_n^{op}}$ generalizing his model structure for complete Segal spaces in the case $n = 1$, in which the fibrant objects are exactly the Θ_n-spaces and in which the cofibrations are the monomorphisms [101]; in particular, all objects are cofibrant. Hence, to induce a cartesian closed ∞-cosmos structure on the full subcategory Θ_n-$\mathcal{S}p$, it suffices to find a Quillen adjunction between this model structure and the model structure for complete Segal spaces whose left adjoint $s\mathcal{S}et^{\Delta^{op}} \to s\mathcal{S}et^{\Theta_n^{op}}$ preserves binary products. We then apply Corollary E.1.4 to the composite of this adjunction with the adjunction (E.2.3).

As in the proof of Proposition E.3.3, we obtain the desired adjunction from Lemma E.3.2 applied to the terminal object $[0] \in \Theta_{n-1}$.

$$\Delta \underset{\Delta\iota[0]}{\overset{\Delta\iota!}{\underset{\perp}{\rightleftarrows}}} \Theta_n \quad \rightsquigarrow \quad s\mathcal{S}et^{\Delta^{op}} \underset{(\Delta\iota[0])^*}{\overset{(\Delta\iota!)^*}{\underset{\perp}{\rightleftarrows}}} s\mathcal{S}et^{\Theta_n^{op}} \qquad (E.3.5)$$

The left adjoint has a further left adjoint given by left Kan extension, and so preserves products.

It remains only to argue that this adjunction is Quillen. The model structure for Θ_n-spaces – and, by specialization, also the model structure for complete

Segal spaces – is defined as a left Bousfield localization of the injective (or, equivalently, Reedy) model structures on simplicial presheaves. In the injective model structure, the cofibrations and trivial cofibrations are defined objectwise in $sSet$, so the left adjoint is manifestly left Quillen with respect to these model structures. Consequently, the adjunction is Quillen for the localized model structures if and only if the right adjoint, which Rezk refers to as the "underlying simplicial space" functor, preserves fibrant objects, because in that case the left adjoint preserves the new trivial cofibrations, which are defined in terms of these. A functor $X \in sSet^{\Theta_n^{op}}$ is fibrant if and only if it satisfies Reedy, Segal, and completeness conditions. Since the adjunction (E.3.5) is Quillen for the injective/Reedy model structure, the Reedy fibrancy condition is preserved, and Rezk proves that the Segal condition is preserved as well [101, 7.2]. The completeness condition for Θ_n-spaces is created from the completeness condition for underlying simplicial spaces [101, §7], so this is preserved as well. Hence, the right adjoint (E.3.5) restricts to a functor $U \colon \Theta_n\text{-}Sp \to \mathcal{CSS}$, which we call the **underlying complete Segal space functor**.

Corollary E.1.4 applies to create a cartesian closed ∞-cosmos structure on $\Theta_n\text{-}Sp$. By Lemma A.7.7, the adjunction (E.3.5) is enriched over bisimplicial sets, and so Proposition E.1.3 can be used to prove that the underlying complete Segal space functor is cosmological. $\square$

There is another model for (∞, n)-categories that generalizes the complete Segal space model for $(\infty, 1)$-categories, which makes use of the notion of a **Rezk object** valued in a model category:

DEFINITION E.3.6. Let $\mathcal{M}$ be a model category.

(i) A simplicial object $X_\bullet \in \mathcal{M}^{\Delta^{op}}$ is **Reedy fibrant** just when the induced map on weighted limits

$$X_m \cong \lim_{\Delta[m]} X \to \lim_{\partial\Delta[m]} X =: M_m X$$

is a fibration for all $m \geq 0$.

(ii) A Reedy fibrant simplicial object $X_\bullet$ is a **Segal object** just when the induced map on weighted limits

$$X_n \cong \lim_{\Delta[n]} X \to \lim_{\Lambda^k[n]} X$$

is a trivial fibration for all $n \geq 2$ and $0 < k < n$.

(iii) A Segal object $X_\bullet$ is a **Rezk object**, just when the induced map on weighted limits

$$\lim_{\mathbb{I}} X \to \lim_{\Delta[0]} X \cong X_0$$

is a trivial fibration.

A map $p : X_{\bullet} \to Y_{\bullet} \in \mathcal{M}^{\Delta^{op}}$ is a **Rezk isofibration** if the relative analogues of the maps appearing in (i), (ii), and (iii) formed by the Leibniz weighted limit of p with the appropriate maps of weights are, respectively, fibrations, trivial fibrations, and a trivial fibration.

Our formulation of Definition E.2.1, which departs slightly from Rezk's framing of his conditions, is intended to make it clear that the complete Segal spaces are precisely the Rezk objects valued in Quillen's model structure for Kan complexes.

PROPOSITION E.3.7. *Suppose* $\mathcal{M}$ *is a Cisinski model category.*[8] *Then the full subcategory* $\mathcal{R}ezk_{\mathcal{M}} \hookrightarrow \mathcal{M}^{\Delta^{op}}$ *of Rezk objects defines an* ∞-*cosmos.*

Proof   We prove this result directly from Proposition E.1.1 by proving that a left Bousfield localization of the Reedy model structure on $\mathcal{M}^{\Delta^{op}}$ defines a Cisinski model structure that is enriched over the model structure for quasi-categories in which the fibrant objects are exactly the Rezk objects.

To begin, observe that the category $\mathcal{M}^{\Delta^{op}}$ is simplicially enriched, tensored, and cotensored, with hom-spaces suggestively denoted by "Fun"

$$(\otimes, \{,\}, \mathsf{Fun}) : s\mathcal{S}et \times \mathcal{M}^{\Delta^{op}} \to \mathcal{M}^{\Delta^{op}}$$

in such a way that the Leibniz tensors of monomorphisms of simplicial sets with (trivial) Reedy cofibrations are (trivial) Reedy cofibrations [35, 4.4]. We apply Jeff Smith's theorem [8] to prove that $\mathcal{M}^{\Delta^{op}}$ admits a model structure in which

- the cofibrations are the monomorphisms,
- the fibrant objects are the Rezk objects,
- the fibrations between fibrant objects are the Rezk isofibrations, and
- weak equivalences are the **Rezk weak equivalences**, those maps $w : U \to V$ that induce equivalences of quasi-categories $w^* : \mathsf{Fun}(V, X) \to \mathsf{Fun}(U, X)$ for all Rezk objects X.

Note that by adjunction, a map $p : X \to Y \in \mathcal{M}^{\Delta^{op}}$ is a Rezk isofibration if and only if for all monomorphisms $m : A \to B \in \mathcal{M}$, the induced map

$$\mathsf{Fun}(B, X) \xrightarrow{\widehat{\mathsf{Fun}(m,p)}} \mathsf{Fun}(A, X) \underset{\mathsf{Fun}(A,Y)}{\times} \mathsf{Fun}(B, Y)$$

of simplicial sets is an isofibration of quasi-categories. By Corollary D.3.11 and

---

[8] A **Cisinski model structure** is a combinatorial model structure on a Grothendieck topos in which the cofibrations are exactly the monomorphisms. It follows that the Reedy model structure on $\mathcal{M}^{\Delta^{op}}$ coincides with the injective model structure, and in particular that all objects are cofibrant [27, 16].

Proposition D.5.2, this is the case if and only if this map has the right lifting property with respect to maps in the set $I \hat{\times} J$ where

$$I := \{\partial\Delta[n] \hookrightarrow \Delta[n]\}_{n \geq 0} \quad \text{and} \quad J := \{\Lambda^k[n] \hookrightarrow \Delta[n]\}_{n \geq 2, 0 < k < n} \cup \{\mathbb{1} \hookrightarrow \mathbb{I}\}.$$

By adjunction again, and Proposition C.2.9(i), p is a Rezk isofibration if and only if it has the right lifting property with respect to the sets of maps $(i \hat{\times} j) \hat{*} m \cong j \hat{\otimes} (i \hat{*} m)$ for all $i \in I$, $j \in J$, and m among the generating cofibrations in $\mathcal{M}$, where $*$ denotes the pointwise tensor $*: s\mathcal{S}et \times \mathcal{M} \to \mathcal{M}^{\Delta^{op}}$. Since by Proposition C.5.8 the Reedy cofibrations in $\mathcal{M}^{\Delta^{op}}$ are generated by the set of maps $i \hat{*} m$ for $i \in I$ and as m ranges over the generating cofibrations in $\mathcal{M}$, we conclude by adjunction that p is a Rezk isofibration between Rezk objects if and only if

$$\text{Fun}(V, X) \xrightarrow{\widehat{\text{Fun}(c, p)}} \text{Fun}(U, X) \underset{\text{Fun}(U, Y)}{\times} \text{Fun}(V, Y)$$

is an isofibration for all monomorphisms $c: U \to V$ in $\mathcal{M}^{\Delta^{op}}$.

Now it is easy to verify the conditions of Jeff Smith's theorem. The Rezk weak equivalences are accessible and satisfy the 2-of-3 property. We argue that the Rezk weak equivalences contain all Reedy weak equivalences and hence the Reedy trivial fibrations, characterized by the right lifting property against the monomorphisms. Transposing the observations already made in [35, 4.4] about the Reedy model structure on $\mathcal{M}^{\Delta^{op}}$, we see that for any Reedy trivial cofibration $w: U \to V$ and Rezk object X, $w^*: \text{Fun}(V, X) \to \text{Fun}(U, X)$ is an equivalence of quasi-categories. By Ken Brown's lemma C.1.10, the same is true when w is a mere Reedy weak equivalence. Note that a map $w: U \to V$ is both a Rezk weak equivalence and a cofibration just when $w^*: \text{Fun}(V, X) \to \text{Fun}(U, X)$ is a trivial fibration between quasi-categories. This characterization proves that the class of Rezk weak equivalences and cofibrations is stable under pushout and transfinite composition. Jeff Smith's theorem now implies that the model structure for Rezk objects exists.

To see that the model structure for Rezk objects is enriched over the model structure for quasi-categories, we must verify the three conditions for

$$(\otimes, \{,\}, \text{Fun}): s\mathcal{S}et \times \mathcal{M}^{\Delta^{op}} \to \mathcal{M}^{\Delta^{op}}$$

to define a Quillen two-variable adjunction (see Definition C.3.8). The cofibrations in the localized model structure for Rezk objects are the same as the cofibrations for the Reedy model structure on $\mathcal{M}^{\Delta^{op}}$, so we already know that Leibniz tensors of cofibrations are cofibrations. To verify the remaining 2/3rds of this axiom, we appeal to a result of Dugger [35, 3.2], which tells us that in the presence of the first 1/3rd, to verify that Leibniz tensors of monomorphisms of simplicial sets with trivial cofibrations are trivial cofibrations, it suffices to show that the simplicial cotensor $(-)^K : \mathcal{M}^{\Delta^{op}} \to \mathcal{M}^{\Delta^{op}}$ preserves fibrations between fibrant objects. For left Bousfield localizations, Rezk isofibrations between Rezk objects coincide with Reedy fibrations between Rezk objects [55, 3.3.16]. It is easy to verify directly that $(-)^K$ preserves Rezk objects, and the preservation of Reedy fibrations is one of the facts we knew already.

For the final 1/3rd of the Quillen two-variable adjunction, we use the second part of Dugger's [35, 3.2], which tells us that in the presence of the first 2/3rds, we need only verify that for all Rezk objects Z and trivial cofibrations of simplicial sets $j : J \to K$, the map $Z^j : Z^K \to Z^J$ is a Rezk weak equivalence (assuming $\mathcal{M}^{\Delta^{op}}$ is left proper, which is the case here since all objects are cofibrant). In fact, we can show that this map is a trivial fibration, by checking the right lifting property against the monomorphisms $c : U \to V \in \mathcal{M}^{\Delta^{op}}$. Transposing, we see that c lifts against Z^j if and only if j lifts against $c^* : \mathsf{Fun}(V, Z) \to \mathsf{Fun}(U, Z)$. But we verified already that c^* is an isofibration between quasi-categories, so the desired lifting property holds. $\square$

The model structure for Rezk objects on $\mathcal{M}^{\Delta^{op}}$ remains a Cisinski model structure, so this construction can be iterated. Barwick's n-**fold complete Segal space** model of (∞, n)-categories is formed by iterating the Rezk objects construction n times [6]. Specializing Proposition E.3.7, we conclude that for all $n \geq 1$, there exist ∞-cosmoi $\mathcal{CSS}_n$ of n-fold complete Segal spaces.

REMARK E.3.8. If $\mathcal{M}$ is a left proper combinatorial model category, the proof just given constructs a model structure on $\mathcal{M}^{\Delta^{op}}$ whose fibrant objects are the Rezk objects that is enriched as a model category over the model structure for quasi-categories. The only hitch is that without the Cisinski condition, it is possible that not all fibrant objects are cofibrant. Nonetheless, this generalization can be understood as defining an ∞-cosmos in a weaker sense developed in [110, §2].

Verity constructs a general family of cartesian model structures on the category of marked simplicial sets whose fibrant objects are complicial sets of various flavors and whose fibrations are the corresponding notions of complicial isofibration [129, §9.3]. One of these model structures presents the ∞-cosmos of Proposition E.2.8. Here, we consider model structures whose fibrant objects, called n-*complicial sets*, model (∞, n)-categories for $0 \leq n \leq \infty$, where the "∞-complicial sets" are the complicial sets of Definition D.1.9 that are saturated in a sense alluded to in Digression D.4.21 and elaborated upon in the proof. The definitions are arranged so that a 0-complicial set is a (maximally marked)

Kan complex, a 1-complicial set is a (naturally marked) quasi-category, and a m-complicial set is an n-complicial set whenever $m < n$.

PROPOSITION E.3.9. *For each $0 \leq n \leq \infty$, the category of n-complicial sets defines a cartesian closed ∞-cosmos n-$\mathcal{C}omp$. Moreover, whenever $m < n$, the functor* core : n-$\mathcal{C}omp \to m$-$\mathcal{C}omp$ *that discards all simplices in dimension $k > m$ that are not marked is cosmological.*

Proof For a suitable class of monomorphisms J, Verity defines a cartesian closed model structure on the category of marked simplicial sets whose fibrant objects and fibrations between them are the J-complicial sets and J-complicial isofibrations, characterized by a right lifting property against J [129, §9.3]. The cofibrations are the monomorphisms so in particular all objects are cofibrant. In more detail, for the n-complicial sets, the class of monomorphisms is defined to be

$$J_n := \left\{\Lambda^k[m] \hookrightarrow_r \Delta^k[m]\right\}_{m \geq 1, k \in [m]} \cup \left\{\Delta^k[m]' \hookrightarrow_e \Delta^k[m]''\right\}_{m \geq 2, k \in [m]} \cup$$
$$\left\{\Delta[r] \hookrightarrow_e \Delta[r]_t\right\}_{r > n} \cup \left\{\Delta[j] \star \Delta[3]_{\mathrm{eq}} \star \Delta[k] \hookrightarrow \Delta[j] \star \Delta[3]^{\sharp} \star \Delta[k]\right\}_{j, k \geq -1}$$

The first set of maps are the complicial horn extensions (D.1.10) while the second set defines the complicial thinness extensions (D.1.11). The third set imposes the condition that all simplices in dimension greater than n are marked (see Notation D.1.4), while the final condition is **saturation**, which in the presence of the other conditions, implies that all equivalences are marked (see Exercise D.4.iii and [105]). To apply Verity's theorem, the sets J_n must satisfy some technical conditions spelled out in [129, 91–92], which have been verified in this case by Ozornova and Rovelli [89, 1.26]. By construction, the n-complicial sets live in the subcategory n-$s\mathcal{S}et^+$ of n-trivial marked simplicial sets, with all simplices in dimension greater than n marked, and we may restrict the cartesian closed model structures to these subcategories.

The ∞-cosmoi 0-$\mathcal{C}omp$ and 1-$\mathcal{C}omp$ are isomorphic to the ∞-cosmoi $\mathcal{K}an$ and $\mathcal{QC}at$, respectively, so for now we consider $2 \leq n \leq \infty$. To define the ∞-cosmos n-$\mathcal{C}omp$, we apply Proposition E.1.3 to convert these self enrichments into an enrichment over quasi-categories via a string of Quillen adjunctions whose left adjoints preserve binary products:

$$s\mathcal{S}et \underset{(-)_0}{\overset{(-)^{\flat}}{\rightleftarrows}} 1\text{-}s\mathcal{S}et^+ \underset{\mathsf{core}_1}{\rightleftarrows} 2\text{-}s\mathcal{S}et^+ \rightleftarrows \cdots \rightleftarrows n\text{-}1\text{-}s\mathcal{S}et^+ \underset{\mathsf{core}_{n-1}}{\rightleftarrows} n\text{-}s\mathcal{S}et^+ \cdots$$

In the limiting case, we also consider adjunctions

$$n\text{-}s\mathcal{S}et^+ \underset{\mathrm{core}_n}{\overset{\subset}{\rightleftarrows}} s\mathcal{S}et^+$$

where $\mathrm{core}_n X \hookrightarrow X$ is the simplicial subset containing only those simplices whose faces in dimension greater than n are marked. By adjunction, these functors carry $(n + 1)$-complicial sets to n-complicial sets (see Exercise D.1.v). Since the left adjoints preserve monomorphisms and products, this is enough to verify that the adjunctions are Quillen. Corollary E.1.4 now induces the desired ∞-cosmoi and Corollary E.1.2 supplies the cosmological core functors.   □

Building on past work of Hirschowitz–Simpson [56] and Pellissier [90], Simpson iterates the construction of the model structure for Segal categories [116, §19.2–4]. When the base model category is taken to be Quillen's model structure for Kan complexes, the n-th iteration defines the **Segal n-categories**, though Simpson also considers more general model categorical bases. Under suitable hypotheses, satisfied in the case of Segal n-categories, the model structure so produced is cartesian closed and has all objects cofibrant, which strongly suggests that there exists an ∞-cosmos spanned by its fibrant objects: the Reedy fibrant Segal n-categories. We leave the confirmation of this as an exercise for the interested reader.

Exercises

EXERCISE E.3.i. Given an explicit formulation of the "relative analogue" of the conditions (i), (ii), and (iii) used in Definition E.3.6 to define the notion of Rezk isofibration.

EXERCISE E.3.ii. Investigate potential ∞-cosmos structures on the Segal n-categories of Hirschowitz and Simpson [56].

EXERCISE E.3.iii. Search for cosmological biequivalences between the ∞-cosmoi constructed in this section (and please share your discoveries with the authors).

Appendix F

The Analytic Theory of Quasi-Categories

The aim in this final appendix is to prove that the synthetic theory of quasi-categories is compatible with the analytic theory pioneered by André Joyal, Jacob Lurie, and many others. In §F.1, we prove an equivalence between the synthetic and analytic definitions of a terminal element in a quasi-category. In §F.2, we extend these results to an equivalence between the synthetic and analytic definitions of limits of diagrams of arbitrary shape valued in a quasi-category.

In §F.3, we provided a new analytic characterization of those isofibrations between quasi-categories that admit a right adjoint right inverse. This is used in §F.4 to compare the synthetic and analytic definitions of cartesian fibrations of quasi-categories and cartesian arrows. Finally, in §F.5, we prove that the synthetic and analytic definitions of an adjunction agree, despite their quite different forms.

F.1 Initial and Terminal Elements

In this section, we complete the argument sketched in Digression 4.3.14 and prove that the synthetic definition of a terminal element in a quasi-category coincides with the analytic definition first introduced by Joyal [61, 4.1]. We prove the equivalence between four synthetic definitions of a terminal element – (i) which appeared first in Definition 2.2.1; (ii) which is Lemma 2.2.2; (iii) which appears as Proposition 4.3.13; and (iv) which appears commonly in the literature (see, e.g., [88]) – and two analytic definitions of a terminal element (v) and (vi), which Joyal proves are equivalent [61, 4.2].

PROPOSITION F.1.1. *For a quasi-category A and element $t : 1 \to A$ the following are equivalent:*

(i) *The element t defines a right adjoint to the unique functor:*

$$1 \underset{t}{\overset{!}{\rightleftarrows}} A$$

(ii) *There exists a natural transformation*

$$A = A$$
$$\underset{!}{\searrow} \underset{1}{\overset{\Downarrow\eta}{}} \underset{t}{\nearrow}$$

so that the component ηt is an isomorphism.

(iii) *The domain projection functor*

$$\mathrm{Hom}_A(A, t) \xrightarrow{\ p_0\ } A$$

defines a trivial fibration.

(iv) *For any element $a : 1 \to A$, the mapping space $\mathrm{Hom}_A(a, t)$ is contractible.*

(v) *The projection functor*

$$A_{/t} \xrightarrow{\ \sim\ } A$$

whose domain is the slice of A over t is a trivial fibration.

(vi) *Any sphere in A whose final vertex is t admits a filler:*

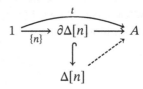

*When these conditions hold, t defines a **terminal element** of A.*

Proof   Unpacking (i), all that is required to define an adjunction $! \dashv t$ is to define a unit natural transformation $\eta : \mathrm{id}_A \Rightarrow t!$ so that the component $\eta t = \mathrm{id}_t$ is an identity. But as proven in Lemma 2.2.2, if ηt is invertible then it is necessarily an identity. This proves the equivalence of (i) and (ii).

Proposition 4.3.13 establishes the equivalence of (i) and (iii) in any ∞-cosmos. By Corollaries 12.2.13 and 5.5.14, the domain projection functor is a trivial fibration if and only if its fibers $\mathrm{Hom}_A(a, t)$ are contractible, proving the equivalence of (iii) and (iv).

By Corollary D.6.6, for any vertex t in a quasi-category, there is an equivalence

$$A_{/t} \xrightarrow{\ \sim\ } \mathrm{Hom}_A(A, t)$$
$$\searrow \underset{A}{} \overset{p_0}{\swarrow}$$

between the canonical projection from the slice construction of Proposition 4.2.5 and the domain projection isofibration. Consequently, by the 2-of-3 property, one isofibration is an equivalence if and only if the other is, proving the equivalence of (iii) and (v).

By Definition 1.1.25, the projection $A_{/t} \twoheadrightarrow A$ is a trivial fibration if and only if the following right lifting property holds for all $n \geq 0$

$$
\begin{array}{ccc}
\partial\Delta[n] & \xrightarrow{\ u\ } & A_{/t} \\
\downarrow & \nearrow_{w} & \downarrow \\
\Delta[n] & \xrightarrow{\ v\ } & A
\end{array}
$$

Via the adjunction

$$
sSet \xrightleftharpoons[{-/-}]{-\star\Delta[0]} \Delta[0]/sSet
$$

the sphere $u : \partial\Delta[n] \to A_{/t}$ transposes into a map $\Lambda^{n+1}[n+1] \to A$ with final vertex t, with the simplex $v : \Delta[n] \to A$ providing a filler for the open face of the horn. Thus, the lifting problem transposes to define a sphere $\partial\Delta[n+1] \to A$ with final vertex t. The desired lift $w : \Delta[n] \to A_{/t}$ exists just when this transposed sphere admits a filler. In this way, we see that the right lifting properties

$$
\forall n \geq 0 \quad
\begin{array}{ccc}
\partial\Delta[n] & \longrightarrow & A_{/t} \\
\downarrow & \nearrow & \downarrow \\
\Delta[n] & \longrightarrow & A
\end{array}
\quad \leftrightsquigarrow \quad
\begin{array}{ccc}
1 \xrightarrow[\{n\}]{} \partial\Delta[n] & \longrightarrow & A \\
\downarrow & \nearrow & \\
\Delta[n] & &
\end{array}
\quad \forall n \geq 1
$$

are transposes, proving the equivalence of (v) and (vi). $\qquad\square$

There is a relative extension of Joyal's characterization (vi):

LEMMA F.1.2. *Suppose E and B are quasi-categories which possess a terminal element and $p : E \twoheadrightarrow B$ is an isofibration which preserves them: if t is terminal in E then pt is terminal in B. Then any lifting problem of the following form has a solution*

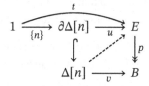

Proof Using the universal property of the terminal object t in E and Proposition

F.1.1(vi), we may extend the sphere u to a map $w : \Delta[n] \to E$. This defines two maps $pw, v : \Delta[n] \to B$ with a common boundary $pu : \partial\Delta[n] \to B$, which we may use to define a sphere $h : \partial\Delta[n+1] \to B$ with $h\delta^{n+1} = pw$ and $h\delta^n = v$ by starting with the degenerate simplex $pw\sigma^n : \Delta[n+1] \to B$, restricting to its boundary, and then replacing the nth face in this sphere with $v : \Delta[n] \to B$. By construction, h maps the vertex $\{n+1\}$ to pt which is terminal in B, so it follows that we may fill this sphere to define a simplex $k : \Delta[n+1] \to B$.

We construct a horn $g : \Lambda^n[n+1] \to E$ by restricting the degenerate simplex $w\sigma^n : \Delta[n+1] \to E$. This pair of maps defines a factorization of the commutative square of the statement:

$$
\begin{array}{ccc}
\partial\Delta[n] \xrightarrow{\ u\ } E & & \partial\Delta[n] \xhookrightarrow{\ \delta^n\ } \Lambda^n[n+1] \xrightarrow{\ g\ } E \\
\downarrow \qquad \downarrow p & = & \downarrow \qquad\qquad \downarrow \quad\nearrow^{\ell}\quad \downarrow p \\
\Delta[n] \xrightarrow{\ v\ } B & & \Delta[n] \xhookrightarrow{\ \delta^n\ } \Delta[n+1] \xrightarrow{\ k\ } B
\end{array}
$$

Since the central vertical map of this commutative rectangle is an inner horn inclusion and its right hand vertical is an isofibration of quasi-categories, it follows that the lifting problem on the right has a solution $\ell : \Delta[n+1] \to E$ as marked, and now it is clear that the map $\ell\delta^n : \Delta[n] \to E$ provides a solution to the original lifting problem.   □

Exercises

EXERCISE F.1.i. Prove that if A and B are quasi-categories which possess a terminal element and $f : A \to B$ is a functor, not necessarily an isofibration, which preserves terminal elements, then given any lifting problem as below-left in which t is terminal in A

$$
\begin{array}{ccc}
1 \xrightarrow[\ \{n\}\ ]{\ \ \overset{t}{\frown}\ \ } \partial\Delta[n] \longrightarrow A & & \partial\Delta[n] \longrightarrow A \\
\downarrow \qquad \downarrow f & \rightsquigarrow & \downarrow \quad\overset{\cong}{\nearrow}\quad \downarrow f \\
\Delta[n] \longrightarrow B & & \Delta[n] \longrightarrow B
\end{array}
$$

there exists a lift as above-right so that the upper-left triangle commutes up to natural isomorphism and the bottom-right triangle commutes on the nose.

EXERCISE F.1.ii. State and prove the equivalence between various synthetic and analytic definitions of an initial element in a quasi-category.

F.2 Limits and Colimits

In this section, we expand Proposition 4.3.2 to prove that the synthetic definition of a limit of a diagram indexed by a simplicial set and taking values in a quasi-category coincides with the analytic definition first introduced by Joyal [61, 4.5]. We prove the equivalence between four synthetic definitions of a limit cone – (i) the original Definition 2.3.8; (ii) appearing in Proposition 4.3.4; (iii) appearing as Definition 9.4.7 and in Proposition 9.4.8; and (iv) from Proposition 4.3.2 – and one analytic one (v) which is Joyal's. In this case, by the results just cited and Proposition F.1.1, there is nothing left to do but state the result and provide references for its components.

PROPOSITION F.2.1. *Consider a functor* $d : J \to A$ *between quasi-categories and a cone* $\lambda : \Delta\ell \Rightarrow d$. *The following are equivalent:*

(i) *The pair* (ℓ, λ) *defines an absolute right lifting diagram*

$$
\begin{array}{ccc}
 & & A \\
 & \overset{\ell}{\nearrow} \Uparrow\lambda & \downarrow \Delta \\
1 & \xrightarrow{\quad d \quad} & A^J
\end{array}
$$

(ii) *The cone* $\ulcorner\lambda\urcorner : 1 \to A^{J^{\triangleleft}}$ *defines an absolute right lifting diagram*

$$
\begin{array}{ccc}
 & & A^{J^{\triangleleft}} \\
 & \overset{\ulcorner\lambda\urcorner}{\nearrow} \| & \downarrow \text{res} \\
1 & \xrightarrow{\quad d \quad} & A^J
\end{array}
$$

(iii) *The cone* $\lambda : \ell! \Rightarrow d$ *defines a pointwise right extension diagram*

$$
\begin{array}{ccc}
J & \xrightarrow{\quad d \quad} & A \\
! \downarrow & \Uparrow\lambda \nearrow & \\
1 & \overset{\ell}{} &
\end{array}
$$

(iv) *The quasi-category of cones* $\mathrm{Hom}_{A^J}(\Delta, d)$ *admits a terminal element* $\ulcorner\lambda\urcorner : 1 \to \mathrm{Hom}_{A^J}(\Delta, d)$, *representing a cone* $\lambda : \Delta\ell \Rightarrow d$.

(v) *The quasi-category* $A_{/d}$ *admits a terminal element* $\ulcorner\lambda\urcorner : 1 \to A_{/d}$, *transposing to define an extension of* d *to a diagram*

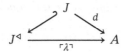

When these conditions hold, the data variously labeled (ℓ, λ) or $\ulcorner \lambda \urcorner$ defines the **limit cone** *over d.*

Note that any diagram $d' : J' \to A$ indexed by a simplicial set and valued in a quasi-category can be extended along an inner anodyne extension to a weakly equivalent diagram $d : J \to A$ between quasi-categories. Alternatively, there is also no cost in the settings of (i), (ii), and (iv), and (v) to working with the original diagram, as the quasi-categories of diagrams $A^J \simeq A^{J'}$ and cones $\mathrm{Hom}_{A^J}(\Delta, d) \simeq \mathrm{Hom}_{A^{J'}}(\Delta, d')$ and $A_{/d} \simeq A_{/d'}$ are equivalent.

Proof   The equivalence of (i) and (ii) is proven in Proposition 4.3.4 for any ∞-category A and simplicial set J; the simplicial set $J^{\triangleleft} := \mathbb{1} \star J$ is the join from Definition 4.2.4. The equivalence of (i) and (iii) is proven in Proposition 9.4.8 for any cartesian closed ∞-cosmos. Proposition 4.3.2 proves the equivalence between (i) and (iv) for any diagram valued in any ∞-category.

Finally, the equivalence between (iv) and (v) is a consequence of the equivalence of quasi-categories $\mathrm{Hom}_{A^J}(\Delta, d) \simeq A_{/d}$ over A of Proposition D.6.4, which provides two models for the quasi-category of cones over d. The final ingredient is Lemma 2.2.7, which proves that if one of these quasi-categories has a terminal element, they both do, as terminal elements are preserved by equivalences. $\square$

Exercises

EXERCISE F.2.i.   State and prove the equivalence between various synthetic and analytic definitions of the colimit of a diagram valued in a quasi-category.

F.3   Right Adjoint Right Inverse Adjunctions

To our knowledge, right adjoint right inverse adjunctions between quasi-categories have not been given much attention. Nonetheless, we pause to establish a useful analytic characterization of such adjunctions, which will help us compare various other synthetic and analytic definitions.

LEMMA F.3.1.   *An isofibration $f : B \twoheadrightarrow A$ of quasi-categories admits a right adjoint right inverse if and only if for every element $a : 1 \to A$, there exists an element $ua : 1 \to B$ with $fua = a$ that has the property that any lifting* *problem of the following form with $n \geq 1$ has a solution.*

$$
\begin{array}{ccc}
1 & \underset{\{n\}}{\overset{ua}{\rightrightarrows}} \partial\Delta[n] \longrightarrow & B \\
& \downarrow \quad\quad\nearrow & \downarrow f \\
& \Delta[n] \longrightarrow & A
\end{array}
\tag{F.3.2}
$$

Proof If $f : B \twoheadrightarrow A$ is an isofibration of quasi-categories admitting a right adjoint right inverse, then by Lemma 3.6.9 its right adjoint u may be chosen to that the counit $\epsilon : fu = \mathrm{id}_A$ is the identity and the adjunction is fibered over A. In this case, the induced fibered equivalence $\ulcorner \epsilon \circ f(-) \urcorner : \mathrm{Hom}_B(B, u) \overset{\sim}{\to} \mathrm{Hom}_A(f, A)$ of Proposition 4.1.1 is represented by the map induced by f between the comma ∞-categories defined in Proposition 3.4.5.

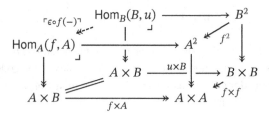

By that result – or alternatively, by Proposition C.1.12, on which its proof relies – we see that the induced map between comma ∞-categories is also an isofibration. Combining these facts, we see that $\ulcorner \epsilon \circ f(-) \urcorner : \mathrm{Hom}_B(B, u) \twoheadrightarrow \mathrm{Hom}_A(f, A)$ is a trivial fibration over $A \times B$. This trivial fibration pulls back over any vertex $a : 1 \to A$ to define a trivial fibration $\mathrm{Hom}_B(B, ua) \twoheadrightarrow \mathrm{Hom}_A(f, a)$ over B. By Corollary D.6.6, the domain and codomain are equivalent to Joyal's slices, so the isofibration $f : B_{/ua} \twoheadrightarrow f_{/a}$ induced by f is also a trivial fibration between quasi-categories. The defining lifting property of Definition 1.1.25

$$
\begin{array}{ccc}
\partial\Delta[n-1] & \longrightarrow & B_{/ua} \\
\downarrow \quad\quad\nearrow & & \downarrow\!\!\downarrow f \\
\Delta[n-1] & \longrightarrow & f_{/a}
\end{array}
$$

for $n \geq 1$ transposes to the lifting property of (F.3.2).

Conversely, the lifting property (F.3.2) can be used to inductively define a section $u : A \to B$ of f extending the choices of elements $ua : 1 \to B$ lifting each $a : 1 \to A$. The inclusion $\mathrm{sk}_0 A \hookrightarrow A$ can be expressed as a countable composite of pushouts of coproducts of maps $\partial\Delta[n] \hookrightarrow \Delta[n]$ with $n \geq 1$, and each intermediate lifting problem required to define a lift

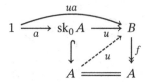

has the form of (F.3.2). To show that u is a right adjoint right inverse to f, it suffices, by Lemma B.4.2 to define a 2-cell $\eta: \ \mathrm{id}_B \Rightarrow uf$ that whiskers with u and with f to isomorphisms. We construct a representative for η by solving the lifting problem

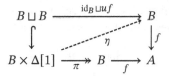

which is again permitted by (F.3.2); note that the inclusion $B \sqcup B \hookrightarrow B \times \Delta[1]$ is bijective on vertices, and note further that every simplex of the codomain that is missing from the domain has its final vertex in the image of u. By construction $f\eta = \mathrm{id}_f$ is certainly invertible.

To show that ηu is an isomorphism it suffices, by Corollary D.4.19, to check that each of its components $\eta u(a): ua \to ufua = ua$ are isomorphisms in A. Inverse isomorphisms to these components can be found by elementary applications of the lifting property (F.3.2), whose details we leave to the reader. □

Exercises

EXERCISE F.3.i. Verify the final statement made in the proof of Lemma F.3.1.

EXERCISE F.3.ii. Formulate an analogous lifting property to characterize those isofibrations $f: B \to A$ that admit left adjoint right inverses.

F.4 Cartesian and Cocartesian Fibrations

The aim in this section is to establish the equivalence between synthetic and analytic characterizations of those isofibrations $p: E \twoheadrightarrow B$ between quasi-categories that define cartesian or cocartesian fibrations. We start by considering p-cartesian arrows, proving the equivalence between the three synthetic definitions of Theorem 5.1.7 and three analytic ones, which appear as [78, 2.4.1.1,

2.4.1.4, 2.4.1.8]. Then we prove a similar comparison between synthetic and analytic characterizations of cartesian fibrations and use this to strengthen the lifting properties associated with cartesian fibrations. We conclude by demonstrating that the discrete cocartesian fibrations coincide with Joyal's left fibrations, while the discrete cartesian fibrations coincide with right fibrations.

In §5.1, the appellation "cartesian arrow" referred to a generalized element $\ulcorner\psi\urcorner\colon X \to E^2$ or equally to the natural transformation $X \underset{e}{\overset{e'}{\rightrightarrows}}\Downarrow\psi\, E$ it

represents. In the ∞-cosmos of quasi-categories, an X-shaped arrow is cartesian if and only if its components $\ulcorner\psi x\urcorner$ at each element $x\colon 1 \to X$ are cartesian (see Proposition 12.2.9), so to simplify the following discussion, we only consider arrows $\ulcorner\psi\urcorner\colon 1 \to E^2$, or equally 1-simplices in E, that we depict with simplified notation as $\psi\colon e' \to e$.

PROPOSITION F.4.1. *Fix an isofibration of quasi-categories* $p\colon E \twoheadrightarrow B$. *The following are equivalent and characterize when a 1-simplex* $\psi\colon e' \to e$ *in E is* **p-cartesian**:

(i) *The isofibration induced by the inclusion* $\delta\colon \lrcorner \hookrightarrow 3$ *with image* $0 \to 2 \leftarrow 1$

$$
\begin{array}{ccc}
E^3 & \xrightarrow{\ \delta\widehat{\pitchfork}p\ } & B^3 \underset{B^{\lrcorner}}{\times} E^{\lrcorner} \\
& \!\!\!\!{}_{p_{12}}\searrow \qquad \swarrow{}_{p_{12}}\!\!\!\! & \\
& E^2 &
\end{array}
$$

pulls back to define a trivial fibration on the fiber over ψ.

(ii) *The commutative triangle defines an absolute right lifting diagram:*

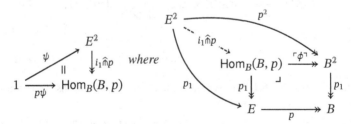

where

(iii) *There is an absolute right lifting diagram with* $p_1\epsilon = \psi$ *and* $p_0\epsilon = \mathrm{id}_{pe'}$

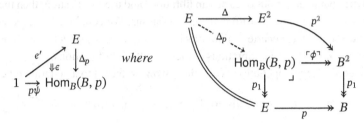

(iv) *The induced map to the pullback in the square*

$$
\begin{array}{ccc}
E_{/\psi} & \longrightarrow\!\!\!\!\! \rightarrow & E_{/e} \\
p\downarrow & & \downarrow p \\
B_{/p\psi} & \longrightarrow\!\!\!\!\! \rightarrow & B_{/pe}
\end{array}
$$

is a trivial fibration $E_{/\psi} \twoheadrightarrow B_{/p\psi} \times_{B_{/pe}} E_{/e}$.

(v) *Any lifting problem of the following form for* $n \geq 2$ *has a solution:*

$$
\begin{array}{ccc}
\Delta[1] \underset{\{n-1,n\}}{\overset{\psi}{\rightrightarrows}} \Lambda^n[n] & \longrightarrow & E \\
\downarrow & \nearrow & \downarrow p \\
\Delta[n] & \longrightarrow & B
\end{array}
\qquad (\mathrm{F}.4.2)
$$

(vi) *Any lifting problem of the following form for* $n \geq 1$ *has a solution:*

$$
\begin{array}{ccc}
\Delta[1] \underset{\{n\}\times\mathrm{id}}{\overset{\psi}{\rightrightarrows}} \partial\Delta[n]\times\Delta[1] \underset{\partial\Delta[n]\times\{1\}}{\cup} \Delta[n]\times\{1\} & \longrightarrow & E \\
\uparrow & & \downarrow p \\
\Delta[n]\times\Delta[1] & \longrightarrow & B
\end{array}
\qquad (\mathrm{F}.4.3)
$$

Proof Theorem 5.1.7 proves the equivalence of conditions (i), (ii), and (iii) in any ∞-cosmos.

By Lemma 5.1.2, the map in (i) is isomorphic to the induced map to the pullback in a square that very similar to the square appearing in (iv).[1] We will demonstrate that these squares are equivalent by constructing a natural

---

[1] Indeed, the initial object in the square (5.1.3) is a quasi-category we denote by $E_{/\psi}$ because it is equivalent (though not isomorphic) to Joyal's slice quasi-category (see Warning 4.2.10). To avoid confusion here, we write $E_{/\!/\psi}$ for that quasi-category and reserve the notation $E_{/\psi}$ for the slice quasi-category of Definition 4.2.4.

equivalence displayed below-left between the top-horizontal maps in each weakly cartesian square:

$$
\begin{array}{ccc}
E_{/\psi} & \longrightarrow\!\!\!\!\rightarrow & E_{/e} \\
\downarrow{\scriptstyle\wr} & & \downarrow{\scriptstyle\wr} \\
E_{/\!/\psi} & \longrightarrow\!\!\!\!\rightarrow & \mathrm{Hom}_E(E, e)
\end{array}
\qquad\longleftrightarrow\!\!\sim\qquad
\begin{array}{ccc}
\Delta[n+2] & \xleftarrow{\ \delta^{n+1}\ } & \Delta[n+1] \\
\pi\uparrow & & \uparrow\rho \\
\Delta[n] \times \Delta[2] & \xleftarrow[\ \mathrm{id}\times\delta^1\ ]{} & \Delta[n] \times \Delta[1]
\end{array}
$$

The maps in this diagram are most easily described in terms of their actions on n-simplices: each map is given by restriction along the corresponding functor in the commutative square displayed above-right, where π and ρ are both defined by the formula:

$$
\rho(i, j) := \pi(i, j) := \begin{cases} i & j = 0 \\ n+j & j > 0 \end{cases}
$$

Propositions D.6.3 and D.6.4 prove that the left-hand vertical map is an equivalence, while Corollary D.6.6 demonstrates that the right-hand vertical map is an equivalence. Thus, the two squares are equivalent and, by Lemma 5.1.2, (i) is equivalent to (iv).

By the adjunction of Proposition 4.2.5, the lifting property that characterizes the trivial fibration of (iv) transposes to the lifting property of (v)

$$
\begin{array}{ccc}
\partial\Delta[n] & \longrightarrow & E_{/\psi} \\
\downarrow & \nearrow & \downarrow \\
\Delta[n] & \longrightarrow & B_{/p\psi} \underset{B_{/pe}}{\times} E_{/e}
\end{array}
\qquad\longleftrightarrow\!\!\sim\qquad
\begin{array}{ccc}
\Delta[1] \xrightarrow[\{n+1,n+2\}]{\ \ \psi\ \ } \Lambda^{n+2}[n+2] & \longrightarrow & E \\
\downarrow & \nearrow & \downarrow{\scriptstyle p} \\
\Delta[n+2] & \longrightarrow & B
\end{array}
\quad \text{for } n \geq 0
$$

proving that (iv) is equivalent to (v).

Now we argue that the lifting properties (F.4.2) and (F.4.3) are equivalent. One implication holds on account of the retract diagram

$$
\begin{array}{ccccc}
\Delta[1] \xrightarrow[\{n,n+1\}]{\ \{n\}\times\mathrm{id}\ } \Lambda^{n+1}[n+1] & \hookrightarrow & \partial\Delta[n]\times\Delta[1] \underset{\partial\Delta[n]\times\{1\}}{\cup} \Delta[n]\times\{1\} & \longrightarrow & \Lambda^{n+1}[n+1] \\
\uparrow & & \uparrow & & \uparrow \\
\Delta[n+1] & \xhookrightarrow{\ \iota\ } & \Delta[n]\times\Delta[1] & \xrightarrow{\ \rho\ } & \Delta[n+1]
\end{array}
$$

$$\text{(F.4.4)}$$

in which ρ is the map defined above and ι is its unique section. By Lemma

C.2.3 it is now clear that the lifting property (F.4.3) implies the lifting property (F.4.2).

For the converse, we show that the lifting property assumed in (v) suffices to solve this lifting problem presented by (vi). Our task is to find lifts along p for each of the $n + 1$ shuffles of $\Delta[n] \times \Delta[1]$. We number these shuffles $0, \ldots, n$ starting from the closed end of the cylinder. Proceeding inductively for $k < n$, we choose a lift for the kth shuffle by filling a $\Lambda^{k+1}[n + 1]$-horn, which can be done since p is an isofibration between quasi-categories. To lift the nth shuffle, we are required to fill a $\Lambda^{n+1}[n + 1]$-horn whose final $\{n, n + 1\}$ edge is ψ, which can be done with the lifting property (F.4.2). This demonstrates that (v)$\Leftrightarrow$(vi). □

Our next result compares the three synthetic definitions proven equivalent in Theorem 5.2.8 and a fourth synthetic definition of Proposition 5.2.11 with two analytic definitions due to Lurie [78, §2.4.1-2].

PROPOSITION F.4.5. *For an isofibration* $p : E \twoheadrightarrow B$ *between quasi-categories, the following are equivalent and define what it means for* p *to be a **cartesian fibration**:*

(i) *Every natural transformation* $\beta : b \Rightarrow pe$ *as below-left admits a lift* $\chi : e' \Rightarrow e$ *as below-right:*

$$
\begin{array}{ccc}
X \xrightarrow{\ e\ } E & & X \overset{e}{\underset{e'}{\Longrightarrow}}{\scriptstyle\Uparrow\chi} E \\
\ \searrow \ {\scriptstyle\Uparrow\beta} \ \downarrow{\scriptstyle p} & = & \qquad\ \downarrow{\scriptstyle p} \\
\quad b \searrow B & & B
\end{array}
$$

with the property that:

- **induction:** *Given any functor* $f : Y \to X$ *and natural transforma-tions* $Y \overset{e''}{\underset{ef}{\Rightarrow}}{\scriptstyle\Downarrow\tau} E$ *and* $Y \overset{pe''}{\underset{pe'f}{\Rightarrow}}{\scriptstyle\Downarrow\gamma} B$ *so that* $p\tau = p\chi f \cdot \gamma$, *there exists a lift* $Y \overset{e''}{\underset{e'f}{\Rightarrow}}{\scriptstyle\Downarrow\bar\gamma} E$ *of* γ *so that* $\tau = \chi f \cdot \bar\gamma$.

- **conservativity:** *Any fibered endomorphism of a restriction of* χ *is invertible: if* $Y \overset{e'f}{\underset{e'f}{\Rightarrow}}{\scriptstyle\Downarrow\zeta} E$ *is any natural transformation so that* $\chi f \cdot \zeta = \chi f$ *and* $p\zeta = \mathrm{id}_{pe'f}$ *then* ζ *is invertible.*

(ii) *Every natural transformation* $\beta : b \Rightarrow pe$ *as below-left admits a lift*

$\chi : e' \Rightarrow e$ as below-right:

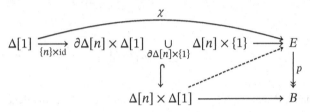

with the property that the induced map is a trivial fibration:

$$E_{/\chi} \twoheadrightarrow B_{/\beta} \underset{B_{/pe}}{\times} E_{/e}$$

(iii) *The functor* $\Delta_p : E \to \mathrm{Hom}_B(B, p)$ *admits a right adjoint over* B.

(iv) *The functor* $i_1 \widehat{\pitchfork} p : E^2 \twoheadrightarrow \mathrm{Hom}_B(B, p)$ *admits a right adjoint right inverse.*

(v) *Any 1-simplex* $\beta : b \to pe$ *in* B *admits a lift* $\chi : e' \to e$ *in* E *so that any lifting problem for* $n \geq 1$

$$\begin{array}{ccc}
\Delta[1] \xrightarrow[\{n\}\times\mathrm{id}]{} \partial\Delta[n] \times \Delta[1] \underset{\partial\Delta[n]\times\{1\}}{\cup} \Delta[n] \times \{1\} \xrightarrow{\chi} & E \\
\Big\downarrow & \Big\downarrow p \\
\Delta[n] \times \Delta[1] \longrightarrow & B
\end{array}$$

has a solution.

(vi) *Any 1-simplex* $\beta : b \to pe$ *in* B *admits a lift* $\chi : e' \to e$ *in* E *so that any lifting problem for* $n \geq 2$

$$\begin{array}{ccc}
\Delta[1] \xrightarrow[\{n-1,n\}]{} \Lambda^n[n] \xrightarrow{\chi} & E \\
\Big\downarrow & \Big\downarrow p \\
\Delta[n] \longrightarrow & B
\end{array}$$

has a solution.

Condition (vi) appears to be mildly stronger than [78, 2.4.2.1], which only requires that p is an inner fibration with the lifting property (F.4.2), but it follows easily that any such p must be an isofibration (see Exercise F.4.i).

Proof The equivalence of (i) and (iii) is proven in Proposition 5.2.11, using the equivalence of (i) is equivalent to (iv) in Proposition F.4.1 to identify the equivalence that characterizes p-cartesian arrows, while the equivalence of (ii), (iii), and (iv) is proven in Theorem 5.2.8.

It remains to verify the equivalence between any of these synthetic conditions and the corresponding analytic ones. We will demonstrate that (iv)⇔(v) and (v)⇔(vi).

By Lemma F.3.1, the isofibration $i_1 \mathbin{\widehat{\pitchfork}} p$ admits a right adjoint right inverse if and only if any 1-simplex $\beta : b \to pe$ in B admits a lift $\chi : e' \to e$ in E with the lifting property

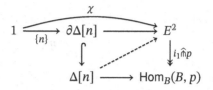

for $n \geq 1$. This lifting property is equivalent to the transposed lifting property

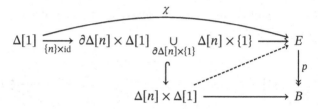

again for $n \geq 1$, proving the equivalence between (iv) and (v).

Finally, Proposition F.4.1(v)⇔(vi) demonstrates the equivalence between (vi) and (v). □

We now extend the notion of cartesian arrow to define a notion of "cartesian cylinder" to describe a variant of the lifting properties appearing in Proposition F.4.1.

DEFINITION F.4.6 (cartesian cylinders). Suppose that $p : E \twoheadrightarrow B$ is a cartesian fibration of quasi-categories and that X is any simplicial set. We say that a cylinder $e : X \times \Delta[1] \to E$ is **pointwise p-cartesian** if and only if for each 0-simplex $x \in X$, e maps the 1-simplex $(x\sigma^0, \mathrm{id}_{[1]}) : (x,0) \to (x,1)$ to a p-cartesian arrow in E.

LEMMA F.4.7. *Let $p : E \twoheadrightarrow B$ be a cartesian fibration of quasi-categories. A cylinder $e : X \times \Delta[1] \to E$ is pointwise p-cartesian if and only if $e : \Delta[1] \to E^X$ defines a p^X-cartesian arrow for the cartesian fibration $p^X : E^X \twoheadrightarrow B^X$.*

Proof   First note that Corollary 5.3.5 implies that $p^X : E^X \twoheadrightarrow B^X$ is a cartesian fibration, while Proposition 5.6.2 proves that restriction along any $f : Y \to X$

defines a cartesian functor of cartesian fibrations:

$$
\begin{array}{ccc}
E^X & \xrightarrow{E^f} & E^Y \\
p^X \downarrow & & \downarrow p^Y \\
B^X & \xrightarrow{B^f} & B^Y
\end{array}
$$

In particular, p^X-cartesian arrows are pointwise p-cartesian. Conversely, any pointwise p-cartesian cylinder factors through a p^X-cartesian arrow via a pointwise isomorphism. By Corollary 1.1.22, a pointwise isomorphism is in fact an isomorphism, so by isomorphism stability of p^X-cartesian arrows, any pointwise p-cartesian cylinder is a p^X-cartesian arrow. $\qquad\square$

The following lifting property is used to prove Proposition 12.2.4.

LEMMA F.4.8. *Let $X \hookrightarrow Y$ be a simplicial subset of a simplicial set Y.*

(i) *Any lifting problem for $n \geq 0$*

$$
\begin{array}{ccc}
X \times \Delta[1] \cup Y \times \{1\} & \xrightarrow{e} & E \\
\downarrow & \nearrow^{\bar{e}} & \downarrow p \\
Y \times \Delta[1] & \xrightarrow{b} & B
\end{array}
$$

with the property that the cylinder $X \times \Delta[1] \subseteq X \times \Delta[1] \cup Y \times \{1\} \xrightarrow{e} E$ is pointwise p-cartesian admits a solution $\bar{e}$ which is also pointwise p-cartesian.

(ii) *Any lifting problem for $n \geq 1$*

$$
\begin{array}{ccc}
X \times \Delta[n] \cup Y \times \Lambda^n[n] & \xrightarrow{e} & E \\
\downarrow & \nearrow^{\bar{e}} & \downarrow p \\
Y \times \Delta[n] & \xrightarrow{b} & B
\end{array}
$$

in which the cylinder $Y \times \Delta^{\{n-1,n\}} \subseteq X \times \Delta[n] \cup Y \times \Lambda^n[n] \xrightarrow{e} E$ is pointwise p-cartesian admits a solution $\bar{e}$.

Proof The Leibniz tensor with $X \hookrightarrow Y$ defines a functor $sSet^2 \to sSet^2$ that preserves the retract diagram (F.4.4), so the lifting property (ii) follows from (i). In turn, the lifting property (i) follows inductively from Proposition F.4.1(vi) combined with the fact that any monomorphism $X \hookrightarrow Y$ can be decomposed as a sequential composite of pushouts of coproducts of inclusions $i_n : \partial\Delta[n] \hookrightarrow \Delta[n]$, and by Proposition C.2.9(vii), the pushout product with $\{1\} \hookrightarrow \Delta[1]$ is then similarly a sequential composite of pushouts of coproducts of the pushout products $\partial\Delta[n] \times \Delta[1] \cup \Delta[n] \times \{1\} \hookrightarrow \Delta[n] \times \Delta[1]$. $\qquad\square$

A final result demonstrates that discrete cocartesian and discrete cartesian fibrations between quasi-categories coincide with the classes of left fibrations and right fibrations introduced by Joyal [61].

PROPOSITION F.4.9. *For an isofibration $p: E \twoheadrightarrow B$ between quasi-categories, the following are equivalent and define what it means for p to be a **discrete cartesian fibration**:*

    (i) *The map $p: E \twoheadrightarrow B$ is a cartesian fibration whose fibers are Kan complexes.*

    (ii) *Every 2-cell $\beta: b \Rightarrow pe$ in the homotopy 2-category of quasi-categories has an essentially unique lift: given $\chi: e' \Rightarrow e$ and $\psi: e'' \Rightarrow e$ so that $p\chi = p\psi = \beta$, then there exists an isomorphism $\gamma: e'' \Rightarrow e'$ with $\chi \cdot \gamma = \psi$ and $p\gamma = \mathrm{id}$.*

    (iii) *The induced functor $i_1 \widehat{\pitchfork} p: E^2 \twoheadrightarrow \mathrm{Hom}_B(B, p)$ is a trivial fibration.*

    (iv) *The functor $p: E \twoheadrightarrow B$ is a right fibration, with the lifting property:*

$$
\begin{array}{ccc}
\Lambda^k[n] & \longrightarrow & E \\
\downarrow & \nearrow & \downarrow{\scriptstyle p} \\
\Delta[n] & \longrightarrow & B
\end{array}
\qquad \text{for} \quad n \geq 0 \quad \text{and} \quad 0 < k \leq n.
$$

Proof The first characterization (i) is a reinterpretation of the original Definition 5.5.3 using Proposition 12.2.3, which says that an isofibration $p: E \twoheadrightarrow B$ defines a discrete object in $\mathcal{QC}at_{/B}$ if and only if its fibers are Kan complexes. The equivalence of (i) with (ii) is proven in Proposition 5.5.6, while the equivalence of (i) with (iii) is proven in Proposition 5.5.8.

We conclude by demonstrating the equivalence of (iii) and (iv). By adjunction, $i_1 \widehat{\pitchfork} p$ is a trivial fibration if and only if the lifting problem below-right has a solution

$$
\begin{array}{ccc}
\partial\Delta[n] & \longrightarrow & E^2 \\
\downarrow & \nearrow & \downarrow{\scriptstyle i_1\widehat{\pitchfork}p} \\
\Delta[n] & \longrightarrow & \mathrm{Hom}_B(B, p)
\end{array}
\quad \leftrightsquigarrow \quad
\begin{array}{ccc}
\partial\Delta[n] \times \Delta[1] \cup \Delta[n] \times \{1\} & \longrightarrow & E \\
\downarrow & \nearrow & \downarrow{\scriptstyle p} \\
\Delta[n] \times \Delta[1] & \longrightarrow & B
\end{array}
$$

By Proposition D.3.8, if $p: E \twoheadrightarrow B$ is a right fibration, satisfying condition (iv), then the lifting problem above-right admits a solution and hence $i_1 \widehat{\pitchfork} p: E^2 \twoheadrightarrow \mathrm{Hom}_B(B, p)$ is a trivial fibration proving (iii). Conversely, for $0 < k \leq n$ the horn inclusion $\Lambda^k[n] \hookrightarrow \Delta[n]$ is a retract

$$\Lambda^k[n] \xrightarrow{\Lambda^k[n] \times \{0\}} \Lambda^k[n] \times \Delta[1] \cup \Delta[n] \times \{1\} \xrightarrow{\ r\ } \Lambda^k[n]$$

$$\Delta[n] \xrightarrow{\Delta[n] \times \{0\}} \Delta[n] \times \Delta[1] \xrightarrow{\qquad r \qquad} \Delta[n]$$

$$\text{where} \quad r(i,0) = i \quad \text{and} \quad r(i,1) = \begin{cases} i & i \neq k-1 \\ k & i = k-1. \end{cases}$$

Thus, to solve the lifting problem postulated by (iv), it suffices to show that p lifts against the pushout products $(\Lambda^k[n] \hookrightarrow \Delta[n]) \hat{\times} (\{1\} \hookrightarrow \Delta[1])$, which transposes to a lifting problem between the monomorphism $\Lambda^k[n] \hookrightarrow \Delta[n]$ and $i_1 \hat{\pitchfork} p : E^2 \twoheadrightarrow \operatorname{Hom}_B(B, p)$. If (iii) holds and $i_1 \hat{\pitchfork} p$ is a trivial fibration, then this constructs the desired lift. $\square$

Exercises

EXERCISE F.4.i. Suppose $p : E \to B$ is an inner fibration between quasi-categories so that any 1-simplex $\beta : b \to pe$ in B admits a lift $\chi : e' \to e$ in E so that any lifting problem for $n \geq 2$

$$
\begin{array}{ccc}
& \xrightarrow{\quad \chi \quad} & \\
\Delta[1] \xrightarrow[\{n-1,n\}]{} \Lambda^n[n] & \longrightarrow & E \\
\downarrow & \nearrow & \downarrow p \\
\Delta[n] & \longrightarrow & B
\end{array}
$$

has a solution. Show that p is an isofibration.

EXERCISE F.4.ii. State and prove the equivalence between various synthetic and analytic definitions of

 (i) a cocartesian fibration between quasi-categories,
 (ii) a cocartesian arrow in a quasi-category, and
 (iii) a discrete cocartesian fibration between quasi-categories.

F.5 Adjunctions

The comparison between the analytic and synthetic definitions of adjunction between quasi-categories is somewhat more subtle than for the other categorical notions considered in this appendix, as these are typically presented with different data. The synthetic definition of an adjunction, originally due to Joyal [63], involves two specified functors $f : B \rightarrow A$ and $u : A \rightarrow B$, together with specified maps $\eta : B \times \Delta[1] \rightarrow B$ and $\eta : A \times \Delta[1] \rightarrow A$ up to homotopy in B^B or A^A, together with 2-simplices in the quasi-categories B^A and A^B that witness the triangle equalities. By contrast, the analytic notion of an adjunction between A and B, due to Lurie [78, 5.2.2.1], is defined to be an isofibration[2] $M \twoheadrightarrow \Delta[1]$ that is both a cartesian fibration and a cocartesian fibration, called a **correspondence**, with equivalences $M_1 \simeq A$ and $M_0 \simeq B$ identifying the quasi-categories A and B with the fibers over the endpoints of the 1-simplex

Since Proposition 2.1.13 demonstrates that the synthetic notion of adjunction is equivalence-invariant, we simplify our notation somewhat and let B and A denote the fibers over 0 and 1, respectively, of the isofibration $M \twoheadrightarrow \Delta[1]$. Our aim in this section is to show that from a cocartesian and cartesian fibration $M \twoheadrightarrow \Delta[1]$, one can extract an adjunction between B and A, with the adjoint functors determined uniquely up to isomorphism, and conversely from a 2-categorical adjunction, one can construct a corresponding correspondence $M \twoheadrightarrow \Delta[1]$, which is unique up to fibered equivalence. As the proof is more involved, we break the argument up into several intermediate steps.

PROPOSITION F.5.1. *Let M be a quasi-category equipped with a map $M \twoheadrightarrow \Delta[1]$ that is both a cocartesian fibration and cartesian fibration. Then*

- *the fibers $B := M_0$ and $A := M_1$ and*
- *the functors $f : B \rightarrow A$ and $u : A \rightarrow B$ defined by the cocartesian and cartesian lift, respectively, of the generic arrow in $\Delta[1]$*

define an adjunction

$$A \underset{u}{\overset{f}{\underset{\perp}{\leftrightarrows}}} B$$

Proof   This is a special case of Proposition 5.4.7. We recall the construction of f and u and leave the rest of the details to that result and Remark 5.2.7. Let χ denote a cocartesian lift of the generic arrow in $\Delta[1]$ whose domain is the inclusion $B \hookrightarrow M$ of the fiber over 0. The codomain of this lifted arrow lands in the fiber over 1 and thus factors uniquely through the inclusion $A \hookrightarrow M$ of that

---

[2]   The nitpicker might note that Lurie only requires an inner fibration, but any inner fibration over $\Delta[1]$ is automatically an isofibration. In fact, so long as M is a quasi-category, any simplicial map $M \rightarrow \Delta[1]$ is automatically an isofibration.

fiber:

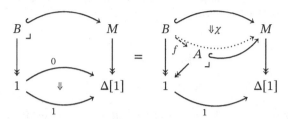

This factorization defines the functor $f : B \to A$. Since cocartesian lifts of a fixed arrow and adjoints to a fixed functor are each unique up to isomorphism, this construction is suitably well-defined. The construction of the right adjoint $u : A \to B$ is dual, involving a cartesian lift of the generic arrow in $\Delta[1]$. $\quad\square$

The converse makes use of something we call the **quasi-categorical collage construction**.

DEFINITION F.5.2 (the quasi-categorical collage construction). For any cospan $B \xrightarrow{g} C \xleftarrow{f} A$ of quasi-categories, define a simplicial set $\mathrm{col}(f, g)$ by

$$\mathrm{col}(f, g)_n = \left\{ (a \in A_i, b \in B_j, c \in C_n) \;\middle|\; \begin{array}{l} c|_{\{0,\dots,i\}} = f(a),\ i, j \geq -1, \\ c|_{\{n-j,\dots,n\}} = g(b),\ i + j = n - 1 \end{array} \right\}$$

with the convention that conditions indexed by $\Delta[-1]$ are empty (or that each simplicial set is terminally augmented). There are evident inclusions that fit into a commutative diagram:

$$
\begin{array}{ccccc}
B & \lhook\joinrel\longrightarrow & \mathrm{col}(f, g) & \longleftarrow\joinrel\rhook & A \\
\downarrow & {}^{\lrcorner} & \downarrow{\scriptstyle\rho} & {}_{\llcorner} & \downarrow \\
\{1\} & \lhook\joinrel\longrightarrow & \Delta[1] & \longleftarrow\joinrel\rhook & \{0\}
\end{array}
$$

The map ρ sends an n-simplex $(a : \Delta[i] \to A, b : \Delta[j] \to B, c : \Delta[n] \to C)$ to the n-simplex $[n] \to [1]$ that carries $0, \dots, i$ to 0 and $i + 1, \dots, n$ to 1. Note that the fiber of ρ over 0 is isomorphic to A while the fiber of ρ over 1 is isomorphic to B

As is our custom for two-sided fibrations and modules, we write $B + A \hookrightarrow \mathrm{col}(f, g)$ for the inclusions of the fibers over 1 and 0 – with the fiber over 1 on the left and the fiber over 0 on the right. This positions the covariantly acting quasi-category on the "left" and the contravariantly acting quasi-category on the "right."

LEMMA F.5.3. *The map* $\rho : \mathrm{col}(f, g) \to \Delta[1]$ *is an inner fibration. In particular, the simplicial set* $\mathrm{col}(f, g)$ *is a quasi-category.*

Proof   Since the fibers of ρ over 0 and 1 are the quasi-categories A and B, it suffices to consider inner horns

$$
\begin{array}{ccc}
\Lambda^k[n] & \longrightarrow & \mathrm{col}(f,g) \\
\downarrow & \nearrow & \downarrow \rho \\
\Delta[n] & \xrightarrow{\quad\alpha\quad} & \Delta[1]
\end{array}
$$

for which $\alpha \colon [n] \to [1]$ is a surjection. Suppose α carries $0, \ldots, i$ to 0 and $i+1, \ldots, n$ to 1. Note that for any $0 < k < n$, the faces $\{0, \ldots, i\}$ and $\{i+1, \ldots, n\}$ of $\Delta[n]$ belong to the horn $\Lambda^k[n]$. In particular, the map $\Lambda^k[n] \to \mathrm{col}(f,g)$ identifies simplices $a \colon \Delta[i] \to A$ and $\Delta[n-i-1] \to B$ together with a horn $\Lambda^k[n] \to C$ whose initial and final faces are the images of these simplices under $f \colon A \to C$ and $g \colon B \to C$. Since C is a quasi-category this horn admits a filler $c \colon \Delta[n] \to C$ and the triple (a, b, c) defines an n-simplex in $\mathrm{col}(f,g)$ solving the lifting problem. $\square$

We write $\mathrm{col}(f, B)$ for the collage of $f \colon A \to B$ with the identity on B.

LEMMA F.5.4. *For any $f \colon A \to B$, the map $\rho \colon \mathrm{col}(f,B) \to \Delta[1]$ is a cocartesian fibration.*

Proof   By Proposition F.4.5(vi), to prove the claim, we need only specify cocartesian lifts of the non-degenerate 1-simplex of $\Delta[1]$ and demonstrate that these edges have the corresponding universal property. To that end, for any vertex $a \in A_0$, let $\chi_a \colon \Delta[1] \to \mathrm{col}(f,B)$ be the 1-simplex

$$
\chi_a := (a \colon \Delta[0] \to A, fa \colon \Delta[0] \to B, fa\sigma^0 \colon \Delta[1] \to B),
$$

defined by the degenerate edge at $fa \in B_0$ lying over the 1-simplex in $\Delta[1]$. To show that χ_a is ρ-cocartesian, we must construct fillers for any left horn

$$
\begin{array}{ccc}
\Delta[1] \xrightarrow[\{0,1\}]{\overset{\chi_a}{\longrightarrow}} \Lambda^0[n] & \longrightarrow & \mathrm{col}(f,B) \\
\downarrow & \nearrow & \downarrow \rho \\
\Delta[n] & \xrightarrow{\quad\beta\quad} & \Delta[1]
\end{array}
$$

whose initial edge is χ_a. Note that this condition implies that the bottom map $\beta \colon [n] \to [1]$ carries 0 to 0 and the remaining vertices to 1. The map $\Lambda^0[n] \to \mathrm{col}(f,B)$ defines a horn $\Lambda^0[n] \to B$ in the quasi-category B whose first edge is degenerate. By Proposition 1.1.14, this "special outer horn" admits a filler $b \colon \Delta[n] \to B$ and the triple

$$
(a \colon \Delta[0] \to A, b\delta^0 \colon \Delta[n-1] \to B, b \colon \Delta[n] \to B)
$$

defines an n-simplex in $\mathrm{col}(f, B)$ that solves the lifting problem. □

PROPOSITION F.5.5. *For any* $f: A \to B$ *between quasi-categories, the collage* $\mathrm{col}(f, B)$ *defines the* **oplax colimit** *of f in* $\mathcal{QC}\mathrm{at}$. *That is* $\mathrm{col}(f, B)$ *defines a cone under the pushout diagram*

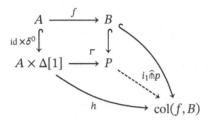

so that the induced map k is inner anodyne, and in particular a weak equivalence in the Joyal model structure.

Proof The map k is a quotient of the map h, which has the following explicit description. For each n-simplex $(a, \alpha): \Delta[n] \to A \times \Delta[1]$ define $i := |\alpha^{-1}(0)| - 1$, so that $-1 \le i \le n$. Then h carries (a, α) to the n-simplex of $\mathrm{col}(f, B)$ corresponding to the triple

$$(a|_{\{0,\dots,i\}}: \Delta[i] \to A, fa|_{\{i+1,\dots,n\}}: \Delta[n-i-1] \to B, fa: \Delta[n] \to B).$$

Note that the composite $\rho h: A \times \Delta[1] \to \Delta[1]$ is the projection.

It remains to present k as a sequential composite of pushouts of coproducts of inner horn inclusions. To do so, first note that

$$\mathrm{col}(f, B)_n = A_n \amalg A_{n-1} \times_{B_{n-1}} B_n \amalg \cdots \amalg A_0 \times_{B_0} B_n \amalg B_n$$

where each map $B_n \to B_i$ is the initial face map corresponding to $\{0, \dots, i\} \hookrightarrow \Delta[n]$. From the perspective of this decomposition, P_n is the subset containing the sets A_n and B_n and the subset of $A_i \times_{B_i} B_n$ whose component in B_n is in the image of f. The n-simplices of $\mathrm{col}(f, B)$ that remain to be attached correspond to elements of $A_i \times_{B_i} B_n$, for $0 \le i < n$, that are not in the image of f in the sense just discussed. Note in particular that the map on vertices $k: P_0 \hookrightarrow \mathrm{col}(f, B)_0$ is an isomorphism and $k: P_n \hookrightarrow \mathrm{col}(f, B)_n$ is an injection for all $n \ge 1$.

To enumerate our attaching maps, we start with the collection of non-degenerate n-simplices of $\mathrm{col}(f, B)$ for $n \ge 1$ that are not in the image of f and remove also those elements of $A_i \times_{B_i} B_n$ whose components $b \in B_n$ are in the image of the degeneracy map $\sigma_i: B_{n-1} \to B_n$. Partially order this set of simplices first in the order of increasing n and then in order of increasing index i; that is we lexicographically order the collection of pairs (n, i) with $n \ge 1$ and $0 \le i < n$.

We filter the inclusion $P \hookrightarrow \mathrm{col}(f, B)$ as

$$P \hookrightarrow P_{(1,0)} \hookrightarrow P_{(2,0)} \hookrightarrow P_{(2,1)} \hookrightarrow P_{(3,0)} \hookrightarrow \cdots \hookrightarrow P_{(n,i)} \hookrightarrow \cdots \hookrightarrow \mathrm{colim} \cong \mathrm{col}(f, B)$$

where the simplicial set $P_{(n,i)}$ is built from the previous one by a pushout of a coproduct of inner horns indexed by the set of n-simplices $(a, b) \in A_i \times_{B_i} B_n$ with b not in the image of f or σ_i. The filler for the horn indexed by (a, b) will attach this n simplex to B_n as the missing face of the horn and also the $n + 1$ simplex $(a, b\sigma^i) \in A_i \times_{B_i} B_{n+1}$.

Consider a simplex $(a, b) \in A_i \times_{B_i} B_n$ with b not in the image of f or σ_i. Define a horn

$$
\begin{array}{ccc}
\Lambda^{i+1}[n+1] & \longrightarrow & P_{(n,i)} \\
\downarrow & & \downarrow \\
\Delta[n+1] & \xrightarrow[(a, b\sigma^i)]{} & \mathrm{col}(f, B)
\end{array}
$$

For each $0 \le j < i + 1$, the δ^j-face of the $n + 1$ simplex $(a, b\sigma^i)$ is the n-simplex $(a\delta^j, b\sigma^i\delta^j)$, which lies in $P_{(n,i-1)}$ or in $B \hookrightarrow P$ in the case $i = 0$. For each $i + 1 < j \le n + 1$, the δ^j-face of the $n + 1$ simplex $(a, b\sigma^i)$ is the n-simplex $(a, b\sigma^i\delta^j) = (a, b\delta^{j-1}\sigma^i) \in A_i \times_{B_i} B_n$, which was previously attached to $P_{(n-1,i)}$. So the horn $\Lambda^{i+1}[n+1]$ indeed maps to $P_{(n,i)}$, permitting an inductive construction of the next simplicial set in this sequence as the pushout

$$
\begin{array}{ccc}
\coprod_{\sim} \Lambda^{i+1}[n+1] & \longrightarrow & P_{(n,i)} \\
\downarrow & & \downarrow \\
\coprod_{\sim} \Delta[n+1] & \longrightarrow & P_{(n,i)+1}
\end{array}
$$

where $P_{(n,i)+1}$ is $P_{(n+1,0)}$ in the case $i = n - 1$ and $P_{(n,i+1)}$ otherwise. $\square$

Putting these results together, we are now able to prove the desired equivalence between the synthetic and analytic notions of adjunction.

PROPOSITION F.5.6. *For a pair of functors between quasi-categories $f : B \to A$ and $u : A \to B$, the following are equivalent and define what it means to have an adjunction*

$$B \underset{u}{\overset{f}{\rightleftarrows}} A$$

(i) *There are natural transformations $\eta : \mathrm{id}_B \Rightarrow uf$ and $\epsilon : fu \Rightarrow \mathrm{id}_A$ satisfying the triangle equalities: $u\epsilon \cdot \eta u = \mathrm{id}_u$ and $\epsilon f \cdot f\eta = \mathrm{id}_f$.*

(ii) The functor f defines an absolute left lifting of id_B *through u:*

$$
\begin{array}{ccc}
 & & A \\
 & {\scriptstyle f}\nearrow & \Big\downarrow{\scriptstyle u} \\
 & {\scriptstyle \Uparrow\eta} & \\
B & =\!=\!= & B
\end{array}
$$

(iii) The functor u defines an absolute right lifting of id_A *through f:*

$$
\begin{array}{ccc}
 & & B \\
 & {\scriptstyle u}\nearrow & \Big\downarrow{\scriptstyle f} \\
 & {\scriptstyle \Downarrow\epsilon} & \\
A & =\!=\!= & A
\end{array}
$$

(iv) There is a pointwise left extension diagram that is absolute:

$$
\begin{array}{ccc}
B & =\!=\!= & B \\
{\scriptstyle f}\Big\downarrow & {\scriptstyle \Downarrow\eta} & \nearrow \\
A & & {\scriptstyle u}
\end{array}
$$

(v) There is a pointwise right extension diagram that is absolute:

$$
\begin{array}{ccc}
A & =\!=\!= & A \\
{\scriptstyle u}\Big\downarrow & {\scriptstyle \Uparrow\epsilon} & \nearrow \\
B & & {\scriptstyle f}
\end{array}
$$

(vi) The modules $\mathrm{Hom}_A(f,A) \simeq_{A\times B} \mathrm{Hom}_B(B,u)$ *are equivalent over* $A \times B$.

(vii) The collages $\mathrm{col}(f,A)$ *and* $\mathrm{col}(B,u)$ *are equivalent under* $A+B$ *and over* $\Delta[1]$, *in which case* $\mathrm{col}(f,A) \twoheadrightarrow \Delta[1]$ *or equivalently* $\mathrm{col}(B,u) \twoheadrightarrow \Delta[1]$ *defines both a cocartesian and a cartesian fibration.*

Proof The equivalence between (i) and (ii) or (iii) is proven in Lemma 2.3.7, while the equivalence with (iv) or (v) is proven in Proposition 9.4.1. The equivalence between (i) and (vi) is proven in Proposition 4.1.1. We conclude by showing that (i) is equivalent to (vii).

First suppose that $\mathrm{col}(f,A) \simeq \mathrm{col}(B,u)$ under $A + B$ and over $\Delta[1]$. By Lemma F.5.4 and Corollary 5.3.1 this means that the map $\mathrm{col}(f,A) \to \Delta[1]$ is both a cocartesian and a cartesian fibration. By Proposition F.5.1 it follows that the 1-arrow in $\Delta[1]$ from 0 to 1 induces an adjunction between the fibers B and A. By inspection of that proof, the left adjoint functor so-constructed in the case of the bifibration $\mathrm{col}(f,A) \to \Delta[1]$ is f; substituting the equivalent bifibration $\mathrm{col}(B,u) \to \Delta[1]$, we see that the right adjoint is equivalent to u. Thus (vii) implies (i).

For the converse, we work in the opposite ∞-cosmos $\mathcal{Q}\mathcal{C}at^{\mathrm{op}}$, an ∞-cosmos in which "not all objects are cofibrant," as described in [110]. In that context,

Proposition F.5.5 proves that $\mathrm{col}(f, A)$ and $\mathrm{col}(B, u)$ construct the contravariant and covariant comma objects associated to the functors f and u. If $f \dashv u$ in $\mathcal{QC}at$ then these functors are also adjoint in $\mathcal{QC}at^{\mathrm{op}}$ and Proposition 4.1.1 then proves that the commas $\mathrm{col}(f, A)$ and $\mathrm{col}(B, u)$ are equivalent under $A + B$. By construction, this equivalence also lies over $\Delta[1]$. Alternatively, if the reader prefers not to dualize, $\mathrm{col}(f, A)$ and $\mathrm{col}(B, u)$ can be shown to define "weak cocomma objects" in the homotopy 2-category $\mathfrak{h}\mathcal{QC}at$, satisfying the 1-categorical duals of the weak universal properties of Proposition 3.4.6. Using these weak universal properties, the proof of Proposition 4.1.1 can be repeated in the dual to construct the desired equivalence $\mathrm{col}(f, A) \simeq \mathrm{col}(B, u)$ under $A + B$ and over $\Delta[1]$. So (i) implies (vii), completing the proof. $\square$

Exercises

EXERCISE F.5.i. Construct a correspondence that encodes the adjunction that expresses the universal property of a terminal element.

References

[1] Adámek, J., and Rosický, J. 1994. *Locally Presentable and Accessible Categories.* Cambridge University Press.

[2] Anel, M., and Lejay, D. 2018. Exponentiable higher toposes. arXiv:1802.10425.

[3] Antolín Camarena, O. 2016. A whirlwind tour of the world of $(\infty, 1)$-categories. Page 15–62 of: Bárcenas, N., Galaz-García, F., and Rocha, M. M. (eds), *Mexican Mathematicians Abroad: Recent Contributions.* Contemporary Mathematics, vol. 657. American Mathematical Society.

[4] Ara, D. 2014. Higher quasi-categories vs higher Rezk spaces. *J. K-Theory*, **14**(3), 701–749.

[5] Ayala, D., and Francis, J. 2020. Fibrations of ∞-categories. *High. Struct.*, **4**(1), 168–265.

[6] Barwick, C. 2005. (∞, n)-*Cat as a closed model category*. Ph.D. thesis, University of Pennsylvania.

[7] Barwick, C., Glasman, S., and Nardin, D. 2018. Dualizing cartesian and cocartesian fibrations. *Theory Appl. Categ.*, **33**(4), 67–94.

[8] Beke, T. 2001. Sheafifiable homotopy model categories. *Math. Proc. Cambridge Philos. Soc.*, **129**(03), 447–475.

[9] Bénabou, J. 1967. Introduction to Bicategories. Page 1–77 of: *Reports of the Midwest Category Seminar.* Lecture Notes in Math., no. 47. Springer-Verlag.

[10] Bénabou, J. 1973. *Les distributeurs.* Tech. rept. Université Catholique de Louvain, Institut de Mathématique Pure et Appliqué.

[11] Berger, C. 2007. Iterated wreath product of the simplex category and iterated loop spaces. *Adv. Math.*, **213**(1), 230–270.

[12] Berger, C., and Moerdijk, I. 2008. On an extension of the notion of Reedy category. *Mathe. Z.*, **269**, 1–28.

[13] Bergner, J. E. 2007a. A characterization of fibrant Segal categories. *Proc. Amer. Math. Soc.*, **135**(12), 4031–4037.

[14] Bergner, J. E. 2007b. Three models for the homotopy theory of homotopy theories. *Topology*, **46**(4), 397–436.

[15] Bergner, J. E. 2018. *An Introduction to $(\infty, 1)$-Categories.* Cambridge University Press.

[16] Bergner, J. E., and Rezk, C. 2013. Reedy categories and the Θ-construction. *Mathe. Z.*, **274**(1–2), 499–514.

[17] Betti, R., Carboni, A., Street, R., and Walters, R. 1983. Variation through enrichment. *J. Pure Appl. Algebra*, **29**, 109–127.

[18] Bird, G. J., Kelly, G. M., Power, A. J., and Street, R. H. 1989. Flexible Limits for 2-Categories. *J. Pure Appl. Algebra*, **61**(1), 1–27.

[19] Blanc, G. 1978. Équivalence naturelle et formules logiques en théorie des catégories. *Archiv math. Logik*, **19**, 131–137.

[20] Blumberg, A., and Mandell, M. 2011. Algebraic K-theory and abstract homotopy theory. *Adv. Math.*, **226**, 3760–3812.

[21] Boardman, J. M., and Vogt, R. 1973. *Homotopy Invariant Algebraic Structures on Topological Spaces*. Lecture Notes in Mathematics, vol. 347. Springer-Verlag.

[22] Bourke, J., Lack, S., and Vokřínek, L. 2020. Adjoint functor theorems for homotopically enriched categories. arXiv:2006.07843.

[23] Brown, K. S. 1973. Abstract homotopy theory and generalized sheaf cohomology. *Trans. Amer. Math. Soc.*, **186**, 419–458.

[24] Burroni, A. 1971. T-catégories (catégories dans un triple). *Cah. Topol. Géom. Différ. Catég.*, **12**, 215–321.

[25] Cartmell, J. 1986. Generalised algebraic theories and contextual categories. *Ann. Pure Appl. Logic*, **32**, 209–243.

[26] Cheng, E., Gurski, N., and Riehl, E. 2014. Cyclic multicategories, multivariable adjunctions and mates. *J. K-Theory*, **13**, 337–396.

[27] Cisinski, D. C. 2006. *Les préfaisceaux comme modèles des types d'homotopie*. Astérisque, vol. 308. Soc. Math. France.

[28] Cisinski, D. C. 2019. *Higher Categories and Homotopical Algebra*. Cambridge Studies in Advanced Mathematics, vol. 180. Cambridge University Press.

[29] Cordier, J. M. 1982. Sur la notion de diagramme homotopiquement cohérent. *Cah. Topol. Géom. Différ. Catég.*, **23**, 93–112.

[30] Cordier, J. M., and Porter, T. 1986. Vogt's theorem on categories of homotopy coherent diagrams. *Math. Proc. Cambridge Philos. Soc.*, **100**, 65–90.

[31] Cruttwell, G. S. H. 2008. *Normed Spaces and the Change of Base for Enriched Categories*. Ph.D. thesis, Dalhousie University.

[32] Cruttwell, G. S. H., and Shulman, M. A. 2010. A unified framework for generalized multicategories. *Theory Appl. Categ.*, **24**(21), 580–655.

[33] Day, B. 1970. On closed categories of functors. Page 1–38 of: *Reports of the Midwest Category Seminar IV*. Lecture Notes in Mathematics, vol. 137. Springer-Verlag.

[34] de Brito, P. B. 2018. Segal objects and the Grothendieck construction. Page 19–44 of: Ausoni, C., Hess, K., Johnson, B., Moerdijk, I., and Scherer, J. (eds), *An Alpine Bouquet of Algebraic Topology*. Contemporary Mathematics, vol. 708. American Mathematical Society.

[35] Dugger, D. 2001. Replacing model categories with simplicial ones. *Trans. Amer. Math. Soc.*, **353**(12), 5003–5028.

[36] Dugger, D., and Spivak, D. I. 2011. Mapping spaces in quasi-categories. *Algebr. Geom. Topol.*, **11**, 263–325.

[37] Dwyer, W. G., and Kan, D. M. 1980. Simplicial localizations of categories. *J. Pure Appl. Algebra*, **17**, 267–284.

[38] Dwyer, W. G., Kan, D. M., and Smith, J. H. 1989. Homotopy commutative diagrams and their realizations. *J. Pure Appl. Algebra*, **57**, 5–24.

[39] Dwyer, W. G., Hirschhorn, P. S., Kan, D. M., and Smith, J. H. 2004. *Homotopy Limit Functors on Model Categories and Homotopical Categories*. Mathematical Surveys and Monographs, vol. 113. American Mathematical Society.

[40] Ehlers, P. J., and Porter, T. 2000. Joins for (augmented) simplicial sets. *J. Pure Appl. Algebra*, **145**, 37–44.

[41] Eilenberg, S., and Kelly, G. M. 1966. Closed categories. Page 421–562 of: *Proceedings of the Conference on Categorical Algebra (La Jolla 1965)*. Springer-Verlag.

[42] Eilenberg, S., and Mac Lane, S. 1945. General theory of natural equivalences. *Trans. Amer. Math. Soc.*, **58**, 231–294.

[43] Freyd, P. J. 1976. Properties invariant within equivalence types of categories. Page 55–61 of: Heller, A., and Tierney, M. (eds), *Algebra, Topology and Category Theories*. Academic Press.

[44] Gabriel, P., and Zisman, M. 1967. *Calculus of Fractions and Homotopy Theory*. Ergebnisse der Math., no. 35. Springer-Verlag.

[45] Gambino, N. 2010. Weighted limits in simplicial homotopy theory. *J. Pure Appl. Algebra*, **214**(7), 1193–1199.

[46] Gepner, D., Haugseng, R., and Kock, J. 2021. ∞-operads as analytic monads. *International Mathematics Research Notices*.

[47] Grandis, M., and Paré, R. 2004. Adjoint for double categories. *Cah. Topol. Géom. Différ. Catég.*, **45**(3), 193–240.

[48] Gray, J. W. 1966. Fibred and cofibred categories. Page 21–83 of: Eilenberg, S., Harrison, D. K., Röhrl, H., and MacLane, S. (eds), *Proceedings of the Conference on Categorical Algebra (La Jolla 1965)*. Springer-Verlag.

[49] Groth, M. 2016. Characterizations of abstract stable homotopy theories. arXiv:1602.07632.

[50] Guillou, B., and May, J. P. 2011. Enriched model categories and presheaf categories. arXiv:1110.3567.

[51] Harpaz, Y. 2017. Introduction to stable ∞-categories. www.math.univ-paris13.fr/~harpaz/stable_infinity_categories.pdf.

[52] Henry, S. 2020. The language of a model category. www.uwo.ca/math/faculty/kapulkin/seminars/hottestfiles/Henry-2020-01-23-HoTTEST.pdf.

[53] Hermida, C. 2000. Representable multicategories. *Adv. Math.*, **151**, 164–225.

[54] Hess, K., Kędziorek, M., Riehl, E., and Shipley, B. 2017. A necessary and sufficient condition for induced model structures. *J. Topol.*, **20**(2), 324–369.

[55] Hirschhorn, P. 2003. *Model Categories and Their Localizations*. Mathematical Surveys and Monographs, no. 99. American Mathematical Society.

[56] Hirschowitz, A., and Simpson, C. 1998. Descente pour les *n*-champs. arXiv:math/9807049.

[57] Hovey, M. 1999. *Model Categories*. Mathematical Surveys and Monographs, vol. 63. American Mathematical Society.

[58] Johnson, N., and Yau, D. 2019. A bicategorical pasting theorem. arXiv:1910.01220.

[59] Johnson, N., and Yau, D. 2021. *2-Dimensional Categories*. Oxford University Press.

[60] Joyal, A. 1997. Disks, duality and Θ-categories. ncatlab.org/nlab/files/JoyalThetaCategories.pdf.

[61] Joyal, A. 2002. Quasi-categories and Kan complexes. *J. Pure Appl. Algebra*, **175**, 207–222.

[62] Joyal, A. 2008a. Notes on quasi-categories. www.math.uchicago.edu/~may/IMA/Joyal.pdf.

[63] Joyal, A. 2008b. *The Theory of Quasi-Categories and its Applications*. Quadern 45 vol II. Centre de Recerca Matemàtica Barcelona.

[64] Joyal, A., and Tierney, M. 2007. Quasi-categories vs Segal spaces. Page 277–325 of: Davydov, A. (ed), *Categories in Algebra, Geometry and Mathematical Physics (StreetFest)*. Contemp. Math., vol. 431. American Mathematical Society.

[65] Kazhdan, D., and Varshavsky, Y. 2014. Yoneda lemma for complete Segal spaces. *Funct Anal Its Appl*, **48**(2), 81–106.

[66] Kelly, G.M. 1964. On Mac Lane's conditions for coherence of natural associativities, commutativities, etc. *J. Algebra*, **1**, 397–402.

[67] Kelly, G. M. 1989. Elementary observations on 2-categorical limits. *Bull. Aust. Math. Soc.*, **49**, 301–317.

[68] Kelly, G. M. 2005. Basic concepts of enriched category theory. *Repr. Theory Appl. Categ.*, **10**.

[69] Kelly, G. M., and Mac Lane, S. 1971. Coherence in Closed Categories. *J. Pure Appl. Algebra*, **1**(1), 97–140.

[70] Kelly, G. M., and Street, R. H. 1974. *Review of the Elements of 2-Categories*. Lecture Notes in Mathematics, vol. 420. Springer-Verlag. Page 75–103.

[71] Lack, S. 2002. A Quillen model structure for 2-categories. *J. K-Theory*, **26**(2), 171–205.

[72] Lack, S. 2007. Homotopy-theoretic aspects of 2-monads. *J. Homotopy Relat. Struct.*, **2**(2), 229–260.

[73] Lack, Stephen. 2010. A 2-Categories Companion. In: May, J. P., and Baez, J. C. (eds), *Towards Higher Categories*. The IMA Volumes in Mathematics and its Applications, vol. 152. Springer.

[74] Lawvere, F. W. 2005. An elementary theory of the category of sets (long version) with commentary. *Repr. Theory Appl. Categ.*, **11**, 1–35.

[75] Leinster, T. 2002. Generalized enrichment of categories. *J. Pure Appl. Algebra*, **168**, 391–406.

[76] Leinster, T. 2004. *Higher Operads, Higher Categories*. London Mathematical Society Lecture Note Series, vol. 298. Cambridge University Press.

[77] Lurie, J. 2006 (May). *Stable Infinity Categories*. arXiv:math/0608228.

[78] Lurie, J. 2009a. *Higher Topos Theory*. Annals of Mathematical Studies, vol. 170. Princeton University Press.

[79] Lurie, J. 2009b. $(\infty, 2)$-*Categories and the Goodwillie Calculus I*. arXiv:0905.0462.

[80] Lurie, J. 2017 (September). *Higher Algebra*. www.math.ias.edu/~lurie/papers/HA.pdf.

[81] Mac Lane, S. 1998. *Categories for the Working Mathematician*. Second edn. Graduate Texts in Mathematics. Springer-Verlag.

[82] Makkai, M. 1995. *First Order Logic with Dependent Sorts, with Applications to Category Theory*. www.math.mcgill.ca/makkai/.

[83] Makkai, M., and Paré, R. 1989. *Accessible Categories: The Foundations of Categorical Model Theory.* Contemp. Math., vol. 104. American Mathematical Society.

[84] May, J. P., and Ponto, K. 2012. *More concise algebraic topology: Localization, Completion, and Model Categories.* Chicago Lectures in Mathematics. University of Chicago Press.

[85] Mazel-Gee, A. 2016. Quillen adjunctions induce adjunctions of quasicategories. *New York J. Math.*, **22**, 57–93.

[86] Mazel-Gee, A. 2019. A user's guide to co/cartesian fibrations. *Graduate Journal of Mathematics*, **4**(1), 42–53.

[87] Myers, D. J. 2018. *String Diagrams For Double Categories and Equipments.* arXiv:1612.02762.

[88] Nguyen, H. K., Raptis, G., and Schrade, C. 2020. Adjoint functor theorems for ∞-categories. *J. Lond. Math. Soc.*, **101**(2), 659–681.

[89] Ozornova, V., and Rovelli, M. 2020. Model structures for (∞, n)-categories on (pre)stratified simplicial sets and prestratified simplicial spaces. *Algebr. Geom. Topol.*, **20**, 1543–1600.

[90] Pellissier, R. 2002. *Catégories enrichies faibles.* Ph.D. thesis, Université de Nice-Sophia Antipolis.

[91] Power, A. J. 1990. A 2-categorical pasting theorem. *J. Algebra*, **129**, 439–445.

[92] Preller, A. 1985. A language for category theory in which natural equivalence implies elementary equivalence of models. *Zeitschrift f. math. Logik und Grundlagen Math.*, **31**, 227–234.

[93] Quillen, D. G. 1967. *Homotopical Algebra.* Lecture Notes in Mathematics, vol. 43. Springer-Verlag.

[94] Radulescu-Banu, A. 2009. Cofibrations in homotopy theory. arXiv:math/0610009.

[95] Rasekh, N. 2017a. Cartesian fibrations and representability. arXiv:1711.03670.

[96] Rasekh, N. 2017b. Yoneda lemma for simplicial spaces. arXiv:1711.03160.

[97] Rasekh, N. 2021a. Quasi-categories vs. Segal spaces: cartesian edition. *J. Homotopy Relat. Struct.*

[98] Rasekh, Nima. 2021b. Cartesian fibrations of complete Segal spaces. arXiv:2102.05190.

[99] Reedy, C. 1974. Homotopy theory of model categories. www-math.mit.edu/~psh/reedy.pdf.

[100] Rezk, C. 2001. A model for the homotopy theory of homotopy theory. *Trans. Amer. Math. Soc.*, **353**(3), 973–1007.

[101] Rezk, C. 2010. A cartesian presentation of weak n-categories. *Geom. Topol.*, **14**(1), 521–571.

[102] Rezk, C. 2019. *Stuff about quasi-categories.* faculty.math.illinois.edu/~rezk/quasicats.pdf.

[103] Riehl, E. 2014. *Categorical Homotopy Theory.* New Mathematical Monographs, vol. 24. Cambridge University Press.

[104] Riehl, E. 2016. *Category Theory in Context.* Aurora Modern Math Originals. Dover Publications.

[105] Riehl, E. 2017. Complicial sets, an overture. In: *2016 MATRIX Annals.* MATRIX Book Series, no. 1–28. Springer.

[106] Riehl, E., and Shulman, M. 2017. A type theory for synthetic ∞-categories. *High. Struct.*, **1**(1), 147–224.

[107] Riehl, E., and Verity, D. 2014. The theory and practice of Reedy categories. *Theory Appl. Categ.*, **29**(9), 256–301.

[108] Riehl, E., and Verity, D. 2015. The 2-category theory of quasi-categories. *Adv. Math.*, **280**, 549–642.

[109] Riehl, E., and Verity, D. 2016. Homotopy coherent adjunctions and the formal theory of monads. *Adv. Math.*, **286**, 802–888.

[110] Riehl, E., and Verity, D. 2017. Fibrations and Yoneda's lemma in an ∞-cosmos. *J. Pure Appl. Algebra*, **221**(3), 499–564.

[111] Riehl, E., and Verity, D. 2018. The comprehension construction. *High. Struct.*, **2**(1), 116–190.

[112] Riehl, E., and Verity, D. 2020a. On the construction of limits and colimits in ∞-categories. *Theory Appl. Categ.*, **35**(30), 1101–1158. arXiv:1808.09835.

[113] Riehl, E., and Verity, D. 2020b. Recognizing quasi-categorical limits and colimits in homotopy coherent nerves. *Appl. Categ. Struct.*, **28**(4), 669–716. arXiv:1808.09834.

[114] Riehl, E., and Verity, D. 2021. Cartesian exponentiation and monadicity. arXiv:2101.09853.

[115] Shah, Jay. 2018. Parametrized higher category theory and higher algebra: Exposé II - Indexed homotopy limits and colimits. arXiv:1809.05892.

[116] Simpson, C. 2012. *Homotopy Theory of Higher Categories*. New Mathematical Monographs, no. 19. Cambridge University Press.

[117] Street, R. H. 1974a. Elementary cosmoi I. Page 134–180 of: *Category Seminar (Proc. Sem., Sydney, 1972/1973)*. Lecture Notes in Math., vol. 420. Springer.

[118] Street, R. H. 1974b. Fibrations and Yoneda's lemma in a 2-category. Page 104–133. of: *Category Seminar (Proc. Sem., Sydney, 1972/1973)*. Lecture Notes in Math., vol. Vol. 420. Springer.

[119] Street, R. H. 1980. Fibrations in Bicategories. *Cah. Topol. Géom. Différ. Catég.*, **XXI-2**, 111–159.

[120] Street, R. H. 1987a. The algebra of oriented simplexes. *J. Pure Appl. Algebra*, **49**, 283–335.

[121] Street, R. H. 1987b. Correction to "Fibrations in bicategories". *Cah. Topol. Géom. Différ. Catég.*, **28**(1), 53–56.

[122] Street, R. H. 2003. Weak omega-categories. Page 207–213 of: *Diagrammatic Morphisms and Applications*. Contemp. Math., vol. 318. American Mathematical Society.

[123] Street, R. H., and Walters, R. 1978. Yoneda structures on 2-categories. *J. Algebra*, **50**, 350–379.

[124] Sulyma, Y. J. F. 2017. ∞-categorical monadicity and descent. *New York Journal of Mathematics*, **23**, 749–777.

[125] The Univalent Foundations Program. 2013. *Homotopy Type Theory: Univalent Foundations of Mathematics*. Institute for Advanced Study: homotopytypetheory.org/book.

[126] Ulmer, F. 1968. Properties of dense and relative adjoint functors. *J. Algebra*, **3**, 77–95.

[127] Verdier, J. L. 1996. Des Catégories Dérivées des Catégories Abéliennes. *Astérisque*, **239**, 1–253+ix.

[128] Verity, D. 2008a. Complicial Sets, Characterising the Simplicial Nerves of Strict ω-Categories. *Mem. Amer. Math. Soc.*, **193**(905).

[129] Verity, D. 2008b. Weak complicial sets I, basic homotopy theory. *Adv. Math.*, **219**, 1081–1149.

[130] Verity, D. 2011. Enriched Categories, Internal Categories and Change of Base. *Repr. Theory Appl. Categ.*, **20**, 1–266.

[131] Weber, M. 2007. Yoneda structures from 2-toposes. *Appl. Categ. Structures*, **15**(3), 259–323.

[132] Wood, R. J. 1982. Abstract proarrows I. *Cah. Topol. Géom. Différ. Catég.*, **XXIII-3**, 279–290.

[133] Zaganidis, D. 2017. *Towards an* $(\infty, 2)$-*category of Homotopy Coherent Monads in an* ∞-*cosmos*. Ph.D. thesis, École Polytechnique Fédérale de Lausanne.

Glossary of Notation

$(-)^{co}$, 544
$(-)^{coop}$, 544
$(-)^{op}$, 36, 461, 463, 544
$(-)^{-}$, 13, 23, 35, 68, 146, 503, 513, 518
$(-)^{\circ}$, 461
$(-)^{\flat}$, 622
$(-)^{\natural}$, 698
$(-)^{\sharp}$, 622
$(-)_0$, 39, 504
$=$, 9
$\Rrightarrow$, 548
$\Rightarrow$, 540
$*$, 512, 540
$\perp$, 56, 141, 549
$\boxtimes$, 582
$\square$, 602
$\cdot$, 512, 540
$\hookrightarrow$, 255
$\cong$, 46
$\dashv$, 56, 549
$\dashv_B$, 125
$\diamond$, 139, 637
$\twoheadrightarrow$, 13, 21, 569, 591
$\dot{\times}$, 248
$\star$, 140, 632, 633, 641

$\widehat{\otimes}$, 585
$\widehat{\pitchfork}$, 15, 181
$\vDash$, 440
$\oplus$, 170, 631
$\otimes$, 146, 323, 518, 585
$\lrcorner$, 77, 146
$\ulcorner$, 146
$\pitchfork_B$, 34
$\rightsquigarrow$, 404, 408
$\approxrightsquigarrow$, 406, 411
$\rightarrowtail$, 591
$\sim$, 625
$\simeq$, 15, 48, 625
$\simeq_B$, 95
$\boxdot$, 147
$\times^B$, 34
$\times$, 320
$\top$, 141
$\twoheadrightarrow$, 16, 22, 569
$\twoheadrightarrow$, 15, 22, 41, 48, 569, 591
$\widehat{\{,\}}$, 585
$\{,\}$, 585

1, 39
$\mathbb{1}$, 7
$\mathbb{2}$, 7
$\mathbb{3}$, 7

Index

Printed in the United States
by Baker & Taylor Publisher Services